# ION IMPLANTATION

*SERIES*    DEFECTS IN CRYSTALLINE SOLIDS

Editors:

S. AMELINCKX
R. GEVERS
J. NIHOUL

Studiecentrum voor kernenergie, Mol,
and
University of Antwerpen Belgium

NORTH-HOLLAND PUBLISHING COMPANY - AMSTERDAM · LONDON
AMERICAN ELSEVIER PUBLISHING COMPANY, INC. - NEW YORK

# ION IMPLANTATION

## G. DEARNALEY, J. H. FREEMAN, R. S. NELSON
### and J. STEPHEN

*Atomic Energy Research Establishment*
*Harwell, England*

1973

NORTH-HOLLAND PUBLISHING COMPANY - AMSTERDAM · LONDON
AMERICAN ELSEVIER PUBLISHING COMPANY, INC. - NEW YORK

252834

Library of Congress Catalog Card Number: 72-93492
North-Holland ISBN for the series: 0 7204 1750 3
North-Holland ISBN for this volume: 0 7204 1758 9
American Elsevier ISBN: 0 444 10488 7

Publishers:

North-Holland Publishing Company - Amsterdam
North-Holland Publishing Company Ltd. - London

Sole distributors for the U.S.A. and Canada:

American Elsevier Publishing Company, Inc.
52 Vanderbilt Avenue
New York, N.Y. 10017

Printed in The Netherlands

# PREVIOUSLY PUBLISHED BOOKS IN THIS SERIES:

(G. Agricola, "De Re Metallica" 1556, emended by M. W. Thompson.)

Professor Thompson has drawn attention to the undue preoccupation with the diffusion furnace (front right) as opposed to the ion implantation equipment (rear left) apparent in this early woodcut, an attitude which persists in a few workshops even to this day.

# PREFACE

*The pace of scientific progress often reaches its maximum at the boundaries and interfaces of the traditional disciplines, because it is here that pre-existing resources and expertise can be devoted to a virgin field. So it has been with regard to ion implantation over the past few years: facilities and ideas derived from the nuclear energy programme have been turned to the investigation of solid state problems. The interaction and mutual feedback between these two branches of physics has proved most fruitful, and a spate of publications and conferences has ensued. Commercial exploitation is well under way, while the general area of work on the interaction of particle beams with solids is fast becoming a discipline for study in its own right and it has been selected by the Science Research Council of the U.K. for special support.*

*When development is so rapid there is naturally a tendency for all but those who are deeply involved to become bewildered by it, and it is easy to lose track of ramifications and of inroads that are being made into neighbouring areas of science. This presents a serious problem for those technologists to whom we must look for the future exploitation of the scientific ideas: they have not the time to search the literature and distil from it the information relevant to their current problems. We hope this book will help them to scan the subject in an orderly fashion and provide them with the theoretical bases for further development. We shall be happy if it leads them to discuss with us their problems for which ion implantation may offer some solution.*

*The growing academic participation in the study of ion implantation which commenced a few years after its early investigation within the government laboratories is leading to a requirement, among students, for a textbook which covers the topic in a broad manner and serves as a reasonably pedagogical introduction to it. We hope that the book may itself attract a few more good students to enter the field, and we look forward to discussions which may thereby be generated. We have purposely presented the subject in an open-ended way since we are aware of the danger of defining prematurely and dogmatically the scope of the field in terms of our own interests.*

*Let us therefore explain here the form of presentation which has been adopted in the succeeding chapters. The basic physical processes are dealt with first of all since they control both what applications are feasible and also the design of equipment for carrying out the process. The physics is divided broadly into two parts: the effect of the target medium upon the incident ion, i.e. its slowing down and the influence upon this of the ordered structure of a crystalline target, and secondly the effect of the implantation of the ion upon the state of the target, i.e. radiation damage and the subsequent migration and interactions of defects. Next we chose to describe the experimental means of performing ion implantation: ion sources, accelerators and target chambers. This section is presented in some detail since there has been no survey of this topic hitherto and the practical experience of producing and handling ion beams is not yet widely disseminated. Augmenting this, an appendix describes many practical points of value in producing different ion beams, and we trust that this will be a useful compilation. One of the needs at the present time is to overcome a barrier of unfamiliarity with accelerators, as compared with conventional tools such as epitaxy reactors, for example. We therefore make no apology for the length of this chapter. Nor in view of the paucity of systematic published data in this field do we apologise for what in consequence may appear to be an undue emphasis on our own experience and equipment. Two chapters on practical applications complete the book, the first dealing with semiconductor devices and the second with other applications. Devices are here described in the form of examples to illustrate the advantages and limitations of ion implantation. Whereas there have been many publications and several conferences devoted to semiconductor device aspects of ion implantation, by contrast there has been relatively little work in other materials openly published. This does not mean that investigations are not being made, but necessarily this final chapter can merely outline in a rather sketchy manner some of the developments now taking place.*

*Finally, some details should be mentioned. After careful consideration, we decided against the adoption of S.I. units in this edition owing to the widespread usage of the older units in the literature we have surveyed. We do not anticipate that this feature will cause any significant inconvenience, since the number of units involved is relatively small.*

*Before concluding this preface, we should like to join in the acclaim which has been awarded to an earlier book on 'Ion Implantation in Semiconductors (Silicon and Germanium)' by our colleagues Jim Mayer, Lennart Eriksson and John Davies. We hope that it will not seem superfluous to publish a second book on this developing topic. The discerning reader will observe, on comparing the two, that, whilst we have tried to make our book complete in itself we have adopted a markedly different emphasis, devoting more attention to aspects such as the means of carrying out ion implantation, device applications and ion implantation outside the semiconductor field. Energy loss theory and radiation damage receive a different treatment. Likewise, we have given less weight to those topics such as atom location by channelling techniques and Hall effect measurements in implanted silicon which are dealt with very fully in the earlier book. We hope, for this and other reasons, that the two volumes will find themselves together on many bookshelves.*

*Finally, we should like to acknowledge the generosity of many colleagues throughout the world in allowing us to make use of their data, in many cases prior to its publication. We are grateful to Geoff Gard and Bill Temple for assistance with Appendix I, to Bernard Smith for his help and the preparation of the tables in Appendices II, III, and IV and to Professor J. F. Gibbons for his kindness in allowing us to make use of his computer program for range calculations for appendix III.*

# CONTENTS

# 1 | INTRODUCTION

## 1.1. Historical introduction

The fact that a flux of energetic atomic particles will produce physical changes in a solid target specimen has been recognized for several decades. It is unlikely that nuclear physicists, during the 1930's, making use of the early ion accelerators then available, could have failed to observe visible modification of certain target materials, but if so these were merely regarded as troublesome side-effects.

With the development of the nuclear reactor in the early 1940's, the situation became very different. Materials were for the first time subjected to intense irradiation by fast neutrons, which cause the atoms they collide with to recoil with energies up to a few hundred keV. Within the fuel, energetic fission fragments are also a potent source of damage. The theoretical aspects of these problems were first considered by E. P. Wigner in about 1942, though not published until later.

The increasing study of nuclear reactions induced by energetic ions led very soon to a need to know the energy an ion would have at different depths in a solid target. This led directly to the classic work of Bohr (1948) and others on the theory of ion penetration through matter.

At just about this time, the electrical properties of elemental semiconductors such as silicon and germanium were rapidly being exploited, and this development led to the first transistor in 1948. This naturally initiated a close study of all means by which electrically-active impurities could be introduced into semiconductors and, at the same time, semiconductor device applications in the nuclear energy field stimulated an investigation of radiation damage in semiconductors, initially at Purdue University and Oak Ridge National Laboratory (Davis et al., 1948).

Hitherto, perhaps without exception, the effects of nuclear radiation on solid objects had been regarded as deleterious. It therefore required imagination to attempt to improve semiconductor devices by ion bombardment, but in 1952 Russell Ohl at Bell Telephone Laboratories bombarded point-contact silicon diodes with helium ions and observed an improvement in the reverse current characteristics. Implantation of silicon at temperatures between 300 and 500 °C produced photosensitive junctions. At the time, the bombardment was believed to induce a densification of the semiconductor surface, leading to a change in the band-gap. In 1955, W. D. Cussins, in Cambridge, bombarded germanium crystals with twelve different types of ion and in each case observed an increase in the surface acceptor concentration. Clearly, a chemical doping effect was being sought but this was overwhelmed by the damage centres produced during the slowing down of the ions. Although unsuccessful, this work by Cussins is perhaps the first example of a deliberate attempt to alter the properties of a solid by introducing dopant impurities in the form of an energetic ion beam. We distinguish this from the technique of merely incorporating ions into a solid target, e.g. in the preparation of electro-magnetically-separated isotopic targets. It should be remarked, however, that at this time the practice had been developed of converting the surface of metal targets to oxide or nitride as a means of holding the separated isotopes of $O^+$ or $N^+$ (Smith, 1956).

The first published work in which a distinction was established between the damaging and doping effects of implantation was by Bredov et al. (1961) in the U.S.S.R. Their experiments demonstrated differing results in germanium bombarded with oxygen and nitrogen ions. Already, by this time, far-reaching patents had been granted on the ion-implantation technique to Ohl (1956), Shockley (1957) and Moyer (1958), and the last two of these reveal that the authors understood the need for thermal annealing after implantation in order to restore the crystalline nature of the bombarded surface. Such a knowledge owed a good deal to the work in Lark–Horovitz' school at Purdue.

A slow but steady series of publications continued to appear. Rourke et al. (1961) implanted ions of group III and group V elements into silicon at energies of about 10 keV. Some advantages of this technique were realized, but the properties of the diodes produced were not reported and no reference was made to annealing: it is difficult to believe that the results could have been very promising. Nevertheless, this publication was widely read and stimulated the interest of scientists who were studying ways of improving junction devices in silicon. Certainly, two of the present authors considered making use of the process, at Harwell, for the fabrication of nuclear particle detectors. This was soon afterwards carried out by Alväger and Hansen (1962), who implanted phosphorus into silicon and, after annealing at 600 °C, obtained diodes with a shallow, large-area junction appropriate for nuclear particle detection. This work revealed that phosphorus, when implanted into silicon and subsequently annealed, can produce a similar doping effect to that introduced by conventional diffusion.

Serious industrial interest in ion implantation dates from about this time, with the early work at Ion Physics Corporation in Massachusetts. Progress was hampered, however, by a lack of close association with the semiconductor device industry which itself was bound to regard the use of ion accelerators as a costly and hazardous development. It was perhaps natural that the next stimulus should come from laboratories which were exploiting the use of ion beams in research.

In 1963 the phenomenon of ion channelling in crystals became recognized and scientists at several major nuclear laboratories (Chalk River Nuclear Laboratories, Canada; Oak Ridge National Laboratory, Tennessee and the Atomic Energy Research Establishment, Harwell) began to study the motion of ions through crystals. Already these centres had built up a considerable experience in the production and handling of ion beams and their use in experiments to simulate radiation damage produced in reactors. The first international meeting to include a session on what would now be termed ion implantation took place at Aarhus University, Denmark in 1965. This occasion aroused interest at Aarhus in ion channelling, the study of which followed naturally on the earlier work of J. Lindhard and his school on the theory of ion penetration, and the development at the same institute of electromagnetic isotope separators capable of ion beam production. There is no doubt that these fundamental studies, begun at several centres in 1963–5, attracted capable scientists to the subject of ion implantation and awakened interest in the technological potential, particularly for the doping of semiconductors, the properties of which are governed by minute concentrations of electrically active material.

At Harwell, for example, a programme of work was initiated in 1965, co-ordinating the activities in four different Divisions, and aimed at exploring the physics and technology of the ion implantation process quite broadly. Besides the basic understanding of ion penetration and the accompanying radiation damage, electrical effects in semiconductor structures were to be examined and designs produced for machines capable of carrying out the process on an industrial scale. It was recognized that close contact with the semiconductor device industry was essential, and this was achieved under the auspices of a government defence organization (C.V.D.*) which itself supported a complementary research programme at SERL, Baldock. Almost all the results of this work, over the succeeding years, have been published in the open literature.

Activity on a world-wide scale has increased very considerably over the past few years, and after constituting a major part of a conference on Applications of Ion Beams to Semiconductor Technology (at Grenoble, 1967), Ion Implantation became the subject of a regular series of international conferences, initiated in 1970 with a meeting in Thousand Oaks, California dealing with the physics and technology of the process. This was shortly afterwards followed by a European meeting held at the University of Reading. The next of the international series took place at Garmisch-Partenkirchen in 1971, followed by another held at Yorktown Heights, N.Y., U.S.A. in December, 1972. In parallel, another series of conferences on basic atomic collision processes in solids has been held at the University of Sussex (1969) and at Gausdal, Norway (1971), while the next of this series will take place at Gatlinburg in September 1973. Industrial interest has quickly grown since the time when ion implanted devices first showed real advantages over conventionally manufactured structures. Bell Telephone Laboratories, Hughes Research Laboratories, and IBM in the U.S.A., Hitachi and Toshiba in Japan, and Mullard and G.E.C. in the U.K. are examples of major semiconductor device companies which have mounted large-scale ion implantation programmes; and in common with a number of smaller firms several of these are now in a position to market ion-implanted devices. In parallel there has grown up a lively market for implantation machines and target chambers and companies in the U.S.A., Britain, Denmark and France are offering a variety of such machines, some of which are capable of large-scale production, i.e. sufficient to meet the daily throughput of a major device manufacturer.

In this sense, ion implantation may claim to have arrived, and yet its full assessment has still to be made. In the semiconductor field, the recent tightening of the economic situation generally has held back the installation of new capital equipment. Some possibly less expensive, though less versatile, processes such as silicon gate technology are being evaluated. The economics

* Department of Components, Valves and Devices Procurement Executive, Ministry of Defence.

of ion implantation in full-scale use on the factory floor have yet to be worked out, though the capabilities of modern equipment allow optimistic estimates to be made. Besides its application in semiconductor device technology there is scope for the exploitation of ion implantation in spheres where properties other than the electrical behaviour require to be modified. Whereverthe surface nature of a high-cost component presents some problem it is worthwhile considering the application of the ion implantation technique, for the same reasons for which it offers advantages in the already highly-developed semiconductor field.

From the historical survey above we have seen that interest in ion implantation and the capability for undertaking research in it sprang up largely in the nuclear energy field. The industrial exploitation necessitates a transfer of this experience into quite different fields of microelectronics and solid state science. There is a problem, therefore, of scientific communication between disciplines which must be solved with the minimum delay.

We believe, therefore, that this book is a timely outline of the physics and technology of ion implantation, not only in semiconductors but also in other potential fields. We have tried to provide the basic understanding from which future technical developments may derive. Moreover, as we shall see in the next section, there is among the universities a new interest in ion implantation as part of a general field of study of atomic collision phenomena in solids. In Britain, the Science Research Council singled out the topic of ion implantation, in this general sense, for particular support and chose three main centres, at the Universities of Salford, Surrey and Sussex, to be fully equipped with accelerators suitable for different aspects of such work. Their work, together with that of the government laboratories already mentioned, has shown the value of ion beams both for modifying and probing solids of many types. This same pattern of activity is matched in other countries, where university groups have been well supported in this new and inter-disciplinary field.

## 1.2. The scope of the subject

It may be helpful at this early stage to outline in more detail the scope of the subject we term ion implantation, since it is a relatively new area of research lying between the more traditional studies of solid state physics, nuclear physics and atomic physics. Its techniques are drawn from each of these fields, which is one reason why it offers particularly good possibilities for the training of versatile research students.

We might define ion implantation to be the process of modifying the physical or chemical properties of a solid by embedding into it appropriate atoms in the form of a beam of ionized particles. These properties may be electrical, optical or mechanical; they may relate to the superconducting behaviour of the material, or its corrosion behaviour. The ions introduced may be radioactive, in which case the subsequent decay processes may reveal information about the solid itself. The solid may be crystalline, polycrystalline or amorphous and need not be homogeneous. Although the early work on ion implantation has been overwhelmingly concerned with semiconductors, it is now clear that the field is far more extensive.

It is important to point out those areas we shall not include. Thus the radiation damage of solids studied in a general way is a well-established topic: only those aspects which are relevant to the implantation of ions will be covered, and readers who wish to broaden their understanding are referred to texts such as those by Strumane et al. (1964) and Thompson (1969). The ejection of material from a solid by ion bombardment, i.e. sputtering, is only briefly mentioned since, although this is an important technological process, the effects are generally negligible in ion implantation as applied so far. We shall simply mention the effect where relevant. Nor do we deal with the release, from an implanted solid, of gaseous material as a result of heating. This and other related topics are dealt with by Carter and Colligon in their textbook (1968).

We do, however, include a number of techniques for the examination of implanted solids, such as electrical conductivity and Hall coefficient measurement, transmission electron microscopy, scanning electron microscopy, and probing by means of a more penetrating ion beam. This last technique receives particular attention because it has developed in parallel with the study of ion channelling in crystals, and may be less familiar to readers than the more conventional methods of examination mentioned above.

As pointed out in section 1.1, we have aimed at describing not only the physics and technology of ion implantation as a process of industrial potential but also to point out its possibilities in research. In many solid-state problems one may wish to introduce, controllably, some particular impurity atoms and study the change produced. Alternative techniques such as diffusion or co-evaporation may often be impracticable, in which case ion implantation offers a straightforward and reproducible means of achieving what is required. Being a non-equilibrium process, one may exceed solubility limits and observe the subsequent precipitation phenomena. One may incorporate any kind of ion without the need for developing a set of diffusion conditions or the appropriate chemical combination method. Several cases have arisen in which a rapid survey of many combinations of dopant and host material were accomplished by ion implantation. Furthermore, the phenomena of ion channelling, energy loss processes in solids, defect migration and clustering are of basic interest for their own sake. They are therefore given treatment in more detail than would be warranted by the purely technico-logical role of ion implantation: we have kept in mind the needs of students in the compilation of these sections of the book.

The interaction of ion beams with solids is relevant also to those experiments which set out to simulate the effects of fast neutron damage in reactors by means of ion bombardment. Whatever species of ions are used, the implantation of material into the target will produce changes which must be understood and distinguished from the results of atomic displacement. Even in space research, scientists must be aware of the effects of ion bombardment resulting from the particles of the solar wind. On the lunar surface, there is evidence of material which has been implanted in this way for extreme periods of time, and knowledge of the effects of such a process will be relevant to interpreting observations made on lunar surface samples. We have in mind the hope that the present book may stimulate the imagination of scientists in other fields and inform them of the possibilities of ion implantation for research and technology. Too often the more specialised research papers published in the literature are read only by specialists within the same field: there is a need, we feel, to disseminate more widely the information regarding ion implantation and its possibilities and limitations.

## References

Alväger, T. and N. J. Hansen, 1962, Rev. Sci. Instr. **33**, 567.

Bohr, N., 1948, Kgl. Danske Vid. Selsk., Matt.-Fys. Medd. **18**, no. 8.

Bredov, M. M., V. A. Lepilin, J. B. Schestakov and A. L. Shakh-Budagov, 1961, Sov. Phys.-Sol. State **3**, 195.

Carter, G. and J. S. Colligon, 1968, 'Ion Bombardment of Solids' (Heinemann, London).

Cussins, W. D., 1955, Proc. Phys. Soc. London B **68**, 213.

Davis, R. E., W. E. Johnson, K. Lark-Horovitz and S. Siegel, 1948, Phys. Rev. **74**, 1255.

Mayer, J. W., L. Eriksson and J. A. Davies, 1970, 'Ion Implantation in Semiconductors (Silicon and Germanium)', (Academic Press, N.Y.).

Moyer, J. W., 1958, U.S. Patent no. 2 842, 466.

Ohl, R., 1952, Bell Syst. Tech. J. **31**, 104.

Ohl, R., 1956, U.S. Patent no. 2 750, 541.

Rourke, F. M., J. C. Sheffield and F. A. White, 1961, Rev. Sci. Instr. **32**, 455.

Shockley, W., 1957, U.S. Patent no. 2 787, 564.

Smith, M. L., 1956, 'Electromagnetically Enriched Isotopes and Mass Spectrometry' p. 97 (Butterworths, London).

Strumane, R., J. Nihoul, R. Gevers and S. Amelinckx (Ed.) 1964, 'The Interaction of Radiation with Solids' (North-Holland, Amsterdam).

Thompson, M. W., 1969, 'Defects and Radiation Damage in Metals' (Cambridge University Press).

Wigner, E. P., 1946, J. Appl. Phys. **17**, 857.

SRC Panel Report on Ion Implantation, 1970 (Science Research Council, London).

Proceedings of International Conference on Electromagnetic Isotope Separators and their Applications, Aarhus June 1965, Nucl. Instr. and Methods **38** (1965).

Proceedings of International Conference on Applications of Ion Beams to Semiconductor Technology, Grenoble May 1967 (Editions Ophrys, Gap, France 1968).

Proceedings of 1st International Conference on Ion Implantation in Semiconductors, Thousand Oaks, Calif., May 1970, Radiation Effects **6** (1970).

Proceedings of European Conference on Ion Implantation, Reading University, September 1970 (Peter Peregrinus Ltd., Stevenage, 1970).

Proceedings of 2nd International Conference on Ion Implantation in Semiconductors, Garmisch-Partenkirchen May 1971 (Springer Verlag).

Proceedings of 3nd International Conference on Ion Implantation in Semiconductors and other materials, Yorktown Heights, December 1972 (to be published).

# 2 | ION PENETRATION AND CHANNELLING

## 2.1. Introduction

An understanding of the mechanisms which govern the slowing down of ions in solids, and particularly those which apply in crystalline media, is of considerable importance in achieving controlled and reproducible dopant distributions in ion implantation. Factors such as the ion energy and ion species, the crystal orientation and temperature all influence the range distributions obtained. There is, therefore, the possibility of a much greater controlled variation of the distribution of material introduced into the specimen than exists in the case of diffusion. If the implantation behaviour is well understood, as a function of ion energy and dose, it is feasible to program these parameters so as to achieve almost any desired distribution, and in principle this processing can be highly automated. We shall return to the technological problems associated with control of the ion implantation process and the appropriate equipment in a later chapter, and here begin by introducing the basic ideas of the energy loss of energetic ions in motion through solids.

These ideas have developed steadily throughout the twentieth century, since the time of Rutherford's early experiments on the transmission of alpha particles through thin metal foils. Although a great deal is now known regarding ion penetration, particularly in amorphous materials, our understanding of the problem is still far from adequate and there is as yet no unified approach to the behaviour in different cases. Some of the models proposed, while successfully accounting for complex phenomena, have not yet been put on a sound quantum-mechanical footing. Our aim will therefore

9

be twofold: to give the reader, so far as is currently possible, a practical capability of estimating the range and distribution of a given ion in a given medium, and in addition, to outline the state of present theories, their assumptions and inadequacies. Since many readers may be unfamiliar with concepts which are as a rule encountered only in atomic and nuclear physics, it will be necessary to introduce certain ideas from first principles and to follow others chronologically to their present form.

## 2.2. Specific energy loss

Before considering the ranges of ions it is convenient to discuss the differential function $(-dE/dx)$, known as the stopping power or specific energy loss. Here $E$ is the ion energy and $x$ is the distance, usually measured along the direction of incidence of the ions, but in other cases it may be along the instantaneous direction of the ion trajectory. It is customary to distinguish between two major processes of energy loss, due on the one hand to elastic Coulomb interactions between the 'screened' nuclear charges of ion and target atom, and on the other to inelastic interactions of the ion with bound or free target electrons. The screening referred to here arises because the positively-charged nucleus is surrounded by a cloud of orbital electrons: as the approaching ion penetrates this cloud it experiences an increasing electrostatic repulsion, which deflects its path. It was of course the observation that a small proportion of alpha particles would recoil through very large angles which led Rutherford to the concept of the nuclear atom. A heavy ion, however, itself carries orbital electrons which partially screen the nuclear field. The energy losses due to nuclear and electronic interactions are correlated, since in close collisions both are taking place. For most purposes, however, it is justifiable to ignore these correlations and regard the electronic stopping as a continuous process (Lindhard, 1954): the very close collisions in which nuclear recoils are important are comparatively rare events. In addition, there may be a contribution to the energy loss because of charge (i.e. electron) exchange between the moving ion and target atom, electrons from one transferring to the other during their close proximity.

The total specific energy loss is thus taken to be the sum of three separable components: nuclear, electronic and charge-exchange

$$dE/dx = (dE/dx)_e + (dE/dx)_{ch} + (dE/dx)_n \qquad (2.1)$$

The variation of the first two of these components with ion energy is shown schematically in fig. 2.1, in which it can be seen that both increase with

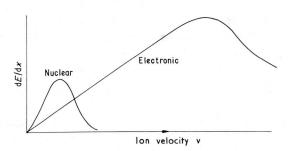

Fig. 2.1. The behaviour of the nuclear and electronic contributions to the specific energy loss $dE/dx$ as a function of ion velocity $v$.

energy, reach a maximum and then decrease again. At the lowest energies, nuclear stopping dominates and is responsible for most of the angular dispersion of an ion beam. At higher energies, electronic collisions are the more important and in slowing down to rest from these energies the bulk of the particle energy is dissipated in the form of electronic, rather than nuclear, motion. On this fact many particle detectors depend. At the highest energies, relativistic factors become significant in the energy loss of charged particles, but this region lies outside the energy range of interest in ion implantation.

### 2.3. Atomic cross-sections

A widely-used concept in atomic and nuclear physics is that of a scattering or reaction cross-section. Consider a parallel beam of $n_0$ particles per second incident on a thin slice of solid material of thickness $dx$ and containing $N$ atomic centres per unit volume. The probability, $P$, of a collision, divided by the number of centres per unit area, $Ndx$, determines an effective cross-section, $\sigma$, ascribed to each centre. The mean free path, $\lambda$, between collisions is defined as $1/N\sigma$. One is sometimes interested in differential cross-sections for scattering into a given solid angle, or for a collision resulting in a given range of energy transfer values: it is then only necessary to redefine the probability $P$.

### 2.4. Electronic stopping

When moving at velocities greater than $v = Z_1e^2/\hbar$, i.e. greater than the K-shell electron velocity, an ion will have a high probability of being fully stripped of its electrons. Here $Z_1$ is the atomic number of the ion and the ratio $e^2/\hbar$, often denoted $v_0$, is known as the Bohr velocity. The theory of energy loss under these circumstances derives from the work of Bohr (1913), who carried out a classical calculation which led to:

$$-dE/dx = \frac{4\pi Z_1^2 e^4 N}{mv^2} B \qquad (2.2)$$

where $m$ is the electronic mass and $B$ is a dimensionless 'stopping number',

$$B = Z_2 \ln\left(p_{max}/p_{min}\right) \qquad (2.3)$$

in which $Z_2$ is the atomic number of the target atom and $p_{max}$ and $p_{min}$ are the upper and lower limits of the ion–electron separation at closest approach. $p$ is usually referred to as the 'impact parameter'. For head-on collisions

$$p_{min} = 2Z_1 e^2/mv^2 \qquad (2.4)$$

while Bohr chose $p_{max}$ to correspond to a 'collision time', $p/v$, equal to the rotational period of the electron in its parent atom:

$$p_{max} = v/w_j \qquad (2.5)$$

where $w_j$ is a characteristic frequency appropriate to the $j$th electronic orbit.

Bethe (1930) gave a quantum-mechanical derivation based on the Born approximation method and therefore valid if

$$\frac{2Z_1 Z_2 e^2}{\hbar v} < 1 \qquad (2.6)$$

or, physically, if the potential energy is less than the kinetic energy of relative motion, so that the incident wave is only slightly perturbed by the presence of the interaction potential. Neglecting relativistic factors, which will be irrelevant to the ion implantation field, Bethe's formula reduces to:

$$B = Z_2 \ln\left(2\,mv^2/I\right) \qquad (2.7)$$

Here $I$ is the average excitation energy of the electrons in the target. Bloch (1933) added correction terms to take account of the perturbation of the atomic wave-functions by the bombarding ion, and also showed that, on the basis of the Thomas–Fermi model of the atom, $I$ is approximately proportional to $Z_2$

$$I = kZ_2, \quad \text{with } k \simeq 11 \text{ eV} \qquad (2.8)$$

Walske (1952) and Bichsel (1963) have considered certain 'shell corrections' which become significant when core electrons fail to contribute to the stopping process:

$$B = Z_2 \ln (2 mv^2/I) - C_K - C_L \tag{2.9}$$

Recent accurate experimental measurements (Andersen et al., 1967) have shown that $k$ varies between 9.5 and 11 eV and shows an oscillatory dependence on $Z_2$, while $C$ is itself energy dependent, reaching a maximum at the value given approximately by $(Z_2/10)$ MeV.

As the ion velocity diminishes towards the value corresponding to maximum electronic energy loss (fig. 2.1) the ion proceeds to capture progressively more electrons from the absorber atoms and its charge falls below $Z_1$. It is still possible to fit the observed dependence of $dE/dx$ by substituting in the above formulae, instead of $Z_1$, an effective charge $\gamma Z_1$, where $\gamma$ is a function of $Z_1$, $Z_2$, $v$ and the density of the absorber medium. The region of validity of the Bethe–Bloch formula (i.e. equations (2.2) and (2.9)) has been considered by Fano (1963). At low ion velocities the expression is invalid for almost all values of $Z_1$ and $Z_2$ except the first few because the fluctuation in ion charge must be taken into account.

As mentioned above, charge-exchange processes between ion and absorber atoms give rise to a significant contribution to the specific energy loss. It is possible to define charge-changing cross-sections for the capture and loss of electrons, denoted by $\sigma_{Z, Z-1}$ and $\sigma_{Z, Z+1}$ respectively. If $\phi_Z$ is the probability, at a given ion velocity $v$, that a given ion has a charge $Z$, then given that the rate-determining process is one of electron capture rather than loss, it is not difficult to see that

$$-(dE/dx)_{ch} = N \sum_Z \phi_Z \sigma_{Z, Z-1}(J + mv^2/2) \tag{2.10}$$

where $J$ is the binding energy of the captured electron in the absorber atom. Experimental data for $\phi_Z$, $\sigma_{Z, Z-1}$ and $\sigma_{Z, Z+1}$, all as a function of ion velocity, now exist for a number of ion species. It is known, for instance, that $\phi_Z$ is influenced by the state of the absorber, i.e. whether it is a solid or a gas. This is because highly-excited electron configurations of the ion have a short lifetime: in a gas there is time between collisions for these states to decay to more stable ones, but in a solid the collisions are much more frequent and the equilibrium charge state distribution centres about higher values of $Z$.

An alternative approach to the problem of electronic stopping at low and intermediate ion velocities was formulated originally by Firsov (1959).

In his model each binary collision of an ion and an atom is viewed as leading to an overlap of their electronic orbitals. Electrons transferred either temporarily or permanently from the ion to the atom, and vice versa, cause a transfer of momentum which slows down the ion. A surprisingly successful yet simple mathematical approach to this problem has been to define a hypothetical interaction plane, which in the symmetrical case of similar particles lies midway between them and defines the regions of the atomic potentials. Whenever an electron crosses this plane its momentum is considered to be transferred abruptly to that appropriate to the other régime (fig. 2.2). If the density of electrons in one of the atomic particle, as a function of

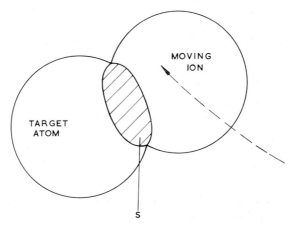

Fig. 2.2. Schematic representation of the model employed in the Firsov treatment of electronic stopping. The interaction is localized at the plane, $S$.

radius, is $n(r)$ and the orbital electron velocity is $u(r)$ then the energy loss due to electrons crossing the interaction plane, $S$, is given by

$$\Delta E = \frac{m}{4} \int_S nu\dot{r}\, dr\, dS \qquad (2.11)$$

Firsov used the Thomas–Fermi statistical model of the atom to calculate with some success the electronic energy loss of ions moving through gases. More recently Cheshire, Dearnaley and Poate (1969) have used the more realistic Hartree–Fock atomic wave functions and a modified surface integral to give

$$\Delta E = \frac{m}{4} \int \dot{r}\, dr \int_S \left\{ \sum_\alpha v_\alpha |\Psi_\alpha|^2 + \sum_\beta v_\beta |\Psi_\beta|^2 \right\} dS \qquad (2.12)$$

where $\Psi_{\alpha, \beta}$ are the Slater orbitals of the projectile and target, and the corresponding velocities $v_{\alpha, \beta}$ are obtained by equating $mv_\alpha^2/2$ to the expectation value of the kinetic energy for each electron orbital. The interaction surface, $S$, is furthermore taken to intersect the line joining the ion to the target nucleus at the point where the electron density is a minimum. Although the Firsov model is a classical one, the use of self-consistent quantum-mechanical atomic wave-functions in this way renders it more trustworthy, and the results, we shall see below, compare well with experiment. There is a need, however, to place the model on a firmer quantum-mechanical basis and to recognize the fact that, owing to the Pauli principle, the self-consistent electron orbitals should be re-calculated at each successive stage as the ion and atom collide. Such a computation, in a quasi-molecular model and incorporating the idea of 'curve-crossing' to manipulate the transfer (with appropriate energy and momentum) of electrons from one orbital to another, has been carried out by Fano and Lichten (1965) and could be applicable here.

So far, we have considered models in which binary collisions between ions and atoms are treated as independent events. An alternative procedure, introduced by Fermi and Teller (1947), is to regard the absorber as a Fermi gas of electrons, the density and energy distributions of which are given by Thomas–Fermi or Hartree–Fock models. Such a gas, in which the positively-charged nuclei are considered to be embedded, is effectively a plasma with a characteristic oscillation frequency $\omega_0$ corresponding to the net relative movement of positive and negative charge. This can be pictured as an oscillatory surging of the electrons with respect to the less mobile nuclei. A fast charged particle entering this system brings about a polarization of the medium, which behaves as if it were a dielectric with a complex dielectric constant, the imaginary part of which relates to the energy absorption processes. Such a model gives a very natural explanation of the upper limit, $p_{max}$, for the interaction distance between ion and electron (eq. 2.5) since in a plasma there is so-called adiabatic cut-off

$$p_{max} = v/\omega_0 \tag{2.13}$$

beyond which the interaction is too weak to excite the plasma. The model was introduced by Fermi and Teller to treat the slowing down of muons but it has since been extensively applied to ionic stopping by Lindhard and Scharff (1953, 1961). The electrons are here no longer considered to be attached to any specific atom so that the concepts of charge-changing collisions disappear: although this is an advantage for computational purposes it is not clear that it is physically wholly justifiable.

Seitz (1949) had earlier assumed that there exists a threshold ion energy, $E_{th}$, below which ionization of the absorber atoms becomes highly unlikely, and this threshold was related to the band gap, $E_g$, of the material by

$$E_{th} = M_1 E_g/8m \qquad (2.14)$$

in which $M_1$ is the incident ion mass. In the electron-gas model of energy loss, however, Lindhard (1954) has pointed out that the problem of momentum change for a point charge moving through the gas is equivalent to the problem of the scattering of an electron gas in a metal due to the presence of charged centres and the calculation of electrical resistance. This argument leads to an electronic energy loss proportional to the relative drift velocity (at least for velocities small compared with the Fermi velocity) and the analogous result is familiar to us in the form of Ohm's law. Similar general considerations enabled Lindhard and Scharff (1961) to derive a linearly velocity-dependent value for the electronic energy loss, valid up to ion velocities of about $Z_1^{\frac{2}{3}} e^2/\hbar$. This corresponds to an energy of about 2.5 MeV for $B^+$ ions. 15 MeV for $Na^+$ ions, 200 MeV for $Kr^+$ and 1400 MeV for $Xe^+$ ions. Taking a Thomas–Fermi atomic model, Lindhard and Scharff obtained

$$-(dE/dx)_e = \xi_e \cdot 8\pi e^2 N a_0 Z_1 Z_2 (Z_1^{\frac{2}{3}} + Z_2^{\frac{2}{3}})^{-\frac{3}{2}} \cdot v/v_0 \qquad (2.15)$$

where $a_0$ is the Bohr radius of the hydrogen atom (approximately 0.53 Å), $v_0$ is the Bohr velocity, and $\xi_e$ is a dimensionless constant of the order $Z_1^{\frac{1}{6}}$. It is convenient to define a reduced energy

$$\epsilon = Ea M_2/Z_1 Z_2 e^2 (M_1 + M_2) \qquad (2.16)$$

in which $M_1$, $M_2$ are the masses of projectile and target atom and $a$ is a commonly-occurring Thomas–Fermi parameter,

$$a = 0.885 a_0 (Z_1^{\frac{2}{3}} + Z_2^{\frac{2}{3}})^{-\frac{1}{2}} \qquad (2.17)$$

and in addition, a reduced range $\rho$ given in terms of range $x$ by

$$\rho = x N M_2 a^2 \, 4\pi M_1/(M_1 + M_2)^2 \qquad (2.18)$$

In terms of these parameters Lindhard, Scharff and Schiøtt (1963) were able to express the specific energy loss as

$$-(d\epsilon/d\rho)_e = K\epsilon^{\frac{1}{2}} \qquad (2.19)$$

in which

$$K = \xi_e \frac{0.0793 Z_1^{\frac{1}{2}} Z_2^{\frac{1}{2}} (M_1 + M_2)^{\frac{3}{2}}}{(Z_1^{\frac{2}{3}} + Z_2^{\frac{2}{3}})^{\frac{3}{4}} M_1^{\frac{3}{2}} A_2^{\frac{1}{2}}} \tag{2.20}$$

(here the atomic weight $A_2$ is substituted for $M_2$ in order to make the equation dimensionally correct). We shall return later to examine the agreement between this formula and the experimental results.

Let us now pursue in greater detail some consequences of the free-electron gas model of atomic stopping at high particle velocities. Suppose that the density of free electrons is $n$: it is easy to derive, an elementary classical model, the oscillation frequency, $\omega_p$, of this plasma. Imagine a displacement of the electrons radially outwards from some fixed point, taken to be the origin of co-ordinates. If the displacement is $\zeta(r)$, the number of electrons leaving a sphere of radius $r$ is $4\pi n r^2 \zeta(r)$, thus leaving the sphere with a net positive charge $4\pi n e r^3 \zeta(r)$. The electric field at $r$ is therefore $4\pi n e \zeta(r)$ and so each electron experiences a force $4\pi n e^2 \zeta(r)$. Therefore we have

$$m\ddot{\zeta} + 4\pi e^2 n \zeta = 0 \tag{2.21}$$

and from this we deduce an angular frequency of plasma oscillation (neglecting the random thermal motion of the electrons) given by

$$\omega_p = (4\pi n e^2 / m)^{\frac{1}{2}} \tag{2.22}$$

which, in a metal, is of the order $10^{16}$ rad/sec. It is customary to refer to the quantum of collective excitation of free electrons, $\hbar\omega_p$, as a 'plasmon'. In silicon and germanium the energies of plasmons are about 15 eV. This is far above average thermal energies and, as a result, plasmon excitation can be observed only when large amounts of energy are supplied to valence electrons from outside the system. This is what occurs when a fast charged particle passes through the solid. In terms of this classical treatment we can now see that there will be a cut-off energy, for a given incident ion, below which plasmon excitation cannot occur. If the maximum energy transfer $4 mE/M_1$ to an electron is less than the plasmon energy, such excitation will be classically forbidden. For protons in silicon this occurs at 7 keV energy, and increases linearly with ion mass. Generally speaking, therefore, the process can be disregarded for most heavy ion implantation energies, but will always be involved in the slowing down of the light, energetic ion beams which are used for probing implanted solids (see section 2.18).

In a quantum-mechanical treatment of plasma oscillations it is necessary to consider the phase factors $e^{-i\mathbf{k}\cdot\mathbf{r}}$ relating to electron waves of different wave-number $k$. Certain complex sums over $k$, of these phase factors, occur in the derivation and it is usual to assume that there is sufficient randomness in the system to justify setting these sums to zero: this is known as the random phase approximation (Bohm and Pines, 1952). With this assumption, one can deduce an upper limit to the wave number of plasma oscillation

$$k_c \sim \omega_p/v_F \qquad (2.23)$$

where $v_F$ is the Fermi velocity, i.e. the maximum velocity of an electron in the material at 0 °K. In a typical metal, $k_c$ is about $10^8$ cm$^{-1}$, corresponding to a minimum wavelength of the order 1 Å. Therefore, taking the case of a moving ion, for impact parameters, $p$, less than $1/k_c$ the electrons behave like a collection of individual particles and not like a plasma. We can therefore separate two contributions to the specific energy loss due to free electrons. Reference to eqs. (2.2), (2.3), (2.13) and (2.23) shows us that, for plasmon excitations:

$$-(dE/dx)_{plas} = \frac{4\pi Z_1^2 e^4 N Z_2^f}{mv^2} \ln\left(\frac{k_c v}{\omega_p}\right) \qquad (\text{for } v \ll v_F) \qquad (2.24)$$

where $Z_2^f$ is the number of *free* electrons per target atom. For single-particle excitations we have:

$$-(dE/dx)_{s.p.} = \frac{4\pi Z_1^2 e^4 N Z_2^f}{mv^2} \ln\left(\frac{2mv}{\hbar k_c}\right) \qquad (2.25)$$

the minimum impact parameter, $p_{min}$, being determined in this case by the uncertainty in position of the electron, $\hbar/2mv$, corresponding to a maximum momentum transfer $2mv$.

It is interesting to compare the sum of these two formulae:

$$-(dE/dx)_{plas} - (dE/dx)_{s.p.} = -(dE/dx)_{free} = \frac{4\pi Z_1^2 e^4 N Z_2^f}{mv^2} \ln\left(\frac{2mv^2}{\hbar\omega_p}\right) \qquad (2.26)$$

with the expression by eqs. (2.2) and (2.7); $I$ in (2.7) is replaced by the plasmon energy $\hbar\omega_p$. Moreover, Lindhard and Winther (1964) have shown that, at least for high particle velocities $v$, there should be an equipartition of the energy loss between collective and single-particle excitations:

$$(dE/dx)_{plas} \sim (dE/dx)_{s.p.} \qquad (2.27)$$

In accordance with this rule, one may observe that for a typical metal $\hbar k_c^2$

is roughly equal to $2\,m\omega_p$ as is to be expected from eqs. (2.24), (2.25) and (2.27).

The question as to what value one should ascribe to $Z_2^f$ in these equations has been considered by Pines (1963) in relation to observed plasmon energies $\hbar\omega_p$ determined by energy loss measurements for electrons transmitted through thin metal foils. For most elements there is agreement between the observed value of what is called the characteristic energy loss $\Delta E_c$ and the values of $\hbar\omega_p$ corresponding to a $Z_2^f$ equal to the number of valence electrons. In the case of Cu, Zn, Ag and Au, however, $\hbar\omega_p < \Delta E_c$, due perhaps to some additional participation by d-shell electrons, while in transition metals $\hbar\omega_p > \Delta E_c$, indicating that not all the electrons outside the closed shell may here be regarded as free.

In order to complete the treatment of electronic energy loss we must, of course, consider the contribution due to bound or core electrons. Lindhard and Scharff (1953) had chosen to ignore the stopping due to inner electrons, i.e. those within a cut-off radius determined by

$$\hbar\omega_p(r) < \sqrt{2}\,mv^2 \qquad (2.28)$$

where the local plasma frequency $\omega_p(r)$ is found from a Thomas–Fermi model. Bonderup (1967) adopted the approach of extrapolating the stopping due to outer electrons, taking over an expression for the slowing down of a low-velocity particle in a dense, but free, electron gas. Incorporating also a first-order correction term to eq. 2.26, due to Lindhard and Winther (1964), he was able to account very successfully for the experimentally-observed behaviour of the shell correction terms in eq. 2.9. In other words, Lindhard and his school prefer not to draw any distinction between core and valence electrons, but treat the whole atom as a localized electron gas. The perturbation treatment adopted then depends merely upon the criterion $v \lesssim v_F$. Other authors have considered it justifiable to treat valence and core electrons separately in stopping calculations, considering the former as an electron gas and the latter in terms of independent two-body collisions for which the energy transfer may be derived classically. Williams (1945) was the first to put forward this method, on the grounds that the incident ion is highly localized and the momentum transfers to inner electrons must be well-defined. Taking Bohr's formula (2.3), one must then define a value of $(p_{max})_j$ for each electronic sub-shell $j$ of the target atom. A criterion for $(p_{max})_j$ has been proposed by Bohm (1951) with the form

$$(p_{max})_j \propto (Z_1 v)^{\frac{1}{2}}/\Delta E_j \qquad (2.29)$$

in which $\Delta E_j$ is the energy of transition from the $j$th subshell to the lowest unoccupied level and is obtainable from some model of the atom. To the extent that the logarithms in eqs. (2.7) and (2.26) are relatively insensitive to the difference between $\hbar\omega_p$ and $I$, the ratio of core excitation to free electron excitation is easily gauged:

$$\frac{(\mathrm{d}E/\mathrm{d}x)_{\text{core}}}{(\mathrm{d}E/\mathrm{d}x)_{\text{free}}} \sim \frac{Z_2 - Z_2^{\text{f}}}{Z_2^{\text{f}}} \tag{2.30}$$

In fact, since $I > \hbar\omega_p$, this naive expression will tend to overestimate the degree of core excitation. While we shall see below that these ideas have been used to estimate the energy loss of ions in crystal channels, it remains true that a complete and valid treatment of the energy loss of heavy particles in amorphous media, in terms of core and valence electron excitation, has yet to be made.

### 2.5. Nuclear stopping

The specific energy loss is here due to collisions of the ion and a target nucleus, and is derived by considering these as independent elastic two-body interactions. At ion energies within the scope of this book, inelastic nuclear excitation, for example by nuclear reactions or Coulomb excitation, is a relatively unlikely process by comparison with elastic scattering and may be ignored for practical purposes. We shall similarly omit consideration of the very lowest ion energies below about 10 eV, when it is no longer permissible to neglect chemical binding forces and the collision must be regarded as taking place quantum-mechanically with the lattice as a whole.

Figure 2.3 illustrates a typical two-body scattering process between an ion and an atom, with an impact parameter $p$. The incident ion is deflected through an angle $\theta_1$, transferring energy $T$ to the struck atom, which recoils at an angle $\theta_2$. The energy transfer $T(E, p)$ may take any value between zero (for $p = \infty$) and a maximum, $T_m$

$$T_m = \frac{4M_1 M_2}{(M_1 + M_2)^2} E_1 \tag{2.31}$$

corresponding to a head-on collision. It is easy to see that the total energy loss, $\Delta E$, in traversing a slab of material of thickness $\Delta x$ is, integrating over $p$, given by

$$\Delta E = -N\Delta x \int_0^\infty T(E, p) 2\pi p \, \mathrm{d}p \tag{2.32}$$

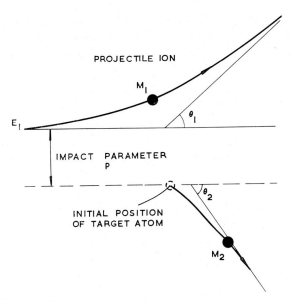

Fig. 2.3. A typical two-body scattering process with an impact parameter $p$.

and therefore

$$-(dE/dx)_n = N \int_0^\infty T(E, p)2\pi p\, dp \qquad (2.33)$$

An alternative description of the problem, due to Bohr (1948), is in terms of the differential scattering cross-section $d\sigma(E, T)$ at energy $E$ involving an energy transfer between $T$ and $T + dT$. We have then, clearly,

$$-(dE/dx)_n = N \int_{T_{min}}^{T_m} T\, d\sigma(E, T) \qquad (2.34)$$

where $T_{min}$ is the minimum energy transfer. The form of $T(E, p)$ and $d\sigma(E, T)$ depend upon the interaction potential $V(r)$ between the two particles. It is appropriate, therefore, to digress at this point in order to introduce the approximations which have been made to the interatomic potential.

## 2.6. The interatomic potential

Unfortunately, there is no one analytical expression for the interatomic potential function $V(r)$ that is valid for all interaction radii. For this reason a large number of approximate formulae have been devised, but these have varied and limited areas of validity. Thus the choice of a potential function is governed by the circumstances of each problem and one must remain aware of the range of applicability of the method chosen in generalising the results.

It is well known that the general form of the interaction potential is as shown in fig. 2.4, being attractive at large radii and strongly repulsive at short distances, the intermediate minimum corresponding to the equilibrium interatomic separation. The attractive part of the potential arises, in ionic solids, almost entirely from Coulomb forces while in covalent solids it results from

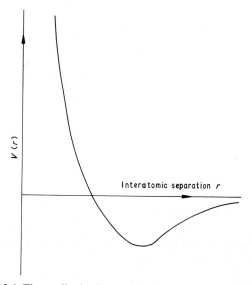

Fig. 2.4. The qualitative form of the interatomic potential $V(r)$.

the electron exchange interaction. The repulsive part of the potential is due to the electrostatic repulsion of the positively-charged nuclei, screened by their surrounding electrons.

Information regarding the strength of the interaction near the equilibrium separation can be obtained from the compressibility of a solid constituted of the two atomic species, and from other elastic constants. The Born–Mayer function

$$V(r) = A \exp(-r/B) \tag{2.35}$$

is relatively successful in describing the potential at these large values of $r$, and $A$ and $B$ are derivable from the elastic moduli. This potential is, however, much too weak at small distances.

When the atoms have only a small separation (up to 0.2 Å, say) the screened Coulomb potential, due to Bohr, is more appropriate:

$$V(r) = (Z_1 Z_2 e^2/r) \exp(-r/a) \tag{2.36}$$

Here $a$ is given by

$$a = a_0 (Z_1^{\frac{2}{3}} + Z_2^{\frac{2}{3}})^{-\frac{1}{2}} \tag{2.37}$$

although some authors have preferred to use empirical values of $a$, up to three times the Bohr value.

It is in the intermediate region of separation that interest lies in most cases, and a useful potential in this region has been obtained from the Thomas–Fermi model of the atom

$$V(r) = Z_1 Z_2 e^2 \phi_{TF}(r/a)/r \tag{2.38}$$

in which $\phi_{TF}$ is the so-called Thomas–Fermi screening function which has been calculated numerically and tabulated by Gombas (1956). Firsov (1958) has given a modified derivation in which $a$ has a somewhat smaller value. When an analytical approximation to the function $\phi_{TF}$ is desired, Lindhard (1965) has proposed the simplified form

$$\phi_{TF}(r/a) \sim \frac{r/a}{[(r/a)^2 + C^2]^{\frac{1}{2}}} \tag{2.39}$$

with $C$ as an adjustable parameter: the value $C = \sqrt{3}$ may be taken as a reasonable average. Empirical relations which approximate the behaviour in the intermediate region of separation have been given by Brinkman (1954), for example:

$$V(r) = \frac{A Z_1 Z_2 e^2 \exp(-Br)}{1 - \exp(-Ar)} \tag{2.40}$$

in which $A$, $B$ are adjustable parameters. It can be seen that for small $r$ this expression tends towards the unscreened Coulomb interaction $Z_1 Z_2 e^2/r$, while as $r$ tends to infinity it approaches the Born–Mayer form (2.35).

Nielsen (1956) suggested an inverse square law potential fitted (in magnitude and slope) to the exponentially screened Coulomb potential

$$V(r) = (Z_1 Z_2 e^2 a/r^2) \exp(-1) \tag{2.41}$$

The relationship between these various potential functions when calculated for two copper atoms, as a function of their separation, is shown in fig. 2.5. In this figure the weakness of the Bohr potential at large $r$ and of the Born–Mayer form at small $r$ can be clearly seen.

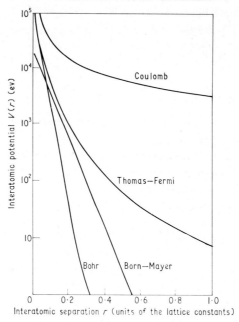

Fig. 2.5. Several commonly-used approximations to the interatomic potential in copper (after Carter and Colligon, 1968).

At low velocities it is often convenient to regard the impact of one atom with another as a collision between two hard spheres of radius $r_0$. In other words, there is no interaction for $r > r_0$, i.e. $V(r) = 0$, but the potential rises to infinity at $r = r_0$ (the region $r < r_0$ obviously being physically unreachable). Some measure of reality is introduced into this by allowing the radius $r_0$ to be energy-dependent, and it is usual to set it equal to the distance of closest approach in a head-on collision. If the energy of the moving atom is $E$, then for a collision between like atoms the energy measured relative to the (symmetrical) centre of mass of the system is $E/2$ for each particle, and

$$V(2 r_0) = E/2 \tag{2.42}$$

An expression for $r_0(E)$ can be obtained by 'matching' to one of the more justifiable potentials such as (since the energies are low) the Born–Mayer, in which case we have

$$r_0(E) = (B/2) \ln (2A/E) \tag{2.43}$$

and the radius decreases slowly as $E$ increases. The hard-sphere potential fails to predict accurately the angular distribution of scattering in atomic collisions, but its simplicity in analytical calculations makes it useful in studying the general features of atomic collisions.

Another simple analytical form which, when matched to one of the potentials already discussed, gives a useful approximation is the inverse power-law potential

$$V(r) = a_p^{S-1} Z_1 Z_2 e^2 / Sr^S \qquad (2.44)$$

in which $a_p$ is related to $a$ in eq. 2.37 by an adjustable dimensionless parameter. It is possible to match the power-law potential, for instance, to the Thomas–Fermi potential at the distance of closest approach, or alternatively to match $\partial V/\partial r$, and so derive an expression for $S$ as a function of $r$. In practice, however, the value of a simple analytical form is lost unless $S = 1$ or 2. Lindhard and Scharff (1961) have shown that, by comparison with the Thomas–Fermi potential, the approximation $S = 1$ is reasonable for close collisions involving large energy transfers, while $S = 2$ is a better approximation when the energy transfer is small.

Nevertheless, it is obvious that at large impact parameters, where the exponential Born–Mayer potential is valid, any power law potential will fall off too slowly. For this reason Leibfried and Oen (1962) and Sigmund and Vajda (1964) suggest cutting off the matched potential at some radius $r_0$, typically at half the equilibrium separation of the atoms (i.e. half the lattice spacing). In recent years the need for tractable analytical approximations to the interatomic potential has lessened to some extent with the development of powerful computers, able to handle the dynamics of collision processes with the more complex potential functions such as Molière's (1947) empirical approximation to the Thomas–Fermi potential.

## 2.7. Nuclear stopping formulae

We are now in a position to resume the calculation of the energy loss due to nuclear collisions, according to equation (2.34). The first step, however, is to transform the problem from the laboratory set of coordinates to a system of coordinates in which the origin is located at, and moves with, the centre of mass of the two colliding particles (fig. 2.6). It is convenient, also, to introduce the parameter $u = r^{-1}$, which has limiting values of zero and $u_{max}$, the reciprocal of the distance of closest approach. If now the incident particle energy $E$ is referred to centre of mass coordinates it transforms to

$$E_{cm} = E\mu/M_1 \qquad (2.45)$$

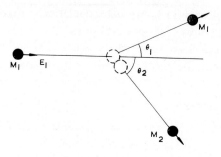

LABORATORY  SYSTEM

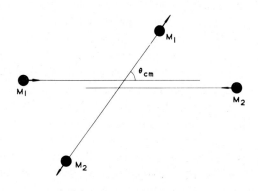

CENTRE  OF  MASS  SYSTEM

Fig. 2.6. A two-body collision in the laboratory and centre-of-mass systems of coordinates.

where $\mu$, called the reduced mass is given by

$$\mu = M_1 M_2 / (M_1 + M_2) \qquad (2.46)$$

The conservation of the angular momentum of the system, $(2M_1 E_{cm})^{\frac{1}{2}} p$, and of energy in an elastic collision lead to

$$E_{cm}(1 - p^2 u_{max}^2) = V(u_{max}) \qquad (2.47)$$

(where $p$ = impact parameter) and

$$\dot{u}^2 + u^2 = \frac{1}{p^2}\left(1 - \frac{V(u)}{E_{cm}}\right) \qquad (2.48)$$

Solution of these equations yields the centre of mass scattering angle $\theta_{cm}$

$$\theta_{cm} = -2p \int_0^{u_{max}} \frac{du}{\left(1 - \frac{V(u)}{E_{cm}} - p^2 u^2\right)^{\frac{1}{2}}} \tag{2.49}$$

Unfortunately, this integral equation can be solved explicitly only for certain simple interaction potentials, $V(u)$. These are the Coulomb potential, $V(u) = Z_1 Z_2 e^2 u$, the inverse square-law potential, $V(u) \propto u^2$, and the hard-sphere potential. For other cases approximation methods must be used.

Having obtained $\theta_{cm}$, the laboratory scattering angle $\theta_1$ is obtained by

$$\tan \theta_1 = \sin \theta_{cm}/(\cos \theta_{cm} + M_1/M_2) \tag{2.50}$$

and the relation between the energy transfer, $T$, and the scattering angle $\theta_1$ is

$$\cos \theta_1 (1 - T/E) = 1 - (M_1 + M_2)T/2M_1 E \tag{2.51}$$

Thus eqs. (2.49), (2.50) and (2.51) allow $T$ to be expressed as a function of $E$ and $p$, as required for solving (2.33). In the simple case of the Coulomb potential,

$$T = E \sin^2 \theta_{cm}/2 \tag{2.52}$$

and

$$p = b \cot \theta_{cm}/2 \tag{2.53}$$

where

$$b = Z_1 Z_2 e^2/E_{cm} \tag{2.54}$$

is the distance of closest approach in a head-on collision.

The differential scattering cross-section for a given energy transfer $T$ under these conditions becomes

$$d\sigma/dT = \tfrac{1}{4}\pi b^2 T_m/T^2 \tag{2.55}$$

showing that small energy transfers and therefore small-angle scattering events are the most probable. Using equation (2.34), we have

$$-(dE/dx)_n = N \int_{T_{min}}^{T_m} T \, d\sigma(E, T) \tag{2.56}$$

where $T_{min}$ is the minimum effective energy transfer, chosen as the energy transfer at some cut-off radius corresponding to the interatomic spacing in the target, to prevent the integral becoming infinite. It follows, for Rutherford scattering

$$-(dE/dx)_n = \tfrac{1}{4}\pi N b^2 T_m \int_{T_{min}}^{T_m} dT/T \tag{2.57}$$

$$= \tfrac{1}{4}\pi N b^2 T_m \ln (T_m/T_{min}) \tag{2.58}$$

Obviously the awkwardness of this treatment reflects the fact that at large radii the use of an unscreened potential is invalid. Nielsen (1956) suggested the potential given in eq. (2.41) and, treating the scattering as isotropic in the centre of mass system, he obtained

$$d\sigma/dT = 4\pi R^2/T_m \tag{2.59}$$

where $R(E)$ is an effective hard-sphere radius derived from the head-on collision relation

$$V(2R) = EM_2/(M_1+M_2) = \frac{Z_1 Z_2 e^2 a}{4R^2} \exp(-1) \tag{2.60}$$

So

$$R^2(E) = \frac{Z_1 Z_2 e^2 a \exp(-1)(M_1+M_2)}{4EM_2} \tag{2.61}$$

and

$$-(dE/dx)_n = \frac{NZ_1 Z_2 e^2 a \exp(-1)(M_1+M_2)}{ET_m M_2} \int_0^{T_m} T \, dT$$

$$= 2\pi N \exp(-1) Z_1 Z_2 e^2 aM_1/(M_1+M_2) \tag{2.62}$$

Note that the integral no longer diverges for $T_{min} = 0$ and, furthermore, that the stopping is energy independent, a feature we know to be incorrect.

In order to handle the more realistic Thomas–Fermi potential (eq. 2.38) Lindhard et al. (1963, 1968) introduced an approximation method. First a new parameter $t$ was introduced, which in terms of the reduced energy $\epsilon$ (eq. 2.16) is simply

$$t = \epsilon^2 \sin^2 \theta_{cm}/2 \tag{2.63}$$

and $t$ is proportional to the product $ET$. Their approximation formula for the differential cross-section

$$d\sigma/dt = \pi a^2 f(t^{\frac{1}{2}})/2t^{\frac{3}{2}} \tag{2.64}$$

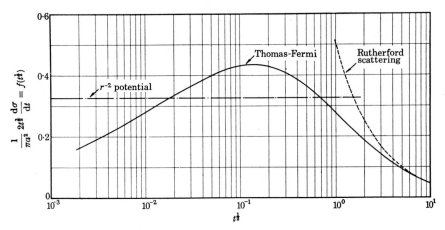

Fig. 2.7. The universal scattering function $f(t^{\frac{1}{2}})$ defined by Lindhard et al. (1963, 1968).

is usually valid to within about 20 per cent. Here $f(t^{\frac{1}{2}})$ is a universal scattering function which has been evaluated numerically and tabulated by Lindhard, Nielsen and Scharff (1968). It is shown as a function of $t^{\frac{1}{2}}$ in fig. 2.7. For large values of $t$, that is for large energy transfers in close collisions, this function merges smoothly with the prediction of the Coulomb formula (i.e. Rutherford scattering), and in fact eq. (2.64) is exact for the Coulomb potential, for which $f(t^{\frac{1}{2}}) = \frac{1}{2}t^{\frac{1}{2}}$. The treatment fails for very low values of $\epsilon < 10^{-2}$, owing to the inadequacy of the Thomas–Fermi potential at the correspondingly large radii of interaction. One may now calculate the specific energy loss by means of eq. (2.34) since both $T$ and $d\sigma/dt$ are known

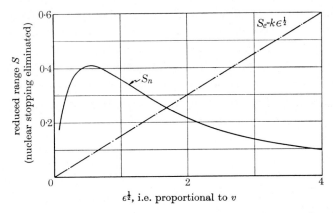

Fig. 2.8. The reduced nuclear and electronic specific energy loss as a function of energy (after Lindhard, Scharff and Schiøtt, 1963).

as functions of $t(E, \theta_{cm})$. The results of this numerical integration, as published by Lindhard, Scharff and Schiøtt (1963), are shown as a function of the
reduced energy $\epsilon$ in fig. 2.8, which also shows the velocity-proportional
electronic energy loss for a typical instance.

One should note that the energy transfer, $T$, in nuclear collisions may
occasionally reach large values, corresponding to $T_m$. There is therefore a
statistical fluctuation in the specific energy loss, leading to a dispersion or
straggling of the ion range. This is very important in determining the
distribution of implanted material. Statistical arguments show that the mean
square fluctuation in energy loss is given by:

$$\frac{\overline{E^2}}{E} = N \int_0^{T_m} T^2 \, d\sigma \tag{2.65}$$

which again can be calculated in the manner we have described, for a given
potential. Straggling is more important, however, in relation to the range
distribution, so we shall return to consider it in the section 2.9 dealing with
range theory.

## 2.8. Experimental measurement of energy loss

In this section we shall compare some experimental measurements of
the specific energy loss $-(dE/dx)$ in amorphous materials with the results
of the theoretical treatments discussed above. It will become clear that there
are deficiencies in both aspects of the topic.

A direct measurement of the specific energy loss is obtained by trans-
mitting a beam of ions of known initial energy $E$ through a thin section of
the absorber, and measuring the energy loss $-\Delta E$ and the thickness of
material $\Delta x$. If $\Delta E < 0.1\, E$ it is usually permissible to set $-\Delta E/\Delta x$ equal to
$-dE/dx$ for the mean energy $E - \Delta E/2$. In other cases the variation of $dE/dx$
with $E$ must be taken into account, and if $dE/dx = f(E)$ one can define an
effective energy $E_{eff}$ such that

$$\Delta E/\Delta x = dE/dx|_{E_{eff}}$$

$$E_{eff} = \frac{\displaystyle\int_{E-\Delta E}^{E} Ef(E)\,dE}{\displaystyle\int_{E-\Delta E}^{E} f(E)\,dE} \tag{2.66}$$

The thickness of material $\Delta x$ must not be so little that it cannot be measured accurately, e.g. by weighing a known area, by optical interferometry, or by $\beta$-particle absorption techniques (see Allison and Warshaw, 1953).

An independent determination of the specific energy loss is feasible from measurements of the equilibrium charge distribution of the transmitted ions, using the r.m.s. charge $\gamma\, Z_1$ to relate the ionic energy loss to the known behaviour of protons. This owes its validity to the fact that when nuclear stopping is negligible the specific energy loss is proportional to the mean square charge (see eq. 2.26). This approach seems to be of greater value in testing the basic assumptions of energy-loss theory than in determining an accurate energy loss. A more generally useful indirect method of measuring specific energy loss is by differentiation of the smoothed range-energy data, obtained by methods we shall consider in sections 2.10.

Most of the experimental data we have on specific energy loss has been gathered for a rather limited number of easily-obtainable ion species, such as the gaseous elements, though the growing availability of versatile ion sources should remedy this deficiency in the future. Also, owing to practical considerations, the data relate to an energy range corresponding to at least 10 keV per mass number.

For the lightest ions, e.g. protons, some highly accurate $dE/dx$ determinations have been made by a calorimetric technique (Andersen et al., 1967) in which the particle energies were measured by the heating they produced in a thermally-insulated target at about 4 °K. A precision approaching 0.1 per cent was achieved in this work. For heavier ions, electrostatic analysis has been applied to the measurement of the incident and transmitted ion energies (Porat and Ramavataram, 1960, 1961).

Despite the considerable precautions against error in the early work on energy loss determination, it is feasible that some results will have been systematically reduced by a degree of ion channelling. Frequently the specimens were in the form of rolled metal foils, in which it is known that the microcrystals assume a preferred orientation normal to the surface. Of course, this effect was unsuspected prior to the discovery of channelling in 1963.

Since the angular aperture of the detector is generally quite small in specific energy loss measurements and it is customary to measure the energy loss for particles transmitted close to the incident beam direction, the wide angular dispersion that accompanies close elastic (nuclear) collisions causes such events to be discriminated against. The measurements are thus dominated by inelastic electronic losses, particularly in that energy range over which accurate measurements are feasible. Therefore the data allow comparison merely with electronic stopping theory and nuclear stopping contributes only a small perturbation. A little work has been published in which transmitted energy spectra were observed for wider angles of scattering (e.g. van Wijngaarden and Duckworth, 1962; Ormrod and Duckworth, 1963) and here it is easy to see the effect of elastic processes. Figure 2.9 shows energy spectra of 35 keV $Ne^{20}$ ions transmitted by a thin carbon foil at four angles of observation. The superimposed histograms were calculated using a Monte Carlo program in which a hard-sphere potential was assumed for the elastic

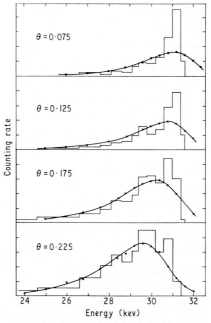

Fig. 2.9. Experimental and computed energy distributions of 35 keV Ne$^+$ ions transmitted through a thin carbon foil (thickness 3.36 $\mu$gm/cm$^2$) at various angles of emergence, $\theta$, measured in radians( after Ormrod and Duckworth, 1963).

scattering, while the inelastic energy loss was taken to be continuous and proportional to the path length between collisions. The electronic stopping power was then deduced and, by measurements taken over a range of ion energies, was shown to be proportional to $E^p$ where $p = 0.47$ for $Ne^{20}$ and between 0.42 and 0.50 for other heavy ions. In more recent work with ions in the energy range 100 to 500 keV, Fastrup et al. (1968) have observed values of $p$ between 0.44 and 0.78. Theoretical expressions such as eq. (2.15) all predict $p = \frac{1}{2}$ in this energy region.

In dealing with the specific energy loss in different materials it is usually convenient to define a stopping cross-section $S$:

$$S = -(1/N)\, dE/dx \qquad\qquad (2.67)$$

where $N$ is the number of absorber atoms per $cm^3$. When the electronic stopping cross-section, $S_e$, for a given material is plotted for a variety of ion species moving at the same velocity, an interesting phenomenon, first observed by Teplova et al. (1962) becomes apparent. $S_e$ exhibits an oscillatory dependence upon $Z_1$ and this was immediately interpreted as due to some kind of electron shell effects. Figure 2.10 shows the more recent data, for amorphous carbon, obtained by Ormrod et al. (1965) and by Hvelplund and Fastrup (1968), and superimposed are the theoretical predictions of Lindhard, Scharff and Schiøtt (1963), which are a monotonically increasing function of $Z_1$ according to eq. (2.20). Also shown in fig. 2.10 is the result of

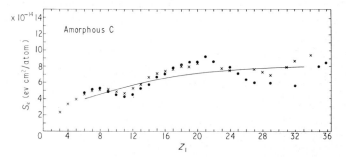

Fig. 2.10. The experimentally-measured electronic stopping cross-section of various ions in amorphous carbon as a function of atomic number $Z_1$ compared with the theoretical predictions of Lindhard and Scharff (1961) (continuous curve) and of Cheshire et al. (1968, 1969) (shown by X) (from Dearnaley, 1969).

a calculation by Cheshire, Dearnaley and Poate (1969) based upon the Firsov model for electronic stopping and incorporating the use of Hartree–Fock atomic wave-functions as in eq. (2.12). The variation in size and distribution of the electron shells leads to an oscillation in the degree of overlap of the electron orbitals and hence, in the Firsov model, correspondingly modulates the stopping cross-section. In the calculation it was necessary to integrate over the range of impact parameters determined by the acceptance angle of the energy analyser used in the experiment. The oscillation is not predicted by the Firsov model using Thomas–Fermi electron distributions, which alter monotonically with $Z_1$ and therefore appears to be a true shell effect, with minimum cross-sections corresponding to the small ions

$Li^+$, $Na^+$ and $Cu^+$. We shall see below that the model described by Cheshire et al. (1968, 1969) is also very successful in explaining the energy loss of ions in crystals, and it can account satisfactorily for the reduction in amplitude of the oscillatory behaviour at higher ion velocities, observed by Hvelplund and Fastrup (1968).

However, Lindhard and Finneman (1969) are developing an alternative explanation of the $Z_1$ oscillations within the framework of the free electron gas model of electronic stopping. One here pictures the electron gas as sweeping past a stationary ion and, just as in the Ramsauer–Townsend phenomenon for an electron beam moving through a gas target, quantum-mechanical oscillations in cross-section are introduced when the size of the ion matches a number of electron wavelengths. Such 'size resonances' can result even from the Thomas–Fermi model and should not be regarded as shell effects. We shall return to this argument in section 2.14 when dealing with the energy loss of channelled ions, but it is clear that further experimental and theoretical work is needed before the $Z_1$ oscillations in $S_e$ can be fully understood.

While, as fig. 2.10 shows, the Lindhard, Scharff, Schiøtt (1963) curve is a good average fit to the behaviour of $S_e$ in solid targets such as carbon, Fastrup et al. (1968) and Ormrod (1968) found that the experimental data for gaseous targets such as argon and nitrogen lie 30 to 40 per cent lower than the theoretical curve. It appears that the collision frequency (which is much lower in a gas) may affect the mean charge state of the ions and hence alter the energy loss.

Teplova et al. (1962) set out to test the symmetry of two-body interactions which are a feature of the Firsov model of atomic stopping. They demonstrated that $N^+$ ions moving in argon suffer the same electronic energy loss as $A^+$ ions in nitrogen, when travelling at the same relative velocity. However, Ormrod, MacDonald and Duckworth (1965) have since found that a significant asymmetry exists between the energy loss of carbon ions in aluminium and of aluminium ions in carbon. The explanation may lie in the fact that the ion-atom collision is not truly symmetrical, in that the moving particle has lost some of its electrons. The results of Cheshire et al. (1969) show that this significantly modifies the energy loss behaviour, and the Firsov model can therefore be refined to take this into account.

The variation of specific energy loss with ion energy in aluminium is shown in fig. 2.11, for $C^+$, $N^+$ and $Ne^+$ ions, and a smoothed curve is drawn through the experimental points, which are principally those of Porat and Ramavataram (1961). In the case of $N^+$ ions the prediction of Lindhard,

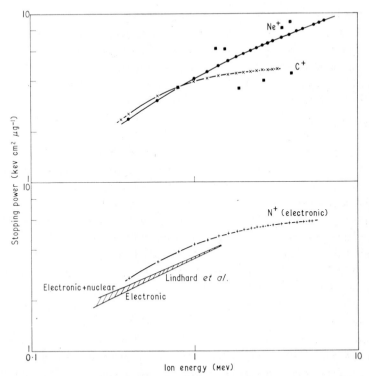

Fig. 2.11. Experimental values of the specific energy loss of $C^+$, $N^+$ and $Ne^+$ ions in aluminium as a function of ion energy. In the lower curve the results are compared with the theory of Lindhard et al. (1963) and the shaded area indicates the contribution due to nuclear stopping. Points show results of charge-distribution measurements (from Dearnaley, 1969; data from Porat and Ramavataram (1961)).

Scharff and Schiøtt (1963) is superimposed: in this instance the theoretical result is systematically about 15 per cent low, and the discrepancy is not removed by including the elastic nuclear stopping, which adds only a small contribution. Also shown, for $C^+$ and $Ne^+$ ions, are the results of stopping power calculations based on the measurement of equilibrium charge distributions of the emerging ions (see section 2.8). These depart in a significant but unsystematic manner from the direct measurements, and there is as yet no explanation of these results. It is generally considered that, within the solid, the moving ion will tend to attract or 'clothe' itself with target electrons but that these will be shed on emerging into the vacuum. A further problem is that an excited ion may undergo 'auto-ionisation' during the flight between target and detector, so that the measured charge state is not representative of that immediately upon emergence.

In summary, the state of energy loss theory cannot be regarded as satisfactory, and the new stimulus which the interest in ion implantation seems bound to arouse, particularly in the interesting energy region around 10 keV per nucleon, may well lead to these problems being tackled more vigorously.

### 2.9. Range theory

From the point of view of ion implantation the range of a particle in a solid is more important than its specific energy loss, and the latter is generally used as an intermediate parameter. First let us define carefully what is meant by the range: when the energy of an ion falls to about 20 eV it ceases to move through the solid, becoming trapped by the cohesive forces of the material, and the total distance travelled from the surface to this point is called the (total) range. In many cases there may be subsequent motion due to electrostatic forces (ion drift) or to thermal diffusion, but these are distinguished by being exceedingly slow mechanisms by comparison with the $10^{-14}$ sec or so involved in the stopping of the ion, so we shall consider these effects separately.

In discussions of the range of an ion we must distinguish between a number of statistically-related concepts which arise because of the fluctuations in energy loss and scattering angle at each collision. Thus we can define, in terms of a probability distribution of penetration depth, a most probable range $R$, a median range $R_m$, a mean range $\bar{R}$, and sometimes, though not always, a maximum range, $R_{max}$, if the distribution shows a definable cut-off. Since the ion will generally follow a devious route with many changes of direction it is also convenient to define, besides a total range $R_{tot}$, a projected range $R_p$ (measured parallel to the incident ion direction) and a perpendicular range $R_\perp$ (measured perpendicular to the incident ion direction). These last two concepts are relevant to device applications of ion implantation.

There are two ways of representing range distributions, one being the differential form as a probability distribution of ion range, and the other being the integral form which shows the number of particles not yet stopped as a function of the penetration depth. The former is usually the more immediately informative representation, which we shall use wherever possible, but for practical reasons many curves have been published in integral form.

The total range $R_{tot}$ is related to the specific energy loss by

$$R_{\text{tot}} = \int_0^E \frac{dE}{-(dE/dx)} \tag{2.68}$$

and this calculation can be performed readily if $dE/dx$ is known as a function of $E$. The integration is particularly simple in the case of isotropic elastic scattering in an inverse square low potential, the case considered by Nielsen (1956). We saw that, neglecting electronic stopping, eq. (2.62) gave then an energy-independent specific energy loss and hence

$$R_{\text{tot}} = \frac{0.6(Z_1^{\frac{2}{3}} + Z_2^{\frac{2}{3}})^{\frac{1}{2}}(M_1 + M_2)M_2 E}{Z_1 Z_2 M_1}, \tag{2.69}$$

proportional to energy. The constant 0.6 in this equation is chosen so that if $E$ is measured in keV, $R_{\text{tot}}$ is given in micrograms of target per $\text{cm}^2$.

The projected range, $R_p$, is much more useful, and Lindhard et al. (1961) have considered the ratio $R_{\text{tot}}/R_p$ for the inverse square law potential and obtained

$$\frac{R_{\text{tot}}}{R_p} = \frac{1}{4}\left[(5+A)\frac{(1+A)}{2A}\cos^{-1}\frac{1-A}{1+A} - 1 - 3A\right] \tag{2.70}$$

where $A = M_2/M_1$. For $M_2 < M_1$ this may be simplified to

$$R_{\text{tot}}/R_p \simeq 1 + M_2/3M_1 \tag{2.71}$$

Lindhard, Scharff and Schiøtt (1963) have made numerical calculations of the range as a function of energy for the more realistic Thomas–Fermi potential, and these remain the most useful set of data for estimating ion penetration in non-crystalline materials. The reduced range and energy parameters $\rho$ and $\epsilon$, defined in eqs. (2.16) and (2.18), allow us to write, from eqs. (2.18) and (2.20)

$$\frac{d\epsilon}{d\rho} = \left(\frac{d\epsilon}{d\rho}\right)_n + K\epsilon^{\frac{1}{2}} \tag{2.72}$$

With the aid of the universal scattering function $f(t^{\frac{1}{2}})$ shown in fig. 2.7, Lindhard et al. were able to obtain 'universal' range-energy plots, valid for all ions and targets. Since however $K$ is a function of $Z_1$, $Z_2$, $M_1$ and $M_2$ (eq. 2.20) it is necessary to plot a series of curves for different $K$ and interpolate as required. Such a series is shown in fig. 2.12. Since it is some-what tedious to manipulate these cumbersome expressions for every estimate

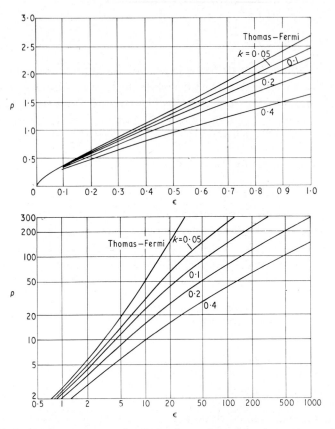

Fig. 2.12. Reduced range-energy plots for various values of the electronic stopping parameter $K$ (from Lindhard et al., 1963).

of ion range, numerous tabulations and graphs of range-energy relations based upon the Lindhard, Scharff, Schiøtt formulae have now appeared (e.g. Channing and Turnbull, 1968; Johnson and Gibbons 1970). For ease of reference, we include in Appendix 2 a series of curves for projected range as a function of ion energy, including cases of particular interest in ion implantation.

Another important quantity is the range straggling or mean square fluctuation in range, and for some purposes (e.g. in junction depth determination) the skewness of the range distribution must be considered in terms of higher moments of the range. The range straggling is obtained from the mean square fluctuation in energy loss, $\overline{\Delta E^2}$, given by eq. (2.65). The mean square fluctuation in range due to each collision is

$$\overline{\Delta R^2} = N\Delta\overline{R}\left(\frac{\mathrm{d}\overline{R}}{\mathrm{d}E}\right)^2 \int_0^{T_m} T^2\,\mathrm{d}\sigma(T) \tag{2.73}$$

$$= N\Omega^2\left(\frac{\mathrm{d}\overline{R}}{\mathrm{d}E}\right)^3 \mathrm{d}E \tag{2.74}$$

where

$$\Omega^2 = \int_0^{T_m} T^2\,\mathrm{d}\sigma(T) \tag{2.75}$$

Thus the mean square fluctuation in the total range is given by the integral

$$\overline{\Delta R_{\mathrm{tot}}^2} = \frac{1}{N^2}\int_0^E \frac{\Omega^2(E)\mathrm{d}E}{(\mathrm{d}\overline{R}/\mathrm{d}E)^3} \tag{2.76}$$

For the case of pure electronic stopping Bohr (1948) was able to show that

$$\overline{\Delta R_{\mathrm{tot}}^2} = \int_0^E 4\pi Ne^4 Z_1^2(\mathrm{d}E/\mathrm{d}x)^{-3}\,\mathrm{d}E \tag{2.77}$$

and, making use of the Bethe–Bloch formula we have, for low-mass absorbers: straggling,

$$S_{\mathrm{tr}} = \frac{\overline{\Delta R_{\mathrm{tot}}^2}}{(\overline{R}_{\mathrm{tot}})^2} = 4m/M_1\ln\left(2mv^2/I\right) \tag{2.78}$$

while for heavy absorbers:

$$S_{\mathrm{tr}} = 3m/4M_1 \tag{2.79}$$

In contrast with these small values ($< 0.01$) for $S_{\mathrm{tr}}$, pure nuclear stopping assuming isotropic scattering can lead to values of the straggling parameter $S_{\mathrm{tr}}$ close to unity. Lindhard, Scharff and Schiøtt (1963) carried out the numerical computation of eq. (2.76) for the Thomas–Fermi potential, including varying fractions of electronic stopping (determined by $K$ in eq. 2.20), and the results are shown in fig. 2.13. Naturally $S_{\mathrm{tr}}$ decreases as $K$ increases. A more recent discussion of range straggling has been given by Schiøtt (1970).

This treatment of straggling assumes, as a first approximation, that the range distribution is Gaussian. In other words, the probability distribution $P(R)$ of range is given by

$$P(R)\mathrm{d}R = \frac{1}{2\pi}(\overline{\Delta R^2})^{\frac{1}{2}}\exp-\left(\frac{R-\overline{R}}{2\Delta R^2}\right) \tag{2.80}$$

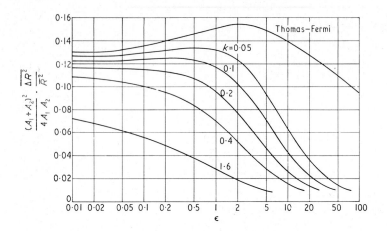

Fig. 2.13. Reduced relative mean-square straggling in range as a function of reduced energy for various values of the electronic stopping parameter, $k$ (from Lindhard et al., 1963).

In fact, experiment shows that at times the range distribution may be quite asymmetric, the degree of asymmetry being associated with the relative importance of nuclear collision processes. Sanders (1968) has therefore generalized the treatment of Lindhard et al. (1963) to include higher-order moments of the range distribution, within the approximation of an inverse power-law potential. From the series of moments he was able to construct a distribution function for the projected range, for comparison with experiment (see next section).

At low enough energies it may be preferable to derive the range by an alternative approach which neglects the inelastic (electronic) energy loss. Such an approach was developed by Holmes and Leibfried (1960) to deal with the behaviour of an ion recoiling in a lattice of similar atoms, a classic problem in radiation damage theory. The colliding atoms are taken to have energy-dependent hard-sphere radii $r(E)$. Since the cross-section

$$\sigma(E) = 4\pi r^2(E) \tag{2.81}$$

a mean free path $\lambda(E)$ can be defined as in section 2.3

$$\lambda(E) = \tfrac{1}{4}\pi N r^2(E) \tag{2.82}$$

It is convenient to define a probability $P(E_1, E_2)$ that a collision will cause an ion with energy $E_1$ to recoil with residual energy between $E_2$ and $E_2 + dE_2$. For the hard sphere model the scattering is isotropic in the centre of mass system and

$$P(E_1, E_2) = dE_2/E_1 \tag{2.83}$$

Averaging over all possible paths after the first collision:

$$\bar{\lambda}_1 = \int \lambda(E_1) P(E_0, E_1) dE_1 \tag{2.84}$$

and after the second

$$\bar{\lambda}_2 = \int\int \lambda(E_2) P(E_0, E_1) dE_1 P(E_1, E_2) dE_2 \tag{2.85}$$

The direction cosines must be taken into account, but in the equal-mass case $M_1 = M_2$ we have:

$$\cos \theta_i = (E_i/E_{i-1})^{\frac{1}{2}} \tag{2.86}$$

Averaging over the direction cosines in each of the preceding collisions yields the vector relation for the velocity $U$

$$U_i = U_0 (E_i/E_0)^{\frac{1}{2}} \tag{2.87}$$

and the above expressions lead to a formula for the mean range $\bar{R}(E_0)$

$$\bar{R}(E_0) = \lambda(E_0) + (E_0)^{-\frac{1}{2}} \int_0^{E_0} \lambda(E) E^{-\frac{1}{2}} dE \tag{2.88}$$

which can be solved since $\lambda(E)$ is given by eq. (2.82). The projected range $R_p(E_0)$ can also be averaged and for a hard-sphere radius $r(E)$ matched to an inverse square law potential, Holmes and Leibfried obtained the ratio $\bar{R}/\bar{R}_p = 1.2$, to be compared with the result (again for $M_1 = M_2$) of $\bar{R}/\bar{R}_p = 1.33$ of the Lindhard et al. formula (2.71).

While this simple hard-sphere model is useful at times, it is difficult to calculate more than the average range $\bar{R}$. A more powerful approach has been to carry out a Monte Carlo calculation of the trajectories of many ions injected into a random lattice of atoms. Binary collision theory with the appropriate interatomic potentials are assumed and impact parameters are selected from a limited distribution of values by means of a set of random numbers. Naturally this method, developed primarily at Oak Ridge National Laboratory by Oen, Holmes and Robinson (1963), demands the use of a large digital computer. Typically about 1000 primary ions are followed until either their energy falls to 25 eV (when they are assumed to stop) or they escape from the surface of the solid due to back-reflection. Range parameters are then obtainable to an accuracy of about 3 per cent, and it is possible to compare the predictions of the hard-sphere, Bohr, inverse square law and Thomas–Fermi potentials with the results of experiment. In this way it was found that the ranges predicted by the Thomas–Fermi potential were much smaller than those deduced from the Bohr potential at low ion energies (a few keV), while the experimental data fall somewhere between the two. This is because, for large interaction distances, the Bohr potential is too weak. Evidence was obtained that a better matching of the hard-sphere to the Bohr potential could be achieved, not by equating the distances of closest approach, as in eq. (2.43), but by equating the corresponding stopping powers.

It is clear that a machine calculation such as this provides a powerful heuristic tool for assessing the various analytic approximations to the inter-atomic potential. Its chief virtue, perhaps, is the fact that it can readily be extended to deal with ion penetration in single-crystal lattices, and it was the result of such a computation, by Robinson and Oen (1963) that led to the identification of the phenomenon of ion channelling, an aspect of ion penetration to which we shall return in section 2.11.

## 2.10. Experimental measurement of ion range

The most useful technique in heavy ion range determination, and one which has been developed extensively during the past few years, is by stripping away known thicknesses of material from a bombarded specimen, and measuring either the amount of implanted material which remains or the amount which is removed. Since the number of implanted ions is usually very small a sensitive method of detection is required, and most of the work has been carried out with radioactive ions and appropriate tracer techniques. In some cases of gaseous ion implantation it has been possible to use mass spectroscopic determination of the ions released. In semi-conductor crystals certain ions give rise to a strong electrical conductivity which can be measured as a function of depth: however, we shall see that this is not an unequivocal measure of the ion range owing to defect inter-actions following implantation and leading to a loss of electrical activity.

Four principal stripping techniques have been employed, of which the most precise is probably the 'anodic stripping' method developed largely by Davies and his collaborators at Chalk River Nuclear Laboratories (Davies et al., 1960a, b). An anodic oxide layer is grown electrochemically to a thickness controlled by the anodizing voltage, or alternatively, the anodizing time at a constant current density. Then the oxide is stripped away by immersion in some reagent which does not attack the underlying material. It is necessary to calibrate the amount of substance removed in each layer, and this may be done by rendering the material uniformly radioactive and measuring the loss of activity, or alternatively by weighing the specimen before and after a number of equivalent layers are removed. If, as in much of the work, inert gas ions are implanted it is essential to measure the radioactivity remaining after removal of successive layers. In other cases, however, it may be feasible to measure the activity of the material dissolved from the surface, and the results yield directly the differential range distribution.

Other stripping techniques have made use of chemical polishing, controlled by the immersion time (Bredov and Okuneva, 1957) or electrochemical polishing, controlled by the transported charge. A two-stage chemical stripping process has been developed by Andersen and Sørensen (1968) in which corrosion films produced by immersion of metal specimens in organic solvents containing iodine or bromine are then dissolved by immersion in a reagent which does not attack the metal. These methods are probably less precise than anodic stripping, since there is more tendency for non-uniform removal and a greater influence of chemical impurities on radiation damage. Nevertheless it is probably true to say that some of the anomalies observed in early experiments using these techniques (Bredov and Okuneva, 1957) were due to the unidentified effects of ion channelling, and not, as was thought for some time, to errors in the technique.

Anodic stripping is restricted in its applicability to those substances, such as Al, W, Ta, Si, which will form an anodic oxide. It is, however, also possible to apply the technique to range measurement in the oxides themselves. This is done by implanting a series of specimens on which anodic oxide films of a variety of thicknesses have previously been grown. Then the radioactivity is measured before and after stripping the oxide, so yielding the integral range distribution.

A less restricted method of stripping owes its development to Whitton (1965) and makes use of mechanical abrasion by an aqueous slurry of 0.05 micron alumina in a vibratory polisher. The removal rate is calibrated by weight-loss measurement or, in appropriate cases, by comparison with the anodic stripping technique. The technique is capable of removing layers as thin as 20 to 50 Å, and the fact that the method has been applied already to range determination in Au, Cu, W, Ta, Al, Cu, Si, $UO_2$, $Ta_2O_5$, $ZrO_2$ and GaAs (Whitton, 1965, 1968) demonstrates its versatility.

Another non-chemical method of stripping is by the controlled sputtering of material from the specimen by low-energy inert gas ion bombardment. Such a method was used by Burtt, Colligon and Leck (1961) for the measurement of range distributions of inert gases in Pt, W and Mo. After sputtering for a measured time the residual inert gas was released by raising the temperature of the specimen almost to the melting-point, and the amount of gas was determined by a calibrated mass spectrometer. Freeman and Latimer (1968) have made use of the same process to study ion ranges in gold and aluminium. Lutz and Sizmann (1963) also employed sputtering to remove known-thickness layers from copper and gold but in their work the release of radioactive $Kr^{85}$ was monitored by measuring the decrease in target activity. This technique proved very valuable when measurements were required following low-temperature bombardment of tungsten, but prior to raising the specimen temperature (Hermann, Lutz and Sizmann, 1966).

The most direct method of stripping, namely by separating, after bombardment, a stack of thin foils has proved the least useful owing to the difficulty of preparing sufficiently uniform foils. The method has been utilised by Thulin (1955) for range measurements in formvar. An added problem is that the density and other properties of extremely thin foils may vary from those of bulk material.

A different principle of range determination involves probing the bombarded specimen with radiation which is sensitive to a change in some property of the implanted layer. Thus refractive index changes have been studied (Hines, 1960) in transparent target materials such as glass and quartz: in this instance the results are related more to the distribution of radiation damage rather than to the true range distribution. In semiconductor crystals certain ions act as donors or acceptors of charge carriers and when implanted into a crystal containing a sufficiently low concentration of the opposite impurity type will, after thermal annealing to minimise the effects of lattice disorder, create a p-n junction. If the material is exposed by bevelling or grooving it is possible to apply a chemical stain which is sensitive to the electrical conductivity type, and thus delineate the junction. By bombarding a series of targets with different but known electrical resistivities it is possible to determine the depths at which the implanted ion density exactly compensates each level of impurity concentration, and in this way the range distribution is obtained. Such a technique has been applied to a variety of ions in silicon (Kleinfelder et al., 1968; Ruth and Eisen, 1967). The method is very simple to use, but is subject to the criticism that the staining process may be influenced by the electrical depletion layer which extends beyond the implanted layer into the lightly-doped substrate. Again, the property measured is not directly the implanted ion concentration but the electrically-active dopant concentration and effects of aggregation and charge neutralization due to defect interactions may distort the behaviour.

Methods which are specific to the implanted ion species frequently make use of a nuclear reaction or scattering process induced by a probing beam of light ions. The total yield may be measured at successive stages during stripping of the specimen, as described above. Alternatively, knowledge of the energy variation of the probing beam as a function of depth may be utilized to deduce the distribution of target nuclei. Rutherford back-scattering of protons or heavier particles, of mass $M_1$, results in a back-scattered energy $E'$ governed by the mass of the target nucleus, $M_2$, and the angle of scattering, $\theta_{cm}$ (in the centre of mass system):

$$E'/E = 1 - \frac{2M_1 M_2(1 - \cos \theta_{cm})}{(M_1 + M_2)^2} \tag{2.89}$$

The energy lost by the bombarding particles in penetrating to a given depth and returning to the surface can next be calculated if $dE/dx$ is known. Hence the measured energy spectrum of scattered particles leads uniquely to a depth distribution (Powers and Whaling, 1962). With the aid of a magnetic spectrometer Bøgh (1968) has achieved a depth resolution of 50 Å in such measurements. The technique is, however, limited to the study of relatively heavy ions in a light matrix owing to the mass dependence of equation (2.89). Alexander et al. (1970a) and Domeij et al. (1970) have used $N^{14}$ and $C^{12}$ probing ions in order to obtain sufficient mass resolution in back-scattering experiments on iron and silicon, respectively. A complementary technique applicable to certain light ions implanted in heavier targets employs a nuclear reaction characterized by a strong, narrow and isolated resonance in the cross-section. Such a resonance occurs, for example, in the $N^{15}(p, \alpha\gamma)$ reaction at a proton energy of 898 keV and provides a prolific yield of 4.43 MeV gamma-rays. Thus, if a specimen is bombarded with protons of slightly higher energy the reaction is excited in the lamina within which the proton energy is degraded through the 898 keV resonance. In this way Barker and Phillips (1965) determined the ranges of energetic $N^{15}$ ions in a number of metals. Similar strong resonances are available in the $B^{11}(p, \alpha)$, $O^{18}(p, \alpha)$ and $F^{19}(p, \alpha\gamma)$ reactions.

Finally, another group of range determination methods, all making use of radioactive decay, is worthy of mention. If alpha-unstable ions, e.g. of $Rn^{222}$, are implanted into a solid, the alpha-particles subsequently emitted will have an energy spectrum which is degraded according to the depth of the emitting atom below the surface, and the mean projected range $\bar{R}_p$ and maximum range $R_{max}$ can be measured in this way (Domeij et al., 1963). A similar technique was also developed for low-energy beta-emitting ions, such as $Xe^{125}$ (Bergstrom et al., 1963). A somewhat different type of measurement involves the observation of the Doppler shift in the spectrum of short-lived gamma-emitters recoiling through a foil. If the slowing-down time of the ions is comparable with, or less than, the mean lifetime there will be a significant shift which, at least for energetic ions, can be measured. If the half-life is known from other experiments it is possible to estimate the slowing-down time and hence the mean total range $\bar{R}_{tot}$. In such work the ions are produced in a short-lived excited state by a nuclear reaction or Coulomb excitation, and the necessary energy is imparted by inducing the reaction with an energetic heavy ion beam from a Tandem electrostatic accelerator, or cyclotron.

We turn now to consider the results of some of the more significant range determinations in amorphous materials, leaving the discussion of the strikingly different behaviour in crystalline solids until the next section. Since these channelling effects are also significant in polycrystalline targets the most reliable results are those obtained in truly amorphous materials such as the anodic oxides $Al_2O_3$, $Ta_2O_5$ and $WO_3$ by Davies and his co-workers, using the technique described above. In this work (Jespersgård and Davies, 1967) the ranges of the radioactive ions $Na^{24}$, $K^{42}$, $Kr^{85}$, $Xe^{125}$

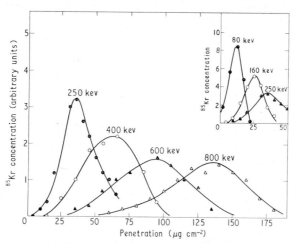

Fig. 2.14. Range distributions of $Kr^{85}$ ions of various energies implanted into amorphous $Al_2O_3$ (from Jespersgaard and Davies, 1967).

and $Xe^{133}$ were investigated. Some of the measured range distributions for $Kr^{85}$ and $Al_2O_3$ are shown in fig. 2.14, and the mean projected ranges $\bar{R}_p$ as a function of energy for various ions are shown in fig. 2.15, together

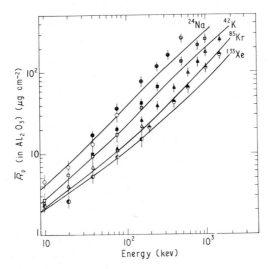

Fig. 2.15. Median projected ranges of various radioactive ions in amorphous alumina as a function of energy, compared with the theoretical curves of Lindhard et al. (1963) (from Jespersgaard and Davies, 1967).

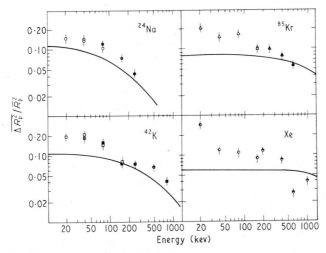

Fig. 2.16. Relative mean-square straggling in range for various ions in amorphous $Al_2O_3$ as a function of energy, compared with theoretical curves of Lindhard et al. (1963) (from Jespersgaard and Davies, 1967).

with theoretical curves derived from the formulae of Lindhard, Scharff and Schiøtt (1963). It is seen that although agreement is good at low energies there are systematic discrepancies at the higher energies, amounting to as much as 50 per cent. Figure 2.16 shows the corresponding range straggling $\overline{\Delta R_p^2}/\overline{R}_p^2$ compared with the theoretical results, and again there are sufficient deviations, particularly at low energies, although these rarely exceed a factor of two. Oen and Robinson (1964) have shown by machine calculations, however, that at low energies the Bohr potential leads to a larger straggling than does the Thomas–Fermi potential.

At low energies, nuclear stopping dominates, and some results obtained by Davies et al. (1961) for various ions in aluminium are shown in fig. 2.17 for reduced energies $\epsilon$ between $10^{-2}$ and 1. The agreement with the theory of Lindhard et al. (1963) is good. For comparison, the results of the simple Nielsen treatment are also shown, and fit well between $\epsilon = 0.1$ and 1.

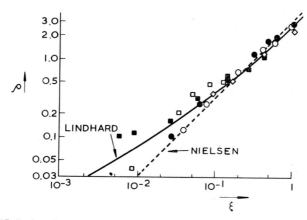

Fig. 2.17. Reduced range-energy data for various ions in aluminium compared with the theoretical predictions of Lindhard et al. (1963) and Nielsen (1956). Data taken from Davies et al. (1961).

Range distributions determined by Domeij et al. (1964) for $Kr^{85}$ ions in $Al_2O_3$ over the energy range 40–160 keV were compared with the results of Sanders' (1968) analytical development of the Lindhard theory, and the agreement is extremely good (fig. 2.18). It would be interesting to compare this theory with the results of experiments at higher energies, where, for instance Barker and Phillips (1965) observed, for $N^{15}$ ions in Ni, Ag and Au, significant departures from the theoretical expectations in the energy region 0.4 to 2.5 MeV. Neglecting nuclear stopping, which is small at these

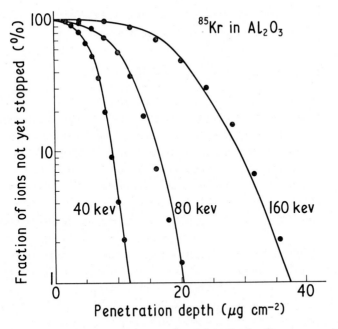

Fig. 2.18. Comparison between the experimental results (continuous curves) of Domeij et al. (1964) for the integral range distribution of Kr $^{85}$ ions in $Al_2O_3$ and the theoretical results calculated by Sanders (1968).

energies,

$$\rho = \int_0^\epsilon \frac{d\epsilon}{-(d\epsilon/d\rho)_e} = -\int_0^\epsilon \epsilon^{-\frac{1}{2}}/k \cdot d\epsilon \qquad (2.90)$$

hence

$$\rho = 2\epsilon^{\frac{1}{2}}k \qquad (2.91)$$

Barker and Phillips (1965) observed stopping cross-sections 20 per cent higher than expected, and showing a more rapid increase with energy than this $E^{\frac{1}{2}}$ dependence, particularly at the higher energies. It appears that there is a systematic tendency for experimental stopping cross-sections to exceed the theoretical values at these higher energies. This was also the energy range in which Gilat and Alexander (1964) reported, from range measurements with dysprosium ions, that the electronic stopping cross-section increased more rapidly than the ion velocity. There is a need for further experimental data to ascertain whether these interesting results are indicative of the general pattern of stopping power behaviour. If so, the theoretical ideas developed in section 2.9 will need a fundamental reconsideration.

### 2.11. Ion penetration in crystal lattices: channelling

Until about 1960 it was tacitly assumed that ion penetration in crystal-line material would not differ substantially from that in amorphous media, and there was little or no experimental evidence to suggest that the regular structure of the crystal lattice was significant in atomic stopping. This was in spite of the prediction by Stark, made as early as 1912, that interatomic fields would play a major part in determining the motion of charged particles through crystals. His ideas were rejected, however, since they were linked with an unsuccessful attempt to explain the Laue X-ray diffraction patterns of crystals.

In 1957, Bredov and Okuneva detected an anomalously deep penetration of radioactive $Cs^{134}$ implanted at 4 keV into single-crystal germanium. Unfortunately, however, the probable cause of this result was not identified. In studies of the sputtering of single-crystal copper Rol et al. (1959) and Almén and Bruce (1961) reported low sputtering ratios wherever the ion beam was incident along a low-index crystal direction. This effect was explained in terms of what was called a 'transparency model'. Low-index crystal axes or planes present large open or transparent avenues for the incident ions and fewer atoms are exposed to the bombarding particles, with the result that the number of atomic collisions which might give rise to atomic ejection, or sputtering, is reduced. Sputtering is, however, essentially a surface phenomenon involving only the first few atomic layers, and so it was not immediately appreciated that similar crystallographic effects might influence deeply-penetrating ion trajectories. At this same time, Davies et al. (1960a), observing the range distribution of heavy ions implanted into metals such as aluminium, found skew distributions with a penetrating tail instead of the nearly Gaussian form anticipated from theory, and were unable to explain them.

A little later, Robinson and Oen (1963) at Oak Ridge National Labora-tory began to find anomalous ion ranges in a computer calculation of the trajectories of low-energy copper ions moving in a simulated copper lattice. At the outset of the work the ions were generated isotropically in all direc-tions, as would be the case in neutron-induced radiation damage, but when certain ions were found to have excessively long ranges, ion injection was simulated along the major crystal directions. This soon revealed that the long-range trajectories were associated with those ions travelling so that their motion was constrained within the open 'channels' between adjacent close-packed rows of atoms. Furthermore, this was *not* simply a transparency phenomenon as the trajectories of these ions were actually steered by a series of glancing collisions with the atomic rows. Clearly, such a trajectory which suffers only glancing collisions will correspond to a reduced rate of energy loss since, as we have seen, it is in the close encounters that large momentum transfers occur. This immediately suggested that if a beam of energetic particles is incident close to a low-index direction in a single-crystal target, enhanced penetration is to be expected.

During 1963, clear experimental evidence in support of this phenomenon of 'ion channelling', as it came to be called, was provided independently by Davies and his collaborators at Chalk River Nuclear Laboratories, by Nelson and Thompson at Harwell, and by Lutz and Sizmann at Munich University. The Oak Ridge work had suggested that the penetrating component observed in range studies in polycrystalline material might have been due to enhanced ion ranges in those crystallites suitably oriented to the ion beam, and this was confirmed by experiments in single-crystal aluminium (Piercy et al., 1963), the results of which are shown in fig. 2.19. The method of anodic stripping was used to measure these ranges, for radioactive $Kr^{85}$ ions. Lutz and Sizmann (1963) employed the sputtering technique to measure

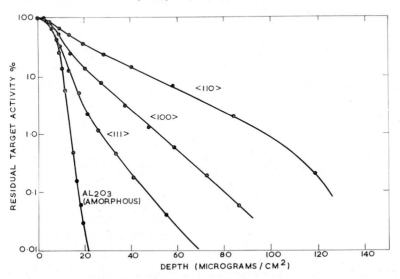

Fig. 2.19. Integral range distribution of 40 keV $Kr^{85}$ ions implanted into single-crystal aluminium along different crystal directions (from Piercy et al., 1963).

the range distribution of 140 keV $Kr^+$ ions in single-crystal copper, and again a deeply-buried component was found in cases when the crystal was oriented with an open channelling direction parallel to the beam. In the work of Nelson and Thompson (1963) the penetration of 75 keV protons through thin gold crystals was studied, for different crystal orientations, with the experimental arrangement shown in fig. 2.20. A collimated ion

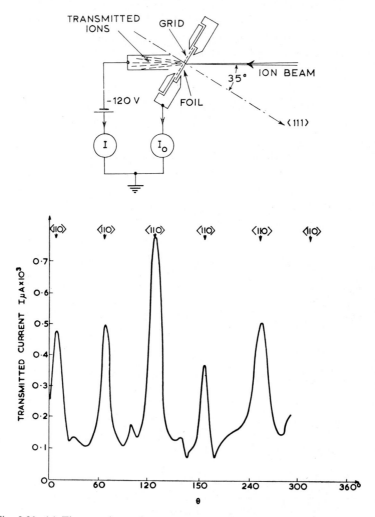

Fig. 2.20. (a) The experimental arrangement used by Nelson and Thompson (1963) for studying the channelling of transmitted ions. (b) The transmitted current observed as a function of the angle $\theta$ of the gold foil.

beam was incident at 35.2° to the normal to the foil, which corresponded to the ⟨111⟩ crystal direction, and the transmitted protons were collected in a Faraday cup. On rotating the foil the transmitted ion current varied in a manner shown in fig. 2.20(b), with maxima at positions

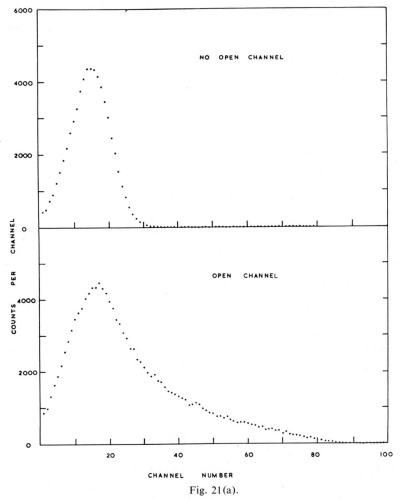

Fig. 21(a).

Fig. 2.21. (a) Energy spectra of 2 MeV protons transmitted through a 33 μm silicon crystal. The top curve shows the spectrum obtained after transmission along a random direction; the lower spectrum shows a tail extending to much higher energies when the beam travelled along the ⟨110⟩ direction (after Dearnaley, 1964). (b) the yield of transmitted protons, of energy above a preset discriminator setting, during one rotation of the silicon crystal. Peaks correspond to transmission along the ⟨110⟩ and ⟨114⟩ channels (from Dearnaley, 1964).

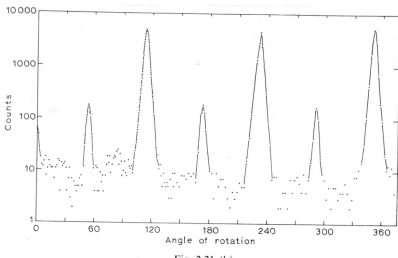

Fig. 2.21 (b).

for which the ions were incident along ⟨110⟩ axial directions (the apparent six-fold symmetry arising from twinning about ⟨111⟩ in the specimen). These experiments were all carried out in energy regions in which nuclear stopping normally dominates the energy loss. Dearnaley (1964) extended the lower energy experiments of Nelson and Thompson by examining the orientation dependence of the energy loss of 2.1 MeV protons transmitted through thin silicon crystals, and fig. 2.21. shows some of the energy spectra obtained. Here the stopping is due almost entirely to electronic collisions. By using a silicon absorber which was, in fact, a fully-depleted silicon particle detector Dearnaley showed that, to a high accuracy, the mean energy loss per ionizing collision was independent of crystal orientation. Hence the number of ionizing events must be reduced as a consequence of the protons, even at these energies, being steered away from the regions along atomic rows where the electron density is high.

These experiments at MeV energies opened the way for many sophisticated studies of the energy loss and angular distribution of particles transmitted through crystals (Gibson et al., 1965; Datz et al., 1965; Dearnaley et al., 1968a; Eisen, 1968a, etc.). Much of the work has been carried out using accelerators and techniques developed for nuclear physics research.

The effects of ion channelling have been observed not only in range measurements but also in the dramatic reduction of cross-sections for nuclear scattering and reaction processes (Thompson, 1964; Bøgh, Davies and Nielsen, 1964; Andersen et al., 1965). This is because the systematic steering of ions away from the rows of nuclei suppresses the close encounters necessary for nuclear reactions or wide-angle scattering to occur. For the same reason characteristic X-ray yields are reduced in channelling (Brandt et al., 1965). Of course, the yield from atoms located in interstitial positions centred on a channel direction, or even somewhat removed from an atomic row, is not expected to be diminished by channelling. We shall see later that this forms the basis for a very useful technique for distinguishing the nature of the sites occupied by implanted atoms (Bøgh and Uggerhøj, 1965).

### 2.12. Channelling at low energies

The earliest analytical treatment of the ion channelling process was given by Lehmann and Leibfried (1963), for the case of 10 keV $Cu^+$ ions in a copper lattice. Nuclear stopping normally dominates at this low energy: we shall return to the question of the relative importance of nuclear and electronic stopping for channelled particles in the next section. Lehmann and Leibfried chose two different interatomic potentials:
(i) an exponentially-screened Coulomb potential

$$V(r) = (99.43 \, a/r) \exp{(-r/a)} \, (\text{keV}) \qquad (2.92)$$

and (ii) a Born–Mayer potential

$$V(r) = 22.5 \exp{(-r/0.197 \, [\text{Å}])} \, (\text{keV}) \qquad (2.93)$$

Since the interatomic distances involved in channelling events at these energies are large, it is expected that the Born–Mayer will prove the more realistic potential. Next, in a 'continuum approximation' it is assumed that the interaction between the moving ion and the atomic row is given by the average potential:

$$\overline{U}(r) = \frac{1}{D} \int_{-\infty}^{\infty} V[(x^2 + r^2)^{\frac{1}{2}}] \, dx \qquad (2.94)$$

where the x-axis coincides with the channel axis and $D$ is the distance between successive atoms in the row, while $r$ is measured radially from the row. Contours of $\overline{U}(r)$ can then be constructed for a given channel in the lattice, and the minima (corresponding to the equilibrium positions of stable trajectories) may be located. The influence of rows other than the nearest ones is neglected. Lehmann and Leibfried next expanded the average potential $\overline{U}(y, z)$ in terms of average components

$$\overline{U}(y, z) = \overline{U}_0 + \tfrac{1}{2}\{\overline{U}_{yy} y^2 + \overline{U}_{zz} z^2\} \tag{2.95}$$

By the impulse approximation of classical mechanics, the change of momentum, $\Delta p$, of the ion is assumed to be given by

$$\Delta p_z = \int_0^\infty - \frac{\partial V(r)}{\partial r} \cdot \frac{z}{r} \cdot dt \qquad \text{etc.} \tag{2.96}$$

It is further assumed that the struck atoms do not recoil. Then the equations of motion of a channelled ion travelling close to the axis are:

$$M\ddot{x} = 0; \qquad dx/dt = v, \text{ to first order}$$

$$M\ddot{y} = -\overline{U}_{yy} y; \; d^2y/dx^2 = -(\overline{U}_{yy}/2E)y \tag{2.97}$$

$$M\ddot{z} = -\overline{U}_{zz} z; \; d^2z/dx^2 = -(\overline{U}_{zz}/2E)z$$

The ion trajectory is therefore sinusoidal and, in general, would describe a Lissajous figure in the $y$-$z$ plane. The wavelength, $\lambda$, of the trajectory is given by

$$\lambda_{y,z} = \frac{2\pi(2E)^{\frac{1}{2}}}{(\overline{U}_{yy,zz})^{\frac{1}{2}}} \tag{2.98}$$

and is typically many times the lattice spacing. As the energy of the ion decreases the wavelength shortens and so does the transverse amplitude of the motion. Lehman and Leibfried (1963) further considered the effects of thermal lattice vibration and showed that its influence would be critically dependent upon the choice of potential, as was the maximum range. The latter was calculated by methods similar to those described in section 2. The momentum transferred to the lattice at each collision causes a damping of the amplitude of the trajectory and the specific energy loss becomes

$$-dE/dx = \frac{G_0}{E(x)} \cdot \frac{M_1}{M_2} \tag{2.99}$$

where $G_0$ is a geometrical factor which was evaluated numerically. This expression leads to a channelled range

$$R_{\max} = (M_2/2G_0 M_1)E_1^2 \tag{2.100}$$

Andersen and Sigmund (1965) have since shown that recoil motion of the atoms in the channel wall will lead to a lower exponent in the energy dependence, and they also considered the requirements which must be satisfied if a maximum channelled range, $R_{\max}$, is to be observed experimen-

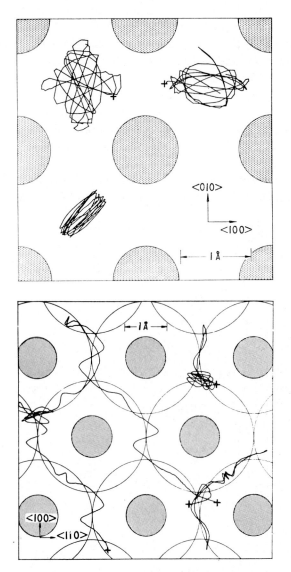

Fig. 2.22. Projection of some $\langle 001 \rangle$ and $\langle 01\bar{1} \rangle$ channelled trajectories calculated by Robinson and Oen (1963) for 1 keV Cu$^+$ ions injected into a copper lattice. The points of entry of the ions, normal to the diagram, are indicated by crosses.

tally. We shall return to consider the experimental measurements in section 2.17.

The calculation of ion trajectories in a computer model, as carried out by Robinson and Oen, has been discussed briefly in section 2.11. The interaction potentials used were similar to those of Lehmann and Leibfried (eqs. 2.92 and 2.93) but were usually truncated (or set equal to zero) beyond a critical radius $R_c$, chosen to be $\frac{1}{2} D^{110}$, the interatomic spacing along the $\langle 110 \rangle$ axis. Figure 2.22 shows the projections of the computed trajectories of 1 keV copper atoms travelling along the $\langle 110 \rangle$ and $\langle 100 \rangle$ directions, after injection normal to the plane shown at the points marked. The circles indicate the cut-off radius $R_c$ of a truncated Bohr potential. The stability of channelled trajectories is clearly shown, in confirmation of Lehmann and Leibfried's analytical result. Note a tendency in the $\langle 110 \rangle$ case for particles to wander from one channel to another. The energy loss of the moving atoms was calculated, as in Lehmann and Leibfried's work, using the im-

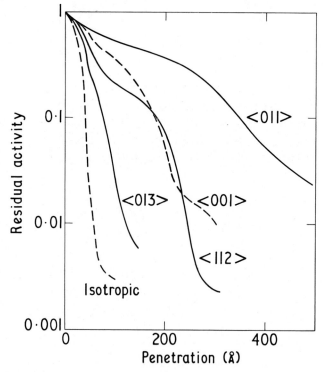

Fig. 2.23. Calculated integral range distributions of 5 keV Cu$^+$ ions slowing down in copper (after Robinson and Oen, 1963).

pulse approximation, and by simulating very many trajectories it was possible to construct the range distribution. Figure 2.23 shows such theoretical integral range distributions for 5 keV copper atoms injected along the five most open channels of the face-centred cubic copper lattice. Also shown is the distribution expected for a random lattice, and it can be seen that channelling has increased the maximum penetration by up to an order of magnitude. This factor depends upon the interaction potential, here taken to be a truncated Born–Mayer type.

Thermal vibration of the lattice plays a major part in determining the range distribution at low ion energies. If the range of a well-channelled particle is $R_0$ for the case of zero vibration, and $R_T$ at temperature $T$, Lehmann and Leibfried (1963) estimated that

$$R_T = R_0 \exp\left(-2\bar{u}^2/a^2\right) \tag{2.101}$$

where $\bar{u}^2$ is the mean square thermal vibration amplitude and $a$ is the screening parameter of the interaction potential. For a given material, values of $\bar{u}^2$ may be obtained from tables (e.g. Lonsdale, 1952) or estimated as follows. At temperatures $T > \Theta_D/4$, say, where $\Theta_D$ is the Debye temperature, we have

$$\bar{u}^2 \sim 2 \cdot 10^{-3} D^2 T/T_m \tag{2.102}$$

where $D$ is the interatomic distance and $T_m$ the melting point. Hence the maximum range is expected to diminish exponentially with temperature, at very low energies. This was the result observed by Howe and Channing (1967) for 40 keV $Xe^{135}$ ions channelled in gold crystals, but the energy loss in this case would be partly by electronic processes.

### 2.13. Channelling at higher energies

So far we have considered cases in which the energy loss even of channelled particles may be considered to take place primarily as a result of elastic collision processes. At higher energies this is not so, although the steering of the particle along the channel does still occur by dint of screened nuclear interactions. It is the nature of this steering process that we shall be concerned with in the present section.

Nelson and Thompson (1963) treated the channelling of protons in gold crystals at energies up to 100 keV in terms of a Bohr potential. The effective potential in the channel was derived, as in eq. (2.94) and since this is a relatively square potential well the proton trajectory will, under channelling conditions, oscillate with almost a zig-zag form. A heavier ion, however, experiences a different potential and would move with a more sinusoidal trajectory (similar to that described by Lehmann and Leibfried above) Dearnaley et al. (1968a). The distance of closest approach $r_{min}$ is then defined by equating the transverse energy component to the average row potential at $r_{min}$, i.e.

$$E\psi^2 = \overline{U}(r_{min}), \qquad \text{for small } \psi \tag{2.103}$$

for a particle of energy $E$ incident at an angle $\psi$ to the row.

Effectively, channelling occurs only for those trajectories which do not penetrate a forbidden region of $r < r_{min}(E)$.

Lindhard (1965) gave a more extended treatment of the problem, in terms of a Thomas–Fermi potential of the form shown in eq. (2.38). The average potential $\overline{U}(r)$ is approximated by the expression

$$\overline{U}(r) = \frac{Z_1 Z_2 e^2}{D} \ln \left\{ \left( \frac{a_{TF} C}{r} \right)^2 + 1 \right\} \tag{2.104}$$

with $C \sim \sqrt{3}$ and $a_{TF} = 0.46 \ (Z_1^{\frac{2}{3}} + Z_2^{\frac{2}{3}})^{-\frac{1}{2}}$ Å. Lindhard then suggested that in the case of ion velocities $v > Z_1^{\frac{2}{3}} e^2/\hbar$ the minimum distance of approach to an atomic row under channelling conditions is likely to be close to $a_{TF}$, and scattering from the channelled or 'aligned' beam into the 'random' beam will occur for closer encounters. The assumption of an energy-independent $r_{min}$, with equations (2.103) and (2.104) then give us a limiting or critical angle for channelling

$$\psi_c \simeq \left( \frac{2Z_1 Z_2 e^2}{ED} \right)^{\frac{1}{2}} \qquad (\text{for } \psi_c < a_{TF}/D) \tag{2.105}$$

while at lower energies, for $\psi_c > a_{TF}/D$ Lindhard gives another approximation

$$\psi_c' \simeq \left( \frac{a_{TF}}{D} \psi_c \right)^{\frac{1}{2}} \tag{2.106}$$

In the higher energy case it is clear that the effective barrier height at the distance of closest approach for channelling is independent of energy:

$$\overline{U}_{max} \simeq 2Z_1 Z_2 e^2/D \tag{2.107}$$

and is typically about 10 $Z_1 Z_2$ eV, while at low energies there is a transition to an energy-dependent potential

$$\overline{U}_{max}'(E) \simeq \left( \frac{3Z_1 Z_2 e^2 E}{D^3} \right)^{\frac{1}{2}} \tag{2.108}$$

In another class of experiments, the moving particle starts effectively at a lattice site rather than being injected into the channel from outside the crystal. Thus in large-angle scattering experiments a close encounter with a nucleus must occur and an ion may recoil at an angle $\phi$ to an atomic row (fig. 2.24). Lindhard considers that the scattering from the averaged potential $\overline{U}$ occurs after the particle has travelled midway to the next scattering centre.

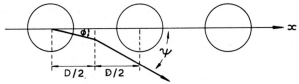

Fig. 2.24. Model of the emission of an atom from a lattice site at an angle $\phi$ to the atomic row. $D$ is the interatomic spacing.

Conservation of transverse energy yields, approximately

$$E\psi^2 \simeq E\phi^2 + \overline{U}(\tfrac{1}{2}\phi D) \tag{2.109}$$

Substitution of the maximum value of $\overline{U}$ gives a minimum value of the angle $\psi$

$$\psi_{\min} = K\psi_2, \qquad \text{for } \psi_c < a_{TF}/D \tag{2.110}$$

where $K$ is a constant of about 1.5. Similar critical angles therefore define the conditions under which an ion can become channelled, on the one hand, or steered away from a nearest neighbour in an atomic row. The latter effect has been termed 'blocking' (Gemmell and Holland, 1965) or 'shadowing' (Tulinov, 1965).

It is easy to understand axial channelling (and blocking) in terms of the potential of atomic rows. Planar channelling of particles injected parallel to a crystal plane results from interaction with planar arrays of atomic rows. Lindhard (1965) derives an average planar potential $\overline{U}_p$ at a distance $z$ from the plane:

$$\overline{U}_p(z) = 2\pi N_p \int_0^\infty \rho V(z^2 + \rho^2)^{\frac{1}{2}}\, \mathrm{d}\rho \tag{2.111}$$

where $N_p$ is the atomic density in the plane and $\rho$ is the average distance parallel to the plane, i.e. $\rho^2 = x^2 + y^2$. In Lindhard's approximation to the Thomas–Fermi potential,

$$\overline{U}_p(z) = 2\pi Z_1 Z_2 e^2 N_p \{(z^2 + C^2 a_{TF})^{\frac{1}{2}} - z\} \tag{2.112}$$

yielding a critical angle for channelling by a planar potential

$$\psi_p = (2\pi N_p Z_1 Z_2 e^2 a_{TF}/E)^{\frac{1}{2}} \tag{2.113}$$

Planar potentials are typically about an order of magnitude lower than the corresponding axial ones, with the result that planar critical angles are three or four times smaller than axial values.

Accepting Lindhard's assumption that if a particle comes to within a distance $a_{TF}$ of a row or plane it is scattered into the random component we see that if the dimensions of an axial or planar channel are less than $2a_{TF}$, no channelling effects will be seen. At low velocities the minimum distance of closest approach (for channelling to be maintained) is energy-dependent and channelling should be observable for only the most open axial channels, while planar effects may essentially disappear.

The atoms of the crystal lattice undergo thermal vibration, and in many materials the r.m.s. vibrational amplitude at room temperature is comparable with $a_{TF}$, so that the influence of temperature on ion channelling may be quite significant. There have been several approaches to this problem and it is instructive to compare them critically. Erginsoy (1965) derived a temperature-dependent average potential, by integrating over a Gaussian probability distribution of vibration amplitudes in directions normal to a row or plane. Thus for a plane

$$\overline{U}_p(z, T) = \overline{U}_p(z, 0) \exp\left(u_{\perp}^2/2a_{TF}\right)$$

where $z$ is the distance from the plane and $\bar{u}_{\perp}^2$ is the mean square thermal vibration amplitude normal to the plane. The potential function was calculated using Molière's (1947) approximation to the Thomas–Fermi potential. For the case of atomic rows, the probability $P(\rho)$ of finding an atom at a distance $\rho$ from a row is

$$P(\rho)\, d\rho = 2\rho/\bar{u}^2 \exp\left(-\rho^2/\bar{u}^2\right) \tag{2.114}$$

and the temperature-dependent row potential is obtained from

$$\overline{U}(r, T) = \int_0^{\infty} d\rho \int_0^{2\pi} P(\rho)\overline{U}\{(r^2+\rho^2-2\rho r \cos\theta)^{\frac{1}{2}}\}\, d\theta \tag{2.115}$$

by numerical integration. The result is shown in fig. 2.25 for protons incident on the $\langle 110 \rangle$ rows in silicon. As the temperature increases the average potentials fall, the distance of closest approach becomes smaller in this model, and the critical angles for channelling decrease (in accordance with experiment).

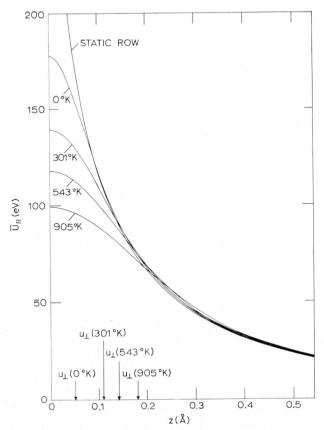

Fig. 2.25. The temperature-dependent average row potential $\bar{U}$ calculated for the $\langle 110 \rangle$ row in silicon by Appleton et al. (1967).

Oen (1965) carried out a numerical computation of the effect of thermal vibrations on the 'blocking' effect of a single neighbour atom, using an exponentially-screened potential. In fact, however, as Andersen (1967) was able to show, the second, third, etc., neighbours play a significant rôle and it is preferable therefore to utilise Lindhard's concept of a string potential. Into Lindhard's expression (2.109), Andersen introduced a Gaussian probability distribution of the position of the emitting atom. Numerical computation then becomes necessary, and the calculations were made for a range of emitted particle energies, crystal temperatures, and for the planar and axial cases. The widths of experimentally-observed dips in back-scattered yield were well-reproduced, but it was not feasible to explain the value of the minimum yield (fig. 2.26).

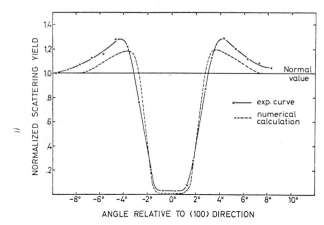

Fig. 2.26. Experimental and calculated dip in Rutherford scattering yield for 480 keV protons incident near a ⟨100⟩ direction in a tungsten crystal at 390 °K (from Andersen, 1967).

Morgan and Van Vliet (1968, 1970a, 1970b) have adopted a different approach. Following the work of Robinson and Oen (1963) they simulated the motion of an ion in a crystal lattice by means of a computer model. Each collision is treated as an isolated two-body interaction, with either a screened Coulomb or Thomas–Fermi potential. Many particle trajectories are simulated, and by inspection of the simulated paths such parameters as critical channelling angles can be identified. Table 2.1 shows a comparison of the critical angles and minimum distance of approach to the rows, compared with the predictions of Lindhard's (1965) analytical treatment: the agreement is very satisfactory. This established, Morgan and Van Vliet (1968) went on to incorporate thermal vibrations of all the atoms, the displacements being chosen from a triangular distribution with a fixed r.m.s. amplitude with the aid of a set of random numbers. Andersen (1967) had earlier shown that it is justifiable to neglect the correlations in the thermal motion of adjacent atoms. The results showed that phenomena involving axial channelling are much more temperature-dependent than those associated with planar channelling. Physically, this is because in a perfect lattice ions approach more closely the atoms in a row than in a planar array so that changes in impact parameter due to thermal vibrations produce correspondingly larger effects, while secondly, the continuum row potential (2.104) decreases more rapidly with distance than does the planar potential (2.112), so that a change in the distance of closest approach effects the critical angle more strongly for rows. An interesting result emerges when the minimum

TABLE 2.1

Critical parameters for rows; protons in copper

| Row | Energy (keV) | $\Psi_{\text{crit}}$ | | | $\rho_{\min}/a$ | |
|---|---|---|---|---|---|---|
| | | Computer | Analytic | Lindhard | Computer | Analytic |
| $\langle 100 \rangle$ | 5 | 0.080 | 0.095 | 0.11 | 1.75 | 1.56 |
| $\langle 100 \rangle$ | 20 | 0.070 | 0.078 | 0.078 | 0.93 | 0.85 |
| $\langle 100 \rangle$ | 100 | 0.047 | 0.052 | 0.051 | 0.40 | 0.38 |
| $\langle 100 \rangle$ | 500 | 0.027 | 0.030 | 0.022 | 0.21 | 0.17 |
| $\langle 110 \rangle$ | 20 | 0.085 | 0.10 | 0.10 | 0.87 | 0.72 |

From: Morgan and Van Vliet (1968)

distance of approach $\rho_{\min}$ to a row or plane is followed as a function of temperature, (Morgan and Van Vliet, 1970a). Figure 2.27 shows the result of a two-dimensional calculation of $\rho_{\min}/a$ for protons at various energies in copper. There is an almost linear dependence upon the r.m.s. thermal vibration amplitude. While Erginsoy's (1965) model would suggest that $\rho_{\min}$ decreases with increasing temperature, because of the behaviour of the continuum potential, eq. (2.115), the computer model shows that there is a dominating effect of an opposite character. There was agreement between

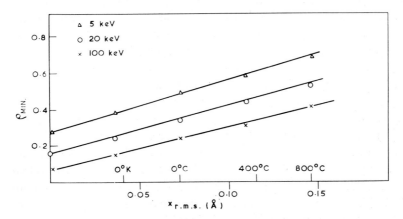

Fig. 2.27. The critical approach distance to $\langle 100 \rangle$ rows as a function of the r.m.s. displacement $x_\perp$ normal to the row, for protons in a 2-dimensional copper lattice. $\triangle = 5$ keV; $\bigcirc = 20$ keV; $\times = 100$ keV (from Morgan and Van Vliet, 1970a).

the values of $\rho_{min}$ derived from the trajectories and from the equation for the conservation of transverse energy (eq. (2.109)), demonstrating that the change in critical angle results primarily from a change in the critical approach distance and *not* to a change in the row potential as proposed by Erginsoy. Atoms vibrating out of the rows compel the channelled ions to remain further and further out in order to avoid close collisions which would violate the conservation of transverse energy. Another way of considering this is to note that, in figure 2.25, the variation in row potential due to thermal vibration is confined to radii of $r < a$: channelled ions do not have an opportunity of approaching this region. More recent calculations by Morgan and Van Vliet (1970b) have confirmed these ideas for the three-dimensional case: here, however, rather than a linear dependence of $\rho_{min}$ upon $x_{r.m.s.}$ they obtain:

$$\left(\frac{\rho_{min}}{a}\right)^2_T = \left(\frac{\rho_{min}}{a}\right)_{static} + C\left(\frac{x_{r.m.s.}}{a}\right)^2 \tag{2.116}$$

where

$$C \propto \left(\frac{Ea}{Z_1 Z_2 e^2}\right) \tag{2.117}$$

While the agreement between computer simulation and Andersen's (1967) analytical treatment was rather good for the case of ions entering a crystal and becoming channelled, the same was not the case for emission (or back-scattering) of ions from a lattice site. The angular distribution of ions emerging from a crystal is then influenced at three distinct stages:

(i) the emission and immediate interaction with an atomic row or plane,

(ii) subsequent passage through the crystal, with channelling and dechannelling,

(iii) transmission through the crystal surface.

The idea of the conservation of transverse energy under a continuum axial or planar potential enable stages (i) and (iii) to be analysed. The abrupt termination of the potential at the crystal boundary thus leads to a refraction of the trajectories at the surface, analogous to an optical situation. However, the effects of lattice vibration are introduced into the analytical treatment solely in terms of the vibration of the emission site, as it introduces too great a complication to vibrate all the atoms. This can, however, be included in

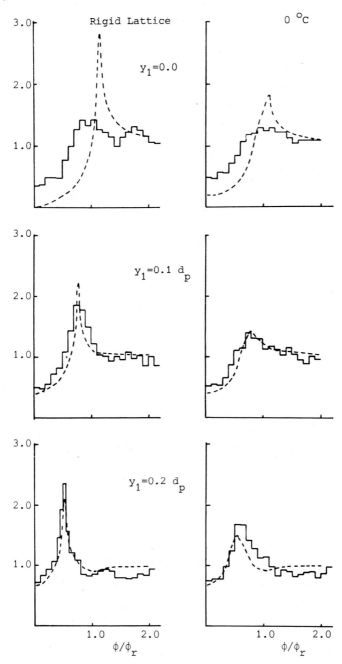

Fig. 2.28. Emission profiles for 20 keV protons in Cu as a function of the distance from the atomic row and the temperature. The results of computer simulations are shown in the histogram; comparable theoretical curves based upon the continuum model are shown as dashed curves (from Morgan and Van Vliet, 1970b).

the computer simulation. Morgan and Van Vliet (1970a) have shown that although there is reasonable agreement for emission (or scattering) from sites away from the atomic row, for the case of 'blocking', i.e. lattice site emission, the analytical treatment differs seriously from the results of a computer calculation (fig. 2.28). The reason is simple: the continuum model breaks down when ions approach very closely on atomic row or plane, and thus it should strictly not be applied to emission from a lattice site. When approached more closely than a critical distance, $\rho_{min}$, crystal rows or planes behave more like a collection of individual scattering centres than a smeared-out potential. The continuum model gives its best results for light ions at high energies, and so may be reasonable for the case of MeV protons.

Thus the computer model, properly used, is able to throw considerable light on the appropriate analytic forms to be used and in this sense is again a heuristic tool. Further results of a similar nature will be discussed in section 2.16.

### 2.14. Experimental measurement of critical channelling angles

There have been many measurements of critical angles for channelling, at both high and low energies. Here we select a few examples to illustrate the different types of experiment and the measure of agreement with equations (2.105) and (2.106).

At the lowest ion energies, Andreen and Hines (1966) have transmitted protons, deuterons and helium ions through thin (200 Å) gold crystals and measured the transmitted intensity of channelled ions as a function of the crystal orientation. Figure 2.29 shows their results, together with the theoretical curve for $\psi'_c$, and it is clear that the agreement is very good.

Brandt et al. (1968) measured the yield of K X-rays from bombarded crystals, as a function of target orientation. Since the X-ray yield increases rapidly with energy, most of the yield is from the surface layer, and the width of the dip in observed yield corresponds to the full incident ion energy. Some results, in aluminium crystals, are shown in fig. 2.30, and again there is excellent agreement with the theoretical predictions.

Andersen and Uggerhøj (1968) measured the widths of the dips in yield of back-scattered ions from crystals such as tungsten, as a function of temperature. In this case either the incoming or outgoing beam may travel near to the direction of an atomic row. An important concept in Lindhard's treatment is that of reversibility: the probability of close collision with a

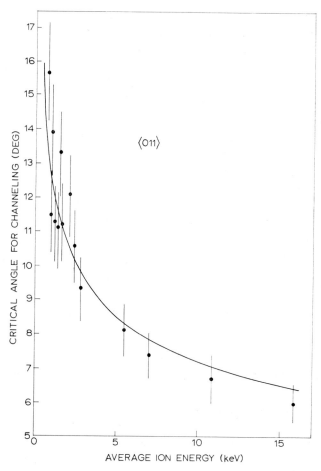

Fig. 2.29. Critical channelling angles of deuterons in gold crystals about 200 Å thick as a function of energy, compared with the predictions of Lindhard (1965) (after Andreen and Hines, 1966).

lattice atom, for a particle entering a crystal in a given direction is equal to the probability for the particle to emerge from the crystal in that same direction when emitted from a lattice site. This will be so if the particle trajectory is invariant to time-reversal, i.e. if the influence of energy loss along the path can be neglected. By selecting particles scattered from near the surface of the crystal, this condition can best be met, and fig. 2.31 shows how closely the minima in scattering yield agree, when 400 keV protons either enter or leave along a $\langle 100 \rangle$ direction in tungsten. The temperature-

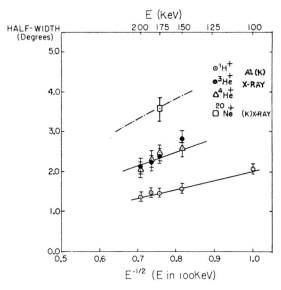

Fig. 2.30. Experimental half-widths of the dip in yield of X-rays about the $\langle 100 \rangle$ direction in single-crystal Al, for the ions indicated. The straight lines show the predictions of equations 2.105 and 2.106 (from Brandt et al., 1968).

dependence of the half-width of such dips is shown in fig. 2.32, at two energies, and compared with the results of Andersen's (1967) calculations.

### 2.15. The energy loss of channelled ions

So far we have considered the scattering mechanisms responsible for the channelling process and the factors which control the critical angle for channelling without regard to the energy loss of the ions. The energy loss of channelled ions can vary considerably from that in amorphous media and depends strongly upon the atomic number of the ion. We shall begin, however, with the case of light, fast-moving ions such as have frequently been used for probing implanted crystals.

Lindhard (1965), in the first treatment of this problem, showed that on the basis of the electron-gas model half the energy loss of a fast, fully-ionized particle will be due to close single-particle collisions, while half will be due to distant, collective excitations. In this model, close collisions are assumed to be proportional to the local electron density and distant collisions to the average density of electrons. Under channelling conditions the former may be drastically reduced while the latter cannot be affected so that Lindhard's

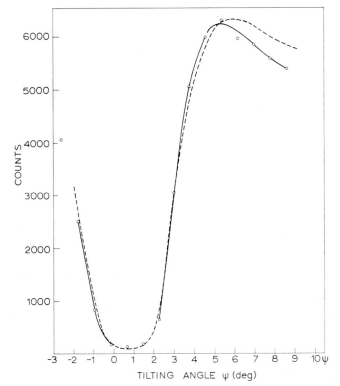

Fig. 2.31. Experimental verification of the reversibility rule governing incoming and out-going beams interacting at lattice sites. In this case 400 keV protons are scattered from a (100) plane in tungsten (after Andersen et al., 1967). (– – – –) outgoing, and (————) incoming beam.

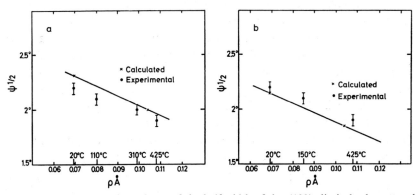

Fig. 2.32. Temperature dependence of the half-width of the $\langle 100 \rangle$ dip in back-scattered yield for protons incident on a tungsten crystal (a) at 400 keV; (b) at 48 keV (from Andersen and Uggerhøj, 1968).

equipartition rule (eq. 2.27) leads to a maximum possible reduction in the electronic stopping cross-section, $(dE/dx)_e$, of a factor of two in a well-channelled ion trajectory. Although this was in agreement with early experimental results in silicon, this argument fails to explain why the energy loss of channelled protons in germanium (Sattler and Dearnaley, 1965) can be only one-third of the non-channelled energy loss.

Erginsoy (1965) put forward a different model, adapted from the treatment of electronic stopping given in section 2.4, in which the contributions of core and valence electrons are separated. In place of equations (2.24) and (2.25) Erginsoy wrote, for channelled ions,

$$-\left(\frac{dE}{dx}\right)_e = \frac{4\pi Z_1^2 e^4 N}{mv^2}\left(Z_{loc} \ln \frac{2mv}{\hbar k_e} + Z_{val} \ln \frac{k_c v}{\omega_0}\right) \qquad (2.118)$$

where $NZ_{loc}$ is the localized electron density in the vicinity of the channel axis, and $NZ_{val}$ is the average density of free, or valence, electrons in the solid. Thus $Z_{val}$ is the same as $Z_2^f$ in eq. (2.24) but $Z_{loc} \neq Z_{val}$. Using the Bethe–Bloch treatment with an adiabatic criterion for core electrons

$$(p_{max})_j = \hbar v / \Delta E_j \qquad (2.119)$$

where $\Delta E_j$ is the binding energy of an electron in the $j$th shell, Erginsoy showed that the energy loss to K- and L-shell electrons may be ignored for protons of a few MeV channelled in silicon. In other words, the maximum impact parameter for these shells is less than the minimum distance of approach of channelled protons. Then, taking $Z_{val} = 4$, the calculated energy loss agreed well with experiment. A further confirmation was claimed from the comparison of the fluctuation in energy loss of channelled and non-channelled protons, but here it was assumed that the width of the spectrum of transmitted proton energies was dominated by statistical effects, i.e. straggling, rather than by variations in the trajectory of the ion: more recent work by Clark et al. (1970) has shown that this argument is probably not justified.

Luntz and Bartram (1968) have taken Erginsoy's approach somewhat further by treating the core excitation produced by channelled alpha-particles (and heavier ions) in more detail. They chose a slightly different adiabatic criterion:

$$(p_{max})_j = (Z_1^* v)^{\frac{1}{2}} / \Delta E_j \qquad (2.120)$$

where $Z_1^*$ is the velocity-dependent effective charge of the ion, given approximately by

$$Z_1^* = Z_1[1 - \exp(\hbar v / Z_1^{\frac{2}{3}} e^2)]$$  (2.121)

By calculating the energy loss due to free electron and core electron excitation, Luntz and Bartram estimated the specific energy loss for various ions channelled in the scintillator crystals NaI and CsI, in order to gauge the influence of channelling on their luminescent response. For channelled alpha particles, in this model, the energy loss along the most open $\langle 111 \rangle$ channel is reduced by about a factor of four, compared with the random energy loss. Experimental measurement of the behaviour of 4 MeV protons channelled in CsI (Clark et al., 1969) has shown a reduction factor of 3.0 or 2.3, depending upon whether the energy of well-channelled protons is taken to be the maximum or the most probable energy, respectively, in the channelled proton spectrum.

This raises the question as to whether the spread in energies of channelled particles arises predominantly by statistical processes, i.e. straggling, or by real differences in the particle trajectories. Only in the latter case would it be justifiable to select the maximum transmitted energy to correspond to the best-channelled particles. Clark et al. (1971) have examined this situation closely, and since they were able to induce marked variations in the shape of the spectrum with small changes in crystal orientation it is concluded that the energy loss of channelled protons is dominated by the precise trajectory followed through the crystal. Clark et al. (1970, 1971) have therefore attempted to correlate the energy loss of best-channelled protons with the physical dimensions and geometry of the various crystal channels, in the hope of distinguishing experimentally between single-particle and collective excitations.

The most recent treatment of channelled energy losses has been given by Bonsignori and Desalvo (1969, 1970), using the complex dielectric contact approach outlined in section 2.4. The imaginary part of this function determines the energy loss. These authors restricted their treatment to nearly-free electrons (i.e. S-electrons) in order to handle the lattice periodicity of the electron gas. As a result, it is difficult to compare their predictions with experiment since the experimental value is nearly five times greater, in the case of random orientation. However, some interesting results emerge. There is no equipartition between close and distant collisions with valence electrons, and the close collisions contribute more than twice as much to the stopping cross-section. Secondly, the effect of channelling on plasmon excitation is about the same as that on single-particle excitation, contrary to the assumptions of Lindhard (1965) and Erginsoy (1965). It will be interesting to see this model extended to core electrons.

We can see, therefore, that equipartition of energy between single-particle and collective modes of excitation 'under asymptotic energy conditions may not be a very useful concept in understanding the energy loss of ions under channelling conditions. Some authors believe that even close collisions between an ion and an electron can excite collective modes of excitation, and the single particle and collective excitations are then held to be strongly-coupled, a situation analogous to that in nuclear theory where short-range interactions between nucleons bring about the collective excitation known as the 'giant dipole resonance'. On such grounds, Klein (1966) believes that, even in the case of non-channelled trajectories, a fast charged particle dissipates most of its energy in plasmon creation. With a model based on this hypothesis he is able (Klein, 1968) to obtain a theoretical value for the Fano factor in germanium in close agreement with the experimental figure of 0.13. (The Fano factor is the ratio between the observed variance of the number of ionizing events in a germanium gamma-ray detector and the value calculated with the naive assumption that all the events are independent.) Furthermore, when protons are channelled through silicon radiation detectors (Dearnaley, 1964) it is found that the mean energy per electron-hole pair created remains the same (3.60 eV) as in the non-channelled case. This figure comprises three roughly equal contributions due to the competing processes of ionization, phonon production and residual kinetic energy of electrons and holes. If there were indeed two distinct processes of single-particle and collective electronic excitation, it would be surprising if each should have the same efficiency for the production of secondary ionization, since the energy distributions of the primary excitation vary widely in the two cases. The observation suggests, therefore, that a single mode of excitation should be considered, and it may be that, at least in crystals, the partition of localized and non-localized excitations is not a useful separation. As further evidence of this, it has been found (Howie, 1967) that the mean free path for the production of plasmons by fast electrons is increased when the electrons are channelled through thin crystal foils. (The channelling of electrons has been considered by Howie et al. (1969).) This suggests that the generation of collective excitations by an electron is affected by conditions which are local to the particle trajectory, but of course the electron beam in such a case must be handled by diffraction theory.

In the case of channelled protons, Brice (1968) has proposed a 'diffraction model' in which it is implicitly assumed that the energy losses of an ion channelling through a crystal are predominantly to collective modes of excitation. Brice considers each atom in the lattice as a heavy, positively-charged core (consisting of the nucleus together with the closely-bound inner electrons) surrounded by a negatively-charged shell of loosely-bound electrons. The shells and cores are considered to move independently, with four possible modes of phonon-like excitation: two of these are the ordinary acoustical and optical phonon bands with core and shell moving in phase, while the other two correspond to modes in which the core and shell move out of phase, giving rise to a local time-dependent polarization of the medium. Brice identifies the low-energy portion of this set of excitations with plasmon oscillations and uses the known properties of plasmons in their description. From experimental measurements of the energy loss of protons and deuterons channelled in germanium (Sattler and Dearnaley, 1965) Brice deduces the effective number $N_e$ of electrons per atom in the loosely-bound shell. This number is found to be a function of the incident ion direction, which governs the average distance of approach of the ion to the channel wall and hence the strength of the local interaction. The values of $N_e$ for germanium varied between 5 and 9. Brice remarks that his model should be valid also for heavy ions, in the electronic stopping region, with the proviso that the appropriate interatomic potential is used.

So far in this section we have been dealing with the relatively simple case of fast, fully-ionized particles, yet it is clear that there is no generally-accepted theory for explaining the energy loss in both the channelled and non-channelled case, and there is a paucity of experimental observations. We next turn to consider the energy loss of channelled particles at lower energies, where $v < Z_1 e^2/\hbar$ and only a few of the electrons are stripped.

The first point to be established is that the energy loss of well-channelled ions is dominated by electronic processes rather than nuclear collisions, simply because these ions only rarely come close enough to a nucleus. Experimentally, this point has been well established even for low-energy $Xe^+$ ions by range measurements in tungsten crystals (Eriksson et al., 1967). By differentiating the maximum observed ion range along the open $\langle 100 \rangle$ channel, the specific energy loss was derived as a function of ion energy. The results, plotted against $E^{\frac{1}{2}}$, are shown in fig. 2.33. Above about $E = 25$ keV the data fits a linear $E^{\frac{1}{2}}$ dependence almost exactly and this line, extrapolated, passes through the origin. This is therefore identified with the electronic stopping, which is expected to be proportional to ion velocity, and the nuclear stopping at the lowest energies is obtained by simple subtraction. Note that the two components are equal at a $Xe^+$ ion energy of only about 4 keV, whereas in the case of non-channelled $Xe^+$ ions this equality occurs at about 2.7 MeV. Except at the lowest ion energies, therefore, we shall ignore nuclear stopping for well-channelled ions. Of course, an ion which approaches the critical channelling angle, or in fact becomes dechannelled at some stage, will experience a much greater degree of elastic energy loss and its range will be shorter.

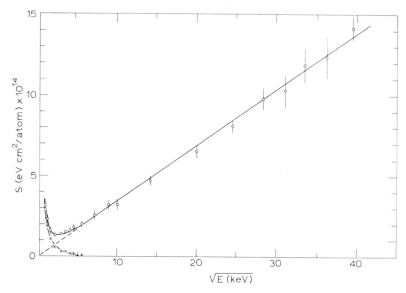

Fig. 2.33. Stopping cross-section values derived from the maximum range of $Xe^+$ ions in tungsten, along the $\langle 100 \rangle$ direction, showing electronic and nuclear stopping (after Eriksson et al., 1967).

An interesting set of data on the specific energy loss of ions channelled through silicon has been presented by Eisen (1968, 1969). A succession of different ion beams was transmitted, at a uniform initial velocity of $1.5 \times 10^8$ cm/sec, through aligned silicon crystals only about 1 $\mu$m in thickness, and the maximum transmitted energy in each case was measured by means of an electrostatic analyser and ion detector. The results are shown in fig. 2.34, which shows that the oscillatory dependence of the electronic stopping cross-section, $S_e$, upon $Z_1$ which is also discernible in amorphous media (fig. 2.10), is greatly enhanced in crystals, with $S_e$ fluctuating by over an order of magnitude.

A number of attempts have been made to explain these striking results. El-Hoshy and Gibbons (1968) adjusted the effective charge of the ion in a semi-empirical manner and so fitted the observed oscillations. The physical justification for this procedure seems, however, inadequate. The method adopted by Cheshire et al. (1968, 1969), making use of the Firsov model of electronic stopping, has been outlined above in section 2.4. Oscillations in $S_e$ then appear naturally as a result of shell effects which influence the size and electron distribution of the ions. The assumption was made that the ions are predominantly singly-charged throughout, and the impact parameter for each collision was set equal to the mean radius of the channel. The results are shown in figs. 2.34 and 2.35 in which it can be seen that the positions of the maxima and minima and the skewness of the peaks are well

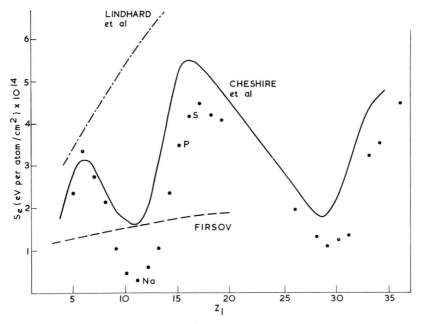

Fig. 2.34. Experimental data of Eisen (1968, 1969) for the minimum electronic stopping cross-section for various ions along the ⟨110⟩ direction in single-crystal silicon as a function of atomic numbers $Z_1$, compared with the theoretical predictions of Lindhard et al. (1963), Firsov (1958) and Cheshire et al. (1968), (from Dearnaley, 1969).

reproduced, but the magnitude of the stopping cross-section is overestimated, particularly near the minima. This could be due to the approximation that the lattice atomic wave-functions are the same as those of free atoms: it is known that covalent bonding in the crystal will modify the distribution in the outer shell. The virtue of this type of approach is that it can be applied equally well to crystalline and amorphous materials, with appropriate choice of impact parameter distribution. Channelling is seen as merely altering this distribution. A rather similar treatment of the problem has been given by Bhalla and Bradford (1968), who also used the Firsov model as a basis for their calculations. They deduced also that, in the limiting case of complete overlap of two atoms described by the Thomas–Fermi model, their expression for $S_e$ reduces to the form developed by Lindhard and Scharff (eq. 2.15), while in the case of little or no overlap the limiting

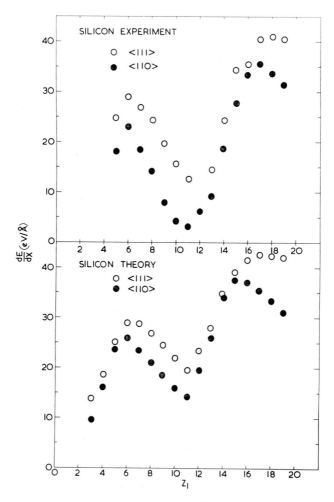

Fig. 2.35. Comparison between the experimental electronic stopping cross-section data of Eisen (1968, 1969) for the $\langle 110 \rangle$ and $\langle 111 \rangle$ channels of silicon, measured at $1.5 \times 10^8$ cm/ sec, and calculations by Cheshire and Poate (1969).

equation corresponds to that derived from the Firsov model by Teplova et al. (1962).

Quite a different approach was mentioned in section 2.8. Lindhard and Finneman (1969) are engaged in constructing a model in which the energy loss of a heavy ion moving through an electron gas is treated quantum-mechanically. Resonances in the cross-section occur in a manner identical to that involved in the Ramsauer–Townsend effect (for electrons transmitted through gases).

Thus the energy loss of channelled ions is becoming better understood, and current work is throwing more light on the physical bases of the Firsov and Lindhard models of electronic stopping. It is not yet clear, however, why the channelled stopping cross-section is often not proportional to ion velocity. Eisen (1968) expressed $S_e$ as a function of energy in the empirical form

$$S_e = kE^p \qquad (2.122)$$

and his experimentally-measured values of the exponent $p$ are plotted in fig. 2.36. It is interesting that $1/p$ shows an oscillation with $Z_1$ which

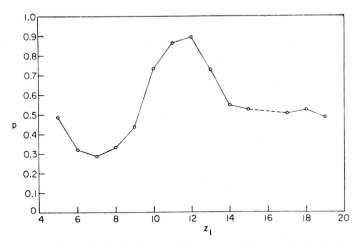

Fig. 2.36. Experimentally derived values of the exponent $p$ in equation (2.122) as a function of the atomic number $Z_1$ of the projectile ion (from Eisen, 1968).

follows that of $S_e$. The energy dependence of $S_e$ is important, of course, if attempts are to be made to estimate the maximum range of channelled ions by integration (as in eq. 2.68), and at the present time this cannot be done with certainty. Eisen (1969) remarks that results obtained in this manner may depart by a factor of two from the experimentally determined range.

Throughout the above we have been concerned with cases in which the absorber thickness is sufficiently large compared with the wavelength, $\lambda$, of the channelled trajectory that a large number of interactions take place with the atomic rows which form the channel. If instead very thin single-crystal absorbers are used, fine structure is observed in the transmitted energy spectrum (Lutz et al., 1966; Gibson et al., 1968c; Datz et al., 1969). Since the energy loss is a function of the electron density, which rises steeply near the atomic rows, Lutz et al. proposed that the structure originates from the transverse oscillation of the particle in a channel and the small number of such oscillations in traversing a thin crystal. Much of the work has been carried out with energetic $Br^+$ and $I^+$ ions accelerated in a Tandem accelerator and transmitted through gold crystals less than 1 $\mu$m in thickness. Figure 2.37 shows a typical energy spectrum for 60 MeV $I^+$ ions passed through a crystal in which the $\{111\}$ plane was inclined at 0.5° to the beam direction. Here the ion energies were determined by magnetic analysis, but in other experiments a time-of-flight method was used to measure the ion velocities.

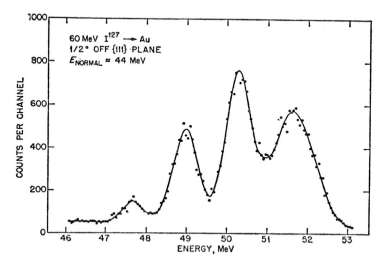

Fig. 2.37. Magnetically analysed energy spectrum of 60 MeV $I^{127}$ ions after transmission through a 0.7 $\mu$m crystal of gold. The detector was in line with the beam and the crystal was tilted $\frac{1}{2}$° from a $\{111\}$ plane. The energy of a particle traversing the crystal in a random direction would be about 44 MeV (from Datz et al., 1969).

Datz et al. (1969) describe a model by which to explain this structure. They consider a planar channel consisting of two adjacent close-packed lattice planes (fig. 2.38) which give rise to an average potential $U(x)$,

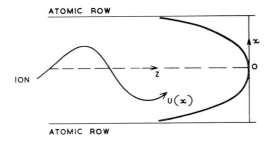

Fig. 2.38. Model showing the average potential $U(x)$ experienced by an ion in a planar channel (after Datz et al., 1969).

where $x$ is the distance from the mid-plane. With axes as shown, the two-dimensional equation of motion (assuming the channel walls to be rigid) is

$$\frac{d^2x}{dz^2} = -\frac{1}{2E}\frac{\partial U(x)}{\partial x} \tag{2.123}$$

and if the potential is expressed in the polynomial form

$$U(x) = U(0) + a_1 x^2 + b_1 x^4 + \cdots \tag{2.124}$$

the equation of motion becomes that of an undamped anharmonic oscillator. Figure 2.39(a) shows ion trajectories for three different impact parameters with respect to the mid-plane. Ions entering close to this plane are deflected only weakly, so the amplitude of their motion is small and the wavelength $\lambda$ is large. Those ions entering close to the edge of the channel, on the other hand, are more strongly repelled and have paths of larger amplitude and shorter wavelength. The energy loss of these particles is greater because they have penetrated the atomic planes more deeply and more often. Hence the energy loss depends upon the impact parameter in a continuous manner. The discrete components observable in the transmitted energy spectrum arise because the detector subtends a small angular aperture and hence accepts only particles emerging from the crystal in a certain direction. This introduces, as can be seen from fig. 2.39(b), two sets of conditions for the wavelength of the particles detected:

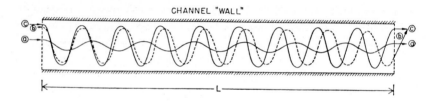

$$\psi_0 = \psi_e \neq 0$$

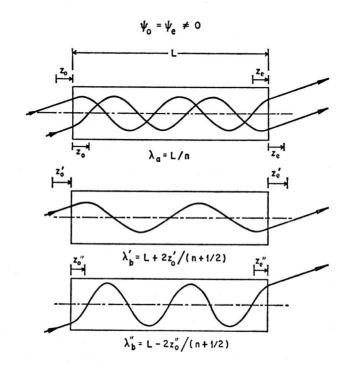

Fig. 2.39. (a) Ion trajectories for three different impact parameters in a planar channel; (b) Detection conditions in terms of the wavelengths of channelled trajectories, i.e. solutions of equation 2.125 (from Datz et al., 1969).

$$\lambda_a = (L \pm z_i \mp z_e)/n$$
$$\lambda_b = (L \pm z_i \pm z_e)/(n + \tfrac{1}{2})$$

$$(2.125)$$

where $n$ is any integer, $L$ is the thickness of the crystal and $z_i$, $z_e$ are the lengths associated with the phase shifts of the incident angle $\psi_i$ and the exit angle $\psi_e$. The experimental results enable the specific energy loss to be plotted as a function of the path length $L$, for various values of $n$. Similar measurements have been made for 3 MeV alpha-particles channelled through gold (Datz et al., 1969). The stopping power, as expected, increases for decreasing wavelength, $\lambda$, of the trajectory, but so far it has not been possible to obtain a precise mapping of the stopping cross-section across the channel, and thereby a determination of the interatomic potentials, owing to excessive mosaic spread in the gold crystals used. Robinson (1969) has developed an interpretation of these results, taking Molière's (1947) approximation to the Thomas–Fermi potential. Electronic stopping was introduced semi-empirically in the form

$$-(\mathrm{d}E/\mathrm{d}x) = S(x)E^p \qquad (2.126)$$

in which $S(x)$ and $p$ were obtained by fitting the experimental data. The values of $p$ for $I^+$ ions in gold varied with the trajectory, but were in all cases significantly lower than the value of $\frac{1}{2}$ which is to be expected on theoretical grounds. Robinson interprets this as indicating that the energy dependence of the stopping cross-section for channelled ions may be different from that appropriate for randomly-directed ones. This is a point that clearly needs further investigation.

### 2.16. Dechannelling

We consider here the various factors which are responsible for the ejection of ions from channels, since it is their magnitude which will determine the form of a channelled range distribution. Dechannelling will in all cases involve a relatively close atomic collision, and therefore an elastic scattering process. The circumstances leading to this collision can, however, be divided into four categories: (i) nuclear collisions, (ii) electronic collisions, (iii) thermal lattice vibration and (iv) crystal defects such as dislocations.

The effect of nuclear scattering is seen very clearly in the two-dimensional computer simulations of ion trajectories carried out by Morgan and Van Vliet (1968). Ions incident at angles slightly greater than the critical channelling angle are initially reflected back by the row potential, but their angle with respect to the channel axis increases at successive collisions until they are scattered out of the channel by a violent collision (fig. 2.40). A well-channelled ion would, however, (in the absence of any other effects) never approach the rows closely enough to suffer this form of dechannelling. This is no longer true in a crystal with interstitial impurities, particularly if these show a tendency to cluster together, as may frequently be the case. Then the channelled ion may have a high probability of collision resulting in dechan-

$\psi = 4.4^{\circ}$                    $\psi = 3.8^{\circ}$

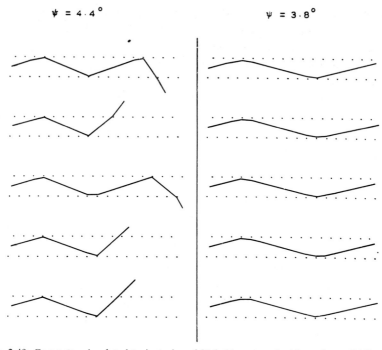

Fig. 2.40. Computer-simulated trajectories of 20 keV protons incident along $\langle 100 \rangle$ rows of copper just above and just below the critical channelling angle. The vertical axis is magnified $\times 4$. Each trajectory corresponds to a slightly different initial position on the mid-channel axis (from Morgan and Van Vliet, 1968).

nelling. Yet another form of nuclear dechannelling occurs in almost every practical situation, since the surface of the crystal is covered with an oxide film which, if even a few monolayers in thickness, can introduce substantial dechannelling. Thus Bøgh (1968) has shown that a 20 Å layer of oxide on tungsten crystals reduces the channelled fraction of a 20 keV beam of $Kr^{85}$ ions by one order of magnitude. In other cases, such as gold, the structure of the surface layers may differ from that of the bulk crystal, producing the same effect as on oxide.

Ions, we have seen, undergo inelastic energy loss processes to the electrons of the lattice atoms, and in each of these events there is a small but finite momentum transfer. The effect of many such collisions is to produce a gradual, statistical fluctuation in the direction of the ion, by multiple scattering. Clearly, this process will bring a well-channelled ion closer to the atomic rows and eventually may result in a dechannelling collision. The effect of multiple inelastic scattering increases as the ion comes closer to the rows, while it is believed to be the dominant process affecting the trajectories of well-channelled ions. Feldman, Appleton and Brown (1967), without identifying the physical mechanisms involved, set up a model for the escape of ions from channels based upon a diffusion equation. They introduced the concept of the 'half-thickness', i.e. that thickness of crystal after traversal of which the channelled fraction of the beam has been reduced by half. For MeV protons this half-thickness was found to be proportional to ion energy, and was influenced by crystal temperature. A full theoretical treatment of dechannelling by multiple inelastic scattering, and including temperature effects, remains to be made.

Thermal lattice vibrations increase the probability of dechannelling by increasing, as we saw in section 2.13, the critical distance of approach, $\rho_{min}$, to an atomic row. Moreover, the mean electron density near the centre of a channel will be increased and so multiple scattering will be enhanced. The magnitude of these effects will depend upon the mass and Debye temperature of the crystal. Morgan and Van Vliet (1970b) have studied the thermal dechannelling of protons in copper crystals, using a computer simulation, and tabulate the non-channelled (or random) fraction $\chi_{min}$ of the incident beam, defined by

$$\chi_{min} = \frac{n_0 - n_{ch}}{n_0} \tag{2.127}$$

where $n_0$ is the number of incident well-aligned ions and $n_{ch}$ is the number successfully channelled.

No real crystal is perfect, and it is important to consider the effect of crystalline faults on the trajectories of channelled ions. The first attempt to do this was made by Quéré (1968) who was interested in utilising the dechannelling of transmitted ion beams as a means of examining defects in thin crystals. He applied an analytical method, arguing that a channel is effectively blocked if, at its point of maximum distortion, the continuum potential cannot provide sufficient centripetal acceleration to an ion at the critical approach distance to keep it from colliding with an atomic row. A dislocation will be surrounded by a cylinder of radius $\lambda/2$ within which dechannelling may occur. Quéré concluded that the total dechannelling probability for an ion beam traversing a dislocation is proportional to the square root of ion energy and the square root of the Burgers' vector of the dislocation, $b$. More recently, Morgan and Van Vliet (1970b) have carried out computer simulations of dechannelling by various types of lattice dislocation, including point defects. The results confirm the energy depen-

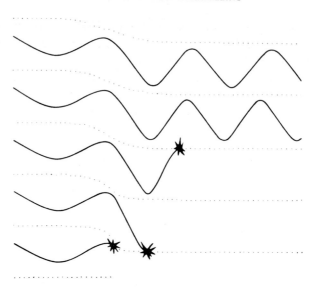

Fig. 2.41. Dechannelling at an edge dislocation, simulated for 20 keV protons along {100} planes in copper. The vertical axis is magnified ×10 (from Morgan and Van Vliet, 1970a).

dence predicted by Quéré, but indicate a dechannelling probability proportional to $b^{\frac{3}{2}}$ rather than $b^{\frac{1}{2}}$. In the case of interstitial defects, Morgan and Van Vliet found that there could be as much dechannelling from the strained lattice, which relaxes around such a defect, as from direct collision with an interstitial atom. One then has in operation both dechannelling categories (i) and (iv) mentioned at the beginning of this section.

Since lattice defects are relatively efficient in dechannelling transmitted ions, this can serve as a means of observing such defects in thin crystals. Some good examples of the possibilities of this technique have been demonstrated by Delsarte et al. (1970) using a radioactive source of alpha-particles, or alternatively a $U^{235}$ fission source. This source is placed next to the foil to be studied, and the transmitted ions are detected in a sheet of cellulose nitrate (for $\alpha$ particles) or mica (for fission fragments). After suitable etching, tracks are made visible, and the presence of any defect causing dechannelling will lead to a decrease in the number of recorded tracks if the sample foil thickness is chosen such that only channelled ions can emerge. Surface defects, grain boundaries, dislocation loops and stacking faults have been shown to be detectable in this way.

In summary, we have seen that dechannelling is a complex process in any real crystal, involving several interrelated physical mechanisms. No complete theoretical understanding of all of these can be given at the present time. The simulation, in a computer, of ion trajectories emerges as a very powerful method of studying this problem, and providing analytical expressions which can be used to extend the analysis still further, e.g. by including the energy loss of the ions, following the path over much greater distances, etc. Until this can be done there is little hope of computing range distributions even when the energy loss processes discussed in earlier sections become better understood.

Closely related to the process of dechannelling is the converse scattering of ions into a channel. This again demands an elastic collision taking place at a site greater than $\rho_{min}$ away from an atomic row. Since there is a reciprocity between the transitions into and out of a channelled trajectory (upon simple time reversal), those factors, such as temperature, which enhance dechannelling will increase the probability of what is sometimes called 'feeding' into channels. This can be a very important process, for example when attempts are made to avoid channelling by misorientation of a target specimen. Scattering into channels will always occur to some extent, and consideration of the factors outlined in this section can assist in estimating the degree to which this may modify the range distribution.

## 2.17. Experimental measurements of range distributions in crystals

We have seen that, owing to some uncertainties regarding the energy loss of channelled ions and a lack of full understanding of dechannelling behaviour, it is not yet possible to calculate the range distribution for ions of a given species injected into a given crystal lattice. It is therefore necessary at the present stage to gather more experimental data under varied conditions in order to build up our understanding of these problems and also because this semi-empirical approach is the only way of obtaining the information by means of which to design implanted structures accurately. In this section we shall therefore gather together the results of a number of experiments on channelled range distributions in metals and semiconductors and use them to illustrate the effects which are taking place.

One point of particular interest will be the maximum range of implanted ions, corresponding to the best-channelled ion trajectories, since this can be compared with the energy-loss measurements discussed in section 2.15. Moreover, this will determine the position of an implanted junction in high-resistivity semiconductors. We shall see that, under some circumstances, the upper limit of the range distribution may be blurred out, a feature which has some important practical consequences (see also section 2.18).

Two principal methods have been used for sectioning implanted crystals for the measurement of range distributions: these are the anodic stripping technique and the mechanical method of vibratory polishing, both described in section 2.10. Some additional evidence has been obtained, in semi-conducting material, by measurement of the junction depth and by capaci-tance–voltage studies of implanted diodes, after annealing, but there are dangers of interpreting wrongly from such data the true distribution profile of the implanted atoms.

Firstly we review an important series of measurements, made by the Chalk River group, on single-crystal tungsten, chosen because the anodic stripping method is easy to apply to this metal, and the channelling effects observed are rather pronounced. Various radioactive species were implanted under a range of conditions, and the results were usually presented as integral range distributions, which show the fractional radioactivity remain-ing in the specimen as a function of the thickness of material removed.

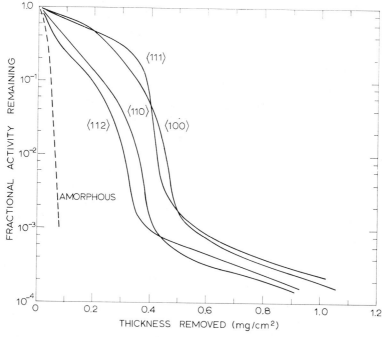

Fig. 2.42. Integral penetration curves for 40 keV $A^{41}$ ions in tungsten crystals (after Kornelsen et al., 1964).

Figure 2.42 shows the distribution of 40 keV $A^{41}$ ions injected along different channel directions in tungsten, and the increase of penetration with the openness of the channel is immediately evident. There is, however, another feature. In this, and nearly every other case of ion distributions in tungsten, a very deeply penetrating component is observed, comprising about 0.1 per cent of the implanted material (Kornelsen et al., 1964). Since, in a subsequent experiment (Davies and Jespersgård, 1966) the same tail to the distribution was observed when $Xe^{125}$ ions were injected into equivalent $\langle 111 \rangle$ channels inclined at 90° and 20° to the crystal surface, it appeared probable that the 'supertail', as it became called, is not an enhanced channelling effect (as suggested by Erginsoy, 1964) but rather that it is caused by a rapid interstitial diffusion, taking place after the particles have been brought to rest. Hermann et al. (1966) confirmed this by low-temperature (77 °K) implantations, showing that the 'supertail' does not appear until the crystal is subsequently warmed up and, moreover, that it can be considerably suppressed by a post-bombardment with a heavier dose of non-radioactive ions before heating the crystal. Alternatively, as Davies et al. (1968) demonstrated, a pre-bombardment with $Ne^{20}$ ions will eliminate the 'supertail' which would otherwise have been observed in a subsequent $K^{42}$ implantation (fig. 2.43). It is presumed that the supertail involves only those ions which

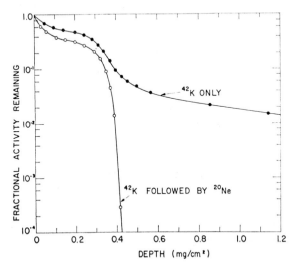

Fig. 2.43. The effect of post bombardment in suppressing the tail of a $K^{42}$ distribution in tungsten at 30 °K: trace bombardment with $K^{42}$ was followed in one case by $3 \times 10^{15}$ $Ne^{20}$ ions/cm², and all bombardments were along the $\langle 100 \rangle$ axis. The range distribution was measured at room temperature (from Davies et al., 1968).

have slowed down without producing local damage to the surrounding lattice. These particularly well-channelled ions may proceed to diffuse interstitially and isotropically until they find a trapping centre such as a vacancy, a dislocation, or a surface. While the fraction of material undergoing such diffusion is small for room temperature implantations, at 30 °K it may involve nearly 10 per cent of the implant (fig. 2.44) indicating that the fraction of the beam coming to rest in interstitial sites is strongly temperature-dependent, presumably owing to the dechannelling mechanisms we discussed in section 2.17. Andersen and Sigmund (1965) have considered under what conditions

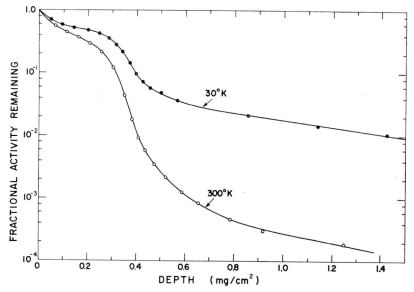

Fig. 2.44. Integral range distributions of 40 keV $K^{42}$ ions implanted along the $\langle 100 \rangle$ direction in tungsten at 30 °K and at 300 °K (from Davies et al., 1968).

an implanted ion may come to rest as an interstitial without creating damage in the surrounding metal lattice. They conclude that an almost perfectly-channelled ion of atomic number $Z_1$ moving through a crystal of atomic number $Z_2$ can lose energy without creating defects provided $Z_1 < Z_2$. This rule appears to be obeyed in the case of tungsten since the only case in which a supertail was not observed was when $Rn^{222}$ was implanted.

   One may therefore separate experimentally the transient effects, occurring during the slowing down of the ion in a period of the order $10^{-14}$ sec, from the non-transient effects, such as subsequent diffusion, which may occupy times of the order of minutes or hours (depending upon the temperature). By using either a pre-bombardment or post-bombardment with 40 keV Ne$^{20}$ ions to suppress the 'supertail', Davies et al. (1968) were able to determine the true maximum range, $R_{max}$, by a linear extrapolation of data such as that in fig. 2.44. The result emerges (fig. 2.45) that to within the experimental accuracy of 3 per cent, $R_{max}$ is independent of temperature over the range 30 to 725 °K. Davies et al. (1968) conclude that the electronic stopping of a well-channelled ion does not depend significantly on the

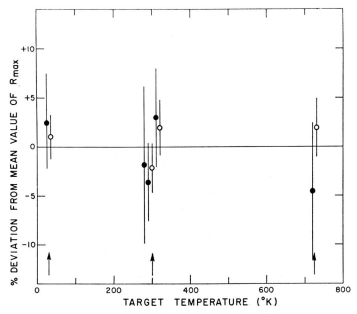

Fig. 2.45. The temperature dependence of the maximum range, $R_{max}$, for 40 keV K$^{42}$ ions implanted along the $\langle 100 \rangle$ direction in tungsten. The more accurate values ($\bigcirc$) were obtained by using either a prebombardment or postbombardment with 40 keV Ne$^{+}$ ions to suppress the tail of the distribution (from Davies et al., 1968).

vibrational amplitude of the lattice atoms, in contradiction of the theoretical prediction of Harrison and Greiling (1967). Indeed, Davies et al. (1968) find that $R_{max}$ is essentially independent of slight misorientation of the specimen, surface contamination or the degree of lattice disorder, and that it is therefore a useful and well-defined experimental parameter. (This conclusion is somewhat hard to reconcile with the results of Howe and Channing (1967) for the penetration of 40 keV Na$^{24}$ ions in tungsten, since in this case it appeared that the maximum range was markedly affected by temperature.)

It is interesting next to follow the energy dependence of the maximum range, and fig. 2.46 shows the results for $Xe^+$ ions channelled along the $\langle 100 \rangle$ axis in tungsten. Up to an energy of about 10 keV the relation is close to an $E^{\frac{3}{2}}$ power law, changing over to an $E^{\frac{1}{2}}$ behaviour, which is followed above about 100 keV (Davies et al., 1965). These results are in agreement with the theoretical ideas presented in section 2.12. At the highest ion energies the range distribution shows two peaks (fig. 2.47), the one nearer the surface occurring at approximately the depth predicted in amorphous tungsten by the Lindhard, Scharff and Schiøtt (1963) formulae. It would be easy to identify this peak with the ranges of ions which are dechannelled at the surface of the crystal, but for the fact that it is sometimes significantly deeper than the amorphous range (Eriksson et al., 1967). It appears that a significant proportion of ions may be 'quasichannelled', i.e. may make a few oscillations within the channel before making a close, dechannelling

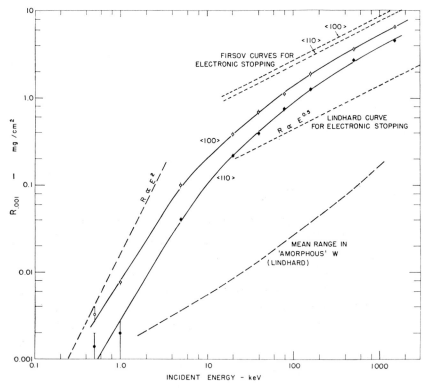

Fig. 2.46. The maximum range of $Xe^+$ ions along the $\langle 100 \rangle$ direction in tungsten, together with some theoretical predictions of the behaviour (from Davies et al., 1965).

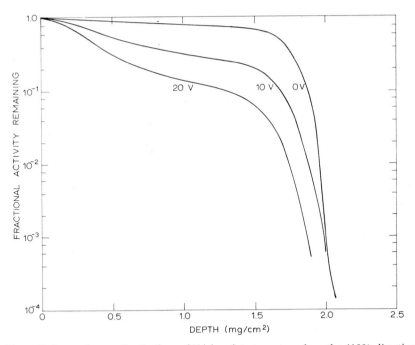

Fig. 2.47. Integral range distributions of $K^+$ ions into tungsten, along the $\langle 100 \rangle$ direction, for different anodic oxide layer thicknesses: the anodic oxide has a thickness $\backsim 10$ Å/volt (after Whitton, 1968).

collision. This phenomenon is evident in many of the computed trajectories of Morgan and Van Vliet (1968). The deeper peak is clearly associated with those particles which enter the crystal with sufficiently large impact parameters with the atomic rows so that they remain channelled throughout most of their path in the crystal. For this to be the case, the energy loss by electronic processes must dominate the multiple inelastic scattering leading to a dechannelling collision. In terms of the critical approach distance, $\rho_{min}$, Eriksson et al. (1967) equate the fraction of ions which fail to be channelled to $n\pi\rho_{2\,min}^2$, where $n$ is the number of rows per $cm^2$. He is then able to plot $\rho_{min}$ as a function of energy for the case of $K^{42}$ ions channelled along the $\langle 100 \rangle$ axis in tungsten. Using eq. (2.103) to estimate the critical channelling angle, Eriksson obtains values roughly in agreement with Lindhard's formulae. The discrepancy suggests that surface effects may have led to an overestimation of $\rho_{min}$ in this example.

By differentiation of the maximum range data obtained at different energies for a variety of radioactive ion species, Eriksson et al. (1967) were able to derive the electronic stopping cross-section as a function of the atomic number, $Z_1$, of the ion. The results, shown for the $\langle 100 \rangle$ channel in tungsten follow an oscillatory curve very similar to that subsequently published by Eisen (1968) for his ion transmission experiments in silicon. The maxima and minima occur at similar values of $Z_1$ in each case, as one would expect on the basis of the theoretical work of Cheshire et al. (1969).

The effect of a thin surface oxide contamination of the tungsten on the range distribution of $K^{42}$ ions was studied by Whitton (1968) who found that even a few atomic layers of oxide can produce significant effects (fig. 2.27). Presumably this comes about by nuclear dechannelling collisions at the crystal surface.

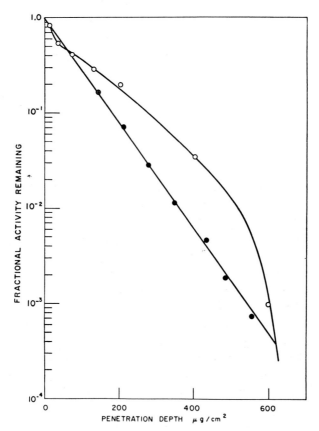

Fig. 2.48. Integral range distributions of 40 keV $Xe^{133}$ ions along the most open channelling directions in gold and tungsten: ● $\langle 110 \rangle$ Au (Whitton 1967); ○ $\langle 111 \rangle$ W (Kornelsen et al., 1964).

A further series of experiments has been carried out in gold crystals (Whitton and Davies, 1964; Lutz et al., 1965; Whitton, 1968), with the interesting result that there are several striking differences in the channelled range distributions, compared with those in tungsten, despite the similarity in the atomic weights of the materials. Thus fig. 2.48 shows the integral range distributions for 40 keV $Xe^{133}$ ions injected at room temperature into the most open channelling directions in tungsten and gold, and it can be seen that the distribution in gold is very nearly exponential, with no indication of a maximum ion range. Measurements at different temperatures (fig. 2.49) show that only at the lowest temperatures is there any bending over of the range distribution, corresponding to a maximum range (Channing, 1967) and this value is at least twice that of the same ions in the most open channel in tungsten. Whitton has also measured the penetration of a number of different ions channelled, at a uniform initial velocity, along the $\langle 110 \rangle$ axial channel in gold. The results, shown in fig. 2.50, show relatively little dependence upon the atomic number of the ion. It seems likely that thermal vibration effects are responsible for masking the true maximum range by producing a good deal more dechannelling in gold than occurs in tungsten. Whitton points out three factors which may be responsible for this:

(a) the lattice structure: Au is f.c.c. while W is b.c.c.;

(b) the thermal vibration amplitude in gold is nearly twice that in tungsten, and

(c) the electronic configuration of the atoms is different, with tungsten having several outer electrons compared to the one in gold. Whitton (1968) and Whitton et al. (1970) consider the first two of these factors to outweigh the third. By a comparison of range distributions of $Xe^{133}$ ions in W, Ta, Au, Cu, Al and Ir crystals, Whitton finds that a discernible maximum range is associated with lower thermal vibration amplitudes and with b.c.c. lattice structure. However, we should expect from what has gone before in section 2.15 that $Na^+$ ions would have a significantly greater penetration than $K^+$ ions, on the basis of a lower electronic stopping cross-section in gold (Bötti-ger and Bason, 1969). Since this is not apparent in the range distribution it would appear that the dechannelling processes are most effective for those ions possessing a small size and a small electronic stopping cross-section. This is not what our present understanding of dechannelling mechanisms (section 2.16) would lead us to anticipate, and so the behaviour of different ions in gold crystals remains an unexplained anomaly.

The earliest channelled range measurements in silicon were made by Davies et al. (1964), for radioactive $Xe^{125}$ ions. The anodic stripping tech-

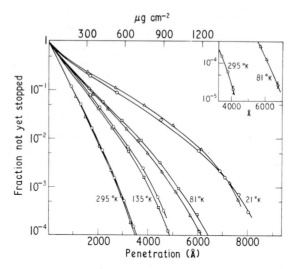

Fig. 2.49. Effect of temperature on the integral range distributions of 40 keV Xe$^{133}$ ions in gold (from Channing, 1967).

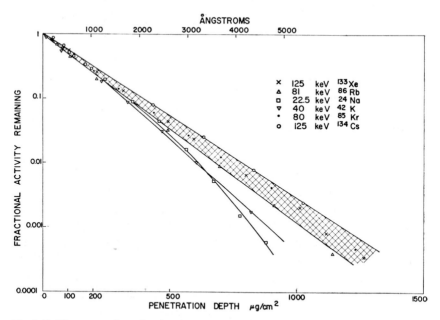

Fig. 2.50. The penetration of various ions channelled along the $\langle 110 \rangle$ direction in gold at a uniform incident velocity of 4250 cm/sec (after Whitton, 1968) ($\times$) 125 keV Xe$^{133}$; ($\triangle$) 81 keV Rb$^{86}$; ($\square$) 22.5 keV Na$^{24}$; ($\triangledown$) 40 keV K$^{42}$; ($\bullet$) 80 keV Kr$^{85}$; ($\bigcirc$) 125 keV Cs$^{134}$.

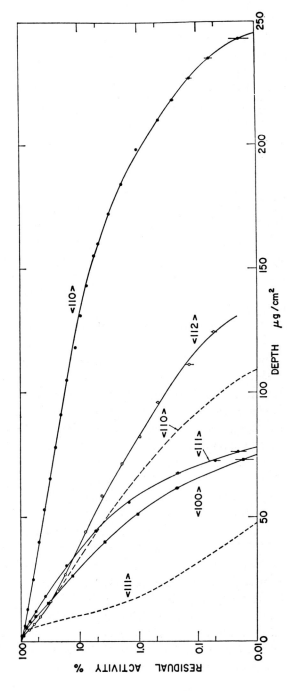

Fig. 2.51. Integral range distributions for 40 keV Xe$^{125}$ ions in silicon crystals compared with (dashed lines) the behaviour in aluminium (after Davies et al., 1964).

nique (see section 2.10) yielded the results shown in fig. 2.51. More extensive measurements have since been carried out by Dearnaley et al. (1968b, 1970) using a somewhat different anodic stripping technique which has been described by Wilkins (1968). In these studies the influence on the range distribution of ion energy, crystal orientation and temperature, crystal quality, ion dose and the type of ion implanted were investigated. A difficulty that arises in silicon to a much greater degree than in metal crystals is the relative ease with which the lattice structure is disrupted by ion bombardment (as we shall see in Chapter 3). It was therefore necessary in this work to ascertain the ion dose which could be allowed without significant modification of the channelled range distributions: typically, a dose of around $10^{12}$ to $10^{13}$ ions/cm$^2$ could be tolerated. By no means all of the implanted ions are radioactive, however, and the presence of numerous complex ion species in the beam can cause problems. Thus in the implantation of P$^{32}$ such species as P$^{31}$ H$^+$ and O$_2^+$ will generally dominate, and it is the total dose of these ions which must be held below the level that introduces serious damage. It is only with considerable care in avoiding source contamination and by skill in machine operation that these experiments can be carried out with good reproducibility and precision (Freeman, 1968, 1970).

Figure 2.52 shows the range distribution of low doses of P$^{32}$ ions implanted at room temperature along the most open channel, the $\langle 110 \rangle$ direction, in silicon. In this and the succeeding figures it should be noted that the range distributions are plotted in the differential form which is more appropriate for semiconductor device applications. The linear plot of the same data in fig. 2.52b shows more clearly how, at the higher energies, a well-defined peak appears, corresponding to the channelled component of the beam. Extending beyond this peak there is seen to be an exponential tail, with a slope which seems to be relatively independent of ion energy. It was suspected that this might be due to an anomalous diffusion mechanism (just as in the case of the 'supertailing' which occurs in tungsten) and in order to test this Dearnaley et al. (1970) channelled 40 keV P$^{32}$ ions at 77 °K into each of two silicon wafers cut from the same crystal. In one case a heavy dose of $10^{15}$ Ne$^{20}$ ions/cm$^2$ was implanted shortly afterwards, the crystal being maintained at 77 °K. The resulting range distributions, when measured as usual at room temperature, showed a significant difference in the slope of the exponential tail (fig. 2.53) and this can be explained only if the migration takes place slowly or not at all at 77 °K. We shall return in the next section (2.18) to consider this and other diffusion processes which may accompany ion

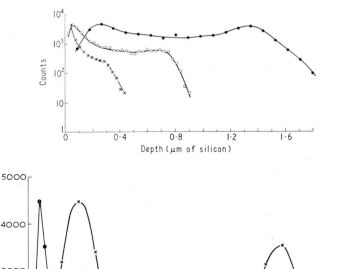

Fig. 2.52. (a) Variation of the range distribution of $P^{32}$ ions implanted along the $\langle 110 \rangle$ direction in silicon, with respect to ion energy: $\times$ 12 keV; $\bigcirc$ 40 keV; $\bullet$ 100 keV (after Dearnaley et al., 1968b). (b) The same data plotted on a linear scale.

implantation, and for the present we shall concentrate upon effect of the channelling process alone in determining ion range distributions.

By subtracting away the exponential 'supertail' in range distributions of the type shown in fig. 2.53 it is possible to derive a fairly well-determined maximum channelled range, which is shown plotted as a function of ion energy in fig. 2.54. Included in this plot are points derived by electrical conductivity measurements and junction-depth studies, and it can be seen that these all agree very well. The maximum range, $R_{\max}$, for channelled phosphorus shows an $E^{0.50}$ dependence on energy, as is expected

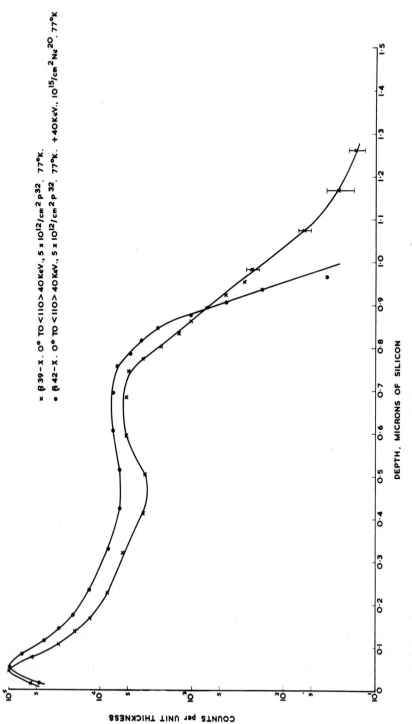

Fig. 2.53. Range distributions of P³² ions channelled at 40 keV along the ⟨110⟩ direction in silicon at 77 °K, followed in one case by a heavy dose of Ne²⁰ ions (from Dearnaley et al., 1970).

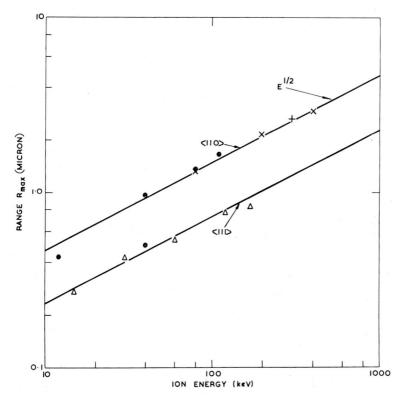

Fig. 2.54. The maximum channelled range of $P^+$ ions in silicon as a function of energy, for the $\langle 110 \rangle$ and $\langle 111 \rangle$ channels. The points ● are obtained from radioactivity measurements; those shown ×, △ and + by electrical measurements (from Goode et al., 1970).

theoretically on the basis of pure electronic stopping. The point at 12 keV in fig. 2.54 may be low either because of nuclear stopping effects or due to the effect of even a very thin surface contaminant film.

Next, fig. 2.55 shows the effect of crystal orientation on the ion penetration. As expected, at lower energies the profile changes less rapidly with the angle of inclination of the beam and channel axis than it does at higher energies. In fact, the rate of decrease of the channelled fraction with respect to angle increases by a factor of about 3.5 over the energy range 12 to 110 keV: this is close to an $E^{\frac{1}{2}}$ dependence. Note that (particularly at the higher energy) the depth of the shallow peak in the distribution varies with crystal orientation and, with good alignment, is a good deal deeper than the calculated median range in amorphous material. This was also observed in tungsten (Eriksson et al., 1967) and is attributed to a large number of

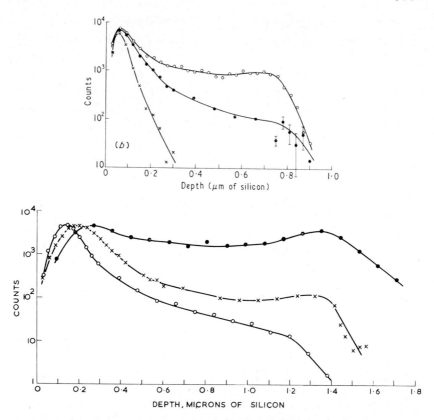

Fig. 2.55. The dependence of the range distribution of $P^{32}$ ions implanted into silicon upon the orientation of the crystal with respect to the $\langle 110 \rangle$ axis. (a) at 40 keV: ○ when aligned; ● 2° and × 8° misalignment. (b) at 110 keV: ● when aligned; ×, 0.43° and ○, 1° misalignment (after Dearnaley et al., 1968b).

'quasi-channelled' trajectories, in which ions are channelled for a little way before making a dechannelling collision. An interesting feature of the orientation dependence is that beyond a certain degree of misalignment the concentration profile does not greatly alter. Even under the least favourable conditions there is still a significant tail. The question as to whether this arises due to ions scattered into channelling directions or due to anomalous diffusion has been considered by Dearnaley (1970) who concluded that, for phosphorus ions in silicon, diffusion is probably the major cause. This point will be taken up in more detail in section 2.18. It is interesting to compare the misaligned distribution in single-crystal silicon with that in silicon which has previously been rendered amorphous by a heavy irradiation with energetic $Ne^{20}$ ions (Dearnaley et al., 1970). The results in this case (fig. 2.56) are very close to those predicted by the Lindhard, Scharff and Schiøtt formulae (1963).

The dose dependence of the distribution profile, referred to above, is illustrated in fig. 2.57 for 40 keV $P^{32}$ ions in silicon. Extrapolation of this and other similar data indicates an threshold for observable damage, by such ions, of about $10^{13}$ ions/cm$^2$ (see also fig. 3.22 below). Nelson and Mazey

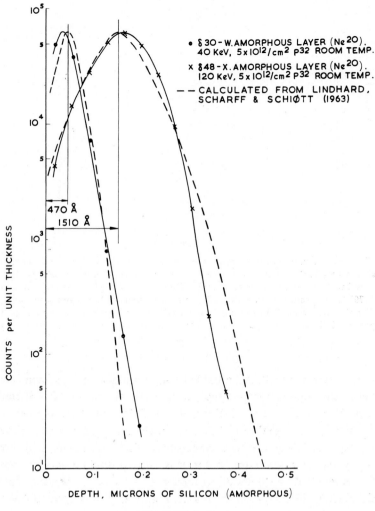

Fig. 2.56(a).

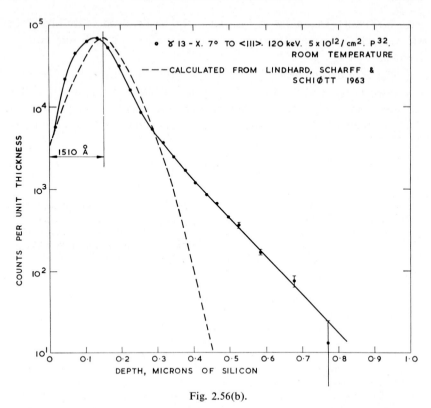

Fig. 2.56(b).

Fig. 2.56. (a) Experimentally-measured range distributions of $P^{32}$ ions implanted into silicon previously rendered amorphous by bombardment with energetic $Ne^+$ ions, compared with the theoretical predictions of Lindhard et al. (1963) (from Dearnaley et al., 1970) (b) The range distribution of $P^{32}$ ions implanted at 120 keV into a silicon crystal in a direction well removed from any major channelling axis or plane. Superimposed is the predicted distribution in amorphous silicon, calculated from the formulae of Lindhard et al. (1963) (from Dearnaley et al., 1970).

(1968) have shown that the stable ion-induced damage in silicon is critically dependent upon crystal temperature (see section 3.1.2.2). Therefore an experiment was performed (Dearnaley et al., 1968b) to test whether, at a dose around the damage threshold, the channelled fraction could be increased by implantation at an elevated temperature of 200 °C. Figure 2.58 shows that though some increase does occur, it is rather a small effect. The reason for this is revealed by studies of the influence of crystal temperature upon (low-dose) channelled range distributions. Figure 2.59 shows that at 400 °C there is a pronounced reduction in the channelled component and a large propor-

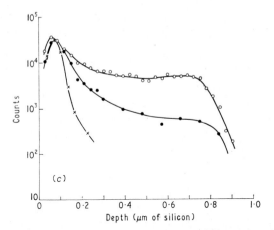

Fig. 2.57. The dose dependence of the range distribution of 40 keV P$^{32}$ ions implanted along the ⟨110⟩ direction into silicon (from Dearnaley et al., 1968b). ○ < 10$^{13}$ ions/cm$^2$; ● 9 × 10$^{13}$ ions/cm$^2$; × 7 × 10$^{14}$ ions/cm$^2$.

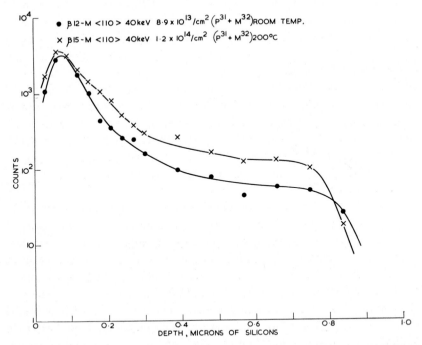

Fig. 2.58. The effect of temperature on 40 keV P$^{32}$ range distributions for a dose corresponding to the threshold of formation of an amorphous surface layer, i.e. about 10$^{14}$ ions/cm$^2$.

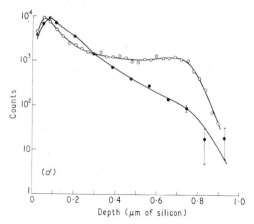

Fig. 2.59. The temperature dependence of 40 keV $P^{32}$ range distributions in silicon ($\bigcirc$) room temperature; ($\bullet$) 400 °C.

tion of the ions come to rest at depths only a little greater than the random or amorphous range. The maximum range, however, is not significantly affected. The range distribution at 400 °C is almost exponential in form, and this is reminiscent of the behaviour of range profiles in tungsten at high temperatures (Kornelsen et al., 1964). These observations are consistent with increased dechannelling as a result of thermal lattice vibrations.

The form of the range distribution obtained under channelling conditions varies considerably among different ion species. This was revealed when Dearnaley et al. (1970) measured the distributions of $Na^{24}$, $P^{32}$, $S^{35}$, $Cu^{64}$ and $Kr^{85}$ channelled along the $\langle 110 \rangle$ direction in silicon (fig. 2.60). In the case of the deeply-penetrating $Na^+$ and $Cu^+$ ions there is no evidence of a maximum range.

Now, it could be argued that such long range distributions, particularly for $Na^+$ and $Cu^+$ ions, might arise by some anomalous diffusion process, analogous to the 'supertailing' discussed above, rather than by a dynamic channelling mechanism. That diffusion is not responsible for the penetration of $Na^+$ and $Cu^+$ in silicon was demonstrated conclusively by Dearnaley et al. (1970) by an ion transmission experiment. In this, a collimated ion beam was directed on to a crystal of silicon, about 4 $\mu$m thick, mounted in a goniometer. Behind the crystal and in line with the beam was a channel electron multiplier plate (Adams and Manley, 1967) and phosphor screen assembly. Ions striking the channel plate release secondary electrons which are multiplied, with a gain of around $10^6$, and on emerging from the other face

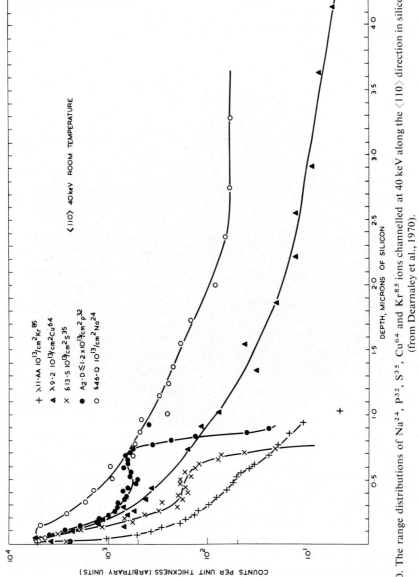

Fig. 2.60. The range distributions of $Na^{24}$, $P^{32}$, $S^{35}$, $Cu^{64}$ and $Kr^{85}$ ions channelled at 40 keV along the $\langle 110 \rangle$ direction in silicon crystals (from Dearnaley et al., 1970).

of the plate these electrons are imaged on the screen. Figure 2.61 shows the pattern of scintillations observed when 400 keV $Cu^+$ ions were directed parallel to the $\langle 111 \rangle$ crystal axis. If the crystal was rotated, this pattern disappeared and the angular width of the visible effects was in reasonable agreement with the prediction of eq. (2.105). It has since been shown by the

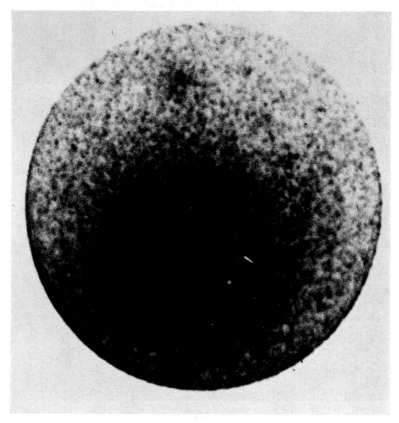

Fig. 2.61. A photograph of the distribution of scintillations observed in a channel electron multiplier plate and phosphor screen assembly detecting $Cu^{63}$ ions channelled through a 4 $\mu$m silicon crystal along the $\langle 111 \rangle$ axis. In this case the channel plate was approximately 9 cm behind the crystal; the screen diameter is 2.5 cm, and the incident ion energy was 400 keV (from Dearnaley et al., 1970).

same group that 150 keV $Cu^+$ ions can be transmitted through some 10 $\mu$m of silicon along the more open $\langle 110 \rangle$ channel.

Although no maximum range is discernible in $Na^{24}$ profiles at room temperature, there is evidence of one if the sodium is implanted at 77 °K (fig. 2.62) and for 40 keV ions this range is over 5 $\mu$m, compared with a Lindhard, Scharff and Schiøtt (1963) range in silicon of 0.07 $\mu$m. The variation in the penetration of different ion species channelled into silicon, as a function of their atomic number, is well correlated with the data of Eisen (1968) on the electronic stopping of these ions (see fig. 2.34). Small ions, such as $Na^+$ and $Cu^+$, can therefore channel to very great depths in crystals and from the device point of view this can be important, particularly

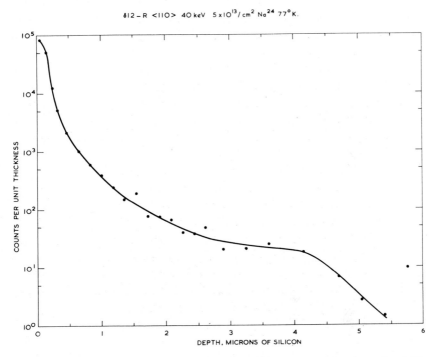

Fig. 2.62. The range distribution of $Na^{24}$ ions implanted at 40 keV along the $\langle 110 \rangle$ direction in silicon at 77 °K (from Dearnaley et al., 1970).

in near-intrinsic semiconductors. Long tails in an implanted dopant distribution will then cause a deep junction which could be troublesome, for example, in a nuclear radiation detector. Such very deep junctions have been observed by Meyer (1969), using very low incident ion energies.

In this case, and in the data shown in figs. 2.60 and 2.62 it is possible that some anomalous diffusion is occurring as well as enhanced channelled penetration. Thus, if a silicon crystal is misaligned the distribution of 40 keV $Na^{24}$ ions shows detectable radioactivity at a depth of no less than

1.4 $\mu$m. It is interesting to note that those ions (e.g. $Li^+$, $Na^+$, $Cu^+$) which channel so readily in silicon are just those which exhibit fast interstitial diffusion. McCaldin (1965) has pointed out that, although the paths followed by the ion are different in these two processes, the controlling factor in each is the energy needed for movement from one interstitial site to the next. We now know (Cheshire et al., 1968) that it is the ionic size, or rather the electron distribution in the outer regions of the atom, which governs the electronic stopping cross-section and the channelled range. It is reasonable that this parameter should strongly affect the activation energy for interstitial migration and, indeed, the probability of the ion favouring an interstitial rather than a substitutional site in the lattice.

The choice of ion species for implantation can thus, by combination of channelling and diffusion, give rise to very different dopant distributions. Depending upon what is required, one may therefore wish to choose $As^+$ rather than $P^+$ in doping silicon, or $Si^+$ rather than $Ge^+$ in doping a III–V compound. For the present, however, it is impossible to calculate what the range distribution will be under a given set of circumstances and so more data is needed for a greater variety of implanted ion species.

So far we have considered the method whereby radioactive ions are implanted, but it is quite feasible, as an alternative, to implant the corresponding non-radioactive species and activate it by neutron irradiation in a reactor. Fairfield and Crowder (1969) have in this way studied the range distributions of $P^{31}$ and $As^{75}$ in silicon, using anodic stripping to section the specimen after activation. In this method the silicon itself becomes radioactive through the $Si^{30} + n \rightarrow Si^{31}$ reaction, but the half-life for the decay of this to $P^{31}$ is only about 2.4 hours so that, after a short time, only the longer-lived activity remains significant. So far this type of experiment has been restricted to fairly substantial ion doses ($> 10^{15}$ ions/$cm^2$), and so has not allowed the study of channelled range distributions below the threshold of severe damage to silicon crystals. The method does, however, extend the range of implantation facilities by means of which range measurements can be made.

The above methods, using as they do radioactive tracer ions of convenient half-lives, are restricted to certain isotopes, though it is fortunate that this list includes such dopant species as $P^{32}$, $As^{76}$, $Sb^{122}$, $In^{114}$, $Bi^{210}$, $S^{35}$, $Zn^{65}$, $Se^{72}$ and $Te^{132}$. A method which is free from this restriction has recently been developed by Cairns and Nelson (1970) and is based upon measurement of the characteristic X-rays produced from an implanted specimen when bombarded with heavy ions such as $Kr^+$. The method utilises the fact that the thresholds for X-ray production depend upon details of the electron shell configuration, and therefore vary from one target element to another (fig. 2.63). It is often feasible to choose a bombardment energy such that the silicon threshold ($\sim 95$ keV for $Kr^+$) is not exceeded. Since the X-ray yield in this region varies so strongly with ion energy, essentially all the yield comes from the surface layer, $\sim 50$ Å deep. Thus, if the yield is measured at each successive stage of sectioning the speci-

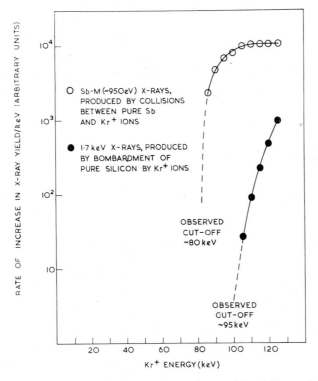

Fig. 2.63. X-rays generated by $Kr^+$ ion bombardment of Si and Sb, as a function of energy (from Cairns et al., 1970).

men (e.g. by anodic stripping) the plot of X-ray yield corresponds to the range distribution of the emitting ions. (Note that in this respect the method differs from the electron-induced X-ray emission technique described by Legrand et al. (1970) for the study of implanted metal specimens.) An example of the use of this method was the determination, by Cairns, Holloway and Nelson (1970a), of the distribution of antimony implanted into silicon by the use of 100 keV $Kr^+$ ions to excite antimony M X-rays in a selective manner (fig. 2.64). The same group has also succeeded in measuring implanted boron concentration profiles in silicon, this time using 100 keV A$^+$ ions. The sensitivity in this case, for such a light element, is not so great and the boron dose required was $10^{17}$ ions/cm$^2$. Once again, this dose produces severe damage to the crystal and it is therefore impracticable to examine channelled range distributions this way in materials such as silicon, though of course this would not be the case in, for example, metal crystals. The

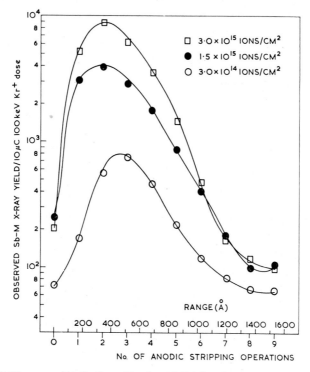

Fig. 2.64. The range distribution of implanted Sb$^+$ in silicon, determined by measurement of characteristic X-rays (from Cairns et al., 1970).

experimental values of median projected range obtained by Cairns, Holloway and Nelson (1970a) agreed very well with the values predicted by the Lindhard, Scharff and Schiøtt (1963) theory.

This result contradicts the finding of Lecrosnier and Pelous (1970) who used very different methods of range determination for boron ions in silicon. These workers employed an ion microprobe analyser developed by Castaing and Slodzian. In this instrument a primary argon ion beam is used to sputter a small hole in the boron-implanted sample, and the secondary ions ejected are mass-analysed and recorded as a function of time. The depth of the hole is measured, for example by interferometric methods, and by assuming a uniform rate of sputtering a plot of ejected atoms of a given atomic weight is obtained as a function of depth. For many ions, including boron, this method has a high sensitivity and Lecrosnier and Pelous (1970) were able to work with samples implanted with only $10^{15}$ boron ions/cm$^2$: this is low enough for channelling studies to have been made, though in the work so far reported only mis-oriented samples were investigated. An interesting feature of the results is the considerable departure from agreement with the Lindhard, Scharff, Schiøtt theory, the observed mean penetration being some 30 per cent less than predicted. However, Lecrosnier and Pelous obtained very similar results by a method involving anodic stripping, and a similar ratio of observed to theoretical stopping was observed by Seidel (1971) in capacitance–voltage measurements on boron-implanted silicon. Further work to determine boron distributions in silicon with the aid of an ion microprobe analyser has been carried out by Gittins et al. (1972), who examined the limitations of the technique besides confirming the enhanced degree of electronic stopping of boron ions first reported by Eisen et al. (1970).

The ion microprobe analyser has greatest sensitivity for the alkali metal ions and Ca, Al, Cr and Ga. It is therefore expected to be particularly valuable for the measurement of channelled range distributions for these elements. The method has the advantage that, in principle, it can be applied to any material so long as the calibration of the sputtering rate can be carried out. Also, it may be feasible to make measurements of implanted samples maintained at low temperatures, though heating by the primary ion beam ($\sim 20$ W/cm$^2$) is clearly a problem. Calibration of the sensitivity of the instrument must be carried out with samples of a known, uniform composition.

Next we turn to a different technique which has often been utilised to obtain information regarding the penetration of ions in crystals. In semiconductors it is possible to study the range distribution of dopant ion species by measurement of the junction depth after annealing the implanted specimen. The position of the junction can be revealed by lapping the crystal at a small angle to its surface, or by grinding a shallow groove of known shape. Then a metallographic staining technique may be used to produce a visible deposit on material of only one conductivity type: Kleinfelder et al. (1968) used a stain consisting of concentrated HF solution to which were added a few drops of either $HNO_3$ or copper nitrate solution, applied under bright illumination. The procedure consists in implanting a specimen of known impurity concentration with a measured dose of ions. It is then assumed (i) that the junction corresponds to the point where two types of impurity compensate each other and (ii) that the staining technique delineates this point. It is possible, however, that the annealing does not leave all the implanted ions in an electrically active state: a comparison between electrical measurements and radioactive tracer studies has enabled this effect to be investigated. Since a small potential always exists between the two sides of the junction, a depletion layer extends a short distance on either side of it, penetrating more deeply into the less highly-doped substrate. It is likely that the stain marks the boundary of this layer rather than the junction. Note that these two effects tend to cancel each other, and it may be for this reason that the few junction depth determinations for well-channelled dopant ions agree with the results of other techniques (see fig. 2.54).

An alternative, but more time-consuming method is to measure the distribution of electrically-active implanted impurities by differential electrical conductivity and Hall mobility measurements, combined with anodic sectioning. These methods will be discussed more fully in Chapter 5, but here we note that the range distributions of channelled ions in silicon measured in this way have agreed well with the radioactive tracer and X-ray excitation methods (Bader and Kalbitzer, 1970). However, there are deviations in that region of the distribution profile corresponding to non-channelled ions. Here the electrical activity, particularly at high ion doses, may fall significantly below 100 per cent, and the concentration of electrically-active ions can decrease with increasing ion dose (Galaktionova et al., 1968). This effect has since been correlated with the distribution of un-annealed defects in the heavily-damaged lamina within which non-channelled trajectories terminate. On the other hand, damage is less severe along a channelled trajectory (see section 3) and anneals more readily.

Another type of electrical measurement of ion ranges in semiconductors has been by capacitance-voltage data in an annealed implanted wafer which is thin and has a surface-barrier contact on the opposite face. Under an applied voltage a depletion layer moves into the silicon, under the surface-barrier, to a depth determined by the impurity profile. In this way Bower et al. (1966) observed a very deeply-penetrating distribution of 20 keV $Sb^+$ ions implanted into silicon. fig. 2.65 (b). What was surprising was that this distribution, extending as deep as 10 $\mu$m, was independent of crystal orientation, temperature, annealing and surface condition. Some question

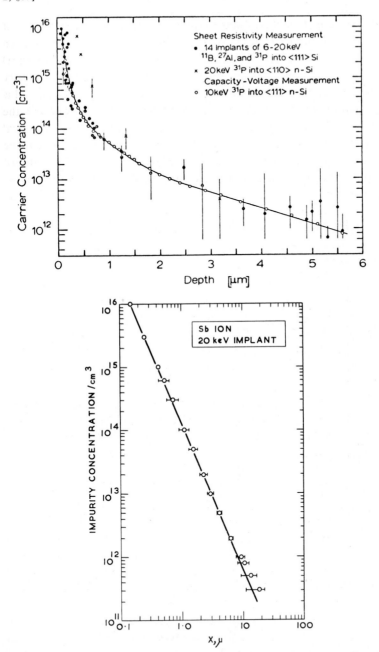

Fig. 2.65. (a) Carrier concentration distributions following B$^{11}$, Al$^{27}$ and P$^{31}$ implantation of high-resistivity silicon, by sheet resistivity measurements (from Bader and Kalbitzer, 1970). (b) Carrier concentration distribution following 20 keV Sb$^+$ ion implantation of silicon, by capacitance–voltage measurements (after Bower et al., 1966).

has since been thrown on the validity of this technique (Kennedy et al., 1968) since what is measured is not the ionised impurity profile but the carrier distribution. If the impurity concentration, $N$, varies rapidly with distance, the majority carrier concentration will not follow it exactly, the condition for quasi-neutrality being that the relative changes of $N$ are small within a distance equal to the Debye length corresponding to a local carrier concentration $n = N$. As Kennedy et al. (1968) show, the consequences of failure of this criterion are complex and unpredictable. They would apply also to the other techniques in which it is the carrier distribution which is measured, and as a result there remains a degree of uncertainty about the results of electrical measurements of channelled range distributions. It is hoped that

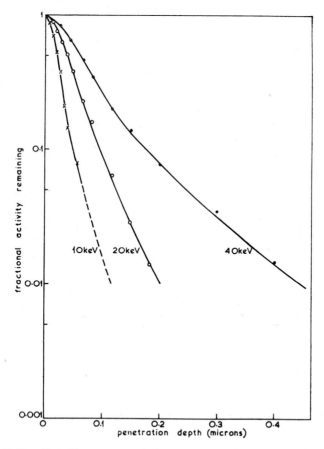

Fig. 2.66. The effect of ion energy on the integral range distribution of $Kr^{85}$ ions implanted in GaAs along the $\langle 100 \rangle$ axis( from Whitton et al., 1969).

further comparisons with the true (e.g. radioactive tracer) determination of impurity concentration profiles will resolve this doubt.

In all of these electrical measurements of ion penetration annealing is necessary before it can be assumed that the results can be interpreted in terms of an ionic concentration, and this demands a reasonable understanding of the damage behaviour. It is obviously impossible to study whether the dopant is migrating during the anneal by this method. From the practical point of view of electrical devices, however, it is the carrier concentration profile after annealing which is of prime importance.

Having concluded this review of range distribution measurements in single-crystal silicon, we next consider the more recent study of ion penetration in compound semiconductors. We shall see that this has already broadened our understanding of the energy loss of channelled ions. A comprehensive study of range distributions in the III–V compound semi-

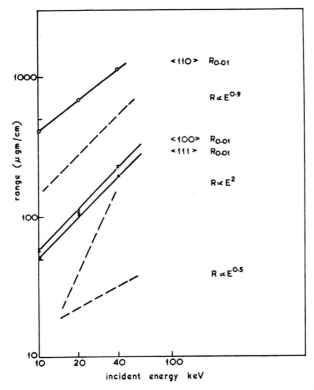

Fig. 2.67. Range-energy relationship at a level of 1 per cent activity remaining, for $Kr^{85}$ ions implanted into GaAs crystals (from Whitton et al., 1969).

conductors GaAs and GaP has been made by Whitton et al. (1969) and Whitton and Carter (1970). Vibratory polishing was employed to section the implanted specimens, since anodic oxidation has so far not been found to be feasible in these materials. In the first experiments, $Kr^{85}$ ions were implanted under varied conditions, and fig. 2.66 shows the effect of ion energy on the integral range distribution for incidence along the $\langle 110 \rangle$ axis. The range–energy relationship for different channels is interesting, since it departs markedly from the $E^{\frac{1}{2}}$ behaviour anticipated for electronic stopping (fig. 2.67). That GaAs is rather easily damaged during ion bombardment was revealed both by the visible appearance of a golden surface film at low doses, and by the dose dependence of the channelled range distribution (fig. 2.68).

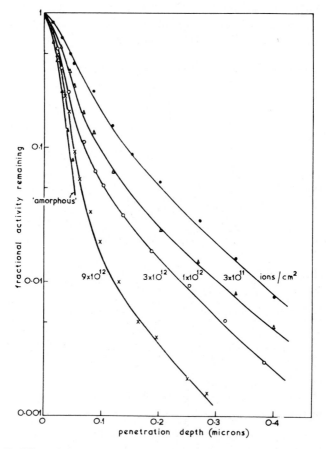

Fig. 2.68. Effect of bombardment dose on the integral range distribution of 40 keV $Kr^{85}$ ions implanted along the $\langle 111 \rangle$ direction in GaAs (from Whitton et al., 1969).

In subsequent experiments, (Whitton and Carter, 1970), $S^{35}$ and $Na^{24}$ ions were implanted, with strikingly different results (fig. 2.69). When injected along the $\langle 110 \rangle$ axis, one sees, just as in silicon (Dearnaley et al., 1970), that while the $S^{35}$ and $Kr^{85}$ range distributions show well-defined maximum ranges, none is discernible for $Na^{24}$. The maximum range of

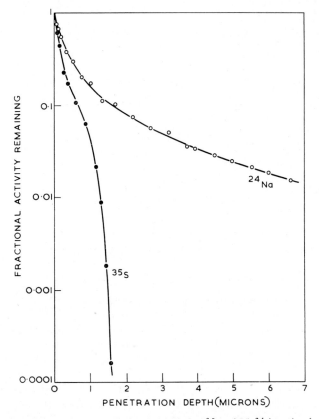

Fig. 2.69. Integral range distributions of 40 keV $S^{35}$ and $Na^{24}$ ions implanted along the $\langle 110 \rangle$ direction in germanium (from Whitton and Carter, 1970).

$S^{35}$ ions, along the $\langle 110 \rangle$ channel varies with energy as $E^{0.43}$–not far from the expected $E^{0.5}$. An interesting behaviour becomes apparent when the channelled range distributions are measured along different axes (fig. 2.70): $S^+$ and $Kr^+$ ions interchange with respect to their degree of penetration. Whitton and Carter (1970) argue that in the $\langle 111 \rangle$ direction where the channels are relatively small and the rate of dechannelling consequently

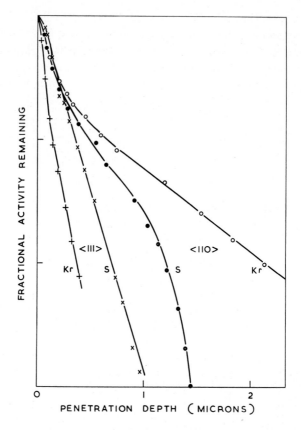

Fig. 2.70. The effect of channel size on the integral range distributions of 40 keV $S^{35}$ and $Kr^{85}$ ions implanted into GaAs (from Whitton and Carter, 1970).

high, nuclear stopping will be the dominant process, as evidenced by the $E^{0.9}$ energy dependence of $R_{max}$. Under these circumstances the penetration of $S^+$ ions exceeds that of the heavier $Kr^+$. In the more open $\langle 110 \rangle$ channels, however, the dechannelling rate is lower, and electronic stopping dominates, hence $R_{max}$ varies as almost $E^{0.5}$ and $S^+$ ions are known (Eisen, 1968) to have a high electronic stopping cross-section. Radiation damage develops at a rather lower rate during bombardment with $S^+$ ions than with $Kr^+$ ions, as shown by the method of Rutherford scattering of helium ions from implanted specimens (fig. 2.71). Whitton and Carter (1970) found that implanted germanium crystals showed ion range distributions identical to those in GaAs, with regard to ion energy, crystal orientation, ion species

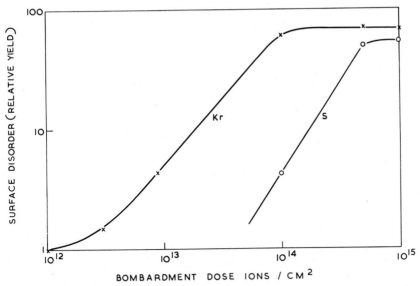

Fig. 2.71. The increase of disorder in GaAs crystals bombarded along the $\langle 110 \rangle$ direction with $Kr^{85}$ and $S^{35}$ ions, as a function of ion dose, and measured by the Rutherford back-scattering technique (from Whitton and Carter, 1970).

and the onset and saturation of damage. This demonstrates that the band gap of the semiconductor has little or no influence on these phenomena.

Channelled range distributions have been measured in indium antimonide by Wilkins and Dearnaley (1970) for radioactive phosphorus and sulphur ions. Anodic stripping proves very straightforward in this material and the dose necessary to disorder the lattice is approximately an order of magnitude greater than in silicon, other things being equal (fig. 2.72). All the range distributions show long penetrating tails and it seems likely that anomalous diffusion is occurring here as in silicon (see section 2.18). There are marked differences between the channelled range distributions of $P^+$ ions in InSb and in Si, as fig. 2.73 shows, and we can perhaps explain this behaviour by recourse to the Firsov model of electronic stopping, as adopted by Cheshire et al. (1968). In this model, each binary collision is treated in a manner which is symmetrical with regard to the moving ion and struck atom, with the sole difference that (in covalently-bonded materials) the latter is not ionized. Thus one expects the strong $Z_1$ dependence of electronic stopping (fig. 2.34) to be reflected in a similar $Z_2$ dependence, other things being equal. Difficulties arise here because factors such as lattice structure, lattice dimensions, and thermal lattice vibration also have

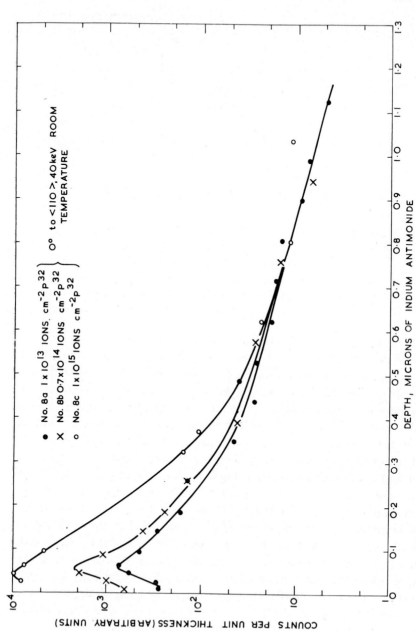

Fig. 2.72. Range distributions of different doses of 40 keV P³² ions implanted along the ⟨110⟩ direction in InSb (from Wilkins and Dearnaley, 1970).

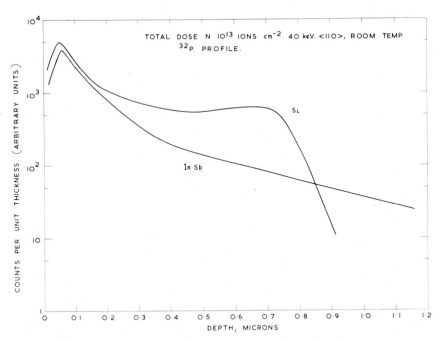

Fig. 2.73. Comparison of range distributions of P[32] ions implanted at 40 keV along the ⟨110⟩ direction in silicon and InSb (from Wilkins and Dearnaley, 1970).

a profound effect on the range distribution. However, the series Si, Ge and InSb offers a set of crystals with a common structure and comparable lattice dimensions. The maximum range of 40 keV S[35] ions channelled along the ⟨110⟩ direction is approximately twice as great in Ge as it is in Si, despite the higher density and atomic number of Ge (Dearnaley et al., 1970; Whitton and Carter, 1969). Ge, with $Z_2 = 32$, lies close to a minimum in the curve of electronic stopping cross-section against atomic number (fig. 2.34), while Si, with $Z_2 = 14$, lies well away from a minimum. Indium antimonide, with Z-values of 49 and 51, is another example of an absorber lying at a minimum in this curve, and we believe that this is why the ion penetration is large. On this hypothesis, Wilkins and Dearnaley (1970) made an interesting test of the symmetry of ion and absorber in determining the form of range distributions. Silicon ($Z = 14$) and phosphorus ($Z = 15$) are well away from a minimum in the electronic stopping cross-section curve (fig. 2.34), while Cu ($Z = 29$), In and As lie at minima. The experiment consisted in comparing the penetration of P[32], in InSb with that of Cu[64] in Si, and as fig. 2.74 shows, there is a remarkably close agreement. Two factors

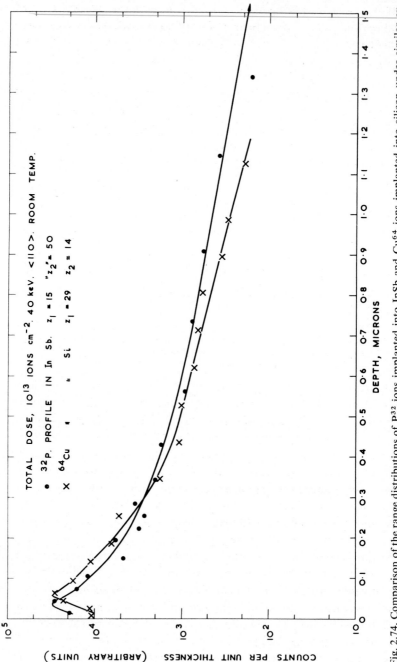

Fig. 2.74. Comparison of the range distributions of $P^{32}$ ions implanted into InSb and $Cu^{64}$ ions implanted into silicon under similar experimental conditions (from Wilkins and Dearnaley, 1970).

must be taken into account: strictly the comparison should be made at equal ion velocities rather than the same energies, and the material densities should be introduced. Wilkins and Dearnaley estimate that these two factors will tend to cancel out, fortuitously, in this case.

Very little work has so far been carried out in the II–VI compound semiconductors, but nevertheless some interesting results have been reported on range distributions in CdS. Eldridge et al. (1970) measured the penetration of 25 keV $Bi^{210}$ ions into CdS platelets using a chemical etching technique for stripping. Figure 2.75 shows their results, in terms of the residual activity

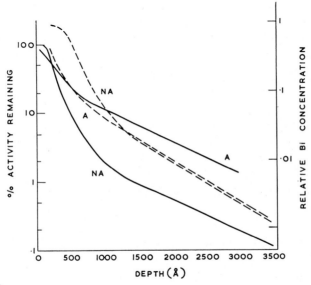

Fig. 2.75. Integral (——) and differential (– – –) range distributions of 25 keV $Bi^{210}$ ions implanted into CdS crystals. In case A, the ions were aligned with the ⟨0001⟩ axis; in case NA the crystals were misaligned by 5° (from Eldridge et al., 1970).

as a function of depth. An exponential tail is apparent in both these curves, highly reminiscent of the anomalous diffusion or 'supertails' in Si and W. Armitage (1970) has obtained confirmation of the very deep penetration of low-energy Bi ions in CdS by a totally different technique, namely the back-scattering of energetic $N^{14}$ ions from the implanted crystal. The spectrum of back-scattered $N^+$ ions exhibited a tail corresponding to Bi centres deeper than 1000 Å, i.e. an order of magnitude greater than the Lindhard, Scharff, Schiøtt (1963) range. Armitage also found that, even after a dose of $10^{16}$ bismuth ions/cm$^2$ at 25 keV the surface of the crystal retained a high degree of order, as judged by the back-scattered yield of nitrogen ions. Presumably the more ionic nature of the CdS lattice renders it less readily damaged than Si or Ge, at least at room temperature. These results are in agreement with the work of Olley et al. (1970) on the optical behaviour of implanted CdS.

## 2.18. Anomalous diffusion accompanying ion implantation

In the course of the previous section, we have learned that both channelled and non-channelled distributions of implanted ions may be modified by the diffusion of atoms after they have been reduced to thermal velocities, resulting in the so-called 'supertails'. In the sense that diffusion, in silicon, of such species as phosphorus and sulphur does not normally occur at room temperature, we have termed this diffusion anomalous, but this does not imply that we are wholly ignorant of its mechanism. It is, furthermore, important to distinguish a second type of 'anomalous' diffusion, properly called radiation-enhanced diffusion, which may have important consequences in the practical application of ion implantation. We shall begin with this second process, the existence of which has been known for the best part of a decade (Pfister and Baruch, 1962; Strack, 1963).

Impurity diffusion in crystals, at least for most electrically-active dopants in semiconductors, is believed to be by interaction between vacancies (or vacancy complexes) and substitutional impurity atoms. This is closely related to the problems of self-diffusion, yet even in the case of that well-studied material, silicon, the mechanisms of self-diffusion have yet to be fully explained (Kendall and de Vries, 1969), and the di-vacancy is probably also an important participant. When a crystal is subjected to ionizing radiation, additional vacancies and interstitials are created, in concentrations which may far exceed those produced thermally. These excess vacancies may therefore enhance the rate of diffusion, particularly at relatively low temperatures. There are two cases to consider: (i) the vacancy enhancement may be brought about by the implantation of the impurity species itself, or (ii) the vacancies may be created by some other ionizing radiation, e.g. protons or electrons, in which case the impurity atoms may have been previously introduced either by diffusion or by implantation.

The first of these two cases has been treated by Tsuchimoto and Tokuyama (1970) in terms of the diffusion equation of the vacancies:

$$\frac{\partial N_v(x, t)}{\partial t} = D_v \frac{\partial^2 N_v(x, t)}{\partial x^2} - \frac{N_v(x, t)}{\tau_v} + g(x) \qquad (2.128)$$

where $N_V(x, t)$ is the vacancy concentration, $D_V$ is their diffusion coefficient, $\tau_V$ is the recombination lifetime of vacancies and $g(x)$ is the generation rate (assumed constant with respect to time). Neglecting thermal generation, at the temperatures of interest, $g(x)$ may be estimated from the ion range, the straggling, the amount of energy released in the form of nuclear recoils, and the displacement energy. The diffusion coefficient $D_i$ of substitutional impurities may normally be written

$$D_i(T) = S(T) \cdot N_V(T) \qquad (2.129)$$

where $S(T)$ is a function of the migration energy for the impurity, the atomic jump frequency and the crystal lattice constants. Assuming that this parameter would be the same in the case of radiation-enhanced and normal diffusion. Tsuchimoto and Tokuyama (1970) chose a value, for phosphorus in silicon, of $S = 2.1 \times 10^{-22} \exp(-1.16/kT)$. The vacancy concentration $N_V(x)$ and diffusion coefficient $D_i$ could then be calculated as a function of depth, and the final ion distribution after a given time could then be deduced. Figure 2.76 shows some results of this calculation.

Enhanced diffusion of antimony in ion implanted silicon has been studied by Namba et al. (1970), who used neutron activation of the crystals together with anodic stripping for determination of concentration profiles. Good agreement was found between the experimental results and a calculation, similar to that outlined above, assuming that diffusion occurs by a vacancy mechanism. Rather than predict the concentration profile from an assumed behaviour of the diffusion coefficient, as did Tsuchimoto and Tokuyama, Namba et al. (1970) chose to extract the effective diffusion coefficient from the experimental data, with the interesting result that it remained constant, at $1.1 \times 10^{-15}$ cm²/sec between 500 and 800 °C (fig. 2.77). It appears possible that the deep penetration of Sb⁺ ions implanted at 300 °C to 500 °C into silicon (Bower et al., 1966) arose from the same mechanism, if indeed the whole effect was not spurious (see below).

Namba et al. (1970) further observed that the enhancement of diffusion of antimony during annealing, after a room temperature ion implantation, was the same as that produced by a high temperature implantation. This they ascribed to the release of vacancies, trapped into complexes and clusters, during the annealing process. In both cases the vacancies diffuse more rapidly than the ion-implanted impurities.

In the second class of experiments, vacancies are created by proton or inert gas ion bombardment of a previously implanted or diffused specimen. The calculation of $N_V(x, t)$ proceeds as above, but the final ion distribution

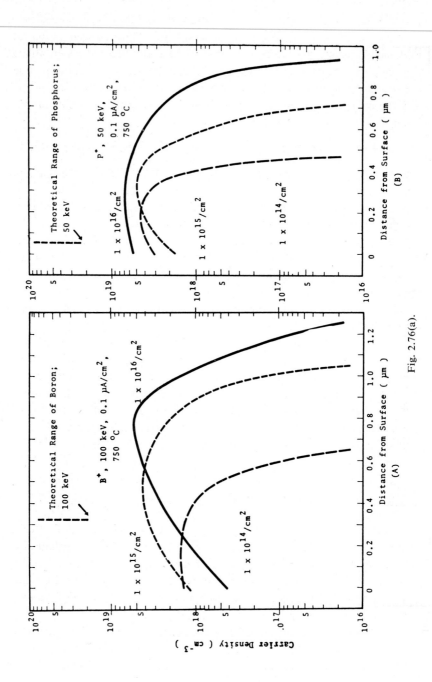

Fig. 2.76(a).

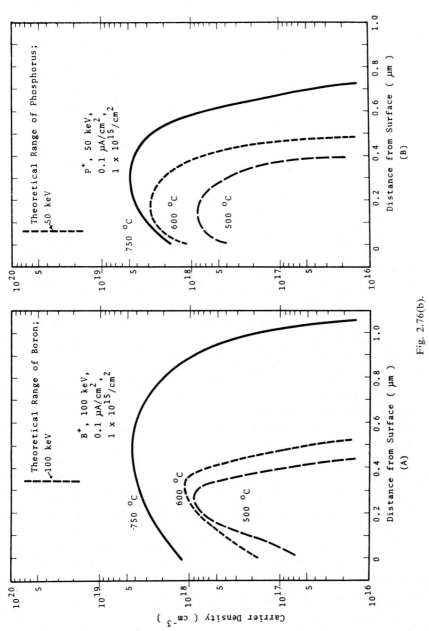

Fig. 2.76(b).

Fig. 2.76. Variation of the carrier distribution, as a function of depth, with (a) ion dose and (b) substrate temperature, as calculated by Tsuchimoto and Tokuyama (1970).

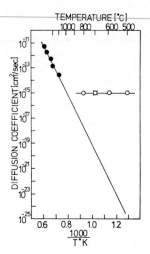

Fig. 2.77. The temperature dependence of the diffusion coefficient of Sb in silicon: ○ indicates the diffusion coefficient of Sb during high-temperature implantations and × indicates the diffusion coefficient during annealing at 700 °C following room temperature implantation. The diffusion coefficient for normal thermal diffusion is shown by ● (from Namba et al., 1970).

will depend upon the initial concentration and distribution of the impurity with respect to the ion range. D. G. Nelson et al. (1969) and Tsuchimoto and Tokuyama (1970) have examined this case and the latter authors find, for example, that proton irradiation, with a 1 $\mu$A beam of 50 keV energy to a dose of $10^{17}$ protons/cm$^2$ produced an increase in the junction depth of a boron-diffused silicon specimen, held at 750 °C, from 0.55 $\mu$m to 1.44 $\mu$m. Gibbons (1970) differs from previous authors in using much smaller proton beam currents, typically $10^{-8}$ A per cm$^2$. One of the advantages of these methods of enhancing ion penetration is that, beyond the ion range, the substrate is essentially perfect and so the junction, in a semiconductor, will be formed by material diffusing substitutionally into this undisturbed lattice. As a result, the electrical activity of the dopant may be expected to be higher than in the case, for example, when Group III dopants are implanted into silicon, since the opportunities for certain defect interactions will probably be reduced. Ion pairing and cluster formation is still feasible, and it will be interesting to compare electron micrographs of material doped by proton-enhanced diffusion with those of material implanted in the conventional manner.

The diffusion mechanisms we have considered so far in this section involve hot implantations, typically at temperatures of 500 to 800 °C, i.e.

in the range immediately below normal diffusion temperatures. The second type of anomalous diffusion may, however, take place at much lower temperatures, and for this reason, may be difficult to avoid. Besides the purely substitutional and interstitial diffusion mechanisms in crystals there exists what has been called the interstitial-substitutional or dissociation diffusion process. Some impurities such as Au in silicon or Zn in GaAs have a finite probability of occupying an interstitial lattice site. The effective diffusion coefficient is controlled by the diffusion coefficient of the interstitial multiplied by the fractional probability of interstitial occupancy. The migrating interstitials interact with vacancies to create substitutional impurities. These vacancies may be entering from surfaces, or may be released from internal sites such as dislocations, vacancy clusters, etc. at high temperatures, and the final impurity distribution will depend upon the relative vacancy concentration and at least two diffusion coefficients (for interstitials and vacancies). Ion implantation is a non-equilibrium process and it is possible, therefore, to introduce impurities into sites which would be highly improbable

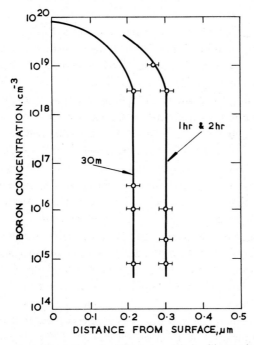

Fig. 2.78. A proton-enhanced boron diffusion profile in silicon, obtained by bombarding a boron-diffused sample at 700 °C with about $5 \times 10^{11}$ protons/cm$^2$/sec at an energy of 10 keV (after D. G. Nelson et al., 1969).

under conditions of thermal equilibrium. This is the mechanism proposed by Davies and Jespersgaard (1966) to explain the so-called 'supertails' in implanted tungsten crystals. In the case of tungsten all species, except $^{222}$Rn, show an exponential 'supertail' in the channelled range distribution. Can such an explanation account for the tails observed in the distribution of phosphorus implanted into silicon? Apart from the fact that the phosphorus interstitial has never been observed in electron spin resonance or any other studies on irradiated phosphorus-doped silicon, this might well be the diffusing species. An alternative possibility, suggested by Dearnaley et al. (1971), is the phosphorus-vacancy pair or $E$-centre, known to be mobile in silicon at temperatures as low as 150 °C. Yet another possibility is the phosphorus-divacancy combination, thought to be even more mobile. Defects such as the $E$-centre can exist in different charge states, each with a different activation energy for migration. Picraux and Vook (1971) have pointed out that the intense ionization which accompanies ion implantation may result in the production of more mobile defects, and suggest that this may explain the difference in temperature for annealing during and after implantation. In the same way, ionization stimulated migration of defects away from the ion track may be involved in enhanced diffusion in silicon. Under such ionizing conditions interstitial phosphorus may migrate by the mechanism involving alternate capture and loss of electrons, as proposed by Bourgoin and Corbett (1972).

Whatever the migrating species, the final distribution of material will depend upon the density of trapping centres, and if we assume a uniform distribution of such sites it is easy to see that an exponential distribution of migrant species will result (Davies and Jespersgaard, 1966). The slope of this exponential will obviously depend upon the trap concentration, and it is not surprising that this may vary considerably from crystal to crystal (fig. 2.75). For this kind of reason, preliminary data (Dearnaley, 1970) erroneously suggested that $S^{35}$ implanted into silicon shows no exponential tail. In fact, $P^{32}$ and $S^{35}$ show very similar tails when implanted into the same crystal of silicon as shown in fig. 2.79. There is no evidence, however, of a tail when $S^{35}$ is implanted into germanium or gallium arsenide (Whitton and Carter, 1970) though one is apparent in indium antimonide (Wilkins and Dearnaley, 1970).

The extent of an exponential tail is characterised by that distance, $\lambda$, over which the impurity concentration falls by the exponential factor, $e$. Whereas some early indication was obtained (Kleinfelder et al., 1968) for an $E^{\frac{1}{2}}$ dependence of $\lambda$ upon ion energy, suggesting that ion channelling is

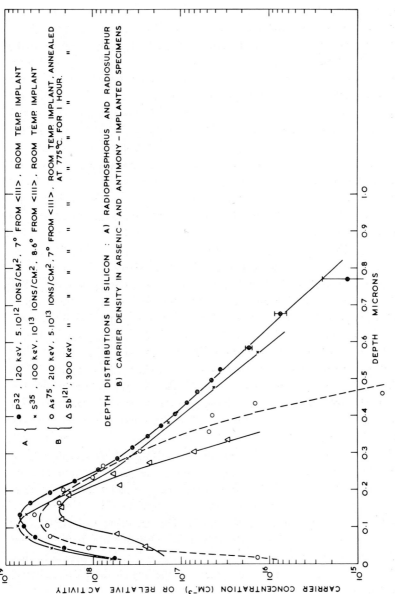

Fig. 2.79. Distribution of 120 keV P³² and 100 keV S³⁵ implanted into misoriented crystals cut from the same ingot of silicon, showing a very similar extent of exponential tail. Also shown are the results of electrical measurements of the donor distribution following implantation of 210 keV As⁷⁵ and 300 keV Sb¹²¹ ions. There is distinctly less tail in these cases.

involved, Moline (1970) finds essentially no variation of $\lambda$ with $E$, between 100 and 600 keV, for phosphorus in silicon. Recent work by Seidel (1971) shows that boron behaves similarly. We have seen, in the previous section, that Bi implanted into CdS shows what is apparently a similar exponential tail (Eldridge et al., 1970), and Wilkins and Dearnaley (1970) reported the same effect in InSb, implanted with $P^{32}$.

It is interesting to enquire whether the characteristic length, $\lambda$, is the same (in a given crystal) in the case of a channelled and a non-channelled implantation. There is insufficient evidence at the present time to answer this question, but certainly the values do not differ by much more than the experimental accuracy. Moreover, crystals which show large values of $\lambda$ in a channelled implantation also do so under mis-aligned conditions. However, factors such as dislocation density which may influence diffusion may also modify the channelling behaviour. Recent attempts (Blood et al., 1973) to determine the diffusion mechanism by implantation of various species, such as $Sn^+$, designed to trap defects such as vacancies, have yielded negative results. Scattering of ions into obliquely-inclined channels is therefore favoured as the cause of the tails. This would demand a series of collisions probably involving misplaced atoms at defects rather than scattering by a single atom on a lattice site. An ion moving on after such an event is likely to be 'blocked' from any channelling direction by nearest-neighbour interaction, and the magnitude of its transverse energy relative to a channel axis will preclude it being steered into such a direction. What channelling can influence, however, is the likelihood of an ion coming to rest in, say an interstitial position in a region of the crystal which has suffered relatively little radiation damage. Hence one may expect the fraction of implanted ions present in the tail to vary with crystal orientation, and this is the case.

We now have the hypothesis that $\lambda$ is controlled by the density of those defects, such as vacancies, which will act as traps for anomalously diffusing ions. A pre-bombardment with an inert ion species should therefore modify $\lambda$, and such an effect is seen in fig. 2.80 for the case of sulphur implanted into silicon (Dearnaley, Wilkins and Goode, 1971). Of course, if the silicon is rendered completely amorphous by previous ion bombardment (fig. 2.56) we have already seen that a true Gaussian range profile results, for $P^{32}$. The exponential tail (just like ion channelling) is therefore a characteristic of implantation into single crystals.

Under some circumstances the tail produced by anomalous diffusion following ion implantation may not be exponential. Sparks (1966) considered

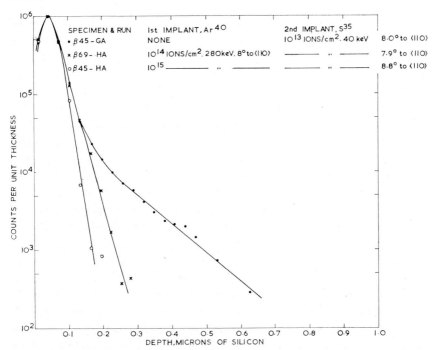

Fig. 2.80. The effect of 280 keV argon pre-bombardment at two different dose levels on the distribution of $S^{35}$ ions subsequently implanted at 40 keV into misoriented crystals of silicon.

a 'two-stream' diffusion of vacancies, self-interstitials and interstitial impurity atoms, with the first two combining on collision, and impurity interstitials interacting with vacancies to yield substitutional impurities. This can, under typical conditions, lead to a power-law dependence of the concentration of immobile substitutional impurities upon depth. Sparks attempted to fit his theory to the experimental results of Bower et al. (1966), but it now seems likely that the capacitance–voltage measurements were mis-interpreted in terms of donor concentration, owing to reasons which have been discussed by Kennedy, Murley and Kleinfelder (1968). This may not rule out the applicability of Sparks' treatment in other cases. However, it seems that the theory would require modification to take account of the vacancy concentration which may be 'frozen' into a crystal and which, as we have seen, can modify the behaviour of a diffusion tail in different specimens of silicon.

To conclude this discussion, we shall try to assess the relevance of these anomalous diffusion mechanisms to the practical application of ion implantation. Perhaps the most difficult of them to judge is proton-enhanced dffusion: it is too early to know whether this will prove economically attractive in device fabrication. In one sense it provides a near-ideal doping profile with a high electrical activity and a control of the penetration depth which is one of the virtues of ion implantation. Since it will do this for dopants which have been introduced to a shallow depth by diffusion, it seems bound to appeal to manufacturers who are nervous of launching into the realms of ion implantation. However, such an approach fails to achieve the uniformity of doping which ion implantation offers, and, particularly in shallow diffusions, it is difficult to obtain a reproducible total dopant concentration. In Chapter 4 we discuss the desirability of this for resistor fabrication. Ideally, therefore, proton-enhanced diffusion should be combined with a low-energy ion implantation and is then seen as a modification to the process of implantation followed by a drive-in diffusion. Its advantages are probably a better control of doping profile and an increase in the effective diffusion rate, which allows much reduced process temperatures.

By comparison, enhanced diffusion induced by a very hot implantation, making use of vacancies created by the implanted ions themselves, seems less attractive. Although only a single operation is required, it is probably difficult to reconcile the throughput necessary for economic utilisation of an ion implantation facility with the time needed for diffusion to occur. In contrast, a proton accelerator, operating at a few tens of keV and delivering the small currents which Gibbons has shown to be adequate, is a less costly piece of equipment, though it remains to be seen how it can best be combined with a diffusion furnace.

This leaves the third, and by far the simplest diffusion mechanism from the point of view of exploitation, i.e. the interstitial or other type of migration at low temperatures of an implanted dopant. By an appropriate choice of dopant and crystal characteristics it is possible, as we have seen, to cause a fraction of the implanted ions to penetrate to depths of nearly half a micron. Again, one can achieve a high substitutional fraction, after trapping by vacancies, and there are no damage complexes to be annealed. The damaged surface layer may be removed, by etching, to leave a doped but near-perfect substrate. Such a technique appears to have been used by Eldridge et al. (1970) to prepare electroluminescent devices in CdS, without any necessity for annealing. In thermally unstable materials this can be very attractive and it opens ways of implanting materials which could not be doped by diffusion. There are, of course, severe limitations to this simple process. Only a relatively small fraction (perhaps a few per cent) of the implanted ions are to be found in the penetrating diffusion tail, though this fraction can be increased by channelling. The exponential profile which generally results is not suitable for many devices. It might be feasible to control the profile by proton irradiation, creating a sheet of vacancies at the depth corresponding to the proton range, but this will add a certain degree of complexity to the equipment. Preferably, the two ion bombardments would be carried out simultaneously and, if only just enough isolated vacancies were created to interact with migrating interstitial dopants, subsequent annealing would again be unnecessary.

In terms of determining the final location of implanted atoms, and from the practical point of view of obtaining, reproducibly, a desirable doping profile it has been realized that the diffusion mechanisms which accompany ion implantation in crystals are of major importance. They have recently become the subject of a good deal of research, the results of which seem certain to have an impact on the methods utilized for ion implantation and the variety of devices which can be fabricated by this process.

### 2.19. Atom location by channelling techniques

The phenomenon of ion channelling has provided a direct and sensitive technique for the location of non-substitutional atoms in crystal lattices. This information is of particular interest in the study of ion implantation owing to the fact that, unlike diffusion, implantation is not an equilibrium process. Dopant atoms which are substitutional in diffused samples may, in implanted crystals, tend to occupy non-substitutional sites with consequently a different electrical behaviour. Precipitation may occur during thermal processing, particularly if, as is quite possible, the solubility limit has been exceeded. During the implantation of a single ion, some $10^3$ lattice atoms may be displaced from their substitutional sites and it is very useful to have a quantitative means of monitoring this radiation damage during annealing. All these processes, which take place in a shallow layer around 1 $\mu$m in thickness, can be investigated by means of a channelled probing beam of energetic ions, and measurement of the yield either of back-scattered particles or alternatively of some reaction product, such as alpha particles, X-rays or $\gamma$-rays. In this section we shall merely introduce the principles and theoretical assumptions of the technique. Specific examples of its application to the study of radiation damage will be given in Chapter 3, and to the electrical behaviour of implanted semiconductors in Chaper 4.

We have seen that when a collimated ion beam enters a perfect crystal lattice parallel to a low-index direction, many of the ions will become channelled. Those which experience an impact parameter with an atomic row less than a critical value $\rho_{min}$ will suffer an elastic dechannelling collision on entering the crystal, and constitute what is usually called the 'random' beam. Since $\rho_{min}$ decreases with ion velocity, for MeV ions of low mass the random beam may be only a few per cent of the total flux. The channelled beam only penetrates closer than $\rho_{min}$ to the atomic rows as a result of dechannelling mechanisms (section 2.16) such as thermal vibrations. In crystals with reasonably low thermal vibration amplitudes (such as silicon, tungsten, etc.) the channelled fraction remains high throughout the depth to which dopant ions would be implanted. This limited approach of the probing ions to the atomic rows suppresses the yield of processes such as large-angle Rutherford scattering, nuclear reactions, or X-ray production, and ideally the attenuation is equal to the ratio of the total channel area $A_{ch}$ to $\pi\rho_{min}^2$, a factor which may reach values of 20 to 100.

If, however, there are impurity atoms which are sited a distance greater than $\rho_{min}$ away from an atomic row, e.g. in an interstitial site or a precipitate cluster, then the attenuation will no longer exist as the atom will be exposed to a flux of channelled particles. It has become clear, over the past year, that the magnitude of this flux may be considerably greater than that which would exist in a mis-aligned or 'random' situation. We shall return to this point in more detail below. On the other hand, impurity atoms which occupy truly substitutional sites are effectively shadowed by their neighbours in the row, and give rise to a scattering or reaction yield no greater than that from the host lattice. That this is so in the case of a well-behaved substitutional system (the copper-gold alloy crystal) has been demonstrated quantitatively by Alexander et al. (1970a, b).

If the surface layers of the crystal are not perfect, but have been dis-ordered to some extent by previous ion bombardment, then it is clear that the scattering or reaction yield will be increased because the probability of channelling is diminished. In the extreme case of total disorder or 'amorphici-ty' as it is sometimes called, the yield must approach that corresponding to a zero channelled fraction, and this is approximately the result of orienting the crystal so that the probing ions enter well away from a channelling direc-tion. Critical channelling angles are small at high energies, and the probabil-ity of scattering into a neighbouring channel is also small. If the surface disordered layer is shallow, many of the ions transmitted through it will become channelled in the undisturbed lattice below, and hence the back-scattered spectrum shows a peak (fig. 2.81), the width of which is related to the depth of the disorder, and the yield, corresponding to the area under the peak, is (to a first order) proportional to the number of displaced atoms.

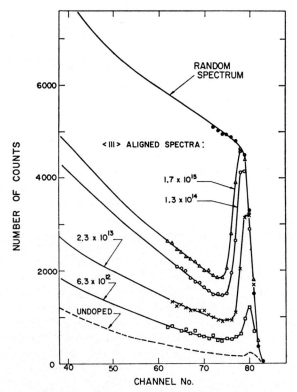

Fig. 2.81. Aligned ⟨111⟩ and random backscattered α-particle spectra from Ge crystals implanted at room temperature with various doses of 40 keV In⁺ ions. The incident alpha-particle energy was 1.0 MeV (from Mayer et al., 1968).

This correspondence breaks down, however, at high doses, since then the displacement clusters overlap and the method cannot tell whether an atom has been displaced more than once. Since the depth and distribution of the displaced atoms is of considerable interest, it is advantageous in such experiments to use heavier particles, such as $N^{14}$ ions which have a greater stopping cross-section. The depth resolution $\Delta t$ is related to the overall experimental resolution $\Delta E$ and the stopping cross-sections $S(E_1)$ for the incident energy, and $S(E_2)$ for the scattered energy, by the expression (Bøgh, 1969):

$$\Delta t \simeq \Delta E [S(E_1)K^2 + S(E_2)\cos \theta_1/\cos \theta_2]^{-1} \qquad (2.130)$$

where

$$K = \frac{M_1 \cos \theta}{M_1 + M_2} + \left[ \left( \frac{M_1 \cos \theta}{M_1 + M_2} \right)^2 + \left( \frac{M_2 - M_1}{M_1 + M_2} \right) \right]^{\frac{1}{2}} \qquad (2.131)$$

and $\theta_1$ is the angle of entry (with respect to the surface normal), $\theta_2$ is the angle of exit, and $\theta$ is the angle of scattering (in the laboratory system). The advantage of using heavier ions to improve the depth resolution is maintained only if the detector has good energy resolution $\Delta E$. With $N^{14}$ ions, it is possible to obtain, in a silicon surface-barrier detector, a resolution of about 50 keV at 2–3 MeV and a corresponding depth resolution of about two hundred Ångstroms. Much better energy resolution is feasible with a magnetic analyser, and in this way Bøgh has achieved a depth resolution of about 50 Å. However, the angular acceptance of such a system is very restricted, and hence it is not possible to gather very much data before the radiation damage induced by the probing beam itself perturbs the damage being measured. This is particularly so if heavy probing ions are used, and there-

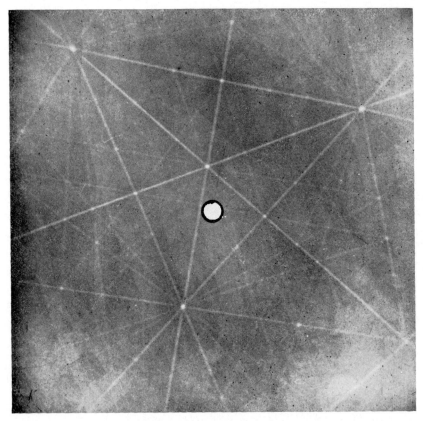

Fig. 2.82. Pattern of tracks produced in cellulose nitrate film by alpha-particles, of incident energy 1.5 MeV, scattered back from a Ge crystal. The ion dose was 3 $\mu$ Coulombs.

fore it is necessary to ensure in each case that this damage can be neglected.

We next consider how the experiments are carried out, for the purpose of determining the lattice location of impurity atoms. The first requirement is to orient the specimen so that the probing ions enter along a major channelling axis. Usually the lattice orientation is known to a few degrees and the specimen is mounted in a two- or three-axis goniometer (see Chapter 5 for a discussion of such apparatus) within the target chamber of the ion accelerator. Two commonly-used methods of orientation themselves utilise the back-scattering process. In one (Marsden et al., 1969) a sheet of cellulose nitrate film is mounted near to the specimen, the ion beam passing through a small hole near its centre. Bombardment with $\sim 1~\mu$ Coulomb of particles will produce sufficient backscattered ions to create developable tracks in the plastic, as a result of soaking for a few minutes in warm NaOH solution. The pattern of blocking due to intersecting crystal planes produces a clearly identifiable record of the crystal orientation (fig. 2.82) from which the appropriate adjustment of the goniometer can be calculated. Often, however, such a degree of bombardment cannot be allowed owing to the amount of radia-

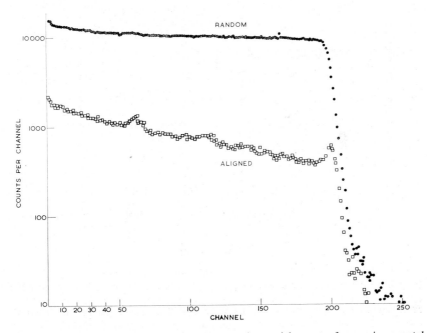

Fig. 2.83. Aligned ⟨111⟩ and random backscattered α-particle spectra from an iron crystal. The incident energy was 1.5 MeV.

tion damage it would cause: then an appropriate method is the one described by Andersen et al. (1965) in which a rotation stereogram of the crystal is obtained by monitoring the scattered yield in a detector during one complete rotation about an axis normal to the crystal. This method, and other techniques of crystal orientation are discussed in Chapter 5.

Once the required orientation of the specimen has been achieved, energy spectra are recorded along two low-index directions and along a selected randomly-oriented direction (fig. 2.83). Ions that are scattered back from shallow implanted impurities which are heavier than the host lattice emerge with energies greater than the maximum energy of the continuum spectrum (fig. 2.84) and as a result, can be readily discriminated. If, however, the number of implanted atoms is small (say $< 10^{13}/cm^2$) the peak is generally obscured by the background due to pile-up of pulses resulting from the

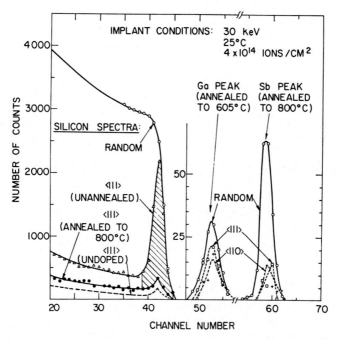

Fig. 2.84. Aligned $\langle 111 \rangle$ and $\langle 110 \rangle$ together with random backscattered α-particle spectra from Si crystals implanted with $Sb^+$ or $Ga^+$ ions (30 keV, 25 °C, $4 \times 10^{14}$ ions/cm²). The spectra on the left, due to scattering from silicon, correspond to the $Sb^+$-implanted specimen: those for $Ga^+$ implantation are almost identical. The shaded area here represents the region of surface disorder. Peaks on the right-hand diagram correspond to scattering from implanted impurities, at the crystal orientations indicated. The analysing beam was of 1.0 MeV alpha-particles (from Eriksson et al., 1968).

many particles scattered by the lattice. Under these circumstances some form of pile-up rejection is essential: one of the simplest generates a timing signal at the commencement of each pulse, and if the interval between successive pulses should fall below a preset value (as would be the case in pile-up), the second pulse is rejected automatically.

Next we examine how, from the energy spectra obtained, the lattice site occupancy may be calculated. Prior to 1970, the number of atoms occupying sites greater than $\rho_{min}$ away from the atomic rows was deduced with the naive assumption that the backscattered (or reaction) yield from such sites is equal to the yield when bombarded under non-channelled conditions. Then one has simply, for the percentage of atoms occupying substitutional sites,

$$f_{subst} = \frac{100(N_A - N_R)}{N_R \left(\dfrac{N_A^h}{N_R^h}\right) - N_A} \tag{2.132}$$

where $N_A$, $N_R$ are the number of counts integrated over the aligned and random impurity peaks respectively, and $N_A^h$, $N_R^h$ are the corresponding figures for the host lattice integrated over an equivalent depth of crystal. This expression neglects the effect of dechannelling of the beam as it passes through the implanted layer. It proves very difficult to incorporate this effect fully: see Bøgh (1968) for a consideration of the error due to neglecting it.

The realization that the channelled ion flux may depart markedly from the random flux value came independently from two investigations, which have been put into perspective recently by Van Vliet (1971). The computer simulation of ion trajectories in crystals has been described in section 2.13 (Morgan and Van Vliet, 1970a, b), and a study of the coordinates of channelled protons simulated in copper crystals revealed a pronounced 'flux peaking' near the centre of the channels. Alexander et al. (1970a, b) applied this approach to problems of the lattice location of implanted atoms in copper crystals, including a treatment of the influence of inelastic collisions on the dispersion of the ion beam. This was followed by a somewhat similar set of computations for Si and Ge by Massa and Clark (1970). Typically, the flux at the centre of a channel may be two to four times the random value, but the ratio is dependent upon many factors, such as ion energy, crystal temperature, channel dimensions, crystal alignment, beam collimation and the depth within the crystal. The effects of all these variables have now been studied (Van Vliet, 1971).

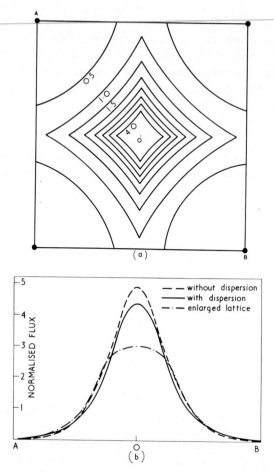

Fig. 2.85. (a) Flux contours across a $\langle 100 \rangle$ channel in Cu at 20 °C for 500 keV He⁺ ions incident along the channel axis, from a computer simulation by Alexander et al. (1970a). (b) Flux distribution along the channel diagonal AOB with and without electronic dispersion, and (–.–.–) with an artificial enlargement of the unit cell from 3.6 to 5 Å (ibid, 1970). In each case the results are averaged over the first 1000 Å of penetration.

The second approach to the problem was stimulated by the experimental observation of a backscattered yield in excess of the random value. Such an effect was reported by Fischer, Sizmann and Bell (1969) in deuterium-implanted copper, and by Andersen et al. (1970) in ytterbium-implanted silicon. The Aarhus group developed an analytical model, based upon Lindhard's (1965) treatment of ion channelling. Two factors are involved: the detailed shape of the continuum potential in the central region of the channel, and the angular dispersion of an initially well-channelled ion beam. It is assumed that these ions have gained sufficient transverse energy to move freely over a central 'flat' region of the continuum potential, of area $A_1$, and then if $A_{ch}$ is the area of the channel the flux peak, $F_{max}$, at the centre of the channel relative to the average flux is given by the simple expression

$$F_{max} = 1 + \ln\left(A_{ch}/A_1\right) \qquad (2.133)$$

Further analysis enables $A_1$ to be derived in terms of the minimum fluctuation in transverse energy due to multiple scattering, and the results agree very well with the computer simulation (fig. 2.86). However, the latter shows an oscillatory behaviour of $F_{max}$ near the surface of the crystal, before the

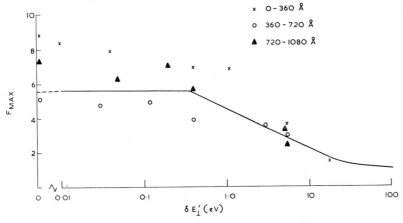

Fig. 2.86. The values of $F_{max}$ (see text) as a function of the minimum fluctuation, $\delta E_1$, in transverse energy at three intervals of depth, obtained by a computer simulation for 1 MeV He$^+$ ions in Cu along the $\langle 100 \rangle$ channel compared with the prediction (solid line) of equation (2.133) (after Van Vliet, 1971).

equilibrium transverse energy is attained. This transient region can be as much as 0.1 $\mu$m in thickness, and may therefore be comparable with the range of those ions for which the lattice site location is required.

    In conclusion, the yield enhancement from non-substitutional atoms means that the interpretation of atom location experiments is by no means as simple as had previously been thought. Additional complications arise in the case of interstitial atoms not in the centre of a channel, or a channel axis not coinciding with the position of minimum potential energy in the channel. It seems necessary to determine the back-scattered yield as a function of angle around the channel direction, and from the shape of this

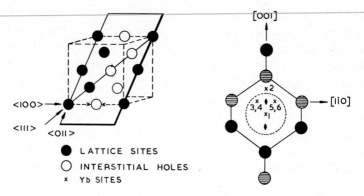

Fig. 2.87. (a) Configuration of the {110} plane of silicon, showing (●) lattice sites and (○) regular interstitial sites and (×) the probable location of implanted Yb atoms deduced by α-backscattering. (b) The six equivalent Yb sites viewed along a ⟨100⟩ channel. The dotted circle indicates the maximum area over which the observed two-fold flux peaking can occur (after Davies, 1970).

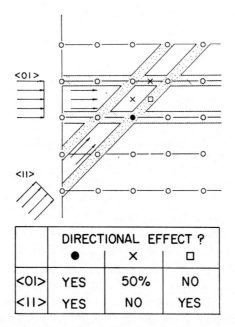

| | DIRECTIONAL EFFECT ? | | |
|---|---|---|---|
| | ● | × | □ |
| ⟨01⟩ | YES | 50% | NO |
| ⟨11⟩ | YES | NO | YES |

Fig. 2.88. A two-dimensional model illustrating how the channelling phenomenon can be used to locate foreign atoms in a crystal. Three typical sites, shown by ●, □ and ×, can be uniquely distinguished by studying the back scattering of a particle beam channelled firstly along the ⟨01⟩ and then along the ⟨11⟩ direction (from Eriksson et al., 1969).

curve to deduce the location of non-substitutional atoms. Andersen et al. (1971) claim a probable accuracy of $\sim 0.1$ Å for the location of Yb atoms implanted into silicon (fig. 2.87).

Having determined the percentage of substitutional atoms along one low-index direction, it is necessary to carry out the experiment along another axis in order to distinguish sites which are not truly substitutional but are shadowed by other atoms of the structure. The choice of axes studied will thus be determined by the structure of the crystal (fig. 2.88).

The method we have so far discussed has made use of a probing beam of ions directed along a channelling direction, while the detector is placed at a random backward angle. However, greater sensitivity is achieved in determining, for example, very small amounts of disorder or low concentrations of non-substitutional impurities, by means of what is called double-alignment geometry. The method involves both introducing and observing ions along low-index directions, and takes advantage of the blocking effect. Since large-angle scattering demands what is essentially the emission of the scattered ion from a lattice site, the yield of such particles may be reduced by an order of magnitude or more when observations are made close to the direction of an atomic row. The direction chosen for observation may also be the one along which the ions enter the crystal: this has been called the 'uniaxial' geometry, and an annular detector subtending a small solid angle at the crystal is mounted with the incident beam passing through its central hole (Appleton and Feldman, 1970). Alternatively, some other axis may be chosen in which case a simple and convenient means of positioning the detector is as follows (Marsden et al., 1969). Cellulose nitrate film is used, as described above, to record the blocking pattern of scattered ions. After developing the tracks, a small hole is punched in the film, of a size just equal to the axial blocking spot, and the particle detector is mounted behind it so that the film acts as a collimator. Of course, it is essential to ensure that the film can be accurately re-mounted in the scattering chamber.

Up to this point we have concentrated upon the Rutherford backscattering technique, since this has been the most widely used channelling method for atom location. It will be clear from energy spectra such as that shown in fig. 2.84 that the method is impracticable for locating species which are close to, or less than, the host lattice in terms of atomic weight. In the case of implants of the lightest atoms (e.g. deuterium, lithium, boron, carbon, oxygen, nitrogen and fluorine) it is often possible to find some nuclear reaction of high yield which is induced by a beam of protons, deuterons or alpha-particles. If so, the procedure is very much as we have described above, with the advantage that the reaction may be highly specific. The lattice location of $Li^7$ and $B^{11}$ in silicon has been investigated in this way by Eriksson et al. (1966) and by Gibson et al. (1968c). The commonest difficulty is that the detector is subjected to a high flux of elastically scattered ions, and these may overload the pulse-handling system: if the incident beam is reduced it may take a long time to gather the necessary data since reaction yields tend to be lower than scattering yields.

Another, less restricted, method of lattice site location for impurities is by observation of the characteristic X-rays produced by a channelled proton bombardment. Cairns and Nelson (1968) have in this manner demonstrated that antimony ions implanted into silicon occupy substitutional positions after annealing.

All the above techniques discussed in this section share a common shortcoming: it is not possible to deduce by them the local configuration of an ion but merely its location with respect to neighbouring atomic rows. Thus it would be wrong to conclude that, if an ion is observed to be non-substitutional, it is necessarily an isolated interstitial. Any form of ion pairing, either with a similar ion or in a complex, will generally result in a relaxation of the surrounding lattice so that both particles are likely to be 0.1 Å or more away from any atomic row. Several impurity atoms may group together in the form of a tiny precipitate and the electron microscope shows that very large clusters of impurities and intrinsic defects are common. In the absence of any other data, e.g. from electron spin resonance or infra-red absorption, it is far from easy to correlate the changes in substitutional lattice site occupancy with the dissociation and migration of specific defects. In our view there has been too much emphasis on the channelling techniques for observing implanted crystals, without the necessary backing from spectroscopic methods.

In semiconductors or insulators there is now the serious question as to whether probing by ionizing radiation will cause a transient migration of non-substitutional species, for example by the mechanism of Bourgoin and Corbett (1972). In such a case the measured lattice locations would not correspond to the equilibrium situation. Lattice location measurements using ion channelling can be fully trusted only in metals.

## References

Adams, J. and B. W. Manley, 1967, Philips Tech. J. **28**, 156.

Alexander, R. B., G. Dearnaley, D. V. Morgan and J. M. Poate, 1970a, Phys. Lett. **32A**, 365.

Alexander, R. B., G. Dearnaley, D. V. Morgan, J. M. Poate and D. Van Vliet, 1970b, Proc. Conf. on Ion Implantation, Reading. p. 181.

Allison, S. K. and S. D. Warshaw, 1953, Rev. Mod. Phys. 25, 779.

Almén, O. and G. Bruce, 1961, Nucl. Instr. and Meth., 11, 257; 279.

Andersen, H. H., C. C. Hanke, H. Sørensen and P. Vajda, 1967, Phys. Rev. 153, 338.

Andersen, H. H. and P. Sigmund, 1965, Nucl. Instr. and Meth. 38, 238.

Andersen, J. U., 1967, Kgl. Danske Vid. Selsk., Matt.-Fys. Medd. 36, No. 7.

Andersen, J. U., O. Andreasen, J. A. Davies and E. Uggerhøj, 1971, Radiation Effects 7, 25.

Andersen, J. U., J. A. Davies, K. O. Nielsen and S. L. Andersen, 1965, Nucl. Instr. and Meth. 38, 210.

Andersen, J. U., W. M. Gibson and E. Uggerhøj, 1967, Proc. Conf. on Ion Beams in Semiconductor Technology, Grenoble.

Andersen, J. U. and E. Uggerhøj, 1968, Can. J. Phys., 46, 517.

Andersen, T. and G. Sørensen, 1968, Can. J. Phys. 46, 483.

Andreen, C. J. and R. L. Hines, 1966, Phys. Rev. 151, 341.

Appleton, B. R. and L. C. Feldman, 1970, Proc. Conf. on Atomic Collision Phenomena in Solids, Brighton (North-Holland, Amsterdam).

Appleton, B. R., C. Erginsoy and W. M. Gibson, 1967, Phys. Rev. 161, 330.

Armitage, S. A., 1970, Proc. Conf. on Ion Implantation, Reading, p. 138.

Bader, R. and S. Kalbitzer, 1970, Appl. Phys. Lett. 16, 13.

Barker, P. H. and W. R. Phillips, 1965, Proc. Phys. Soc., London, 86, 379.

Bergstrom, I., F. Brown, J. A. Davies, J. S. Geiger, R. L. Graham and R. Kelly, 1963, Nucl. Instr. Methods, 21, 249.

Bethe, H. A., 1930, Ann. Physik 5, 325.

Bhalla, C. P. and J. N. Bradford, 1968, Phys. Lett. 27A, 318.

Bichsel, H., 1963, Handbook of Physics, Section 8C (McGraw-Hill, N.Y.).

Bloch, F., 1933, Ann. Physik, 16, 285.

Blood, J., G. Dearnaley and M. A. Wilkins, 1973, to be published.

Bøgh, E., 1968, Can. J. Phys. 46, 653.

Bøgh, E., 1969, Proc. Roy. Soc., A311, 35.

Bøgh, E. and E. Uggerhøj, 1965, Nucl. Instr. and Meth. 38, 216.

Bøgh, E., J. A. Davies and K. O. Nielsen, 1964, Phys. Lett. 12, 129.

Bohm, D., 1951, 'Quantum Theory' (Prentice-Hall, N.Y.).

Bohm, D. and D. Pines, 1952, Phys. Rev. 85, 338 and 82, 625.

Bohr, N., 1913, Phil. Mag., 25, 10.

Bohr, N., 1948, Kgl. Danske Vid. Selsk., Matt.-Fys. Medd., 18, No. 8.

Bonderup, E., 1967, Kgl. Danske Vid. Selsk., Matt.-Fys. Medd., 35, No. 17.

Bonsignori, F. and A. Desalvo, 1969, Nuovo Cim. Lett. 1, 589.

Bonsignori, F. and A. Desalvo, 1970, J. Phys. Chem. Sol. 31, 2191.

Bourgoin, J. and J. W. Corbett, 1972, Phys. Lett. 38A, 135.

Bøttiger, J. and F. Bason, 1969, Radiation Effects, 2, 105.

Bower, R. W., R. Baron, J. W. Mayer and O. J. Marsh, 1966, Appl. Phys. Lett. 9, 203.

Brandt, W., J. M. Khan, D. L. Potter, R. D. Worley and H. P. Smith, 1965, Phys. Rev. Lett. 14, 42.

Brandt, W., R. Dobrin, H. Jack, R. Lambert and S. Roth, 1968, Can. J. Phys. 46, 537.

Bredov, M. M. and N. M. Okuneva, 1957, Doklady Akad. Nauk SSSR 113, 795.

Bredov, M. M., V. A. Lepilin, I. B. Schestakov and A. L. Shakh-Budagov, 1961, Sov. Phys. – Solid State 3, 195.

Brice, D. K., 1968, Phys. Rev. 165, 475.

Brinkman, J. A., 1954, J. Appl. Phys. 25, 961.

Burtt, R. B., J. A. Colligon and J. H. Leck, 1961, Brit. J. Appl. Phys. 12, 396.

Cairns, J. A. and R. S. Nelson, 1968, Phys. Lett. **27A**, 14.

Cairns, J. A., D. F. Holloway and R. S. Nelson, 1970a, Radiation Effects **7**, 163, 167.

Cairns, J. A., D. F. Holloway and R. S. Nelson, 1970b, Proc. Conf. on Ion Implantation, Reading, p. 203.

Carter, G. and J. S. Colligon, 1968, 'Ion Bombardment of Solids' (Heinemann).

Channing, D. A., 1967, Can. J. Phys. **45**, 2455.

Channing, D. A. and J. A. Turnbull, 1968, C.E.G.B., Berkeley, Nucl. Laboratories Report RD/B/N 1114.

Cheshire, I. and J. M. Poate, 1669, Proc. Conf. on Atomic Collision Phenomena in Solids, Sussex (North-Holland, Amsterdam, 1970) p. 351.

Cheshire, I., G. Dearnaley and J. M. Poate, 1968, Phys. Lett., **27A**, 304.

Cheshire, I., G. Dearnaley and J. M. Poate, 1969, Proc. Roy. Soc. **A311**, 47.

Clark, G. J., D. V. Morgan and J. M. Poate, 1969, Proc. Conf. on Atomic Collision Phenomena on Solids, p. 388 (North-Holland 1970).

Clark, G. J., G. Dearnaley, D. V. Morgan and J. M. Poate, 1971, (unpublished).

Datz, S., C. D. Moak, T. S. Noggle, B. R. Appleton and H. O. Lutz, 1969, Phys. Rev. **179**, 315.

Datz, S., T. S. Noggle and C. D. Moak, 1965, Phys. Rev. Lett. **15**, 254.

Davies, J. A. and P. Jespersgård, 1966, Can. J. Phys. **44**, 1631.

Davies, J. A., J. Friesen and J. D. McIntyre, 1960a, Can. J. Chem. **38**, 1526.

Davies, J. A., J. D. McIntyre, R. L. Cushing and M. Lounsbury, 1960b, Can. J. Chem. **38**, 1535.

Davies, J. A., J. D. McIntyre and G. A. Sims, 1961, Can. J. Chem. **39**, 611.

Davies, J. A., G. C. Ball, F. Brown and B. Domeij, 1964, Can. J. Phys. **42**, 1070.

Davies, J. A., L. Eriksson and P. Jespergård, 1965, Nucl. Instr. and Meth. **38**, 245.

Davies, J. A., L. Eriksson and J. L. Whitton, 1968, Can. J. Phys. **46**, 573.

Dearnaley, G., 1964, IEEE, Trans. Nucl. Sci., **NS-11**, No. 3, 249.

Dearnaley, G., 1969, Reports on Progress in Physics, **32**, 405.

Dearnaley, G., 1970, Proc. Conf. on Ion Implantation, Reading, p. 162.

Dearnaley, G., B. W. Farmery, I. V. Mitchell, R. S. Nelson and M. W. Thompson, 1968a, Phil. Mag., **18**, 985.

Dearnaley, G., J. H. Freeman, G. A. Gard and M. A. Wilkins, 1968b, Can. J. Phys. **46**, 587.

Dearnaley, G., M. A. Wilkins, P. D. Goode, J. H. Freeman and G. A. Gard, 1970, Proc. Conf. on Atomic Collision Phenomena in Solids, Sussex, p. 633.

Dearnaley, G., M. A. Wilkins and P. D. Goode, (1971) Proc. Conf. on Ion Implantation in Semiconductors, Garmisch (Springer-Verlag, Berlin, 1971), p. 439.

Delsarte, G., J. C. Jousset, J. Mory and Y. Quéré, 1970, Proc. Conf. on Atomic Collision Phenomena in Solids, Sussex, p. 501.

Domeij, B., I. Bergstrom, J. A. Davies and J. Uhler, 1963, Arkiv f. Fysik. **24**, 399.

Domeij, B., F. Brown, J. A. Davies and M. McCargo, 1964, Can. J. Phys. **42**, 1624.

Domeij, B., G. Fladda and N. G. E. Johansson, 1970, Radiation Effects, **6**, 155.

Eisen, F. H., 1968, Can. J. Phys. **46**, 561.

Eisen, F. H., 1969 (priv. comm.).

Eisen, F. H., B. Welch, J. E. Westmoreland and J. W. Mayer, 1970, Atomic Collision Phenomena in Solids, 111 (North-Holland).

Eldridge, G., P. K. Govind, D. Nieman and F. Chernow, 1970, Proc. Conf. on Ion Implantation, Reading, p. 143.

El-Hoshy, A. A. and J. F. Gibbons, 1968, Phys. Rev. **173**, 454.

Erginsoy, C., 1964, Phys. Rev. Lett. **12**, 366.

Erginsoy, C., 1965, Phys. Rev. Lett. **15**, 360.

Eriksson, L., J. A. Davies, J. Denhartog, H. J. Matzke and J. L. Whitton, 1966, Can. Nucl. Tech. **5**, No. 6, 40.

Eriksson, L., J. A. Davies and P. Jespersgård, 1967, Phys. Rev. **161**, 219.
Eriksson, L., J. A. Davies and J. W. Mayer, 1968, 'Radiation Effects in Semiconductors', ed. F. Vook, pp. 398–405 (Plenum Press, N.Y.).
Fairfield, J. M. and B. L. Crowder, 1969, Trans. Met. Soc., AIME, **245**, 469.
Fano, U., 1963, Ann. Rev. Nuc. Sci., **13**, 1.
Fano, U. and W. Lichten, 1965, Phys. Rev. Lett. **14**, 627.
Fastrup, B., A. Borup and P. Hvelplund, 1968, Can. J. Phys. **46**, 489.
Feldman, L. C., B. R. Appleton and W. L. Brown, 1967, Proc. Conf. on Solid State Physics Research with Accelerators, Brookhaven (Brookhaven National Lab. Report BNL 50083 (C-52)).
Fermi, E. and E. Teller, 1947, Phys. Rev. **72**, 399.
Firsov, O. B., 1958, Sov. Phys., J.E.T.P. **7**, 308.
Firsov, O. B., 1959, Sov. Phys. J.E.T.P. **9**, 1076.
Fischer, H., R. Sizmann and F. Bell, 1969, Z. Physik **224**, 135.
Freeman, J. H., 1968, Proc. Conf. on Applications of Ion Beams to Semiconductor Technology, Grenoble, May 1967, pp. 75–90.
Freeman, J. H., 1970, Proc. Conf. on Ion Implantation, Reading, p. 1.
Freeman, N. J. and I. D. Latimer, 1968, Can. J. Phys. **46**, 467.
Galaktionova, I. A., V. M. Gusev, V. G. Naumenko and V. V. Titov, 1968, Proc. Conf. on Applications of Ion Beams to Semicond. Tech., Grenoble, p. 619 (Editions Ophrys.).
Gemmell, D. S. and R. E. Holland, 1965, Phys. Rev. Lett. **14**, 945.
Gibbons, J. F., 1970, (priv. comm.).
Gibson, W. M., C. Erginsoy, H. E. Wegner and B. R. Appleton, 1965, Phys. Rev. Lett. **15**, 357.
Gibson, W. M., F. W. Martin, R. Stensgaard, F. Palmgren-Jensen, N. I. Meyer, G. Galster, A. Johansen and J. S. Olsen, 1968a, Can. J. Phys. **46**, 675.
Gibson, W. M., F. W. Martin, R. Stensgaard, F. Palmgren-Jensen, N. I. Meyer, G. Galster, A. Johanssen and J. S. Olsen, 1968b, Proc. Conf. on Applications of Ion Beams to Semicond. Tech., Grenoble, p. 449 (Editions Ophrys).
Gibson, W. M., J. B. Rasmussen, P. Ambrosius-Olesen and C. J. Andreen, 1968c, Can. J. Phys. **46**, 551.
Gilat, J. and J. M. Alexander, 1964, Phys. Rev. **136**, 1298.
Gittins, R. P., D. V. Morgan and G. Dearnaley 1972, J. Phys. D. **5**, 1654.
Gombas, P., 1956, Handbuch der Physik **36**, 109, (Springer, Berlin).
Goode, P. D., M. A. Wilkins and G. Dearnaley, 1970, Radiation Effects **6**, 237.
Harrison, D. E. and D. S. Greiling, 1967, J. Appl. Phys. **38**, 3200.
Hermann, H., H. Lutz and R. Sizmann, 1966, Z. Naturforschung **21a**, 365.
Hines, R. L., 1960, Phys. Rev. **120**, 1626.
Holmes, D. K. and G. Leibfried, 1960, J. Appl. Phys. **31**, 1046.
Howe, L. M. and D. A. Channing, 1967, Can. J. Phys. **45**, 2467.
Howie, A., 1967 (priv. comm.).
Howie, A., M. S. Spring and P. N. Tomlinson, 1969, Proc. Conf. on Atomic Collision Phenomena in Solids, Sussex (North-Holland, Amsterdam, 1970) p. 34.
Hvelplund, P. and B. Fastrup, 1968, Phys. Rev. **165**, 408.
Jespersgård, P. and J. A. Davies, 1967, Can. J. Phys. **45**, 2983.
Johnson, W. S. and J. F. Gibbons, 1970, 'Projected Range Statistics in Semiconductors', distributed by Stanford University Bookstore.
Kendall, D. L. and D. B. De Vries, 1969, 'Semiconductor Silicon' Ed. R. R. Haberecht and E. L. Kern (Electrochem. Soc., N.Y.).
Kennedy, D. P., P. C. Murley and W. J. Kleinfelder, 1968, IBM J. Res. and Devel. 399.
Klein, C. A., 1966, J. Phys. Soc. Japan **21**, Suppl., 307.
Klein, C. A., 1968, IEEE Trans. on Nucl. Sci. **NS-15**, No. 3, 214.

Kleinfelder, W. J., W. S. Johnson and J. F. Gibbons, 1968, Can. J. Phys. **46**, 597.

Kornelsen, E. V., F. Brown, J. A. Davies and G. R. Piercy, 1964, Phys. Rev. **136A**, 849.

Lecrosnier, D. P. and G. Pelous, 1970, Proc. Conf. on Ion Implantation, Reading, p. 102.

LeGrand, C., C. Bahezre, J. Le Duigou and J.-J. Trillat, 1970, Comptes Rend. **271**, 88.

Lehmann, C. and G. Leibfried, 1963, J. Appl. Phys. **34**, 2821.

Leibfried, G. and D. S. Oen, 1962, J. Appl. Phys. **33**, 2257.

Lindhard, J., 1954, Kgl. Danske Vid. Selsk., Matt.-Fys. Medd. **28**, No. 8.

Lindhard, J., 1965, Kgl. Danske Vid. Selsk., Matt.-Fys. Medd. **34**, No. 14.

Lindhard, J. and C. Finneman, 1969, Priv. Comm.

Lindhard, J. and M. Scharff, 1953, Kgl. Danske Vid. Selsk., Matt.-Fys. Medd. **27**, No. 15.

Lindhard, J. and M. Scharff, 1961, Phys. Rev. **124**, 128.

Lindhard, J. and A. Winther, 1964, Kgl. Danske Vid. Selsk., Matt.-Fys. Medd. **34**, No. 4.

Lindhard, J., M. Scharff and H. E. Schiøtt, 1963, Kgl. Danske Vid. Selsk., Matt.-Fys. Medd. **33**, No. 14.

Lindhard, J., V. Nielsen and M. Scharff, 1968, Kgl. Danske Vid. Selsk., Matt.-Fys. Medd. **36**, No. 10.

Lonsdale, K., 1948, Acta Crystl. **1**, 142.

Lonsdale, K. (Ed.) International Tables for X-ray Crystallography, 1952 (Kynoch Press, Birmingham).

Luntz, M. and R. H. Bartram, 1968, Phys. Rev. **175**, 468.

Lutz, H. O., S. Datz, C. D. Moak and T. S. Noggle, 1966, Phys. Rev. Lett. **17**, 285.

Lutz, H. O., R. Schuckert and R. Sizmann, 1965, Nucl. Instr. and Meth. **38**, 241.

Lutz, H. and R. Sizmann, 1963, Phys. Lett. **5**, 113.

Marsden, D. A., N. G. E. Johanssen and G. R. Bellavance, 1969, Nucl. Instr. and Meth. **70**, 291.

Massa, I. and G. J. Clark, 1970, Proc. Conf. on Ion Implantation, Reading, p. 207.

Mayer, J. W., L. Eriksson, S. T. Picraux and J. A. Davies, 1968, Can. J. Phys. **46**, 663.

McCaldin, J. O., 1965, Nucl. Instr. and Meth. **38**, 153.

Meyer, O., 1969, Proc. Conf. on Special Techniques and Materials for Semiconductor. Detectors, (Ispra), p. 161.

Molière, G., 1947, Zeits. f. Nat. **20**, 133.

Morgan, D. V. and D. Van Vliet, 1968, Can. J. Phys. **46**, 503.

Morgan, D. V. and D. Van Vliet, 1970a, Proc. Conf. on Atomic Collision Phenomena in Solids, Sussex, p. 476.

Morgan, D. V. and D. Van Vliet, 1970b, Radiation Effects **5**, 157.

Namba, S., K. Masuda, K. Gamo, A. Doi, S. Ishihara and I. Kimura, 1970, Radiation Effects **6**, 115.

Nelson, D. G., J. F. Gibbons and W. S. Johnson, 1969, Appl. Phys. Lett. **15**, 246.

Nelson, R. S., 1968, 'The Observation of Atomic Collisions in Crystalline Solids', (North-Holland, Amsterdam).

Nelson, R. S. and D. J. Mazey, 1968, Can. J. Phys. **46**, 689.

Nelson, R. S. and M. W. Thompson, 1963, Phil. Mag. **8**, 1677.

Nielsen, K. O., 1956, 'Electromagnetically Enriched Isotopes and Mass Spectrometry' (Academic Press, N.Y.).

Oen, O. S., 1965, Phys. Lett. **19**, 358.

Oen, O. S., D. K. Holmes and M. T. Robinson, 1963, J. Appl. Phys. **34**, 302.

Oen, O. S. and M. T. Robinson, 1963, Appl. Phys. Lett. **2**, 83.

Oen, O. S. and M. T. Robinson, 1964, J. Appl. Phys. **35**, 2515.

Olley, J. A., P. M. Williams and A. D. Yoffe, 1970, Proc. Conf. on Ion Implantation, Reading, p. 148.

Ormrod, J. G., 1968, Can. J. Phys. **46**, 497.

Ormrod, J. H. and H. E. Duckworth, 1963, Can. J. Phys. **41**, 1424.

Ormrod, J. H., J. R. MacDonald and H. E. Duckworth, 1965, Can. J. Phys. **43**, 275.
Pfister, J. C. and P. Baruch, 1962, J. Phys. Soc., Japan **18**, Suppl. 251.
Picraux, S. T. and F. L. Vook, 1971, Radiation Effects **11**, 179.
Piercy, G. R., F. Brown, J. A. Davies and M. McCargo, 1963, Phys. Rev. Lett. **10**, 399.
Pines, D., 1963, 'Elementary Excitations in Solids' (Benjamin, N.Y.).
Porat, D. I. and K. Ramavataram, 1960, Proc. Phys. Soc., London, **77**, 97.
Porat, D. I. and K. Ramavataram, 1961, Proc. Phys. Soc., London **78**, 1135.
Powers, D. and W. Whaling, 1962, Phys. Rev. **126**, 61.
Quéré, Y., 1968, Phys. Stat. Sol. **30**, 713.
Robinson, M. T., 1969, Phys. Rev. **179**, 327.
Robinson, M. T. and O. S. Oen, 1963, Phys. Rev. **132**, 2385.
Rol, P. K., J. M. Fluit, F. P. Viehböck and M. de Jong, 1959, Proc. 4th Conf. on Ionization Phenomena in Gases, Uppsala.
Ruth, R. P. and F. H. Eisen, 1967, Proc. Conf. on Applications of Ion Beams to Semiconductor Technology, Grenoble, May 1967, pp. 539–558.
Sanders, J. B., 1968, Can. J. Phys. **46**, 455.
Sattler, A. R. and G. Dearnaley, 1965, Phys. Rev. Lett. **15**, 59.
Schiøtt, H. E., 1970, Radiation Effects **6**, 107.
Seidel, T. E., 1971, Proc. Conf. on Ion Implantation in Semiconductors, Garmisch, 47 (Springer-Verlag, Berlin).
Seitz, F., 1949, Disc. Faraday Soc. **5**, 271.
Sigmund, P. and P. Vajda, 1964, Danish AEC reports No. 83, 84.
Sparks, M., 1966, Phys. Rev. Lett. **17**, 1247.
Stark, J., 1912, Phys. Zeits. **13**, 973.
Strack, H., 1963, J. Appl. Phys. **34**, 3405.
Teplova, Y. A., V. S. Nikolaev, I. S. Dmitriev and L. N. Fateeva, 1962, Sov. Phys., JETP **15**, 31.
Thompson, M. W., 1964, Phys. Rev. Lett. **13**, 756.
Thulin, S., 1955, Arkiv f. Fysik **9**, 107.
Tsuchimoto, T. and T. Tokuyama, 1970, Radiation Effects **6**, 121.
Tulinov, 1965, Dokl. Akad. Naut., USSR **162**, 546.
Uggerhøj, E., 1969, Priv. Comm. via J. A. Davies.
Van Vliet, D., 1971, Radiation Effects **10**, 137.
Van Wijngaarden, A. and H. E. Duckworth, 1962, Can. J. Phys. **40**, 1749.
Walske, M. C., 1952, Phys. Rev. **88**, 1283.
Whitton, J. L., 1965, J. Appl. Phys. **36**, 3917.
Whitton, J. L., 1967, Can. J. Phys. **45**, 1947.
Whitton, J. L., 1968, Can. J. Phys. **46**, 581.
Whitton, J. L. and G. Carter, 1970, Proc. Conf. on Atomic Collision Phenomena in Solids, p. 615.
Whitton, J. L., G. Carter, J. H. Freeman and G. A. Gard, 1970, J. Mat. Sci. **4**, 208.
Whitton, J. L. and J. A. Davies, 1964, J. Electrochem. Soc. **111**, 1347.
Wilkins, M. A., 1968, AERE report R.5875.
Wilkins, M. A. and G. Dearnaley, 1970, Proc. Conf. on Ion Implantation, Reading, p. 193.
Williams, E. J., 1945, Rev. Mod. Phys. **17**, 217.

# 3

# THE PHYSICAL STATE
# OF ION IMPLANTED SOLIDS

The preceding chapter has provided the basic background for our understanding of the atomic collision phenomena pertinent to ion implantation. We have seen how energy is lost from an energetic ion during its passage through a solid and we have gained a knowledge of the ion penetration depths to be expected. Further, we have seen the dramatic influence of the regular nature of the crystal lattice on these processes. However, we must now consider the important problems relating to the physical state of the implanted solid; for example, is the solid damaged during implantation and what is the fate of the implanted ions? In this chapter we will consider these problems in some detail in an attempt to provide the reader with a fairly comprehensive understanding of present day knowledge.

## 3.1. Radiation damage produced during implantation

### 3.1.1. ATOMIC DISPLACEMENT

The elastic collisions which occur between the incident ions and the atoms of the solid result in energy transfers ranging from fractions of an eV to many tens of keV. The exact energy spectrum of those atoms recoiling from such primary encounters depends intimately on the scattering cross-section which in turn depends on the interatomic potential between the colliding particles. However, in this instance we will not dwell on this point as we will be mainly concerned with subsequent events.

Provided a recoiling lattice atom receives an energy in excess of a minimum value, about 25 eV, called the displacement energy, $E_d$, it can leave its lattice site to become permanently displaced within the solid. The exact

154

magnitude of this energy not only depends on the solid in question but also on the recoil direction within the lattice. In other words, it is somewhat easier to displace an atom in certain well defined directions than in others. This problem has been considered both theoretically and experimentally for a variety of materials (Banbury et al., 1966; Erginsoy et al., 1964; Lomer and Pepper, 1968). For instance fig. 3.1. shows a displacement threshold map for b.c.c. iron as calculated in a computer simulation study. It is readily apparent that in this case the easiest displacement direction is the ⟨100⟩ at around 17 eV whilst atoms recoiling in the region near the centre of the unit

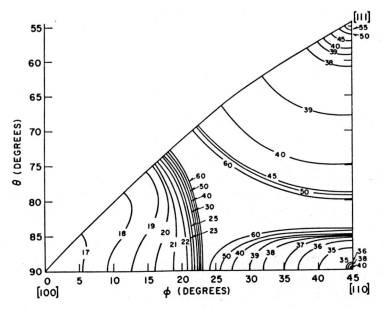

Fig. 3.1. Contours of displacement threshold within the unit triangle (after Erginsoy et al. 1964).

triangle require many times this energy. In general, however, and especially in the context of ion implantation, it will suffice to take 25 eV as a good average value for virtually all materials.

In most cases of interest lattice atoms recoil from primary collisions with kinetic energies far in excess of the displacement energy, and as such are capable of penetrating many atomic distances into the surrounding lattice. To a good approximation, as these recoil atoms start from lattice sites, their penetration into the solid is expected to result from a succession of uncorrelated collisions and is therefore amenable to calculation. As the mean energy falls, the scattering cross-section will become so large that the mean free path between successive collisions will be essentially equal to the distance between adjacent atoms. In such a situation it is quite clear that a cascade of secondary collisions will be initiated very close to the original primary event, see fig. 3.2. It is the spreading of such collision cascades which is therefore important to radiation damage, as those regions remaining in the wake of the cascades will contain many displaced atoms.

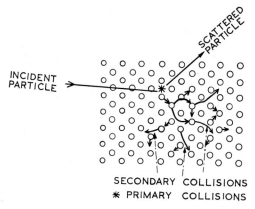

SECONDARY COLLISIONS
* PRIMARY COLLISIONS

Fig. 3.2. The interaction of an incident particle with a solid showing the primary collision and subsequent secondary collisions.

From the foregoing it is readily apparent that in order to understand radiation damage we must first focus our attention on the details of the collision cascade and the initial configuration of displaced atoms. The first of these has been covered extensively by Volume 1 in this series of monographs and, in this instance, it will be sufficient to present just an outline of the more important points. The majority of our knowledge has been gained from both theoretical and experimental studies on metals; although recently, due to the interest in ion implantation in semiconductors, a substantial effort has been directed towards Si, Ge and GaAs. In this monograph we are not exclusively concerned with implantation into semiconductors, consequently we shall discuss the production of radiation damage in metals in some detail, as it is such materials which have provided the basis for our present understanding. Originally, theories of the collision cascade assumed the solid to have no structure and the multiplication of collisions was assumed to be completely random. However, it is now known that the regular nature of the crystal lattice plays a vital part in the spreading of the collision cascade. It was Silsbee (1957) who first pointed out that the ordered atomic array would impose a directional correlation between successive collisions and that energy and momentum would be focused into those directions consisting of close-packed rows of atoms. To illustrate this effect consider the simple collision illustrated in fig. 3.3 between two identical atoms, represented by two perfectly elastic hard spheres of radius $R$, initially

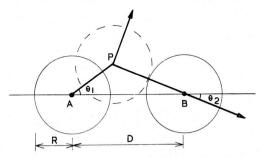

Fig. 3.3. Simple collision between two atoms of radius $R$ initially separated by $D$.

separated by a distance $D$. Suppose the first atom is given momentum in the direction AP, making an angle $\theta_1$, with the line of centres AB. It will move along AP until a collision occurs when its centre has reached P. The second atom will then move off along PB at an angle $\theta_2$ to AB. For the case of small angles a simple geometrical argument shows that

if $D > 4R$, then $\theta_2 > \theta_1$

or                                                                                          (3.1)

if $D < 4R$, then $\theta_2 < \theta_1$

Thus, in a row of such atoms, provided the atomic radius is greater than one quarter the interatomic spacing, successive angles will converge towards zero until all collisions are effectively 'head-on'. Under these conditions, therefore, momentum has been focused into the line, and for this reason such sequences of correlated collision events have been called *focused collision sequences* or *focusons*. In a real crystal it is necessary to relate the effective radius of an atom to the interatomic potential. At the atomic separations of interest the nuclear charges of the colliding atoms are heavily screened by their orbital electrons and a purely exponential potential provides a reasonable approximation to the interaction, for example a Born–Mayer potential of the form

$$V(r) = A \exp\left(-r/a\right)$$                                                      (3.2)

where $A$ and $a$ are constants for a particular atom. Under these conditions, to a first approximation, near head-on collisions can be treated as if the atoms were perfectly elastic hard-spheres. Such an approximation clearly over-simplifies the problem but has proved very successful in obtaining order of magnitude estimates. The interaction is likened to that between two billiard balls of radius $R$ such that in any collision their centres are separated by a distance $2R$. Then transforming the collision co-ordinates into the centre of mass system, the total kinetic energy of the system is equal to $\frac{1}{2}E$, where $E$ is the kinetic energy of the moving atom in the laboratory system. In a head-on collision this energy is then simply equal to the potential energy of the interaction i.e.

$$V(2R) = \tfrac{1}{2}E \tag{3.3}$$

whence using $(3.2)$

$$R = \tfrac{1}{2}a \ln (2A/E) \tag{3.4}$$

Using this expression we can relate the geometric focusing condition contained in $(3.1)$ to a practical case. The first point to notice is that the hard-sphere radius is inversely proportional to the logarithm of energy so that at high energy $R < \frac{1}{4}D$ and focusing cannot occur. Then as the energy decreases the radius becomes slowly larger until at a critical energy corresponding to $R = \frac{1}{4}D$ the correlation between successive collisions permits a reduction in angle. This energy is known as the *focusing energy* and is given simply by

$$E_f = 2A \exp (-D/2a) \tag{3.5}$$

This hard-sphere approach has provided us with a simple basis to discuss focused collision sequences. However, in practice atoms are not rigid spheres and we must allow for the 'soft' nature of the interaction. In a realistic collision, the true scattering angle corresponding to a given impact parameter is somewhat less than that predicted by the hard-sphere model. This means that the hard-sphere approach overestimates the focusing and so in practice focusing will occur at an energy somewhat less than that given by $(3.5)$. On the other hand, with a realistic potential, atoms ahead of the main energy packet will start moving before its arrival; this leads to a shortening of $D$ and a consequent increase in focusing. Lehmann and Leibfried $(1961)$ have considered these effects in some detail, but their treatment is too sophisticated for our present purposes. However, a simpler version due to Sigmund (1965) gives the following reasonably simple analytical expression:

$$E_f = 2A \exp\left[-\frac{D}{4a}\left\{1+\left(1+\frac{8a}{D}\right)^{\frac{1}{2}}\right\}\right] \qquad (3.6)$$

In a real crystal it is not only sufficient to satisfy the above conditions for the propagation of focused collision sequences but also that the surrounding atoms do not interfere to such an extent as to destroy the sequence. A simple consideration of both f.c.c. and b.c.c. lattices shows that in each case their close-packed rows, i.e. the $\langle 110 \rangle$ and $\langle 111 \rangle$ directions respectively, are situated in an environment which allows the unhindered propagation of collision sequences. But on the other hand, in the diamond lattice, such isolated rows do not exist and the propagation of such sequences is unlikely.

In order to put the foregoing discussion on a quantitative basis let us consider a row of copper atoms equi-spaced by a distance equal to the close-packed $\langle 110 \rangle$ direction of the f.c.c. copper lattice, 2.55 Å. Using the Born–Mayer constants used by Vineyard et al. (1960) of $A = 2.2 \times 10^4$ eV and $a = \frac{1}{13} D^{110}$, we obtain a focusing energy using (3.6) of $E_f^{110} = 35$ eV. The next step is to consider how far such focused collision sequences are likely to travel. As already pointed out atoms are not just hard-spheres, with the consequence that as sequences propagate through the lattice, both potential and kinetic energy will be dissipated. Further, as in reality the atoms of a crystal are in a continual state of thermal vibration, perfect focusing can never be attained and energy is inevitably lost to neighbouring atoms as a consequence of lateral displacement. A full discussion may be found in Volume 1 of this series; the net result being that for copper at room temperature the mean energy lost per collision is roughly constant at 3 eV, and so a sequence starting off at the focusing energy has a maximum range of some 10 collisions, in other words about 25 Å.

So far we have only considered simple focused collision sequences which propagate along the close-packed rows of the crystal lattice. If we also consider the possibility of such sequences propagating along the next close-packed rows of atoms, i.e. the $\langle 100 \rangle$ directions in both the f.c.c. and b.c.c. structures, their focusing energies as calculated by (3.6) turn out to be too low for practical importance. However, in both cases, atoms which recoil close to these $\langle 100 \rangle$ directions have their trajectories essentially focused by the surrounding atoms, see fig. 3.4. Once again, it is possible to focus energy into the line by a succession of such collisions. This effect is called *assisted focusing* and has been treated theoretically by Nelson and Thompson (1961)

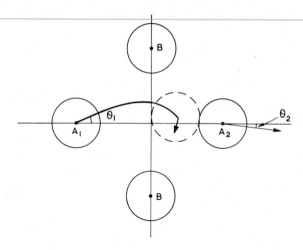

Fig. 3.4. Plan view showing the atomic trajectory of a $\langle 100 \rangle$ collision sequence.

and by Leibfried and Dederichs (1962). The range of such assisted focused collision sequences is generally less than expected for simple sequences.

In the context of radiation damage we must assess the influence of focused collision sequences both on the numbers of atoms displaced and on their initial distribution. From the foregoing we know that whenever the energy of the cascade falls below the focusing energy, the random multiplication process ceases and successive collisions are strongly correlated so that pulses of energy are focused into certain low index directions. We can therefore assume that once this situation is reached the number of further displacements is strictly limited. In the random collision model we must first of all estimate the total fraction of energy within the collision cascade that is dissipated via nuclear rather than electronic losses. From the previous chapter, using the Lindhard expressions for energy loss, we can simply estimate the relative fractional energy loss of the primary recoil atom itself as a function of energy. Then by integration we can obtain the total fractional energy loss to nuclear collisions for the primary from its starting energy $E_p$ until it comes to rest. However, we must remember that the secondary and subsequent recoils also lose a fraction of their energy to electronic excitation and so the total energy dissipated in nuclear collisions within the cascade is still further reduced from that fraction lost by the primary alone. Figure 3.5 shows an estimate of the net total energy lost to such nuclear collisions $(\Delta E)_n$ within a collision cascade in Cu initiated by primaries of given energy plotted as a function of primary energy. Then in the hard sphere approxi-

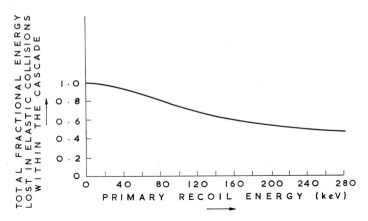

Fig. 3.5. The net total energy lost to elastic collisions within a collision cascade in copper plotted as a function of primary recoil energy.

mation without focusing, the final number of displaced atoms is simply given by $(\Delta E)_n/2E_d$. On the other hand, a more realistic treatment which accounts for the inadequacy of the hard sphere approximation would give $0.8(\Delta E)_n/2E_d$. However, if we include correlation effects it would seem more appropriate to use an expression for the initial number of displacements such as:

$$n_d = 0.8(\Delta E)_n/2\bar{E}_f \qquad (3.7)$$

where $\bar{E}_f$ is a mean focusing energy averaged over all possible focusing directions.

With regards to the spatial distribution of displaced atoms it is well known (see Volume 1) that in some instances focused collision sequences propagate via a series of replacement collisions such that interstitial-vacancy pairs can be separated by many atomic distances. Further, at the extremity of the cascade the majority of collision sequences will be directed radially outwards from the cascade centre with the inevitable result that the damaged region will in the first instance have a vacancy rich core surrounded by a region rich in interstitials, see fig. 3.6. This will be especially true for the heavier metals where, due to the large scattering cross-section, the central core region of the collision cascade is rather small. However, in lighter metals such as aluminium the collision cascade is spread over a very much larger volume and also, because focusing is expected to be rather weak in aluminium, the net result is that the collision cascade contains a relatively uniform distribution of interstitials and vacancies.

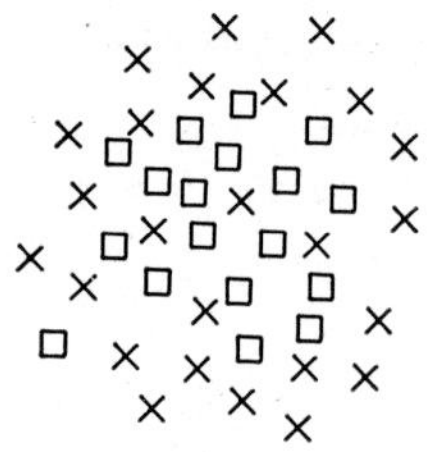

☐ Vacancy　　　　　× Interstitial

Fig. 3.6. The initial distribution of vacancies and interstitials left in the wake of a collision cascade.

From the creation of the primary recoil the high energy secondary collisions are completed in about $10^{-13}$ sec. However, this leaves a highly damaged region with an accumulation of excess energy which can only be dissipated by thermal vibration. This results in the creation of a local hot-spot, called a *thermal spike*. The life-time of such a spike depends on the efficiency of heat conduction to the surrounding lattice. In this case, however, the energy is initially given to the atomic system, and, as the transfer of energy between the atomic and electronic systems is relatively slow, we must consider the cooling of the spike to proceed by atomic processes rather than by the usual equilibrium electronic conduction processes. After about 10 atomic vibration periods, the velocity distribution of the atoms in the heated region approximates to that expected from Maxwell–Boltzmann statistics and as such becomes amenable to calculation. Experiments by Thompson

TABLE 3.1

The temperature, size and lifetime of thermal spikes produced by 45 keV $Xe^+$, $A^+$ and $Ne^+$ ions

| Metal | Ion | $T_s$ (°K) | $\tau_s$ ($10^{-12}$ sec) |
|---|---|---|---|
| Au | Xe | 910 | 3 |
| | A | 600 | 6 |
| Ag | Xe | 530 | 6 |
| | A | 523 | 4 |
| | Ne | 518 | 7 |
| Cu | Xe | 490 | 5 |
| Zn | Xe | 150 | 1 |
| Bi | Xe | 600 | 9 |
| Ge | Xe | 1060 | 3 |

and Nelson (1962) and Nelson (1965) have specifically studied the sizes, temperatures and life-times of thermal spikes in a variety of solids; a few of these are listed in table 3.1. In general, as the temperatures of such spikes are only a few hundred degrees above that of the surrounding lattice and their life-times are only a few times $10^{-12}$ sec, it is unlikely that significant defect rearrangement will occur as a result of thermal spikes, except perhaps for the recombination of some close interstitial-vacancy pairs.

### 3.1.2. The physical configuration of damage

*3.1.2.1. Metals.* In the preceding section we outlined the mechanisms responsible for the production of radiation damage. However, in reality the target is maintained at some temperature above 0 °K, with the result that the individual defects created within the collision cascade can migrate under the influence of thermal activation. On the other hand, even at 0 °K some defect rearrangement is possible due to the mutual elastic interaction between defects. For instance, computer calculations have shown that there is a critical separation between an interstitial and a vacancy which must be attained before the defects do not spontaneously recombine. This sets an ultimate limit on the density of damage that can be produced even at temperatures too low for thermally activated migration. Experimentally this is manifest in the saturation in the change of some physical property, such as electrical resistivity on irradiation; see fig. 3.7 for instance. From an analysis of such results it is possible to make estimates of the average volume around a point defect which is unstable against spontaneous recombination. Typically

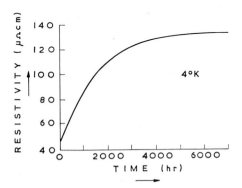

Fig. 3.7. The saturation of resistivity of α Pu plotted as a function of irradiation time held in liquid helium (after King et al., 1965).

this is of the order of 50–100 atomic volumes, which therefore sets an absolute limit on the number of individual point defects that can exist at $10^{-2}$.

However, in most circumstances, at temperatures typical of the conditions pertaining during ion implantation both individual vacancies and interstitials can migrate through the lattice. The first question to answer is, what is the configuration of damage in the vicinity of displacement cascades under these conditions, e.g. room temperatures in copper, and when the ion dose is sufficiently low that such cascades are well separated? Clearly, a substantial degree of mutual recombination will occur, thus reducing the total number of defects from that originally created. But perhaps even more important, due to the dynamic segregation of defects within the collision cascade, which occurs predominantly in heavy materials, the vacancy rich core can rearrange to form a compact defect cluster, so leaving the interstitials and residual vacancies free to migrate away from the damaged region.

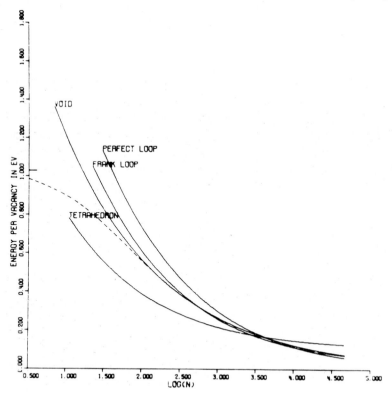

Fig. 3.8. Specific energy per vacancy plotted as a function of number of vacancies for different defect agglomerates in Cu (after Sigler and Kuhlmann-Wilsdorf, 1967).

The precise morphology of these clusters, i.e. whether they form three dimensional voids, faulted dislocation loops, unfaulted dislocation loops or stacking fault tetrahedra, depends on a variety of parameters such as stacking fault energy and surface energy. For instance, Sigler and Kuhlmann-Wilsdorf (1967) have made a number of predictions for a variety of pure metals depending on the number of vacancies within the agglomerates, see fig. 3.8. However, due to the inadequacy of our knowledge of the parameters involved, we must treat the quantitative results of such computations with some reservation. In light materials, especially those associated with poor focusing such as aluminium or graphite, where the segregation of vacancy and interstitial defects is somewhat limited, the heterogeneous nucleation of defect clusters is less likely, with the result that perhaps every vacancy and interstitial is free to migrate through the lattice in equal numbers. Some experimental evidence for these ideas has been provided by field-ion and electron microscopy. Both techniques suggest that in heavy metals such as copper, gold, silver, iridium and tungsten, at temperatures where defects are mobile, the individual collision cascades leave regions which give contrast consistent with some

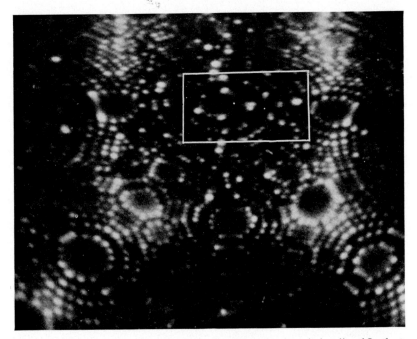

Fig. 3.9. A field ion microscope photograph of a dislocation loop in irradiated Ir, the two white dots define the ends of the dislocation (after Hudson, 1969).

Fig. 3.10. Block spot defects in Cu irradiated with 30 keV Cu⁺ ions, each spot corresponds approximately to one incident ion. Analysis of the diffraction contrast suggests that such defects are vacancy type dislocation loops (after Wilson, 1970).

sort of vacancy agglomerate, generally a small dislocation loop, figs. 3.9 and 3.10 (Hudson, 1969; Wilson, 1970). However, in light metals bombarded to low doses, no such evidence for cluster formation within the collision cascade has been forthcoming as expected.

Next we must discuss the general behaviour as the bombardment dose steadily builds up to where the individual collision cascades overlap. Whilst the individual cascades are well separated a large fraction of the freely migrating defects will be lost to the surface, which in most cases of interest acts as the most dominant sink. However, in heavy metals, as the dose builds up the isolated vacancy agglomerates will build up linearly until eventually overlap and saturation prevails, fig. 3.11. Concurrent with this, the interstitial concentration will build up to a sufficiently high level that

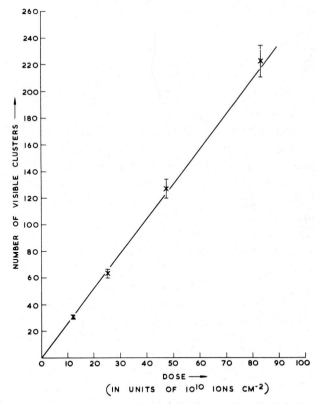

Fig. 3.11. The number of clusters in 30 keV Cu$^+$ ion bombarded Cu with diameters greater then 25 Å plotted as a function of dose showing the linear relationship (after Wilson, 1970).

the homogeneous nucleation of interstitial clusters occurs most probably in the form of loops as in fig. 3.12. As the dose is increased still further those new vacancies and interstitials which escape recombination will be captured by the existing defect clusters rather than sustain further nucleation. The clusters will therefore steadily grow until they interact and eventually form a complicated dislocation entanglement as illustrated in fig. 3.13. This situation represents an ultimate saturation in configuration of damage as further irradiation will simply result in the rearrangement of this entanglement by the processes of slip and climb.

On the other hand, in light metals such as aluminium, where the heterogeneous nucleation of vacancy clusters within cascades is considered unlikely, substantial self-annihilation occurs in the early stages. However, as the damage level increases, it is thought that homogeneous nucleation of both interstitial and vacancy clusters occurs simultaneously (Beevers and Nelson, 1963).

It is of interest to estimate the actual number of atoms which remain displaced from their lattice sites, especially in the vicinity of the surface. As we have already stated, under normal conditions, e.g. implantation of say copper at 20 °C, the number of free vacancies and interstitials is essentially zero, these having clustered to form a complex dislocation array. It is well known that dislocations exhibit strong elastic interaction both with each other and with the surface. The surface interaction is manifest in the creation of an image force which acts so as to remove the dislocation from the solid. The net result is that even under saturation damage conditions the number of atoms in the vicinity of the surface which are not situated in equilibrium lattice positions is only about $10^{-3}$.

Let us next consider the situation at slightly higher temperatures too low for the thermal activation of vacancies from extended sources, but where the discrete vacancy rich regions created within displacement cascades spontaneously dissociate, e.g. 200 °C in copper. Under these conditions vacancy cluster formation is unlikely, however providing some vacancies can escape to nearby sinks, such as the surface, or become trapped as part of small gas bubble nuclei, some interstitial clusters will in fact form and survive. For instance, the large intrinsic faulted loops which occur during proton irradiation of copper foils at $\sim$ 250 °C, fig. 3.14 (Mazey and Barnes, 1968; Tunstall et al., 1970). In such an environment, if the irradiation dose is continued to very high doses, the excess vacancies which remain in supersaturation can condense to form voids (Nelson et al., 1970), fig. 3.15. At higher irradiation temperatures still the thermal activation of vacancies from

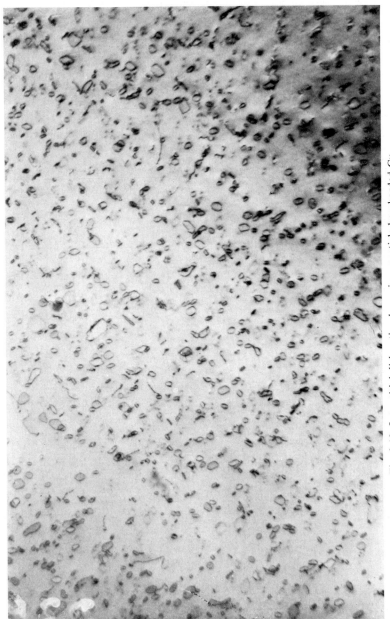

Fig. 3.12. Interstitial dislocation loops in $\alpha$-particle bombarded Cu.

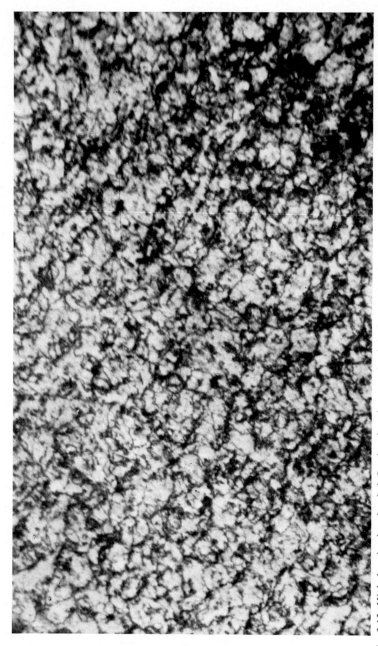

Fig. 3.13. High dose ion bombarded Cu showing the growth of the interstitial dislocation loops into a complex dislocation entanglement.

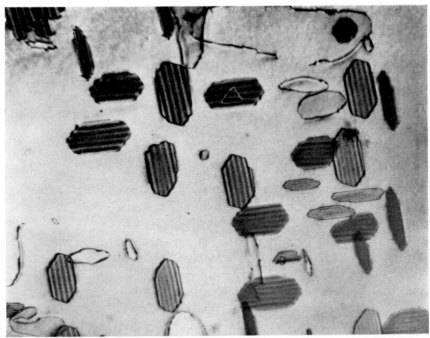

Fig. 3.14. Faulted interstitial loops in proton irradiated Cu at 250 °C (after Mazey and Barnes, 1968).

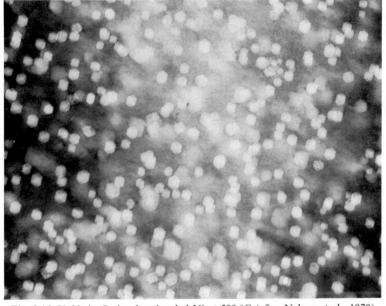

Fig. 3.15. Voids in Cu ion bombarded Ni at 500 °C (after Nelson et al., 1970).

the extended surface source will generally preserve thermodynamical equilibrium and prevent cluster formation. However, as we shall see later, even under these circumstances some clustering can in fact occur, especially during bombardment with inert gas ions.

*3.1.2.2. Semiconductors.* So far we have considered the nature of radiation damage in metallic solids where to all intents and purposes the solid remains essentially crystalline, even after excessive bombardment to where each atom is displaced many hundreds of times. However, it will be readily apparent from the following that the class of solids exhibiting strong covalent bonding, including both semiconductors and insulators, behave quite differently under irradiation.

Electron microscope studies of ion bombarded Ge (Parsons, 1965), Si (Mazey et al., 1968) and GaAs (Mazey and Nelson, 1969) have shown that radiation damage is manifest in the creation of essentially amorphous regions $\sim 30$ Å diameter. It is thought that these regions result from the disorder produced in the vicinity of the primary recoils as a consequence of displacement cascades. The precise mechanism for their creation and why such a phenomenon does not occur in metals is not yet understood; but as we will see, there is an apparent relationship between the temperature of the solid and the size of the collision cascade, which strongly influences their formation. Furthermore, the actual number of atoms not occupying proper lattice sites within these regions may be substantially in excess of those dynamically displaced during the spreading of the collision cascade. But clearly not all displaced atoms are contained within the disordered zones as those interstitials and vacancies created at the outermost of the collision cascades will be free to migrate randomly through the lattice either to recombine or to cluster and create the well known defect centres described for example in Corbett (1966) and Watkins (1967).

In this instance we shall not discuss such isolated defect centres at length, for, as we shall see, in most cases of interest to ion implantation the fraction of displacements which remain as isolated point defects is very small, especially for high dose implantations which result in the complete destruction of the crystalline lattice. Moreover, to obtain reasonable electrical activity of the implanted species within the semiconductor, it is usually necessary to heat the sample to a temperature well above that which results in the complete annealing of such isolated defects. On the other hand, a study of, for instance, the depth distribution of isolated point defects within a solid implanted to low doses, can provide valuable information to compare with theoretical predictions of such distributions. The most significant data on simple point defects produced in semiconductors has been provided for Si by electron paramagnetic resonance (EPR) and local mode infra-red adsorption techniques. As yet the self interstitial has not been identified, but the vacancy and certain vacancy complexes are thought to have been identified unequivocally. For instance, the doubly charged vacancy in n-type Si is thought to migrate with an activation energy of $0.18 \pm 0.02$ eV, and the neutral vacancy in p-type Si is thought to migrate with an activation energy of $0.33 \pm 0.03$ eV. Another defect that has been studied is the di-vacancy, this has an activation energy for migration of about 1.2 eV.

As the bombardment dose builds up the disordered zones steadily increase in number until eventually overlap occurs. At this stage the bombarded regions will have to all intents and purposes degenerated into a non-crystalline amorphous phase. The depth below the surface at which overlap first occurs will depend on the spatial distribution of primary recoils along the ion track. Sigmund and Sanders (1967) have calculated the depth distribution of radiation damage expected during ion bombardment using a realistic scattering model. They obtained a Gaussian-type distribution curve very similar to that expected for the projected range curve of the implanted ion. Overlap will first occur at the peak of the distribution and will gradually spread to either side as the integrated dose builds up. In general, for low bombardment energies the side of this distribution nearest to the surface is 'cut off' by the surface at say half-peak value; on the other hand the distribution on the far side of the surface decreases gradually to zero. Thus, at a dose corresponding to overlap at the surface, a substantial fraction of zones will still remain isolated on the far side of the peak. In this instance ion bombardment will result in a completely amorphous phase to a depth of approximately 1.5 to 2 times the projected range, together with a gradually decreasing concentration of isolated amorphous zones. In the case of rather more energetic ions (e.g. > 200 keV), the projected range distribution forms a peak somewhat deeper into the solid and in this case it is possible to form an amorphous layer completely within the target. The existence of the amorphous phase is illustrated in fig. 3.16, where a pre-thinned ( $\sim$ 1000 Å thick) specimen of GaAs has been irradiated with 80 keV $Ne^+$ ions. The top area of light contrast was shielded from the ion beam with a grid, whilst the lower area suffered bombardment with $4.0 \times 10^{15}$ ions $cm^{-2}$ at 20 °C. Selected area electron diffraction of these regions clearly demonstrates the degeneration from a crystalline to the amorphous state.

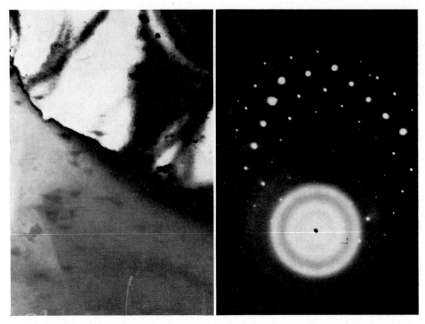

Fig. 3.16. Transmission electron micrograph of Ne ion bombarded GaAs at 20 °C together with diffraction patterns. The light contrast area has been shielded during irradiation (after Mazey and Nelson, 1969).

During bombardment of Si, that part of the surface on which the ions impinge undergoes visible colour changes as the dose builds up (Nelson and Mazey, 1968). At threshold, faint blue or pink hues are observed, the actual colour depending on the ion and its energy. However, as the dose increases these hues eventually saturate into a characteristic 'milky white' appearance. Figure 3.17 shows a photograph of a silicon slice bombarded through a grid so as to provide a series of demarkation lines; the milky appearance is readily seen. It is throught that these visual effects arise because of a Rayleigh type scattering of the incident white light by the small disordered zones and can be qualitatively understood as follows.

At low ion doses the individual disordered zones act as isolated scattering centres, and provided their radii are very much less than the wavelength of the illumination, the scattered energy per unit volume can be described by the Rayleigh relation:

$$E \propto nv^4 \sin^{-2} \gamma \qquad (3.8)$$

where $\gamma$ is the frequency of the radiation, $n$ is the number of scattering

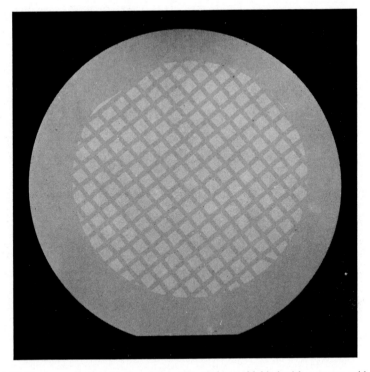

Fig. 3.17. Photograph of an ion bombarded Si specimen shielded with a coarse grid. The 'milky' effect is clearly visible.

centres per unit volume, and $\gamma$ is the angle between the direction of polarization of the incoming beam and the direction of propagation of the scattered beam (see Ditchburn, 1952). Then

$$\sin^2\gamma = 1 - \sin^2\alpha \cos\beta \qquad (3.9)$$

where $\alpha$ is the included angle between the direction of propagation of the incoming and scattered beam, and $\beta$ is the angle between the direction of polarization of the incoming beam and the plane defined by propagation vectors of the incoming and scattered beams. For the purposes of the present discussion, we shall suppose that the scattering centres are uniformly distributed throughout the surface layers to a depth equal to the range of the incident particle $(R)$. In reality, of course, this is not so, as these centres are more likely to be distributed in some Gaussian form.

Consider the scattering from a layer thickness $dx$ at $x$ below the surface in the direction specified by spherical co-ordinates $\theta = \theta_s$ and $\phi = 0$;

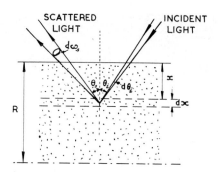

Fig. 3.18. Diagrammatic representation of scattering in $d\omega_s$ at an angle $\theta_s$, from the amorphous zone in a layer thickness $dx$ at a depth $x$ below the surface; $R$ defines the maximum depth of the scattering centres, which are assumed to be uniformly distributed throughout the irradiated volume.

fig. 3.18. If the incident beam is specified by $\theta = \theta_i$ and $\phi = \phi_i$, $\alpha$ is given by

$$\cos \alpha = \cos \theta_s \cos \theta_i + \sin \theta_s \sin \theta_i \cos \phi_i \qquad (3.10)$$

then, for a particular frequency, the intensity remaining after absorption in travelling to the layer is

$$I(\nu, x, d\theta_i) = I_0 \exp\left(-a_\nu x \sec \theta_i\right) \sin \theta_i d\theta_i \frac{d\theta_i}{2\pi} \frac{d\beta}{2\pi} \qquad (3.11)$$

where $a$ is the absorption coefficient. The fraction scattered per unit area in $dx$ is proportional to $n\nu^4 \sin^2\gamma dx$, and so the intensity scattered into the solid angle $d\omega_s$, subtended by the viewer's eye, at $\theta_s$, is

$$I_s(\nu, x, d\theta_i, \theta_s, d\omega_s) = CI_0 n\nu^4 d\omega_s \sin \gamma \exp\left(-a_\nu x \sec \theta_i\right)$$

$$\times \exp\left(-a_\nu x \sec \theta_s\right) \sin \theta_i d\theta_i \frac{d\phi_i}{2\pi} \frac{d\beta}{2\pi} dx \qquad (3.12)$$

where $C$ is a constant. The total scattered intensity from a depth $R$ is then

$$I_s(\nu, \theta_s, d\omega_s) = \frac{C}{4} I_0 n\nu^4 d\omega_s \int\int_0^R \left[3 - \cos^2 \theta_s + (3 \cos^2 \theta_s - 1) \cos^2 \theta_i\right]$$

$$\times \exp\left[-a_\nu x(\sec \theta_s + \sec \theta_i)\right] \sin \theta_i dx d\theta_i \qquad (3.13)$$

The relative intensity can thus be calculated for scattering into $d\omega_s$ at $\theta_s$ as a function of frequency. Figure 3.19 shows $I_s(\nu, \theta_s, d\omega_s)$ plotted as a function of photon energy ($h\nu$) for different values of $R$, together with the appropriate limits to the visible spectrum. Values of $a_\nu$ for silicon have been taken from Dash and Newman (1955). Such spectral distributions, which cover the

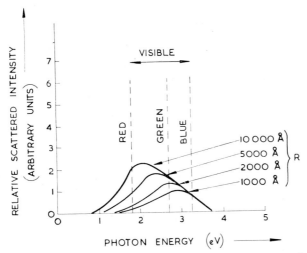

Fig. 3.19. Relative scattered intensity at $\theta_s = 45°$ plotted as a function of photon energy, for different scattering thicknesses $R$. The limits of the visible spectrum are included for convenience.

whole of the visible range, exhibit a hue, the colour of which depends on the position of their maximum (Wright, 1946). For instance, in the case of a shallow scattering region < 1000 Å, the scattered light will exhibit a blue hue, whereas, as the scattering region moves farther and farther into the specimen, the hue will appear pink. With increasing ion dose, the density of scattering centres increases; appreciable multiple scattering will then occur, with the result that the coloured hues will be replaced by a milky-white appearance typical of diffuse scattering. This final state will appear similar independent of ion energy or mass.

The onset of the milky appearance in Si has been used as a useful criterion for the existence of an amorphous surface phase. For instance in the ion implantation of semiconductors, it is of major importance to know the threshold dose for the production of an amorphous surface phase under a variety of conditions of ion species, ion energy, target temperature and target orientation. We will now outline a selection of such experiments together with their results. Firstly, the influence of bombardment temperature on the formation of the amorphous layer has been readily studied by observing the sample continuously whilst it was under irradiation on a heated support (Nelson and Mazey, 1968). The results are illustrated in fig. 3.20 for 80 keV Ne+ ions bombarding the {111} face of a Si sample, which shows the threshold dose for the milky appearance plotted logarithm-

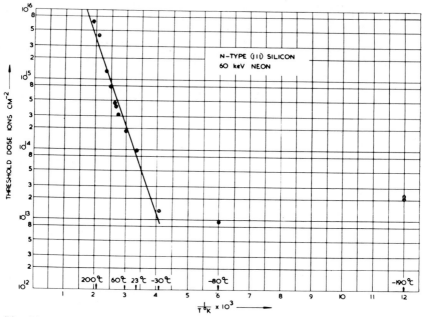

Fig. 3.20. Graph of the threshold dose required to produce a visible amorphous film, plotted as a function of reciprocal temperature.

ically as a function of the inverse of the absolute temperature. The most striking feature of this curve is that an increase in temperature by as little as 200 °C results in the retention of the crystalline nature of the Si to doses more than two orders of magnitude above those for room temperature. The result was apparently independent of dose rate as this was varied by more than two orders of magnitude, and was also independent of the type or resistivity of the material. Similar effects are seen for both Ge and GaAs. In this context it is worth noting that Parsons (1965) observed an apparent temperature dependence on the size distribution of disordered zones in Ge (fig. 3.21), the medium size becoming smaller the higher the temperature. Analysis of the threshold curves in the case of Si suggests an activation energy of $0.3 \pm 0.05$ eV. It is interesting to note that the measured activation energy for migration for the neutral vacancy in Si is quoted at 0.33 eV (Watkins, 1963), and this immediately suggests that the formation of the disordered zones depends on temperature via the migration of such vacancies.

The effects can be qualitatively understood as follows. The continuous amorphous phase results when the individual disordered zones overlap, and this depends on the size of the zones and their production rate per unit

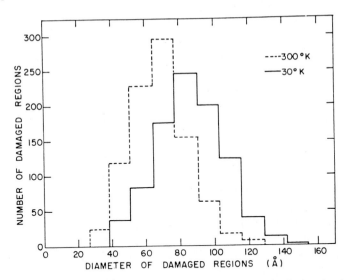

Fig. 3.21. Histogram size distribution of disordered regions in Ge bombarded at room temperature and at 30 °K (after Parsons, 1965).

volume. The initial creation of displacement cascades depends on the scattering cross-section and to a first approximation we can use the hard-sphere expression resulting from a screened Coulomb interaction, i.e.

$$\sigma(E) = \frac{2\pi a_B^2 E_R Z_1 Z_2 (M_1 + M_2)}{eE(Z_1^{\frac{2}{3}} - Z_2^{\frac{2}{3}})^{\frac{1}{2}} M_2} \tag{3.14}$$

where $a_B$ is the Bohr radius, $E_R$ the Rydberg energy and $e$ is the base of natural logarithms. Averaging this from the initial energy $E_0$ to some lower energy limit $E_1$ we have the average cross-section

$$\sigma = \frac{2\pi a_B^2 E_R Z_1 Z_2 (M_1 + M_2)}{e(Z_1^{\frac{2}{3}} + Z_2^{\frac{2}{3}})^{\frac{1}{2}} M_2} \cdot \frac{\ln (E_0/E_1)}{(E_0 - E_1)} \tag{3.15}$$

The density of collision cascades is then simply $\phi t N_0 \sigma$, where $\phi$ is the incident ion flux, $t$ the irradiation time and $N_0$ the atomic density. In the hard sphere approximation all primary recoil energies are equally probable, and so to a first approximation we have a uniform size distribution of collision cascades within the irradiated volume. Then whether or not these individual damaged regions subsequently form stable disordered zones, apparently depends on the temperature. For at high temperatures ($\sim 200$ °C) only the largest survive, whereas as the irradiation temperature is lowered smaller and smaller regions spontaneously rearrange to form stable zones.

It is possible that the controlling process is the rate at which the neutral vacancies can migrate away from the damaged regions before they degenerate into stable disordered zones. Thus at very low temperatures where the vacancy mobility is low, even the smallest disordered zones are stable and zone overlap occurs rapidly. However, as the temperature increases fewer and fewer damaged regions form stable zones with the result that the ion dose for overlap becomes steadily larger. The horizontal line in fig. 3.20 simply results from the fact that in order to see the milkyness, we must have overlap, and at these temperatures the vacancy mobility is so slow that even the smallest (say 5 Å) zones are stable, zones smaller than this are meaningless inasmuch as they contain too few atoms to significantly influence the result.

To a first approximation we can say that overlap occurs when the average sized zones of radius $r$ just touch. Then in the temperature independent region, where all zones are stable, this occurs when the mean density of zones equals the inverse of the volume of the average zone, i.e. when

$$\phi t N_0 \sigma = 1/\tfrac{4}{3}\pi r^3 \tag{3.16}$$

In other words the dose for overlap at low temperature is simply

$$\phi t = 3/4\pi r^3 N_0 \sigma \tag{3.17}$$

Using the observed results this yields an average value for the average zone size of $\sim 40$ Å which is in reasonable agreement with that found by electron microscopy (Parsons, 1965; Mazey et al., 1968).

A similar experiment measured the threshold dose for amorphicity as a function of bombarding ion mass at constant temperature (Nelson, 1969). The results for Si bombarded at room temperature with 80 keV ions are illustrated in fig. 3.22. From the above discussion we expect this threshold to vary to a first approximation via an inverse proportionality to $r^3\sigma$. However, within the accuracy of the calculations we can ignore the small variations in $r$ which result from the mass dependence of the maximum recoil energy, and the threshold dose should be roughly inversely proportional to the average scattering cross-section. In fig. 3.22 this variation is shown as a dotted line fitted to the experimental points for comparison.

So far, in our discussion of the formation of the amorphous phases which occur on ion bombardment of covalently bonded solids we have neglected any possible influence of the crystalline nature of the target with respect to channelling effects. If the crystal is bombarded with an aligned beam under conditions where the incident ions are channelled, the probabili-

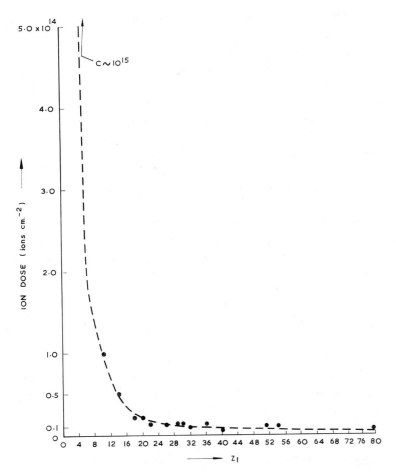

Fig. 3.22. The threshold dose for a visible 'milky' appearance of ion bombarded Si plotted as a function of atomic number of the bombarding ion. The dashed line shows the result of a simple theoretical treatment fitted to the experimental points.

ty of large-angle elastic collisions with the lattice atoms is reduced. The milkyness resulting from amorphicity should therefore appear after a dose greater than that corresponding to random trajectories of the incident ions. A simple experiment which clearly illustrates the effect has been described by Nelson and Mazey (1967). Single crystals of Si 2.5 cm in diameter and 0.03 mm thick, having either {111}, {100} or {110} planes parallel to the surface, were mounted in turn on a rotatable target holder which allowed simultaneous rotation about two orthogonal axes, fig. 3.23. The vertical

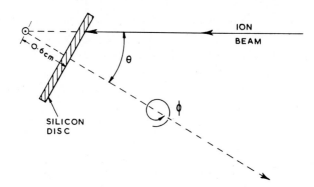

Fig. 3.23. Schematic experimental arrangement indicating the sense of the two orthogonal rotations of the crystal relative to the ion beam.

axis provided changes in inclination between the crystal and the ion-beam direction ($\theta$), and the horizontal axis provided continuous rotation about the normal to the crystal surface ($\phi$). The face of the crystal was held at a distance of 6 mm from the vertical axis, so that as $\theta$ was increased by 1° intervals the beam would describe a series of concentric circles on its surface. After a dose of 80 keV Ne$^+$ ions just sufficient to form an amorphous layer under nonchannelling conditions, as determined by the onset of the milky appearance viewed through the vacuum window, the angle of inclination was reduced by a succession of 1° intervals until the whole sample had been scanned with the ion beam. The irradiation time was carefully adjusted at each angular setting so as to produce a uniform dose over the bombarded surface. Figure 3.24 is a direct photographic reproduction of the {100}crystal surfaces after bombardment, together with a simple radial projection of the corresponding crystal face for comparison. Similar patterns were also recorded for the {111} and {110} faces. It is evident from the regular patterns mapped out on the crystal surface that those regions that do not exhibit a milky appearance correspond to angles of incidence close to the major axial and planar channelling directions. It should be noted that the apparent milkyness at the centre of the larger spots is an optical effect believed to be a consequence of photographing the coloured hues around these regions with sodium light. The results are consistent with a reduction in radiation damage and hence the rate of amorphous-phase formation, whenever the incident ion is channelled.

As the bombardment was continued, the milkyness gradually spread into those parts of the patterns which correspond to channelling, until eventually the whole silicon surface had become amorphous. Then to provide a measure of the reduction in radiation damage under conditions of channelling, the dose required to obliterate the patterns was compared with the dose necessary to just turn the surface milky under conditions when the incident ions were not channelled. For the major channelling direction, namely the $\langle 110 \rangle$, this occurred after a dose increment of about $8 \times$ and, therefore, infers a reduction radiation damage by a factor of 8 when the crystal is oriented so as to present its $\langle 110 \rangle$ channels to the incident ion beam.

The general features of the radiation damage produced in Si during ion implantation as described above have been verified by the more sophisticated techniques of Rutherford Scattering as outlined in Chapter 2. For instance Davies et al. (1967) have made a comprehensive study of the damage ensuing on 40 keV $Sb^+$ implanted Si. Samples having a $\{111\}$ orientation were

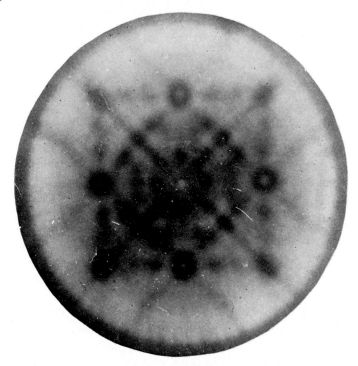

Fig. 3.24(a).

Fig. 3.24. (a) A photographic reproduction of a $\{100\}$ crystal surface showing the regular crystallographic patterns corresponding to a reduction in radiation damage whenever the ion beam is channelled. (b) A simple radial projection of the $\{100\}$ face of a f.c.c. crystal showing the low index planes and poles for comparison.

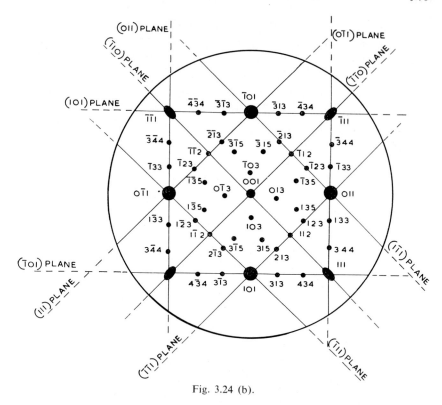

Fig. 3.24 (b).

implanted at room temperature with $Sb^+$ ion doses ranging from $10^{12}$ to $10^{16}$ ions $cm^{-2}$. An analysing beam of 1 MeV $He^+$ ions was then used to probe the lattice disorder blocking the $\langle 111 \rangle$ channels by observing the changes in backscattered yield for different $Sb^+$ ion doses. The results are shown in fig. 3.25(a). The peaks which result from increased scattering close to the surface are caused by backscattering from those Si atoms displaced from their equilibrium lattice sites. The area under the peaks is therefore a measure of the number of such atoms which remain displaced after implantation. However, in order to obtain this measurement the inherent peak due to the thin surface oxide film must first be subtracted (the dashed curve in fig. 3.25). The net results are illustrated in fig. 3.25(b) where it can be seen that as the implanted dose builds up the lattice disorder initially increases proportionally to the dose until after a dose of about $5 \times 10^{13}$, 40 keV $Sb^+$ ions $cm^{-2}$ the disorder quickly saturates. This behaviour is entirely consistent with the electron microscope observations previously discussed. The

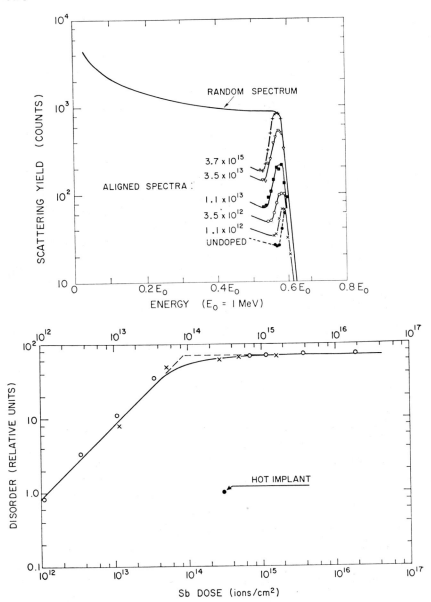

Fig. 3.25. (a) Aligned ($\langle 111 \rangle$) and random backscattering spectra for Si crystals implanted at 25 °C with various doses of 40 keV Sb$^+$/cm$^2$. An undoped crystal is included for comparison. Random spectra coincide within the statistical counting errors. The analyzing beam was 1.0 MeV He$^+$ (after Davies et al., 1967). (b) Lattice disorder in 40 keV Sb implanted Si at room temperature as a function of dose (after Davies et al., 1967).

slope of the linear region allows a determination of the magnitude of the disorder associated with each ion; in the case cited this yields a value of 3000 displaced atoms per ion. If we assume the disordered regions to be roughly spherical, this gives an average size of about 18 atoms diameter which is of the same order as that found by electron microscopy. The technique also allows us to obtain an estimate of the depth of the amorphous phase. For example, the results such as those illustrated in fig. 3.26 can be used directly from a knowledge of $dE/dx$ for $He^+$ ions in Si to give the effective thickness of the disordered layer. A typical set of values is given in table 3.2, and these compare favourably with measurements of amorphous layer thickness estimated by other techniques such as successive anodic stripping together with electron diffraction. More recent experiments (Eriksson et al., 1969) have used Rutherford scattering techniques to study the disorder produced accompanying hot implantation of Si with 40 keV $Sb^+$ ions. For example, fig. 3.27 shows the normalized degree of disorder after a dose of $2 \times 10^{14}$ ions $cm^{-2}$ at different bombardment temperatures. Once again the results bear out the rather crude optical measurements of Nelson and Mazey (1968) inasmuch as they show a strong temperature dependence of

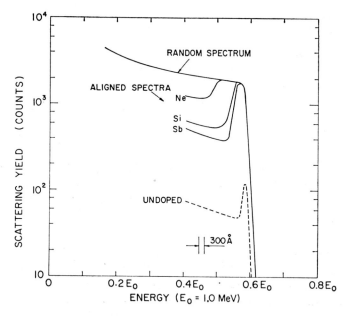

Fig. 3.26. A Rutherford scattering determination of the thickness of the damaged layer produced in Si with various bombarding ions (after Davies et al., 1967).

TABLE 3.2

Effective amorphous layer thicknesses at saturation

| Implanted ion | Energy (keV) | Thickness (Å) |
|---|---|---|
| Ne | 40 | 1600 |
| Si | 20 | 720 |
| Sb | 40 | 530 |
|  | 30 | 370 |

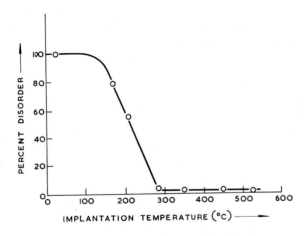

Fig. 3.27. Radiation damage produced in Si during Sb bombardment as a function of temperature (after Eriksson et al., 1969).

the production of disordered regions; furthermore, the activation energy calculated from fig. 3.27, 0.35 eV, agrees rather well with that deduced from the dose for the onset of amorphicity as a function of bombardment temperature.

Another useful technique for the study of damage produced in ion implanted semiconductors has been pioneered by Crowder et al. (1970). In this, electron spin resonance (ESR) signals of Si implanted at room temperature with $Si^+$, $P^+$, or $As^+$ ions were used to provide information both on the nature and the total density of radiation damage. The ESR signals were measured with the irradiated samples immersed in liquid nitrogen using an x-band microwave spectrometer. The number of spins,

TABLE 3.3

ESR data for ion bombarded Si compared with amorphous Si

| Material | $g$ value | Line width | Density of spins |
|---|---|---|---|
| Si bombarded with $10^{16}$ P$^+$ ions cm$^{-2}$ at 280 keV | $2.0059 \pm 0.0005$ | $5.2 \pm 0.4$ | $2 \times 10^{20}$ |
| Amorphous Si prepared by sputtering | $2.0055 \pm 0.0005$ | $4.7 \pm 0.4$ | $2 \times 10^{20}$ |

the line width and the $g$ factor of the signal were determined. Table 3.3 shows their results for $10^{16}$ 280 keV P$^+$ cm$^{-2}$ implantation compared with those for amorphous Si films produced by sputtering (e.g. Brodsky and Title, 1969). The agreement verifies that amorphous Si produced by ion implantation is very similar to amorphous Si prepared by other techniques, as was in fact previously inferred from the electron diffraction data of Mazey et al. (1968) and Large and Bicknell (1967). Crowder et al. (1970) also studied the ESR signal from Si at ion doses far below that necessary to create a continuous amorphous layer. The agreement with sputtered Si suggests that the individual disordered zones observed by electron micro-scopy are in fact amorphous Si. To determine the dose dependence of damage the ESR signal was measured from samples having received diffe-rent doses of 280 keV P$^+$ ions. The result is given in fig. 3.28, which shows,

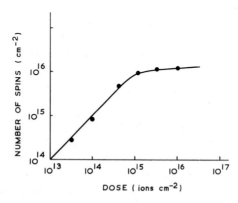

Fig. 3.28. The number of spins in room-temperature ion-damaged Si as determined from ESR experiments as a function of dose of 280 keV P ions.

as expected, a linear increase until saturation occurs at about $10^{15}$ ions cm$^{-2}$ when the bombarded region has been rendered completely amorphous.

Yet another method of detecting amorphous Si has been developed by Davidson and Booker (1970). They made use of the fact that electron channelling patterns, seen in the scanning electron microscope, disappear when crystalline Si has been rendered amorphous. Furthermore, as electron channelling patterns are developed only within the last 100 Å of the surface, by using conventional anodic stripping procedures, Davidson and Booker were also able to determine the depth distribution of amorphous material within their implanted Si.

It is important to point out that all of the above techniques for observing amorphous Si, also detect the small disordered zones which are produced within the crystalline matrix. Consequently, it is not necessarily an easy matter to determine the exact point at which these disordered zones overlap to form a continuous amorphous layer. For instance, in some cases, when Rutherford scattering shows complete disorder, electron diffraction shows a pattern composed of both diffuse rings (indicative of amorphous material), and single crystal spots, suggesting that the crystal structure has not been completely destroyed. Perhaps the most useful technique for determining the point at which overlap occurs is ESR; as in this case, the signal relates unambiguously to amorphous Si. As we shall see in Chapter 4, a knowledge of this overlap point is vital from the viewpoint of maximising the electrical activity of some implanted species.

*3.1.2.2.1. Damage in Semiconductors with Ions of low Z.* Up until now in discussing the production of radiation damage in semiconductors we have concentrated on the build up of small disordered zones which lead to the eventual creation of a continuous amorphous phase. Furthermore, we pointed out the strong dependence of bombardment temperature on the formation of this phase and suggested that this might bear a relationship to the initial size of the most densely damaged regions left in the wake of the collision cascade. With ions of large atomic number, the stopping cross-section, in say Si, is sufficiently large that the damage produced by successive primary recoils from just one ion overlaps to form essentially one highly disordered zone. Then whether this zone degenerates into an amorphous character depends on the temperature of the lattice. However, in the case of light ions such as B, the damage is more spread out with the consequence that only occasionally does one obtain localised regions of high disorder. In the main, displacements occur as well separated Frenkel pairs which can subsequently migrate through the lattice as isolated defects.

We must now consider the fate of these isolated displaced atoms, as both in the case of warm implants ($>$ 100 °C) with ions of large atomic number and in the case of room temperature implants with light ions such as B, such defects constitute a predominant fraction of the damage until eventual overlap of the amorphous zones occurs. At the temperatures of interest individual interstitials and vacancies in Si, Ge and GaAs are thought to be mobile, with the result that during bombardment defect annihilation and agglomeration, similar to that observed in metals, can readily occur. Some recent work on 200 keV $O^+$ ion bombarded Si by Stein et al. (1969) using local-mode infra red adsorption technique has provided valuable data on such agglomerates. It is shown conclusively that for doses which are significantly below that necessary for the production of a continuous amorphous phase at room temperature, a predominant defect within the crystalline Si lattice is the divacancy. Divacancies are readily detectable by a characteristic 1.8 $\mu$m absorption band and have been identified to exist in crystalline Si at room temperature. Furthermore, by combining the infra red measurements with anodization and stripping the same authors (Stein et al., 1970) were able to determine the integral depth distribution of divacancies in 400 keV $O^+$ ion implanted Si. Their results are shown in fig. 3.29 together with a theoretical calculation due to Brice (1970). Further evidence for agglomeration is provided by electron microscopy; for instance, figs. 3.30(a), (b) show electron micrographs of Si and GaAs bombarded at elevated temperatures to doses insufficient to create a continuous amorphous phase at that temperature. Both micrographs show dense dislocation entanglements, thought to result from the agglomeration of interstitials, but exhibit diffraction patterns (shown inset) which are typical of essentially crystalline material; note there is no evidence of the characteristic diffuse ring patterns associated with gross amorphicity which would be present in both cases if the bombardment was carried out at room temperature. Just as in metals, the formation of such networks proceeds by the formation and growth of dislocation loops. Such loops are generally nucleated homogeneously within the damaged region by the random agglomeration of like defects. Many theories for such a nucleation process have been proposed, perhaps the most pertinent is that of Brown et al. (1969) which discusses the formation of interstitial loops during Frenkel pair formation under irradiation. The main conclusion is that unlike amorphous zone formation, which proceeds proportional to dose

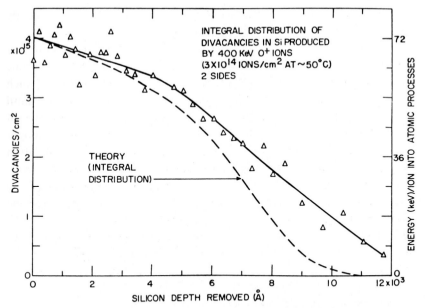

Fig. 3.29. Integral distribution of divacancies produced in 400 keV O⁺ ion bombarded Si together with a theoretical calculation (after Stein et al., 1970).

and independent of dose rate, the density of dislocation loops increases in a rather complicated way with dose and depends intimately on dose rate and temperature. In the case of semiconductors bombarded with light ions the heterogeneous formation of amorphous zones proceeds concurrently with the homogeneous nucleation and growth of dislocation loops. The precise dependence of total damage on dose and dose rate therefore depends crucially on the fraction of defects contained in amorphous zones to those left free to migrate throughout the lattice, and this in turn depends on the bombardment temperature. In the case of very heavy ions bombarding Si near room temperature, amorphous zone formation dominates and the damage is expected to increase proportionally with dose and be independent of dose rate. In fact Hart (1970) has shown that this is certainly time for 40 keV $Sb^+$ bombardment of Si. On the other hand, in the case of light ion bombardment of Si at room temperature, where, at least in the early stages of irradiation, the majority of defects are not contained within amorphous zones, we would expect that the total radiation damage does not increase linearly with dose and is dose rate dependent. However, at large doses where the amorphous zones occupy a substantial fraction of the solid, or at low

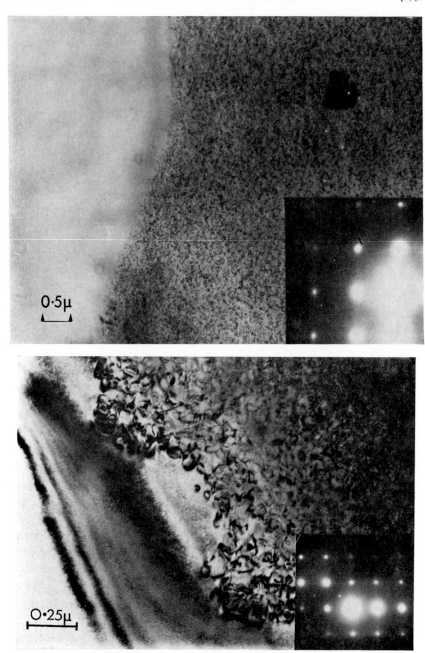

Fig. 3.30. (a) Dislocation entanglement in Ne bombarded Si at 200 °C. The clear region was shielded during radiation, the diffraction pattern corresponds to the irradiated region. (b) Dislocation entanglement in Ne bombarded GaAs at 35 °C. The clear region was shielded during radiation, the diffraction pattern corresponds to the irradiated region.

temperatures where amorphous zone formation is more efficient, the dependence of damage on dose and dose rate should be similar to the very heavy ion case. Such a behaviour has in fact been observed by Eisen (1970) during 200 keV $B^+$ ion irradiation of Si, where only at low temperature was there no dose rate dependence to the damage production. Furthermore, Chadderton and Eisen (1970) have demonstrated, using electron microscopy, that the amorphous zone formation in $B^+$ implanted Si at 25 °C is significantly reduced compared to similar implants at liquid nitrogen temperatures.

### 3.1.3. ANNEALING OF DAMAGE

Annealing is the thermal treatment of a damaged solid which eventually results in the recovery of crystal structure and physical properties. In line with the rest of this chapter we will primarily concern ourselves with implantation at about room temperature and assume that for practical cases both isolated interstitials and vacancies are free to migrate through the lattice by thermal activation. This assumption is of course not generally true, for in the high melting point materials such as W, Mo, Pt and Ir, only the interstitial is thought to be freely mobile at 20 °C. However for Cu, Au, Ag, Al, Si and Ge this assumption is certainly valid. Consequently we will limit our discussion to the annealing of defect clusters and assume that the individual vacancies and interstitials have precipitated into such agglomerates.

*3.1.3.1. Metals.* In most cases of interest dislocation loops are stable at room temperature, but some loss to the surface of the bombarded specimen can occur as a result of slip under the influence of the surface image force. As the temperature is raised the thermal activation of vacancies from a variety of sources occurs. Such vacancy sources are dislocation lines including vacancy loops, grain boundaries with the exception of coherent twin boundaries, and the free surfaces. In normal circumstances, under these conditions vacancy clusters and loops will steadily evolve their vacancies and shrink. However, in some special cases where the surface is prevented from acting as a sink for vacancies due to thermally induced chemical reactions which inject copious supplies of vacancies into the solid, the high vacancy supersaturations can in fact result in the growth of vacancy loops. Such a situation is generally to be found during oxidation which proceeds via a vacancy controlled diffusion process, e.g. during the formation of MgO,

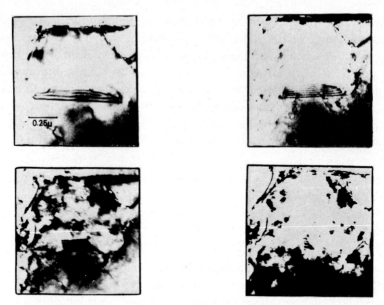

Fig. 3.31. The shrinkage of a fault in Si after annealing in vacuum at 1100 °C for (a) 0 min, (b) 10 min, (c) 20 min, (d) 30 min (after Sanders and Dobson, 1969).

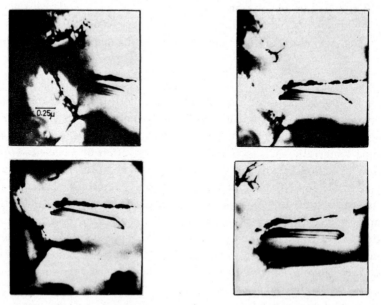

Fig. 3.32. Growth of a fault in Si after annealing in air at 1100 °C for (a) 0 min, (b) 10 min, (c) 15 min, (d) 45 min (after Sanders and Dobson, 1969).

$Al_2O_3$, $ZnO$ and $SiO_2$. This particular phenomena has been studied in some detail by Dobson and Smallman (1966) and Sanders and Dobson (1969). For instance, fig. 3.31 shows the shrinkage of a fault in Si during a vacuum anneal at 1100 °C, whereas fig. 3.32 shows the growth of a similar fault in Si during an anneal in air at the same temperature. These faults are clearly vacancy in nature which when annealed in vacuum release their vacancies but when annealed in air capture vacancies form the supersaturation which prevails on oxidation.

Unlike vacancies, the energy required to create an interstitial atom is so great that the thermally activated release of interstitials from interstitial loops is out of the question. However the steady acquisition of thermal vacancies will result in the gradual shrinkage of such loops. For instance, fig. 3.33 shows an annealing sequence of a number of faulted interstitial loops having been produced in copper during proton irradiation at 200 °C, the gradual shrinkage is readily apparent (Mazey and Barnes, 1968). In general, as thermally produced vacancies become available both vacancy and interstitial loops will shrink and dislocation networks will climb out to the surface; the temperature at which this occurs in any particular material can be compared with the temperature at which thermally activated self-diffusion becomes appreciable. We must point out that the annealing processes just described can be significantly hindered by effects such as precipitation, and these will be discussed in detail in section 3.2.3.

*3.1.3.2. Semiconductors.* Once again the covalent semiconductors fall into a different category, as is only to be expected in view of the fact that damage results in the complete destruction of the crystal lattice. It has been shown (Mazey et al., 1968) that for not too large doses the milky appearance in Si, which results from the overlap of disordered zones, quite suddenly vanishes at a temperature of about 650 °C. This immediately suggests a recovery of crystalline structures at this temperature. Furthermore, electron microscope studies (Mazey et al., 1968) show that at this same temperature the amorphous layer anneals by epitaxial recrystallization onto the undamaged Si below. On the other hand, where the ion range was greater than the specimen thickness (e.g. near the edge of a thin tapering section) epitaxial recrystallization could not occur and the region was found to contain small randomly oriented crystallites $\sim 1000$ Å diameter. Figure 3.34 shows two electron micrographs of Si bombarded with $10^{15}$ 40 keV $Ne^+$ ions $cm^{-2}$. After annealing at 650 for 1 hour in vacuo, typical of these two conditions, their

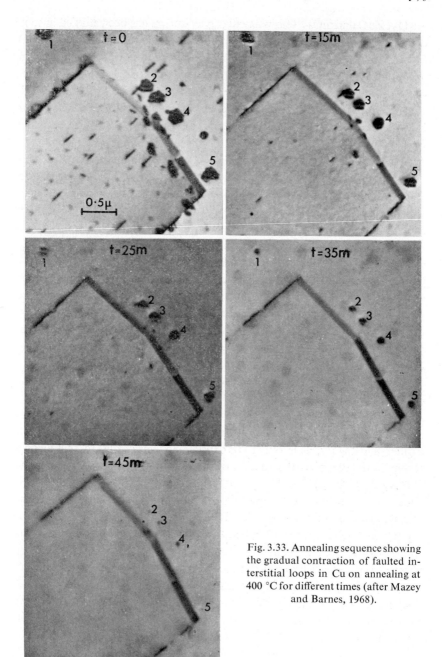

Fig. 3.33. Annealing sequence showing the gradual contraction of faulted interstitial loops in Cu on annealing at 400 °C for different times (after Mazey and Barnes, 1968).

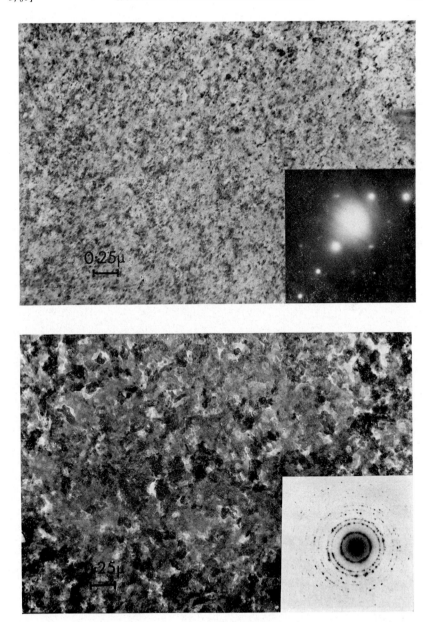

Fig. 3.34. (a) Electron micrograph and diffraction pattern of recrystallized amorphous S having grown epitaxially onto a single crystal substrate. (b) Recrystallization of an isolated thin film (1000 Å) of Si at 650 °C together with diffraction pattern showing fine grain polycrystalline regrowth.

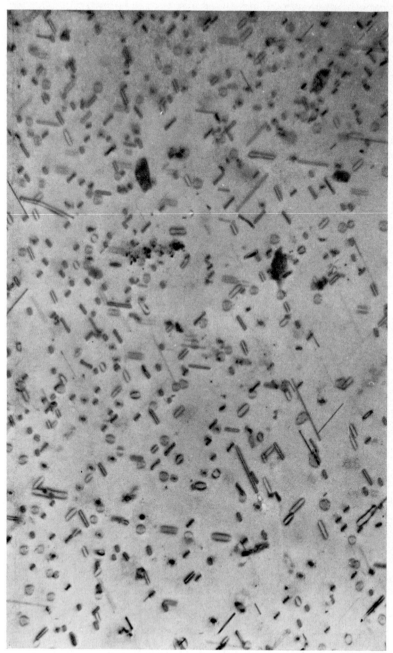

Fig. 3.35. Residual dislocation loops and dipoles in Si bombarded with $3 \times 10^{15}$ 40 keV B ions cm$^{-2}$ after recrystallization and annealing at 800 °C for 4 hours.

electron diffraction patterns are shown inset. It should be pointed out that in contrast the small disordered zones recrystallise epitaxially onto the surrounding Si at a temperature between 200–500 °C, but see later. An important feature of fig. 3.34(a) is the small regions of dark contrast which are thought to result from small clusters of defects within the essentially crystalline material. Numerous observations have shown that on annealing at higher temperatures, i.e. between 700 °C and 850 °C, these defect clusters grow by processes of slip and climb into larger clearly resolvable dislocation loops and dipoles. A typical example is shown in fig. 3.35 which is an electron micrograph of a Si specimen bombarded with $3 \times 10^{15}$ 40 keV $B^+$ ions cm$^{-2}$ which has been subsequently recrystallized by an 800 °C anneal for 4 hours. The detailed nature, size and distribution of these dislocation structures depends to some extent on the particular bombarding ion and on the dose as we will see later on. The same general annealing characteristics are to be found in both Ge and GaAs, however the recrystallization temperatures are somewhat different, see table 3.4.

TABLE 3.4

Fully amorphous phase recrystallization temperatures

| Material | Temperature (°C) | Reference |
|---|---|---|
| Si | 630 | Mazey et al. (1968) |
| Ge | 400 | Parsons (1965) |
| GaAs | 250 | Mazey and Nelson (1969) |

The Rutherford scattering techniques (Mayer et al., 1968) confirm the general nature and temperatures of the recovery processes, however these can give no information on the nature of the residual dislocation damage as can electron microscopy. Figure 3.36(a), (b) show some typical results for Sb$^+$ implanted Si and In$^+$ implanted Ge at different doses. In both cases, the low dose curves corresponding to isolated disordered zones prior to overlap show low temperature recrystallization, whereas the higher dose curves corresponding to saturation amorphicity show rather sharp epitaxial recrystallization temperatures as also found by electron microscopy.

It is thought that the activation energy for epitaxial recrystallization depends on the radius of the amorphous region, and to a very rough approximation the ratio of the recrystallisation temperature for an amorphous

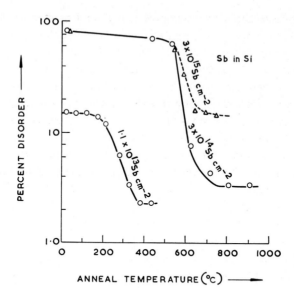

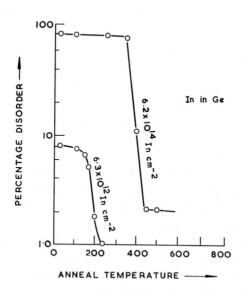

Fig. 3.36. (a) The recrystallization of Sb implanted Si as a function of annealing temperature for different ion doses (after Mayer et al., 1968). (b) The recrystallization of In implanted Ge as a function of annealing temperature for different ion doses (after Mayer et al., 1968).

region of radius $r$ to that for an infinite volume (e.g. a flat interface) is given by

$$\frac{T_r}{T_\infty} = 1 - \frac{2\gamma_i\,\omega}{rE_\infty} \tag{3.18}$$

where $\gamma_i$ is the interfacial free-energy between crystalline and amorphous Si, $\omega$ is the activation volume for recrystallization (i.e. the volume of the 5 atom agglomerations which constitute amorphous Si, and $E_\infty$ is the activation energy for an infinite interface. Choosing reasonable values for $\gamma_i = 500$ erg cm$^{-2}$, $\omega = 2 \times 10^{-22}$ cm$^{-3}$ and $E_\infty = 5$ eV, and approximate annealing temperature curve as a function of zone size can be constructed, fig. 3.37.

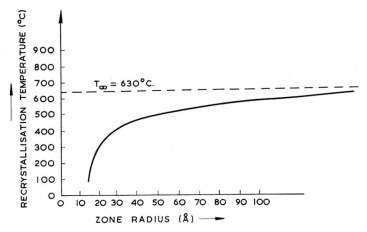

Fig. 3.37. The recrystallization temperature of amorphous zones in Si plotted as a function of zone radius.

From this it is readily apparent that the small isolated amorphous zones anneal at substantially lower temperatures than the complete amorphous phase. Furthermore, due to variation in primary recoil energies, an implantation to a dose insufficient to produce overlap (e.g. $\sim 10^{15}$ B cm$^{-2}$ in Si at 20 °C), will result in a size distribution of disordered zones. Consequently, in such a case, recrystallization will occur over a range of temperatures, but will probably be complete after annealing at about 500 °C where zones of up to $\sim 50$ Å will have recrystallized.

It is interesting to note that fig. 3.36(a) suggests that for rather high doses, e.g. $2.6 \times 10^{15}$ Sb$^+$ ions cm$^{-2}$ in Si, recovery is far from complete even at 800 °C. However, as we will see later, this is a consequence of exceeding the Sb solubility limit in Si which in turn hinders perfect epitaxial recrystallization. On the other hand, providing the bombarded solid does not contain so many impurities that the solubility limit is exceeded, the dislocation stuctures can generally be annealed out completely at higher temperatures, e.g. of the order of 1000 °C in Si, in just the same way that damage is finally removed in metals.

*3.1.3.3. Defect Structures in Recrystallized Semiconductors.* In section 3.1.3.2 we pointed out that amorphous layers created on the surface of ion bombarded silicon, germanium and gallium arsenide underwent epitaxial recrystallization on annealing to certain specified temperatures. In this section we will discuss the nature of the residual structures which remain as observable defects within the lattice after recrystallization. However, in the present context we will simply present an outline of the results rather than pursue a highly sophisticated electron microscopical account which requires detailed knowledge of diffraction contrast theory. As we will see, we must distinguish between the situations corresponding to ion doses which do and do not result in complete amorphicity (Mazey et al., 1968; Davidson and Booker, 1970).

In the first instance, let us suppose that the isolated disordered zones have not overlapped but are of sufficient density to produce strong diffuse Debye-Scherrer rings in the diffraction pattern. The damage then consists of small disordered regions together with isolated or clustered point defects in essentially crystalline material. Further, as the disordered zones are some-what less dense than the surrounding matrix, it is reasonable to assume that this surrounding matrix contains an excess of both clustered and trapped interstitials. On annealing, even at temperatures below that at which recrystallization of a complete amorphous layer occurs, vacancies boil off from the disordered zones either as singles or di-vacancies. Some of these will recombine with interstitials and, some may cluster to form vacancy loops but, due to the close proximity of the surface sink, a large fraction will be lost to the surface so leaving a net surplus of clustered interstitials. On further annealing, the interstitial clusters will grow either by the capture of previously trapped single interstitials, or by the usual dislocation growth processes of slip and conservative climb, the net result being that the diffraction pattern shows only single crystal reflections with perhaps some slight streaking due to planar defects; whilst electron microscopical examination reveals dislocation loops and dipoles, fig. 3.38. The precise nature of the dislocation loops has been found to depend on the particular bombarding

Fig. 3.38. Dislocation loops and dipoles in recrystallized B implanted Si to a dose corresponding to incomplete overlap of individual disordered zones.

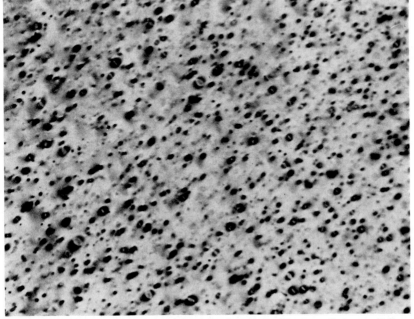

Fig. 3.39. Residual dislocation loops in P implanted Si after annealing a partially a-morphous layer at 800 °C.

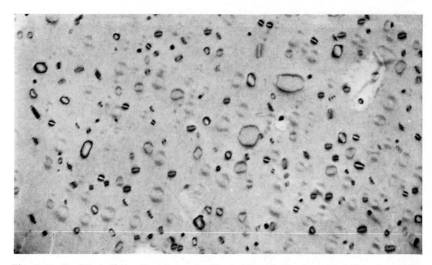

Fig. 3.40. Residual dislocation loops in Ne implanted Si after annealing a partially amorphous layer at 800 °C.

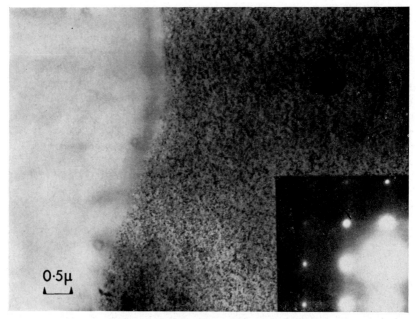

Fig. 3.41. Implantation of Si to a high dose at 200 °C showing complex dislocation entanglement.

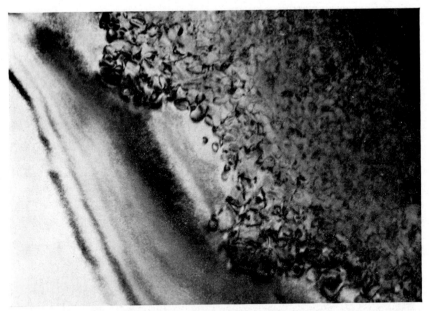

Fig. 3.42. Implantation of GaAs to a high dose at 35 °C showing complex dislocation entanglement.

ion species. For instance, figs. 3.39 and 3.40 show typical dislocation structures produced in Si after annealing samples bombarded with P and Ne ions respectively. It is thought that some impurity atoms become trapped within the dislocation structures so influencing their morphology and annealing behaviour. Furthermore, in such samples which have not been rendered completely amorphous during bombardment, the final defect density is dependent on the variable bombardment parameters such as dose, temperature and crystal orientation. For instance, bombardment of Si and GaAs to high doses at temperatures ∼ 200 °C results in the growth of individual dislocation loops into complex dislocating entanglements, figs. 3.41 and 3.42.

Let us now turn to bombardments which correspond to doses and temperatures sufficient to render the surface layers completely amorphous. In general, provided the amorphous layer is supported by the parent crystalline lattice then recrystallization during elevated temperature anneals is essentially epitaxial. However, both the diffraction patterns and electron micrographs show slight signs of micro-twinning and other defects typical of imperfect epitaxial growth, e.g. fig. 3.43. Detailed studies in Si (Davidson

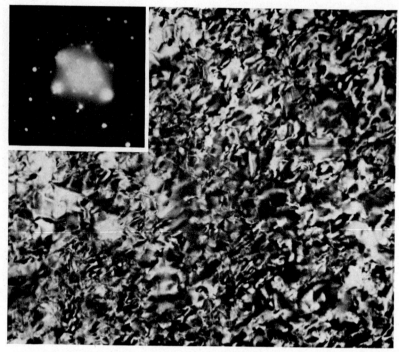

Fig. 3.43. The imperfect recrystallization of Sb implanted Si to a dose sufficient to render the surface layer completely amorphous. The micrograph shows clear evidence of fine microtwins and the diffraction pattern gives some indication of the degree of misorientation (after Davidson and Booker, 1970).

and Booker, 1970) show that these defects become more dense towards the surface; in other words, as the recrystallization proceeds from the amorphous crystalline interface the epitaxial regrowth becomes steadily worse. As we will see later, this can have a detrimental effect on the electrical properties of the implanted sample.

In summary, therefore, we may make the following general conclusions concerning the final configuration of residual defects which persist after annealing ion implanted semiconductors. The annealing of implants which correspond to doses and temperatures insufficient to render the surface layers completely amorphous results in a reasonably uniform array of discrete dislocation loops and dipoles throughout the bombarded region, except for a small defect free zone at the surface $\sim$ 100 Å deep. On the other hand, the annealing of implants which have completely destroyed the crystalline structures result in epitaxial recrystallization with a dense array of fine misoriented regions such as micro-twins and planar faults; the degree of misorientation becoming steadily worse as the recrystallization proceeds.

## 3.2. Surface effects

### 3.2.1. SPUTTERING

So far we have discussed in some detail the formation, configuration and annealing of radiation damage produced within the bulk of crystalline solids during ion implantation. However, we must not neglect the damage to the surface itself, as whenever the collision cascade intersects the surface atomic ejection occurs if the surface atom recoils with an energy in excess of the surface binding energy (typically 2–5 eV). This phenomenon is known as sputtering and in most circumstances ejection will initially result in the formation of a surface vacancy, that is except in the special case of ejection via a replacement collision.

In the present context it is not our intention to present an all embracing review of sputtering, but just to point out the more important implications pertinent to ion implantation. We shall discuss just two aspects of sputtering; firstly, the number of surface atoms ejected per incident ion – called the sputtering ratio or sputtering yield, and secondly, the energy spectrum of sputtered atoms. We shall then go on to discuss the state of the sputtered surface and its possible effect on the implanted solid.

*3.2.1.1. Sputtering Yield.* Measurement of the sputtering yield has occupied the experimentalist for many years, in fact since the late eighteen-hundreds when the phenomenon was first discovered as the result of the ion bombardment of cathodes in electrical discharge devices. During the following years, right up to the present day, a host of information has been acquired for a wide variety of ion-target combinations and energies. Recent books and review articles which include a comprehensive collection of data include the following; Wehner (1955), Kaminsky (1965), Behrisch (1964), Carter and Colligon (1968), and Pleshivtsev (1968).

The most comprehensive data on high energy sputtering yields is that of the Swedish group under Almén (e.g. Almén and Bruce, 1961a, b) and the Dutch group under Kistemaker (e.g. Rol et al., 1960). We shall reproduce just a few of their results to establish the general trend of the variation sputtering ratio with bombarding ion, energy and angle of incidence. Figures 3.44 and 3.45 illustrate the variation in sputtering ratio of polycrystalline Cu and Mo at normal incidence with incident ion energy for a variety of

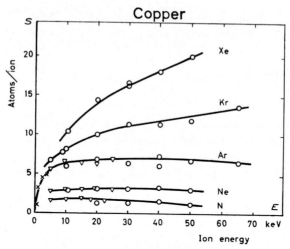

Fig. 3.44. The sputtering ratio of polycrystalline Cu as a function of energy for different bombarding ions (after Almén and Bruce, 1961a, b).

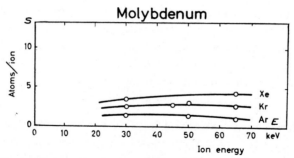

Fig. 3.45. The sputtering ratio of polycrystalline molybdenum as a function of energy for different bombarding ions (after Almén and Bruce, 1961a, b).

incident ion species. The general trend is for the sputtering ratio to increase rapidly from zero and then to level off towards an approximately constant plateau. The heavier the ion the greater the yield and the higher the energy at which the sputtering ratio levels off. However, the plateau does not continue indefinitely as measurements made well above 100 keV (e.g. Perović and Cobić, 1962) show a steady decrease with increasing energy. Two more important results from Almén and Bruce (1961a, b) on polycrystalline targets were the self-sputtering ratio plotted as a function of atomic number, and the variation of sputtering ratio of a particular target (e.g. Ta, Cu or Ag) as a function of increasing atomic number of the bombarding ions. These results are illustrated in figs. 3.46 and 3.47 respectively.

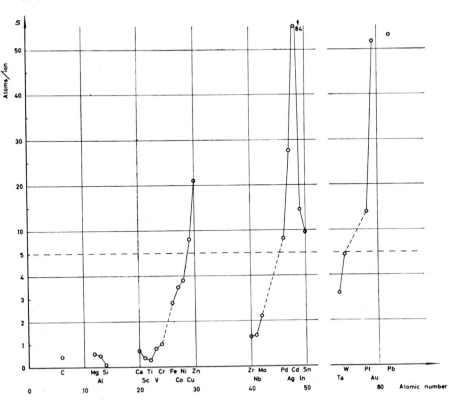

Fig. 3.46. The self-sputtering ratio as a function of atomic number for bombardment with 45 keV ions (after Almén and Bruce, 1961a, b).

The self-sputtering ratio result shows, as expected, a general increase with increasing atomic number but with systematic variations peaking at elements such as Zn and Cd. A cursory glance at the variation in sublimation energy with atomic number immediately suggests a strong inverse correlation with sputtering ratio. We shall see later that the energy spectrum (the number with energy $E$ in the interval $dE$) of atoms crossing a surface due to many collision cascades within a solid is proportional to $1/E^2$, then if we suppose that sputtering occurs for all atoms which intersect the surface with energy greater than the binding energy $E_b$, the total sputtering yield is proportional to the integral of the energy spectrum from $E_b$ to $E_{max}$, which for $E_{max} \gg E_b$ gives a sputtering ratio proportional to $1/E_b$. We are not completely justified in assuming that the binding energy is identical to the sublimation energy, for evaporation no doubt occurs more readily from

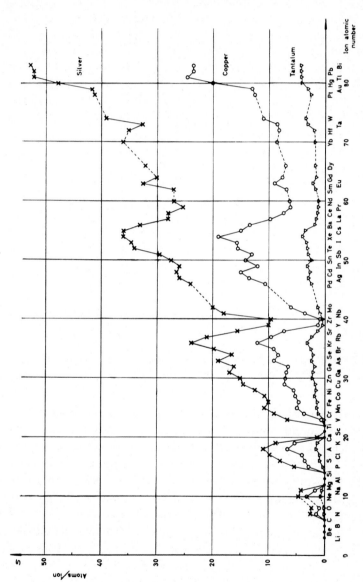

Fig. 3.47. The self-sputtering ratio of Si, Cu, and Ta plotted as a function of the atomic number of the bombarding ions at an energy of 45 keV (after Almén and Bruce, 1961a, b).

specific sites, such as steps. However, it is a reasonable assumption to expect that variations in binding energy will follow the same general pattern as that known to occur for the variations in sublimation energy. Accepting this, we can readily understand the systematic variations observed in the self-sputtering ratio shown in fig. 3.46.

On the other hand, the results of fig. 3.47 are not so readily explained. The general trend is to increase with increasing atomic number but with rather sudden dips at specific elements. Two possible effects can give rise to variations in yield; these are the variation in interatomic potential due to the systematic variation in electron density screening the Coulomb field of the interacting nuclei (Nelson, 1969a), and precipitation effects due to exceeding the solid solubility limit of the bombarding ions within the target. Precipitation effects will be discussed in detail later on when dealing with the fate of the implanted atoms. The first effect is thought to give rise to somewhat smaller and smoother variations than suggested in fig. 3.47, whereas the second effect can be very large and from a consideration of the phase diagrams of binary alloys (e.g. Hansen, 1958) can clearly be the underlying cause of the observed variations.

Another important feature in polycrystalline sputtering is the influence of the angle of incidence of the bombarding ions. Figure 3.48 shows the result of such an experiment by Molchanov and Tel'kovski (1962) from polycrystalline Cu. This result is not surprising since the energy dissipation

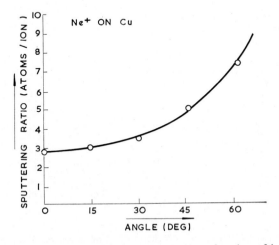

Fig. 3.48. Sputtering ratio of Ne bombarded Cu as a function of bombardment angle (after Molchanov and Tel'kovski, 1962).

from the incident beam occurs closer to the surface as the angle of incidence
is increased. In general, for angles reasonably close to normal incidence, the
angular dependence in sputtering ratio follows a relationship of the form

$$S_\theta = S_0 \sec \theta \qquad (3.19)$$

where $S_0$ is the yield at normal incidence. However, at larger angles the
yield reaches a maximum and falls due to the occurrence of increased ion
scattering from the surface.

In many implantation experiments we use single crystal targets and for
this reason alone we must discuss the special case of sputtering from such targets. The first phenomenon to point out is that due to the channelling of the
incident ions in the target lattice, to a first approximation only those ions
which enter the so-called random component can be responsible for sputtering. For it is the recoils from such ions which are capable of producing
sufficiently energetic collision cascades which ultimately intersect the surface
and result in sputtering. It is not surprising, therefore, that numerous
observations of the variation in sputtering yield with crystal orientation
have been reported, even before the channelling phenomenon was in fact
discovered, e.g. Rol et al. (1959), Almén and Bruce (1961a, b). More recent
results due to Magnuson and Carlston (1963) and Onderdelinden (1968) are
shown in fig. 3.49 for $A^+$ ion bombardment of Cu for the {111}, {100} and
{110} faces as a function of energy. As expected, the more open or better
channelling directions have the lowest sputtering yield.

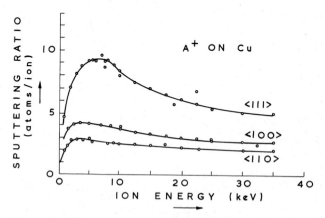

Fig. 3.49. The experimental sputtering ratio curves for A on single crystal Cu as a function
of energy plotted for different crystal orientations. The experimental points are taken
from Magnuson and Carlston, 1963; Onderdelinden, 1968.

Although it does not directly affect implantation it is worth briefly commenting on the spatial distribution of atoms sputtered from targets. Clearly, for polycrystalline targets ejection is spatially random, that is except for the basic cosine type distribution due to the binding energy effect. However, the spatial distribution of atoms sputtered from single crystal specimens is strongly preferred in certain well defined directions. This effect was first observed by Wehner (1955) during low energy ~ 1 keV bombardments; but following this, numerous authors have reported similar observations for a variety of ion-target combinations at both low and high ion energies, e.g. Yurasova et al. (1959), Nelson and Thompson (1961, 1962a, b), Perović (1961) and Nelson (1963). A selection of typical ejection patterns obtained by collecting a visible deposit of the sputtered atoms on a glass plate is shown in figs. 3.50 and 3.51. In general, ejection was found to occur near to the simple close-packed crystal directions; i.e. the $\langle 110 \rangle$ and $\langle 100 \rangle$ in f.c.c. metals and the $\langle 111 \rangle$ and $\langle 100 \rangle$ in b.c.c. metals. Originally, these observations were taken as unequivocal evidence for the existence of long range focused collision sequences. Recently, however, this interpretation has come under serious criticism, both experimentally and theoretically. For

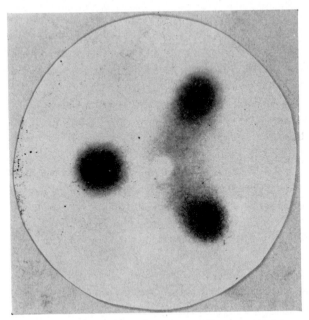

Fig. 3.50. Spot pattern collected on a flat plate held parallel to the {111} crystal surface of Au showing three $\langle 110 \rangle$ spots corresponding to ejection at 35° to the normal.

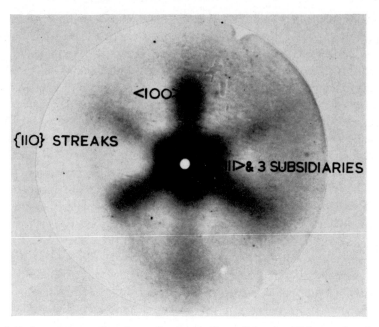

Fig. 3.51. Spot pattern collected on a flat plate held parallel to the {111} crystal surface of W showing ejection in the ⟨111⟩ and ⟨100⟩ directions.

instance, in zinc and α-uranium ejection occurs from directions other than the expected focusing directions; further, Lehmann and Sigmund (1966) have suggested that the majority of preferential ejection can be accounted for by just pairs of atoms in close proximity. In an attempt to clarify the position, Von Jan and Nelson (1968) have developed a model in which the angular distribution and energy spectrum of sputtered atoms is calculated as the cumulative effect of random collisions and simple collision sequences of different range. Their conclusions were that focused collision sequences do play a part in sputtering, but only in the case of high focusing energies is there a significant contribution from long range sequences. Even in the most favourable cases, such as gold, preferential ejection from the first two layers contributes nearly 50 % of the total.

*3.2.1.2. The Energy of Sputtered Atoms.* In the context of ion implantation it is useful to discuss the energies of sputtered atoms for a variety of reasons, if not only to give us the underlying data required for a better understanding of the technique known as 'Recoil Implantation' to be discussed later. It is well known that the majority of atoms sputtered from solid surfaces are ejected as neutral particles and that any that are charged tend to have the highest energy. For this reason rather specialised techniques must be developed in order to obtain worthwhile measurements. Average energies have been obtained by the momentum transferred to a micro-balance (Kopitzki and Stier, 1962) and by the temperature rise of a small thermo-couple (Weijsenfield, 1966). However, these suffer from the drawback that it is impossible to separate out the back scattered high energy incident particles. The best method is undoubtedly a time of flight technique, and to date a variety of attempts using different techniques have been used. Perhaps the most sophisticated is the 'spinning rotor' technique devised by Thompson (1963) and Thompson et al. (1968); however, it is not our intention to pursue the experimental details this time.

To date, measurements have been made on both polycrystalline and single crystal targets, but here we shall just briefly outline a few of the most important results. Figure 3.52 shows the energy spectrum of sputtered gold atoms ejected from a polycrystalline target under 45 keV $A^+$ and $Xe^+$ ion bombardment due to Thompson (1968). In this case the results are plotted as the probability $P(E)$ of detecting an ejected atom with energy $E$ in the interval $dE$, a line of slope corresponding to $1/E^2$ drawn for comparison. For ejection normal to the surface we expect the theoretical form of the energy spectrum outside the surface of a target to have the form

$$\Phi(E) = \Phi_i(E + E_b)/(1 + E_b/E) \tag{3.20}$$

where $\Phi_i(E + E_b)$ is the recoil spectrum within the solid proportional to $1/(E + E_b)^2$ (Thompson, 1968). For $E \gg E_b$, $\Phi(E)$ is simply proportional to $1/E^2$, but for $E \lesssim E_b$ the surface exerts a major influence and causes the spectrum to turn over; both features agree well with the experimental result shown above. The average energy of sputtered atoms is simply obtained from an integration of energy over the energy spectrum from zero to the maximum, which for $E_{max} \gg E_b$ leads to

$$\bar{E} \simeq \tfrac{1}{2}E_b \ln\left(\frac{E_{max}}{4E_b}\right) \tag{3.21}$$

This expression shows that the average energy is strongly dependent on the binding energy but increases rather slowly with bombardment energy. Both features have been qualitatively substantiated by the measurements of Kopitzki and Stier (1962), e.g. see fig. 3.53.

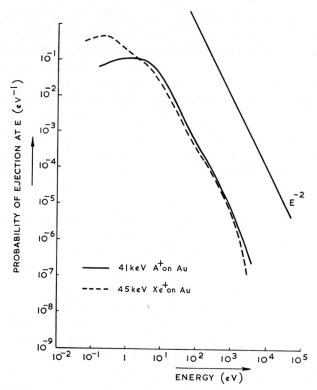

Fig. 3.52. The energy spectra of atoms ejected from a polycrystalline gold target, showing the $E^{-2}$ dependence (after Thompson, 1968).

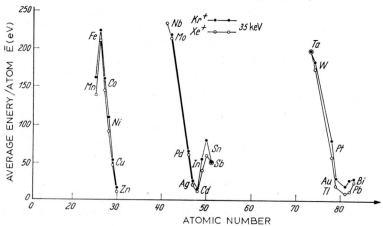

Fig. 3.53. The average energy of atoms sputtered from polycrystalline targets as a function of atomic number (after Kopitzki and Stier, 1962).

*3.2.1.3. The State of the Implanted Surface.* In this sub-section we will concern ourselves with the physical state of the surface of a solid after implantation. However, we shall exclude surface precipitation effects as these will be discussed later on.

Generally, the microscopic surface of an implanted solid is only substantially modified by the bombardment after high doses, i.e. $10^{14}$–$10^{15}$ ions $cm^{-2}$; as for doses below this, sputtering removes less than a single monolayer of atoms from the target surface. The ejection of atoms inevitably results in the formation of surface vacancies. Whether or not these remain as isolated defects depends on the temperature of the lattice, inasmuch as thermally activated migration over the surface and into the bulk can lead to agglomeration and loss. In many materials single vacancies are thought to migrate within the bulk just below room temperature with activation energies $\sim 1$ eV. However, at the surface these activation energies are perhaps lower, both for migration within the surface plane and into the bulk. In such cases, therefore, the dynamic surface vacancy concentration will be virtually insignificant. At lower temperatures, or in high melting point solids such as tungsten and iridium, we might expect the surface vacancy concentration to become important. However, even then, elastic interaction will probably result in spontaneous local rearrangement and agglomeration. We must also remember that for every incident ion something like 50 highly mobile interstitials will find their way to the surface, where immediate annihilation will remove virtually all of the surface vacancies. In fact, such interstitials are likely to build up on the surface and smooth out irregularities; such an effect is easily demonstrated by the fact that etch pits or pits remaining after the leaching of surface precipitation are very rapidly filled in during ion bombardment. We may conclude, therefore, that the dynamic concentration of isolated surface vacancies can be neglected, even in high melting point materials.

In those circumstances requiring a high-ion dose, i.e. if the implanted impurity concentration is to exceed a few per cent, many monolayers inevitably become sputtered from the surface. This in itself imposes a limit to the maximum concentration of impurity atoms that can be implanted into a solid, for when the sputtered surface reaches a point approximately equal to the initial penetration depth a steady state is attained and the implanted concentration is at a maximum. It is easily shown that, provided diffusion processes can be neglected, the maximum atomic concentration that can be attained under such steady state conditions is approximately equal to $1/S$, the inverse of the sputtering ratio. However, during such an implantation, as a consequence of the removal of many hundreds of monolayers well defined structures can develop on the bombarded surface. Due to the orientation dependence of sputtering which arises due to channelling, adjacent grains of a polycrystalline target will be eroded at different rates, which in turn produces an overall relief structure on the surface. Apart from this, a marked surface structure can occur within each grain, e.g. fig. 3.54 (Haymann and Waldburger, 1964). A detailed investigation has recently been carried out on thin Cu foils using transmission electron microscopy (Mazey et al., 1967). Figures 3.55(a) and (b) show typical micrographs illustrating the general nature of the structure which develops after removal of only $\sim$ 300 Å from the surface. In the particular case of grains oriented with their {110} or {310} planes parallel to the surface, aligned bands of light and dark contrast were readily apparent; but in the case of {111} and {112} oriented grains the dark contrast exhibited a 'cellular' type structure. On the other hand, grains oriented with their {100} planes parallel to the surface exhibited no detectable contrast structure even after extended bombardment. Figure 3.56(a) shows that whenever an overlaid electron-microscope grid-bar had been removed after bombardment, its profile is delineated at the boundary between the sputtered and the thicker shielded

Fig. 3.54. Replica of a {100} surface of a silver single crystal after bombardment with 8 keV $A^+$ ions at 75° angle of incidence (after Haymann and Waldburger, 1964).

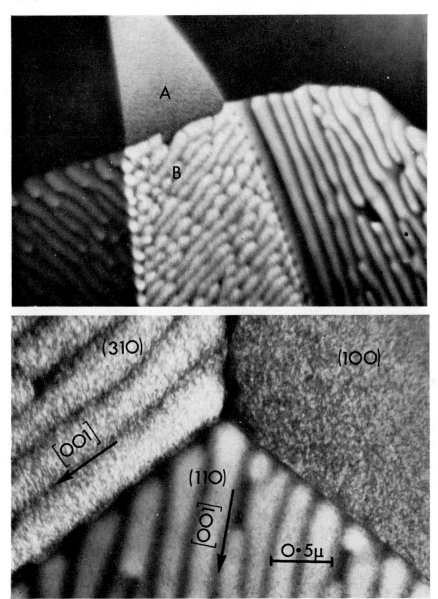

Fig. 3.55. Electron micrographs of the surface structure developed on a copper polycrystal during 60 keV $Xe^+$ ion bombardment at normal incidence. (a) Region A shows the un-sputtered surface caused by an electron microscope grid bar placed so as to shield the specimen from the incoming beam. In region B the unshielded specimen has developed a characteristic surface structure. (b) The intersection of three differently oriented grains showing that no sputtering structure has developed on the (100) surface.

Fig. 3.56. Electron micrograph showing the aligned dislocation arrays in copper bombarded with 60 keV Xe$^+$ ions to a dose of $10^{16}$ ions cm$^{-2}$.

areas. The variation of contrast is a consequence of greater electron absorption and scattering in the thicker regions and suggests the presence of an undulating surface structure. Figure 3.55 also shows that the structure bears some relation to the relative orientation of the grains with respect to the ion beams. Selected area electron diffraction patterns taken from {110} and {310} grains show that the regular 'hill and valley' structure is aligned in the ⟨100⟩ directions parallel to the foil surface, this is illustrated in fig. 3.55(b) where three grains having {110}, {310} and {100} orientations are visible in the same field of view, note the {100} grains have no detectable preferential sputtering structure. Closer examination and geometrical analysis of many different grains leads to the general conclusion that it is the {100} surfaces which develop preferentially by selective sputtering as the surface is steadily eroded.

To investigate the evolution of the sputtering structure, a copper film was given successive bombardments with 60 keV Xe$^+$ ions and examined after each bombardment. Initially at doses ∼ $10^{16}$ ions per cm$^2$ only heavy dislocation damage was observed, and in some grains these tended to become aligned in the ⟨100⟩ directions lying in the plane of the specimen; an example is shown in fig. 3.56. The 'furrowed' structure was first detected at a dose of $6 \times 10^{16}$ ions per cm$^2$, and at this threshold dose exhibited the same periodicity as that of the well developed structure seen after prolonged sputtering. It was found that the sputtering structure was not influenced by pre-existing chemical or electrochemical etching structures, or by the influence of Xe gas bubbles or by lead precipitates, as similar effects were produced by Cu$^+$ ion bombardments. The similarity in direction and in periodicity between the sputtering structure and the dislocation arrays suggested that these were responsible for the nucleation of the structure, and the fact that the {100} faces are expected to have the highest sputtering yield was proposed as the reason for their development.

As in the case of radiation damage the covalent semiconductors again present a special case. At temperature below and near to room temperature, due to the degeneration of the crystalline structure into a non-crystalline amorphous state, no regular crystallographic sputtering structures form on the surface of a pure material. However, if the temperature is raised, either externally or due to the power dissipation of the ion beam, so as to maintain the essentially crystalline nature of the surface layers, then crystallographic structures can in fact form. It should be pointed out that the sputtering ratio of such materials depends critically on the bombardment temperature, inasmuch as whether or not the target remains crystalline in turn determines whether channelling occurs; also the magnitude of the binding energy could well depend on the precise nature of the target.

### 3.2.2. RECOIL IMPLANTATION

The technique of 'Recoil Implantation' is the introduction of impurities into a solid by recoil from, say, a thin layer evaporated onto its surface. The procedure is illustrated in fig. 3.57. The target with its evaporated layer

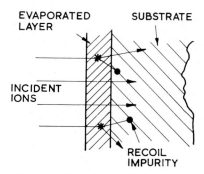

Fig. 3.57. Diagrammatic illustration of the technique of recoil implantation.

intact is irradiated either with energetic protons or heavy ions which pass through the layer into the substrate. Some atoms within the evaporated layer suffer elastic collisions with the bombarding particles and recoil with sufficient energy to penetrate into the substrate where they come to rest as an impurity. Clearly the technique is not so controllable as the more conventional method of ion implantation, but can be quite effective in the absence of a versatile particle accelerator.

In order to make the most of the technique we must attempt to formulate the physical processes involved so as to provide the relevant quantitative data. In the first instance we shall restrict ourselves to the first of the two alternatives, namely proton bombardment right through the evaporated layer. We have to calculate the flux and energy spectrum of recoil atoms crossing the interface between the layer and substrate for a particular proton bombardment.

From our knowledge of the energy density with collision cascades (Robinson, 1965; Thompson, 1968) we know that the flux of recoils across any plane, with energy $E$ in the interval $dE$ at $E$, initiated by a primary recoil spectrum $S(E_r)dE_r$ having a maximum energy of $E_r$ is

$$\Phi(E, E_1) \simeq \frac{D}{E^2} \int_E^{\hat{E}_r} S(E_r) E_r \, dE_r \qquad (3.22)$$

where $E_1$ is the energy of the incident protons assumed to be sufficiently large that we can ignore the loss in passing through the evaporated layer, $\hat{E}_r = (4/M_2)E_1$, $M_2$ is the mass of the target atom and $D$ is the mean interatomic spacing. For energetic protons we know that the differential cross-section, $d\sigma$, can be approximated by a Rutherford type of the form

$$d\sigma(E_r, E_1) = 4\pi a_0^2 (M_1/M_2) Z_1^2 Z_2^2 (E_R^2/E_1)(1/E_r^2) dE_r \qquad (3.23)$$

where $a_0$ is the radius of the hydrogen atom, $E_R$ is the Rydberg energy (13.6 eV) and $M_1$, $Z_1$, $M_2$ and $Z_2$ refer to the masses and charge numbers of the moving and struck atoms respectively. The recoil spectrum is then simply

$$S(E_r)dE_r = N_0 \, \phi_p \, d\sigma(E_r, E_1) \qquad (3.24)$$

where $N_0$ is the atomic density and $\phi_p$ is the flux of the bombarding protons. Substitution, putting $M_1$ and $Z_1 = 1$ for protons and integration gives

$$\Phi(E, E_1)dE = 4\pi a_0^2 D N_0 \, \phi_p \left( \frac{Z_2^2 E_R^2}{M_2 E^2 E_1} \right) \ln (4E_1/M_2 E) dE \qquad (3.25)$$

Neglecting the logarithmic term, this expression shows that recoil spectrum crossing the interface varies as $1/E^2$ just as expected from the previous discussion on the energy spectrum of sputtered atoms. For instance, fig. 3.58 shows $\Phi(E, E_1)$ plotted as a function of $E$ for the particular case of 500 keV protons passing through a thin layer of boron evaporated into a silicon substrate.

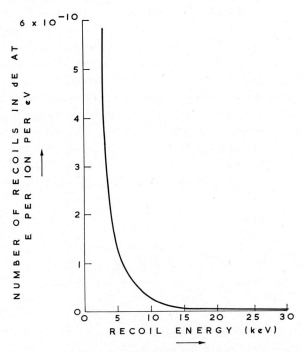

Fig. 3.58. The recoil spectrum of B passing into a Si substrate during irradiation of a thin B layer with 500 keV protons.

However, we are mainly concerned with transforming the information contained in eq. (3.25) into a depth distribution of implanted atoms within the substrate. From our knowledge of the range of heavy particles in solids as discussed in Chapter 2, e.g. Lindhard et al. (1963), this is simply achieved by converting recoil energy into penetration depth. Of course, in reality we must take into account both the straggling in range of the recoil atoms and the fact that by no means all of the recoils cross the interface at right angles. Both effects will modify the resulting penetration distribution calculated using eq. (3.25) and existing range-energy data; however their direct application serves to illustrate the general nature of the implanted profile.

The total number of recoils that penetrate into the substrate can be calculated from eq. (3.25) by integration from some minimum energy corresponding to the threshold for passing from the evaporated layer into the substrate, say 1 keV, to the maximum energy transferable in a head-on collision, i.e. $\hat{E}_r = (4/M_2)E_1$. For 500 keV proton bombardment of a thin boron layer on silicon the total number of impurities crossing into the substrate is $3 \times 10^{-5}$ per proton. However, the number crossing into the substrate with an energy greater than 10 keV is only $3 \times 10^{-7}$ per proton. Such a fraction is of course very small and in order to inject a reasonable concentration of impurities into a semiconductor we would have to accumulate a total proton dose in excess of about $10^{21}$ protons cm$^{-2}$.

In order to reduce the total ion dose delivered to the specimen we must choose a more massive ion such that both the maximum energy transfer and the scattering cross-section are increased. Once again let us suppose that the evaporated layer is sufficiently thin that, to a first approximation, we can ignore the energy loss in the layer. For massive particles the differential cross-section is virtually isotropic and can be approximated by the expression

$$d\sigma(E_r, E_1) = \frac{\sigma(E_1)\,dE_r}{\hat{E}_r} \tag{3.26}$$

where in this case $\hat{E}_r = 4\,M_1 M_2 E/(M_1 + M_2)^2$ and the total cross section is approximately

$$\sigma(E_1) = \frac{2\pi a_0^2 E_R Z_1 Z_2 (M_1 + M_2)}{(\exp 1)E_1 (Z_1^{\frac{2}{3}} + Z_2^{\frac{2}{3}})^{\frac{1}{2}} M_2} \tag{3.27}$$

substituting into eqs. (3.26) (3.27) and integrating we obtain

$$\Phi(E, E_1) = \frac{4\pi a_0^2 D N_0 \phi_i E_R Z_1 Z_2 M_1}{(\exp 1)(M_1 + M_2)(Z_1 + Z_2^{\frac{2}{3}})^{\frac{1}{2}}} \left( \frac{1}{E^2} - \frac{1}{E_r^2} \right) \tag{3.28}$$

where $\phi_i$ is the ion flux. For $\hat{E}_r \gg E$ this expression is independent of $E_1$ and is again proportional to $1/E^2$. However, for $E \lesssim \hat{E}_r$ the dependence on $E_1$ becomes noticeable and it falls even more rapidly than $1/E^2$. In this case for 100 keV Ne$^+$ ion bombardment of boron on silicon the total number of boron recoils above 1 keV that penetrate the substrate is as high as 0.02 per incident ion, and the useful number in excess of 10 keV is $\sim 2 \times 10^{-3}$ per incident ion. Figure 3.59 shows the actual penetration distribution compared with that expected for a direct B$^+$ implant into Si at 40 keV. The relative ion doses have been adjusted to give the same total number of implanted

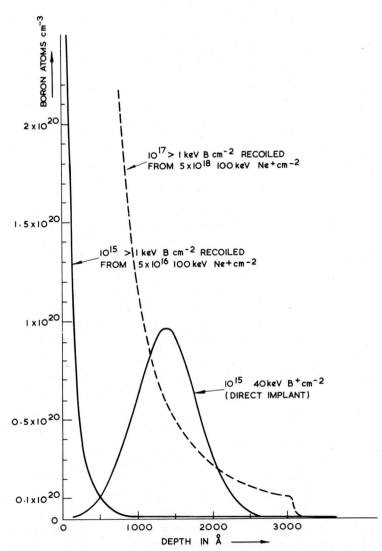

Fig. 3.59. The penetration distribution of recoil implanted B into Si from $5 \times 10^{16}$ 100 keV $Ne^+$ $cm^{-2}$ compared with a direct implantation of 40 keV $B^+$. The dotted curve shows the recoil implanted profile for a higher dose.

atoms within the Si, at a depth greater than about 50 Å below the surface. The difference in the two distributions is readily apparent and illustrates the very strong tendency for small penetration depths in the case of recoil implantation. Due to the independence of the low energy recoil spectrum on

$E_1$, and to the relatively small difference in mass between B, O and Ne, we can, to a first approximation, use the curves in fig. 3.64 to estimate the concentration and depth distribution of contaminant oxygen, which inevitably suffers recoil implantation when directly implanting a Si target with 40 keV $B^+$ ions through an oxide layer. Firstly, we notice that the total number of impurity atoms which recoil into the substrate with an energy greater than 1 keV is about 50 times smaller than the total number of bombarding ions. However, of these, only $\sim 10\%$ penetrate deeper than $\sim 500$ Å below the surface. The fraction of oxygen contaminant to boron impurity in the vicinity of the peak of the implanted boron is then less than about 0.2 %, which for practical purposes can be neglected.

In summary, recoil implantation can be a useful alternative to conventional ion implantation; however, due to the preference for very low energy recoils, whether these be produced by protons or by more massive lower energy primary particles, penetration depth will generally be very small. Further, radiation damage effects will not only be produced by the implanted atoms during their passage into the substrate but also by the primary irradiation, when even the most favourable heavy ion case the total dose to the substrate will be in excess of $10^{17}$ ions $cm^{-2}$. In the case of a semiconductor substrate such a dose will have rendered the material amorphous, which, as we will see later may in fact be advantageous; but on the other hand, the high concentration of implanted bombarding ions may seriously affect the perfection of the final recrystallization on post bombardment annealing.

### 3.3. The fate of ion implanted atoms

So far we have discussed the penetration of ions into a solid, the inevitable radiation damage which ensues, the recovery of such damage, and some special problems relating to ion bombarded surfaces. The outstanding and perhaps most important question still to be answered relates to the eventual fate of the implanted atoms within the host lattice.

3.3.1. GENERAL CONSIDERATIONS

At very low implantation doses where the atomic concentration of impurity atoms is so low that even in thermal equilibrium at elevated temperatures the solubility limit is orders of magnitude greater, the implanted atoms will remain as isolated impurity atoms within the host lattice. In principle these atoms can occupy either substitutional or interstitial sites depending on which corresponds to the lowest free energy state, this in turn depends on parameters such as the size and electronic configuration of the atoms. In an irradiation environment, however, the situation can be quite different and often a steady state different from that expected from purely thermodynamical grounds prevails. For instance, at the commonly used implantation temperature around 20 °C the configuration of radiation damage may drastically influence the situation. In practice, the majority of implanted atoms come to rest in an environment where initially both lattice interstitials and vacancies are plentiful; and in the case of semiconductors, sometimes but not necessarily, within the disordered zones produced whilst the ion is slowing down. Calculation suggests that within a random collision cascade the incident ion has a high probability of finally coming to rest as a consequence of a replacement collision and therefore remains in a substitutional site. If the surrounding lattice interstitials and vacancies produced by the irradiation are mobile, this atom may subsequently move by vacancy exchange, become paired or trapped with a lattice interstitial, be thermally activated into a free interstitial configuration leaving a vacancy, or in some cases can be directly replaced by a lattice interstitial. Generally speaking, free interstitials have low activation energies for migration and move off rapidly through the lattice until they become trapped or cluster. However, this is not always the case; for instance, small impurity atoms such as carbon and nitrogen quite happily occupy stable interstitial positions within the more open b.c.c. lattices such as iron and molybdenum at room temperature.

At higher implantation temperatures or during annealing, thermal activation can result in a variety of modifications. For instance, the break-up of pairs and the release from traps, as well as the diffusion of isolated substitutional impurities to extended defects such as dislocation lines and grain boundaries, where the well established segregation effects can play an important role. However, as we are more concerned with somewhat higher implanted impurity concentrations where the mutual interaction between impurity atoms can, but not necessarily, dominate their final configuration, we will move onto higher dose implants.

3.3.2. RADIATION ENHANCED DIFFUSION

*3.3.2.1. General Considerations.*   Under normal conditions thermally activated diffusion is controlled by an activation energy which is the sum of a formation energy and an activation energy for migration. However, under the particular conditions of irradiation both vacancies and interstitials are created continually and their diffusion will depend only on the activation energies for their migration. Thus, provided there is sufficient thermal energy to maintain migration, but not to allow defect formation, diffusion during irradiation will be controlled primarily by the production of radiation damage and is therefore limited by the damage rate. Both experiment and theory suggest that under conditions where the migrating defects or impurity atoms are lost to fixed sinks, such as dislocation lines or precipitates, 'irradiation enhanced diffusion' should be approximately independent of the damage rate; this implicitly suggest thats, to a first approximation, enhanced diffusion is also independent of the activation energy for migration.

During typical implantation conditions, say with heavy ions having energies in the vicinity of 50 keV, the irradiation enhanced diffusion coefficient, $D^1$, has been estimated at about $10^{-15}$ cm$^2$ sec$^{-1}$ corresponding to a bombardment rate of $10^{13}$ ions cm$^{-2}$ sec$^{-1}$ (P. A. Thackery, 1968, private communication). Such a diffusion coefficient is typically equivalent to a temperature of about 450 °C in copper, 600 °C in iron, 200 °C in aluminium and about 1000 °C in silicon with respect to vacancy controlled diffusion. Thus ion implantation at temperatures below these values, but in excess of that at which vacancy mobility becomes significant, will provide substitutional enhanced diffusion of the implanted species in all these materials at a temperature which would normally not support significant thermally activated diffusion, i.e. at room temperature in aluminium and copper and about 200 °C in iron.

It should be realised that in the main, enhanced diffusion only occurs in that part of the target undergoing radiation damage. However, some diffusion beyond this region will occur due to the migration of vacancies and interstitial atoms to greater depths and this will result in the addition of a slight tail onto the normal penetration distribution. Another point to remember is that, when the irradiation ceases, the formation of the defects responsible for diffusion will immediately stop; however, as the temperature is generally sufficiently high to support the migration of already existing defects, enhanced diffusion will still persist until all these have been lost to sinks, the decay time depending on the thermally activated jump rate for migration.

*3.3.2.2. Enhanced Diffusion during Proton Irradiation.* A specific application of enhanced diffusion is the creation of useful impurity profiles in semiconductors during irradiation with protons. The nature of such profiles has been discussed in Chapter 2 and their application will be described in

Chapter 5. In this chapter we will discuss the theoretical concepts behind such enhanced diffusion during proton irradiation.

The simplest theoretical situation is to use protons of sufficiently high energy that to all intents and purposes the radiation enhanced diffusion coefficient is constant over the region of interest. For instance if we are interested in producing profiles up to 1 $\mu$m below the surface, then irradiation with, say, 1 MeV protons will result in an almost constant damage rate over this depth. The damage rate is readily calculable from an integration of the Rutherford scattering cross-section over the range of recoil energies which result in atomic displacement, and we obtain a displacement rate of

$$N_d = \phi \sigma_d (\overline{T}/2E_d) \text{ displacements/atom/sec}$$

where $\phi$ is the proton flux (particles/cm$^2$/sec), $\sigma_d$ is the displacement cross-section, $\overline{T}$ is the average energy transferred per collision and $E_d$ is the displacement energy. For Rutherford collisions

$$\sigma_d = 4\pi a_0^2 \left(\frac{M_1}{M_2}\right) Z_1^2 Z_2^2 \left(\frac{E_R^2}{E}\right)$$

where $a_0 = 0.53$ Å, $E$ is the proton energy, $E_R = 13.6$ eV and $M_1$, $M_2$ and $Z_1$ and $Z_2$ and the masses and atomic numbers of the protons and target respectively; $\overline{T}$ is given by

$$\overline{T} = \frac{E_d T_m}{(T_m - E_d)} \log_e \left(\frac{T_m}{E_d}\right) \quad \text{where } T_m = \frac{4M_1 M_2}{(M_1 M_2)^2} E.$$

The enhanced diffusion coefficient depends critically on the degree of recombination between the vacancies and interstitials produced by the irradiation in their migration through the lattice. If the temperature is sufficiently high, the damage rate is sufficiently low and defects are lost only to fixed sinks, then to a good approximation we can neglect recombination and the enhanced diffusion coefficient is directly proportional to the damage rate. Furthermore under this situation the enhanced diffusion coefficient is temperature independent. However, if the temperature is sufficiently low that either the vacancies or interstitials migrate only very slowly, and/or the damage rate is very high, the dynamic equilibrium concentration of defects will build up such that appreciable recombination occurs. Under these circumstances the enhanced diffusion coefficient will depend on the damage rate to the half power and will be strongly temperature dependent.

If we are primarily concerned with the diffusion of substitutional impurities, then we must calculate the enhanced vacancy diffusion coefficient.

This can be defined as $D_v = v_v v \lambda^2$, where $v_v$ is the vacancy jump frequency, $v$ is the dynamic vacancy concentration and $\lambda$ is the jump distance ($\lambda^2 \sim 10^{-15} \text{ cm}^2$). In the case of no recombination and loss only to fixed sinks it is simple to show that $v = K/(\alpha v_v \lambda^2)$ where $K$ is the vacancy production rate which equals $N_d$ and $\alpha$ is the fixed sink density. This gives

$$D_v = N_d/\alpha.$$

In order to provide some idea of the magnitude of the enhanced diffusion coefficient during the irradiation of Si with 1 MeV protons we will choose a current density of 1 $\mu$A cm$^{-2}$ and a typical sink density equivalent to $10^{10}$ dislocation lines cm$^{-2}$. Then by simple substitution into the formula we obtain an enhanced diffusion coefficient of $D_v \sim 10^{-16} \text{ cm}^2 \text{ sec}^{-1}$.

If on the other hand mutual recombination dominates then $v = (K/Zv_i)^{\frac{1}{2}}$ where $v_i$ is the interstitial jump frequency and $Z$ is the number of sites around a defect where recombination can spontaneously occur, whence

$$D_v = v_v \left( \frac{N_d}{Zv_i} \right)^{\frac{1}{2}} \lambda^2$$

In these circumstances the enhanced diffusion coefficient is usually at least an order of magnitude less than for loss to fixed sinks.

Which one of these regimes dominates depends largely on the irradiation temperature and on the damage rate as it is these parameters which determine the fraction of irradiation produced interstitials and vacancies which suffer recombination relative to the loss at fixed sinks. From the point of view of using proton enhanced diffusion to tailor impurity profiles it would therefore seem advisable to use the lowest damage rates together with high temperatures such that the enhanced diffusion coefficient is independent of temperature variations during the irradiation. However, the damage rate should not be so low that the absolute value of the enhanced diffusion coefficient is too small to cause any effects in a reasonable time, and the temperature should not be so high that thermal diffusion predominates.

A useful application of proton enhanced diffusion has been discussed in Chapter 2, where the proton energy was chosen to be sufficiently low that the proton's range approximately coincides with the depth to which the profile must be diffused. In this case the enhanced diffusion coefficient will vary with depth due to the fact that the damage rate increases towards the end of the proton's range. However, the precise variation will depend on whether the experiment is carried out under conditions where recombination is dominant or loss to fixed sinks is dominant. Using simple computational techniques it is then possible to ascertain the approximate shape of the diffused impurity profiles. One important point is that the maximum diffusion depth will be rather well defined and will be determined by the extent of enhanced diffusion beyond the limiting damage depth.

### 3.3.3. PRECIPITATION EFFECTS

*3.3.3.1. General Considerations.* As the ion dose is increased more and more atoms are introduced into the lattice and it is of considerable interest to determine whether or not the alloy so formed remains as a solid solution indefinitely. In comparison with conventionally prepared alloys, if the concentration of impurity exceeds the maximum solubility limit in the target material, and the diffusion conditions are suitable, then precipitation of a second phase may result. However, unlike conventional systems where precipitation effects can only be observed in a limiting number of alloy systems, due to the supersaturation which occurs on the reduction in temperature of a solid solution; there is, in principle, no such limitation in the case of ion implantation, as ions of any element can be injected into any solid irrespective of solubility considerations.

At low temperatures where thermally activated diffusion of the implanted atoms is virtually prohibited we might readily jump to the conclusion that the alloy so formed would in fact remain as a supersaturated solid solution. But, on the other hand, if the implantation temperature was sufficiently high that impurity diffusion became significant, we might expect agglomeration and precipitation. For instance, the inert gases are known to have vanishingly low solubilities in solids and provided their mobility is adequate precipitation will occur for even the smallest concentrations. In this case, due to the inability of the inert gases to form compounds, precipitation takes the form of an agglomerate of gas atoms which can best be described as a bubble; we will discuss this in some detail later on. However, less inert impurities precipitate out to form a compound on attaining their solubility limit, for example Cu–Al forms $CuAl_2$. But, on the other hand, in some instances such as Ag–Au the one species is quite stable within the other and remains in substitutional solid solution at all concentrations. For a review of conventionally prepared alloys the reader is referred to Kelly and Nicholson (1963). However, as previously pointed out, the radiation damage and associated effects which inevitably occur during implantation can significantly modify the conditions pertaining to thermodynamical equilibrium. In dealing with such a complicated phenomenon, which depends on a particular case, it is very difficult to present a general understanding. However we will attempt to discuss the underlying considerations and illustrate the conclusions with a few specific examples (see Thackery and Nelson, 1969).

*3.3.3.2. The Formation and Stability of Precipitates during Implantation.* It is an experimentally proven fact that small precipitates within a solid can be an dissolved by irradiation. For instance, fig. 3.60 illustrates the dissolution of $\eta$-phase $MgZn_2$ precipitates in Al-6(wt %), Zn-2(wt%), Mg alloy after 50 keV $A^+$ ion bombardment at room temperature. It is clear that precipitates have been dissolved only in that part of the specimen not shielded with an electron microscope support grid. This dynamic re-solution process results from the passage of energetic collision cascades through the interface between the precipitate and the host lattice, as if an atom within a precipitate recoils with sufficient energy to penetrate into the surrounding lattice it is to all intents and purposes dissolved (Nelson, 1969b). The minimum energy for such a mechanism to occur can be estimated from our knowledge of

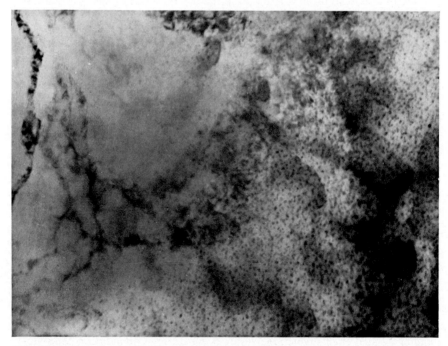

Fig. 3.60. Micrograph showing the dissolution of fine $\eta$-phase $MgZn_2$ precipitates in Al-6 (wt %), Zn-2 (wt %) Mg alloy after a 50 keV $A^+$ ion bombardment at room temperature. The region still showing precipitates was shielded during the bombardment.

radiation damage processes. For instance we know that the energy necessary to separate an interstitial from its vacancy so that their spontaneous recombination does not occur is of the order of 25–50 eV. Thus an atom within the surface layers of the precipitate which recoils with an energy in excess of such a value into the surrounding lattice has a good chance of escaping into solution. However, if this atom is thermally mobile within the surrounding lattice, it has a finite chance of returning to the same precipitate rather than diffusing through the lattice.

In order to estimate the kinetic dissolution rate, we must calculate the flux of recoils with energy in excess of say 50 eV which cross the precipitate surface. In this instance, we will be concerned with displacement cascades initiated by primary recoil atoms of about 50 keV, and from our knowledge of the numbers and energy spectra of atoms ejected from crystals under ion bombardment (Thompson, 1968) we can estimate the flux of such recoils, $\psi$, to be $\sim 5 \times 10^{12}$ cm$^{-2}$ sec$^{-1}$ corresponding to a bombarding flux, $\phi$, of $10^{13}$ ions cm$^{-2}$ sec$^{-1}$. Then, if we assume for convenience that the implanted solid contains $n$ spherical precipitates of radius $r$, the kinetic dissolution rate per unit volume within the irradiated region is

$$d_k = 4\pi r^2 f \psi n \text{ cm}^{-3} \text{ sec}^{-1} \tag{3.29}$$

where $f$ is the probability of a kinetically dissolved atom not jumping back into the same precipitate. Further, the thermal dissolution rate per unit volume is given approximately as

$$d_t = 4\pi r^2 \Omega n f^1 v \exp\left(-E_s/kT\right) \tag{3.30}$$

where $\Omega$ is the atomic surface density at the precipitate, $f^1$ is the probability of escape, $v$ is the thermal vibration frequency and $E_s$ is the thermal activation energy for solution.

On the other hand, the capture rate per unit volume from a uniform concentration $c$ in solution by these precipitates is approximately

$$K = (D + D^1) cnr/3a^3 \text{ cm}^{-3} \text{ sec}^{-1} \tag{3.31}$$

where $(D + D^1)$ is the net solute diffusion coefficient being the sum of the thermal and irradiation enhanced values, and $a$ is the nearest neighbour spacing in the host lattice. In a steady state, the capture rate per unit volume equals the solute atom capture rate per unit volume and equating (3.29), (3.30) and (3.31) we obtain

$$c = \frac{12\pi a^3 r}{(D + D^1)} \left\{ f\psi + f^1 \Omega v \exp\left(-E_s/kT\right) \right\} \tag{3.32}$$

As the implantation dose builds up we also know that if we neglect those ions scattered back from the surface the fraction of implanted atoms in solution and in precipitates must equal the total number implanted in time $t$, thus

$$\phi t/R = N_c + \tfrac{4}{3}\pi r^3 n N \tag{3.33}$$

where $R$ is the incoming particles mean range in the solid, and the atomic density $N$ is taken to a first approximation to be the same in both the precipitate and in the surrounding matrix.

Equating (3.32) and (3.33) we obtain the relation

$$c^3 = \left[\frac{12\pi a^3 \{f\psi + f^1 \Omega v \exp\left(-E_s/kT\right)\}}{D + D^1}\right]^3 \left(\frac{3}{4\pi n}\right)\left(\frac{\phi t}{NR} - c\right) \tag{3.34}$$

Furthermore from homogeneous nucleation theory (Greenwood and Speight, 1957) it is readily shown that the nucleation density is given by

$$n = \frac{1}{10a^2}\left\{\frac{\phi}{(D + D^1)RN}\right\}^{\frac{1}{4}} \tag{3.35}$$

Let us initially assume that kinetic solution dominates thermal solution, i.e. the thermal solution energy $E_s$ is sufficiently large so that at temperatures of interest $f\psi \gg f^1 \Omega v \exp(-E_s/kT)$. Then choosing typical values, i.e. $R = 1000$ Å, $\phi = 10^{13}$ ions cm$^{-2}$ sec$^{-1}$, $N = 2.5 \times 10^{22}$ atoms cm$^{-3}$, $a = 2$ Å, $f = 0.1$, we can calculate $c$ as a function of $t$ for different values of $(D + D^1)$. This is shown in fig. 3.61 where it is readily apparent that for a total diffusion coefficient of $\gtrsim 10^{-16}$ cm$^2$ sec$^{-1}$ virtually all the implanted atoms will be in solution, even up to concentrations approaching 50 %. However, if as we have stated the temperature is sufficiently high to support vacancy migration an incident flux of $10^{13}$ ions cm$^{-2}$ sec$^{-1}$ produces an enhanced diffusion coefficient of $\sim 10^{15}$ cm$^2$ sec$^{-1}$, and so even at room temperature in metals such as aluminium and copper substantial precipitation of the implanted atoms can occur leaving less than a quarter in dynamic solution after irradiation times in excess of $10^3$ sec. If, on the other hand, the total number of implanted atoms is sufficiently small to produce a concentration below about 0.5 %, then during an implantation at a rate of $10^{13}$ ions cm$^{-2}$ sec$^{-1}$ the solute atoms will always be in solution. At the instant the irradiation ceases, however, the mobile irradiation produced vacancies will result in some fraction of the implanted atoms forming precipitate nuclei. At higher temperatures the situation will depend on the relative importance of the two thermally activated terms $D$ and $\exp(-E_s/kT)$. For instance if $D > D^1$ but $E_s$ is so large that thermal solution is small

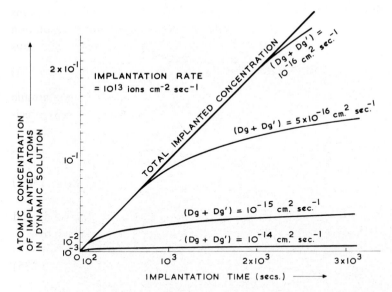

Fig. 3.61. The atomic concentration of implanted solute atoms in dynamic solution plotted as a function of irradiation time for different fusion coefficients.

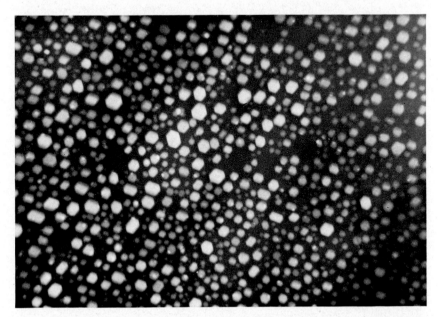

Fig. 3.62. Argon bubbles in Cu, note the crystallographic equilibrium shape defined by the criterion of minimising free surface energy.

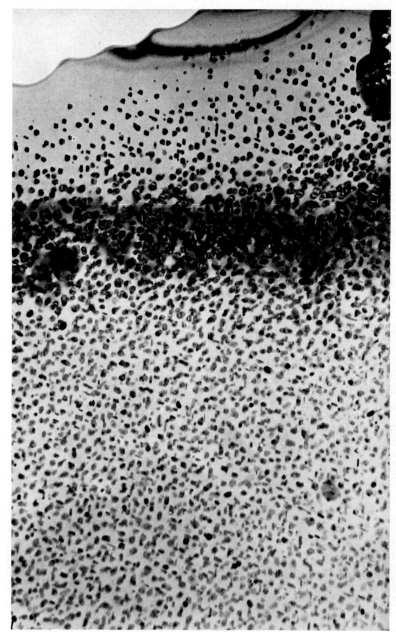

Fig. 3.63. Micrograph of 80 keV Sb$^+$ implanted into Al at 330 °C to a dose of $5 \times 10^{16}$ ions cm$^{-2}$ showing AlSb precipitates.

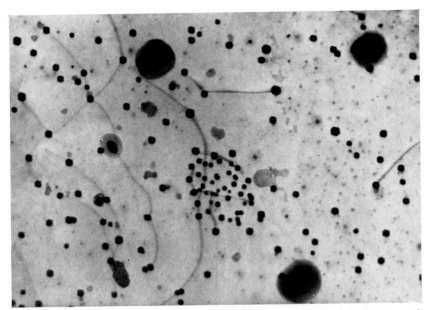

Fig. 3.64. Al implanted with 70 keV Pb$^+$ ions at 400 °C to a dose of $10^{16}$ ions cm$^{-2}$. Pure lead precipitates have well defined shape and are all oriented parallel to each other and bear a f.c.c. epitaxial relationship to the substrate Al.

compared to kinetic solution, as is the case for the inert gases, then virtually all the implanted atoms will precipitate leaving only a small fraction in solution. But, on the other hand, if $D > D^1$ and $E_s$ is sufficiently small that thermal solution dominates, as say for copper in aluminium at temperatures > 500 °C, then the alloy so formed will be virtually identical to that expected from purely thermodynamical arguments. Figures 3.62, 3.63, 3.64, 3.65 and 3.66 show a selection of precipitates produced during ion implantation of copper aluminium and silicon targets.

In the foregoing discussion, on precipitate effects, we have assumed that the implanted region was uniformly irradiated, and that the solute atoms were uniformly distributed in depth. Further, we ignored both the surface and the non-irradiated region beyond. Of these, the surface is likely to play the major role in modifying our general conclusions; the fact that the implanted atoms are deposited in a non-uniform way will simply introduce a tail which corresponds to lower solute concentrations and lower enhanced diffusion rates. In nearly every circumstance the surface will contain a higher solute concentration than within the implanted layer and consequently precipitation will occur more readily (as confirmed by the

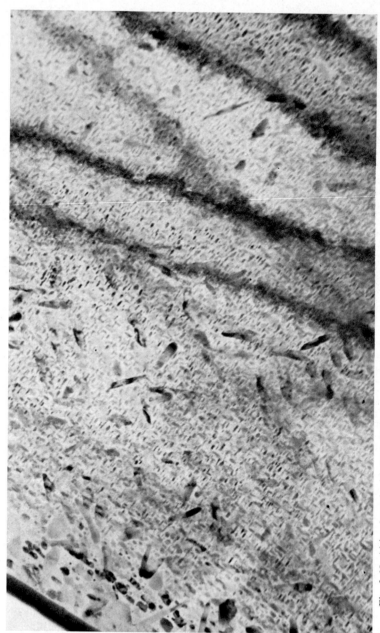

Fig. 3.65. Al implanted with 70 keV $Cu^+$ ions at 40 °C to a dose of $4 \times 10^{16}$ ions $cm^{-2}$ showing $CuAl_2$ precipitates.

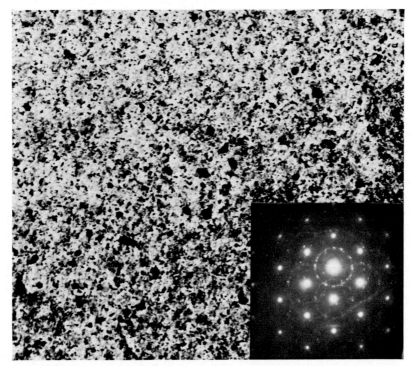

Fig. 3.66. Micrograph showing Pd precipitates in Si after recrystallization at 650 °C. Note that although the Si is essentially single crystal the Pd precipitates have a random orientation.

larger scale precipitates in fig. 3.65). The stability of these precipitates ultimately depends on the erosion of the surface by sputtering, as during the implantation processes, especially to high doses, the surface is being continuously removed. If, then, the surface precipitates have a higher sputtering rate than the surrounding matrix their growth will be somewhat restricted however, if they sputter less readily than the matrix they will grow.

### 3.3.4. EXPERIMENTAL LOCATION OF IMPURITY ATOMS USING CHANNELLING TECHNIQUES

In Chapter 2 the use of channelling techniques as a tool for the investigation of atomic position was discussed in general terms. In this section we will describe the specific application of this technique to the investigation of impurity atom location in ion implanted semiconductors. The majority of this work has been carried out either by or in association with J. A. Davies and his group at Chalk River (Davies et al., 1967; Mayer et al., 1968; Eriksson et al., 1969) and in this instance it will suffice to present an overall picture of the results obtained to date.

A general requirement of the ion implantation of Si is that, after annealing, the impurity atoms must reside as isolated impurities occupying substitutional sites within the lattice. In the case of diffusion, where processes are controlled by thermodynamical considerations, the impurity atoms have a fair chance of finding such sites. However, after the nonequilibrium process of ion implantation, it is impossible to predict their final state for a specific case. For instance, after high doses, it is often possible to exceed the solubility limit with the inevitable consequence that precipitation occurs. On the other hand, the interaction between the implanted impurities and the residual radiation damage can significantly influence their final configuration. We are principally concerned with the fate of certain specific elements in Si; for example, the group III and group V elements. Unfortunately, because the Rutherford scattering technique is only useful in the case of heavy impurity atoms, the atomic position of the most commonly used dopants in Si, namely B and P, cannot be studied. On the other hand, in principal the characteristic X-ray technique can be used, but at the time of writing such experiments have not been performed. However, using Rutherford scattering, a detailed study has been carried out for Ga, In and Tl on the one hand, and As, Sb and Bi on the other.

If we recall Chapter 2, we can say that, to a first approximation, the percentage attenuation of back scattering during proton or $He^+$ bombardment along a channelling direction relative to that from a random direction, is a measure of the substitutional impurity component. To date the majority of data has been acquired for hot implantation rather than recrystallized room temperature implants, in an attempt to minimise the effects of radiation damage. Table 3.5 is a resume of the results and indicates the percentage attenuation along the major channelling directions after 40 keV bombardment with $1.5 \times 10^{14}$ ions $cm^{-2}$ at 450 °C. It is clear that, because of the large percentage attenuation in the case of group V elements, especially Sb and Bi, these elements exhibit large substitutional components after implantation under these conditions. Such elements should therefore provide suitable dopants from the viewpoint of their electrical activity. On the other hand the percentage attenuation in the case of the group III elements is quite different. For instance, the attenuation in the $\langle 111 \rangle$ direction is approximately double that for the $\langle 110 \rangle$, indicating that roughly equal numbers of impurity atoms are located at substitutional and interstitial sites.

TABLE 3.5

The hot implant behaviour of group III and V elements in Si

| Implanted atom | % attenuation | |
| --- | --- | --- |
| | $\langle 111 \rangle$ | $\langle 110 \rangle$ |
| As ⎫ | 56 | 55 |
| Sb ⎬ V | 90 | 88 |
| Bi ⎭ | 87 | 86 |
| Ga ⎫ | <10 | <10 |
| In ⎬ III | 61 | 34 |
| Tl ⎭ | 53 | 27 |

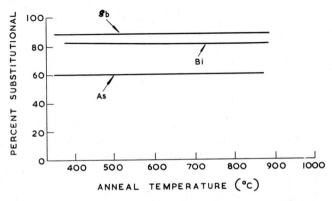

Fig. 3.67. Annealing behaviour of the substitutional component of the group V implantations at as temperature of about 400 °C (after Eriksson et al., 1969).

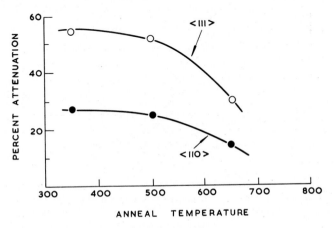

Fig. 3.68(a).

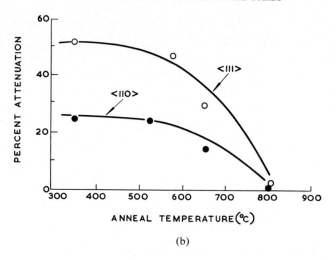

(b)

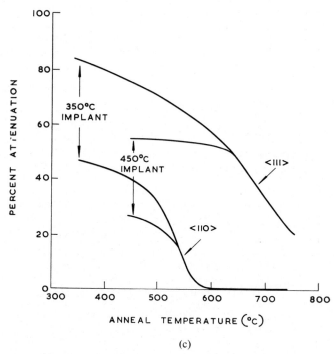

(c)

Fig. 3.68. (a) Annealing behaviour of 40 keV Ga implanted into Si at 350 °C (after Eriksson et al., 1969). (b) Annealing behaviour of 40 keV In implanted into Si at 350 °C (after Eriksson et al., 1969). (c) Annealing behaviour of 40 keV Tl implanted into Si at 350 °C and 450 °C (after Eriksson et al., 1969).

The subsequent annealing behaviour of these hot implanted specimens shows up some further discrepancies between the group V and group III elements. Figure 3.67 shows that even up until nearly 900 °C the substitional component for As, Sb and Bi in Si remains approximately constant, whereas, on the other hand, fig. 3.68(a), (b), (c) shows quite different behaviour for Ga, In and Tl. For instance, in all cases, both the substitutional and interstitial component fall as the temperature is increased, until at temperatures of about 800 °C virtually all the impurity atoms have moved into some configuration which is neither substitutional or interstitial, probably in the form of minute precipitates as discussed previously.

An important observation which suggests that in some cases it is perhaps beneficial to implant the impurity atoms at sufficiently low temperatures such that the Si can in fact degenerate into an amorphous state, has also been reported. Figure 3.69 shows the substitutional and interstitial component of Ga implanted into Si at room temperature to a dose of $4 \times 10^{14}$ ions cm$^{-2}$, following an anneal to cause epitaxial recrystallization. It is readily seen that on recrystallization at 600 °C a substantial fraction of the Ga atoms are forced to occupy symmetrical sites, such that the sum of the substitutional and interstitial is equal to about 60 % of the total number of implanted atoms. However, on subsequent annealing to higher temperatures, where thermally activated diffusion processes occur, the substitutional and interstitial components fall as precipitation occurs. The technique of forcing impurities into substitutional sites on recrystallization, at temperatures too

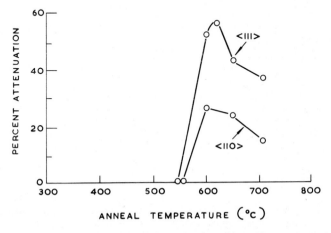

Fig. 3.69. Annealing behaviour of 40 keV Ga implanted into Si at room temperature (after Eriksson et al., 1969).

low for thermal diffusion, has important consequences in the electrical behaviour of such impurities as we will see in Chapter 5.

It must be pointed out that the above interpretation of Rutherford scattering data from implanted impurities must be viewed with caution, for a variety of reasons. The first point, as discussed in detail in Chapter 2, is that isolated interstitial atoms with a regular lattice can give rise to an increased scattering yield along certain crystallographic directions. But perhaps even more important, no account has been taken of the fact that very small impurity clusters can give rise to attenuation in scattering yields which are virtually identical to those expected from isolated impurities. For instance, it is well known that at the nucleation stage, precipitates often bear a coherency relationship to the host matrix, see for example fig. 3.70(a), (b); the mismatch in lattice spacing then gives rise to coherency stains in the lattice. Then, if the precipitate nuclei is of the type shown in fig. 3.70(a) the attenuation in Rutherford scattering yield along any channelling direction could suggest an interpretation in terms of isolated impurities. On the other hand, the result of the experiment shown in fig. 3.70(b) could suggest an interpretation in terms of both isolated substitutional and interstitial impurities corresponding to channelling along the directions $X$ and $Y$ respectively. In either case, as the precipitate grows and loses coherency with the surrounding matrix, the attenuation in Rutherford scattering along any channelling direction would disappear. It is quite possible that some of the results described above could be interpreted in terms of the formation and growth of such precipitates.

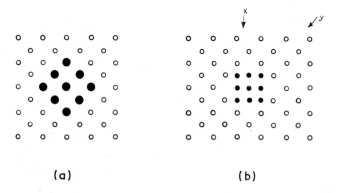

(a)                                        (b)

Fig. 3.70. Small precipitate nuclei bearing a coherency relationship to the host lattice. (a) Rutherford scattering data might suggest an interpretation in terms of isolated impurities. (b) Rutherford scattering data might suggest an interpretation in terms of both isolated substitutional and interstitial impurities.

3.3.5. EFFECT OF IMPURITIES ON THE RECOVERY OF RADIATION DAMAGE

As we have hinted on numerous occasions throughout the preceding sections of this chapter, the thermally induced recovery of radiation damage in solids containing high concentrations of impurity atoms can be quite different from that encountered in pure materials. In line with our previous discussions of damage, we will again differentiate between metals and semiconductors.

*3.3.5.1. Metals.* We have seen that in practical cases radiation damage in metals during ion bombardment degenerates into an array of point defect clusters such as dislocation loops, which, at high doses, eventually interact to form an entangled dislocation network. Such networks anneal at self diffusion temperatures by the processes of climb and slip and are ultimately lost to the free surface. However, if the solid contains a high concentration of impurities which are insoluble at the annealing temperature and form small precipitates, these can pin the dislocation network and prevent its movement to the surface.

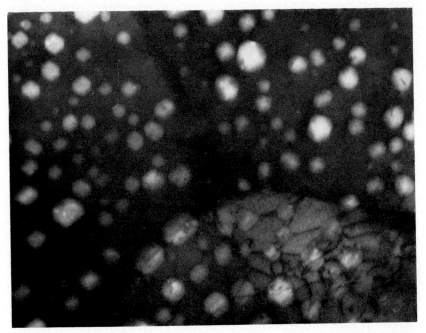

Fig. 3.71. Bubbles of A in Cu predominantly attached to dislocation lines after annealing at 700 °C.

A classic example of this is to be found in the case of copper implanted with inert gas ions such as argon. It is well known that the inert gases readily precipitate out as gas bubbles within the solids and are generally to be found lying on the residual dislocation entanglement that has grown from the radiation damage (e.g. Nelson and Mazey, 1966), for an example see fig. 3.71. In such a situation, if the bubbles are mobile, the network will be essentially pinned irrespective of the supply of thermal vacancies. On the other hand, it is known that such gas bubbles can in fact move in the solid by the migration of atoms over their internal surfaces by surface diffusion (Barnes and Mazey, 1963). However, bubble migration is rather slow and only becomes significant close to the melting point. Figure 3.72 is an electron micrograph of the same area shown in fig. 3.71 after a subsequent anneal at a higher temperature. It is readily apparent that with the progressive loss of bubbles to the surface the residual dislocation network gradually moves out from the solid. The persistence of the damage network depends of the density and size of the precipitates; in other words on the impurity concentration.

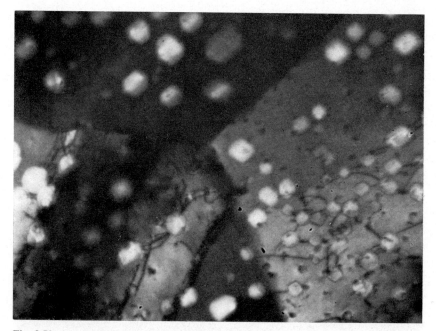

Fig. 3.72. An electron micrograph showing the same area as shown in fig. 3.71 but after a further anneal at 800 °C. Note the loss of bubbles to the free surface especially from the grain boundaries.

Clearly it is not only the inert gases which can pin the residual dislocation structure but any precipitate existing within the solid. For instance,
in the case of Al irradiated with oxygen ions to a sufficiently high dose to
create $Al_2O_3$ precipitates within the metal, the dislocation structure will
persist until the melting point of the Al. On the other hand, in the case of
Al bombarded with say $Cu^+$ ions, the $CuAl_2$ precipitates will only pin the
dislocations until the temperature is sufficiently high for the precipitates to
dissolve in the lattice.

The annealing of radiation damage in ion bombarded metals therefore
depends intimately on the solubility of the bombarding ion in the solid, on
the ion dose, and in the case where insoluble precipitates form, on the
mobility of precipitates to the free surface.

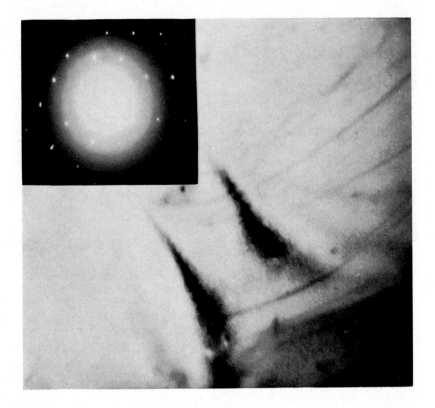

Fig. 3.73. Si implanted with Pb at room temperature. The diffuse rings of the diffraction
pattern indicate that the Si is amorphous and there is no evidence for Pb precipitation.

*3.3.5.2. Semiconductors.* The general comments made in the preceding section on metals regarding solubility are also pertinent to semiconductors. However, due to the fact that semiconductors can become amorphous during ion bombardment and recrystalline on annealing some special problems arise. For instance, within an amorphous semiconductor the concepts of diffusion and precipitation are somewhat different from materials possessing a regular crystal structure. In most cases, experiment has shown that even after excessive bombardments, whilst the solid remains

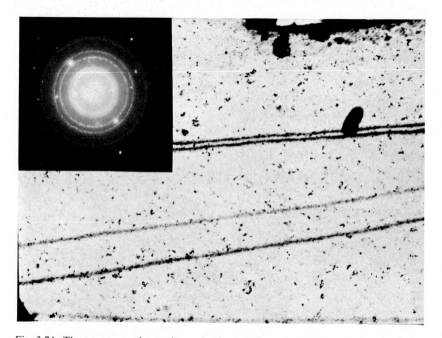

Fig. 3.74. The same sample as shown in fig. 3.73 but after a recrystallization anneal at 800 °C. In this case the Pb precipitates have formed in a random manner within the essentially single crystal Si.

amorphous, there is little or no tendency to precipitation, for example see fig. 3.73 (Matthews, 1969). However, on recrystallization, even at temperatures well below the accepted impurity diffusion temperatures precipitation of the impurity occurs rather than its remaining in supersaturated solution within the newly formed lattice. For example, fig. 3.74 shows small Pb precipitates which have formed within a recrystallized Si sample after annealing to 800 °C.

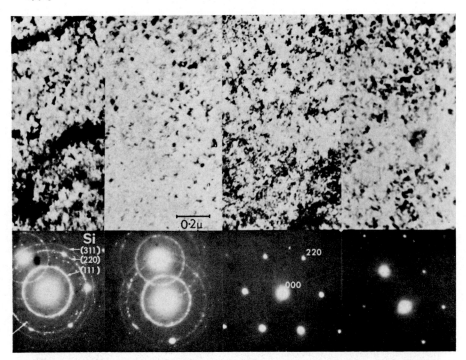

Fig. 3.75. The recrystallization of 50 keV $Sb^+$ implanted Si as a function of dose after annealing at 680 °C for 30 min. At the low doses shown on the right hand side recrystallization was epitaxial, however at the high doses shown on the left hand side small metallic precipitates of Sb have caused the Si to recrystallize in a polycrystalline way (after Matthews, 1969).

The question arises; can the precipitation of implanted impurities affect the epitaxial regrowth of the recrystallized lattice? This has been studied in detail for a number of impurities in Si by Matthews (1969). As an example, fig. 3.75 shows the recrystallization of 50 keV $Sb^+$ implanted Si as a function of dose after annealing at 680 °C for 30 min. At doses up to about $3 \times 10^{15}$ $Sb^+$ ions $cm^{-2}$ recrystallization was epitaxial, as shown by the diffraction patterns together with the usual residual defect structure. However, at higher doses, where the concentration of Sb exceeded the solubility limit, a substantial fraction of the Si recrystallized in a polycrystalline manner, and small metallic Sb precipitates could be identified. This clearly suggests that the precipitation of Sb influences the recrystallization process sufficiently to interfere with the process of epitaxial regrowth. A similar effect can also be seen after the implantation of inert gas ions, fig. 3.77(a), (b). In this case, although the solubility limit is exceeded at even

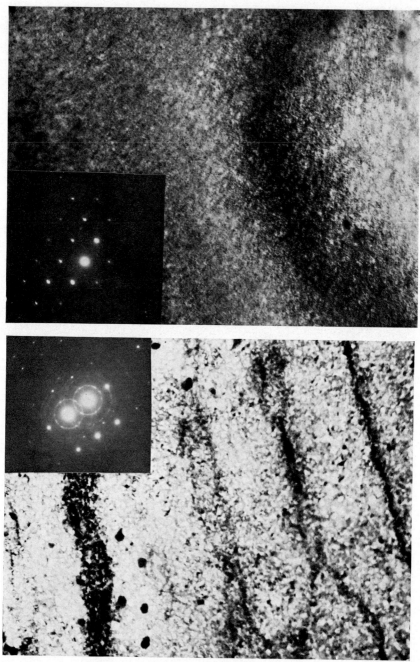

Fig. 3.76. The recrystallization of $Xe^+$ implanted Si. (a) to a relatively low dose showing good epitaxial regrowth, (b) to a high dose showing that the formation of Xe gas bubbles has resulted in a polycrystalline recrystallization.

the lowest doses, the precipitation of gas into bubbles on recrystallization only hinders epitaxial growth after high dose implantation.

A further example of the effect of implanted impurity atoms influencing the recrystallization process which must be mentioned, is the case of Au in Si. Due to the large atomic number of Au, the Si lattice is quickly rendered amorphous during ion bombardment. At low doses $< 10^{16}$ ions cm$^{-2}$ the same general pattern as described above was found to occur, i.e. epitaxial recrystallization and internal precipitation of Au at temperatures about 650 °C. However, for doses in excess of $10^{17}$ Au$^+$ ions cm$^{-2}$, on annealing at temperatures as low as 500 °C a liquid Au–Si amalgam is produced in the heavily implanted regions. On subsequent cooling the amalgam solidified from the solid/liquid interface when the temperature drops to about 370 °C.

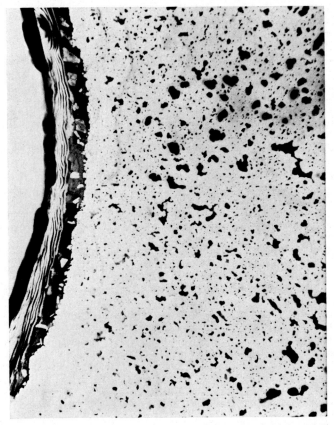

Fig. 3.77. Surface precipitates of Au formed after annealing Au$^+$ implanted Si at 525 °C (after Matthews and James, 1969).

Experiment shows that the solidification processes occur epitaxially onto the underlying crystal lattice and that the majority of the Au which is in excess of the solid solubility limit at that temperature precipitates as metallic Au. However, as is only to be expected a significant fraction of these precipitates form as large particles on the surface of the Si, see fig. 3.77. It is interesting to note that, due to the formation of the liquid phase, epitaxial recrystallization occurs without the excessive twinning which prevails after recrystallization from the usual amorphous layer.

## References

Almén, O. and G. Bruce, 1961a, Nucl. Instr. Methods **11**, 279.
Almén, O. and G. Bruce, 1961b, Nucl. Instr. Methods **11**, 257.
Banbury, P. C., and I. N. Haddad, 1966, Phil. Mag. **14**, 841.
Barnes, R. S. and D. J. Mazey, 1963, Proc. Roy. Soc. **A275**, 47.
Beevers, C. J. and R. S. Nelson, 1963, Phil. Mag. **8**, 1189.
Behrisch, R., 1964, Ergeb. Exakt. Naturw. **35**, 295.
Brice, D. K., 1970, Proc. Int. Conf. on Ion Implantation in Semiconductors, Thousand Oaks, California, p. 101.
Brodsky, M. H. and R. S. Title, 1969, Phys. Rev. Lett. **23**, 581.
Brown, L. M., A. Kelly and R. M. Mayer, 1969, Phil. Mag. **19**, 721.
Carter, G. and J. S. Colligon, 1968, 'Ion Bombardment of Solids' (Heinemann, London).
Chadderton, L. T. and F. H. Eisen, 1970, Int. Conf. on Ion Implantation in Semiconductors, Thousand Oaks, paper PD7.
Corbett, J. W., 1966, 'Electron Radiation Damage in Semiconductors and Metals', Solid State Physics, Sup. 7 (Academic Press. New York).
Crowder, B. L., R. S. Title, M. H. Brodsky and G. D. Pettit, 1970, Appl. Phys. Lett. **16**, 205.
Dash, W. C. and R. Newman, 1955, Phys. Rev. **99**, 1131.
Davidson, S. J. and R. Booker, 1970, Int. Conf. in Ion Implantation, Thousand Oaks.
Davies, J. A., J. Denhartog, L. Eriksson and J. W. Mayer, 1967, Can. J. Phys. **45**, 4053.
Ditchburn, R. W., 1952, 'Light' (Blackie, London).
Dobson, P. S. and R. E. Smallman, 1966, Proc .Roy. Soc. **A293**, 423.
Eisen, F. H., 1970, European Conf. on Ion Implantation, Reading.
Erginsoy, C., A. Englert and G. Vineyard, 1964, Phys. Rev. **133**, 595.
Eriksson, L., J. A. Davies, N. G. E. Johansson and J. W. Mayer, 1969, J. Appl. Phys. **40**, 842.
Greenwood, G. W. and M. V. Speight, 1957, J. Nucl. Mater. **16**, 327.
Hansen, M., 1958, Constitution of Binary Alloys (McGraw Hill, N.Y.).
Hart, R. R., 1970, Int. Conf. on Ion Implantation in Semiconductors, Thousand Oaks, paper B.4.
Haymann, P. and C. Waldburger, 1964, Ionic Bombardment, Theory and Applications (C.N.R.S. Symposium Science Publ. Inc.).
Hudson, J. A., 1969, Thesis, University of Cambridge.
Kaminsky, M., 1965, Atomic and Ionic Impact Phenomena on Metal Surfaces (Springer-Verlag, Berlin).
Kelly, A. and R. B. Nicholson, 1963, Prog. Mat. Sci. **10**, No. 3.
King, E., J. A. Lee, K. Mendlessohn and D. A. Wigley, 1965, Proc. Roy. Soc. **A284**, 325.

Kopitzki, K. and H. E. Stier, 1962, Z. Naturforsch. **17a**, 346.

Large, L. N., and R. W. Bicknell, 1967, J. Mater. Sci. **2**, 589.

Lehmann, C. and G. Leibfried, 1961, Z. Physik **162**, 2, 203.

Lehmann, Ch. and P. Sigmund, 1966, Phys. Status. Solid. **16**, 507.

Leibfried, G. and P. H. Dederichs, 1962, Z. Physik **120**, 320.

Lindhard, J., M. Scharff and H. Schiøtt, 1963, Kgl. Danske Videnskab. Selsk., Mat.-Fys. Medd. **33**, no. 14.

Lomer, J. N. and M. Pepper, 1967, Phil. Mag. **16**, 1119.

Magnuson, G. D. and C. E. Carlston, 1963, J. Appl. Phys. **34**, 3267.

Matthews, M. D., 1969, J. Mat. Sci. **4**, 997.

Matthews, M. D. and P. James, 1969, Phil. Mag. **19**, 1179.

Mayer, J. W., L. Eriksson, S. T. Picraux and J. A. Davies, 1968, Can. J. Phys. **46**, 663.

Mazey, D. J. and R. S. Barnes, 1968, Phil. Mag. **17**, 387.

Mazey, D. J. and R. S. Nelson, 1969, Rad. Effects **1**, 229.

Mazey, D. J., R. S. Nelson and R. S. Barnes, 1968, Phil. Mag. **17**, 1145.

Mazey, D. J., R. S. Nelson and P. A. Thackery, 1967, J. Mater. Sci. **3**, 26.

Molchanov, V. A. and V. G. Tel'kovski, 1962, Zh. Tekh. Fiz. **32**, 1032.

Nelson, R. S., 1963, Phil. Mag. **8**, 693.

Nelson, R. S., 1965, Phil. Mag. **11**, 291.

Nelson, R. S., 1969, Unpublished.

Nelson, R. S., 1969a, Phys. Lett. **28A**, 676.

Nelson, R. S. 1969b, Proc. Roy. Soc. **A311**, 53.

Nelson, R. S. and D. J. Mazey, 1967, J. Mater. Sci. **2**, 211.

Nelson, R. S. and D. J. Mazey, 1968, Can. J. Phys. **46**, 689.

Nelson, R. S. and D. J. Mazey, 1966, unpublished.

Nelson, R. S., D. J. Mazey and J. A. Hudson, 1970, J. Nucl. Mater. **31**, 1.

Nelson, R. S. and M. W. Thompson, 1961, Proc. Roy. Soc. (London) **A259**, 458.

Nelson, R. S. and M. W. Thompson, 1962a, Phil. Mag. **7**, 1425.

Nelson, R. S. and M. W. Thompson, 1962b, Phys. Lett. **2**, 124.

Onderdelinden, D., 1968, Can. J. Phys. **46**, 729.

Parsons, J. R., 1965, Phil. Mag. **12**, 1159.

Perović, B., 1961, Bull. Inst. Nucl. Sci. Bonis Kidrich **11**, 226.

Perović, B. and B. Cobić, 1962, in Proc. Vth Int. Conf. Ionization Phenomena in Gases, Munich (North-Holland Pub. Co., Amsterdam).

Pleshivtsev, N. V., 1968, 'Cathode Sputtering' (Atomizdat, Moscow).

Robinson, M. T., 1965, Phil. Mag. **12**, 145.

Rol, P. K., J. M. Flint and J. Kistemaker, 1960, Physica **26**, 1009.

Rol, P. K., J. M. Flint, F. P. Viehbock and M. De Jong, 1959, in Proc. IVth Int. Conf. on Ion Phenomena in Gases, Uppsala, p. 257.

Sanders, I. R. and P. S. Dobson, 1969.

Sigler, J. A. and D. Kuhlmann-Wilsdorf, 1967, Phys. Status Solidi **21**, 545.

Sigmund, P., 1965, Risø, (Denmark) Report No. 103.

Sigmund, P. and J. Sanders, 1967, Proc. Conf. on Ion Beams in Semiconductors, Grenoble.

Silsbee, R. H., 1957, J. Appl. Phys. **28**, 1246.

Stein, H. J., F. L. Vook and J. A. Borders, 1969, Appl. Phys. Lett.

Stein, H. J., F. L. Vook, D. K. Brice, J. A. Borders and S. T. Picreau, 1970, Proc. Int. Conf. on Ion Implantation in Semiconductors, Thowsand Oaks, California, p. 17.

Thackery, P. A. and R. S. Nelson, 1969, Phil. Mag. **19**, 169.

Thompson, M. W., 1963, Phys. Lett. **6**, 24.

Thompson, M. W., 1968, Phil. Mag. **18**, 377.

Thompson, M. W., B. W. Farmery and P. Newson, 1968, Phil. Mag. **18**, 361.

Thompson, M. W. and R. S. Nelson, 1962, Phil. Mag. **7**, 2015.

Tunstall, W. J., L. Eriksson and D. J. Mazey, 1970, Phil. Mag. 21, 617.
Vineyard, G. H., J. B. Gibson, A. N. Goland and M. Milgran, 1960, Phys. Rev. 130, 1229.
Von Jan, R. and R. S. Nelson, 1968, Phil. Mag. 17, 149.
Watkins, G. D., 1963, J. Phys. Soc. Japan, Suppl. II 18, 22.
Watkins, G. D., 1967, Conf. on Radiation Effects in Semiconductors, Santa Fe, p. 67.
Wehner, G. K., 1955, Phys. Rev. 102, 690.
Weijsenfeld, C. H., 1966, Thesis, Rijksuniversiteit, Utrecht.
Wilson, M. M., 1970, Radiation Effects 1, 207.
Wright, W. D., 1946, 'The Measurement of Colour' (Adam Hilger Ltd. London).
Yurasova, V. E., N. V. Pleshivtsev and I. V. Orfanov, 1959, Zh. Eksperim. i Theor. Fiz.
37, 966.

# 4 | THE PRODUCTION AND MANIPULATION OF ION BEAMS FOR IMPLANTATION

## 4.1. Introduction

The rapid growth of interest in recent years in doping solids with ions has brought about a synthesis of the requirements of different accelerator disciplines. It has created a particular need for versatile machines capable of producing and resolving a range of intense heavy ion beams over wide energy limits. In certain cases existing accelerators have been uprated to meet these requirements, in others some quite novel concepts have gone into the construction of new machines.

In this chapter we shall be concerned with the techniques of ion implantation. We shall either omit, or only briefly discuss, those aspects of accelerator design and operation such as high voltage generation, ion beam handling lenses and vacuum techniques, which are already adequately treated in standard texts and for which the equipment is usually available commercially. We shall however consider at some length those factors, such as the need for heavy ions, the special target chamber requirements, and the problems of beam analysis, which may distinguish implantation from other ion beam studies. We shall, in particular, pay considerable attention to the various methods of producing suitable ion beams since the versatility and success of an implantation facility is particularly dependent upon the quality of the ion source. We shall also describe in some detail the constructional and operational problems which are an unfortunate concomitant of the need in certain heavy ion sources to operate at high temperatures with corrosive or reactive vapours.

It is not our intention to give either an historical or a comprehensive

255

account of the very large number of accelerators and heavy ion sources which have been described. We shall however consider several typical examples of machines and sources, which illustrate the important features of the various possible types of implantation facility.

In the most general terms ion implantation can be carried out in a remarkably wide diversity of machines. These can range from an elementary glow discharge tube, or a simple exploding wire (4.2.6.6), to the very high energy nuclear physics accelerators. For certain experiments the sub-microampere beams of the mass spectrometer may be adequate, for others the hundred milliampere capability of production scale isotope separators may be required.

We shall not attempt to review in detail this very wide range of possible facilities and for practical purposes we shall restrict ourselves to ion energies of about one to five hundred kiloelectronvolts, and to beam intensities of microamperes to milliamperes. These limits, in fact, encompass a substantial fraction of the studies which have been carried out, and it seems probable that the majority of future accelerators designed specifically for ion doping will fall somewhere within this range.

At energies below one kiloelectronvolt the penetration of ions is largely restricted to the first few atomic layers of the target material. At these depths the doping behaviour is likely to reflect the condition and properties of the surface itself rather than the bulk properties of the host material. Such doping may be of interest in certain research problems or in the formation of contacts to surfaces, but most implantations are in fact carried out at substantially higher energies. The choice of the higher energy limit is somewhat more arbitrary and several important studies have been carried out with MeV ions. However at these energies we are moving into the realms of conventional or modified nuclear physics accelerators and these are beyond the scope of this book. In general, at energies above a few hundred kiloelectronvolts, the cost of the equipment begins to rise rapidly and the problems of producing and analysing the heavy ions become severe. Also at very high energies the implanted ions may be deposited well below the surface to form a buried layer. While such doped regions are of considerable importance in certain applications, the more usual objective in implantation is to modify the bulk properties of solids in layers of adequate thickness lying close to the surface.

It is important to note too, that the performance of certain accelerators deteriorates as the energy is reduced; both the quality of focus and the resolved beam intensity may fall significantly. If a machine is required for doping over a broad energy range, this factor may set an upper limit on the practicable maximum voltage.

The required beam intensities in implantation experiments may vary over very wide limits indeed. For radioactive tracer studies the ion current may be almost undetectable, and even for certain semiconductor doping experiments doses below $10^{10}$ ions/cm$^2$ ($\sim 10^{-3}$ $\mu$A sec/cm$^2$) have been used. At the other extreme, the chemical conversion of a surface with reactive ions (e.g. $Si + 2O^+ \rightarrow SiO_2$) may require doses of around $10^{19}$ ions/cm$^2$ ($\sim 10^3$ mA sec/cm$^2$) to attain stoichiometry. The somewhat arbitrary beam requirements ($\mu$A–mA) which we have set, in fact, include most of the reported experimental results. Although a high current availability clearly offers the maximum versatility, the use of milliampere beams of a wide range of heavy ions has so far been largely restricted to certain types of accelerator. In high energy machines a limitation may be set by the restricted availability of suitable ion sources capable of producing the required intensity with a sufficiently wide range of ions, and also by the additional constraints which these may pose on the power, cooling and pumping requirements at the high voltage terminal. As we shall see below, this problem may in certain circumstances be circumvented by the use of low energy accelerators combined with high voltage targets.

We shall also see that in practical applications the temperature rise due to the ion beam energy deposited in the irradiated sample may set an upper limit to the intensities which can be usefully employed. However in many experiments, doping at the atomic per cent level can be obtained in acceptable times even with microampere beam intensities, whilst the milliampere heavy ion beams which are well within the scope of certain accelerators are adequate for the large scale industrial production of semiconductor devices.

Because of the broad scope of ion doping, and its application in several scientific disciplines, it is not possible, even within the performance limits set out above, to define a general purpose accelerator. A varied range of facilities has been used and we shall consider several machine configurations which satisfy the needs of particular areas in ion implantation.

For convenience, after a general consideration of ion beam and accelerator concepts, we shall consider implantation machines as being made up of three main components: the ion source; the beam analyser and focusing lenses; and the target chamber. We would stress however that this artifice must be employed with discretion, and that in a complete machine the separate parts must be carefully matched.

## 4.2. General accelerator concepts

Before proceeding to a detailed treatment of the main accelerator components we must first consider some of the general features of implantation facilities. We shall also define those qualities of the ion beam and the analysing system which essentially determine the overall performance of the equipment and include a brief description of a number of machines which illustrate the general concepts.

### 4.2.1. THE DESIGN OF IMPLANTATION ACCELERATORS

The performance limits which we set out above encompass a variety of conventional types of accelerator each of which satisfies some, but not all, of the requirements for ion implantation. The needs in certain applications for intense beams at high voltages, or for profile control by energy programming, are best satisfied by somewhat unconventional machine designs.

In certain restricted cases implantation can be carried out with a simple ion gun directed at a target but since most sources deliver a spectrum of ions which must be resolved, we shall consider the more general situation of accelerators with some form of mass analysis. Beam handling lenses may also be required to steer and focus the ions, and some form of scanning facility may be included to provide beam coverage over the target area.

A detailed description of the several methods of generating high voltages, or of the design of suitable pumping equipment, is beyond the scope of this chapter and these subjects are adequately covered in standard texts. Since most of the requirements are conventional and can be satisfied by using commercially available equipment, we shall simply comment briefly on some of the particular design aspects which distinguish implantation from other accelerator applications.

The requirement for versatile heavy ion sources may create a number of particular problems at the high voltage terminal. For example, as we shall see below, it is normally essential that the available E.H.T. current should be considerably in excess of the required beam intensity at the target stage, and the accelerating voltage must also be adequately stabilised to give the required degree of resolution after analysis. In general too, the ion source auxiliary power requirements at high voltage will be in excess of those associated with simple conventional gas sources. This higher power dissipation may in turn increase the need for cooling in the source terminal. The use of such sources may also increase the pumping requirements and the system must be designed to cope with corrosive vapours if compounds such as volatile halides are to be ionised. Careful design of the accelerating lenses and tubes may be required to minimise electrical breakdowns due to the probable condensation of evaporated source feed materials. If such breakdowns are likely it is also desirable that the high voltage generator should have a fast recovery time.

As in the case of conventional accelerators a variety of high voltage generating systems may be employed. It is notable however that open air high voltage terminals are more commonly used for implantation than the more compact enclosed pressurised accelerators. Although these latter have proved to be reliable and versatile in other applications, their flexibility for experimental ion doping is somewhat reduced by the restricted accessibility to the ion source.

Whilst it is axiomatic that undesirable pressure dependent effects such as electrical breakdowns and beam scattering can be minimised by the use of the highest possible vacuum in any accelerator, the final design is usually a compromise between cost and convenience. In many cases pumping speed is of greater importance than the ultimate vacuum which may be obtained under static conditions. The operating pressure in the high voltage terminal, for example, will in general be largely determined by the outgassing of the hot units and by the unionised vapour escaping from the discharge chamber. The detrimental effects of these sources of localised high pressures can be minimised by differential pumping of the ion source region, whilst cryogenic pumping is particularly effective in removing hydrocarbons, water vapour and volatile halides.

Conventional oil diffusion pumps are commonly used for the ion source. Turbomolecular pumps may be equally satisfactory, but their long term behaviour with corrosive vapours is a somewhat uncertain factor. Ion pumps have relatively low pumping speeds for the inert gases which are frequently used in heavy ion sources.

In the beam transport system from the ion source to the target chamber the pressure determining factor is frequently the outgassing produced by the ion beam and the leakage of gases or vapours from the source chamber. In this case also, the pumping speed is often of more importance than the ultimate static vacuum. Operating pressures of around $10^{-6}$ torr are generally adequate to minimise ion collisions with residual gases. Such vacua can readily be achieved by simple conventional techniques which permit ready access to, and easy modification of, the various flight tubes, where beam collimators, lenses or probes have to be installed or adjusted. Whenever possible cryogenic pumping should also be used to remove residual hydrocarbons which may otherwise be decomposed by the ion beam and lead to the build-up of deleterious carbonaceous layers on the walls of the vacuum system. Silicone oils should generally be avoided because of the probable formation of highly insulating films which may adversely affect the ion optics of the accelerator.

A particular vacuum problem arises in those regions of the ion beam transport system (normally the section between the analyser and the target chamber) which have a direct line of sight with the sample to be implanted. Both doping level errors and a loss of implantation uniformity may result from the directed flux of stripped ions and fast neutrals which are produced by charge-exchange reactions of the beam with gas molecules.

The magnitude of such effects depends upon the ion species and energy as well as the pressure and length of the flight tubes. The major uncertainty in calculating the likely errors arises from the lack of available data on the appropriate cross-sections in the energy range of interest. It seems probable, however, that for the most common dopants the values will generally be below $10^{-15}$ cm$^2$/molecule. In practical terms this corresponds to beam contamination effects of 1 % for a pressure/path length product of approximately $3 \times 10^{-6}$ torr metres.

In the case of the target chamber the pumping requirements cannot be simply defined since they depend largely upon the particular experimental conditions. For example, ultra high vacuum techniques are essential in studies which depend critically upon the quality and state of the sample surface during implantation. These include measurements of secondary electron emission, or determination of the charge state and energy of transmitted, reflected or sputtered particles.

In more conventional simple doping experiments the most probable consequences of implanting in an inadequate vacuum are the formation of carbonaceous layers on the sample surface and the possibility of impurity doping caused by knock-on collision between the ions and adsorbed gases and vapours. Both of these phenomena could lead to spurious results, particularly in the important case of electrical measurements on implanted semiconductors, and it might appear that the most rigorous vacuum conditions should be used in such experiments. The target chamber design must generally, however, also be compatible with several other experimental requirements. These may include specimen heating and cooling, target orientation, precise current measurement, and a capability of handling multiple samples. Whilst all of these techniques can be employed in ultra high vacuum assemblies there would inevitably be a significant loss of adaptability and speed of operation.

Although, clearly, the aim in all doping studies should be to work at the lowest possible pressure there is, in fact, little evidence to indicate that the careful use of quite conventional vacuum procedures is inadequate for most implantation applications. If necessary the partial pressure of oil vapours, which are undoubtedly the most important contaminant, can be very effectively reduced by the use of cold traps in the target chamber. Turbo-molecular pumps, or mercury diffusion pumps may also be used to minimise hydrocarbon contamination. Additionally, differential pumping and cryogenic trapping techniques can often be employed in the final beam transport tube between the accelerator and the target chamber.

It is worth observing finally that the target vacuum requirements depend not only upon the type of implantation experiment but also upon the available ion beam intensity, since the rate of formation of surface layers is a function of both the hydrocarbon partial pressure and the rate of arrival of ions.

*4.2.1.1. Implantation Accelerator Configurations.* The three principal implantation accelerator configurations incorporating beam analysis are shown diagrammatically in fig. 4.1. Each layout has advantages and drawbacks for ion doping and we shall consider them separately at some length.

The simplest and most conventional form of accelerator is shown in fig. 4.1(a). This is a machine with the ion source at high voltage and the beam analysis and target chamber at earth potential. The advantages of such a facility are that the source is electrically isolated from the rest of the system and that there is free access to the target. This is particularly useful for experiments which may involve, for example, back scattering, beam transmission or electron microscope observations on the bombarded sample.

The main limitations of this simple accelerator layout are the probable loss of beam quality when it is used at low energies and, in the case of the higher energy machines, the need for large and powerful deflecting magnets to analyse the beam. For many research applications their convenient layout is attractive and their versatility can be increased by the use of multiple beam lines as shown in fig. 4.6.

For experiments in which the dopant profile is to be controlled by energy programming of the ion beam they are somewhat inflexible. Each adjustment of the high voltage requires a corresponding correction to the analysing system and continuous energy variation is difficult. This is however a somewhat specialised application of ion implantation and Allen (1969) has demonstrated that on such a machine a good measure of profile control can

CONVENTIONAL ACCELERATOR.

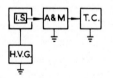

Convenient layout, access restricted
to ion source only. Large analyser
required for high energy beams,
machine performance commonly deteriorates
at low energies.. Energy programming
of ion beam requires analyser adjustment.

ACCELERATOR WITH BEAM ANALYSIS AT HIGH VOLTAGE.

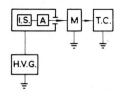

Analysis at constant low energy.
Small analyser adequate.
No deterioration of performance
over wide energy range.
Beam energy programming by H.T.
adjustment only.
Analyser or beam analysis flight
tube must be at high voltage.

ACCELERATOR WITH TARGET CHAMBER AT HIGH VOLTAGE.

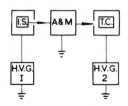

Restricted access to target but convenient
method of increasing accelerator energy
range. Target does not require high
stability H.T. supply.
Beam energy programming by target H.T.
adjustment only.
Wide energy range with small analyser.
High resolution and intense beam currents
readily obtained.

Fig. 4.1. Accelerator configurations. (I.S. – Ion source; A – Analysis; M – Manipulation;
T.C. – Target chamber; H.V.G. – High voltage generator).

be obtained by carrying out a number of successive implantations at different
energies.

Because of their importance and since they illustrate many of the general
features of implantation accelerator design we shall consider a number of
variants of this conventional configuration. An excellent example of such an
accelerator is the 600 kV heavy ion facility at Aarhus. This was designed by
Nielsen and his colleagues (Nielsen, 1967) on the basis of considerable
experience obtained with lower energy isotope separators and it largely com-
bines the flexibility of these latter with the higher energy requirements of

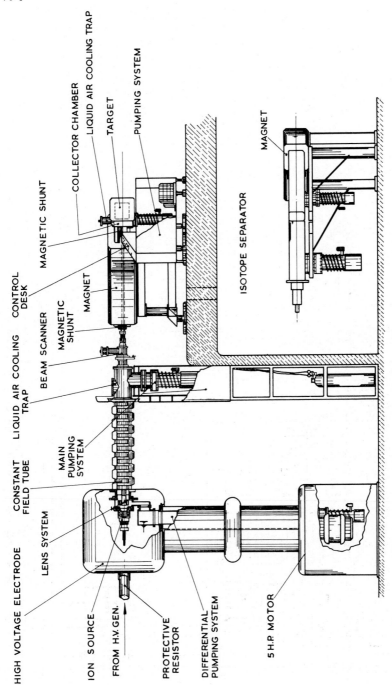

Fig. 4.2. Aarhus 600 kV accelerator (from Nielsen, 1967).

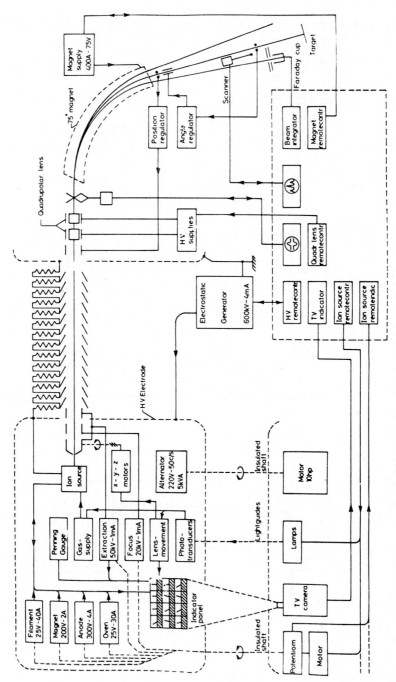

Fig. 4.3. Schematic illustration of supplies for the Aarhus 600 kV accelerator (from Nielsen, 1967).

more general ion beam studies. The layout of the machine which uses a Sames electrostatic generator, a constant field acceleration tube and which is equipped with a general purpose discharge heavy ion source, is shown in figs. 4.2 and 4.3.

Figure 4.4 shows a more compact modern accelerator, the 200 kV machine produced by Accelerators Incorporated. This is designed for industrial scale implantation with dopants such as boron in applications where the limited beam current capability is not important.

Fig. 4.4. Accelerators Incorporated 200 kV accelerator.

Another conventional accelerator, the modified 500 kV Cockroft Walton of Dearnaley and his colleagues at A.E.R.E., Harwell (Goode, 1971) is shown in fig. 4.5.

This machine uses a sputtering source to produce the required range of heavy ions and is characterised by the long flight tube to the principal target chamber which permits accurate beam collimation for channelling experiments. The vertical constant field acceleration tube combined with a rotatable magnet permits the use of several alternative beam lines as shown in fig. 4.6. This increases the flexibility of the facility and also permits maximum machine utilisation.

The semi-industrial implantation machine of Tsuchimoto et al. (1970) is shown in fig. 4.7. The full 200 kV acceleration can be used up to about mass 40. For heavier ions the maximum energy is restricted by the deflecting capability of the analysing magnet and falls to about 50 keV at mass 120. Up to four two inch diameter wafers can be implanted simultaneously using simple $XY$ beam deflecting electrodes and an ion beam intensity of about 1 $\mu A/cm^2$ can be obtained for dopants such as phosphorus.

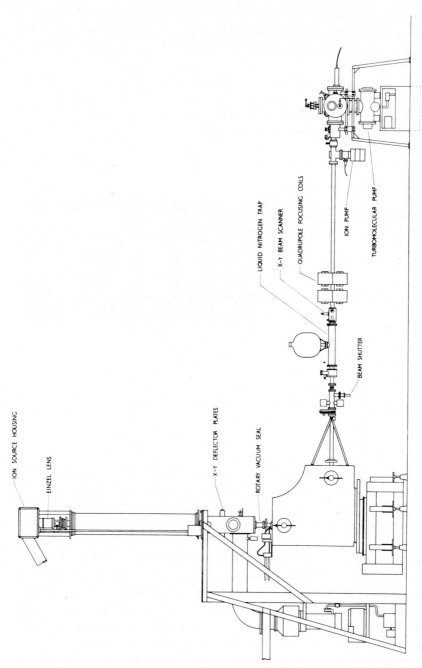

Fig. 4.5. A.E.R.E., Harwell 500 kV Cockcroft Walton accelerator (from Goode, 1971).

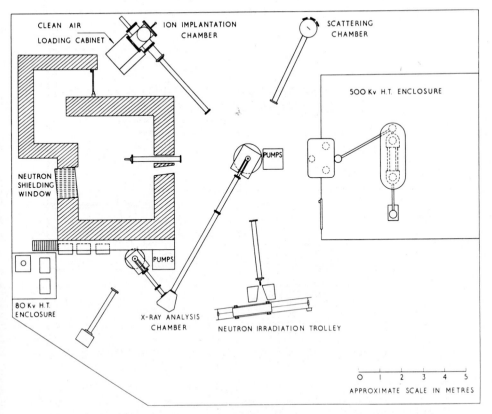

Fig. 4.6. Multiple beam line arrangement of A.E.R.E. Cockcroft Walton accelerator (from Goode, 1971).

Van de Graaff accelerators with the ion source enclosed within the high voltage terminal have proved to be compact and reliable machines for many beam interaction studies. The 400 kV implantation facility of this type at Ion Physics Corporation is shown diagrammatically in fig. 4.8. An interesting feature in the illustration is the proposed technique of profile control by programming the inclination of the target stage to the ion beam during implantation. The principal limitation of this type of accelerator for general purpose ion doping studies lies in the restricted accessibility to the high voltage source terminal. It seems likely that their more widespread use for implantation will depend upon the development of suitable heavy ion source techniques.

For lower energy applications the versatile and small laboratory isotope

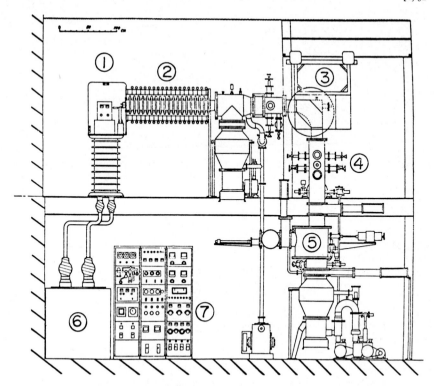

Fig. 4.7. Hitachi 200 kV accelerator (from Tsuchimoto et al., 1970). (1 – Ion source, 2 – Acceleration tube, 3 – Analysing magnet, 4 – Beam scanning, 5 – Target chamber, 6 – High voltage supply, 7 – Controls.). Schematic drawing of newly developed implantation system.

separators are particularly useful. As we shall see below, their energy range can often be extended by the use of multiply charged ions or by employing high voltage targets to post-accelerate the ions after separation. They are characterised by the excellent resolution which can be obtained with a very wide range of dopant species. The first type of machine commonly known as Scandinavian separators (see for example, Nielsen, 1967) is shown diagrammatically in fig. 4.9(a). The principal difference from the previous accelerators is that the primary acceleration generally takes place across a simplified single gap electrode instead of the constant field tube used at higher energies. This simplifies the ion optics, reduces space-charge blow up effects and eases the vacuum pumping problems at the high voltage terminal. In general the ion source is more accessible and it is relatively simple on such machines to introduce additional facilities such as vacuum locks for fast dopant changes

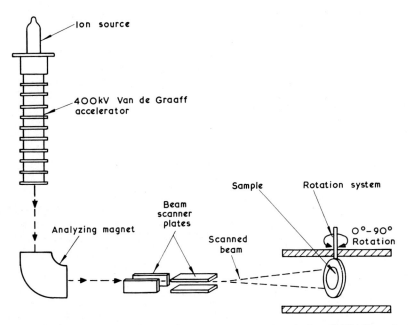

Fig. 4.8. Schematic illustration of Ion Physics Corporation Van de Graaff 400 kV accelerator layout for ion implantation (after Davies et al., 1968).

at the high voltage terminal. Typical accelerating voltages for this type of accelerator are in the range of 50–80 kV. The manipulation of the ion beam is carried out using conventional electrostatic lenses and deflector plates. The general size and layout of such a separator can be seen in fig. 4.9(b).

The beam current on target, of the various conventional accelerators described above, is typically microamperes or tens of microamperes. Such intensities are adequate for many experimental requirements but much higher beam currents ($\sim$mA) can be obtained on the second type of isotope separator shown diagrammatically in fig. 4.10. In this type of machine the ions are extracted from a slit instead of the circular aperture used on the accelerators described above. This results in the formation of a wedge shaped beam. The increased intensity arises largely from the larger extraction area. Somewhat paradoxically the ion optics of these high output machines are considerably simpler than those of conventional accelerators because of the parallel nature of the beam in the plane of the extraction slit. This feature permits the construction of compact versatile machines. The accelerating voltage in this case also is normally limited to around 50 kV. Figure 4.11 shows such a separator designed at A.E.R.E. Harwell (Freeman, 1969a).

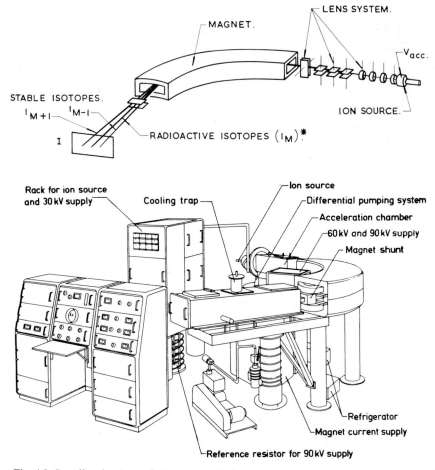

Fig. 4.9. Scandinavian type of electromagnetic isotope separator with axial extraction of ion beam (from Nielsen, 1967). (a) beam manipulation, (b) machine layout.

The quality of focusing which can be obtained with milliampere heavy ion beams can be seen in fig. 4.12.

In the second accelerator configuration with the analyser at high voltage (fig.4.1(b)), the mass separation is carried out at low energy, typically only tens of kilovolts and a small analyser can be used. The energy of the final beam can be readily controlled, and programmed if required, by continuous adjustment of the secondary acceleration voltage. The extracted beam current is largely independent of the total accelerating voltage and the system has the same flexibility at the target stage as the conventional layout. There

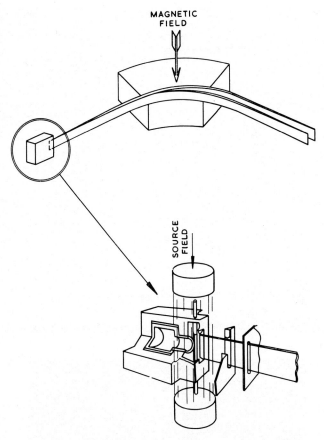

Fig. 4.10. Beam handling in medium output electromagnetic isotope separator with slit
extraction of ion beam.

is however the complication that the analyser with its supplies must be at H.T.
or alternatively the vacuum flight tube must be electrically isolated from the
analyser. A typical example of such a machine, the 160 kV accelerator used at
Grenoble for implantation (Guernet et al., 1970) is shown in fig. 4.13.

The third configuration with the target chamber at high voltage (fig.
4.1(c)) may in its simplest form consist of a small conventional accelerator
with a biassed target stage to extend the energy range, or it may be a machine
specifically designed for implantation. It has the same advantage as the
previous configuration, that a small analyser can be used, but the immediate
drawback to the use of high voltages on the target is that it restricts access
and limits the experimental flexibility. It may thus, in certain circumstances,

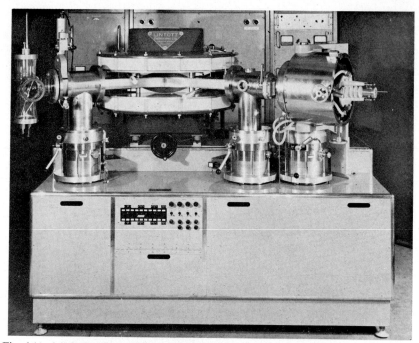

Fig. 4.11. A.E.R.E., Harwell electromagnetic isotope separator (from Freeman, 1969a).

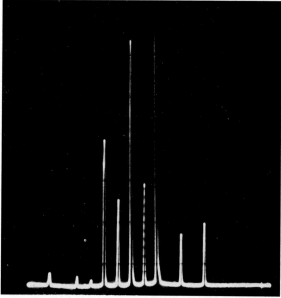

Fig. 4.12. Spectrum of tin isotopes (from Freeman, 1969b).

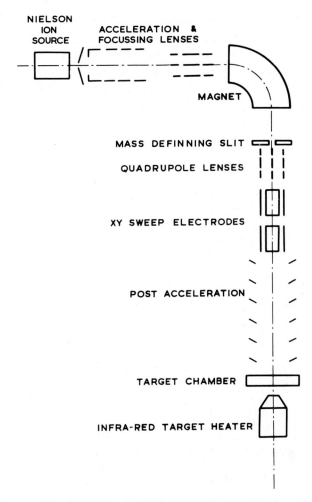

Fig. 4.13. Schematic outline of Grenoble 160 keV accelerator layout, with analysis at high voltage (from Guernet et al., 1970).

preclude back scattering, transmission or electron microscope observations during bombardment. It does not however seriously interfere with the ion doping in routine implantation applications and it offers certain important advantages. Since the beam is accelerated (or decelerated) after analysis, a relatively low cost high voltage power supply is adequate and energy programming can be used to control the implanted profile. The ion source and the associated power supplies need only be at a potential of a few tens of kilovolts. At this energy it is relatively simple to obtain intense beam currents and a considerable degree of source flexibility can be achieved. An additional feature of dividing the total accelerating voltage into two elements in this way is that X-ray generation is reduced.

This method of operation has been used in a number of simple laboratory assemblies. A typical small accelerator designed specifically to operate in this configuration (Wilson, 1967) is shown in fig. 4.14. High voltage target stages have also been employed for the production of solar cells by

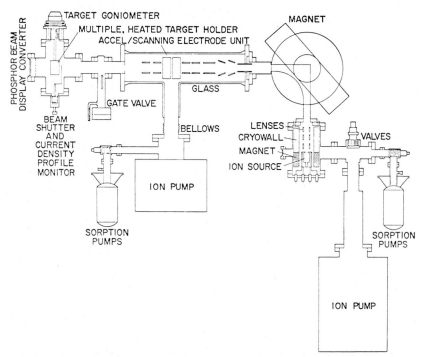

Fig. 4.14. Hughes experimental accelerator with target at high voltage (from Wilson, 1967).

ion implantation at Ion Physics Corporation and the technique has proved particularly useful in the use of small and versatile isotope separators for ion doping. Freeman (1967), (also Freeman and Gard, 1970b) has described the use of such a machine with milliampere beams at energies from a few eV up to several hundred kilovolts, using a simple target stage. Using a more sophisticated high voltage target assembly (fig. 4.122) implantation can be carried out on an industrial scale on this type of accelerator.

## 4.2.2. ION BEAM REQUIREMENTS

The performance of an implantation facility is determined by the quality of the required dopant ion beam *at the target stage*. Although beam intensity and energy are clearly important, several other parameters must be considered. These include the resolution and shape of the ion beam, as well as its physical separation, or dispersion, from other ion species. The current stability, the duration of continuous operation and the ease with which the dopant species can be changed are also important.

Some of these factors will be treated in more detail in later sections of this chapter, but we shall first consider in general terms the question of the beam intensity and energy, and define the important concepts of dispersion, resolution and emittance.

## 4.2.3. BEAM INTENSITY

Although beam transmission losses can be minimised in well matched, high vacuum accelerator assemblies, it is generally essential that the total available current, from the ion source, should be considerably in excess of the required intensity at the target. Many elements are conveniently ionised in a discharge from the vapour of a suitable compound. Although such sources are relatively efficient in breaking down molecules to produce elemental ions the extracted beams normally contain several ion species which must be resolved. In a few exceptional but important cases, the required ion may be only a minor component of the total current. Figure 4.15 shows the spectrum obtained using dysprosium trichloride as the source feed material. Although a range of ions is produced the elemental beams of $Dy^+$ and $Cl^+$ are the most intense. In contrast in the spectrum of boron trichloride shown in fig. 4.16 it can be seen that the boron is only a minor component. Thus in order to obtain a 1 mA $B^+$ doping beam, a total source drain current of 12–14 mA was required (Freeman, 1970a). Additionally in any accelerator using a beam analysis system there must inevitably be a measure of isotopic separation. This may result in the clear dispersion of the peaks shown in the figures above, or in a machine with a lower resolution it will result in an extended blurring of the focused beam. In either case some effective loss of intensity will occur if the whole isotopic spectrum cannot be utilised in the implantation step. This effect is particularly important in the case of elements such as tin, molybdenum or gadolinium which have large numbers of low abundance isotopes*. Finally if the accelerator is to be used for channelled

* A schematic representation of the isotopic abundance of all elements is given in Appendix 1.

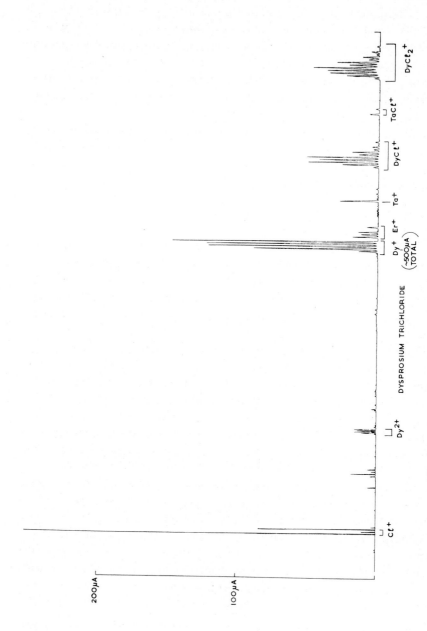

Fig. 4.15. Dysprosium trichloride spectrum (from Freeman, 1970a).

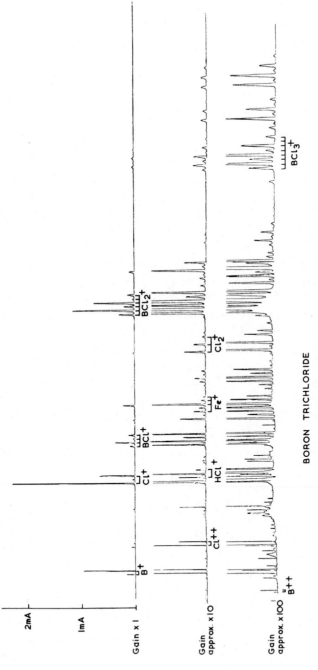

BORON TRICHLORIDE

Fig. 4.16. Boron trichloride spectrum ($\sim$1 mA B$^+$) (from Freeman, 1970a).

implantations there may be a further loss of intensity because of the need for precise collimation to define the beam direction and angular divergence.

The transport of ions to the target can be readily calculated from the electronic charge ($1.6 \times 10^{-19}$ coulombs) and Avagadro's number ($6.02 \times 10^{23}$ mole$^{-1}$), or from Faraday's law, which states that one faraday (96,500 coulombs) is the amount of electricity carried by one gram equivalent of any ion having a unit charge. More conveniently the rate at which ions arrive at the target is:

$$\frac{I_+ N}{nF} \text{ ions/sec,} \quad \text{or} \quad \frac{I_+ A_+}{nF} \text{ g/sec}$$

where $I_+$ is the beam current in amperes, $N$ is Avagadro's number, $n$ is the charge state of the ion, $F$ is the faraday, and $A_+$ is the atomic weight of the ion.

Thus a one milliampere beam current corresponds to a transport of $6.24 \times 10^{15}$ ions/sec or about $A_+/27$ mg/hour (for singly charged ions).

In the case of an ion beam intensity of 1 mA/cm$^2$ bombarding a silicon target, each atom on the surface of the sample is hit very approximately once a second, and although the currents we are considering are small compared with the electrolytic transport of ions, the physical mass which may be implanted is quite significant. Since the penetration of ions in doping experiments is commonly less than one micron, high impurity densities can be rapidly achieved. It is this factor which makes implantation a viable industrial process.

For non-channelled doping the ion penetration profile obtained is approximately gaussian and if the projected range is $R_p$ and the standard deviation $\sigma$, then for a surface dose of $N_+$ ions/cm$^2$ the maximum or peak concentration is

$$C_p = \frac{0.3989}{\sigma_{cm}} N_+/\text{ions/cm}^3$$

and the mean concentration between $\pm \sigma$ is:

$$\bar{C} = \frac{0.6827}{2\sigma_{cm}} N_+/\text{ions/cm}^3 = 0.845 \cdot C_p$$

The doping concentration thus depends on the values of $R_p$ and $\sigma$ in a particular bombardment, but it is instructive to consider a typical semiconductor application requiring the implantation of silicon with 100 keV phosphorus ions. In this case $R_p$ is $\sim 1200$ Å and $\sigma$ is $\sim 350$ Å. Thus if $N_+ =$

$5 \times 10^{15}$ ions/cm$^2$, the maximum peak concentration is $5.7 \times 10^{20}$ ions/cm$^3$ and the mean concentration in the layer lying between $\pm \sigma$ is $4.8 \times 10^{20}$ ions/cm$^3$. This high doping level can be achieved in approximately one second with a beam intensity of 1 mA/cm$^2$.

### 4.2.4. BEAM ENERGY

*4.2.4.1. The Use of Multiply Charged Ions to Increase the Energy Range of an Accelerator.* An important feature of any conventional heavy ion accelerator which incorporates beam analysis is that its maximum energy may be increased by the use of multiply charged ions. In section 4.3 some of the techniques which may be used to obtain high charge states are described but it is worth noting that significant beams of doubly, and in certain instances triply, charged ions can be readily obtained from most discharge sources. Since such species are more readily deflected in magnetic analysers than the corresponding singly charged ions, they provide a convenient means of extending the upper range of available implantation energies. It is also fortuitous that the probability of multiple ionisation tends to increase with mass. The use of such beams is thus particularly applicable to heavy dopant species which may otherwise have an inadequate implantation range at the normal maximum accelerating potential of the accelerator. Figure 4.17 shows the singly, doubly and triply charged ions obtained from a discharge of gallium vapour (Freeman, 1970a). The beam intensities of Ga$^+$, Ga$^{2+}$ and Ga$^{3+}$ respectively were approximately 1 mA, 80 $\mu$A and 3 $\mu$A. (It should be noted that the fractional yield of multiply charged impurity chlorine ions is significantly lower.) The spectrum of the doubly charged ions also gives a clear demonstration of one of the major problems which may be encountered in the use of multiply charged species, namely, that the required low intensity beams may coincide with, or be adjacent to, significant impurity peaks*. Although in this case the resolution is sufficiently high to permit the use of one of the pure isotopic Ga$^{2+}$ beams and thus avoid the contamination from the neighbouring Cl$^+$, this would not necessarily be the case in an accelerator with a lower resolving power. In this particular instance, the Cl$^+$ beam which arose from prior operation with a volatile chloride could be reduced, or even eliminated, by careful cleaning procedures, but trace impurity beams from source constructional materials are almost always present. For example, the silicon and nitrogen beams in the spectrum arise from the use of silicon nitride filament insulators.

* Note also the contamination of the Kr$^{3+}$ beam in fig. 4.18.

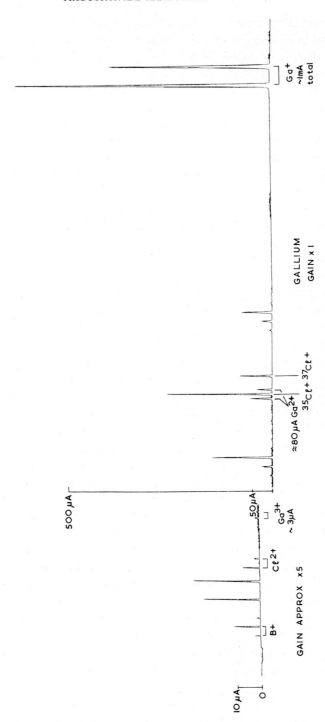

Fig. 4.17. Gallium spectrum showing multiply charged ions (from Freeman, 1970a).

Although the small yields of quadruply charged ions which can be obtained from conventional discharge sources could in principle be used to obtain a four-fold increase in effective energy we have deliberately omitted them in the above discussion. This is because such beams are frequently seriously contaminated with much lower energy ions of the same characteristic spectrum which are produced in the region between the ion source and the analyser by inelastic charge-exchange collisions of the type $M^+ + X \rightarrow M^{++} + X + e$. These doubly charged ions behave during deflection as though they had the mass to charge ratio, $1 : 4$ and thus coincide after analysis with the quadruply charged beam, $M^{4+}$. This effect can readily be demonstrated by raising the pressure in the flight tube before the analyser and observing the increase in the peak intensity due to the higher probability of the charge-exchange reaction. Such a change in an apparent $Kr^{4+}$ beam as a function of pressure is shown in fig. 4.18. It can be seen that the singly charged impurity beam at mass 18 decreases steadily as a function of pressure, as would be expected, because of scattering and neutralisation effects. The krypton beam on the other hand increases significantly at first and only begins to decrease at very high pressures. A characteristic of such beams produced by charge-exchange is that the resolution is low because of scattering during the inelastic collision. There is also normally an asymmetric tail on one side of the spectrum which is produced by charge-exchange occuring in the analysing region. In the example quoted above when the pressure was further reduced by lowering the gas flow into the ion source discharge the spectrum shown in fig. 4.18 was obtained. It seems probable that the more sharply resolved but lower intensity beam, obtained under these conditions was mainly $Kr^{4+}$. It was also found that this particular beam intensity increased, as would be expected, with the use of high discharge voltages.

Although the reaction described above is probably the most important in ion implantation, a wide variety of similar impurity beams, known as Aston bands, can be found in the mass spectrum.

For a simple charge-exchange reaction leading to a change in the degree of ionisation, $M^{n+} \rightarrow M^{m+}$, it can readily be shown that the product ion will be focused by a magnetic analyser at the mass position $M'$ where

$$M' = M \cdot \frac{n}{m^2}$$

In the more complex case of an inelastic molecular ion collision which

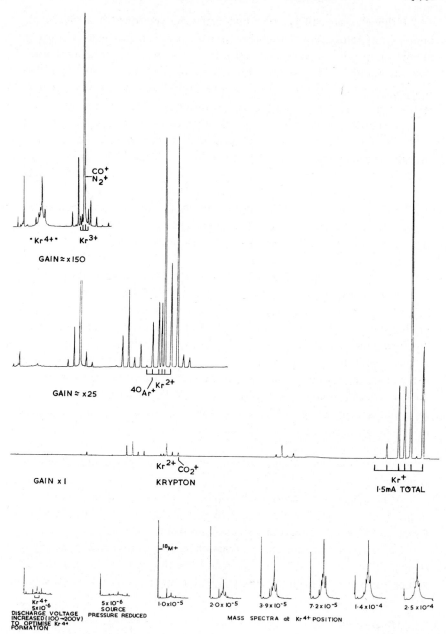

Fig. 4.18. Krypton spectrum, a – showing multiply charged ions, b – effect of pressure and ion source parameters on beam at $Kr^{4+}$ mass position (from Freeman, 1970a).

may result in a change of both mass and charge $(M_1^{n+} \rightarrow M_2^{m+})$ then;

$$M' = \frac{M_2^2}{M_1} \cdot \frac{n}{m^2}$$

*4.2.4.2. Energy Resolution Requirements.* In most implantation experiments there is no particular requirement for very precise control of the ion energy and it would normally be quite difficult, for example, to distinguish a change in the doping effect produced by an energy spread of a few per cent. It is this factor which allows the use of simple, low cost, high voltage power supplies when post-acceleration of the ion beam after mass analysis is employed. The situation is somewhat different in the case of the primary ion accelerating voltage, since low energy resolution before analysis will inevitably lead to an increase in the image size and a commensurate reduction in the mass separating capability of the accelerator. Thus in the typical case of a symmetric homogeneous magnetic analyser of one metre radius an energy spread of $\pm 1\%$ would result in an image broadening of two centimetres. In doping applications where such a loss of definition would be unacceptable the high voltage supply must be adequately stablished. In very high resolution machines, which exploit the low energy spread of certain ion sources both the extraction voltage and the magnet supplies may be stabilised to one part in ten thousand, but for many less stringent accelerator applications a stability of about one part in a thousand is normally adequate. Alternatively the required dopant beam may be spatially stabilised after analysis by the use of position sensing probes which provide correction signals to the magnetic or high voltage supply.

## 4.2.5. DISPERSION AND RESOLUTION

In order to obtain adequately pure ion beams for implantation some measure of mass separation is generally essential. The performance of the analyser is defined by the dispersion and the resolution of the ion beams,

The first factor, the dispersion, is simply the physical separation, $\Delta x$, between two adjacent beams of mass $M$ and $M + \Delta M$, and is solely dependant upon the geometry of the accelerator. Thus in the case of a symmetric homogeneous magnetic analyser, the dispersion factor $D_{m\perp}$, measured perpendicular to the central beam, for $\Delta M = 1$, is given by:

$$D_{m\perp} \approx \frac{R}{M}$$

where $R$ is the radius of deflection of the beam $M$. The dispersion which will be required in a particular machine design depends upon the mode of application. If an analysed and separated beam of given mass is to be transmitted through an aperture before being scanned over the target, a physical separation of a few millimetres between adjacent beams is generally adequate. If on the other hand, as is sometimes the case, the spectrum of ion beams is to be directly scanned, the separation between the required dopant beam and the nearest contaminant species must be greater than the dimensions of the target. The dispersion factor for different types of analyser will be considered in more detail in section 4.4.

Since the dispersion of an instrument is inversely proportional to the mass of the ions, the resolving power (R.P.) could be defined in terms of the mass at which two adjacent beams of image size $B$ and with unit mass difference, could just be distinguished. In practice, however, such a definition would be imprecise because of the ill defined low intensity spread of most focused ion beams, and it is more usual to define the resolving power in terms of the beam width at half height, $B_{\frac{1}{2}}$, so that,

$$\text{R.P.} = \frac{D_{m\perp} M}{B_{\frac{1}{2}}}$$

Thus, if the focused beam shapes are approximately triangular a resolving power of 100, as defined above, would permit the separation of ions only up to mass 50. This is illustrated in fig. 4.19, which shows a typical mass spectrum obtained at the target stage of an implantation accelerator (Allen, 1970). The resolving power of about 70 is adequate to separate the

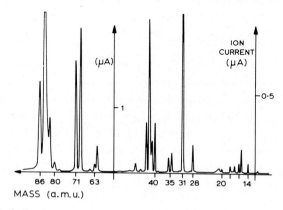

Fig. 4.19. Typical spectrum using sputtering source (Kr+Ga) on S.E.R.L., Baldock implantation accelerator (from Allen, 1970).

isotopic beams of $^{69}Ga^+$ and $^{71}Ga^+$ since in this case $\Delta M = 2$. The beams which were obtained by sputtering gallium phosphide with krypton in the ion source, demonstrate the need for adequate mass separation in such machines.

The resolving power, unlike the dispersion, is a function both of the instrument and the operating conditions*. Several factors may contribute to the image broadening, and the size and form of the focused beams may vary significantly from run to run. It is important to note that a large dispersion factor does not necessarily correspond to a proportionately high resolution since many of the image aberrations increase with the size of the analyser.

Whilst the resolving power is a useful figure of merit to describe the performance of a particular accelerator it does not fully define the degree of separation between resolved ion beams. This is because, even in the highest resolution machines the beams have extended tails of the form shown in fig. 4.20. These arise from instrumental aberrations, gas scattering, charge-exchange and a number of other processes. In order to describe the detailed performance of an instrument, it is necessary to define the contamination factor, $C_{M_2M_1}$, which is a measure of the fraction of a beam $M_2$ arriving at the focal position of the required beam $M_1$,

$$C_{M_2M_1} = \frac{N_{M_2}}{N_{M_1}} \cdot \frac{I_{M_1}}{I_{M_2}} = \frac{(\Delta I_{M_2})_{M_1}}{I_{M_2}}$$

where $N_{M_1}$ and $N_{M_2}$ are the number of ions of $M_1$ and $M_2$ arriving on the target, $I_{M_1}$ and $I_{M_2}$ are the respective total beam intensities of $M_1$ and $M_2$, and $(\Delta I_{M_2})_{M_1}$ is the current of $M_2$ at the focal position of $M_1$.

This global contamination factor, $C_{M_2M'_1}$, which is independent of the relative beam intensities and which is the sum of the various contaminating factors is of particular importance in implantations where the required dopant beam is accompanied by neighbouring ion beams of much higher intensity. This is a typical situation in the case of radioactive tracer doping (see section 4.7), and it may also occur when low yield multiply charged ions are used to extend the energy range of an accelerator.

We can also define an enhancement factor which is the reciprocal of the contamination. In sharp focusing instruments, such as high resolution isotope separators, enhancements of over a thousand can be obtained for a single mass separation, with beams of milliampere intensity. A reduction in

* The effects of the analyser optics and the quality of the stabilised power supplies on the resolving power are considered in section 4.4.1.

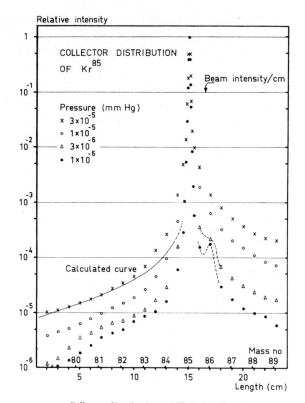

Collector distributions of Kr⁸⁵ at different pressures.

Fig. 4.20. High resolution krypton spectrum showing extended beam tailing (from Uhler, 1963).

the enhancement is, however, inevitable in applications where such beams are defocused or swept at the focal plane to obtain uniform implantations over large areas.

*4.2.5.1. Measurement of Resolving Power.* A corollary of the requirement for mass separation is the need to identify ion beams and to measure the resolving power. In the simplest case the analysing magnet can be scanned over the required mass range and the various ion peaks identified or recorded as the beams pass over a suitably positioned probe. For more precise identification and ease of adjustment mass-meters can be used (Dropesky et al., 1967; Lloyd and Hodson, 1968). A useful feature of these instruments which sample the accelerating voltage, $V$, and magnetic field, $H$, is that the field term is squared to satisfy the analyser relationship $H = k\sqrt{MV}$. If the output of the squaring circuit is taken as the $X$ signal of an $XY$ recorder with the beam probe current for the $Y$ axis the resultant beam scan which is obtained has a linear mass base. The low mass broadening and high mass cramping of normal mass spectra is thus avoided and beam identification is much simpler. Most of the ion beam spectra illustrated in this chapter were obtained in this way (for example figs. 4.15, 4.16, 4.17.).

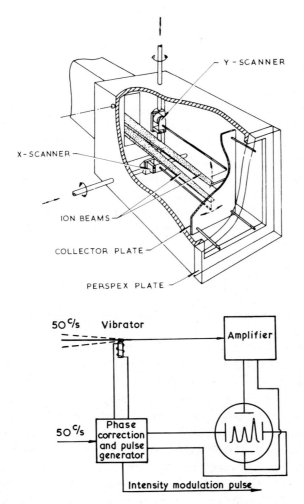

Fig. 4.21. (a) Schematic representation of $X$ and $Y$ beam scanners in a target chamber, (from Nielsen and Skilbreid, 1957), (b) Block diagram of a beam scanner vibrator arrangement (from Alväger and Uhler, 1957).

Additional information on the quality of the ion beams can be obtained by much faster scanning over a shorter mass range, with oscilloscope presentation of the spectrum. Figure 4.21 illustrates a mechanical scanner which operates at 20 cps (Nielsen and Skilbreid, 1957). Alternatively by modulating the H.T. with a suitable superimposed voltage signal the beam can be swept across a fixed probe or slit. Figure 4.12 shows the resolved spectrum of an intense tin beam measured in this way (Freeman, 1969b). In section 4.5.1 we show how this technique of sweeping may also be used for implanting over extended areas.

### 4.2.6. EMITTANCE AND BRIGHTNESS

In the computation of charged particle transport in simple accelerators it is often sufficient to assume a limited divergence point source of ions and to use standard lens formulae or conventional geometric ray tracing techniques to obtain a first order approximation of the ion beam trajectories. Corrections may then be introduced, if necessary, to account both for the finite object size and also the fringing fields and aberrations of the dispersive and focusing lenses.

The validity of the description can be confirmed by measurements in the accelerator of the dispersion, the resolution, and the efficiency of transmission of the ion beam.

In more exacting beam handling situations a fuller treatment of the ion beam characteristics is required and it is necessary to introduce the concepts of emittance and brightness.

Let us consider an idealised beam whose collective motion is directed principally along the axis of propagation ($p_z \gg p_x$ and $p_y$), and in which there is no particle scattering and no mutual interaction between the ions. The state of any ion in the beam can be fully defined momentarily by three position, and three momentum co-ordinates ($x, y, z; p_x, p_y, p_z$). If the beam is cylindrically symmetric the ion motion in the plane perpendicular to the axis can be more simply defined by the radial position $r$, and the transverse angular momentum $p_r$, and depicted as a point on a momentum-displacement phase diagram. If all the particles in the same perpendicular plane are similarly plotted the area which encloses them is known as the emittance. If the momentum of the particles remains constant it can be shown from Liouville's theorem that the emittance is an invariant along the beam axis (Hereward et al., 1956).

In the more general case of beams which are not axially symmetric it can also be shown that the sub phase-space areas $x$, $p_x$ and $y$, $p_y$ are also invariant and these may be used to define the beam emittance in the $x$ and $y$ directions respectively.

Although the shape of the emittance diagrams may be bent, rotated or distorted by beam transport elements the areas remain constant. For convenience the emittance is normally defined not by the transverse momentum co-ordinate but by the more readily measurable slopes of the particle trajectory $(x', y'$ or $r')$ to the direction of beam propagation. At the non-relativistic energies with which we are principally concerned,

$$p_x = x'p_z = x'(2MVe)^{\frac{1}{2}}$$

$$p_y = y'p_z = y'(2MVe)^{\frac{1}{2}}$$

and

$$p_r = r'p_z = r'(2MVe)^{\frac{1}{2}}$$

Typical emittance distributions for a number of beam configurations are shown in fig. 4.22.

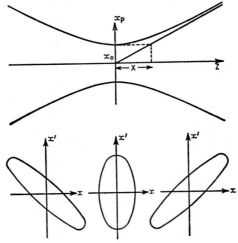

Fig. 4.22. Emittance distributions around a beam waist (from Banford, 1966). Hyperbolic profile of a normal beam in a field free region and the associated phase plots.

In the hypothetical case of particles diverging from a point source in a field free region the displacement-transverse momentum, phase-space diagram would be a straight line of zero emittance. In a more practical situation, however, the ions always have random motions superimposed upon the collective beam motion. These may arise, for example, from ion temperature distributions or from imperfections in the ion extraction surface, and they lead to a finite emittance area. It follows that such a beam can never be made truly parallel, nor can it be focused down to a perfect cross-over. The emittance is thus a measure of the beam quality.

These limitations on ion beam manipulation are of singular importance in the complex or very long beam transport systems which are commonly used in high energy nuclear physics accelerators. In ion implantation they establish the degree of fine focusing which can ultimately be achieved. They are also undoubtedly responsible for some of the large discrepancies between the high extracted currents reported for simple ion source test assemblies, and the lower beam currents commonly obtained at the target stages of suitable accelerators.

A variety of units and definitions of emittance have been used in the literature. Since the approximate form of the phase space distribution is commonly found to be elliptical, Lapostolle (1969) has suggested that the emittance $E$ should be defined as

$$E = \pi \cdot X \text{ radian metres}$$

where $X$ is the product of the semiaxes of the ellipse which circumscribes the area. In situations where the extremities of the beam are diffuse it may be more practicable to define the emittance by the area which contains, for example, 90 % of the total current density ($E_{90\%}$). This is illustrated in fig. 4.23.

So far we have only considered beams of constant momentum. If the particles are further accelerated the transverse momentum $p_r$ will be unaffected, but the slope $r'$, ($= p_r/p_z$), will decrease and the corresponding phase space area will be reduced.

For such beams the normalised emittance, $E_n = E \cdot p_z$ is still invariant and at non-relativistic energies this results (for a given ion mass) in the momentum normalised emittance having the units radian–metre–$(\text{MeV})^{\frac{1}{2}}$ (certain workers use the alternative units mrad–cm–$(\text{MeV})^{\frac{1}{2}}$).

Since in most ion doping experiments we are concerned not only with the spatial description of the beam but also with the useful intensity we can define a final figure of merit, the brightness $B$, which relates the current $I$ to the transverse phase-space volume of the ion beam (Septier, 1967).

$$B = \frac{2I}{E^2} \approx \frac{2I}{E_x \cdot E_y} \qquad \text{(for beams which are not axially symmetric)}$$

The emittance and brightness of ion beams are normally determined

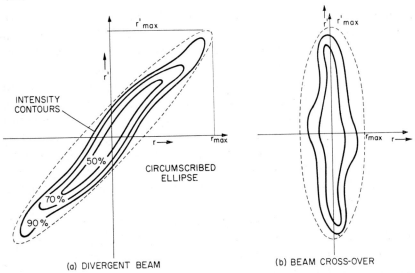

(a) DIVERGENT BEAM                    (b) BEAM CROSS-OVER

Fig. 4.23. Emittance diagram for a beam with a diffuse boundary (from Rose and Galejs, 1965). A typical distribution of the particle slope $r'$ in a diverging beam as a function of radius. The dotted elliptical contour shows the circumscribing ellipse with the smallest area. (b) The emittance at a beam cross-over.

experimentally by scanning the ion beam with slits or apertures and measuring the angular direction, the divergence and the intensity of the transmitted fraction in each position (see, for example, Van Steenbergen, 1967; Septier, 1967). It is worth noting, however, that precise determinations on low emittance beams present certain experimental difficulties. The defining system may result in collimation or scattering of the beam, and Freeman and Bell (1963) have observed that the interaction of apertures with intense, low energy, space-charge neutralised beams of heavy ions may result in significant perturbations of the ion trajectories. In accelerators used for implantation a sufficiently accurate estimate of the emittance and brightness can often be made directly if the focused image size and the beam convergence can be measured.

In this elementary and simplified description of the nature of an idealized accelerated beam in a perfect vacuum, we have assumed the absence of mutual interactions between the particles. The effective emittance of real beams is increased by factors such as scattering and lens aberrations as well as by non-linear space charge forces. More rigorous and detailed treatments of the concept of emittance and its importance in high energy accelerators, can be found in Lichtenberg (1969), Lapostolle (1969), Banford (1966), and Rose and Galejs (1965).

## 4.3. Heavy ion sources

In this section we shall be concerned with the production and extraction of heavy ions. The successful operation of any implantation facility is probably more dependent upon the capability and the performance of the ion source than on any other single factor. For this reason we shall consider the various methods of ion production at some length.

We shall not review the historical development if ion guns, nor shall we attempt to give a comprehensive account of the very large number of source variants which have been reported. We shall however describe in some detail examples of the different kinds of sources which best illustrate the particular problems of heavy ion production.

For convenience, in the absence of a recognised nomenclature, we shall simply characterise particular ion sources by the names of the authors referred to in the text.

For most ion implantation experiments the paramount requirement for adequate beam intensities ($\mu$A–mA) excludes simple, low pressure, electron bombardment sources such as are commonly used in mass-spectrometry. The most successful and widely used method of producing heavy ions is by means of a confined electric discharge or arc, which is wholly or partially sustained by the vapour of the material to be ionised. Very intense ion beams can be extracted from suitably designed discharge sources, and the method has the important advantage that all elements can be ionised in a discharge.

The only limitation to the versatility of many arc sources lies in the technical difficulty of operating over a sufficiently wide temperature range to volatilize the required range of materials. As we shall see below, this problem can be circumvented in certain ion sources by the use of sputtering techniques.

Ionising discharges can be produced in a variety of ways and we shall consider hot cathode, radiofrequency and Penning sources. We shall pay particular attention to the various types of heated filament are sources because of their suitability for a wide range of ion doping requirements.

We shall also describe other methods of heavy ion production which, although more limited in scope, may be important in certain applications.

An examination of the copious literature on heavy ion sources indicates that the most important advances have generally been made experimentally. In many cases the detailed behaviour of the ionising mechanism is complex and is still incompletely understood. It seems likely that, with a few notable exceptions, further development will continue to be largely of a technological and empirical nature. We shall only consider the detailed theory of ionisation and beam formation insofar as it is essential to an understanding of ion source behaviour.

Because of their inevitable complexity, the successful operation of versatile heavy ion sources is to some extent dependent upon the competence of the experimenter. Nevertheless the large, and often inexplicable, discrepancies quoted in the literature between the performances of quite similar sources, highlight the problems of heavy ion production and the inadequacy of the existing ion source theory.

It is noteworthy also, that the extracted beam currents reported for simple ion source test assemblies are only rarely duplicated at the target stage of more complex accelerator systems. In certain cases the discrepancies may be very large. The primary ion source measurements may be optimistic insofar as they may include multiply charged or impurity ion species. They may also correspond to a value of the beam emittance which is unacceptably large for the ion optics of the accelerator system. As we shall see below, an additional factor, particularly important with intense low energy heavy ion beams is the possibility of beam 'blow up' because of space-charge repulsion effects. This may impose constraints upon the optimisation of the ion source during operation in an accelerator. Transmission losses will also occur because of scattering and charge-exchange processes in the beam acceleration and transport system.

It seems somewhat paradoxical that whilst beam currents of tens of milliamperes of heavy ions are commonplace in ion propulsion systems and in production isotope separators, and that although similar currents of protons and certain gas ions can be extracted in many nuclear accelerators, progress in ion implantation may be restricted by the problems experienced in certain machines in obtaining even microampere intensities of suitable ions such as boron.

Whilst there is no fundamental problem in producing intense ion beams of any element, it is not always possible to adapt existing and proven ion sources to a particular accelerator. There may, for example, be difficulties because of space, power or pumping restrictions at the high voltage terminal: There may be inadequate access for the more frequent and complete adjustments associated with the production of heavy ions: The beam extraction system of the source may be incompatible with the ion optics of the accelerator – thus whilst milliampere heavy ion beams are commonly obtained from slit extraction sources, most machines are designed to handle beams of cylindrical geometry.

Because of the wide variety of accelerating systems which may be utilised for ion doping it is not possible to define a universal ion source. Many restricted and generalised approaches to the problems of producing suitable beams of heavy ions have been made. In any particular situation the preferred method must ultimately be decided by considerations such as the type of machine and the experimental requirements.

The most common requirement is nevertheless for general purpose positive ion sources capable of producing suitably intense* beams of a wide range of elements. A high ionisation efficiency may be required because of pumping limitations; it is likely to be essential in applications requiring the use of radioactive ions. The beam acceptance and transmission requirements of the machine may impose emittance limitations on the ion source. For the most efficient use of the accelerator, the source should be simple to install and operate; for experimental convenience it may be desirable to have a facility which permits the rapid change of dopant ions. It should also be possible to operate the source continuously for periods of several hours.

It is these general requirements, and in particular the need for versatility, which impose a number of important constraints upon the source design.

Thus in order to handle a wide variety of elements the source may need to be capable of operation at precisely controlled high temperatures. This requirement creates both heating and cooling problems at the source potential. The use of heat shielding to minimise power losses may influence the geometry of the ionisation chamber. The condensation of source feed vapours on the extraction electrodes may lead to severe electrical breakdowns if the conventional small spacings and high field gradients of gas sources are used. Provision for the compensation of thermal expansion and misalignment effects may be required, and the choice of constructional materials must be compatible with the wide variety of corrosive compounds and elements which have to be vaporized. Finally the source design and operation may be influenced by the need to minimise space-charge effects which can cause serious and irreproducible defocusing of the extracted ion beam.

* If the source is to be used with a wide range of elements it is generally essential that the available ion output should be considerably in excess of the required current at the target. This is because the required dopant beam may in many cases represent only a fraction of the total extracted current. Boron trichloride is, for example, a convenient source feed vapour for the production of boron ions. Recent measurements (Freeman, 1970a) showed, however, that in a typical doping experiment a total source drain current of 12–14 mA was required to obtain a one milliampere beam of boron ions. The analysis of the extracted beam is shown in fig. 4.16. It can be seen that the spectrum is made of several ion species, and that the $B^+$ is a relatively minor component.

Although as we shall see below these various problems have been previously encountered and largely resolved in certain cases, there is no simple solution to the problem of heavy ion production for ion implantation. Nor in view of the wide variety of experiments and the different types of accelerator is this to be expected.

It seems likely however that the rapidly increasing interest both in ion doping and in other ion interaction studies, will lead to a welcome development of ion sources of improved performance and versatility, designed specifically for such applications.

Although the production of suitable beams by novel techniques such as laser ionisation may eventually make a significant contribution, there is considerable scope for improvement using conventional methods. Certain of the problems of heavy ion production are technological, whilst others are associated with our incomplete understanding of the complex ionisation phenomena in electrical discharges. It is likely that, whilst much of the progress in ion source development will continue to be of an empirical nature, techniques such as sputtering and charge-exchange will play an increasingly important role in the ionisation process. The range of accelerators which can be used, and the general scope of implantation research will certainly increase as compact, low power sources, negative ion sources, and sources capable of producing high yields of multiply charged ions become more generally available.

Whilst the requirement for versatility will generally dominate in research, the industrial application of ion implantation will create a separate need for simple and robust sources capable of continuous operation at higher outputs but with a more limited range of ion species.

### 4.3.1. HOT CATHODE ARC SOURCES

The common feature of the sources described below is that an ionising arc or discharge is sustained by electrons which are emitted from a heated filament.

Figure 4.24(b) illustrates a typical arc source which is similar in layout to the familiar mass spectrometry electron-bombardment ion source (see for example Elliot, 1963). In both types of source, electrons extracted and accelerated from a heated cathode are collimated by defining slits and by a suitably oriented magnetic field. Ions produced in the chamber are extracted through a narrow slit and accelerated. Gas or vapour is introduced through a controlled leak from an external reservoir, or, in the case of low vapour pressure materials it may be admitted from a furnace.

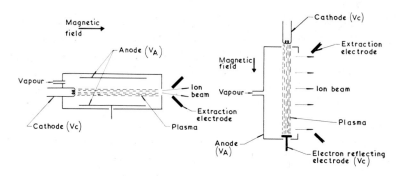

(a). Axial beam extraction from a circular aperature    (b). Lateral beam extraction from a slit in
in the end of the discharge chamber.                         the side of the discharge chamber.

Fig. 4.24. Oscillating electron discharge sources.

In the mass spectrometry source the pressure is normally low and ioni-sation is largely effected by single collisions between the gas molecules and the monoenergetic electrons. The essential and important difference in the discharge source is that the pressure is raised (typically to $10^{-4}$–$10^{-2}$ torr) until a stable arc is struck between the filament and the anode (in this case the walls of the arc chamber). The nature of the ionising discharge dominates the behaviour of the source. The precise and selective ionisation character-istics of the electron-impact source are lost. Because of the formation of a plasma sheath close to the cathode, much larger electron currents can be extracted from the filament. The increased possibility of ionisation in the intense arc leads to higher efficiencies and a considerable increase in the ion beam intensity. As we shall see below, the shape and position of the plasma surface plays an important role in the ion extraction process, whilst funda-mental plasma instabilities may under certain conditions result in beam defocusing.

A variety of ionisation chamber geometries and cathode configurations have been used but the sources fall broadly into two categories. The ions may be extracted in the direction of the plasma column and the electron flow, or the extraction can take place perpendicular to the discharge. Kiste-maker and Zilverschoon (1951) describe these as Finklestein (F), and Heil (H), geometries respectively, but we will simply define the sources as axial (end), or lateral (side) extraction. The two modes of extraction are shown diagrammatically in fig. 4.24.

This classification is somewhat arbitrary since there is no fundamental difference between the two types of source. The same constructional and operational techniques are used in each case and the general mode of behaviour is quite similar. Nevertheless, for practical purposes, the distinction is a very real one since both the ion optics and the beam intensities are quite different in the two geometries.

In general, axial extraction sources have a circular extraction orifice and produce cone shaped beams, which can be subsequently manipulated with conventional electrostatic or magnetic lenses. Although milliampere beams have been reported, more typical intensities of heavy ions at the target stages of accelerators are tens of microamperes. The sources can be quite compact and are readily adaptable to different types of machine.

In lateral extraction sources a slit is normally used to give a wedge shaped beam geometry. Whilst this may have advantages in the simplest machines, where the parallelism in one plane eases the subsequent focusing requirements, such beams are not so readily amenable to conventional accelerator optics. The problem may be aggravated since the milliampere intensity beams which are typical of such sources are more prone to space-charge defocusing effects in electrostatic lenses. The sources are inevitably larger with higher pumping and power requirements. Despite these drawbacks, in many cases they offer the simplest available solution for implantation requirements in which high beam intensities are required. Although it is possible in principle to use them in high energy accelerators they have generally been employed at voltages below 100 kV. The use of high voltage target assemblies and multiply charged ions to extend the energy range is described in section 4.2.

Several of the sources described below have been specifically developed for various types of electromagnetic isotope separator. The requirements for ion doping are similar although not identical. Thus whilst the focusing requirements for implantation will in general be less critical than those of a high resolution separator, the possible additional need for more complex beam manipulation may nevertheless impose stringent brightness requirements. Since most sources produce a complex spectrum of ions, a high resolving power may be necessary to resolve and identify the required dopant beam. It is unlikely, however, that the most exacting requirements of isotope separation, the simultaneous optimisation of efficiency, output and resolution, will be required for most ion implantation experiments. It is thus generally possible to simplify the source operation and in certain circumstances to obtain higher beam currents. An important exception is the special case of low dose radioactive ion implantation. In the most stringent experiments both the requirements for precise ion doping and for rigorous isotope separation have to be reconciled. This type of application is considered in detail in section 4.6.

The basis of much of our current understanding of discharge ion sources is to be found in the comprehensive study of ion beam production carried out in the United States during the early days of the development of production electromagnetic isotope separators (e.g. Guthrie and Wakerling, 1949*). Although these machines known as 'calutrons' were originally intended solely for the separation of the isotopes of uranium they have since been used with only minor modifications, for almost all of the polyisotopic elements, and are still the only accelerators which are capable of producing really high intensity (10–200 mA) focused beams of a wide range of heavy ions.

Although calutrons may be appropriate for certain industrial ion doping requirements, and although they have been used for certain experimental implantation applications, their size, complexity and cost, exclude their general use as research facilities.

The calutron sources have nevertheless provided an important foundation for much of the subsequent work on the production and extraction of heavy ions in lower output machines. We shall consider them briefly before proceeding to a more detailed consideration of the several types of both lateral and axial extraction ion sources.

We shall first outline some of the general features of the ionising discharge in hot cathode sources.

*4.3.1.1. Ionisation in a Discharge.* A number of the elementary processes which may occur in a gas discharge are shown diagrammatically in fig. 4.25 (Penning, 1957). Even in its simplest form in a semi-idealised laboratory apparatus it is a complex phenomenon.

The discharge in heavy ion sources is further complicated by a number of factors. The geometry of the discharge chamber, the choice of constructional materials and the wide variety of vapors to be ionised are largely determined by the experimental accelerator requirements. The problems of attempting a fundamental description of ion source behaviour are also compounded by the inevitable interaction of the beam extraction field with the

---

* A great deal of valuable information on the separator and ion source studies is contained in the official U.S., T.I.D. and N.N.E.S. reports of the programme. These are listed after the Guthrie and Wakerling reference given above.

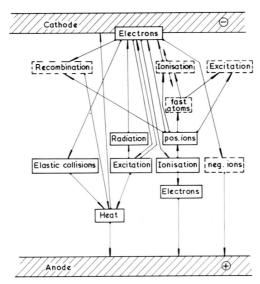

Fig. 4.25. Elementary processes in a gas discharge (from Penning, 1957). The diagram above summarizes a number of the elementary processes occurring in gas discharges. Each process or transition is symbolized by an arrow, the black spot representing the start of the process or transition and the arrow-head its termination. The names of processes and particles that play a particularly important part in inert gases are enclosed by a black line, the others by a broken one. Liberation of electrons from the cathode arrows terminate at 'electrons' on the cathode.

plasma and by the frequent use of magnetic fields to constrain the discharge and increase the ionisation efficiency.

Although a number of detailed studies of ion source discharges have been made (e.g. Bohm, 1949; Almèn and Nielsen, 1957; Chavet, 1965) they have unavoidably been restrictive because of the large number of variable parameters. In particular they have largely concentrated on the relatively simple case of source operation with pure inert gases. They have nevertheless confirmed certain fundamental postulates and the work has led to a considerable measure of optimisation and understanding of source behaviour.

The overall picture in general purpose heavy ion sources is clearly very involved and, while more fundamental studies of a number of aspects of the discharge are desirable, it is far from evident that they will be more successful than empiricism and continued technical development in establishing further significant improvements in performance.

Because of this we shall give only an elementary and approximate description of those fundamental characteristics of the discharge which are essential to an understanding of the general behaviour of ion sources. We shall be primarily concerned with sources in which the discharge is sustained by electrons produced by a hot filament, and we will begin by considering the factors which control the cathode emission. It should be noted however that the mechanisms of ionisation and beam formation in other types of discharge (Penning, radiofrequency) are generally quite similar.

The maximum current density which can be obtained from a metal surface of work function $\phi$ heated to a temperature $T$ is

$$j_e = AT^2 e^{-e\phi/kT} \qquad (4.1)$$

where $A$ and $k$ are constants and $j_e$ is the current per cm$^2$. In spite of the obvious advantages of working with low work function activated filaments these are only rarely used in heavy ion sources because of the intense cathode sputtering which occurs in the discharge, and because of the surface poisoning effects which obtain with certain source feed materials. We shall return to this problem in section 4.3.6.

The maximum, or saturation, current of eq. (4.1) cannot normally be extracted from an incandescent cathode in a vacuum tube. This is because the electrostatic mutual repulsion effect of electrons leaving the filament inhibits the emission to a value known as the space-charge limited current. For a plane cathode and anode separated by a distance $d$, with a potential $V$ this is given by Langmuir's well known formula:

$$j_e = \frac{V^{\frac{3}{2}}}{9\pi d^2 \sqrt{M/2e}} = 2.3 \times 10^{-6} V^{\frac{3}{2}}/d^2 \text{ amperes/cm}^2/\text{sec.} \qquad (4.2)$$

If, however, a small amount of gas is present whose ionisation potential is lower than the anode voltage, the primary electrons will produce positive ions which will tend to neutralise the negative space-charge and thus allow the electron current to increase. With sufficient gas a stable discharge will be created and under suitable conditions the filament emission may rise to the temperature saturation current value (eq. (4.1)).

The effect of space-charge on the lines of electric force and on the distribution of potential between the cathode and the anode is diagrammatically shown in fig. 4.26. (Penning, 1957). The effect of electron emission (fig. 4.26(b)) reduces the effective extraction field at the filament, whereas the presence of positive ions (fig. 4.26(c)) enhances it. When a discharge is formed the potential takes the approximate form shown in fig. 4.26(d). The full voltage drop then occurs over the short distance $CA'$ and the extraction field for electron production is considerably increased. As we shall see below

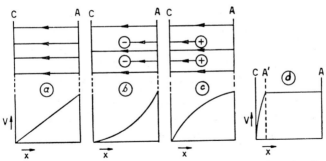

Fig. 4.26. Effect of space charge on the lines of electric force and on the distribution of potential between the cathode and the anode (from Penning, 1957). (a) In the absence of a space charge, (b) with a negative space charge between the electrodes, (c) with a positive space charge and (d) in a discharge.

however, the plasma boundary at $A'$ is not quite analogous to the simple plane anode at $A$ and the maximum electron emission is also related to the flow of positive ion current from the discharge to the filament.

The region $A'A$ corresponds to the plasma of the discharge. This contains roughly equal densities of electrons and positive ions and is approximately field free. The density of ionised particles in ion source discharges depends upon the nature of the plasma but usually lies between $\simeq 10^{10}$ and $10^{14}$ ions/cm$^3$. The neutrality of charge in the plasma is self-regulating since the positive ions and electrons are able to distribute themselves rapidly to compensate any localised field gradients. In the steady state condition the electrical balance must be maintained, and the fluxes of ions and electrons leaving the plasma must be equal to their rate of production. A most important characteristic of the plasma, which dominates its behaviour, is its tendency to maintain its electrical neutrality in spite of external disturbing factors. It thus accommodates any electrodes or surfaces which are not at plasma potential by enveloping them with a charged sheath and effectively shields itself from the electrical fields. (The thickness of such plasma sheaths depends upon the discharge conditions and the probe voltage but is typically less than one millimetre.)

Figure 4.27 illustrates the formation of a double sheath of electrons and ions at the filament surface. At the plasma boundary the electric field must be approximately zero to satisfy the requirement for neutrality. If the electron emission from the filament is space-charge limited the field must also be zero at its surface. Since the plasma is approximately at anode potential, the full voltage drop of the discharge occurs across the ion-electron double layer. At equilibrium, electrons from the filament pass through the sheath

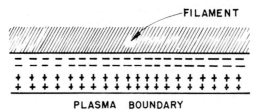

Fig. 4.27. Formation of a double sheath of electrons and ions at filament surface (from Bohm, 1949).

to the plasma, whilst positive ions move in the opposite direction. The concentration of electrons is highest near the cathode surface where they have the lowest velocity, and similarly the ions are concentrated near the plasma boundary. Since the ions move more slowly than the electrons they produce the same charge density with a much smaller current.

In a similar way the neutrality of the plasma is maintained by the formation of a sheath at the anode and around any other probe surfaces in the discharge chamber. If we ignore secondary electron production induced by ion bombardment, and also the effects of negative ions, the treatment and description of such sheaths is simpler since we can assume that they contain mainly positive ions and that the flow of charged particles is monodirectional.

We thus have an approximate qualitative description of the discharge as an electrically neutral plasma surrounded by a sheath, and within which ionisation is primarily induced by electrons accelerated from the filament.

The rate of ion production is determined approximately by the electron current, the pressure, the geometry of the discharge and the ionisation cross section of the neutral atoms. The nature of the dependence of the cross section $\sigma$ on the electron energy for argon and uranium tetrachloride can be seen in fig. 4.28 (maximum cross sections for elements are listed in Appendix 1). Secondary electrons liberated in the ionisation process have appreciable energies and can also contribute significantly to further excitation and ionisation. In simple inert gas discharges recombination processes in the plasma may be neglected and the loss of ions is primarily by diffusion to the walls.

The potential of the plasma normally adjusts itself to a value several volts above that of the anode. In this way the imbalance of electrical neutrality which would result from the loss of the fast moving electrons to the walls of the discharge chamber is prevented. The ions diffuse through the plasma and fall to the walls whilst the majority of electrons are reflected by the potential barrier which is formed at the boundary. The ion temperature

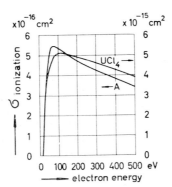

Fig. 4.28. Total ionisation cross section versus electron energy for argon and uranium tetrachloride (from Walcher, 1958).

in the discharge is in general not much higher than that of the neutral gas, whereas the plasma electrons undergo frequent collisions and attain an approximately Maxwellian distribution with a temperature of about two to four volts (Bohm, 1949). At equilibrium the number of electrons energetic enough to escape through the surface potential barrier is equal to the number produced.

According to Bohm a necessary condition for the formation of a stable plasma sheath is that the ions must reach the sheath with a kinetic energy corresponding to approximately half the mean electron temperature. The necessary ion acceleration $(T_+ \to T_e/2)$ occurs because the electrostatic shielding effect of the plasma is not quite perfect at the boundary and there is a small penetration of the potential. The form of the voltage gradient between the plasma and the wall is shown in fig. 4.29.

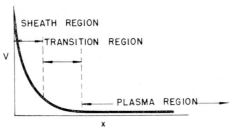

Fig. 4.29. Form of the voltage gradient between the plasma and the wall (from Bohm, 1949).

This effect is important since it follows that the flow of ions from the discharge is directed normally to the surface and is determined by the electron temperature. The current density of singly charged ions $j_+$ through the plasma boundary is approximately given by

$$j_+ = n_+ \bar{v}_+ \approx n_+ \sqrt{\frac{kT_e}{M_+}} \text{ ions/cm}^2/\text{sec.} \quad \text{(Bohm, 1949)} \quad (4.3)$$

where $k$ is Boltzmann's constant, $n_+$ is the density of positive charges in the plasma, $\bar{v}_+$ is the mean value of the velocities of the positive particles normal to the plasma boundary and $T_e$ is the electron temperature.

The precise value of the current density depends upon the geometry of the system and the nature of the discharge, but a commonly used value due to Kamke and Rose (1956) is:

$$j_+ \approx n_+ \sqrt{\frac{kT_e}{2\pi\varepsilon M_+}} \approx 3.5 \times 10^{-13} n_+ \sqrt{\frac{T_e}{M_+}} \text{ mA/cm}^2 \quad (4.4)$$

(where $\varepsilon$ is 2.718, the base of natural logarithms).

Thus for an ion of mass 100, an electron temperature of about 50,000 °K and a plasma ion density of $10^{12}$ ions/cm³, the current density would be approximately 8 mA/cm².

Furthermore if the initial velocity of the positive ions at the plasma boundary is disregarded then the positive current which flows is also a function of the sheath thickness $d$ and the voltage V between the plasma and the wall, and is given by the familiar 'space-charge limited' form of expression:

$$j_+ = \frac{V^{\frac{3}{2}}}{9\pi d^2} \cdot \sqrt{\frac{2e}{M_+}} = 5.45 \times 10^{-8} \frac{V^{\frac{3}{2}}}{d^2\sqrt{M_+}} \text{ amp/cm}^2 \quad (4.5)$$

where $M_+$ is the mass of the ion and $e$ the electronic charge.

The thickness of the plasma sheath which is typically a fraction of a millimetre can be calculated from eqs. (4.4) and (4.5) and is:

$$d = 2.36 \times 10^{-4} \left( \frac{V^{\frac{3}{2}}}{j_+ \sqrt{M_+}} \right)^{\frac{1}{2}} \text{ cm.} \quad (4.6)$$

At the filament, which may be regarded as a negative probe, currents pass through the sheath in both directions: the ions of the plasma moving towards the cathode whilst the electrons move from the cathode through the boundary (fig. 4.27). The maximum space charge limited electron current which can flow is given by (Langmuir, 1929; Bohm, 1949):

$$j_e = j_+ Y \sqrt{\frac{M_+}{M_e}} \quad (4.7)$$

where $Y$ is a correction factor lying between $\frac{1}{3}$ and $\frac{2}{3}$ introduced by Langmuir to account for the condition of the filament.

The sheath thickness in this case is somewhat greater than that of a sheath containing only positive ions because the effective space-charge density of the ions is reduced by the presence of the electrons.

Since the current leaving the filament may be temperature limited (eq. (4.1)) and thus have a value below the space-charge limited value eq. (4.7) is normally written

$$j_e \leqslant j_+ Y \sqrt{\frac{M_+}{M_e}} \tag{4.8}$$

[Carlston and Magnuson (1962) have noted that in the temperature limited case eq. (4.1) should be modified to take account of the secondary electron emission arising from the bombardment of the cathode with positive ions, thus $j_{total} = j_e + k j_+$, where $k$ is the secondary electron coefficient. This factor is of particular importance in certain types of Penning source.]

Finally from the inequality of eq. (4.8) Bohm showed that there must be a minimum threshold pressure for stable operation of the discharge. The relationship which he deduced is somewhat complex since it includes several physical parameters of the discharge but it can readily be seen that if eq. (4.8) is written in the form $j_+/j_e \geqslant (1/Y)\sqrt{(M_e/M_+)}$, since $j_+$ is proportional to the electron flow $j_e$, and to the density of neutral atoms, then

$$p \geqslant K \sqrt{\frac{M_e}{M_+}} \tag{4.9}$$

The existence of threshold operating pressures which usually lie in the range $10^{-4}$–$10^{-2}$ torr can be readily observed in ion source discharges. Source operation is generally particularly unstable close to the minimum pressure and frequently results in beam defocusing.

In the simplified account of ionisation in a discharge given above we have considered the idealised and approximate situation of a relatively homogeneous and quiescent plasma in a simple gas. In many ion sources the presence of magnetic fields to constrain the primary discharge and increase the degree of ionisation, as well as the use of reflecting cathodes to induce electron oscillation introduces a considerable degree of complexity. In such plasmas the electrons are no longer free to move to the anode walls and significant spatial and temporal non-uniformities may result. The treatment of discharge instabilities and oscillations is complex (Bohm, 1949; Kistemaker et al., 1965; Delcroix, 1969) and the problem is generally intractable in real ion source situations. The effects are principally of importance in the extraction of high intensity, low energy ion beams since the

instabilities of the discharge (commonly referred to as 'hash') may be transmitted to the beam with a consequential degradation of both the energy
homogeneity and the space-charge compensation. At lower currents ($<1mA$)
the effects are observable (Alväger and Uhler, 1968) but are usually less
important.

Figure 4.30 shows pictorially the effect of magnetic confinement and
electron reflection on the intense discharge. In even fairly weak magnetic
fields the electrons are constrained to follow tight spiral paths along the
magnetic lines of force. The Larmor radius $R_L$ (cm) of charged particles in
a magnetic field $H$ can be calculated from the relation:

$$R_L = \frac{144}{H} \cdot \sqrt{MV} \tag{4.10}$$

Thus for 100 eV electrons in a field of 300 gauss the radius is only about
one millimetre. This would be the maximum radius of the spiral if all of the
electron energy was in a direction normal to the direction of the magnetic
field. The lower energy secondary electrons are even more tightly constrained.

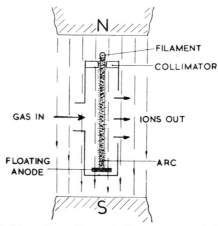

Fig. 4.30. Magnetic confinement of an ion source discharge.

In the design of an ion source based on this principle, considerable care
must be taken to ensure that the magnetic field has the required configuration. The position of the discharge with respect to the ion extraction slit is
normally critical and it follows that the field must be free of curvature and
precisely aligned over the whole length of the discharge chamber. Serious
perturbations may arise from the use of electrical heating elements in the
source and it has also been shown (Freeman et al., 1961) that the self-induced

Fig. 4.31. Displacement of ion source discharge arising from the interaction of the source magnetic field and the self-induced magnetic field of the filament (cf. fig. 4.30) (from Freeman et al., 1961).

field of the filament may produce a significant displacement of the discharge (fig. 4.31).

In spite of the approximations the generalised description given above of the plasma is useful in understanding the basic processes involved and it leads directly to the mechanism of the extraction and formation of ion beams from discharge sources.

*4.3.1.2. Ion Extraction from a Discharge.* In the previous section we considered the transport of ions from a thermalised plasma to the walls of a discharge chamber. We saw that the current density was controlled by the relationships

$$J_+ = 5.45 \times 10^{-8} \frac{V^{\frac{3}{2}}}{d^2 \sqrt{M_+}} \approx 3.5 \times 10^{-16} n_+ \sqrt{\frac{T_e}{M_+}} \cdot \text{A/cm}^2$$

Similar criteria govern the formation of the ion beam in discharge sources. A representation, due to Walcher (1958), for the mechanism of ion extraction, from a plasma is shown in fig. 4.32.

The wall of the discharge chamber is provided with a hole of area $A_0$ for the extraction of ions. These are collected on an extraction electrode which may be negatively biased and which for simplicity is shown as a Faraday cage. If, as in fig. 4.32(a), the probe is first insulated, the plasma will diffuse into the cage and a space charge sheath will be formed in accordance with eqs. (4.1) and (4.4). The probe will acquire a negative voltage with respect to the plasma.

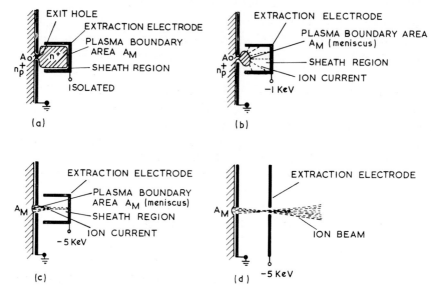

Fig. 4.32. Mechanism of ion extraction from a discharge (after Walcher, 1958).

If a higher negative voltage is now applied to the probe, the plasma will be pushed back towards the aperture as shown in fig. 4.32(b) and will finally take the form shown in fig. 4.32(c). If the Faraday cage is replaced by a suitable electrode, fig. 4.32(d), an ion beam can be extracted from the meniscus of the plasma. The current flow through the discharge boundary is governed by eq. (4.1) but it can be seen that both the position and the area, $A_m$, of the meniscus can vary according to the extraction conditions and the intensity of the discharge. A detailed theoretical treatment of the formation of ion beams has been given by Walcher. Rautenbach (1961) has shown experimentally that the beam parameters are connected by the equation

$$j_+ = K(e/M_+)^{\frac{1}{2}}V^{\frac{3}{2}}$$  (4.11)

where $K$ is a constant depending upon the geometry of the system, and similar results which confirm the general 'space-charge limited' form of the current dependence $(j_+ \propto V^{\frac{3}{2}}/d^2 M^{\frac{1}{2}})$ have been obtained by Almén and Nielsen (1957) and Chavet (1965).

In all of these measurements, care was taken to ensure that the plasma boundary conditions were well defined. It must be stressed that quite erroneous conclusions may be drawn from simple measurements of the ion beam current as a function of the extraction parameters. The flow of ions and hence the beam intensity is determined *primarily* by the degree of ionisation in the plasma and the electron temperature. If these are fixed, variations in the extraction conditions will modify the geometry of the plasma boundary until the two relationships which govern the beam current are simultaneously satisfied.

Although for simplicity we have considered the space-charge limited current flow between plane electrodes, in a more detailed treatment the expressions must be modified if the meniscus is curved (Chavet, 1965). It should be noted also, that in the fairly common case of beam formation from a plasma boundary inside the discharge chamber the true extraction field may be reduced because of the electrostatic shielding effect of the aperture.

The important feature of ion extraction from a discharge is thus that the mobility of the plasma meniscus plays a dominant role in defining the beam quality. The nature of the discharge and the position, the shape and the size of the extraction surface determine the ion beam current.

The curvature of the meniscus plays an equally important role in determining the subsequent angular divergence and emittance of the ion beam. This arises because the ions leave the surface normally under the influence of the weak accelerating field ($\approx kT_e/2e$) close to the plasma boundary (random transverse velocities are generally negligible since they are governed by the plasma ion temperature $T_+$). Some typical extracted beam configurations are shown in fig. 4.33. In the case of converging intense beams, space-charge repulsion leads to the formation of a waist with subsequent divergence. It can be seen that a virtual image which is smaller than the extraction aperture may be formed, from well defined concave or convex plasma boundaries. This effect is exploited in high resolution beam handling, as can be seen in fig. 4.34 which shows a sharply resolved mass doublet ob-

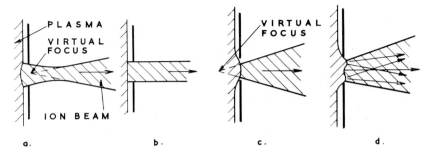

Fig. 4.33. Simplified representation of typical extracted beam configurations from discharge sources (a) concave plasma meniscus, (b) flat plasma meniscus, (c) convex plasma meniscus, (d) distorted plasma meniscus.

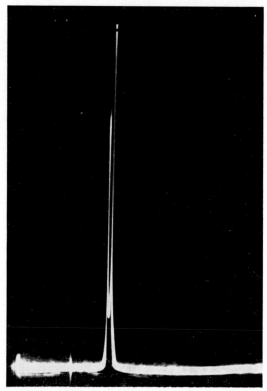

Fig. 4.34. Mass doublet, $^{40}Ar^{2+}$–$^{20}Ne^+$ obtained using an isotope separator ion source (from Freeman, 1963).

tained from an isotope separator heavy ion source (Freeman, 1963). Since the curvature is a function of the discharge parameters and the acceleration geometry, the beam divergence may be precisely controlled by suitable adjustment of the ion source conditions (Chavet, 1965). The shaping of the plasma meniscus can also lead to deleterious effects of the kind shown in fig. 4.33(d). Such plasma inhomogeneities may result in the formation of a beam emittance which is greater than the acceptance of an accelerator.

Another important consequence of the mechanism of extraction of ions from a plasma boundary is that the ion source efficiency $\eta$, may be significantly higher than the degree of ionisation $n_+/n^0$ in the discharge;

$$\eta = \frac{v_+}{v_+ + v_0} \tag{4.12}$$

where $v_+$ and $v_0$ are the fluxes of ions and neutrals leaving the source.

As we have seen above the ion current is proportional to the electron temperature ($E_0 \approx 50{,}000\ ^\circ\mathrm{K}$) whilst the flow of neutral particles is determined approximately by the source temperature which in most cases lies in the range 300–2000 $^\circ\mathrm{K}$. The efficiency for a source with a homogeneous thermalised plasma is approximately given by the expression (Sidenius, 1969):

$$\eta \approx \left[ 1 + \frac{n_0}{kn_+} \sqrt{\frac{T_0}{T_e}} \right]^{-1} \tag{4.13}$$

An additional increase in efficiency is obtained if, as in fig. 4.33(a), the area of the plasma meniscus $A_m$ is greater than the size of the aperture $A_0$ in which case eq. (4.13) becomes

$$\eta \approx \frac{A_m}{A_0} \left[ 1 + \frac{n_0}{kn_+} \sqrt{\frac{T_0}{T_e}} \right]^{-1} \tag{4.14}$$

It should be noted that the use of high source temperatures leads to a decrease in efficiency. For $n_0 \gg n_+$ the efficiency is roughly proportional to $1/T_0^{\frac{1}{2}}$.

Finally an additional important feature of ionisation in a discharge is that, if the plasma is thermalised, the energy spread of the extracted ions is very low ($\approx \mathrm{eV}$) and such sources can produce intense beams which may be sharply focused after mass separation. If however the source is subject to plasma instabilities ('hash') the extracted ions may have a significantly broader energy spectrum.

*4.3.1.3. Ion Beam Formation and Space-charge Effects.* So far we have simply considered the extraction of an ion beam by a single plane electrode with a suitable aperture. In such a system an initially parallel beam would tend to expand during acceleration because of space-charge effects. This problem has been considered in detail by Walcher (1958), (see also Wakerling and Guthrie, 1949), who has shown that the treatment of space charge limited beams developed by Pierce for electron guns is applicable. Figure 4.35 shows the approximate calculated shaping of electrodes to give a rectilinear flow of the charged particles. Although the theory appears to be applicable in sources in which the ions leave a fixed surface (e.g. surface ionisation), the situation in the discharge source is complicated by the facility with which the plasma meniscus can change both its shape and its position as a function of the operating parameters. Thus although a measure of electrode shaping is used in most heavy ion arc sources, the geometries are often empirical and

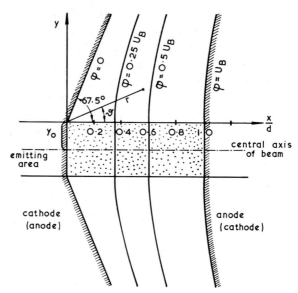

Fig. 4.35. Calculated electrode shaping to give rectilinear charged particle flow (from Pierce, 1949).

are largely determined by practical operating requirements. In a comprehensive study of the factors affecting beam divergence from a slit, Chavet (1965) showed that the curvature of the plasma meniscus played the most important role, and that this could be precisely controlled using a fixed acceleration voltage by varying the distance, $d$, between the source and the accelerating electrode. He found that the divergence could be expressed by the simple relationship:

$$2\alpha = 2.35f\frac{(d - d_0)}{d_0^2},\tag{4.15}$$

where $2\alpha$ is the total beam divergence, $f$ is the emission slit width, $d$ is the acceleration gap distance, and $d_0$ is the corresponding value for emission of the same current from a plane plasma boundary ($\alpha = 0$).

An alternative method of controlling the divergence is to maintain fixed electrode spacings and to vary the accelerating voltage. The final ion energy can be maintained constant by introducing an intermediate accelerating electrode with a variable negative bias between the source and the final electrode. As we shall see below the use of 'accel–decel' systems of this type also reduces the electron drain to the source and helps to maintain the space charge neutralisation of the ion beam.

For lower intensity beams accelerated from circular extraction aperture sources conventional electrostatic focusing lenses can also be used to control the beam divergence. Two such systems described by Alväger and Uhler (1968) are shown in fig. 4.36. In both cases the distance from the source to the first element of the extraction assembly could be varied to give additional control of the ion beam. In the lens shown in fig. 4.36(a) the intermediate focusing electrode is biased whilst the first and last electrodes are at earth potential, in fig. 4.36(b) the outer electrode is earthed and the internal electrode is used for focusing. (Cylindrical magnetic lenses are normally too weak to be used with heavy ion beams.)

Two typical slit geometry extraction systems due to Chavet and Bernas (1967) and Freeman (1963), are shown in fig. 4.37. It is usual in such sources to use relatively large acceleration gaps ( > 1 cm). The highly polished small gap electrode systems which are sometimes used in electron guns, and certain gas sources, are not readily applicable to the production of heavy ions, since the deposition of condensable vapours on the extraction electrodes may lead to electrical breakdown and severe back-streaming of electrons. The

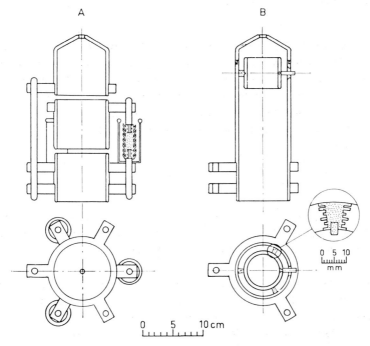

Fig. 4.36. Extraction and focusing lenses for circular aperture source (from Alväger and Uhler, 1968).

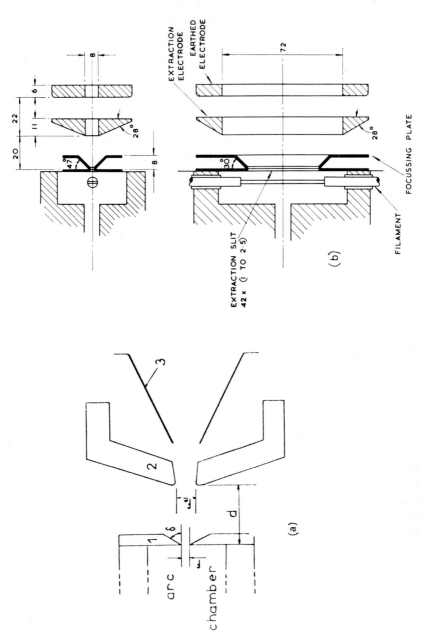

Fig. 4.37. Typical extraction systems for slit geometry sources (a) Chavet and Bernas (1967), (b) Freeman (1963). (In both cases the intermediate electrode is held at a negative potential to improve the space-charge neutralisation of the ion beam.)

shaping of the extraction aperture in high temperature sources may also be determined to some extent by the practical need for heat shielding on the front face of the discharge chamber, whilst the form of the extraction electrode is usually arranged to minimise high voltage sparking, rather than to satisfy the idealised Pierce geometry.

In the acceleration region of the ion source the repulsive forces between the ions result in some degree of beam broadening. There is virtually no compensation of the space-charge, since any electrons which may be present are rapidly accelerated towards the source. The situation is very different however in the field-free region beyond the final electrode. In this volume any slow electrons or negative ions produced by collisions of the primary beam with residual gas molecules are trapped in the positive space-charge potential well of the beam. Low energy secondary positive ions which are also produced in such collisions are accelerated out of the beam volume. If the ion current is free from intensity or spatial fluctuations the build up of electrons continues until the electron density fully compensates the positive field. In this state the beam volume becomes a plasma with equal densities of positive and negative charged particles. Whilst it can be shown that the effects of space-charge may be relatively insignificant for very low intensity beams, or for very high energy ions, this is not the case for the high current, low energy heavy ion beams which may be required in ion implantation. Here the compensation of the space charge is essential for the transmission and focusing of such beams through accelerators. It should be noted that if the ion current is not pulsed and if the beam volume is free of electrostatic fields the neutralisation process occurs automatically even in very high vacuum conditions. The relaxation time, (the time taken to achieve neutralisation), has been measured by Bernas, Kaluszyner and Druaux (1954) as a function of both pressure and ion energy for a number of systems. Typical results obtained for magnesium ions in argon and nitrogen are shown in fig. 4.38. Neutralisation can also be effected by the secondary electrons produced in the collision of ions and fast neutral atoms with surfaces of the vacuum system and electrons may also be artificially introduced into the beam from heated cathodes or by photo-electric effects (Peters, Helmholtz and Parkins, 1949). Detailed accounts of various aspects of the space-charge neutralisation of heavy ion beams have been given by Walcher (1958), Nezlin (1958) and Anastasevich (1957). Walcher has also calculated the effects of repulsion broadening on the focal quality for uncompensated idealised beam forms.

Thus for a fully compensated, or for an uncompensated beam the ion

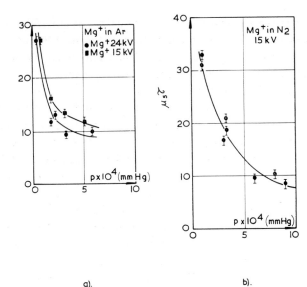

a).                                              b).

Fig. 4.38. Space-charge neutralisation of magnesium ion beams in (a) argon, (b) nitrogen. Relaxation time ($\mu$s) as function of pressure and ion energy (from Bernas et al., 1954).

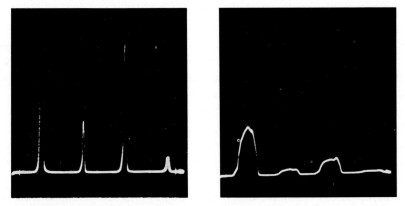

Fig. 4.39. Effect of grounding negative extraction electrode (fig. 4.37b) on focus (from Freeman, 1970a). (a) normal operation – electrode negative, (b) electrode grounded.

trajectories can be computed. In many real situations however the degree of compensation may be variable and difficult to control. The delicate charge balance may be upset if the compensating electrons are extracted locally from the beam by electrostatic fields due to accelerating or focusing lenses. Figure 4.39 shows the deterioration of the focal pattern in a separator when the negative lens of the extraction electrode system in fig. 4.37(b) is grounded. This arises because of the penetration of the source accelerating field into the plasma of the ion beam. If the intermediate lens of the accel–decel arrangement is simply used as a potential barrier to prevent electron extraction, quite low voltages may be adequate. (In the lens system shown in fig. 4.37(a), a negative potential of only a few hundred volts is normally used for positive accelerating voltages of 30 kV.) Beam defocusing may also occur from the charging up of insulating layers formed by cracked hydrocarbon layers on the walls of the vacuum chamber (Freeman and Bell, 1963), but the most significant defocusing effects arise in the ion source itself. Discharge instabilities ('hash') may be transmitted to the beam plasma as coupled ion intensity oscillations. The effect of such fluctuations is to reduce the degree of electron compensation, since, as the ion current falls the excess electrons are rapidly expelled to the walls, then when it rises there is a net positive charge on the beam until the electrical balance is re-established. The degree of defocusing is particularly serious if the characteristic frequency of the plasma oscillations is in the tens to hundreds of kilocycles range, and is thus comparable with the relaxation time. At very high beam intensities ($\approx$ tens of milliamperes) obtained from calutron sources, operating in strong magnetic fields, the beam 'blow up' may be catastrophic.

Space-charge effects in low energy heavy ion beams tend to be inconsistent and unpredictable but they can be minimised by careful design of the accelerator. In high resolution instruments considerable attention has been paid to the minimisation of plasma instabilities in the ion source discharge (see for example, Freeman, 1963). It should be noted however that the operation of a source in a suitably quiescent condition may correspond to a reduction in both beam current and efficiency.

*4.3.1.4. Thermal Evaporation Arc Sources.* The important feature which distinguishes simple gas sources from general purpose heavy ion sources is the need, in the latter, to introduce low vapour pressure materials into the discharge. The technique of sputtering is described below in section 4.3.1.5. We will be concerned here with the several types of ion source which use thermal evaporation of the source feed material from an oven. Some of the problems associated with high temperature operation have been outlined in 4.3. In section 4.3.7 and Appendix 1 we will separately consider the detailed operational techniques for a variety of elements. We will simply stress here that such sources may have to operate over a wide temperature range and that considerable care must be taken in the source design to prevent condensation of the source feed material on the walls of the discharge chamber. In certain cases where suitable volatile compounds may be prepared by chemical reaction in the heated ion source (see section 4.3.6) close temperature control is not necessary; in general however, the oven must be capable of precise and preferably rapid adjustment.

*4.3.1.4.1. Lateral Extraction Sources.* The calutron sources, (Guthrie and Wakerling, 1949; Morozov et al., 1958; Koch, 1958) designed for the large scale separation of isotopes have the unique capability of producing ion beams of hundred milliampere intensity of almost any element. This full ionising potential is only rarely realised in normal operation because of the additional requirement for very sharp focusing to obtain high isotope purities. Nevertheless resolved heavy ion beams in the 10–100 mA range are commonplace in production separators. By conventional accelerator standards the sources are very large and they are quite inappropriate to experimental studies. They may, however, become of importance in certain in-

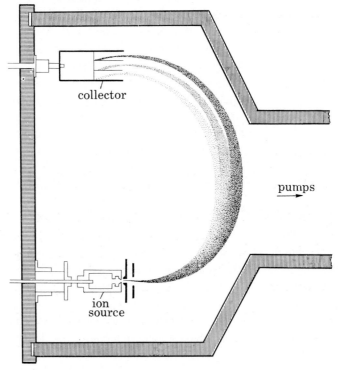

Fig. 4.40. Schematic illustration of a 'calutron' (high current electromagnetic isotope separator).

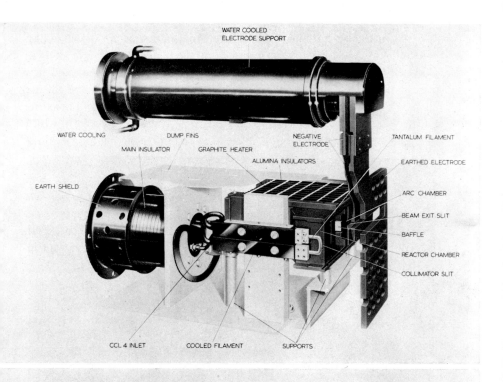

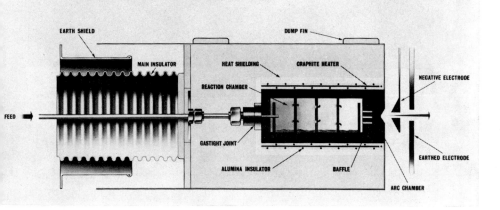

Fig. 4.41. Schematic illustration of a calutron ion source.

dustrial implantation applications and they illustrate clearly the general principles of ionisation in a magnetically constrained discharge.

Schematic illustrations of the calutron and the source are shown in figs. 4.40 and 4.41. The ion source is inserted into the high magnetic field ($> 1000$ gauss) of the separator. The discharge is thus tightly constrained and it is also spatially defined by a collimating slot between the filament and the arc chamber. Such sources are effective in producing a high degree of ionisation but they are also particularly subject to plasma instabilities ('hash'), which may lead to serious defocusing of the extracted ion beam. Figure 4.42 illus-

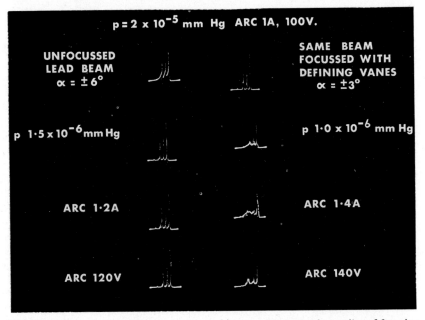

Fig. 4.42. Effect of minor adjustments to machine parameters on the quality of focusing of a 10 mA lead beam in a calutron.

trates the effect of minor adjustments to the discharge parameters on the quality of focusing of a lead ion beam. In many instances similar changes can result in the complete degradation of the beam quality. The performance of the source is very dependent upon the precise positioning of the discharge with respect to the ion extraction slit. This is determined primarily by the filament and collimating slot geometries, but is maintained along the full length of the arc chamber by the strong homogeneous magnetic field. In smaller ion sources using weaker, externally applied, magnetic fields this degree of spatial control may be difficult to achieve. Also as we have seen above, the interaction of the current induced field from the filament in such sources may also have important effects. An additional important feature of the very large calutron sources is that the length of the extraction slit ($\sim 20$ cm) is such that end effects are generally unimportant and the extracted beam is very parallel in the longitudinal direction. This is not the case in lower output sources which typically have extraction slits of one to four centimetres in length. In such cases the extraction geometry must be precisely shaped to obtain adequate control of the beam shapes (Freeman et al., 1961; Chavet, 1965).

A scaled down version of the calutron source was designed by Dawton (1958) for the production of milliampere beams of heavy elements in a sector field isotope separator. Although the source which is shown in fig. 4.43 was generally satisfactory with a range of both gases and solid charge materials, its behaviour was somewhat irreproducible and it was found that optimum operation could only be obtained under a restricted range of discharge conditions. Since in this case the source was not used inside the main beam analysing magnet an external variable magnet field ($\leqslant 1000$ gauss) was applied to collimate the discharge. In a detailed study of the source behaviour Freeman et al. (1961) showed that the interaction of this field with the self induced field of the heavy current filament ($\approx 200$ A) resulted in deleterious spatial displacements of the ribbon discharge (see fig. 4.31). By utilising the maximum available source field the effect could be minimised, but this mode of operation also corresponded to the condition for maximum discharge instability and loss of focus. The use of such high fields also leads to a significant, mass-dependent, angular displacement of the extracted beam and contributes to electrical breakdowns around the ion source. To eliminate these several problems a new source configuration was adopted in which the discharge was spatially fixed in the discharge chamber and which required the use of much lower magnetic fields (see Freeman below).

Although for very prolonged source operation the use of thick filaments is desirable they have a number of disadvantages. In addition to the effects of the self-induced magnetic field described above they impose a requirement for high current supplies at the source voltage and their geometry is relatively inflexible. Experimental ion sources of this type using thinner cathodes have proved more successful and a particularly comprehensive study of their behaviour has been carried out by Chavet (1965).

The source of Chavet and Bernas (1967) shown in fig. 4.44 is a modified version of the conventional Orsay separator source (Bernas, 1954). The resemblance to the calutron source described above can readily be seen. In

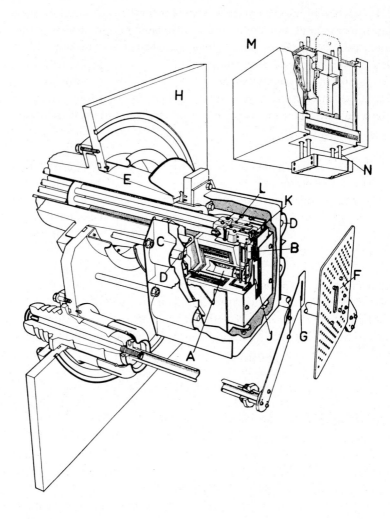

A.  Source furnace.
B.  Front of main graphite block.
C.  Side earth shield.
D.  Electron 'dumps'.
E.  Main insulator.
F.  Front (earthed) electrode.
G.  Accelerating electrode.

H.  Back plate with vacuum seal.
J.  Arc slit.
K.  Tantalum filament.
L.  Water-cooled leads to filament.
M.  Detail of front heater.
N.  Tantalum rods of front heater showing method of support.

Fig. 4.43. Heavy ion discharge source (from Dawton, 1958).

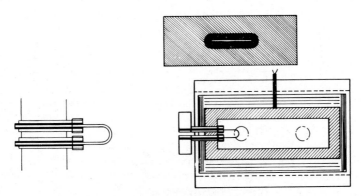

Fig. 4.44. Schematic illustration of heavy ion discharge source (from Chavet and Bernas, 1967).

this case the collimating slot is omitted and the filament is brought directly into the discharge chamber. The maximum externally applied magnetic field is about one kilogauss. Although an insulated anticathode, or electron reflector, may be fitted through the opposite face to the filament it is not normally used. A common feature of this type of source is that although electron reflection should result in an improved performance, no real experimental advantage is found in practice. Since the insulation of the anticathode introduces additional complexity into the source construction it is frequently omitted.

Although most of the source studies were carried out with inert gases the source has been successfully used for a wide variety of both stable and

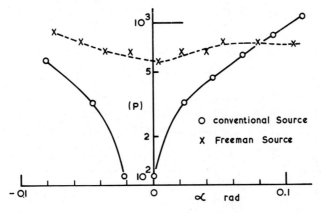

Fig. 4.45. Relationship between beam divergence, $\alpha$, and resolution, $P$ (from Chavet and Bernas, 1967).

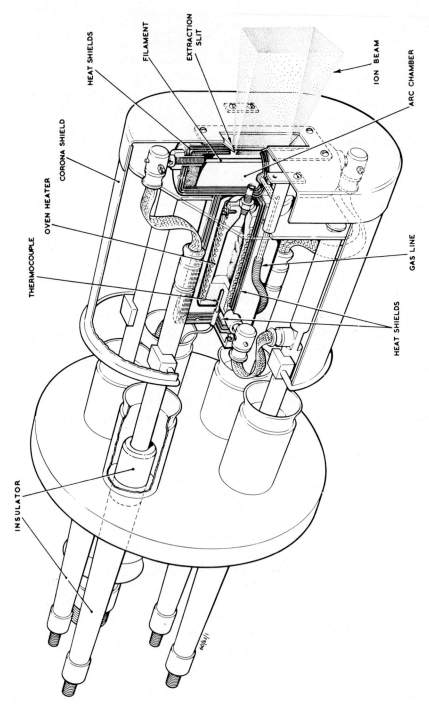

(a) Separator ion source.

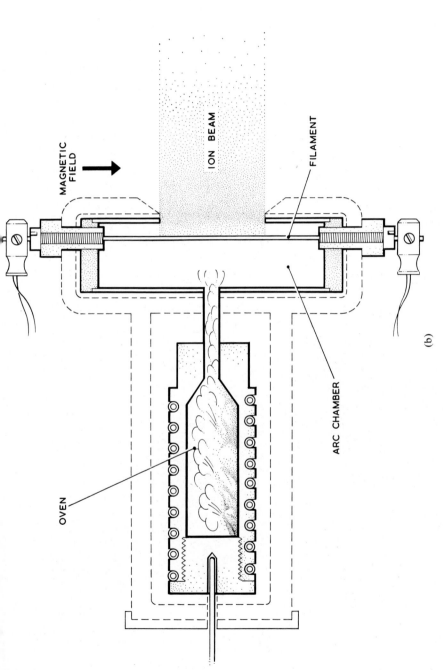

Fig. 4.46. (a) Schematic illustration of heavy ion discharge source. (b) details of discharge chamber and oven assembly from (Freeman, 1963, 1969b).

MAGNETIC FIELD

ION BEAM

FILAMENT

ARC CHAMBER

OVEN

(b)

radioactive isotope separations. Under optimum conditions both the efficiency and the maximum extracted currents are high ($j_+ \approx (0.25 \text{ A/cm}^2)/M^{\frac{1}{2}}$), but it should be noted that for optimum focus such beams, which are obtained from a convex plasma meniscus, have a large angular divergence, ($\sim 0.1$ rad.). The relationship between the beam divergence and the resolving power is shown in fig. 4.45. The broken line shows similar measurements on the same apparatus using a source approximating to that described by Freeman (see below). For certain implantation requirements, particularly those involving the doping of single crystals, a large degree of angular spread is unacceptable, in such applications it would be necessary to apply additional focusing, to collimate the beam, or to operate the ion source in a more restrictive regime. The construction of the source and the use of multiple oven arrangements has been described by Sarrouy (1969).

The lateral extraction heavy ion source described by Freeman (1963, 1969b) is shown in figs. 4.46 and 4.47.

The source had its origins in the difficulties described above in obtaining reproducible behaviour in a scaled down calutron type ion source. By placing the filament parallel with and close to the ion extraction slit, the intense dis-

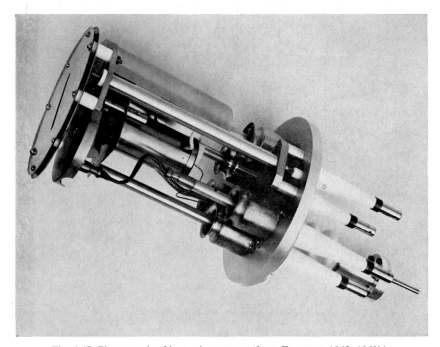

Fig. 4.47. Photograph of heavy ion source (from Freeman, 1963, 1969b).

charge is precisely located in the most advantageous position, and in this case the compound magnetic fields of the filament and the source magnet produce a beneficial increase in the ionising efficiency.

The electrons are fully accelerated as they enter the plasma sheath close to the filament surface. Their subsequent trajectories in the essentially electrostatically field free plasma are determined by the interaction of the magnetic fields of the filament and the external source magnet. A particular characteristic of the source is that optimum operation is normally obtained with a low magnetic field ($< 100$ gauss). A small magnet can thus be used and the undesirable effects which are sometimes associated with more intense fields (sparking, beam deflection, plasma instabilities) are noticeably absent. Within the normal operating limits of the source the quality of the extracted beam is largely independent of the various discharge parameters. These must be optimised for maximum efficiency, or output, but for routine ion doping experiments the operation of the source is generally simple.

The source unit is constructed largely of tantalum and molybdenum to reduce the thermal mass and to minimise outgassing and chemical reactions with corrosive vapours. There is no requirement for high voltage cooling and separate heating of the discharge chamber is not needed for normal opera-

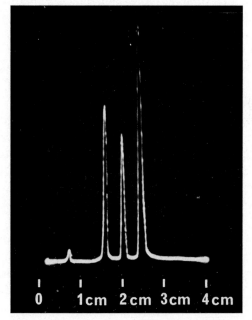

Fig. 4.48. Focal quality of 1 mA lead beam (from Freeman, 1969a, b).

tion ($\leqslant 1000\ °C$). The source may be used with a sputtering probe for the ionisation of platinum and other refractory metals. It can also be fitted with a simple charge loading facility with a vacuum lock. This permits the rapid introduction of solid source feed samples during operation.

Figure 4.48 shows the high resolution which can routinely be obtained with milliampere beams at the target. The source is notable for its ease of operation and with most elements and beams of this intensity can be obtained continuously for periods of tens of hours. It has been extensively used for a wide range of implantations with both stable and radioactive ion beams.

The high temperature ion source of Bouriant (1968) and Bouriant et al. (1969) is shown diagrammatically in fig. 4.49. For normal operation the ionisation chamber, which is a thin walled tungsten tube is also the cathode. The tube can be heated directly to a temperature of around 2700 °C using an alternating current supply and an arc is struck between the cathode and the internal anode when gas or vapour is introduced into the source (no external magnet is used). In this mode of operation very refractory elements such as hafnium, rhenium and tantalum can be directly vaporized into the discharge. Alternatively, the disposition of the cathode and the anode can be inverted (fig. 4.49(b)) in which case the cathode is indirectly heated by the thermal radiation from the heated tube anode. This arrangement is used for more volatile elements. In both cases high efficiencies can be obtained with milliampere beam intensities.

Finally by operating the source without an anode and with the heated tungsten tube simply at source potential it can be used for thermal ionisation (see section 4.3.4). Intense beams of europium, indium, potassium and lithium have been obtained in this way.

*4.3.1.4.2. Axial Extraction Sources.* We have seen above how the intense ion beam requirements of production separators have influenced the subsequent development of experimental lateral extraction heavy ion sources. In a similar way many axial extraction ion sources have been developed largely to meet the requirements of low current laboratory isotope separators of the Scandinavian type (see, for example, Nielsen, 1969; Alväger and Uhler, 1968).

Although typical heavy ion beam currents of tens of microamperes are commonly reported it should be noted that these generally refer to high resolution isotope separations, often with very small samples. For less exacting ion doping requirements considerably higher currents can frequently be obtained.

The most detailed study of this type of ion source is due to Nielsen

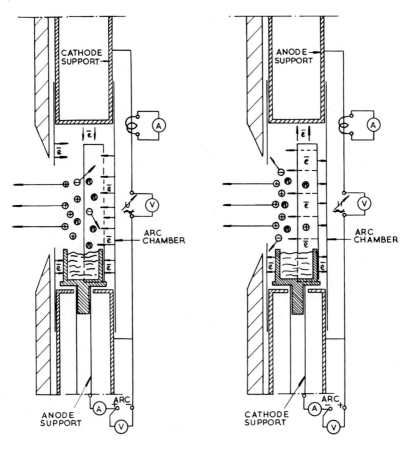

Fig. 4.49. Heavy ion discharge source (from Bouriant, 1968; Bouriant et al., 1969) (a) external cathode, (b) internal cathode.

(1957), and Almén and Nielsen (1957). Most of the subsequent development has been of a technical nature.

In this type of ion gun (fig. 4.50) the discharge is constrained axially by a magnetic field of a solenoid magnet placed concentrically around the source body. An important difference from the lateral extraction sources described above is that, in this case, the end of the discharge chamber is normally at the cathode potential and acts as an electron reflector. The ionising probability is thus considerably increased since the electrons oscillate repeatedly along the magnetic lines of force until they eventually reach the anode side walls.

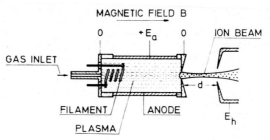

Fig. 4.50. Principle of operation of axial extraction heavy ion discharge source (from Nielsen, 1957).

The source may also be used with the end wall of the chamber at anode potential. In such a chamber the electrons do not oscillate and, according to Almén and Nielsen, in this mode of operation the heating current to the filament is higher and the anode current increases substantially. This may be useful for high temperature operation but the advantage is achieved at the expense of filament life time. As we shall see below the high temperatures required for certain source feed materials may also be obtained by the use of an additional heating element in the arc chamber.

The source assembly shown in more detail in fig. 4.51 (Nielsen, 1969) consists of three main parts; an outer housing with the magnet coil windings, a discharge chamber with thermal radiation shields, and a sublimation furnace.

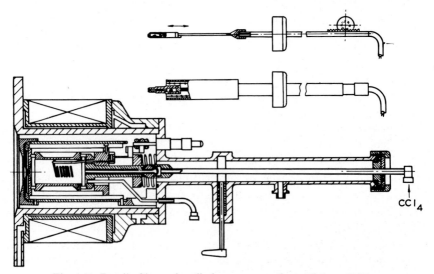

Fig. 4.51. Detail of heavy ion discharge source (from Nielsen, 1969).

Two types of furnace are used. The first of these shown in fig. 4.51(b) has a separate heating element to allow precise temperature control of the charge oven. Alternatively the thermal gradient from the filament may be used. In this case the oven assembly is simpler but some form of drive mechanism is required to locate the sample in the required position. Figure 4.52 shows details of such a furnace, and the temperature as a function of distance from the discharge chamber in a typical source of this type (Lerner, 1965).

In addition to the two types of furnace for use with volatile solids a third modification incorporating a carbon tetrachloride vapour feed can also be used for internal chlorination (see section 4.3.7.3).

In a modified design (Uhler and Alväger, 1958), two independently controlled furnaces are used and an additional heating element may be installed in the discharge chamber (fig. 4.53). Both the furnaces can be replaced rapidly through vacuum locks. This feature permits the separation of trace amounts of short lived radioactive materials. It is also convenient in implantation studies since it allows a rapid change of dopant ions.

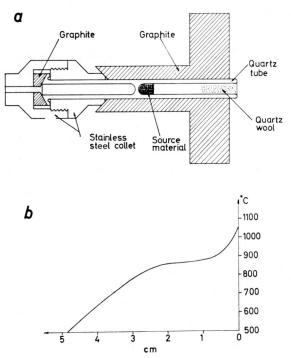

Fig. 4.52. Use of thermal gradient from the discharge chamber to control charge oven temperature (from Lerner, 1965).

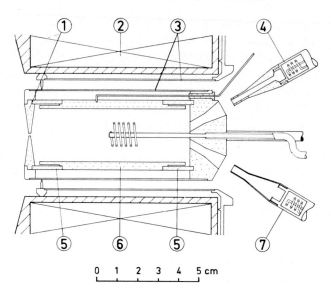

Fig. 4.53. Heavy ion discharge source with two independently controlled furnaces (from Uhler and Alväger, 1958).

In the source of Magnuson et al. (1965) the oven is eliminated and the charge container is placed directly into the discharge chamber (see fig. 4.54). In this case also the temperature is controlled by an additional heating filament inside the source body. Pure metals were used as source feed materials and ions of $Zn^+$, $Ag^+$, $Al^+$, $Fe^+$, $Cu^+$, and $Au^+$ were obtained.

Although, in this section, we are primarily concerned with sources using thermal evaporation techniques the gas source of Crowder and Penebre (1969) is worth noting because of its basic similarity to the sources described above, and because it has been used to produce ion beams of $^{11}B^+$ (from $BF_3$), $^{31}P^+$ (from $PF_5$), and $^{75}As^+$ (from $AsH_3$) for implantation studies. The total power consumption of the source (see fig. 4.55) is stated to be only 50 W. This factor and the simple compact design are attractive for restricted ion doping experiments in conventional accelerators.

It can thus be seen that the inherently simple design of the oscillating electron axial extraction source can be readily modified to meet differing experimental requirements. The technical development has arisen largely from the close collaboration between the several Scandinavian separator groups and this type of source has been extensively used for almost all elements on a very wide range of machines. The sources are well suited to conventional accelerator beam handling systems. Using the normal source

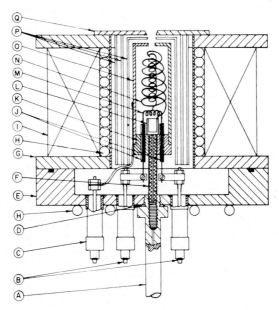

Metal ion source showing details of construction. (A) Gas feed, (B) filament feedthroughs (4), (C) thermocouple feedthroughs (2), (D) metal gasket, (E) top plate, (F) tantalum support rod, (G) vacuum casing, (H) water-cooling coil, (I) magnet coil, (J) insulating sleeves, (K) molybdenum anode base, (L) charge container, (M) emission filament, (N) heating filament, (O) molybdenum anode, (P) radiation shields, (Q) stainless steel bottom plate.

Fig. 4.54. Heavy ion discharge source utilising heating filament in discharge chamber to evaporate source feed materials (from Magnuson et al., 1965).

constructional materials (stainless steel, graphite, quartz) the maximum operating temperature is around 1100 °C. This is adequate for most requirements. Higher temperatures can be achieved by the use of more refractory materials, and additional heating elements.

An end extraction heavy ion source of the unoplasmatron type (Von Ardenne, 1956) has been described by N. J. Freeman et al., (1961), fig. 4.56. The unoplasmatron does not require a magnetic field. The discharge is constricted by the intermediate electrode and is highly concentrated at the outer anode which is also the ion beam extraction aperture of the source. The probability of ionisation is high and efficiencies (without beam analysis) of up to 20 % for argon were obtained. With an extraction gap of 5 mm, currents of 100–200 $\mu$A could be obtained at an acceleration voltage of 60 kV. With solid source feed materials however, condensation on the ex-

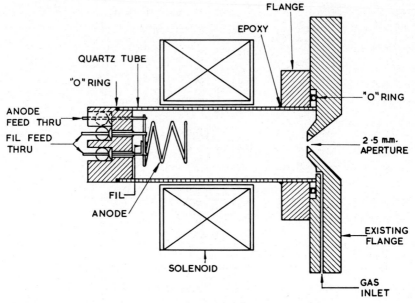

Fig. 4.55. Low power heavy ion discharge source without oven for use with gases and vapour (from Crowder and Penebre, 1969).

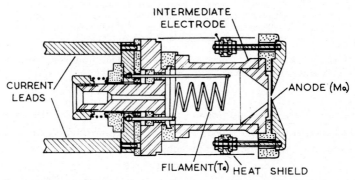

Fig. 4.56. Unoplasmatron heavy ion discharge source (from N. J. Freeman et al., 1961).

traction electrode can lead to electrical breakdown at such fields. Higher resolving powers were obtained with extraction distances of 4–6 cm.

For practical operation on an isotope separator it was stated that a source of the type described by Nielsen was preferred because of its simpler construction, higher reliability, and a reduction in insulator failure. The resolving power was also much less dependent upon the extracted beam intensity.

*4.3.1.5. Sputtering Ion Sources.*   The general principles of discharge formation in sputtering sources are usually quite similar to those described above.
Where this is the case we shall only be concerned with the design and operational features of the sources.

The sputtering of atoms from metallic cathodes as a result of ion bombardment was observed in some of the earliest 19th century experiments
with discharge tubes. In conventional ion sources, where it may contribute
significantly to filament wear and result in the metallisation of insulators,
the phenomenon is generally deleterious.

It can however be utilised to good effect in sputtering sources where it
provides a convenient and controllable means of introducing materials into
the discharge. In the simplest case the cathode can be coated with the compound or element to be ionised (Perović, 1957), but the more usual method
is to introduce a negatively biased probe into the arc chamber of a simple
inert gas ion source. Ions from the plasma are accelerated onto the surface
and, at a sufficiently high voltage, material from the probe is sputtered
(mainly as neutral atoms) into the discharge. The ejected particles have a
finite probability of being ionised before condensing on the walls of the
chamber. The beam extracted from the source under these conditions consists of a mixture of gas and metal ions which can be resolved in the usual
way. The current of metal ions can be readily controlled by varying either
the probe voltage, or the discharge parameters. In certain cases with elements
which have a high sputtering coefficient the support gas can be gradually
turned off, and the discharge maintained on the probe material alone. This
method of operation is however inherently unstable and is only rarely used.

Since sputtering probes can, in many instances, be directly fabricated
from the required metal, and since for other elements, alloys or powder com-
pacts can also be used, such sources can be simple and versatile. They can
also be used in a conventional mode with gases and compounds which have
an adequate vapour pressure at room temperature. In certain cases too the
heating effect due to the ion bombardment can be used to vaporize suitable
materials from the probe.

The problems associated with the use of corrosive materials and with
the high temperatures which may be required in thermal evaporation discharge sources are largely obviated in sputtering sources. The power requirements are quite low and it is possible to obtain beams of refractory
elements which may present serious difficulties in other systems.

Considerable care must however be taken in the design of the source
to avoid the metallisation and electrical breakdown of the insulators. After

prolonged operation the condensation of unionised sputtered materials on the walls of the discharge chamber may cause partial or total blockage of the extraction aperture. This problem may also be minimised by careful source design.

Although the threshold bombardment energy for sputtering is quite low, typically only tens of electron volts, it is usually necessary to use probe potentials of several hundred volts for practical sources. These may be necessary to obtain an adequate rate of sputtering, or to remove surface impurity layers of oxides, carbonaceous deposits, or evaporated or sputtered filament material. As the voltage is raised the bombardment current and the power dissipation in the probe surface also increase. Provision for cooling of the probe is normally required.

The efficiency of sputtering sources is usually low because of the loss, by condensation, of the unionised material. This is not generally important for experimental implantation requirements. In spite of the general use of a support gas (typically krypton or xenon) to sustain the discharge quite intense beams of sputtered ions can be obtained. For example fig. 4.57 shows a 0.5 mA Pt$^+$ beam obtained by inserting a simple uncooled sputtering probe in a conventional separator ion source (Freeman, 1969b).

The sputtering source of Hill and Nelson (1965) is of particular interest since it was designed specifically for this mode of operation and has been successfully used for a wide range of ion interaction studies in a number of accelerators. Because of its relevance to experimental implantation applications we shall consider it below in some detail.

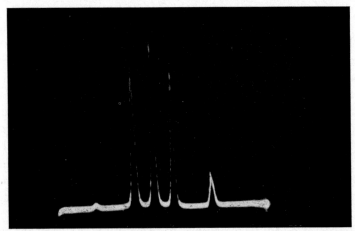

Fig. 4.57. 0.5 mA sputtered platinum beam obtained from modified heavy ion discharge source (from Freeman, 1969b).

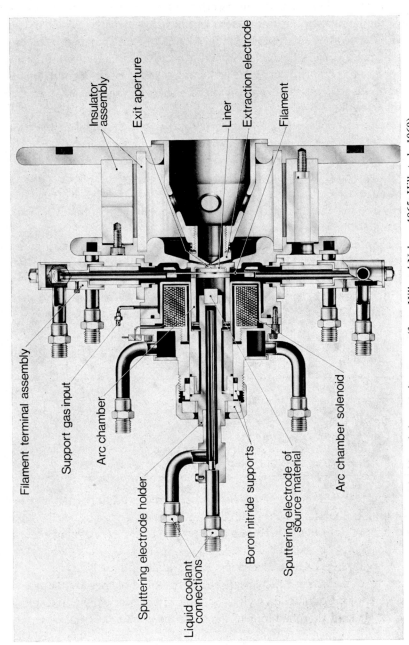

Fig. 4.58. Liquid cooled sputtering source (from Hill and Nelson, 1965; Hill et al., 1969).

In this section we shall not describe sputtering ion sources which involve the bombardment of a surface with a focused energetic primary ion beam followed by the acceleration and focusing of the sputtered secondary ions (e.g. Liebl and Herzog, 1963). Both positive and negative ion beams can be produced in this way but the intensities obtained are normally more appropriate to mass spectrometric studies than to ion implantation (see however section 4.3.5).

A liquid cooled version of the sputtering source of Hill and Nelson (1965) and Hill et al. (1969) is shown diagrammatically in fig. 4.58. The molybdenum arc chamber body has a removable tantalum insert at the front which contains the circular extraction aperture. The material to be sputtered is mounted on an insulated copper or aluminium electrode. The discharge is confined along the axis of the source by the externally applied solenoid magnetic field.

The total power consumption of the gun is about 500 W; with an extraction voltage of 20 kV a total extracted beam current of about 1 mA of positive ions has been reported.

More typical experimental values of the beam current on target after separation, collimation and focusing are given in table 4.1.

TABLE 4.1

Experimentally measured beam currents from a sputtering source (Hill, Nelson and Francis, 1969)

| Ion | Current | Ion | Current | Ion | Current |
|-----|---------|-----|---------|-----|---------|
| $H^+$ | 30 $\mu$A | $P^+$ | 10 $\mu$A | $Sb^+$ | 50 $\mu$A |
| $H_2^+$ | 100 $\mu$A | $Ar^+$ | $>$100 $\mu$A | $Xe^+$ | $>$100 $\mu$A |
| $He^+$ | 50 $\mu$A | $Ar^{++}$ | 15 $\mu$A | $Au^+$ | 25 $\mu$A |
| $B^+$ | 5 $\mu$A | $Cu^+$ | 30 $\mu$A | $Ir^+$ | 5 $\mu$A |
| $Al^+$ | $>$10 $\mu$A | $Ag^+$ | 50 $\mu$A | $Pt^+$ | 5 $\mu$A |

Figure 4.59 shows the relationship between the beam current and the sputtering probe bias for the production of a vanadium ion beam. It can be seen that in this instance there is no advantage in operating at voltages above one kilovolt.

A wide range of pure metals has been sputtered directly in the source. Beams of elements such as phosphorus have been obtained by the use of suitable alloys. Compounds such as gallium arsenide or gallium phosphide can also be used to obtain ions of the constituent materials, and in certain cases where a range of low intensity beams were required copper rods with several different inserts have been used as sputtering probes. Beams of

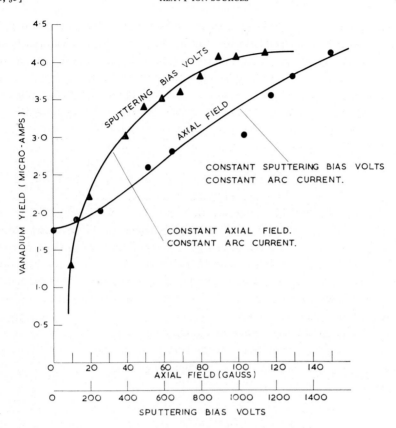

Fig. 4.59. Relationship between the vanadium beam current and probe bias in a sputtering source (from Hill et al., 1969).

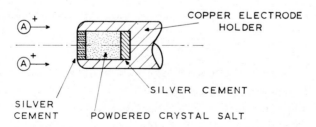

Fig. 4.60. Sputtering source modification for materials in powder form (from Hill et al., 1969).

barium, potassium and sodium ions have been obtained from the powdered halides using the arrangement shown in fig. 4.60. Finally materials such as zinc which vaporize readily have been evaporated from suitable probes by utilising the heating effect of the positive ion bombardment.

An ion source of somewhat similar configuration used both for sputtering and ion bombardment thermal evaporation has been described by Renskaya and Abramycheva (1969). The source has been used for the production and study of finely focused ($<10$ $\mu$m) heavy ion beams.

The lateral extraction sputtering source of Druaux and Bernas (1956) was constructed to overcome the problems of ionising the platinum elements and other low vapour pressure materials in an isotope separator.

A feature of particular interest is that the source shown in fig. 4.61 gives both high beam currents ($>$mA) and high efficiencies. These result from the way in which the ionising discharge is magnetically constrained to run the full length of the sputtering chamber. The use of a tube in place of the more usual probe gives a very large area for sputtering and has the advantage that material which is not ionised has a high probability of usefully recondensing on the chamber walls.

The axial sputtering source of Rautenbach (1960) shown in fig. 4.62 was designed to avoid the need for the precise temperature control for the ionisation of materials which would normally be thermally evaporated from an oven. The discharge chamber was maintained at a sufficiently high tem-

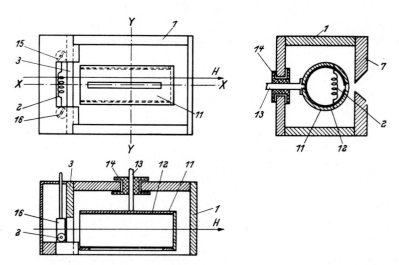

Fig. 4.61. Lateral extraction sputtering source (from Druaux and Bernas, 1956).

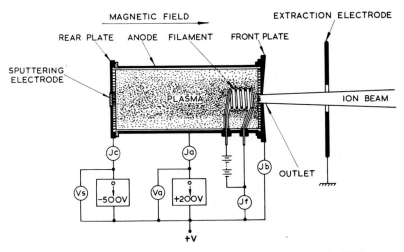

Fig. 4.62. Axial extraction sputtering source (from Rautenbach, 1960).

perature to prevent condensation whilst the sputtering probe was carefully cooled to minimise evaporation. Useful currents of lead, bismuth, cadmium and other fairly volatile elements were obtained with source efficiencies of several per cent.

The sputtering magnetron source of Perović (1957) and Cobić, Tosik and Perović (1963) is shown diagrammatically in fig. 4.63. In spite of the superficial similarity with the side extraction furnace source of Freeman

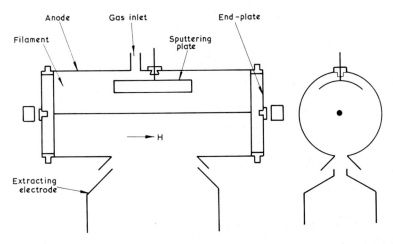

Fig. 4.63. Magnetron discharge sputtering source (from Perović, 1957; Cobić et al., 1963).

described above (4.3.1.4.1) the mode of behaviour and operation appear to be quite different. In this case the filament of the source is placed coaxially in the discharge chamber and much higher magnetic fields are used to exploit the magnetron type of discharge and to obtain a high degree of ionisation. At low operating pressures the electrons from the filament are constrained to follow spiral paths in the crossed electric and magnetic fields. Their path length and hence their ionising efficiency is further increased by reflection at the ends of the chamber, which are at cathode potential. A particular feature of this type of source is the sharp magnetron 'cut off' of the emission at a characteristic high magnetic field (see fig. 4.64). Optimum operation is stated to be obtained near to the 'cut off' condition. The self induced magnetic field of the cathode provides an additional constraint on the electron trajectories and the use of a.c. filament heating currents is recommended to increase the ionising efficiency.

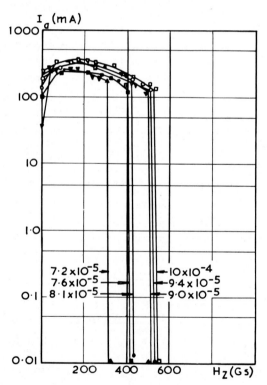

Fig. 4.64. Characteristic 'cut off' of emission in magnetron sputtering source (from Cobić et al., 1963). Anode current as a function of $H_z$ for different values of the gas pressure ($U_a = 50$ V, $I_a = 50$ A, $p_a \times$ the pressure in the vacuum chamber of the separator).

When the source is operated at higher pressures a normal ionising discharge is formed. Under these conditions the simple magnetron theory must be modified since the electrostatic field gradient can no longer exist in the plasma. Hull (1921) has shown however that in such a discharge a magnetron 'cut off' will still be exhibited and this is confirmed by the behaviour of the source.

The source has been used in a dual sputtering mode. In certain applications a biased sputtering probe is inserted into the discharge chamber of the ion source, and ion currents at the target range from 100–500 $\mu$A for a variety of metals. Alternatively the filament is coated with a suspension of the material to be ionised which is then sputtered directly by operating with a high cathode voltage ($\sim$200 V). Although this method is convenient for the ionisation of small samples the source operation is limited to a few hours because of the rapid filament wear.

It is worth noting in this respect however that significant beams of elements such as molybdenum, tantalum and tungsten can readily be obtained from *most discharge sources* by simply using the appropriate metal for the filament and operating with a high arc voltage.

Although the technique of sputtering has also been used in other types of discharge sources (for example Dawton–radiofrequency, 1956, and mercury arc, 1969; Collins–duoplasmatron, 1970) the applications have been of more limited scope than those described above.

*4.3.1.6. Hollow Cathode Heavy Ion Sources.*   Although the hollow cathode discharge has been used extensively for both spectroscopic studies as a high intensity light source, and also to produce highly ionised plasmas (Lidsky et al., 1952) it has only recently attracted serious attention as a source of ions for accelerators.

Sidenius (1969) has described two compact hollow cathode, heavy ion sources. The preliminary results which have been reported indicate that both versions are versatile and capable of playing an important role in experimental ion implantation studies. The small size of the units and the facilities for extracting and replacing the major source components through a vacuum lock are attractive for open air accelerators and may permit their use in pressurised ion source terminals. The high temperature capability (2000 °C) allows the direct ionisation of a number of refractory elements. A striking feature of the source is the current densities and efficiencies which are obtained from plasma volumes as small as 0.3 cm$^3$.

In the first version of the hollow cathode source shown in fig. 4.65 the

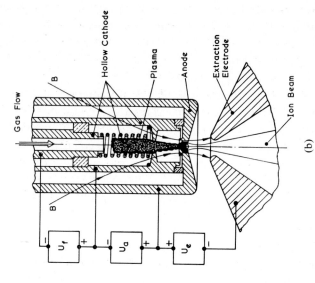

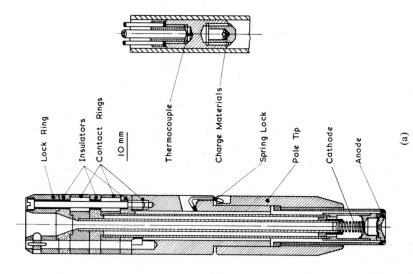

Fig. 4.65. Hollow cathode heavy ion source, version I (a) detail of construction, (b) principle of operation (from Sidenius, 1965, 1969).

ion beam is extracted from the anode region. Although in the original source (Sidenius, 1965) a cold cathode self-sustaining discharge was used, a hot filament was subsequently introduced to allow more control over the discharge parameters. The plasma is constricted before it reaches the anode by a strong inhomogeneous magnetic field and both the ions and the electrons are focused through the aperture. The ions are then extracted from the plasma boundary formed in the expansion cup below the outlet. The required field configuration is obtained by the use of the shaped pole tips shown in fig. 4.65(b). The working temperature of the source normally exceeds 2000 °C. For use with solid charge materials the oven is introduced through the gas inlet position and is slid into position in the thermal gradient (100–2000 °C) of the inner cathode tube.

With argon a total current of over 1.3 mA was obtained from a 0.4 mm diameter outlet and a 1.5 mm diameter expansion cup. The efficiency was over 40 %.

In the second version of the hollow cathode source the ion beam is extracted from the cathode. The principle of operation is shown in fig. 4.66 and a schematic illustration of the detailed construction in fig. 4.66(b). In this case too, a hot emitting filament is used to control the discharge. When a plasma is formed the full cathode voltage drop occurs across the sheath between the cylindrical electron emitter and in the absence of a magnetic field the electrons are accelerated normally to the cathode surface. They are reflected from the opposite face and are only able to move up to the anode by scattering diffusion or weak fields in the plasma. The essential feature of the source is that in the first approximation primary electrons pass through the central area of the discharge at each reflection and that a region of high plasma density will be formed on the axis, in the region from which ion extraction takes place. The weak solenoid magnetic field of the filament will deflect the electrons and assist their movement towards the anode and in certain source configurations the field is compensated by a secondary heating coil outside the cathode (fig. 4.66). The table below shows the extracted beam currents and efficiencies obtained with the source during preliminary trials.

| Ion | $H^+$ | $Ar^+$ | $Ag^+$ | $Au^+$ | $Pt^+$ | $Eu^+$ |
|---|---|---|---|---|---|---|
| Beam current $\mu A$ | 750 | 500 | 250 | 120 | 5 | 120 |
| Efficiency | 2 | 25 | 10 | 7 | – | 77 |
| Method of operation | gas | gas | oven | oven | oven | oven+$CCl_4$ (see section 4.3.6) |

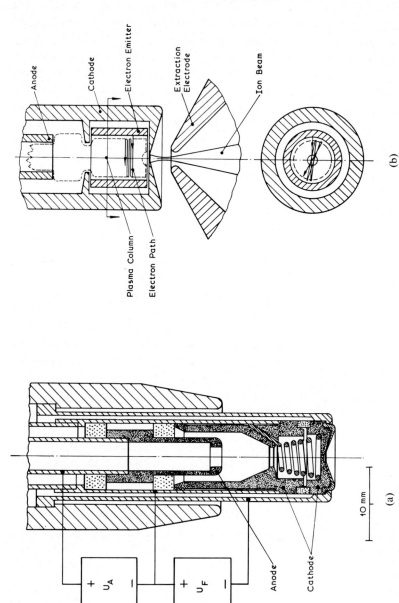

Fig. 4.66. Hollow cathode heavy ion source version II, from (a) detail of construction, (b) principle of operation (from Sidenius, 1969).

*4.3.1.7. Duoplasmatron Heavy Ion Sources.* The duoplasmatron source (Von Ardenne, 1956), shown diagrammatically in fig. 4.67, has been extensively used for many years in conventional accelerators for the production of intense ion beams of hydrogen and helium, and to a lesser extent other gases. The behaviour of the source, which also has a high efficiency and brightness as well as a long operating life, has been studied and described in detail by a number of experimenters (see for example Septier, 1967) and we will restrict ourselves to a description of the use of the duoplasmatron for heavy ion production.

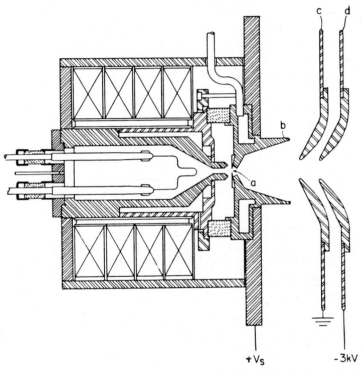

Fig. 4.67. Duoplasmatron ion source and extraction system (after Rose and Galejs, 1965). (a) anode aperture, (b) plasma expansion cone, (c) extractor electrode, (d) suppression electrode.

In spite of its many attractive features the restricted range of gas ions which can be obtained in the conventional mode of operation limits its direct applicability to ion implantation. The use of the duoplasmatron source for the ionisation of other elements presents a number of technical problems, (see for example Klapisch, 1960). This is particularly true for solid source feed materials. The technique described by Masic, Warnecke and Sautter (1969), in which the vapour of the element to be ionised is injected into the discharge expansion cup of the source appears to overcome these difficulties. Although only preliminary results have been reported it is probable that a wide range of heavy ion beams can be produced in this way.

The source shown diagrammatically in fig. 4.68 is conventional except that the anode expansion cup is lengthened and fitted with a separate vapour feed line. The duoplasmatron is operated in the normal way with a gas such as helium, and the gas or vapour to be ionised is introduced directly into the intense plasma in the expansion chamber where it is post-ionised by charge exchange with the primary gas ions and also by electrons in the plasma. The process of ionisation appears to be efficient, and intense beams ($\sim$ mA) of lithium, copper and oxygen have been obtained. For solid ion source charge materials the expansion cup is heated and the element is

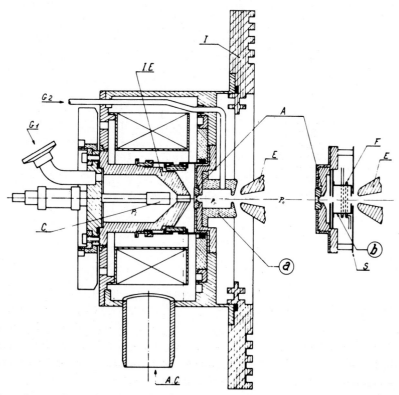

Fig. 4.68. Heavy ion duoplasmatron source with vapour feed into anode expansion cup, (b) detail of anode cup (from Masic et al., 1969).

vaporized in a separate oven. In this mode of operation the filament is protected from the corrosive or reactive vapours, and the problems of heating are significantly reduced since only the expansion cup and the oven need to operate at high temperature.

A somewhat similar principle of secondary ionisation was described by Dawton (1958, 1962) in the source shown diagrammatically in fig. 4.69. Large discharge currents were drawn from the primary low voltage discharge

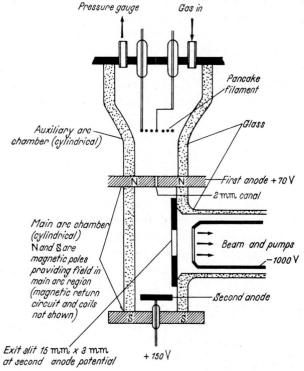

Fig. 4.69. Principle of operation of heavy ion source using secondary ionisation and separate cathode chamber (from Dawton, 1958, 1962).

in the upper chamber through the small hole in the dividing plate into the secondary plasma below. The primary discharge was maintained with a suitable gas while the element or compound to be ionised was introduced directly into the second chamber. Little filament wear was observed and beams of milliampere intensity were extracted from the aperture.

In the conventional duoplasmatron, Collins (1970) has observed that microampere ion beams of certain metallic elements can be obtained by inserting a metal insert into the intermediate electrode aperture. If the electrode is biased negatively, ions from the primary discharge sputter material from the insert into the plasma.

4.3.2. RADIOFREQUENCY DISCHARGE SOURCES

In spite of their low power requirements, and their simplicity and reliability, R.F. sources have only a restricted applicability to implantation. Although they can readily be used for the production of very intense beams of certain gases and vapours, they lack the versatility of conventional hot cathode discharge sources. With the notable exception of the brief report by Komarov et al. (see below) most workers have obtained only a limited success in the production of metallic ions from sputtering or high temperature R.F. sources, and whilst this may be a profitable field for further development, their present interest for heavy ion accelerators is somewhat marginal. It is also worth noting that the energy spread of the extracted ions from such sources (10–200 eV, see for example Valyi, 1970) is generally higher than that from a conventional thermalised discharge. This will normally result in a deterioration of the mass resolution.

A detailed comparative survey of high frequency sources has been given by Blanc and Degeilh (1961). Basically the R.F. source consists of a glass envelope with two internal electrodes, an external high frequency oscillator to excite the discharge and an aperture for beam extraction. The oscillator can be coupled inductively or capacitively. The frequency is normally in the range ∼ 10–100 megahertz with a power of a few hundred watts.

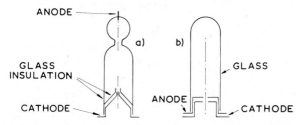

Fig. 4.70. Illustration of two R.F. source geometries (after Blanc and Degeilh, 1961).

Two simple source geometries are shown diagrammatically in fig. 4.70. The intensity of the discharge may be increased by the use of magnetic fields, and according to Gabovich (1958), for inductive coupling the field should be normal to the excitation coil whereas in the capacitive case longitudinal fields should be used.

As in the case of hot cathode sources the potential of the discharge is essentially determined by the anode voltage but in this case the ions are extracted directly from the plasma by the cathode. An extraction voltage of several kilovolts between the cathode and the anode is normally used and the cathode is generally in the form of a metal canal in an insulated tube as shown in fig. 4.71, Moak et al. (1951). This reduces the gas consumption of the source significantly. It does however also result in the neutralisation by charge-exchange of a fraction of the ion current, and the consequent formation of a directed beam of energetic neutrals from the source. The electrode geometry shown in fig. 4.70(a) was used by Thonemann (1946) in one of the earliest R.F. sources and has since been employed by a large number of workers. According to Blanc and Degeilh it gives better practical

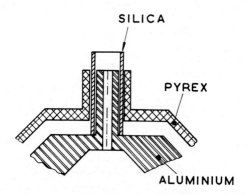

Fig. 4.71. Extraction canal in R.F. source (after Blanc and Degeilh, 1961).

results than the alternative layout of fig. 4.70(b) which was described by Ward (1950). However Valyi (1970) has remarked that since ion sputtering effects are reduced in the Ward configuration, it is preferable for sources intended to operate for long periods. Metallic layers deposited by sputtering or condensation on the glass or silica walls seriously affect R.F. source operation. They result in increased recombination on the walls and the potential distribution may be adversely affected. In extreme cases they are heated by the R.F. power and fuse into the walls. It is the formation of such layers which restrict the use of radiofrequency sources for the production of heavy ions. Dawton (1956) has described the production of silver beams by using a sputtering probe in the high frequency discharge but the life of the source was limited to a few hours. A high temperature silica source has been used by Kozlov et al. (1961) to obtain sodium and nickel ions from the vapours of the chlorides. The source has a separate heated bulb to allow the controlled evaporation of the solid feed material into the discharge. Kozlov et al (1962) have also obtained silicon and germanium beams by feeding the source with the volatile halides but the resolved beam current at the collector was only about one microampere. More recently Komarov et al. (1969) have also briefly described a heated R.F. source, shown in fig. 4.72, and state that stable milliampere intensity beams of a range of heavy ions, including $B^+$, $P^+$, $Si^+$, $Ag^+$, etc., can be extracted for periods of 20–100 hours.

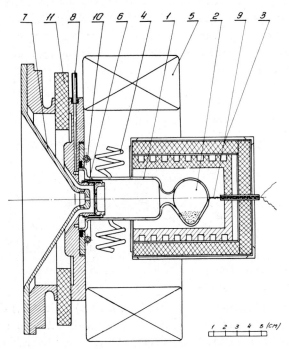

Fig. 4.72. Heated heavy ion R.F. source (from Komarov et al., 1969).

Microampere beams of $Fe^+$ have been produced by placing ferrocene (dicyclopentadienyl iron) powder in the ion source bottle (Hall et al., 1963). The compound which has a vapour pressure of approximately $10^{-2}$ Torr at 20 °C evaporates into the discharge when the source is operated. After some time iron, deposited on the walls of the vessel and the canal sleeve, degrades the focusing of the plasma boundary, and the source has to be cleaned. Although low intensity beams of a wide range of other heavy ions could probably be produced by similar techniques the use of unmodified R.F.

sources in this way for general purpose implantation is likely to remain restricted to accelerators in which other methods of ionisation cannot readily be used.

### 4.3.3. PENNING DISCHARGE SOURCES

The principle of the Penning (or Philips) mechanism of ionisation is well known and is widely used both in vacuum gauges and as a source of gas ions for high energy accelerators. The ionising discharge may be initiated by applying a high voltage between the cathode and anode, by the use of a supplementary heated filament, by indirect heating of the cathode, or by ionisation induced by cosmic rays or a radioactive source. Electrons in the discharge are constrained to oscillate between the cathodes and thus ionise the gas in the chamber. Ions bombarding the cathode produce secondary electrons which are accelerated into the discharge and in certain cases the cathodes become sufficiently hot to emit thermionic electrons. Penning ion sources can thus be relatively simple since a heated filament is not generally required and the cathode elements can be made robust to give a long life in the intense sputtering discharge. A characteristic of such sources is that they are generally used with higher discharge voltages and in stronger magnetic fields than the hot cathode sources described in section 4.2.1. The power dissipation in the high voltage discharge may reach several kilowatts and certain sources are operated in a pulsed mode to minimise the heating effects. Extraction of ions can be effected either axially or laterally from the discharge.

Although these sources are normally used for the ionisation of gases they can also be designed to operate with solid charge feed materials (see for example Arianer et al., 1969). Their relevance to implantation lies mainly in their capability of producing significant beam intensities of multiply charged ions for high energy studies or alternatively in more routine applications where the high beam currents and the robust and simple construction outweighs the somewhat restricted flexibility and the probable need for large power supplies.

We will describe only one source of this type, that of Bennett (1969), since it typifies the general features of this method of ion production and its performance has been studied in detail with a range of heavy gas ions.

The ion source layout is shown in fig. 4.73. The anode is held at ground potential and a negative voltage is applied to both cathodes. The cathodes are larger than the anode chamber diameter to maintain the intense primary discharge as close to the extraction slit as possible. In the magnetic spectrom-

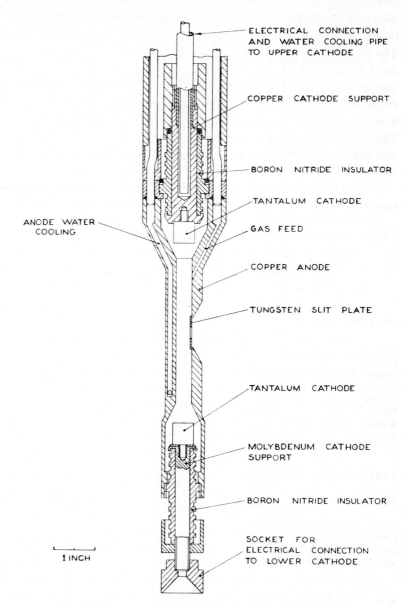

Fig. 4.73. Heavy ion Penning gas source (from Bennett, 1969).

eter used for laboratory tests, beam currents of up to 20 mA were obtained with a 20 kV extraction voltage from a slit approximately 5 mm by 1 mm in size. When the source is used in the cyclotron rather larger slits are used. The discharge is initiated with a voltage of 1.5 to 3 kV. A small current flows which gradually increases as the cathodes become heated by the discharge. To minimise the initiating voltage the gas pressure is also increased in the initial stages. Table 4.2 shows the operating discharge parameters for a series of gases and also the percentage of ions in the different charge states. The life of the tantalum or tungsten cathodes varies greatly with the gas and with the arc conditions. With hydrogen it may be several hundred hours whereas with argon it may be as low as eight hours.

Table 4.3 shows typical beam currents for a series of multiply charged ions obtained *after* acceleration in a variable energy cyclotron.

TABLE 4.2

| Gas | Arc voltage Volts | Arc current Amps | Gas flow st.cc/min | Ion | Percentage of ion current in charge state | | | | | | | |
|---|---|---|---|---|---|---|---|---|---|---|---|---|
| | | | | | 1 | 2 | 3 | 4 | 5 | 6 | 7 | 8 |
| Hydrogen | 160 | 9.5 | 1.5 | $H_1$ | 98.8 | | | | | | | |
| | | | | $H_2$ | 1.2 | | | | | | | |
| | | | | $H_3$ | 0 | | | | | | | |
| Helium | 720 | 5.5 | 0.4 | He | 85 | 15 | | | | | | |
| Carbon monoxide | 1600 | 3.5 | 0.45 | C | 11.1 | 22.9 | 7.9 | 1.2 | 0.003 | | | |
| | | | | O | 11.2 | 20.8 | 16.4 | 8.2 | 0.3 | 0.007 | | |
| Nitrogen | 730 | 8 | 0.15 | N | 15.8 | 37.0 | 37.0 | 9.6 | 0.6 | 0.006 | | |
| Oxygen | 500 | 4 | 0.5 | O | 25.7 | 35.4 | 30.1 | 8.0 | 0.8 | 0.01 | | |
| Neon | 350 | 9 | ⌣0.01 | Ne | 11.4 | 43.0 | 36.1 | 8.0 | 1.6 | 0.02 | | |
| Argon | 800 | 2 | ⌣0.01 | A | 6.4 | 16.5 | 33.0 | 33.0 | 7.9 | 2.4 | 0.6 | 0.1 |
| Xenon | 660 | 2.5 | ⌣0.01 | Xe | 0.8 | 7.1 | 20.9 | 21.0 | 19.4 | 14.6 | 11.2 | 4.9 |

TABLE 4.3

| Particle | Energy, MeV | Internal beam current $\mu$A | Extracted beam current $\mu$A |
|---|---|---|---|
| $C^{4+}$ | 118 | 10 | 5 |
| $C^{5+}$ | 175 | 0.150 | 0.08 |
| $N^{4+}$ | 98 | over 100 (Current limited by permissible dissipation of probe) | 30 (Current limited by permissible dissipation of septum) |
| $N^{5+}$ | 150 | 4 | 2 |
| $Ne^{4+}$ (3rd harmonic) | 65 | 13 | 7.5 |
| $A^{8+}$ (3rd harmonic) | 130 | 0.04 | 0.02 |
| $Kr^{5+}$ (5th harmonic) | 27 | 20 | — |

4.3.4. SURFACE IONISATION SOURCES

The phenomenon of surface ionisation, in which atoms of a metal vapour impinging on a hot metal surface may lose an electron and evaporate as ions is well known (e.g., Langmuir and Kingdom, 1923).

Under certain conditions a high degree of ionisation efficiency can be obtained and intense ion beams can be produced in this way.

Surface ionisation has been widely used in mass spectrometry because of its simplicity and convenience. In recent years it has attracted attention because of its potential use in the ion propulsion of space vehicles (e.g. Forrester and Speiser, 1959). It has also been used to a limited extent for ion bombardment and isotope separator ion sources.

A detailed treatment of the application of surface ionisation to the production of intense ion beams of a variety of elements has been given by Harrison (1958).

By considering the various factors which can influence the ionisation in a practical case he concluded that the ideal equation:

$$\frac{\nu_+}{\nu_0} = \frac{B_+(T)}{B_0(T)} \exp \left[ \frac{11600}{T} (\phi - V_1) \right] \tag{4.16}$$

(in which $\nu_+$ and $\nu_0$ are the fluxes of ions and atoms evaporating from the ionising surface; $B_+(T)$ and $B_0(T)$ the partition functions of the ions and atoms; and $\phi$ and $V_1$ are respectively the work function of the surface, and the first ionisation potential of the atom) can be replaced by the more conventional form of the Saha–Langmuir equation for surface ionisation:

$$\frac{\nu_+}{\nu_0} = \frac{\omega_+}{\omega_0} \exp \left[ \frac{11600}{T} (\phi - V_1) \right] \tag{4.17}$$

(where $\omega_+$ and $\omega_0$ are the statistical weights of the ground states of the ions and atoms).

It can be seen from eq. (4.17) that the probability of ionisation is highest when $(\phi - V_1)$ is positive and large, and that in this case it is inversely proportional to the temperature. When $(\phi - V_1)$ is negative the probability of ionisation increases with temperature.

The attraction of surface ionisation lies in its simplicity: an oven, a hot metal surface and an extraction electrode suffice for the production of quite intense beams of monoenergetic ions. The pumping requirements are modest, and since, with careful operation, using pure materials, only one ion species is produced, subsequent beam analysis is not essential. The beam current can be readily controlled over wide limits and the source does not suffer from the beam defocusing effects which are a characteristic of certain discharge sources.

The technique is however limited to elements having suitable ionisation potentials, and the more versatile surface ionisation sources must be capable of operation at very high temperatures. It is generally necessary to vaporize the element to be ionised rather than one of its compounds and there may thus be handling problems with materials which oxidise rapidly on exposure to air. The source life may be severely restricted by reaction between the vapour and the hot metal surface. An additional problem is that the intensity of the extracted ion current may be limited by electron emission arising from the condensation of the low work function charge material on the extraction system.

Nevertheless, for a number of applications this method of ion production deserves serious consideration. We shall describe in some detail four different sources; those of Harrison (1958), Daley, Perel and Vernon (1966), Wilson (1969) and Raiko, Ioffe and Zolotarev (1961) which illustrate the versatility and simplicity of the method, and the relative ease with which high intensities can be obtained. The application of the multipurpose Grenoble source (Bouriant, 1968) in a thermal ionisation mode has previously been described in section 4.3.1.4.

The source described by Harrison (1958) is particularly interesting because of the variety of elements which were ionised and also because of the increase in the efficiency of ionisation which he was able to obtain by the use of an enclosed ionising surface.

For a conventional ionising surface the efficiency $\eta_1$ can be defined as the ratio of the number of ions, to the total number of ions and atoms which leave the surface in unit time

$$\eta_1 = \frac{v_+}{v_+ + v_0} \tag{4.18}$$

By using an enclosed ionising surface with an exit slit as shown in fig. 4.74 the efficiency can be considerably enhanced in certain cases. Following Harrison's terminology the ionising chamber has a diameter of $\rho$ cm and is heated to $T\,°K$. The exit slit is $L$ cm long and has an area $\Delta$ cm$^2$. An extracting field penetrates the slit and removes all the ions formed on area $\kappa$ cm$^2$. Since the slit impedes the flow of neutrals, whilst allowing the extraction of ions the fraction of neutrals which are ionised, and hence the efficiency is increased.

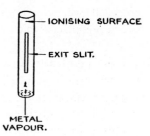

Fig. 4.74. Schematic diagram of ionising chamber of enclosed geometry surface ionisation source (from Harrison, 1958).

If it is assumed that the surface above and below the slit does not contribute significantly to the ion beam then the effective ionising area:

$$\kappa = 2L\rho - \varDelta. \tag{4.19}$$

If $\mu_0$ is the flux of neutral atoms impinging upon the surface and if the atom and ion mean free paths are large compared to the chamber dimensions:

$$\mu_0 = v_+ + v_0 \tag{4.20}$$

and

$$\eta_2 = \frac{\kappa v_+}{\kappa v_+ + \varDelta\mu_0} = \frac{1}{1 + (\varDelta/\kappa)\alpha} \tag{4.21}$$

where $\eta_2$ is the source efficiency, and

$$\alpha = 1 + \frac{v_0}{v_+} = 1 + \frac{\omega_0}{\omega_+} \exp\left[\frac{11600}{T}(V_1 - \phi)\right] \tag{4.22}$$

Thus the amplification factor $(F)$ of the enclosed surface

$$F = \frac{\eta_2}{\eta_1} = \eta_2\alpha = \frac{\alpha}{1 - (\varDelta/\kappa)\alpha} \tag{4.23}$$

It can be seen that the area of the extraction slit, $\varDelta$, should be small compared to $\kappa$; it must not however prevent the penetration of the extraction field to the walls of the chamber.

It should be stressed that for a given ionising surface, $\kappa$ cm$^2$, the amplification refers to the efficiency and not to the ion beam intensity, the gain being most important in situations where the efficiency would otherwise be low. An additional important feature of this type of source arises from the fact that the ions produced over a large ionising surface ($\approx L\pi\rho$) are finally extracted from a narrow slit.

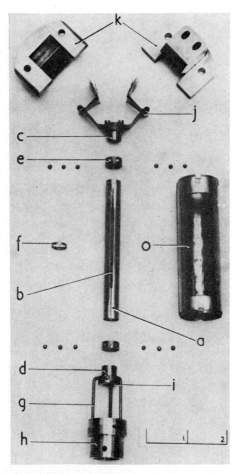

Fig. 4.75. Surface ionisation source – ionising chamber components (from Harrison, 1958). (a) Tungsten tube, (b) slit, (c) top cap, (d) bottom cap, (e) clamping band, (f) slit support, (g) legs, (h) copper connector, (i) heat shields on bottom cap, (j) laminated leaf spring, (k) copper clamps, (o) heat shields for chamber. All components made from tantalum unless otherwise stated.

Figures 4.75 and 4.76 illustrate the source used by Harrison.

Tungsten was chosen as the ionising surface and the source was designed to operate at a maximum temperature of around 2600 °C. Table 4.4 summarises the maximum beam currents and the corresponding efficiencies observed for a number of elements.

The thermal ionisation source of Daley, Perel and Vernon (1966) which was modified specifically for the production of indium ion beams for ion

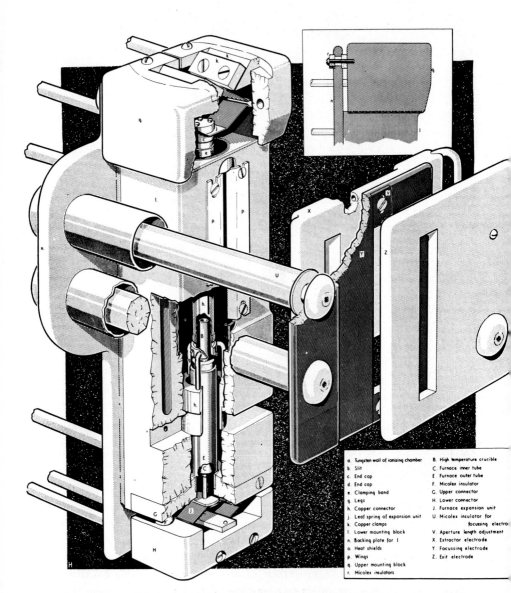

a. Tungsten wall of ionising chamber
b. Slit
c. End cap
d. End cap
e. Clamping band
g. Legs
h. Copper connector
j. Leaf spring of expansion unit
k. Copper clamps
l. Lower mounting block
n. Backing plate for l
o. Heat shields
p. Wings
q. Upper mounting block
r. Micolex insulators

B. High temperature crucible
C. Furnace inner tube
E. Furnace outer tube
F. Micolex insulator
G. Upper connector
H. Lower connector
J. Furnace expansion unit
U. Micolex insulator for
      focussing electro-
V. Aperture length adjustment
X. Extractor electrode
Y. Focussing electrode
Z. Exit electrode

Fig. 4.76. Surface ionisation source, detail of assembly and extraction system (from Harrison, 1958).

TABLE 4.4
Experimental results of a thermal ionisation source (Harrison, 1958)

| Element | K | Ba | Sr | Ca | Al | La |
|---|---|---|---|---|---|---|
| Beam current $\mu A$ | 540 | 220 | 225 | 83 | 29 | 200 |
| Efficiency % | 68.8 | 48.5 | 13.8 | 4.5 | 0.68 | 17.4 |

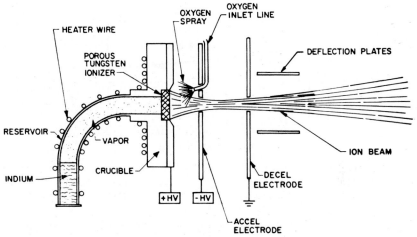

Fig. 4.77. Oxygen spray surface ionisation source (from Daley et al., 1966).

implantation experiments is shown in fig. 4.77. The source which is simple in layout illustrates a quite different approach to that of Harrison described above.

As can be seen from the schematic diagram the indium is evaporated from a reservoir and diffuses through a heated porous tungsten button, 4.6 mm diameter with a density approximately 80 % that of solid tungsten. At the front surface a substantial fraction of the indium is ionised and extracted through an accel–decel electrode system.

The work function for clean tungsten is about 4.6 eV whilst the ionisation potential for indium is 5.8 eV. Thus in this case the exponential term $(\phi - V_1)$ in eq. (4.16) would be negative and this combination of materials would be inefficient.

A particular feature of the source which overcomes this problem is an oxygen spray directed onto the ioniser surface. This increases the work function of the tungsten (see for example Eisinger, 1959), and hence both the efficiency and the ion beam intensity. Values of over 6 eV have been reported for the work function of oxygenated tungsten surfaces. Calculated indium

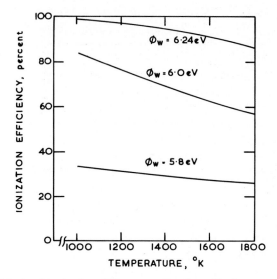

Fig. 4.78. Calculated ionization efficiency as a function of ionizer temperature for indium as an oxygenated tungsten surface. Efficiencies were calculated using the Saha–Langmuir equation for the surface work functions shown (from Daley et al., 1966).

ionisation efficiency using assumed work function values of 5.8, 6.0 and 6.2 eV are shown in fig. 4.78 for a temperature range of 1000 to 1800 °K.

Whilst most of the source studies were carried out with extracted beams of around 50 $\mu$A In$^+$ the highest beam current was 0.57 mA at an operating temperature above 1350 °C. Under normal conditions the volatility of tungsten oxide at this temperature would probably lead to a rapid reduction in the work function. The successful operation of the source is probably due to the continuous flow of oxygen over the surface.

Wilson (1967) has observed however, that the general use of oxygenated surfaces poses several problems. These include the enhancement of emission of impurity ions from the ioniser surface, the deterioration of the surface by oxygen attack and the possible deposition of a layer of the oxide of the element being ionised. He recommends the use of other materials with higher work functions than tungsten (Os 4.83 eV, Re 4.96, Ir 4.86 and Pt 5.6 to 5.8 eV).

Because of the low melting point of platinum Wilson used ionising surfaces of iridium and osmium to obtain beams of aluminium, germanium, indium and the alkali elements. The source shown diagrammatically in fig. 4.79 has a 'curvilinear geometry' to separate the ion beam from the flux of neutral atoms.

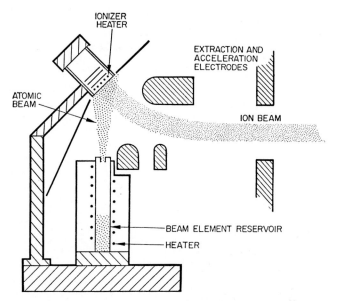

Fig. 4.79. Curvilinear geometry surface ionisation source (from Wilson, 1967).

Raiko, Ioffe and Zolotarev (1961) have used surface ionisation for the production of very high intensity beams of potassium and rubidium in an electromagnetic isotope separator. They choose nickel as the ionising surface because of its high work function (5.03 eV) and its excellent machining properties. The melting point of nickel is only 1728 °K but this is quite adequate for an ion source intended for use with the alkali elements. In the source the alkali vapour from an oven passed between a heated nickel surface and the front face of the ionisation chamber. Ions were extracted from an extraction slit 8 mm × 18 cm. The beam was further focused by using a concave ionising surface. A schematic representation of the source is shown in fig. 4.80. The front surface of the ionisation chamber was made of a lower work function material and was deliberately cooled to prevent surface ionisation from its surface.

Figure 4.81 shows the extracted beam current of potassium as a function of the nickel surface temperature and the vaporizing oven temperature.

Finally it is worth noting that the very high work functions reported for certain single crystal surfaces offer interesting possibilities for high efficiency surface ionisation if the technical problems of fabricating such sources could be overcome. According to Young and Clark (1966) the work function of the 110 face of tungsten is as high as 7 or 8 eV.

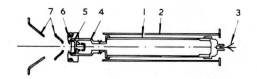

SCHEMATIC DIAGRAM OF THE SOURCE TOP VIEW
1.) CRUCIBLE:  2.) HEATER:  3.) THERMOCOUPLE
4.) VAPOR DISTRIBUTOR:  5). IONISER:  6). LID:
7.) ION-OPTICAL ELECTRODES.

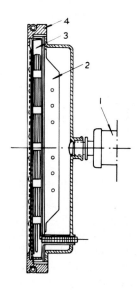

SOURCE IONIZATION UNIT
1). CRUCIBLE:  2). VAPOR
DISTRIBUTOR:  3). IONIZER:
4.) LID.

Fig. 4.80. High intensity surface ionisation source (from Raiko et al., 1961).

The sources described above are of the conventional type in which thermal ionisation is effected on a heated metal surface. As a simple alternative quite intense currents of alkali metal ions can be obtained by heating suitable chemical compounds. Although this is one of the oldest and most successful methods of producing such ion beams (e.g. Kunsman, 1925; Jones, 1933; Blewett and Jones, 1936) it has attracted only limited attention.

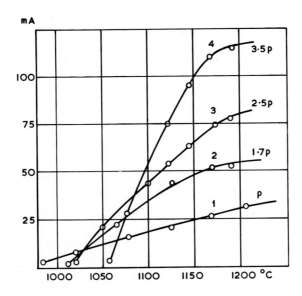

Fig. 4.81. Extracted $K^+$ beam current as a function of temperature and potassium pressure (b) in a high intensity surface ionisation source (from Raiko et al., 1961).

For lithium ions, $\beta$ eucryptite (lithium aluminium silicate), can be used. This is a naturally-occurring compound but for ion source requirements it is normally synthesised by heating alumina and silica with lithium carbonate, at around 1450 °C (Couchet, 1954)

$$Li_2CO_3 + 2SiO_2 + Al_2O_3 \rightarrow Li_2O, Al_2O_3, 2SiO_2 + CO_2.$$

The compound is subsequently melted onto a platinum support foil which is then used as the heated ion emitting surface in a simple ion gun. Using an extraction field of 9 kV Septier and Leal (1964) were able to obtain continuous operation for prolonged periods with $Li^+$ currents of several milliamperes. Mass analysis of the beam indicated that only singly charged ions of lithium were emitted and that the energy spread was very low. Only trace amounts of impurity ions ($Na^+$, $K^+$) could be detected and the flux of these rapidly dropped after a short period of operation.

Other alkali metals can be ionised in a similar way and Dawton (1956) has reported 100 $\mu A$ stable caesium beams for up to fifty hours from a caesium oxide, alumina, ferric oxide mixture.

4.3.5. NEGATIVE HEAVY ION SOURCES

Negative ions are of no particular importance per se in implantation*, and since they are more difficult to produce than positive ions, their practical application is largely restricted to injectors for tandem Van der Graaffs. The production of negative ions for such accelerators has been reviewed by Rose and Galejs (1965) and we shall confine ourselves to a brief account of negative ion sources.

Whilst most of the negative ion studies have been carried out on simple gas sources there is an increasing interest in the acceleration of a range of heavy ions to very high energies for nuclear physics experiments. The tandem accelerators for such requirements are large, but it is worth noting that compact, lower energy laboratory machines could be constructed using the same principles (Rose, 1969).

Although certain negative ions can be extracted directly from discharge sources (Lawrence et al., 1965; Moak et al., 1959; Wittkower et al., 1963), more attention has been given to collisional attachment processes in which a fraction of a beam of fast positive ions will pick up two electrons in successive collisions;

$$\overrightarrow{A^+} + M \rightarrow \overrightarrow{A} + M^+, \quad \overrightarrow{A} + M \rightarrow \overrightarrow{A^-} + M^+$$

In certain favourable cases negative ions may also be formed directly by the process:

$$\overrightarrow{A^+} + M \rightarrow \overrightarrow{A^-} + M^{++}$$

A typical collision attachment ion gun consists of a combination of a conventional positive ion source fitted with a gas target canal as illustrated in fig. 4.82. Unfortunately the attachment processes are always in competition with electron loss processes and it does not follow that a high value for the attachment cross section will necessarily lead to a correspondingly large value for the extracted negative ion beam.

The cross section is energy dependent and reaches a maximum value which may be estimated from the Massey adiabatic criterion (Massey and Burhop, 1952). The maximum is reached approximately when the velocity of the ions is comparable with the velocity of the electrons concerned in the transition. The energy dependence for the production of negative carbon and oxygen ions in a hydrogen target can be seen in fig. 4.83. For certain require-

* Although in principle negative ions could be used advantageously to minimise surface charging effects in the implantation of insulators, the technique does not appear to have been used. The build up of a significant negative potential on such a target surface would be prevented by the secondary electron emission.

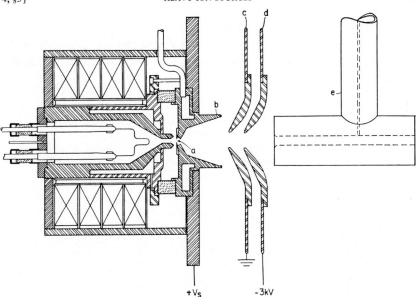

+Vs                    -3kV

Fig. 4.82. Schematic diagram of a duoplasmatron ion source, extraction system and charge-changing target: (a) anode aperture, (b) plasma expansion cone, (c) extractor electrode, (d) suppression electrode, (e) charge-changing canal (from Rose and Galejs, 1965).

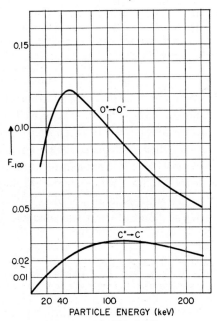

Fig. 4.83. The equilibrium fraction of negative ions of carbon and oxygen after passing through a target of hydrogen (from Rose and Galejs, 1965).

ments compound negative ions may also be used in the accelerator. Table 4.5 lists a variety of negative ions obtained by Rose and his colleagues. Both radiofrequency and duoplasmatron sources were used to provide the beam of primary ions.

TABLE 4.5

Negative ion beams other than hydrogen obtained using the HVEC collisional attachment sources (Rose et al., 1962)

| Particle | Type of positive ion source * | Source material | Attachment gas | Attachment energy (keV) | Negative ion current ($\mu$A) |
|---|---|---|---|---|---|
| $He^-$ | d.p. | He | $H_2$ | 45 | 0.3 |
| $He^-$ | d.p. | He | $H_2$ | 80 | 3.1 |
| $B^-$ | r.f. | $BF_3$ | – | 40 | <1 |
| $C^-$ | r.f. | $CO_2$ | – | 40 | 4 |
| $(NH_n)^-$ | r.f. | $NH_3$ | $H_2$ | 40 | – |
| $N^-$ | d.p. | $N_2$ | $H_2$ | | |
| $O^-$ | r.f. | $CO_2$ | $H_2$ | 40 | 25 |
| $F^-$ | r.f. | $SiF_4$ | | | 25 |
| $Si^-$ | r.f. | $SO_2$ | $NH_3$ | 40 | 2–3 |
| $Cl^-$ | r.f. | $ClF_3$ | $H_2$ | | 25 |
| | | $ClF_3, Cl_2$ | $NH_3$ | | 25 |
| $Br^-$ | r.f. | $BrF_5, Br_2$ | $H_2, A$ | | 10 |
| $I^-$ ** | r.f. | $H_2$ | I | | 2–3 |
| $U^-$ ** | d.p. | $H_2$ | $UF_6$ | 45 | – |

* d.p. – duoplasmatron; r.f. – radio frequency.
** Gentner and Hortig (1963).

The use of alkali metal vapours (particularly lithium) in the collision chamber, although technically more difficult, has proved to be a powerful technique for the production of relatively intense negative ion beams (Donnally et al., 1964; Chapman, 1969; Delauney, 1969). The method has been successfully used by Dawton (1969) in combination with his mercury pool cathode source (see section 4.3.6) for a wide range of elements. The primary positive ion beams of metals were generally produced by sputtering techniques in the source discharge. Some of the beams obtained are listed below.

TABLE 4.6

| Ion | $Be^-$ | $B^-$ | $Al^-$ | $P^-$ | $Cu^-$ | $Ga^-$ | As | $Ag^-$ | $In^-$ | $Sn^-$ | $Sb^-$ | $Ti^-$ | $Bi^-$ | $U^-$ |
|---|---|---|---|---|---|---|---|---|---|---|---|---|---|---|
| Focused ($\mu$A) beam current | 0.05 | 0.3 | 1.2 | 1.2 | 2.3 | 0.6 | 0.7 | 1.5 | 0.4 | 1.1 | 0.7 | 0.5 | 3.0 | 0.12 |

Using a similar technique on a small isotope separator (see section 4.2.1) Freeman et al. (1971) have obtained sharply focused negative ion beams of a variety of elements ($\geqslant 200\ \mu$A Te$^-$, $\geqslant 25\ \mu$A Fe, $\geqslant 50\ \mu$A Fe$^-$).

Finally an interesting and simple source utilising the secondary emission of negative ions from a surface bombarded with positive inert gas ions has been described by Müller and Hortig (1969). The source has a metal disk target which is rotated during bombardment to allow the continuous reactivation of the surface with oxygen or caesium. (If a stationary target is bombarded with gas ions the negative ion yield falls away rapidly.) A wide range of elemental and compound negative ions was obtained from the source.

### 4.3.6. OTHER TYPES OF HEAVY ION SOURCES

In the previous sections we have described a variety of heavy ion sources which satisfy a number of the requirements for ion doping experiments. For more restricted implantation other methods of ion production can be used. We shall consider some of these briefly below, and also describe one or two techniques which could be of importance in future work. We shall not however include methods such as field ionisation which normally produce very low intensity beams.

*4.3.6.1. Sources of Highly Stripped Heavy Ions.* We have previously commented on the advantages of using multiply charged ions as a simple technique to extend the energy range of an accelerator. We have also noted that, since for a given accelerating voltage such ions are more readily magnetically deflected than the singly-charged species, they present no problems in the beam analysis system. The main experimental difficulty in conventional sources arises from the reduced yield as the charge state increases and the increased possibility of doping contamination from Aston bands and impurity ion beams. This problem is partly resolved in Penning type sources (section 4.3.3) but even here the final charge state is generally modest.

In controlled thermonuclear research much higher degrees of ionisation are achieved in certain of the constrained plasma configurations which are used. The problem in many cases is one of containment of the highly stripped ions, and it seems that this may be turned to good effect by extracting the ions and using the device as an ion source.

Although the main aim of such studies is to produce very high energy heavy ions for nuclear physics research the same principle could clearly be used in more modest accelerators for ion implantation.

An ion gun of toroidal configuration known as the 'Hipac' source has been proposed by Dougherty et al. (1969). This is shown in fig. 4.84. The major radius of the torus is 20 cm and it has a circular cross section of 3 cm. High energy electrons ($\sim 20$ keV) are constrained to rotate continuously with cycloidal trajectories around a closed circular path. This gives an effective ionising path length which is very long with low power requirements.

The source is aimed at the production of very highly stripped ions (e.g. $U^{40+}$–$U^{60+}$) and it is worth noting that the design parameters would be much less restrictive for the production of say $Kr^{20+}$. Such a degree of ionisation would be more than adequate for most implantation requirements.

The source will operate in a pulsed mode and a typical operating cycle is shown in fig. 4.85. The length of the cycle depends upon the charge state which is required. The final extraction of the ions is effected in about one millisecond. Figure 4.86(a) shows the estimated ionisation times as a function of the expected charge state and the atomic number of the ionised species. Figure 4.86(b) shows the estimated number of particles per second, or mean current for the same parameters.

In a quite different linear configuration (E.B.I.S.) described by Donets et al. (1969) and shown in fig. 4.87 preliminary experiments have shown the formation of gold ions up to charge state $Au^{19+}$. The aim of this experiment

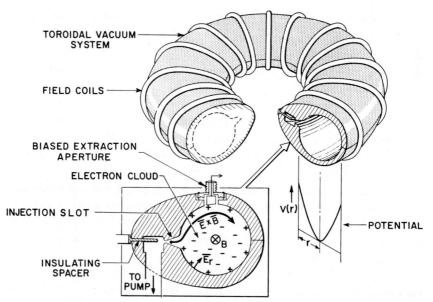

Fig. 4.84. 'Hipac' source of highly stripped ions (from Dougherty et al., 1969).

TYPICAL OPERATING CYCLE

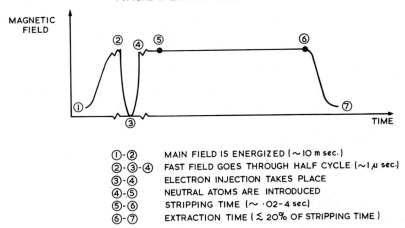

| | |
|---|---|
| ①-② | MAIN FIELD IS ENERGIZED (~10 m sec.) |
| ②-③-④ | FAST FIELD GOES THROUGH HALF CYCLE (~1 μ sec.) |
| ③-④ | ELECTRON INJECTION TAKES PLACE |
| ④-⑤ | NEUTRAL ATOMS ARE INTRODUCED |
| ⑤-⑥ | STRIPPING TIME (~ ·02-4 sec.) |
| ⑥-⑦ | EXTRACTION TIME ( ≲ 20% OF STRIPPING TIME ) |

Fig. 4.85. Typical operating cycle of a Hipac ion source (from Dougherty et al., 1969).

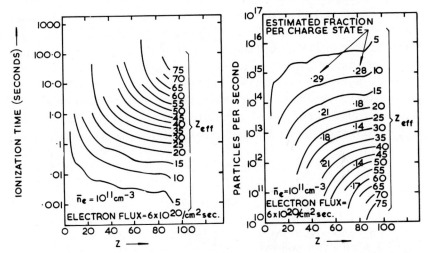

Calculated electron impact ionization times versus $Z$. The parameter $Z_{eff}$ is the number of electrons stripped from the atom. Note that the assumed electron beam carries a current density of 100 amps/cm² at 10 kV corresponding to a power density of 1 MW/cm².

Estimated average ion currents from a HIPAC Ion Source for atoms of all elements stripped to any level $Z_{eff}$.

Fig. 4.86. Estimated ionisation times (a) and mean current (b) as a function of charge state and atomic number in Hipac source (from Dougherty et al., 1969).

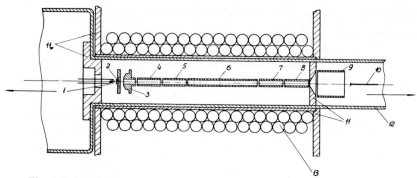

Fig. 4.87. 'E.B.I.S.' source of highly stripped ions (from Donets et al., 1969).

is to produce microampere beams of highly stripped uranium for cyclotron injection.

*4.3.6.2. Laser Sources.* The energy deposition of a finely focused laser beam on a solid sample can lead to the evaporation of the surface at the point of impact. Under certain conditions the vapour cloud is formed as a plasma and can be used as a source of ions as shown in fig. 4.88. Such sources have been described by Eloy (1969) and by Tonon and Rabeau (1969). Although both singly and multiply charged ions of a variety of elements have been produced in this way the efficiency and the ion yields are quite low. At the present stage of development the sources appear to be more appropriate to mass spectrometric and neutron production studies than ion implantation. It seems possible however that the technique might be applied with some

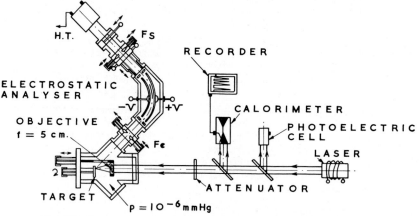

Fig. 4.88. Laser source of heavy ions (after Tonon and Rabeau, 1969).

success to the special case of the production of ion beams of difficult refractory elements such as boron.

*4.3.6.3. Vibrating Contact Arc Sources.* Spark sources and intermittent arc
sources provide a very simple means of ion production. They are of somewhat limited application in general ion doping studies because of the low
current intensities, the energy spread of the ions, and the fluctuations of the
ion beam. However their ease of construction and operation as well as their
low pumping and power requirements make them attractive for certain implantation requirements.

Sources of this type have been described by a number of experimenters
(see for example Wilson and Jamba, 1967; Venkatasubramanian and Duckworth, 1963) but we shall restrict ourselves to a brief description of the
vibrating contact source of Stroud (1969) shown in fig. 4.89 which illustrates
the general principles of the technique. The arc is struck between the vibrating electrode of the element to be ionised and a fixed pure carbon
counter electrode. The electrode separation is about 1.5 mm and the exciting
coil is driven with a 50 Hz supply. Best operation was obtained with an arc

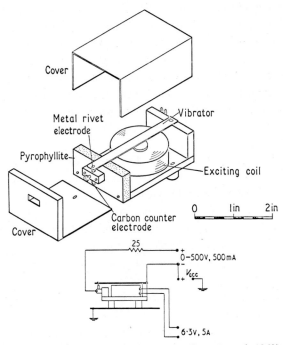

Fig. 4.89. Vibrating contact arc source (from Stroud, 1969).

voltage of 200–300 V and a mean current of 150–300 mA. Ion extraction takes place perpendicular to the arc column. The metal probe is made negative to assist the discharge by self sputtering of the cathode surface. After extraction and magnetic analysis, beam intensities of a fraction of a microampere of $CO^+$, $Al^+$, $Cu^+$ and $Au^+$ were detected. The total ion energy spread was about 90 eV. The table below shows the spectrum of ion species obtained from a variety of cathodes.

TABLE 4.7

Analysis of ion beams (Stroud, 1969)

| Cathode | $M^+$ % | $M^{2+}$ % | $M^{3+}$ % | $M_2^+$ % | $C^+$ % | $C^{2+}$ % |
|---------|---------|------------|------------|-----------|---------|------------|
| Al      | 58.5    | 19         | Trace      | 12.5      | 10      | 0          |
| Cu      | 26.2    | 18.4       | 3.2        | 39.2      | 9.3     | 3.7        |
| Cd      | 96      | 4          | 0          | Trace     | 0       | 0          |
| Au      | 77      | 15         | 0          | –         | 8       | –          |

*4.3.6.4. The Glow Discharge.* A negatively biased probe placed in a glow discharge will attract ions from the plasma. The use of this technique for ion bombardment cleaning of surfaces has attracted considerable attention and is well documented. Although very large fluxes of ions can be obtained from glow discharges the ion bombarding energy is normally ill defined. It provides nevertheless a simple technique for ion implantation. However since only a small number of elements can readily be used to produce pure discharges and since the depth of penetration will generally be severely restricted by the ion energies available the method clearly has a somewhat limited scope. Its major role is likely to be in areas such as contacting, where only superficial surface doping may be required.

*4.3.6.5. Mercury Cathode Sources.* The use of mercury pool cathodes to replace the more conventional heated filaments has been extensively used by Dawton (1969; Allen and Dawton, 1967) in a variety of both positive and negative ion source configurations. Apart from the use of the mercury primary discharge the sources are largely conventional and thermal evaporation, sputtering, and charge-exchange techniques are used to obtain a wide range of heavy ions. The advantage of the mercury pool technique is that the cathode life is virtually unlimited and such sources can run continuously for periods of several months. The cathode can be exposed during operation to air at atmospheric pressure without damage. The source operation does however present problems with certain compounds (particularly chlorides) and the mercury cathode is bulky and requires refrigeration to reduce the mass transfer of the mercury from the pool to other parts of the equipment. The use of the mercury pool in a duoplasmatron charge-exchange source is shown in fig. 4.90.

*4.3.6.6. Exploding Wire Source.* Implantation using the ions produced in the explosive evaporation of a wire has been described by Takahashi, Tuno, Oshima and Kobayisha (1968). The technique is of very limited application because of the uncertainty in the dose, the ion energies, and in the case of coated wires, the complex ion spectrum, but it deserves mention as a very simple implantation apparatus. The equipment consists of a wire of copper, gold or aluminium coated with the impurity to be implanted (Sb, Bi, etc.) stretched between disk electrodes. The specimens to be bombarded are fixed to the negative disk and the wire is exploded by discharging a 10 kV, 150 $\mu$F condenser through it. The ions produced in the explosive evaporation are accelerated onto the samples.

### 4.3.7. ION SOURCE DESIGN AND OPERATION

In the previous sections we have considered the general principles of ion beam production and described a variety of different ion sources. We have also outlined the general nature of the problems of producing heavy ions in high temperature discharges. We shall now briefly consider some of the practical features associated with the design and operation of such sources (a detailed list of recommended source feed materials for most elements is separately given in Appendix 1).

*4.3.7.1. Constructional Materials.* As we have previously commented there is no ideal universal heavy ion source. The choice of constructional materials for a particular design or application is determined by a number of considerations such as availability, machinability, and compatibility with the source feed materials. In certain cases, problems of corrosion or alloying may be best resolved by designing simple low cost replaceable components, in others it may be essential to use suitably resistant materials.

To some extent the problem of high temperature source construction has been eased in recent years by the introduction of new insulating materials such as boron nitride or silicon nitride, and also by the availability of improved quality refractory metals and graphite.

There is unfortunately no single material which is completely resistant to the very wide range of vapours which may be handled in a versatile ion source, nor is it always possible to predict the consequences of possible

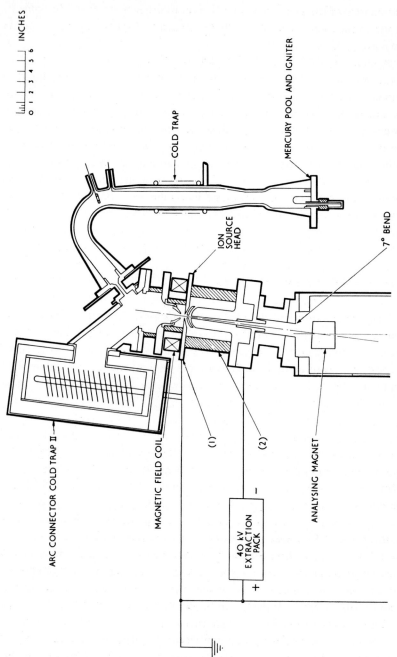

Fig. 4.90. Mercury pool heavy ion discharge source (from Dawton, 1969).

metallurgical or chemical reactions in the discharge chamber. For example tungsten appears to be suitable for the very high temperature ionisation of rare earth elements, but in contact with plutonium it readily alloys. Tantalum filaments, which are useful because of their malleability and which operate well with most elements, cannot readily be used with boron compounds because of the severe deactivation which occurs.

For many applications high density graphite is excellent because of its low cost and ease of machining. It has been extensively used in the construction of a wide variety of ion sources. It is however fragile and it outgasses badly after standing in air. This effect can be troublesome since, unlike metals, the last traces of air and moisture can only be removed at very high temperatures. A number of elements and compounds diffuse readily into the graphite surface under running conditions and this may result in significant persistent memory effect during subsequent operation with other materials. The alkali metals and the alkaline earths are particularly troublesome in this respect; the degree of diffusion may be such that on subsequent exposure to moist air the elements may oxidise and swell under the graphite surface, breaking up the structure.

The range of metals which can be used for source construction is somewhat limited. For low or medium temperature operation (approximately 1000 °C) stainless steel is often preferred, for higher temperature general purpose sources tantalum, molybdenum and tungsten are frequently used. Tantalum can be readily machined and is particularly useful for complex components. It does however embrittle readily under typical operational conditions and must subsequently be handled with care. Molybdenum and tungsten are commonly used for simpler components such as heat shields. It is fortunate that the tendency of these metals to react chemically with source feed vapours is significantly reduced in most ion sources by the formation of protective layers on the exposed surfaces. These films, which are tenacious and hard, are probably produced by a variety of effects such as surface oxidation, or the decomposition of trace amounts of hydrocarbons, or by sputtering of the filament onto the walls of the discharge chamber. If in fact metals are used in the construction it is normally worthwhile operating new ion sources for several hours with inert gases to 'passivate' the surfaces before introducing reactive vapours or gases. Figure 4.91, which is a photograph of a thick layer removed from the inside of an ion source discharge chamber, shows an extreme case of such build up.

A variety of insulating materials can also be used in ion source construction. In this respect silica, pyrophyllite and alumina are being gradually

Fig. 4.91. Illustration of layers formation on wall of an ion source discharge chamber (from Freeman, 1970a).

Fig. 4.92. Effect of vacuum heating moist boron nitride (from Freeman, 1970a).

replaced in many instances by boron nitride and more recently silicon nitride. These last materials, although more costly, have the advantage that
they can be precisely machined, they are not particularly brittle, and they
can be used over a wide temperature range with a considerable range of
reactive vapours. At very high temperatures (approximately 2000 °C) boron
nitride decomposes slowly, but the effect is negligible in most source applications. In the presence of chlorine or chloride vapours some reaction takes
place but the life of components is still quite long. (The reaction is in fact
useful for the production of low intensity boron beams.) Although boron
nitride has an adequate strength for most applications, it is somewhat
variable in quality and the surface, like talc, is slightly friable (because of
this effect contaminated or metalised insulators can be easily cleaned after
use). A particular disadvantage of the material is that it picks up moisture
and if it is subsequently outgassed rapidly at high temperatures under vacuum
the surface decomposes. Figure 4.92 shows a boron nitride component after
such maltreatment. The effect can be almost eliminated by storing boron
nitride components in a laboratory oven until required, or by preheating at
atmospheric pressure before use, and by outgassing carefully under vacuum.
Silicon nitride is much superior in this respect, it is also extremely strong
and can be machined to very fine tolerance. It has however the disadvantage
that it must be machined in the 'green' state before firing. Freeman and
Gard (1970a) have observed that when it is used at high temperatures as a
filament insulator, it tends to break down electrically after several hours
continuous source operation with certain elements.

The choice of source constructional materials is thus largely a question
of expediency, but it also depends to some extent upon the proposed application. Figure 4.93 shows two versions of the same heavy ion source (Freeman, 1969a). The very simple graphite construction is used for repetitive
operations with gases or high vapour pressure source feed materials such as
the boron, phosphorus or silicon chlorides. The much more complex metal
construction is used both for a very wide range of solid elements or compounds as well as for gases and vapours.

The filament shown in the illustration is standard tungsten rod, but
filaments of tantalum or molybdenum are also commonly used in heavy ion
sources. Coated or activated cathodes do not generally present an advantage
because of poisoning and sputtering effects in the discharge.

Filament operation may be seriously affected by activation, deactivation
or alloying with the source vapour, and the choice of filament material in a
practical case may be determined by such considerations. Elements such as

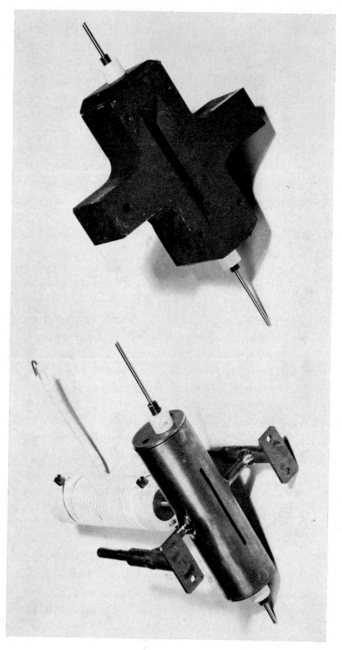

Fig. 4.93. Comparison of graphite and tantalum ion source bodies (from Freeman, 1970a).

boron are particularly troublesome, and the prolonged operation of a discharge source with a tantalum filament in boron trichloride is difficult because of the swelling and loss of emission which occurs when the vapour reacts with the hot cathode. Tungsten is much better, but even in this case if an excess of $BCl_3$ is used in the discharge a rapid deterioration of the filament takes place. The effect can be seen very clearly in fig. 4.94 (Freeman and Gard, 1970a) which shows an unused filament; a filament which has operated for thirty five hours with a steady 1 mA $B^+$ beam; and an identical filament which has only been used for a few hours in an excess of vapour. Similar effects are observed when an excess of carbon tetrachloride is used in internal chlorination source operation (see below).

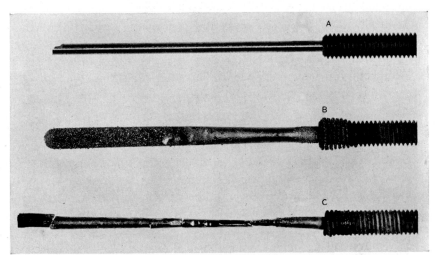

Fig. 4.94. Effect of boron trichloride on tungsten ion source filaments (a) unused filament, (b) filament after brief operation with excess $BCl_3$ in discharge chamber, (c) filament after thirty five hours continuous operation (at 1 mA $B^+$) with correct vapour pressure of $BCl_3$ (from Freeman, 1970a).

Activation of the filament which occurs with low work function materials can be advantageous since it results in lower temperature operation of the cathode and reduced wear. Unfortunately the activation effects are sometimes inconsistent and erratic. This presumably arises because the thin surface layer of contaminant can be removed by ion sputtering or evaporation and cause the emission to oscillate between two extremes. In this respect the common use of carbon tetrachloride to chlorinate such elements (particularly the lanthanides and actinides) can compensate the activation and result In much more stable source operation.

In rarer cases reaction of the source feed vapour with the hot filament to form low melting alloys may also result in premature failure of the cathode. Freeman (1970) found it advantageous to use rhenium filaments to minimise this effect when ionising thorium tetrachloride. He also used carbon cathodes successfully in certain cases but found them too fragile for routine operation (1963).

It is generally desirable to use thick filaments to obtain long periods of continuous operation. The 2 mm diameter cathodes shown in fig. 4.94 normally run for several tens of hours and even larger filaments ($\geqslant 4$ mm diam.) are used in production separators. It must be noted however that the use of such thick wires or rods may be difficult in the case of complex cathode shapes and it inevitably leads to increased heating current requirements. Additionally if the total electron emitting area is large it may be impracticable to operate the filament in a space-charge limited regime (see section 4.3.1.1). Whilst temperature limited electron emission (eq. (4.1)) provides a useful control parameter for the discharge it requires stable current supplies, and is particularly sensitive to filament activation effects.

*4.3.7.2. Source Feed Materials for High Temperature Operation.* Only a limited number of elements are themselves sufficiently volatile, or have compounds with an adequate vapour pressure, for direct use in a low temperature gas source. The use of sputtering techniques described in section 4.3.1.5 overcomes the problem of low vapour pressure with many materials, particularly metals. The alternative which is widely used is to evaporate the element, or a suitable compound, from an oven into the heated discharge chamber.

A convenient temperature range for operation is 500–1000 °C. Lower temperatures may present difficulties because of the heating effect of the discharge and the probable need for rapid and precise control. Higher temperatures are in fact often used but they necessitate higher power dissipation and the use of refractory materials. It is fortuitous that with the notable exception of the precious metals, most elements, or suitable compounds, can be ionised in this range (see Appendix 1). Compounds are generally used for elements with a low pressure but in certain cases they may also be useful for volatile elements which present difficulties at the lowest convenient oven temperature. For example Hg as $Hg_2F_2$, As as GaAs, Zn and S as ZnS. Alternatively particularly volatile materials can frequently be controllably evaporated by using a crucible with a restrictive orifice. This technique has the advantage that the whole source is operated at a relatively high temperature and condensation of vapour on cold spots is prevented. This particular problem of temperature variations is important in the source design. Care must be taken both to ensure that the arc chamber is always hotter over its whole surface than the furnace, and also that the oven is uniformly heated. It is also necessary to ensure that the design of both oven and discharge chamber heaters is such that their induced magnetic fields do not interact with the ion source discharge (see for example, Freeman et al., 1961).

An additional important requirement is that the temperature control should be rapid and precise. The need for temperature control is less important in evaporation sources which make use of the thermal gradient, from the hot discharge, along a tube, (4.3.1), but in this case precise and controllable positioning of the sample must be used. This method of operation may present difficulties with elements which activate the filament since the decrease in cathode power requirements will result in a reduction of the temperature gradient along the tube.

The most commonly used source feed compounds are the halides, and in particular the chlorides. Less frequently, volatile oxides or sulphides may be used and in certain cases thermally unstable alloys or compounds may be employed (e.g. I from AgI, Zn from brass).

Because of their particular importance we shall consider the chlorides in some detail. It is usually essential to use completely anhydrous chlorides for source operation. Even small traces of water will prolong the source outgassing and in many cases will result in the conversion of some of the chloride to the involatile oxychloride or oxide on heating in vacuum. The preparation of anhydrous monovalent and divalent chlorides is generally simple and most of them are readily available. The polyvalent chlorides present a more serious problem because of the difficulties of chemical synthesis and because many of them are extremely hygroscopic. Simple preparative techniques such as the vacuum or low temperature dehydration of the hydrated chloride, hydrochlorination of the oxides, or hydrated chloride (Kleinheksel and Kremers, 1928), heating the oxide with ammonium chloride (Reed et al., 1935) rarely result in pure products. In certain cases the direct chlorination of the elements is useful, and Freeman and Smith (1958) have reviewed a number of preparative methods and described the successful preparation of a wide range of anhydrous chlorides by dehydration of the hydrated chlorides (or chloride solutions) by refluxing with thionyl chloride. The preparation of $PuCl_3$ by hydrochlorination or phosgene treatment of the carbonate or oxalate has been described by Boreham et al. (1960). Boreham and Freeman (1960) have also used the phosgene technique extensively with the lanthanide elements.

In all of these preparations the danger of subsequent hydration can be minimised by fusing the anhydrous compound to reduce its surface area.

*4.3.7.3. Chemical Synthesis of Source Feed Materials in the Ion Source.* The direct preparation of volatile compounds by chemical reaction in the ion source itself is a powerful technique which largely overcomes many of the difficult chemical problems outlined above.

By far the most useful example is the chlorination of heated oxides using carbon tetrachloride vapour first described by Sidenius and Skilbreid (1961). The method was devised for the rapid handling of small radioactive samples but has subsequently been successfully used for a wide range of elements (see Appendix 1) on a very large scale. Kilograms of oxides are commonly treated in this way in high output production separator ion sources. The vapour is corrosive and care must be taken to prevent the erosion of the gas-feed line and metal source components. The large tungsten beam in fig. 4.95 is due to the chlorination of the hot filament by the vapour. To ensure efficient reaction it is normally essential to provide an impedance between the oven and the discharge to maintain an adequate pressure of vapour over the oxide surface. Care must also be taken to prevent volatile chloride diffusing back along the vapour feed line and condensing.

This very simple method of chlorination has a number of very important advantages:

(a) Since the reaction is normally carried out at a relatively high temperature, to ensure the rapid removal of the reaction products from the solid-gas interface, precise temperature control is not required.

(b) The extracted ion current can be readily controlled by adjusting the flow of $CCl_4$ vapour.

(c) The $CCl_4$ tends to suppress uncontrollable filament activation with low work function elements. It also reduces the electron back streaming and sparking from the extraction electrodes which often occurs with such elements.

(d) The method can be used with the oxides of elements such as molybdenum or zirconium, whose chlorides are particularly difficult to prepare or handle by conventional means.

There are also certain problems which may militate against the technique in certain instances. As we have remarked previously the use of an excess of carbon tetrachloride may seriously deactivate the filament. With some oxides there is an induction period before reaction occurs. This may be particularly long in the case of certain difficult elements such as zirconium,

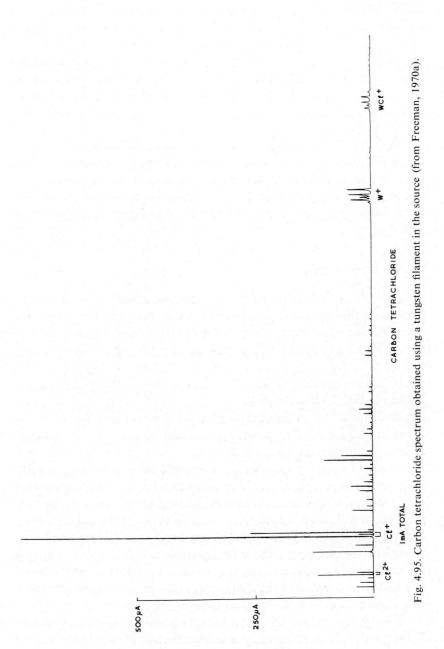

Fig. 4.95. Carbon tetrachloride spectrum obtained using a tungsten filament in the source (from Freeman, 1970a).

gadolinium or uranium, or with materials which have been prepared at high temperature. Finally the ionisation of the carbon tetrachloride itself may result in both a severe current drain on the ion source and in the production of a very complex spectrum of impurity beams which may lead to undesirable doping effects. Figure 4.95 shows the resolved spectrum obtained using only carbon tetrachloride vapour in the discharge. When chlorination occurs there are additional peaks from the element and its halides and from oxygenated carbon ions.

Although the carbon tetrachloride method is undoubtedly the most useful for heavy ion production other chemical conversions can also be carried out in the source. For example the volatile oxides of tungsten and osmium can be prepared by passing oxygen over the heated metal, and chlorine can be used directly to produce volatile chlorides of a range of metals (Freeman, 1970a).

### 4.4. Ion beam analysis

The primary role of the analysing element of a heavy ion accelerator is to separate the required dopant beams from associated impurity ions. The degree of mass discrimination required depends upon the particular application. Whereas a very elementary analyser may be adequate in low mass doping with, for example, boron ions, the problem may be much more severe in more versatile machines in which it is necessary to resolve adjacent beams of much heavier elements.

In addition to providing the essential dispersion or mass separation the use of an analyser may have other important consequences in both the design and the performance of the accelerator. As we have remarked elsewhere, the separation of impurity species must inevitably be accompanied by some measure of isotope separation. If the resolution is low this may only manifest itself for polyisotopic elements as a broadening of the image in the focal plane but it will also frequently result in a reduction in the utilisable beam current. Magnetic analysers exhibit strong focusing behaviour which may be exploited in suitable geometric arrangements to reduce or even eliminate the need for additional beam handling lenses. The quality of field stabilisation and the ion optics of the analyser can also play an important role in determining the overall resolution of the accelerator.

Finally, it is worth noting that in analysis systems involving beam deflection, a useful separation of the ions from the flux of fast directed neutral atoms usually formed in the source region is also obtained.

Since both the theory and the techniques of charged particle analysis are exhaustively covered in standard works on mass spectrometers and accelerators we shall restrict ourselves to a quite generalised treatment of the subject. In particular, only those mass separation techniques which have proved to be of practical consequence in ion doping will be considered.

### 4.4.1. THE HOMOGENEOUS FIELD MAGNETIC ANALYSER

An accelerated charged particle travelling through a homogeneous magnetic field is constrained to a circular trajectory whose radius is given by the well known expression:

$$R = \frac{143.95}{H} \sqrt{\frac{MV}{n}} \tag{4.24}$$

where $M$ is the ion mass (a.m.u.), $V$ is the accelerating voltage (volts), $R$ is the radius of the ion trajectory (cm), $H$ is the magnetic field (gauss), $n$ is the charge state of the ion.

Thus particles of the same charge state and energy but of differing mass originating from a common source will follow paths of different radii and can be separated as shown in fig. 4.96(a). (More strictly the magnetic field is a momentum filter which acts as a mass separator under the conditions of constant ion acceleration which exist in conventional accelerators.)

The magnetic analyser also acts as a first order lens on the divergent beams and in the simple case illustrated the ions are brought to a focus after a deflection of 180°. It can readily be seen from eq. (4.24) that at the focal plane the dispersion for unit mass separation is given by:

$$D_{\mathrm{m}} \approx \frac{R}{M} \tag{4.25}$$

It can also be seen from the geometric construction of fig. 4.96 (b) that the images would have a finite width, $W_\alpha$, where:

$$W_\alpha = 2R(1 - \cos \alpha) = R\alpha^2 \text{ (for small } \alpha) \tag{4.26}$$

and $\alpha$ is the half angle of beam divergence from the ion source in radians.

Thus even in the idealised case in which no other factors contributed to the quality of focusing a separation of adjacent masses in such an uncorrected analyser would only be effected if $W_\alpha \leqslant D_{\mathrm{m}}$, (i.e. $\alpha^2 \leqslant 1/M$). This requires that for high mass operation ($M > 200$) the beam divergence should be less than about $\pm 4°$.

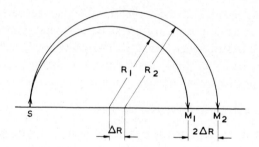

a. DISPERSIVE PROPERTY OF HOMOGEN−
EOUS MAGNET.

(DISPERSION, $D_M = 2\Delta R = \dfrac{\Delta M}{M} R$ )

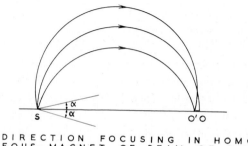

b. DIRECTION FOCUSING IN HOMOGEN−
EOUS MAGNET OF BEAM OF DIVER−
GENCE $\pm\ \alpha$.

(FOCUSED BEAM WIDTH, $W_\alpha = OO' =$
$2 R\ [\ 1 - \cos\alpha ]$)

Fig. 4.96. Dispersive and direction focusing inside a homogenous magnet (ion source and focal plane in the magnetic field).

In a more practical situation the total image width would be made up of a number of contributions and a smaller divergence would be required to obtain an adequate resolution. The magnetic field configuration, both in this elementary type of analyser and in the more common sector field magnets that we shall consider shortly, can be corrected to eliminate such aberrations. However this solution is normally only applied in instruments requiring a high mass resolution because of the difficulty of applying the idealised theoretical treatment to real analysers.

In the context of ion implantation facilities it is usually sufficient to note that a low divergence is desirable since it reduces the image size, improves the resolution, and minimises the ion beam volume in the analysing region.

It can readily be seen from the simple geometry of the analyser considered above that the magnification of the system is unity and that the final
object width will thus include a contribution equal to the size of the ion
source object. Although this may in some cases approximate to the width
of the extraction aperture of the ion source, we have seen in section 4.3
that beam extraction from a plasma can result in the formation of virtual
objects which may in certain circumstances be either smaller or larger than
the source aperture. For high mass separation this contribution to the image
width, which is independent of the deflecting radius, may be important in
determining the required dimensions of a suitable analyser.

Additional image broadening results from instabilities in the power supplies. Differentiation of eq. (4.24) gives:

$$\frac{2\,\mathrm{d}R}{R} = \frac{\mathrm{d}M}{M} + \frac{\mathrm{d}V}{V} - \frac{2\,\mathrm{d}H}{H} \tag{4.27}$$

and thus, for ions of constant mass, the total image width $W_T{}^*$ (excluding
broadening due to beam scattering and space charge effects) is given by:

$$W_T = R\frac{\mathrm{d}V}{V} + 2R\frac{\mathrm{d}H}{H} + R\alpha^2 + S \tag{4.28}$$

where $S$ is the width of the virtual source.

Equation (4.28) permits the minimum stabilisation requirements for the
high voltage and magnet supplies to be calculated in terms of both the image
size and the mass resolution. For example, in the case of an analyser of
radius 50 cm, with power supply stabilities of $1:10^3$, a beam divergence of
$\pm 2°$, and an assumed object size of 2 mm, the focused beam would be
approximately 4 mm wide. It would be possible in such a system to resolve
adjacent beams to around mass 100.

For high resolution work better power supply stabilities ($\sim 1:10^4$) are
normally used and care is taken to minimise the virtual object size.

Although the elementary 180° analyser configuration which we have
considered above is only rarely employed in ion implantation we have described it in some detail. This is because its dispersive and focusing behaviour
can be readily deduced from simple geometric considerations. Also, as we
shall see below, the expressions which we have derived are applicable to the
important case of the symmetric sector analyser with normal entrance and
exit of the beam to the magnet pole edges.

---

* The negative term in eq. (4.27) simply indicates the direction of the broadening effects.
$W_T$ is obtained by summing the numerical values of all the terms.

In the more general case of the homogeneous sector magnet in which both the object and image are situated outside the magnet the analytical treatment is more involved and we shall simply state the results following Herzog (1954).

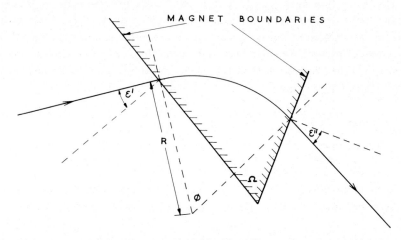

Fig. 4.97. Homogeneous sector magnetic field with oblique entry and exit of ion beam. (In the illustration $\varepsilon'$ is negative and $\varepsilon''$ is positive.)

The first order direction focusing behaviour of any sectorial magnet field in which the boundaries are inclined to the ion trajectory as shown in fig. 4.97 is described by Herzog's general lens equations:

$$f^2 = (l'-g')(l''-g'') \tag{4.29}$$

and

$$\frac{1}{f} = \frac{1}{(l'-h')} + \frac{1}{(l''-h'')} \tag{4.30}$$

where:

$$f = \frac{R \cos \varepsilon' \cos \varepsilon''}{\sin (\phi - \varepsilon' - \varepsilon'')} = \text{focal length}$$

$l'$ and $l''$ are the object and image distances respectively from the entrance and exit field boundaries.

$$g' = \frac{R \cos \varepsilon' \cos (\phi - \varepsilon'')}{\sin (\phi - \varepsilon' - \varepsilon'')} = \text{first focal point measured to the field boundary, object at infinity}$$

$$g'' = \frac{R \cos \varepsilon'' \cos (\phi - \varepsilon')}{\sin (\phi - \varepsilon' - \varepsilon'')} = \begin{array}{l} \text{second focal point measured to the field} \\ \text{boundary, image at infinity} \end{array}$$

$$h' = \frac{R[\cos \varepsilon' \cos (\phi - \varepsilon'') - \cos \varepsilon' \cos \varepsilon'']}{\sin (\phi - \varepsilon' - \varepsilon'')}$$

$$h'' = \frac{R[\cos \varepsilon'' \cos (\phi - \varepsilon') - \cos \varepsilon' \cos \varepsilon'']}{\sin (\phi - \varepsilon' - \varepsilon'')}$$

$h'$ and $h''$ are the principal planes (planes of unit magnification) of the lens. In the diagram shown (fig. 4.97) $\varepsilon''$ is positive and $\varepsilon'$ is negative.

The close analogy of eqs. (4.29) and (4.30) with the conventional optical behaviour of a thick lens is apparent.

Herzog also showed that the lateral image shift $b''$ in terms of the lateral object shift $b'$ is given by:

$$b'' = b' \frac{(g'' - l'')}{f} + \delta[a(1 - \cos \phi) + l''\{\sin \phi + \tan \varepsilon''(1 - \cos \phi)\}]$$

$$(4.31)$$

$\delta = \beta + \gamma$ where the incident velocity of the ions is $v = v_0(1 + \beta)$ and the ions have mass $M = M_0(1 + \gamma)$.

Both the magnification $G = b''/b'$, $(\delta = 0)$ and the dispersion coefficient $D$, $(b' = 0)$ of the magnetic lens can be calculated from eq. (4.31).

In spite of the simplicity of the general lens equations (4.29) and (4.30) their evaluation is numerically tedious. For the rapid evaluation and comparison of magnetic analyser configurations Cartan's (1937) graphical construction which provides a quite general method of locating the image position is useful. This is illustrated in fig. 4.98.

In many analysers the ion beams enter and leave the magnetic field normal to the boundaries as shown in fig. 4.99. In this case the focusing behaviour is much simpler (Bainbridge, 1953; Kerwin, 1963) and is described by Barber's rule which states that $\lambda + \theta + \gamma = \pi$, that is to say, the source, magnet apex and image lie on a straight line. The magnification, $G$, is given by:

$$G = \frac{W}{S} \left( = \frac{b''}{b'} \right) = \frac{q}{p}$$

where the image width $W$, $(b'')$, and the source width $S$, $(b')$, are measured perpendicularly to the central beam.

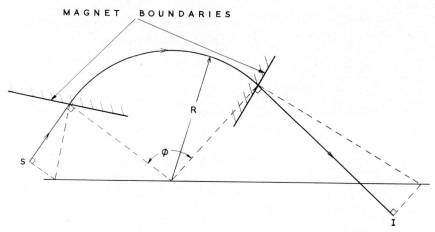

Fig. 4.98. Cartan's general graphical solution for the location of the image position $I$ for any homogeneous sector magnetic analyser when the source position and the radius of deflection are known.

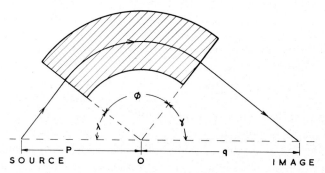

Fig. 4.99. Barber's focusing rule $\phi + \lambda + \gamma = \pi$ for ions which enter and leave homogeneous sector field normally to the magnet boundary.

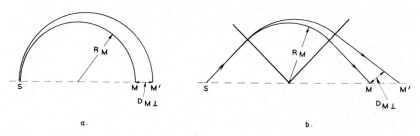

Fig. 4.100. Comparison of dispersion in a 180° (a) and a sector (b) analyser. The separation of the masses $M$ and $M'$ measured perpendicularly to $M$ ($D_{M\perp}$) is independent of the deflection angle for the same radius of deflection ($R_M$).

The dispersion, measured perpendicularly to the central beam, for unit mass separation, is given by:

$$D_{m\perp} \approx \frac{R}{2M} \left(1 + \frac{q}{p}\right)$$

and the image broadening, $W$, due to the beam divergence $(2\alpha)$ is:

$$W_{\alpha} = \frac{\alpha^2}{2} R \left(\frac{p^2}{q^2} + \frac{q}{p}\right)$$

It can readily be seen that in the case of a symmetric layout ($p=q$ and $\lambda=\gamma$) the expressions are identical with those derived earlier for the elementary 180° analyser. Thus the general focusing and dispersion of such systems are independent of the angle of deflection in the analyser. This is illustrated in fig. 4.100.

### 4.4.2. FRINGING FIELD EFFECTS

In the general treatment of beam analysis in homogeneous magnets we have so far been solely concerned with the ion optics in the median plane parallel to the pole faces ($\alpha$ focusing) and we have ignored the effects of the fringing field of the magnet. In any real analyser these cannot be neglected and they may in extreme cases play an important role in determining the beam transmission and the quality of the image. The two main consequences of the fringing field are to extend the effective volume of the real magnet and also to produce vertical, ($\beta$ or Z), focusing of charged particles which do not enter or leave normally to the pole boundaries. Whilst the first effect is minimised by the use of magnetic shields which restrict the fringing field the $\beta$ focusing is usually enhanced under these conditions. The shape and extent of the fringing field which depend upon the design parameters of a particular magnet also affect the second order aberrations which contribute to the image broadening.

Detailed accounts of fringing field focusing effects have been given by Wollnick (1967), Enge (1964), Kerwin (1963) and Bainbridge (1953) and we shall simply outline the main features.

Because of the curvature of the fringing field (see fig. 4.101) ions travelling in a plane parallel to the median plane cross the field lines at an angle. If additionally the ion trajectories are inclined at an angle $\varepsilon$ to the magnet boundary they experience a vertical focusing force which arises from the interaction of the component of the beam direction parallel to the magnet boundary and the horizontal component of the magnetic field. The focusing

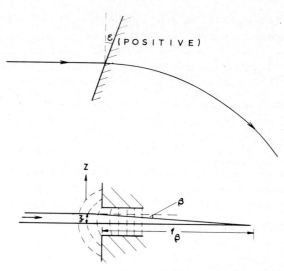

Fig. 4.101. Vertical ($\beta$) focusing when ion trajectories are inclined at an angle $\varepsilon$ to the magnet boundary ($\varepsilon$ positive as in illustration results in convergent focusing).

effect is positive (i.e. converging), for $\varepsilon$ positive as shown in fig. 4.101. The focal length $f_\beta$ in the vertical (or $Z$) plane is given by (Cotte, 1938):

$$f_\beta = \frac{R}{\tan \varepsilon}$$

and the particle is deflected through an angle $\beta$ where:

$$\beta = z \frac{\tan \varepsilon}{R}$$

and $z$ is the vertical displacement of the particle from the median plane.

In all situations involving non-normal beam trajectories at the analyser boundaries the effect of the vertical focusing on the final image height must be taken into consideration. By appropriately choosing the boundary conditions of the magnetic analyser this effect may sometimes be exploited to obtain stigmatic focusing and thus reduce the need for additional beam handling lenses in implantation applications in which a small image is required. A number of stigmatic focusing systems have been described by Cross (1951). It is generally found however that the full predicted vertical focusing is only obtained in real analysers when magnetic shielding is used to define the fringing field (see for example, Andersson et al., 1964). Freeman (1970b) has also shown that even in this case the quality of two directional focusing may be limited by the overall variation in the direction of incidence, $\varepsilon$, arising from the beam divergence $\pm\alpha$.

To a large extent the effects of fringing field focusing in the plane parallel to the pole faces ($\alpha$ focusing) can be minimised by the use of suitably positioned magnet clamps or shields. Under these conditions the theoretical focusing behaviour which was derived in section 4.4.1 and which assumes that the effective field boundary coincides with the physical magnet provides a reasonably valid approximation for practical ion implantation systems. When, as is often the case, such shields are not employed, the focusing behaviour is more complex. As a first approximation it can simply be assumed that the effect of the fringing field is to extend the magnetic induction to a virtual field boundary located about one pole gap width or less from the edges of the analyser. Calculations based on this premise of a sharp cut-off fringing field are commonly referred to as SCOFF. In more rigorous treatments the focusing effect of the extended fringing field (EFF)

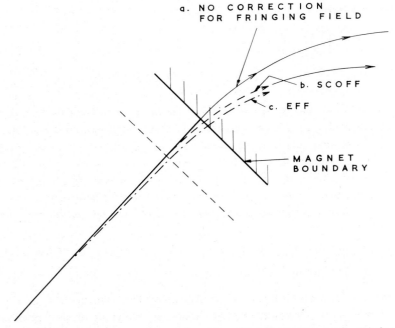

Fig. 4.102. Simplified representation of charged particle trajectories assuming (a) that the physical magnet boundary coincides with the field boundary, (b) that the effect of the fringing field is to extend the magnetic induction to a sharp cut off virtual field boundary located outside the analyser (SCOFF treatment) and (c) the normal extended fringing field distribution (EFF treatment).

is considered (see for example Kerwin, 1963; Bainbridge, 1953). A detailed comparison of the two methods has been given by Enge (1964). A diagrammatic representation of the two field distributions and the approximate charged particle trajectories which result from them is shown in fig. 4.102. The most important point to note is that for both fringing field distributions in a simple symmetric analyser the beam trajectory is significantly displaced from the corresponding position calculated for a magnet whose field terminates sharply at the physical boundary. The differences between the SCOFF and the EFF trajectories although less marked are nevertheless significant.

### 4.4.3. INHOMOGENEOUS FIELD MAGNETIC ANALYSERS

The homogeneous field momentum analyser considered in detail in section 4.4.1 is commonly employed. It is simple to design and construct and has adequate dispersive properties for most applications. It can however be considered as the limiting case ($n = 0$) for the more general class of cylindrically symmetric analysers whose magnetic field is given by:

$$H = H_0 \left( \frac{R_0}{R} \right)^n$$

where $H$ is the field at radius $R$ and where the field gradient is $n$ ($0 \leqq n \leqq 1$). $R_0$ is the mean radius of deflection of mass $M_0$ and satisfies the general equation (4.24). The general focusing properties of such magnets have been described by Judd (1950) and Siegbahn (1965), and several high resolution sector mass analysers utilising inhomogeneous fields have been described (see for example, Bernas et al., 1960; Viehbock, 1960; Fabricius et al., 1965).

The most important feature of such analysers is that for a given mean radius, $R_0$, the dispersion is increased. For unit mass difference the separation of adjacent beams (measured perpendicular to the central beam) is given by:

$$D_\perp \approx \frac{R_0}{2M} \cdot \frac{1+G}{1-n}$$

where $G$ is the magnification (for symmetric focusing arrangements $G = 1$).

Without considering the detailed focusing behaviour of such inhomogeneous magnets it can be seen qualitatively that the improvement in dispersion arises from the effect of the field gradient on the trajectories of ions of different mass. Heavy ions which travel through the weaker field region are deflected less than in the equivalent homogeneous analyser ($H_R = H_0$), whilst the light ions which traverse the stronger field are more strongly bent.

An additional feature of inhomogeneous magnet analysers is that they also have $\beta$, or $Z$, focusing properties. For $n = 0.5$ a twofold increase of dispersion over a conventional analyser of the same radius is obtained and in such a system the axial and radial focal points coincide. Thus a point image of a point source can be obtained without the use of additional focusing lenses.

As a practical consideration the realisation of the required field gradient presents certain difficulties and, in spite of their advantages, the use of inhomogeneous sector magnets for the analysis of heavy ion beams has remained somewhat restricted. It should be noted too that the fringing field effects described in section 4.4.2 also apply to inhomogeneous sector magnets.

### 4.4.4. THE WIEN VELOCITY FILTER

The use of a velocity filter (Wien, 1902) to separate ions of different mass was employed as early as 1934 by Oliphant et al. Wahlin (1964, 1965) has described a flexible, small size, isotope separator based on this principle and more recently the method has been adopted in a number of implantation machines (see for example, the accelerator illustrated in fig. 4.4).

The velocity, $v$ of singly charged ions of mass $M$ accelerated across a potential difference $V$ is given by:

$$v = \left(\frac{2eV}{M}\right)^{\frac{1}{2}} \tag{4.32}$$

If such an ion traverses a velocity filter as depicted in fig. 4.103 it will not be deflected if the opposed magnetic and electric forces from the crossed fields balance one another, i.e. if:

$$Hev = eE \tag{4.33}$$

when $E$ is the static electric field. Also from eq. (4.32) and (4.33):

$$\frac{E}{H} = v = \left(\frac{2eV}{M}\right)^{\frac{1}{2}} \tag{4.34}$$

An ion of the same energy but of mass $M + \Delta M$ travelling at velocity $v + \Delta v$ would experience a net deflecting force, $F$, and would follow a circular trajectory of radius $R$, where:

$$F = He(v + \Delta v) - eE = \frac{(M + \Delta M)(v + \Delta v)^2}{R} \tag{4.35}$$

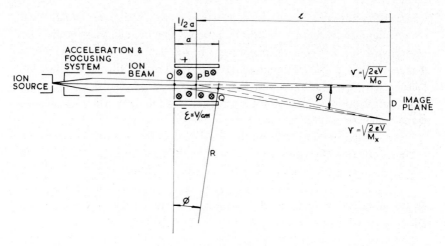

Fig. 4.103. Schematic representation of a velocity filter employing crossed electrostatic and magnetic fields (from Wahlin, 1964).

which from eq. (4.32) can be written:

$$F = \frac{2eV}{R} = eE \left[ \frac{H}{E} \left( \frac{2eV}{M + \Delta M} \right)^{\frac{1}{2}} - 1 \right] \tag{4.36}$$

and substituting for $H/E \cdot (2eV)^{\frac{1}{2}}$ from eq. (4.34) gives:

$$\frac{2eV}{R} = eE \left[ \left( \frac{M}{M + \Delta M} \right)^{\frac{1}{2}} - 1 \right] \tag{4.37}$$

thus:

$$R = \frac{2V}{E} \left[ \left( \frac{M}{M + \Delta M} \right)^{\frac{1}{2}} - 1 \right]^{-1} \tag{4.38}$$

If $\Delta M$ is small the ion will be deflected through an angle $\phi = S/R$ (where $S$ is the length of the filter). The separation, $D$, of the ion beams $M$ and $M + \Delta M$, at a distance $l$ from the centre of the analyser, is then given by

$$D = \frac{lS}{R} = \frac{lSE}{2V} \left[ \left( \frac{M}{M + \Delta M} \right)^{\frac{1}{2}} - 1 \right] \approx \frac{lSE}{4VM} \quad \text{if } \Delta M = 1 \tag{4.39}$$

Unlike the magnetic analysers which were considered previously the focusing action of the velocity filter as described above is relatively weak and a stigmatic image is obtained if the injection optics are also stigmatic (Wahlin, 1964).

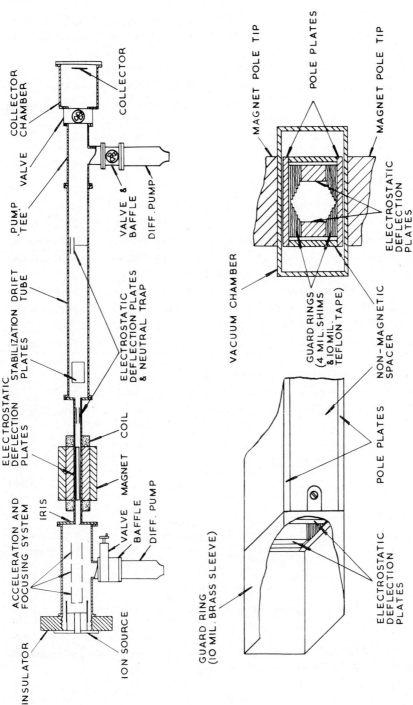

Fig. 4.104. Isotope separator of Wahlin (1965) employing a velocity filter for mass analysis.

A useful feature of the velocity filter is that the dispersion can be altered, for fixed mass selection, by appropriately varying the magnitudes of the magnetic and electric fields. Figure 4.104 shows the layout of the system used by Wahlin. Considerable care was taken in the design of the analyser to optimise the ion optics for best performance. It was also found necessary to introduce an additional electrostatic deflection of the beams to obtain an adequate separation of the ions from the directed flux of fast neutrals produced by charge-exchange in the source region. This is shown in the illustration.

### 4.4.5. THE ELECTRIC QUADRUPOLE ANALYSER

Although high frequency quadrupole analysers have been extensively used in mass spectrometry (an excellent review of the subject has been given by Dawson and Whetton (1969)), they have only attracted limited attention for the separation of higher intensity beams. Paul and his co-workers who carried out extensive studies of quadrupole systems (Paul et al., 1955; Busch et al., 1961) have however described experiments in which a separation of magnesium isotopes was effected with a total beam current of approximately 1 mA. More recently Bahrami et al. (1970) have reported preliminary experiments on a quadrupole mass filter for use on an ion implantation accelerator.

The required electric field gradient of the quadrupole analyser is produced by four hyperbolic electrodes* with a separation $2r_0$ (fig. 4.105). To obtain mass separation a high frequency electric field is superimposed upon

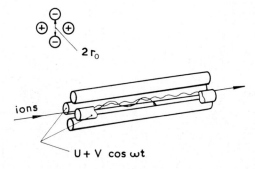

Fig. 4.105. Principle of operation of the electric quadrupole analyser (after Brunnée and Voshage, 1964).

---

* In the practical case circular rods or wires are frequently used to provide an approximation to the theoretical hyperbolic field.

a static field. Ions injected axially into the quadrupole oscillate under the influence of the r.f. field. It can be shown that the oscillations which may be 'stable or unstable' are solely a function of the field parameters $U$, $V$, $\omega$, $r_0$ and the ion mass. In the 'unstable' case the oscillations grow exponentially with time and the ions strike the electrodes. 'Stable' ions are transmitted if the amplitude of the oscillations is smaller than the internal dimensions of the analyser. The stability behaviour is given by the parameters $a$ and $q$ where:

$$a = \frac{8U}{M\omega^2 r_0^2} \quad \text{and} \quad q = \frac{4V}{M\omega^2 r_0^2}$$

The ion is only transmitted through the filter if the $(a, q)$ values lie within certain limits. This is shown in fig. 4.106. By increasing the ratio $a/q$ ($=2U/V$) the stable mass interval may be made small and the resolution increased.

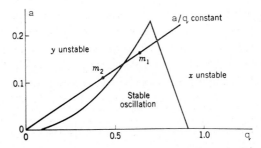

Fig. 4.106. Stability diagram for a quadrupole analyser.

This type of analyser which normally operates at low ion velocities is relatively insensitive to the initial beam energy distribution. Also because of the strong focusing action, gas scattering and space-charge repulsion effects are less significant than in other types of mass separator. Nevertheless at high beam intensities both the transmission and the resolution may be affected by the positive space-charge of the ions and this factor has been considered by Busch et al. (1961). The transmission of the quadrupole analyser is also to some extent dependent upon the lateral positions and the radial velocities of the ions entering the field and some measure of beam collimation is usually required.

The high beam current analyser described by Busch had a length of 3 m and a diameter of 3 cm. The frequency of the alternating field ($\omega/2\pi$) was 2.6 MHz at a voltage ($V$) of 2.6 kV. The dc voltage ($U$) was 175 V. The ion energy was 1.5 kV. It was found that the requirement for high transmission was not compatible with the required resolving power. The performance of the instrument for $^{25}$Mg separations was considerably improved by the use of small auxiliary alternating fields which were in resonance with the frequencies of the unwanted $^{24}$Mg and $^{26}$Mg isotopes and which caused the amplitudes of their oscillations to increase. A maximum relative enhancement of several hundred was obtained with a total Mg$^+$ beam of about 700 $\mu$A.

The mass filter of Bahrami et al. is much shorter ($\sim$28 cm) and is intended to operate with ions of considerably lower energy ($\sim$50 eV). A 2 MHz power supply capable of delivering about 12 kV is used. The transmission is stated to approach 100 per cent below mass 100 and is approximately 50 % at mass 210. The ion implanter which will provide beam currents of up to 1 mA in the energy range 50–300 keV has been designed to have a mass resolution ($\Delta M/M$ measured at half height) of 510 with an enhancement of $10^7$.

In order to obtain high beam currents at the required low energy the extraction lens is operated at a negative voltage while maintaining the ion source at 50 volts positive with respect to the mass filter. After mass analysis the required dopant beam is then accelerated to the required potential.

A more precise comparison of the performance of the two high current mass analysing systems will be possible when the ion implantation equipment is complete.

## 4.5. Target chambers

In view of the multiplicity of ion implantation applications it would be quite impracticable to attempt to define a general purpose target chamber. In experimental studies the wide diversity of requirements and the need for flexibility may lead to complex high vacuum assemblies equipped with detectors and precise target alignment facilities. In chambers designed to exploit the production potential of implantation, the requirements of reliability and reproducible process control will dominate the design. Even here, however, the inevitable vacuum mechanical problems of repetitively manipulating and aligning large numbers of fragile single crystal wafers may lead to a complicated overall design.

In spite of the broad range of requirements it is still possible to identify certain general features of the target stage which are common to many implantation applications. These include the need for uniform doping over extended surfaces, with, in certain cases, precise control of the ion beam alignment and the sample orientation. The integrated ion current must be carefully monitored and in some applications the doping rate and the target temperature may have to be precisely measured and controlled.

We shall consider a number of these factors in more detail below, and describe some typical target stages which illustrate the general principles of ion implantation on both a laboratory and a production scale.

In section 4.2.1 we briefly discussed the vacuum requirements of both the accelerator and the implantation target stage. We stressed that the quality of vacuum which is required depends upon both the application *and the available ion beam intensity*. The time required for the formation of a mono-molecular layer of perfectly adsorbing gas on a clean surface is given by, $t$ (secs) $\simeq 10^{-6}/p$ (torr) (Carter and Colligon, 1968); whilst a beam intensity of 1 mA/cm$^2$ corresponds very approximately to the bombardment of each surface atom once a second. It is evident that if the available beam current is low an improved vacuum will be required to minimise surface contamination effects.

Finally in this general consideration of target handling procedures, we would also stress that, because of the shallow ion penetration in many experiments, meaningful implantation results are only likely to be realised if painstaking attention is paid to the preparation and maintenance of a suitable surface finish on the sample before bombardment. A detailed description of the several preparative and assessment techniques required for single crystal specimens has been given by Whitton (1969). Care must also be taken to ensure that the surface is not contaminated, during implantation, by ion beam sputtering of the vacuum tank walls, or beam defining plates, and in the case of semiconductor implantation studies it is normally essential to take additional stringent precautions to avoid dust particles on the surface. These may result in localised shadowing and cause 'pipes' of undoped material extending to the surface. Dust free loading chambers or cabinets are essential in serious work. Care should be taken too, when admitting air or nitrogen to the vacuum chamber, to ensure that dust particles are not swept in. Porous filters are effective in preventing this. It is also advisable in certain cases to maintain a stream of pressurised and filtered dry nitrogen or argon through the vacuum chamber while it is open. This reduces the dust problems and leads to shorter pump-down times.

4.5.1. SCANNING REQUIREMENTS

An important distinction between the use of ion beams for implantation and the more conventional applications of accelerators is the general need for uniform doping over extended areas. This involves the use of defocused beams or the need for scanning of the ion beam or the sample itself. For certain applications a very high degree of uniformity is required, and for channelling studies the precise alignment of the sample and the ion beam is essential. These requirements can introduce a considerable degree of complexity into both the beam and target handling equipment. They may also conflict with other essential experimental needs.

Although certain assessments of implanted specimens can be carried out on small doped areas, most types of experimental measurement require large surface samples to obtain the most meaningful results. In certain applications the use of such targets may provide the only means of overcoming the limited penetration (and hence the surface loading of heavy ions), and thus providing an adequate amount of the dopant species for subsequent measurements. This is particularly the case for methods of assessment which are sensitive to the total number of impurity atoms rather than to the doping density.

In semiconductor implantations the requirement for uniform coverage of large samples is even more important. We shall consider the use of finely focussed ion beams for the production of micro-circuits in section 4.7. In spite of its potential advantages the likelihood of use of this method of ion doping on an industrial scale in the near future is problematical. On the other hand one of the immediate major attractions of implantation as a commercially viable production process is the degree of spatial impurity control which can be readily achieved by combining ion beam scanning with the conventional photo-mechanical masking procedures which are routinely employed in device manufacture. This requirement to uniformly and controllably flood large target surfaces (typically five centimetre diameter silicon wafers), with suitable heavy ion beams, can be realised in a number of ways.

In many accelerators the simplest and most convenient method of obtaining uniform sample coverage is to use electrostatic $X$ and $Y$ deflection plates to sweep the ion beam over the surface. If the implantation is carried out at the focal plane of the analyser, several ion masses may be simultaneously swept and the dispersion must be sufficiently large to ensure that impurity beams do not contaminate the target. Care must be taken too, with polyisotopic elements, to avoid non-uniform doping which may arise because of the lateral isotopic beam spread. Alternatively the required dopant beam may be transmitted through a defining aperture at the focal plane and then subsequently swept to any required extent. Since the dispersion of a conventional analyser is inversely proportional to the ion mass this technique is

preferable for the implantation of large samples with heavier ions. The two
methods are illustrated in a simplified form in fig. 4.107. It should be noted
however that when post analysis scanning is used, an additional focussing
electrode may be required to bring the diverging beam back to a spot focus.

The disadvantage of elementary sweep systems of this type is that there
is inevitably a loss of angular definition of the incident beam across the

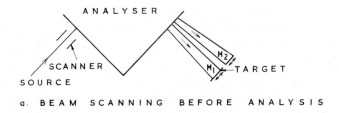

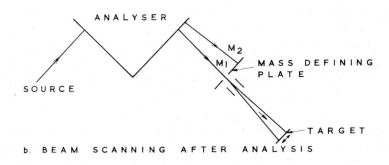

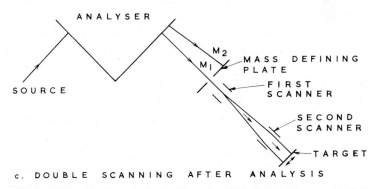

Fig. 4.107. Methods of beam scanning to give uniform implantation over extended targets
(the vertical scanning plates which may also be required for two dimensional coverage are
not shown in the illustration.

target surface. If channelling is to be studied, the maximum angle of the beam deflection must be within the critical acceptance angle, $\theta_c$. For low mass ions at high energy this requirement may be quite stringent, for example, for 500 keV $B^+$ ions, $\theta_c \simeq 1$ mrad in silicon, and a ten metre flight path would thus be required to cover a two centimetre diameter target. In such an exacting experimental situation it would generally be more acceptable to restrict the implanted area and to use a shorter scanning length. Even in doping studies which do not invoke channeling it is still desirable to maintain a reasonable degree of beam alignment for large single crystal targets. Five centimetre diameter silicon wafers are now commonplace in semiconductor applications. In a typical one metre scanning tube the loss of angular definition across such a slice would amount to about three degrees. This may be adequate to produce small but significant changes in the dopant profile in different areas. In such applications a useful alternative technique is to employ a second coupled set of scanning electrodes arranged so that the initial angular deflection is subsequently corrected as shown in fig. 4.107(c). Alternatively a simple converging lens may be employed to produce a parallel beam after the primary deflection system. Although in both cases the requirement for relatively strong focussing at maximum off-axis displacement may result in aberrations, the primary loss of angular resolution can be substantially corrected.

In the use of such electrostatic deflection systems it cannot be simply assumed that the use of relatively high frequency $X$ and $Y$ scanning will ensure adequate uniformity with large area samples and small beam spot sizes. For exacting requirements high quality 'sawtooth' or triangular scanning wave forms are essential and considerable care must be taken to ensure that these do not interact to form coupled, standing wave, beam patterns on the sample. When asynchronous $X$ and $Y$ scanning frequencies are used the duration of the doping must also be adequate to allow uniform coverage. It is important to note too that the beams extracted from certain ion sources may have significant high frequency amplitude modulations, arising from plasma oscillations or as a result of ripple on the arc power supplies. Since these latter will normally be some harmonic of the mains supply they may couple with the applied deflection frequencies and result in non-uniform doping. Care must also be taken to minimise the flux of undeflected fast neutral atoms to the target. These may be produced by charge-exchange reactions in the beam transport system and a significant loss of uniformity arising from this cause has been reported by McGinty et al. (1970).

A useful property of electrostatic deflection systems is that they are mass independent and that the angular deflection is simply a function of the geometry of the plates, the applied voltage and the beam energy. For a parallel plate scanner of length, $l$, and electrode separation, $w$, the deflection $\alpha$ is given by:

$$\alpha = \frac{lE}{2V}$$

where $V$ is the energy of the beam in electron volts and $E$ is the field strength (volts/cm) between the deflector plates.

Although this very simple form of scanning is the most commonly employed in ion implantation it is by no means the only method of achieving uniform doping over extended samples. It has been successfully and extensively used with relatively low beam currents, but it is possible that with space-charge neutralised, higher intensity, heavy ion beams the drain of compensating electrons in the electrostatic field gradient could result in beam defocusing and possibly contribute to electrical breakdown of the electrodes and insulators.

An alternative technique, which has the attraction that it can readily accommodate a wide range of sample sizes, is to mechanically scan the target holder through a fixed beam. Although this can be most simply achieved with samples mounted on rotating drums the loss of angular definition for large single crystals in such a system would generally be unacceptable. This difficulty may be obviated by the use of coupled $X$ and $Y$ linear drive systems. The quality of the beam alignment is maintained over the full distance of travel by this technique but a requirement for uniform doping is that each area of the sample must pass through the beam several times on one axis as it progressively moves along the other direction. It follows that the beam should be large compared to the stepping distance between consecutive scans. A chamber designed on this principle for large scale semiconductor doping is shown in fig. 4.122. This is described in more detail in section 4.5.6, and we shall simply note the mode of scanning at this stage. The beam height in the implantation facility is approximately 3 cm whilst the distance between successive scans can be automatically adjusted to satisfy the particular doping requirements. Thus, a 0.3 mm step will result in a uniformity of about ±1 %. By simultaneously doping several wafers the beam utilisation is improved and the target plate can accommodate a wide variation in sample sizes.

In machines such as medium output isotopes separators in which the ion beam is normally focussed as a line rather than as a spot, single axis scanning only is required to give a rectangular coverage. This technique has been extensively used by Freeman (1967; Freeman and Gard, 1970b) for a range of implantation applications. In this case the beam scanning is obtained by simply superimposing a sawtooth voltage signal on the stabilised ion source accelerating potential. In the mass analyser the ions of different energy are deflected on different trajectories with the magnet acting as the axis of deflection. The loss of angular deflection depends upon the geometry of the accelerator but in this case was approximately $0.2°$ over a 1 cm diameter sample. The slight loss of energy resolution would be negligible in most implantation applications. This simple scanning technique can be used with intense ion beams and by paying careful attention to the ion source and accelerator optics a reasonable degree of uniformity can be obtained in the direction normal to the scanning axis.

The techniques described above are of fairly general application in ion implantation. For particular special requirements other methods such as beam defocussing, or combinations of mechanical and electrostatic scanning, may be simpler or more appropriate.

4.5.2. TARGET AND BEAM ALIGNMENT

In the previous section we briefly outlined the scanning angular definition requirements for implantation channelling studies in single crystals. It is clearly also important that the beam angular spread and the alignment of the required crystal axis to the direction of propagation of the ions should be equally well defined. For the most precise studies of channelling behaviour it is desirable that the *total* uncertainty in the angular definition of all three factors should be significantly less than the critical acceptance angle $\theta_c$. The stringent requirements for 500 keV $B^+$ have been described previously but even for heavier ions at much lower energies it is preferable to maintain an overall angular definition of a fraction of a degree.

Although the angular convergence, or divergence, of the dopant beam can be controlled by the use of low aberration lenses, it is frequently simpler to use mechanical collimation to obtain the required definition. This last method inevitably reduces the available beam intensity but this is not usually of primary concern in such studies.

A number of techniques may be used to determine the crystal orientation and six methods, with varying degrees of accuracy and simplicity will be described. It should be noted that even in implantation studies where channelling is to be avoided, the orientation of the crystal lattice with respect to the ion beam may have to be determined, since it is important, in obtaining reproducible results, to avoid systematically any open direction or plane.

(a) Optical scattering by etch pits in the crystal surface. Certain chemical etches preferentially expose particular crystal faces. In this way small etch pits can be created in the surface of the specimen, each with a number of plane facets. These reflect a collimated parallel beam of light so as to produce a reflected spot on a screen. A low-power laser is well suited to this purpose. The angle between the crystal planes can thus be determined. The accuracy is not, however, very great but may be adequate for some requirements. At best, it gives about $\pm 1°$ precision, and has the merit of requiring very little specialized equipment or skill.

(b) X-ray diffraction is a standard procedure for studying crystals and normally makes use of an appropriately-designed X-ray camera. The crystal orientation, relative to the crystal normal, is determined by solving the pattern of Laue diffraction spots obtained on the exposed X-ray film. The techniques have been described in detail in texts such as that of Nuffield (1966). Typically, an accuracy of about $\pm 0.5°$ can be achieved, though with more sophisticated X-ray cameras it is possible to do better than this.

(c) Detection of the back-scattered yield of particles measured in an ion detector as a function of the crystal setting. This method is applicable if it is known that the crystal normal is near to a prominent channelling axis. The crystal is off-set by about $3°$ in a goniometer and rotated about the crystal normal. The yield of back-scattered particles, for a given incident flux, is plotted for one complete rotation and the angles corresponding to minimum yield are plotted on polar graph paper (fig. 4.108). By constructing the intersection of lines joining these points, the pole corresponding to the crystal axis can be determined to an accuracy of about $0.02°$.

(d) A convenient method, though with lower accuracy, makes use of the proton scattering microscope, Nelson (1967). A fluorescent screen is used to record 20 keV protons scattered from the crystal as shown in fig. 4.109. From the setting of the goniometer and the appearance of the pattern on the screen, the crystal orientation can be found to about $\pm 0.3°$.

(e) Greater accuracy is obtained with higher energy ion beams, since the blocking pattern observed in back-scattering becomes better defined. Marsden et al. (1969) have exploited the sensitivity of cellulose nitrate film to MeV alpha-particles for this work. A finely-collimated beam of 1 MeV alpha-particles is directed on to the crystal specimen though a hole in the centre of a sheet of the plastic, set about 5 cm from the crystal. After an exposure of about 0.5 $\mu$ Coulomb the film is removed, etched in 6 N sodium

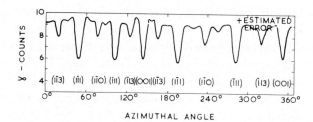

SCAN AROUND THE $(110)$ AXIS WITH A TILTING ANGLE $\phi$ OF 3·8° INCIDENT PROTON
ENERGY : 510 KeV.

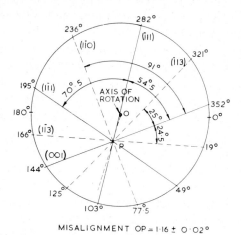

MISALIGNMENT OP = 1·16 ± 0·02°

Fig. 4.108. Use of back scattering for crystal alignment (from Andersen, 1965). Stereo-graphic representation of the rotation azimuthal angles corresponding to the planar dips in A are marked off on the circumference.

hydroxide solution at 70° for 2–5 minutes, when a fine matt pattern appears, corresponding to the etched damage tracks of scattered alpha-particles. In this way the crystal orientation can be determined to about ±0.02°.

(f) At even higher ion energies it may be possible to make use of particles which are transmitted through the crystal specimen, depending upon its thickness. For 0.5 mm silicon crystals, for instance, it is necessary to use 10 MeV protons for this purpose. Dearnaley and Wilkins (1967) made use of the Harwell Tandem accelerator in combination with an optical technique for obtaining the orientation of such crystals to an accuracy of ±0.02°. As in method (e), a permanent record is obtained, in this case in X-ray emulsion, of the orientation of the crystal. An advantage of the technique is that the exposure time can be very brief and, since the particles all penetrate the crystal, the radiation damage it suffers is very slight.

Fig. 4.109. Proton scattering microscope image (20 keV protons on ⟨110⟩ tungsten) (from Nelson, 1967).

In certain accelerators the alignment of the sample to the ion beam can be carried out *in situ* using one of the particle back-scattering techniques described above. For such studies the target chamber must be equipped with suitable ion detectors (typically silicon surface barrier detectors). Two types are commonly used, one the normal plane variety set at some back angle, e.g. 150–160° to the beam direction, the other with annular geometry so that the ion beam can pass through a central hole. This accepts particles very close to 180° to the beam direction. Other effects which are sensitive to ion beam channelling such as electron or X-ray emission or sputtered particle yields may also be used for such alignment.

When proton beams are used in this way to set up the crystal, prior to implantation with a heavier ion species, care must be taken to ensure that the collimated proton trajectory is identical with that of the higher mass beam. Small errors may arise because of perturbations in the field distribution of the analyser at the widely different magnet settings, and with low energy protons in long beam flight tubes even apparently trivial stray fields may result in significant angular deflections. Thus a 100 keV proton beam would be deflected though approximately 0.1° and displaced about 4 mm over a distance of two metres in a field of only one gauss.

The advantage of the *in situ* technique is that it avoids the need to precisely predetermine either the ion beam direction or crystal orientation. It is, however, a time consuming operation and necessitates the use of both a goniometer and a detector in the target chamber. An alternative method is to position a sample of predetermined orientation in a precisely located target holder which has been previously aligned to the ion beam. This mode of operation has been extensively used by Freeman and Gard (1970b) who employ a rigid target stage into which accurately machined target holders can be rapidly inserted through a vacuum lock. The samples are fixed to the face of the holder with shims to correct for misalignment of the crystals. Goode and Dearnaley (1970) employ a similar system but shim the sample in position by means of a single axis goniometer on the target stage. A laser beam reflected from a mirror on the sample mounting is used to obtain a zero setting before adjustment of the goniometer.

In addition to making provision for the precise adjustment of angular definitions for channelling, allowance should be made on the target stage for misorientation through much larger angles ($\sim 10°$) in applications where channelling is to be avoided. It is not sufficient in such cases to simply tilt the specimen randomly at an angle of several degrees to a major crystal axis since the sense of rotation may be along a crystal plane, or it may bring a minor crystal axis into line with the beam. The crystal orientation must be predetermined and the target shimmed to align the required blocking direction with the ion beam axis. For example, according to Ergensoy and Wegner (1965) the best blocking direction in silicon is 20° off the [100] axis in the direction [110] followed by 8° off the [110] axis in the direction of the [111] axis.

*4.5.2.1. Goniometers.* The normal requirements of a goniometer are that it should allow rotation about at least two orthogonal axes intersecting in the plane of the bombarded crystal and on the beam axis. The motion should be accurately controllable to a precision of about $10^{-3}$ rad and reproducible. Ideally the movement should be remotely controlled so that, for instance, the crystal may be oriented by reference to some other monitor such as a scaler or phosphor-coated viewing screen. It should be possible to make measurements at high or low temperatures (say 500 °C down to $-200$ °C). The more versatile goniometer designs allow transmission experiments to be carried out with thin target specimens, as well as implantation and backscattering in thick samples.

Figure 4.110 shows the goniometer used at Aarhus University by Bøgh

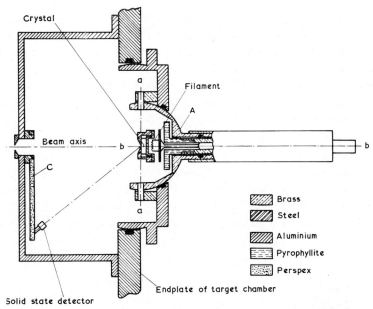

Crystal

a

Filament

A

Beam axis ⋯⋯ b

C

a

b

Brass

Steel

Aluminium

Pyrophyllite

Perspex

Endplate of target chamber

Solid state detector

Fig. 4.110. Goniometer and target assembly (from Andersen and Uggerhøj, 1968).

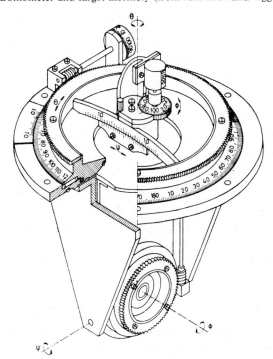

Fig. 4.111. Three-axis goniometer used to control thin crystal orientation relative to the incident-particle beam (from Gibson et al., 1968).

(Andersen and Uggerhøj, 1968) for the study of the effect of temperature and ion energy on back-scattering from crystals. The crystal may be mounted for rotation about axes aa or bb to an accuracy of $\pm 0.05°$. Electron bombardment from a filament enables temperature up to 600 °C to be achieved.

Another goniometer which has been used at Aarhus by Gibson et al. (1968) possesses three axes of rotation (fig. 4.111) and is designed for transmission experiments in which the energy loss of ions is measured in passing through thin crystals. Figure 4.112 shows a design used for similar work by Eisen (1968).

Figure 4.113 shows a three-axis goniometer for remote operation. It has been in use for some years at Harwell (Dearnaley et al., 1968) and enables crystals to be set with an accuracy of $\pm 0.02°$. With appropriate inserts it is suitable for either back scattering or transmission experiments and provision

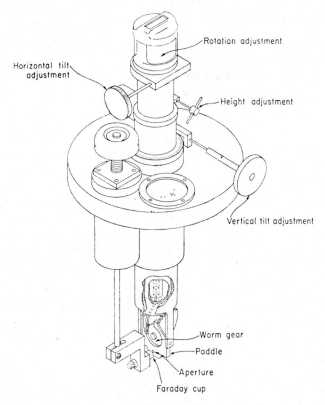

Fig. 4.112. Three axis goniometer used for channeling studies (the sample is mounted in the centre of the worn gear) (from Eisen, 1968).

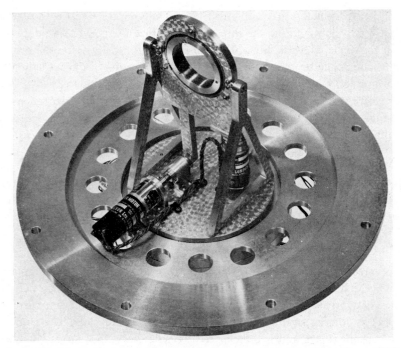

Fig. 4.113. Three axis goniometer designed for remote operation (from Knox, 1970).

has also been made for heating or cooling specimens over the range −150 to 400 °C.

*4.5.2.2. Profile control by target orientation.*   In the fabrication of certain semiconductor devices, such as varactor diodes, a precise tailoring of the dopant profile is desirable. Whereas only a limited measure of control can be achieved by conventional diffusion, ion implantation offers three distinct techniques which, subject to the limitation of the energy restricted penetration depth, may be used to obtain required dopant distributions.

The control of profile by energy programming of the ion beam has already been discussed in section 4.2. The use of channelling also offers both a means of extending the penetration range and obtaining a considerable degree of profile control, insofar as the dopant distribution is sensitive to both temperature and orientation. The limitations on the application of channelling for device fabrication lie in the relatively small dose which can be introduced in this way and the need for very precise and reproducible alignment of the ion beam and the target.

The third method (Ion Physics, 1965; Duffy et al., 1969) is to tilt the target controllably during the implantation. Since the penetration depends upon the angle of inclination, $\theta$, of the sample to the ion beam the profile can be controlled by continuously adjusting the rotation of the target holder according to a predetermined programme using a constant accelerating voltage (see fig. 4.8). If it is assumed that the mean range $\bar{R}$ is independent of the angle of inclination, as in amorphous solids, the penetration depth at an angle $\theta$ is given simply by:

$$R_0 = \bar{R} \sin \theta$$

In the case of implantation in single crystal targets the final profile will be somewhat modified by channelling effects which occur as the various crystal axes and planes pass through the beam direction.

### 4.5.3. ION BEAM CURRENT MEASUREMENTS

A singular advantage of implantation over diffusion, in certain applications, is that the amount of dopant introduced into the specimen can be precisely controlled and measured by integration of the ion beam current. In practice, this apparently simple measurement is complicated by a number of ion beam effects and is difficult to reconcile with other target chamber requirements. Precise dose measurements can only be made if suitable precautions are taken.

Most reported implantations have been carried out with beam intensities in the range $10^{-9}$–$10^{-3}$ A. Such currents can be readily measured and suitable commercial integrators can be used to estimate the total dose. However, the monitored current may be made up of positive and negative ions, and electrons arriving on the target, as well as similar secondary charged particles being emitted as a result of ion bombardment processes. These effects are by no means trivial and if no precautions are taken to suppress them the apparent measured beam current may easily be high or low by a factor of two. The situation is aggravated by the fact that the errors are not consistent and depend upon effects such as the pressure in the chamber and the state of the target surface.

The major cause of inaccuracy in most measurements is usually due to secondary electrons leaving the target and *increasing* the apparent beam current. However, as we have seen in section 4.3.1.3, most ion beams are space-charge neutralised with electrons produced by ion–gas collisions in the vacuum system. If a high impedance integrator or meter, is used to monitor the ion current, the target becomes positively biased, thus reducing the secondary electron yield and also attracting electrons from the beam which result in a *reduced* reading.

These effects are illustrated in fig. 4.114 which shows the monitored current of a constant intensity ion beam on a variety of unshielded target materials as a function of the specimen voltage. The beneficial effect of quite weak magnetic fields is also shown. We shall return to this point below.

An additional, and often unsuspected source of error, arises from fast directed neutrals. These are produced in inelastic charge-exchange collisions in which the ions are neutralised without serious deflection or loss of velocity. Their penetration profile and doping behaviour in the target are similar to those of the ions but their arrival is not monitored with the beam. Similar doping errors also arise from charge-exchange stripping reactions which

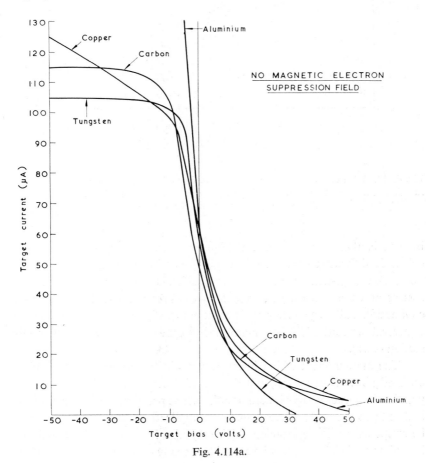

Fig. 4.114a.

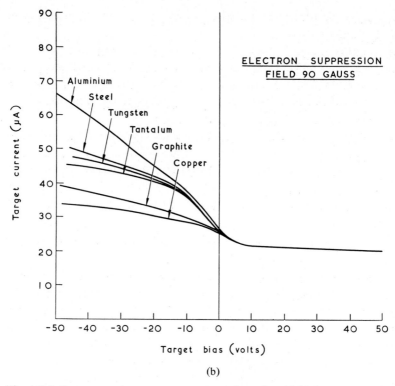

Fig. 4.114. Beam current measurement on a variety of unshielded target materials as a function of the target voltage, (a) with no magnetic suppression field, (b) with a 90 gauss field to suppress secondary electron effects.

become increasingly important at high energies. The importance of these effects is difficult to estimate in a particular instance since they depend upon the charge-exchange cross-section, which is a function of both the ion species and the residual gas, as well as the pressure, the degree of beam collimation and the geometry of the accelerator. They may, however, amount to several per cent of the beam current in the most unfavourable cases (see section 4.2.1).

The measurement of ion beam currents in the presence of the effects outlined above is not novel, and the use of electrostatic and magnetic suppression in Faraday cages is commonplace. The problem in ion implantation is that these precautions are required in applications where the beam may be swept over a large sample, which may also have to be orientated and heated or cooled. There may be a need to visually observe the sample during bombardment and there may also be requirements for detectors placed close to the specimen.

For uniform doping it is the ion beam density rather than total current which is important. This means that the total area and uniformity of the swept beam must also be accurately known, and if beam defining apertures are used in front of the target, allowance must be made for the subsequent change in beam area if it is convergent or divergent.

It is thus apparent that the precise estimation of ion doping dose as a function of measured target currents is far from easy. Curiously, the accurate and relatively simple method of monitoring the total beam current, by electrically isolating the whole target chamber from the adjoining beam lines and pumps does not appear to have been used to any significant extent in implantation studies. In such systems an important advantage is that the target and the support assembly do not require separate insulation. If they are situated a reasonable distance from the beam entry port, and if both the chamber and the adjoining beam line are at approximately the same potential, the net drift of charged particles, other than the ion beam, across the insulated boundary will normally be small. The additional use of a weak externally applied magnetic field across the interface will reduce it to an inconsequential level. Since the secondary processes occurring in the chamber do not result in a net charge transfer the total current between the chamber and earth is an accurate measure of the beam intensity.

In more conventional target assemblies the specimen holder must be electrically isolated from the ancilliary equipment and suitable precautions must be taken to suppress the sources of error in the beam current measurement. The simplest technique is to use a negatively biased suppressor plate in front of the target to reduce the emission of secondary electrons. The impedance of the beam monitoring circuit should be kept low to minimise the build up of positive charge on the bombarded specimen. The flow of secondary electrons may be further reduced by placing the sample at the end of a Faraday cage. The suppressor plate should itself be guarded from primary ion bombardment by an earthed shield with a smaller aperture as shown in fig. 4.115. In this way the principal causes of error can be significantly reduced, although with large targets the secondary electron suppression may be incomplete even with an applied bias of several hundred volts. Certain minor effects are not corrected in this way. For example, secondary electrons produced as a result of collisions of fast sputtered atoms with the suppressor plate are attracted to the target. The loss of sputtered ions from the bombarded specimen also leads to errors in the current measurement,

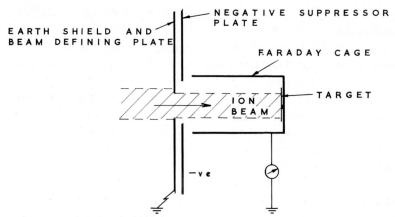

Fig. 4.115. Faraday cage arrangement with suppressor plate for beam current measurements.

and these ions are particularly effective in producing secondary electrons if they are accelerated to the suppressor plate. The additional use of a transverse magnetic field ($\sim 100$ gauss) is very effective in reducing all of these effects. In fact if such a field can be readily applied to the target stage it is often possible to obtain reasonably accurate results without the use of Faraday cages or suppressor plates. The maximum effect is obtained by using a curved fringing field as shown in fig. 4.116(a) or by using a linear field with end shields to collect electrons which spiral along the lines of force (fig. 4.116(b)).

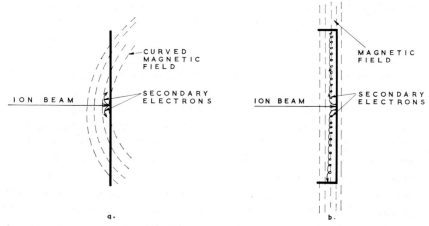

Fig. 4.116. Suppression of secondary electron effects using a magnetic field (a) plane target in curved magnetic field (b) target plate with end guards in linear magnetic field.

Although it is preferable to use an electromagnet whose position and field strength can be adjusted until the beam current reading has reached a plateau (fig. 4.114(b)), smaller permanent magnets can often be used inside the chamber if space is restricted.

It is finally worth noting that calorimetry is one of the most precise methods of measuring ion beam intensity. It has the additional advantage that it also monitors the neutral beam component of the bombarding flux. A number of calorimetric methods have been described by Gunn (1964) and also by Harrison (1957). Some of the precautions which are necessary are discussed by Hara (1968).

The technique does not lend itself directly to ion implantation because of the problems of making good thermal contact between the specimen and the calorimeter. It may, however, be useful for beam calibration purposes.

### 4.5.4. ION BEAM HEATING EFFECTS

Since many of the phenomena induced by ion implantation are temperature sensitive, and since certain target materials are thermally unstable, it is frequently desirable to maintain precise control of the specimen temperature during bombardment. In the next section we shall describe methods of heating and cooling samples for particular implantation requirements, but here we are concerned with the heating effects of the ion beam itself. These are of particular importance in semiconductor applications where it has been well established that the electrical dopant behaviour after annealing is often a function of the implantation temperature. It seems probable in fact that, in certain industrial electronic applications, ion beam heating may set an upper limit on the use of high intensity beams to increase the production rate.

We can reasonably assume, for the purposes of calculating thermal effects, that all of the beam energy is deposited as heat in the bombarded sample. Thus a beam density of $10 \, \mu A/cm^2$ at an ion energy of $100 \, keV$ corresponds to one watt of power for each square centimetre of the target whilst a typical implantation dose requirement of $5 \times 10^{15}$ ion/cm$^2$ at the same bombarding energy corresponds to a total energy deposition of about $100 \, W \, sec/cm^2$ ($10^2 \, J/cm^2$). Much higher current densities and doses may be used in certain applications and unless suitable precautions are taken these may result in temperature excursions of several hundred degrees. The problem is aggravated by the low thermal capacity of many targets and by the acute difficulty of obtaining good reproducible thermal contacts between surfaces in vacuum. Simple pressure contact, such as commonly obtains when, for example, a flat semiconductor wafer is held onto a machined metal holder by clips, is particularly inefficient. This can readily be seen in high dose implantations into silicon which should result in the production of an amorphous surface layer. The effects of such radiation damage can be readily seen as a contrasting milky appearance in the surface of the bombarded region. The dose required to produce amorphicity is strongly temperature dependent and it is notable that when large area specimens are uniformly implanted with appropriate doses of heavy ions the final surface appearance is normally quite non-uniform indicating that the temperature rise due to the bombarding ion energy has resulted in annealing of the damage in certain areas of the target. This can be confirmed by implanting at a very low rate, or by careful thermal bonding of the specimen to the holder, when quite uniform amorphicity can be obtained.

We shall return to techniques to reduce the temperature excursion below, but we shall first consider in more detail the important case of heating effects in bombarded silicon wafers. The precise temperature rise in a particular instance will depend upon a number of factors. These include the dimensions of the specimen and the ion beam, the emissivity of the surface, and the losses by conduction to the target holder. Since these may all vary we can only make a very approximate calculation of the thermal effects, but even this is adequate to show the need for stringent precautions at dose levels which are routinely used for device fabrication.

For the uniform doping of low thermal capacity samples care should be taken to either ensure good bonding to the target holder, or alternatively to minimise the conduction contact between the surfaces. The common compromise in which only part of the sample is in good thermal contact will inevitably lead to undesirable temperature gradients across the surface.

Although for experimental applications good thermal bonding can be obtained, the techniques present certain problems and do not lend themselves as readily to repetitive doping. Because of this we shall begin by considering the case in which heat transfer by conduction has been minimised and assume that the thermal losses arise solely from radiation. The emissivity, $E$, of silicon depends upon both the surface finish and in a practical case may be modified by insulating or metallic layers and by the effects of the bombardment itself. For the purposes of calculation, we shall assume a mean emissivity of 0.5 erg/cm$^2$/sec, and since the wafer has two radiating sides this conveniently corresponds to a total emissivity of 1 erg/cm$^2$/sec of implanted surface. The familiar radiation temperature relation can then be written:

$$E = \sigma(T^4 - T_0^4)$$

where $E$ is the beam energy in ergs, $\sigma$ is the Stefan–Boltzmann constant ($5.7 \times 10^{-5}$ ergs cm$^{-2}$ deg$^{-4}$ sec$^{-1}$), $T$ is the temperature of the bombarded specimen at equilibrium and $T_0$ is the temperature of the surrounds.

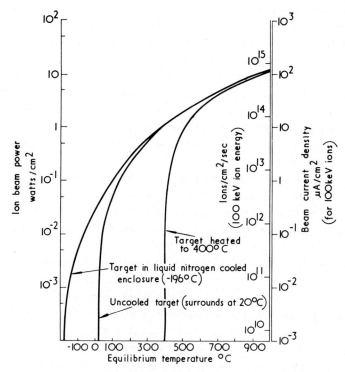

Fig. 4.117. Ion beam heating – equilibrium target temperature as a function of beam power (assuming radiative heat losses only) (from Freeman, 1970a).

In fig. 4.117 the equilibrium target temperature is shown as a function of beam power over a wide range of ion doping rates ($10^{10}$–$10^{15}$ ions/cm$^2$/sec, for a 100 keV beam energy). The calculation has been carried out for three important implantation situations.

(a) A sample in a liquid nitrogen cooled enclosure.

(b) A sample at ambient temperature.

(c) A sample preheated to 400 °C.

In all cases it is assumed that the sample is in poor thermal contact with the target support and that conduction losses are negligible. It is worth stressing however that the thermal mass of the target does not enter into the calculation and that even in the case of a wafer which is well bonded to a metal plate large temperature rises may occur if the ion bombardment is continued long enough and if, as is probable, the target support is itself in poor thermal contact with an adequate heat sink. We shall consider the rate of heating below.

In spite of the approximation involved in assuming the emissivity, and the absence of conduction losses, the general form of the temperature curves bring out some important points which are of importance in semiconductor implantation. In the following discussion we will assume an ion energy of 100 keV, but the results can readily be corrected for other beam energies. Firstly it can be seen that with high, but attainable, beam intensities ($\leqslant 1$ mA/cm$^2$) the temperature excursion may be very large indeed ($\sim 1000$ °C) and the final temperature is quite insensitive to the prebombardment conditions. At somewhat lower beam currents the preheated sample is not significantly affected by the implantation. The room temperature sample begins to show significant heating effects at current densities above 0.1 $\mu$A/cm$^2$ and the temperature rises rapidly above 1 $\mu$A/cm$^2$. Finally in the case of the precooled sample in a liquid nitrogen enclosure the temperature rise is already large at 0.1 $\mu$A/cm$^2$ and at what would normally be considered a modest beam intensity of 0.5 $\mu$A/cm$^2$ the wafer would have risen above the normal ambient temperature. One of the more important applications of low temperature semiconductor bombardments is to produce high electrical activity (after subsequent annealing) in high dose boron implantations ($> 10^{15}$ ions/cm$^2$). It can readily be seen that if the doping is not to take an inordinate time ($10^3$–$10^4$ secs) considerable care must be taken to ensure that the wafer is in good thermal contact with the cooled heat sink.

The precise measurement of the heating effects of ion beams on thin silicon wafers is experimentally difficult because of the problems of ensuring reproducible thermal insulation and of measuring the precise temperature of the bombarded face, but preliminary measurements by Freeman and Gard (1970a) over the temperature range 50–500 °C agree with the calculated room temperature curve to within about 25 %.

In certain experimental implantations in which only a fraction of the slice area is bombarded the temperature rise is smaller since the remainder of the wafer acts as a heat sink. However even in this case when high beam intensities are used ($\sim$ mA/cm$^2$) it can readily be seen that the rate of thermal conduction across the thin wafer is low since the bombarded region becomes significantly hotter than the surrounds.

So far we have only considered the final equilibrium temperature of the bombarded target at which the thermal dissipation of the ion beam energy is just equal to the radiated heat losses. Clearly an equally important factor is the rate of rise of temperature, since for a given implantation dose the bombardment may be complete before the maximum temperature is reached. Alternatively it may be essential to choose a doping rate at which the thermal effects are acceptable, or to take precautions to ensure adequate thermal bonding to the target mount.

To calculate the rate of temperature rise we shall again assume the typical case of a silicon wafer of thickness 0.25 mm ($\kappa$), density 2.32 g/cm$^3$ ($\rho$), and specific heat 0.836 joules/g/°C, ($S$).

It can be seen that the instantaneous rate of temperature rise at a temperature $T$, ($T_0 = 20$ °C) and time $t$ after the commencement of the bombardment arises from the difference between the beam power ($P$) and the radiative power $\sigma \, (T^4 - T_0^4)$ and is given by:

$$\frac{dT}{dt} = \frac{P - (T^4 - T_0^4)}{S\kappa\rho} \tag{4.40}$$

and thus

$$t = \int_{T_0}^{T} \frac{S\kappa\rho}{P - (T^4 - T_0^4)} \, dT \tag{4.41}$$

The evaluation of expression (4.41) for a range of beam intensities ($\mu$A/cm$^2$ at 100 keV) is shown in fig. 4.118. Also shown are the temperature excursions for a series of doping levels (ions/cm$^2$ at 100 keV) as a function of doping rate. (The results can readily be recalculated for samples of different thermal mass or for ions of different energy, but it should be noted that in the case of low temperature implantations an appropriate correction must be applied to the specific heat. The effect of this is to further raise the rate of heating.)

It can be seen that at high doping rates ($> 10$ watts/cm$^2$) the temperature rise is almost instantaneous but that at much lower beam intensities the slice may take several minutes to reach the equilibrium temperature. The curves also show that at doping levels below $10^{14}$ ions/cm$^2$ (at 100 keV) the maximum slice temperature is largely independent of the doping rate whilst for higher doses the beam heating effects can be significantly reduced by increasing the time taken for the implantation.

Although the calculations above of temperature rise and rate of sample heating are necessarily approximate, they are sufficiently accurate to indicate

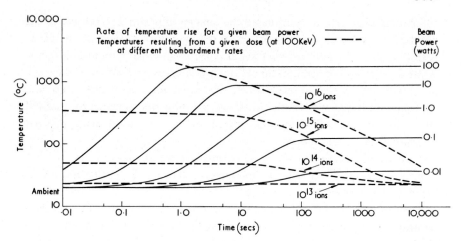

Fig. 4.118. Ion beam heating – rate of temperature rise for silicon wafers as a function of beam power and the temperatures arising from a given dose (at 100 keV) at different bombardment rates (assuming radiative heat losses only) (from Freeman, 1970a).

that these effects may be serious in many typical semiconductor implantation applications. We have stressed that the thermal conduction between contacting surfaces in vacuum is poor and may also lead to non-uniform temperature excursions. It is thus essential for many requirements to ensure both that a good heat conducting contact is achieved between the specimen and the support, and also that the total thermal mass of the target assembly is adequate to minimise the thermal effects. If these conditions are fully satisfied it can readily be shown that the temperature rise of a thin silicon wafer (thermal conductivity 0.2 cals/cm/sec/°C) would be only a few degrees, even with relatively high beam currents ($\leqslant$ watt/cm$^2$).

A number of techniques may be used to bond semiconductor wafers to metal surfaces. The most widely used is silver 'dag' (a colloidal silver suspension) which is sandwiched between the surfaces and allowed to harden. Unless it is very carefully dried before use it shows some tendency to outgas (particularly when warm) in the vacuum system but it is convenient for many requirements. After implantation the wafer can be removed by wetting the surface with acetone. Liquid gallium–indium mixtures ($\sim 80\%$, 20 %) wet both silicon and metals and can be used to provide an excellent thermal bond. Care must be taken however with low temperature applications where the contact is less satisfactory. Finally a very simple technique which is quite suitable for applications where the temperature rise will not be excessive is to use a layer of vacuum grease between the two surfaces. Best results are obtained if the grease is melted and outgassed immediately before application. If it is carefully applied the hydrocarbon surface exposed to the vacuum can be minimised and it can be readily removed after implantation by suspending the sample in a suitable solvent vapour bath.

In these and other thermal contacting techniques considerable care must normally be taken at the end of the experiment to ensure that the bonding material is completely removed from the wafer. This is particularly the case in semiconductor implantations which are frequently followed by a high temperature annealing step.

A quite different approach to the problem of ion beam heating, for large scale production, is to reduce the *effective* beam intensity, but not the overall doping rate by scanning over a larger area. In the case of electrostatic beam scanning this may present difficulties if high uniformity and good alignment are required. The solution is however attractive in applications when mechanical scanning of the sample is employed, since the beam utilisation factor can also normally be increased if several wafers are simultaneously doped. This method has been adopted by Freeman et al. (1970) in the implantation chamber described in section 4.5.6.

### 4.5.5. SAMPLE TEMPERATURE CONTROL

In the previous section we considered heating effects induced by the deposition of ion beam energy in the bombarded target. We saw the importance of these at doping levels and beam intensities which are commonly used in ion implantation. In any experiments requiring precise temperature control, suitable precautions must be taken to correct for, or to minimise, such effects.

The techniques which may be used to heat or cool samples in implantation target chambers are quite conventional and we shall simply briefly discuss some of the particular problems which may arise in certain types of ion doping experiments. We have previously commented on the difficulty of reconciling the requirement for precise current measurements, on large area samples, with the other experimental needs. This is particularly true when it is necessary to heat or cool the specimen and also measure its temperature. The probable need for efficient heat shielding may conflict with requirements for accessibility to the target, whilst the problem of electrical insulation may be difficult to reconcile with the necessity of maintaining good thermal contact to a suitable heat sink. At low temperatures in conventional vacuum systems efficient cryogenic vapour trapping close to the target will be required to prevent surface contamination of the specimen. There may also be problems due to condensation on the outside or inside of the electrical insulators which support coolant feed-throughs into the chamber. At very high temperatures the electrical resistance of the insulators may deteriorate and additional problems with resistive radiant heaters may arise both from thermionic emission from the heater to the target stage and because of metal evaporation from the filament surface. Focussed beam heating from a radiant lamp may be used in certain instances but the equipment may be bulky if large wafers have to be uniformly heated to high temperatures (> 500 °C). If the heater is used outside the chamber it may impose constraints on the design. In certain instances electron bombardment of the specimen may provide a convenient and simple means of heating, but this also conflicts with direct ion beam current measurements and presents problems with large samples.

The most appropriate techniques and precautions which must be taken in a particular situation clearly depend upon the precise experimental requirements. It is worth noting however that the most serious difficulties arise because of the requirement to measure the integrated beam current during the implantation. In certain cases it is simpler to separately monitor a known fraction of the total beam, or to sample the beam with a measuring probe at suitable intervals. The difficulties of such measurements are considerably reduced if the accelerator power supplies and the ion source are well stabilised and deliver a sufficiently constant beam current.

4.5.6. DESCRIPTION OF TARGET CHAMBERS

In this final section we shall briefly describe three target chambers which have been used for ion implantation. These are chosen simply to illustrate the different techniques which have been used to resolve some of the general problems outlined above. We would stress again that in view of the variety of doping requirements, as well as the range of accelerators which may be used, this feature of the overall machine design will normally be a matter of personal preference in any particular application.

A particularly versatile chamber used on implantation accelerators at both A.E.R.E., Harwell (Goode, 1971) and S.E.R.L. Baldock is shown in figs. 4.119 and 4.120. Up to 36 samples can be loaded into the magazine holder. These may be transferred individually onto the target stage in an analogous manner to a conventional slide projector. The stage is equipped both with liquid nitrogen cooling and with radiation heaters which permit temperature control of the specimen during implantation. A biased shield is also incorporated in front of the target to minimise secondary electron

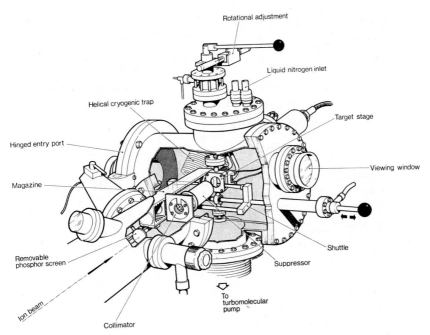

Fig. 4.119. Ion implantation target chamber as used at A.E.R.E., Harwell and S.E.R.L.,
Baldock (from Goode, 1971).

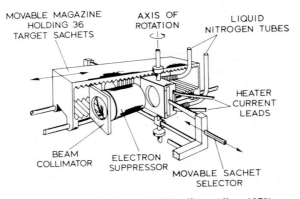

Fig. 4.120. Detail of target holder (from Allen, 1970).

emission. A single axis goniometer permits precise alignment of samples
whose crystallographic orientation has been previously determined.

A simple high voltage target stage as used on an isotope separator at
Harwell is illustrated in fig. 4.121. The outer shield is earthed and the inner

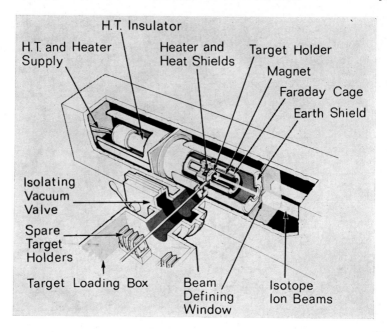

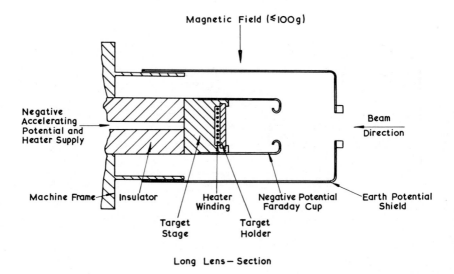

Fig. 4.121. (a) Isotope separator high voltage implantation stage with target loading
facility (b) detail of lens assembly (from Freeman and Gard, 1970b).

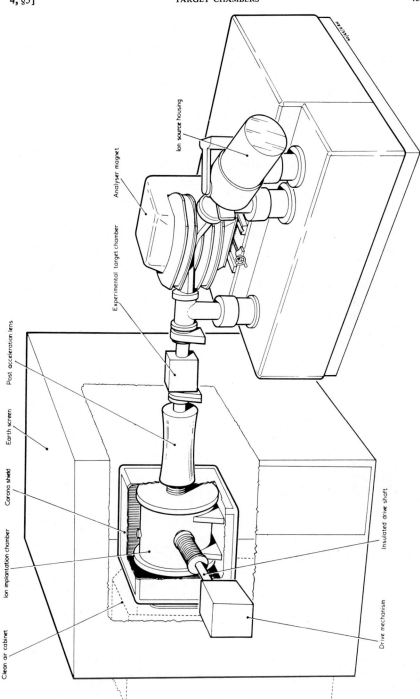

Fig. 4.122. Industrial scale implantation facility employing a high voltage automated target chamber (from Freeman et al., 1970).

target stage can be raised to 100 kV to increase the energy range of the facility. Provision for sample heating is included and targets are changed through a vacuum lock. In this case the target stage is carefully aligned to the ion beam and samples on machined metal holders are shimmed before insertion to obtain the correct orientation for channelling studies. A weak magnetic field minimises secondary electron emission and X-ray production. Such an accelerating system acts as a weak convergent lens. This results in some loss of both beam area and angular resolution at high voltage, but by restricting the doped area the system has been successfully used for channelling studies over a considerable energy range.

Finally figs. 4.122 and 4.123 show an industrial scale implantation facility (Freeman et al., 1970) which uses coupled mechanical $X$ and $Y$ scanning to obtain uniform coverage of large area samples. In this way the problems associated with the electrostatic deflection of intense ion beams is avoided and since several semiconductor wafers are normally doped simultaneously ion beam heating effects are also reduced. The chamber is designed to operate at voltages up to 150 kV and allows the possibility of dopant profile control by continuous energy programming. The 'post acceleration' of the

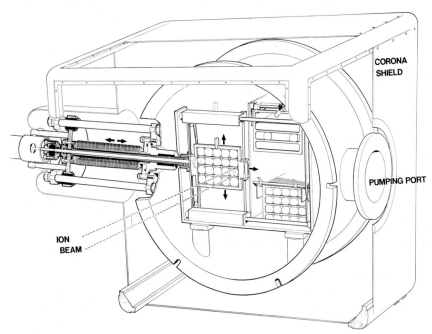

Fig. 4.123. Detail of high voltage automated target chamber (from Freeman et al., 1970).

ion to the target chamber takes place across a single gap electrode system. The lens is designed to maintain angular definition for channelling requirements and also to minimise X-ray production and space-charge 'blow up' of the intense ($\sim$ mA) beams required for production scale implantation at high doping levels. The chamber is normally loaded with twenty plates each of which holds several wafers. The high voltage operation presents no particular problems since the chamber enclosure is fully interlocked and the entire scanning process is activated by one reciprocating and one rotary motion. These are both introduced into the chamber through a single insulated shaft with the electrical drive motors and the electronic controls at earth potential.

### 4.6. The techniques of ion implantation with radioactive dopants

In this section we shall be principally concerned with the particular technical problems which arise from the use of radioactive materials for ion implantation.

The use of tracers for the study of ion penetration in solids has been described in section 2. The technique is however quite general since, as in the case of conventional ion doping, any radioactive species can be introduced into any host with the same precise control of depth, uniformity and doping level. Implanted radioactive tracers have been used in studies of hyperfine interactions, 'hot atom' chemistry, radiation damage, and diffusion as well as in the more technological fields of tribology and temperature measurement.

The normal handling and health physics precautions associated with radioactive materials are beyond the scope of this book and we shall not consider them further.

For most isotopes the accelerator provides an excellent shielded container for moderate amounts of activity. In the case of short lived radioactive materials which will decay rapidly in the system after implantation the experimental precautions can be minimised. With longer-lived activities the problems can be significantly reduced by careful design of the experiment. The use of removable liners or foils, polished metal surfaces and, for volatile materials, removable cold traps in the areas where the activity will be concentrated can greatly ease the subsequent decontamination. Particular care must however be taken with radioactive inert gases such as $^{85}$Kr because of the danger that when implanted into solids they may subsequently be ingested in stable particulate matter. The problem is particularly acute in the case of graphite, which is commonly used for beam stops or defining plates, because of the ease with which the active surface layers may be removed as fine dust.

So far we have only briefly considered the radioactive handling implications of tracer implantation. In the simplest case the only distinction from conventional ion doping lies in the special precautions and the extra degree of care which must be taken.

In many real situations however the use of radioactive isotopes may have much more profound experimental consequences. As we shall see below both the type of accelerator which can be used and the operational conditions may be severely restricted. A characteristic of a number of tracer applications is that the total bombarding ion dose must be kept to a minimum to reduce radiation damage effects, or to prevent the agglomeration of implanted ions into precipitates or bubbles. Thus in the case of channelled penetration measurements in most semiconductors it would be desirable to keep the ion dose below $10^{12}$ ions/$cm^2$ to prevent the modification of the profile by ion-induced disorder in the surface of the aligned single crystal target.

Unless the specific activity of the source feed material is sufficiently high or unless the radioisotope is available in a carrier-free form, a measure of isotopic resolution will be required to separate the tracer ion beam from the inactive parent or carrier.

The degree of separation required in a particular case can be calculated from the specific activity of the radioactive sample, the maximum total ion dose which can be tolerated during the implantation, and the number of counts per unit time required in the target to give the required experimental accuracy. In measurements of ion ranges involving sectioning techniques this last figure may be quite high if the fine structure of the profile is to be examined in detail. The total activity required may be high also in instances where the time taken to carry out the experimental measurements after implantation is long compared to the half-life of the tracer. In fig. 4.124 the number of disintegrations per second is shown as a function of the total number of atoms in the unseparated sample for varying specific activities.

Thus from a knowledge of the counting efficiency of the detector system and the sample area to be implanted, the required isotopic enhancement can be determined. (The self dose due to the radioactive isotope itself, which cannot be avoided, and which may be significant if the half-life is long, is considered separately below.)

In certain cases the mass discrimination of a conventional accelerator

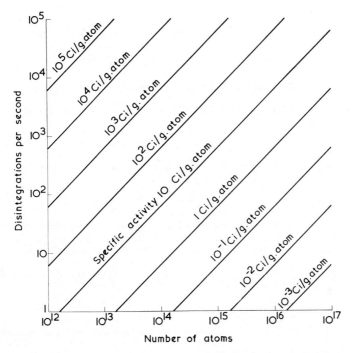

Fig. 4.124. Number of disintegrations per second as a function of the total number of atoms in an *unseparated* sample for varying specific activities.

may be adequate, but for more precise experiments the optimum performance of a high resolution isotope separator is required (see for example Freeman, 1969c). We shall consider this type of application in some detail; we shall see that although radiation damage to the sample may be reduced it cannot be eliminated and that meaningful and reproducible results are only likely to be obtained under exacting operational conditions.

4.6.1. THE USE OF ISOTOPE SEPARATORS FOR RADIOACTIVE IMPLANTATION

Laboratory separators have been successfully used for several years for the resolution of mixtures of radioactive isotopes in nuclear spectroscopy studies (see for example Nielsen, 1969). They are also well suited for certain conventional implantation applications because of their well-established capability of producing high-intensity beams of a wide range of elements. The quite disparate requirements of these two different areas of application, which are listed and compared in table 4.8, are largely combined in precise radioactive ion doping experiments.

TABLE 4.8

Separator requirements for nuclear spectroscopy and inactive ion implantation

|  | Radioactive isotope separation | Conventional ion implantation |
|---|---|---|
| Beam shape | Sharp line or spot | Large uniform area |
| Ion beams | Wide choice of radioactive ions | Wide choice of stable ions |
| Isotope purity | Highest possible enrichment | Not important since isotope effects are normally small |
| Chemical purity of beam | Isobaric impurities normally unimportant unless radioactive | Chemical purity as high as possible |
| Energy requirements | Normally unimportant although deceleration to low energies may be required to deposit thin targets | Wide energy range required to control depth of implantation, energy must be precisely measured. Requirement for ion beam post-acceleration in target chamber |
| Beam intensity | Normally unimportant unless specific activity is low or half-life short | Heating effects may be important with intense beams |
| Beam angular definition | Unimportant | Precise beam alignment and collimation ($\sim$0.1°) for channelling |
| Ion source efficiency | As high as possible | Not generally important |
| Doping level | Normally unimportant | Wide range of doses $\approx 10^9$–$10^{18}$ ions/cm$^2$; must be precisely measured |

In general a wide range of radioactive ions is required. The beam may have to be aligned and precisely collimated for channelling studies. Additional acceleration at the target may be necessary to obtain the required depth of penetration and in many cases the total ion dose may have to be kept low to minimise radiation damage effects. The operational problems associated with this type of rigorous implantation are described below.

*4.6.1.1. Ion Source Efficiency.* The efficiency of an electromagnetic separator is usually several per cent for most elements. Thus for a routine radioactive separation requiring, say, 1 $\mu$Ci of the separated isotope for decay scheme measurements it is sufficient to charge the ion source with a few tens of microcuries of activity. For radioactive implantation experiments the overall efficiency can be much lower and it may be necessary to take several millicuries in order to prepare a few specimens doped to microcurie level. These amounts of activity may be associated with stringent and time-consuming health physics controls, and in the case of long-lived isotopes can lead to an embarrassing build-up of activity in the machine.

The reduction in efficiency arises from the need to collimate the ion beam to meet channelling requirements, and the further need to optimize the quality of the separation. In the case of the medium output type of separator there may be an additional loss, since the target rarely utilizes the full height of the line-focused ion beam.

### 4.6.1.2. Dose Effects.

The total ion dose to the specimens will normally be made up of three distinct components.

(i) Self-dose due to the ions of the radioactive dopant. This can be calculated from the simple expression $N = 1.44T_{\frac{1}{2}}A$ where $N$ is the total number of radioactive ions, $T_{\frac{1}{2}}$ the half-life in seconds and $A$ the activity (d.p.s.) in the sample. Thus 1 $\mu$Ci of activity in the sample would correspond approximately to $10^{10}$ dopant atoms for an isotope with a three day half-life. The dose effect becomes significant ($> 10^{12}$) for half-lives greater than one year.

(ii) Isotopic dose. Although modern separators can give a very high isotopic enhancement in a single stage, the separation is never complete. Because of gas scattering and other effects the well-defined beams have low intensity tails in both the high and low mass directions (see fig. 4.20). By utilizing the very sharp focus of the separator to collect over the smallest possible area, this contamination effect can be minimized and an enrichment of several thousand of one isotope with respect to another can often be achieved. For implantation experiments however it may be necessary to scan the ion beams over the whole sample area with a consequent degradation in the quality of the separation. In many cases the radioactive tracer will have been produced by an (n, $\gamma$) reaction and will thus be associated with an intense ion beam of the parent at the adjacent lower mass position. In a typical case in which the specific activity is 1 mCi/mg for a source feed material of mass 100, a separation giving an isotopic enhancement of 1000 will result in a contaminant dose of about $10^{13}$ ions of the parent for each microcurie of the daughter implanted in the target.

A further cause of isotopic contamination which may dominate in many separations, is that due to the formation of ion-molecules of the form $XH^{+}$. These hydride ions, which are particularly important in separations of inert gases and electronegative elements, can be produced in significant yield from traces of water vapour or hydrocarbons in the ion-source discharge. In the example quoted above, the formation of only one part per thousand of the hydride would also lead to a contaminant dose of $10^{13}$ ions.

Although contamination effects of this order are very difficult to detect, such a dose may be very significant in terms of the damage produced. It is therefore essential that the separation be carried out with the most stringent care under the best possible vacuum conditions.

(iii) *Isobaric and adventitious contamination.* A careful inspection of the mass spectrum of a conventional separator or accelerator during normal operating conditions will reveal the presence of small impurity peaks at almost every mass position. These trace ion beams are normally inconsequential in the separation of stable isotopes. They are usually trivial, too, in conventional ion-doping experiments but they can provide a major fraction of the total dose of ions bombarding the target in radioactive implantation experiments. In many cases the peaks can be attributed to elemental and compound ions of impurities or of ion source constructional materials, but the situation is complicated by the production of multiply-charged ions and Aston bands.

For example, in implantations of radioactive $^{32}$P probable contamination will arise from beams of $^{31}$PH$^+$ and $^{16}$O$_2^+$. Similarly, although implantations of $^{35}$S are subject to only minor $^{34}$SH$^+$ effects from the low

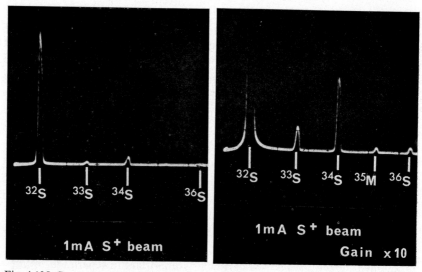

Fig. 4.125. Spectrum obtained during a radioactive sulphur ($^{35}$S) implantation showing stable sulphur isotopic beams and isobaric contamination at mass position 35 ($^{35}$Cl$^+$) (from Freeman 1969c).

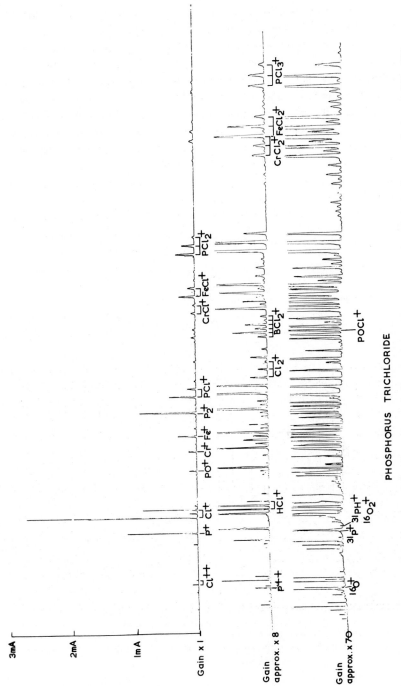

Fig. 4.126. Phosphorus trichloride spectrum over a wide mass range, note low value of contaminant peak at mass 32 position (from Freeman, 1970a).

abundance sulphur-34 isotope, there is frequently a significant beam of $^{35}Cl^+$. Chlorine is commonly present in ion sources as a residual trace impurity from previous separations in which chlorides have been used. Figure 4.125 shows the spectrum of a typical radioactive $^{35}S^+$ separation. It is evident that there is a significant contaminant beam at the mass 35 position. Figures 4.126 and 4.127 show the spectra obtained over a wide mass range for both inactive and radioactive phosphorus trichloride. In both cases, at high magnification the presence of large numbers of impurity peaks can be seen. However in spite of the complexity of the spectrum,

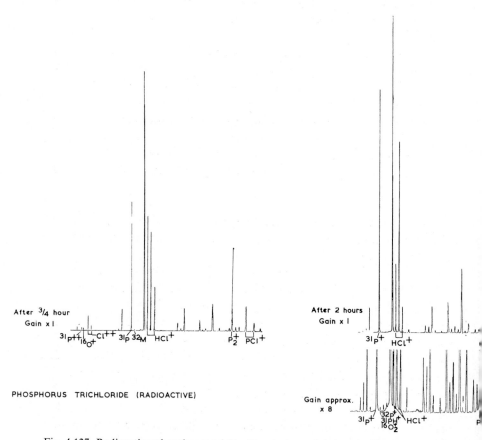

Fig. 4.127. Radioactive phosphorus trichloride spectrum showing significant contamination (PH$^+$) due to moisture in PCl$_3$ sample at mass 32 position. After two hours operation the contamination is reduced but not eliminated. The presence of water in the sample is confirmed by the magnitude of the HCl$^+$ peaks compared with the inactive spectrum of fig. 4.126 (from Freeman, 1970a).

$PCl_3$ is a suitable and convenient source feed material and the contaminant peak at $^{32}M^+$ can be reduced to an extremely low value by taking rigorous precautions to exclude water vapour and air. The spectra of fig. 4.126 were recorded during the setting-up operation for an active $^{32}P$ implantation. In this experiment the precautions were vitiated because the radioactive sample had been contaminated, presumably by moisture. The spectra obtained with the active phosphorus trichloride were quite different and even after pumping away over half the sample much more significant $PH^+$ and $HCl^+$ beams could be observed (fig. 4.127).

Such impurities can only be identified and estimated if the accelerator has a high resolving power and is equipped with a precise high gain beam scanning facility.

The examples of the adventitious doping quoted above are typical of many radioactive implantations. In general the total dose can be reduced or even eliminated by very careful attention to the separator techniques.

The identification of elemental impurity ions is assisted by the use of isotope abundance charts such as that shown in Appendix 1. For compound ions, and in particular those in which the constituents are polyisotopic, the spectra are much more complex. For example the range of ion species obtained when carbon tetrachloride is ionised in a discharge source is shown in fig. 4.95.

*4.6.1.3. Ion Beam Collimation.* Although a variety of focusing techniques exist for the production of low divergence ion beams, these are difficult to reconcile with some of the other separator requirements, such as sharpness of focus, dispersion and beam scanning. This is particularly the case for the medium output type of machine with its wedge-shaped beam. The milli-ampere intensity ion beams also tend to be defocused by electrostatic lenses. On the other hand, such separators are well suited for routine stable implantations, since simple defining window collimation procedures can be readily used to reduce the divergence to a fraction of a degree. The available beam ($\sim 10^{16}$ ions/second) is greatly in excess of normal requirements for channelled applications. The same solution has been adopted for radioactive implantation experiments but as was noted above, this leads to a very significant reduction in the separator efficiency, with a consequent increase in the amount of activity to be handled in the ion source.

**4.7. Finely focused ion beams**

One of the major applications of ion implantation is in the study and production of semiconductor devices. The doped structures may vary from simple diodes or resistive patterns to extremely complex integrated circuits which contain several thousand transistors on an area of only a few square millimetres. In almost all of the work reported to date conventional photo-mechanical alignment procedures have been used to define the implanted regions. Photo engraved masks of metals, oxides and photo resist have all been employed. This technique is particularly suitable since it exploits established techniques and also permits a direct comparison between the electrical performance and yield of diffused and implanted structures. The only accelerator requirement is that the required dopant beam of adequate intensity, purity and alignment should be uniformly implanted over the whole target surface. In section 4.5.1, we described the use of both beam and sample scanning to achieve the required degree of flooding uniformity.

An attractive but, at the present time, more speculative alternative method of implantation would be to use finely focused ion beams to produce the precisely defined structures. The possible production of transistors in this way was foreseen, in a foresighted patent by Shockley as early as 1954.

In principle, beam writing could replace at least some of the exacting and troublesome photolithographic and etching stages which are currently required to maintain the essential definition in diffused, or conventionally implanted microcircuits. A combination of the technique with, for example, focused electron beam processing could possibly lead ultimately to the automated production of complete complex microcircuits in a single vacuum assembly. This would eliminate the need for wet chemistry, high temperature processing and photo mechanical alignment. The development of such a method, if it were reproducible and had an adequate throughput would clearly be of major industrial importance and as such deserves serious consideration. However, it should be noted that such an application would require substantial improvements in ion beam technology and would involve considerable changes in the established circuit and device design procedures. The quality of silicon oxide layers which can currently be produced by electron beams is, for example, quite inadequate for MOST structures and yet it is with such transistors that implantation has so far been most successful. In view of the probable time scales involved the process must also necessarily compete with the continuing improvement in the established semiconductor manufacturing techniques as well as with the significant gains which are likely to result from the more conventional methods of ion implantation.

It seems likely that, as in the case of electron beam technology, the technique will develop gradually and that in the first instance it will be used as a complementary process for relatively simple structures, or for the preparation of small quantities of very specialised devices. A realistic economic assessment will only be possible when the basic development has been more extensively pursued. Nevertheless, the increasing interest in the use of finely focused beams is welcome since at the very least it should lead to the development of a powerful implantation research technique with widespread applications.

### 4.7.1. THE CURRENT STATUS OF FOCUSED ION BEAMS

Finely focused beams have been studied in considerable detail for applications such as ion microprobe analysis, scanning ion microscopy and ion beam micro machining (Gabbay et al., 1965; Long, 1965; Drummond and Long, 1969; Liebl, 1967; Hill, 1970). Although the overall objectives of these studies differ from those of implantation many of the requirements are similar. In particular there is a common need for high intensity heavy ion beams of micron or sub-micron dimensions with some form of beam scanning.

In the ion microprobe analyser (Liebl, 1967) shown diagrammatically in fig. 4.128 the mass analysed and focused spot beam of micron dimensions is scanned across a predetermined area of the target surface. A range of ions may be employed and both the secondary electrons and the sputtered ions which are emitted under bombardment can be used to characterise the specimen. The focused current densities obtained in such devices are typically tens of milliamperes/cm$^2$. Considerably higher beam intensities ($\sim 0.5$ A/cm$^2$) of gaseous ions with spot sizes below one micron have been obtained by Drummond and Long (1969) in the equipment shown in fig. 4.129. In this case, however, the beam transport system is much simpler since there is no mass analysis stage and the very high brightness and intensity were obtained by accepting only the most sharply defined fraction of the extracted beam from an optimised duoplasmatron source. The actual current on a one micron spot was $\sim 5 \times 10^{-9}$ A. Using a somewhat similar device incorporating a Wien filter for mass analysis Hill (1970) has reported the direct implantation of resistors in silicon using beams of nitrogen ions.

Whilst the achieved microfocused beam intensities of 0.1–1 A/cm$^2$ are adequate for most of the applications mentioned above, and are consider-

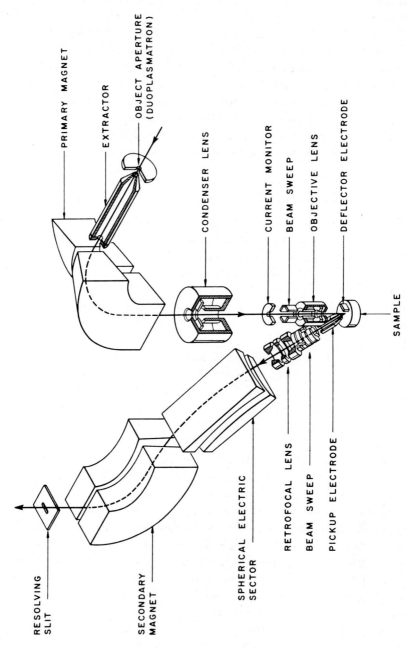

PRIMARY MAGNET

EXTRACTOR

OBJECT APERTURE
(DUOPLASMATRON)

CONDENSER LENS

CURRENT MONITOR

BEAM SWEEP

OBJECTIVE LENS

DEFLECTOR ELECTRODE

SAMPLE

RESOLVING
SLIT

SECONDARY
MAGNET

SPHERICAL ELECTRIC
SECTOR

RETROFOCAL LENS

BEAM SWEEP

PICKUP ELECTRODE

Fig. 4.128. Ion microprobe analyser (from Liebl, 1967).

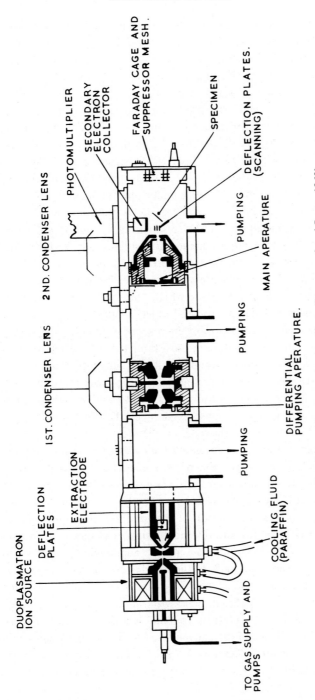

Fig. 4.129. Microfocused beam accelerator (from Drummond and Long, 1969).

ably higher than those normally used for ion doping it is instructive to consider their general relevance for integrated circuit implantation requirements. If we assume, for example, a one micron beam size with an intensity of 1 A/cm$^2$ and a required doping level of $6 \times 10^{14}$ ions/cm$^2$ the doping rate would be $10^3$ secs/cm$^2$. Clearly whilst this could be quite acceptable for experimental requirements the production potential would be severely restricted unless much higher intensities could be obtained. We shall not consider the detailed arguments which indicate that microfocused heavy ion beams of greater brightness are theoretically possible since the results described above represent the limits which have currently been achieved using the highest available beam intensities under optimum conditions. It is worth noting, however, that both space-charge defocusing and localised heating at the point of beam impact could also set an upper limit to the useful beam intensity. The focusing requirements for implantation would additionally be more stringent since for most purposes the spot size would have to be precisely defined. This would present difficulties because of the aberrations of the electrostatic focusing lenses which tend to produce a low intensity penumbra around the sharply focused object. Whilst a small loss of resolution may be inconsequential in ion microscopy it would correspond in implantation to non-uniform doping outside of the delineated structure, and in high dose applications in a high resistivity substrate could result in the formation of a broadened junction. The beam scanning and positioning requirements for precisely defined microcircuit implantation would also be much more exacting than the relatively simple $XY$ deflection techniques used in the equipment described above.

A possible method of overcoming the inherently slow doping of a programmed single beam is to use a pattern of identical microfocused beams to simultaneously implant a large number of devices. Such a technique is directly compatible with the replication methods currently in use and it effectively utilises a larger fraction of the total available ion beam current. It clearly poses a need for an even more complex structure of focusing and scanning lenses but it also presents a more fundamental problem. This arises from the requirement that each of the focused beam elements should be of the same intensity. Since these would normally be formed before demagnification by means of a defining plate with suitable apertures, the primary beam must be sufficiently uniform over the whole mask area. This requirement is far from trivial and the conventional technique of obtaining uniform coverage over an extended area by $XY$ electrostatic scanning of the primary beam provides only a partial solution since at any given time only fractional

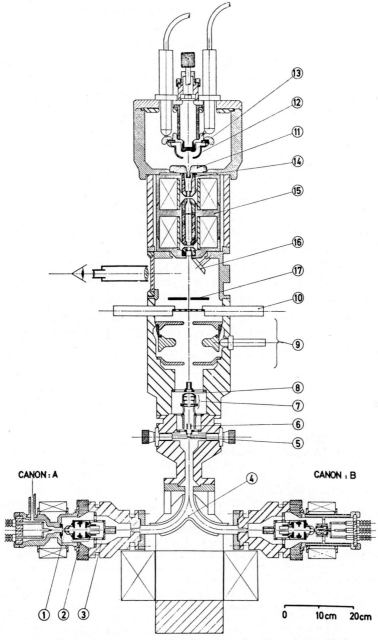

Fig. 4.130. Accelerator for simultaneous implantation of replicate demagnified structures
(from Bernheim, 1970).

coverage of the aperture is obtained. Nevertheless the method does present the possibility of the quasi-simultaneous implantation of the replicated pattern of circuits over the whole surface of a semiconductor wafer and this approach is being studied on the implantation accelerator at Grenoble (Guernet, 1970, see fig. 4.13). The narrow pencil beams produced by sweeping the primary beam over a suitably perforated plate are passed through an identically spaced assembly of miniature focusing lens which produce a pattern of identical beam spots on the sample. In a preliminary experiment it was found that a matrix of thirty six beams could be well focused to a diameter of about thirty microns. The required dopant structure will be obtained by appropriate movement of the defining mask and lens assembly on a mechanical $XY$ table rather than by electrostatic deflection of the beam themselves.

In the facility described by Bernheim (1970) a mask with suitably shaped apertures defines the primary beam into a number of replicate patterns which are then demagnified to permit the simultaneous implantation of several identical structures. The accelerator which is illustrated in fig. 4.130 is equipped with two ion sources to facilitate the change of dopant species. In this case because of the demagnification of the beam envelope only a limited area can be doped in a single operation but the need for scanning is eliminated.

## References

Allen, R. M., 1969, Elect. Letters 5, 111.
Allen, R. M., 1970, private communication.
Allen, W. D. and R. H. Dawton, 1967, Nucl. Inst. and Methods 55, 61.
Almén, O. and K. O. Nielsen, 1957, Nucl. Inst and Methods 1, 302.
Alväger, T. and J. Uhler, 1957, Arkiv för Fysik 13, 145.
Alväger, T. and J. Uhler, 1968, Prog. Nucl. Techniques and Inst. 3, 159.
Anastasevich, V. S., 1957, Soviet Phys., Tech. Phys. 1, 1448.
Andersen, J. U., 1965, Nucl. Inst. and Methods 38, 210.
Andersen, J. U. and E. Uggerhøj, 1968, Can. J. Phys. 46, 517.
Andersson, G., B. Hedin and G. Rudstam, 1964, Nucl. Inst. and Methods 28, 245.
Arianer, J., J. Baixas, P. Debray, A. Serafini and A. Steinegger, 1969, Proc. Int. Conf. on
    Ion Sources, 587 (I.N.S.T.N.-Saclay, France).
Bahrami, H., F. Chernow, D. Denison and G. Eldridge, 1970, Proc. European Conf. on
    Ion Implantation, Reading, England (Peter Peregrinus Ltd. Stevenage, England).
Bainbridge, K. T., 1953, Experimental Nuclear Physics, ed. Segrè (John Wiley, New York).
Banford, A. P., 1966, The transport of charged particle beams (Spon, London).
Bennett, J. R. J., 1969, Proc. Int. Conf. on Ion Sources, 571 (I.N.S.T.N.-Saclay, France).
Bernas, R., 1954, Thesis, University of Paris.

Bernas, R., L. Kaluszyner and J. Druaux, 1954, J. Phys., Radium 15, 273.
Bernas, R., J. L. Sarrouy and J. Camplan, 1960, Electromag, Sep. of Radioactive isotopes (Springer- Verlag, Vienna).
Bernheim, M. 1970, 4th Int. conf. on Electron and Ion Beam Science and Technology, Los Angeles.
Blanc, D. and A. Degeilh, 1961, J. de Physique et le Radium 22, 230.
Blewett, J. P. and E. J. Jones, 1936, Phys. Rev. 50, 464.
Bohm, D., 1949, The characteristics of electrical discharges in magnetic fields, ed. Guthrie and Wakerling, N.N.E.S. 1–5 (McGraw-Hill, New York).
Boreham, D. and J. H. Freeman, 1960, Unpublished data.
Boreham, D., J. H. Freeman, E. W. Hooper, I. L. Jenkins and J. L. Woodhead, 1960, J. Inorg, Nucl. Chem. 16, 154.
Bouriant, M., 1968, C.E.A.-R 3456 (France), also Bouriant, M., M. Boge, S. Dousson, P. Griboval and R. Bouchez, 1969, Proc. Int. Conf. on Ion Sources, 329 (I.N.S.T.N.-Saclay, France).
Brunnée, C. and H. Voshage, 1964, Massenspektrometri (K. Thiemig, K. G. Munich).
Busch, F. V., W. Paul, H. P. Reinhard and U. V. Zahn, 1961, Electromagnetic Separation of Radioactive Isotopes, ed. Higatsberger and Viehbock (Springer-Verlag, Vienna).
Carlston, C. E. and G. D. Magnuson, 1962, Rev. Sci. Inst. 33, 905.
Cartan, L., 1937, J. Phys. Radium 8, 453.
Carter, G. and J. Colligon, 1968, Ion Bombardment of Solids (Heinemann, London).
Chapman, K. R., 1969, Nucl. Inst. and Methods 73, 255.
Chavet, I., 1965, Thesis, University of Paris, also Chavet, I. and R. Bernas, 1967, Nucl. Inst. and Methods 47, 77 and 51, 77.
Cobić, B., D. Tosik and B. Perović, 1963, Nucl. Inst. and Methods 24, 358.
Collins, L., 1970, A.W.R.E., Aldermaston, U.K., private communication.
Cotte, M., 1938, Ann. Phys. 10, 333.
Couchet, G., 1954, Ann. Phys. 9, 731.
Cross, W., 1951, Rev. Sci. Inst. 22, 717.
Crowder, B. L. and N. A. Penebre, 1969, Rev. Sci. Inst. 40, 170.
Daley, H. L., J. Perel and R. H. Vernon, 1966, Rev. Sci. Inst. 37, 473.
Davies, D. E., T. C. Smith and R. N. Cheever, 1968, Solid State Technology, Oct., 33.
Dawson, P. H. and N. R. Whetton, 1969, Advances in electronics and electron physics (Academic Press, U.S.A.).
Dawton, R. H., 1956, Electromagnetically Enriched Isotopes, 37 ed. M. L. Smith (Butterworth, London).
Dawton, R. H., 1958, Electromagnetic Isotope Separators, 156, ed. Koch (North-Holland, Amsterdam).
Dawton, R. H., 1962, Prog. Rep. NIRL/R/23, Rutherford High Energy Lab., U.K.
Dawton, R. H., 1969, Proc. Int. Ion Source Conf. 303 (I.N.S.T.N.-Saclay, France), also Nucl. Inst. and Methods 67, 341.
Dearnaley, G. and M. A. Wilkins, 1967, J. Sci. Inst. 44, 880.
Dearnaley, G., B. W. Farmery, J. V. Mitchell, R. S. Nelson and M. W. Thompson, 1968, Phil. Mag. 18, 155.
Delauney, B., 1969, Proc. Int. Ion Source Conf., 149 (I.N.S.T.N.-Saclay, France).
Delcroix, J. L., 1969, Proc. Int. Ion Source Conf., 1 (I.N.S.T.N.-Saclay, France).
Dearnaley, G. and P. D. Goode, 1970, private communication.
Donets, E. D., V. I. Ilyuschenko and V. A. Alpert, 1969, Proc. Int. Ion Source Conf. 635 (I.N.S.T.N.-Saclay, France).
Donnally, B. L. et al., 1964, Phys. Rev. Letters 12, 502.
Dougherty, J. D., J. E. Eniger, G. S. Janes and R. H. Levy, 1969, Avco Research Report, 333 and Proc. Int. Ion Source Conf. 643 (I.N.S.T.N.-Saclay, France).

Dropesky, B. J., J. B. Deal, J. F. Buchen and G. M. Kelley, 1967, Nucl. Inst. and Methods 48, 329.

Druaux, J. and R. Bernas, 1956, Electromagnetically Enriched Isotopes ed. M. L. Smith (Butterworth, London).

Drummond, I. and J. V. P. Long, 1969, Proc. Int. Conf. on Ion Sources (I.N.S.T.N.-Saclay, France).

Duffy, M. C., P. A. Schumann and T. E. Yeh, 1969, I.B.M. Technical Disclosure Bulletin 12, No. 1.

Eisen, F. H., 1968, Can. J. Phys. 46, 561.

Eisinger, J., 1959, J. Chem. Phys. 27, 1363.

Elliot, R. M., 1963, 69 Mass Spectrometry ed. McDowell (McGraw-Hill, U.S.A.).

Eloy, J. F., 1969, Proc. Int. Ion Source Conf. 619 (I.N.S.T.N.-Saclay, France).

Enge, H. A., 1964, Rev. Sci. Inst. 35, 278.

Ergensoy, C. and H. E. Wegner, 1965, I.E.E.E. Transactions on Nucl. Science 12, 540.

Fabricius, H., K. Freitag and S. Goring, 1965, Nucl. Inst. and Methods 38, 64.

Forrester, A. T. and R. C. Speiser, 1959, Astronautics 4, 34.

Freeman, J. H., 1963, Nucl. Inst. and Methods 22, 306.

Freeman, J. H., 1967, Proc. Int. Conf. Applications of Ion Beams to Semiconductor Technology, ed. Glotin (C.E.N. Grenoble, France).

Freeman, J. H., 1969a, Proc. Int. Mass Spec. Conf. (Kyoto, Japan) also A.E.R.E. Report R 6254.

Freeman, J. H., 1969b, Proc. Int. Ion Source Conf. 369 (I.N.S.T.N.-Saclay, France), also A.E.R.E. Report R 6138.

Freeman, J. H., 1969c, Proc. Royal Soc. A311, 123.

Freeman, J. H., 1970a, Unpublished data.

Freeman, J. H., 1970b, Proc. Int. Conf. on Electromagnetic Isotope Separators, ed. Wagner and Walcher (Marburg, Germany, BMBW-FB K 70-28), also A.E.R.E. Report 6497.

Freeman, J. H. and W. A. Bell, 1963, Nucl. Inst. and Methods 22, 317.

Freeman, J. H., L. R. Caldercourt, K. C. W. Done and R. J. Francis, 1970, Proc. European Conf. on Ion Implantation, Reading, England (Peregrinus Ltd., Stevenage, England) also A.E.R.E. Report R 6496.

Freeman, J. H. and G. A. Gard, 1970a, Unpublished data.

Freeman, J. H. and G. A. Gard, 1970b, A.E.R.E. Report R 6330.

Freeman, J. H., R. W. McIlroy and K. J. Hill, 1961, E. M. Separation of Radioactive Isotopes, 155 (Springer-Verlag, Vienna).

Freeman, J. H. and M. L. Smith, 1958, J. Inorg. Nucl. Chem. 7, 224.

Freeman, J. H., W. Temple and D. Chivers, 1971, Nucl. Inst. and Methods. In press, also A.E.R.E. Report M 2419.

Freeman, N. J., W. A. P. Young, R. W. D. Hardy and H. W. Wilson, 1961, E.M. Separation of Radioactive Isotopes 83 (Springer-Verlag, Vienna).

Gabbay, M., R. Goutte, C. Guillaud and C. Monllor, 1965, C. R. Acad. Sc. Paris 261, 3325.

Gabovich, M. D., 1958, Zh. Tekh. Fiz. 28, 872.

Gabovich, M. D., 1963, Pribory i Tekhnika Eksperimenta 2, 5.

Getner, W. and G. Hortrig, 1963, Z. Physik 172, 353.

Gibson, W. M., J. B. Rasmussen, P. Ambrosius-Olesen and C. J. Andreen, 1968, Can. J. Phys. 46, 551.

Goode, P. D., 1971, Nucl. Inst. and Methods, in press, also A.E.R.E. Report 6401.

Guernet, G., P. Esteve, and P. Glotin, 1970, Note technique LETI/ME, 659, C.E.A., C.E.N.G. Report, Grenoble, France.

Gunn, S. R., 1964, Nucl. Inst. and Methods 29, 1.

Guthrie, A. and R. K. Wakerling, editors, 1949,
   (a) Electromagnetic separation of isotopes in commercial quantities. N.N.E.S. 1–4, T.I.D. 5217.
   (b) The characteristics of electrical discharges in magnetic fields, N.N.E.S. 1–5.
   (c) Sources and collectors for use in calutrons N.N.E.S. 1–6, T.I.D. 5218.
   (d) Problems of physics in the ion source T.I.D. 5219.
Hall, R. S., D. H. Poole and A. E. Souch, 1963, C.E.G.B. Report RD/B/N. 138, Berkeley Nucl. Laboratories.
Hara, E., 1968, J. Sci. Inst. 1, 1032.
Harrison, E. R., 1957, J. Sci. Inst. 34, 242.
Harrison, M. F., 1958, A.E.R.E. Report, GP/R 2505.
Hereward, H. G., K. Johnson and P. Lapostolle, 1956, Proc. CERN Accelerator Conf. 179.
Herzog, R., 1954, Z. f. Physik 89, 447.
Hill, A. R., 1970, Proc. European Conf. on Ion Implantation, Reading, England (Peter Peregrinus Ltd., Stevenage, England).
Hill, K. J. and R. S. Nelson, 1965, Nucl. Inst. and Methods 38, 15.
Hill, K.J., R. S. Nelson and R. J. Francis, 1969, Proc. Int. Ion Source Conf., 413 (I.N.S.T. N.-Saclay, France).
Hull, A. W., 1921, Phys. Rev. 18, 31.
Ion Physics, 1965, U.S. Patent Application 466218.
Jones, E. J., 1933, Phys. Rev. 44, 707.
Judd, D. L., 1950, Rev. Sci. Inst. 21, 213.
Kamke, D. and H. J. Rose, 1956, Z. für Physik 145, 83.
Kerwin, L., 1963, Mass Spectrometry, ed. McDowell (McGraw-Hill, New York).
Kistemaker, J., P. K. Rol and J. Politiek, 1965, Nucl. Inst. and Methods 38, 1.
Kistemaker, J. and C. J. Zilverschoon, 1951, Physica 17, 43.
Klapisch, R., 1960, Thesis, University of Paris.
Kleinheksel, J. H. and H. C. Kremers, 1928, J. Amer. Chem. Soc. 50, 959.
Knox, K. C., 1970, Nucl. Inst. and Methods 81, 202.
Koch, J., 1958, Electromagnetic Isotope Separators and Applications of Electromagnetically Enriched Isotopes (North-Holland, Amsterdam).
Komarov, V. L., S. G. Tsepakin and G. V. Chemyakin, 1969, Proc. Int. Conf. on Ion Sources, 383 (I.N.S.T.N.-Saclay, France).
Kozlov, V. F., V. Ya. Kolot and Sun-Chzhei-tszin, 1962, Prebori i Tekh, Eksperimenta 6, 116.
Kozlov, V. F., V. L. Marchenko and Ya. M. Fogel, 1961, Pribori i Tekh, Eksperimenta 1, 25.
Kunsman, C. H., 1925, Science 62, 269.
Langmuir, I., 1929, Phys. Rev. 33, 954, and Tonks, L. and I. Langmuir, 1929, Phys. Rev. 34, 876.
Langmuir, I. and K. H. Kingdom, 1923, Science 57, 58; Phys. Rev. 21, 380.
Lapostolle, P., 1969, Proc. Int. Ion Source Conf. 165 (I.N.S.T.N.-Saclay, France).
Lawrence, A. P., R. K. Beauchamp and J. L. McKibben, 1965, Nucl. Inst. and Methods 32, 357.
Lerner, J., 1965, Nucl. Inst. and Methods 38, 116.
Lichtenberg, A. J., 1969, Phase-space dynamics of particles (J. Wiley, New York).
Lidsky, L. M., S. D. Rothleder, D. J. Rose, S. Yoshikawa, C. Michelson and R. J. Mackin, 1962, J. Appl. Phys. 33, 2490.
Liebl, H., 1967, J. Appl. Phys. 38, 5277.
Liebl, H. J. and R. F. K. Herzog, 1963, J. App. Phys. 44, 2893.
Lloyd, R. A. and A. E. M. Hodson, 1968, Nucl. Inst. and Methods 58, 298.
Long, J. V. P., 1965, Brit. J. Appl. Physics 16, 1277.

Magnuson, G. D., C. E. Carlston, P. Mahaderan and A. R. Comeaux, 1965, Rev. Sci. Inst. 36, 136.

Marsden, D. A., N. G. E. Johansson and G. R. Bellevance, 1969, Nucl. Inst. and Methods 70, 291.

Masic, R., R. J. Warnecke and J. M. Sautter, 1969, Proc. Int. Conf. on Ion Sources, 387 (I.N.S.T.N.-Saclay, France).

Massey, H. S. W. and E. H. S. Burhop, 1952, Electronic and Ionic Impact Phenomena (Oxford University Press, London).

McGinty, G. K., B. J. Goldsmith and R. A. Thomas, 1970, 41 Proc. European Conf. on Ion Implantation, Reading, England (Peter Peregrinus Ltd., Stevenage, England).

Moak, C. D., H. E. Banta, J. N. Thurston, J. W. Johnson and R. F. King, 1959, Rev. Sci. Inst. 34, 853.

Moak, C. D., H. Reese and W. M. Good, 1951, Nucleonics 9, 18.

Morozov, P. M., B. N. Makov, M. S. Ioffe, B. G. Brezhnev and G. N. Fradkin, 1958, Second Int. Conf. on Peaceful Uses of At. En. (Geneva).

Müller, M. and G. Hortig, 1969, Proc. Int. Ion Source Conf. 159 (I.N.S.T.N.-Saclay, France).

Nelson, R. S., 1967, Phil. Mag. 15, 845.

Nezlin, M. V., 1958, Zhurnal Tekh, Fiz. 30, 168.

Nielsen, K. O., 1957, Nucl. Inst. and Methods 1, 289.

Nielsen, K. O., 1967, Proc. Int. Conf. Applications of Ion Beams to Semiconductor Technology, ed. Glotin (C.E.N., Grenoble).

Nielsen, K. O., 1969, Proc. Int. Conf. on Mass Spec. (Kyoto, Japan).

Nielsen, K. O. and O. Skilbreid, 1957, Nucl. Instr. and Methods 1, 159.

Nuffield, E. W., 1966, X-ray diffraction methods (Wiley, New York).

Oliphant, M. L., E. S. Shire and B. M. Crowther, 1934, Proc. Roy. Soc. A146, 922.

Paul, W. and M. Raether, 1955, Z. Physik 140, 262, also F. von Busch and W. Paul, 1961, Z. Physik 164, 581, 588.

Penning, F. M., 1957, Electrical discharges in gases (Philips Technical Library).

Perović, B., 1957, Proc. 3rd Int. Conf. Ion Phen. in gases 813.

Peters, B., A. C. Helmholtz and W. E. Parkins, 1949, T.I.D. 5217, 11.

Pierce, J. R., 1949, Theory and design of electron beams (D. van Nostrand Inc., New York).

Raiko, V. I., M. S. Ioffe and V. S. Zolotarev, 1961, Pribori i Tekhnika Eksperimenta 1, 29.

Rautenbach, W. L., 1960, Nucl. Inst. and Methods 9, 199.

Rautenbach, W. L., 1961, Nucl. Inst. and Methods 12, 196.

Reed, J. B., B. S. Hopkins and F. L. Audrieth, 1935, J. Amer. Chem. Soc. 57, 1159.

Renskaya, I. V. and O. A. Abramycheva, 1969, Pribory i Tekhnika Eksperimenta 4, 142.

Rose, P. H., private communication.

Rose, P. H., A. J. Gale and R. J. Connor, 1962, Kerntechnik 4, 246.

Rose, P. H. and A. Galejs, 1965, Prog. in Nucl. Techniques and Instrumentation.

Sarrouy, J. L., 1969, Proc. Int. Conf. on Ion Sources, 341 (I.N.S.T.N.-Saclay, France).

Septier, A., 1967, Focusing of charged particles, Vol. II (Academic Press, New York).

Septier, A. and H. Leal, 1964, Nucl. Inst. and Methods 29, 257.

Shockley, W., 1954, U.S. Patent 2,787,564 (published 1957).

Sidenius, G., 1965, Nucl. Inst. and Methods 38, 19.

Sidenius, G., 1969, Proc. Int. Conf. on Ion Sources, 401 (I.N.S.T.N.-Saclay, France).

Sidenius, G. and O. Skilbreid, 1961, Electromagnetic separation of radioactive isotopes, 243, ed. Higatsberger and Viehböck (Springer-Verlag, Vienna).

Siegbahn, K., 1965, Alpha, beta and gamma spectroscopy, Vol. 1 (North-Holland, Amsterdam), also Svarthholm, K. and K. Siegbahn, 1956, Ark. Mat. Astr. Fys. 33A, 21.

Stroud, P. T., 1969, J. Sci. Inst. 2, 452.

Takahashi, R., T. Tuno, M. Oshima and A. Kobayisha, 1969, Japan J. Appl. Phys. **8**, 284.
Thonemann, P. C., 1946, Nature **158**, 61, also, 1948, Proc. Royal Soc. **61**, 483.
Tonon, J. F. and M. Rabeau, 1969, Proc. Int. Ion Source Conf. 605 (I.N.S.T.N.-Saclay, France).
Tsuchimoto, T., T. Tokuyama and T. Ikeda, 1970, 4th Int. Conf. Electron and Ion Beam Science and Technology (Los Angeles, U.S.A.).
Uhler, J., 1963, Arkiv för Fysik **24**, 329.
Uhler, J. and T. Alväger, 1958, Arkiv för Fysik **14**, 473.
Valyi, L., 1970, Nucl. Inst. and Methods **79**, 315.
Van Steenbergen, A., 1967, Nucl. Inst. and Methods **51**, 245.
Venkatasubramanian, V. S. and H. E. Duckworth, 1963, Can. J. Phys. **41**, 234.
Viehbock, F., 1960, Electromag. Sep. of radioactive isotopes (Springer-Verlag, Vienna).
Von Ardenne, M., 1956, Tabellen für Elektronen, Ionenphysik und Ubermikroskopie (V.E.B. Deutscher Verlag der Wissenschaften, Berlin).
Wahlin, L., 1964, Nucl. Inst. and Methods **27**, 55, also 1965, **38**, 133.
Wakerling, R. K. and A. Guthrie, 1949, U.S.A.E.C. report T.I.D. 5217.
Wakerling, R. K. and A. C. Helmholtz, 1949, T.I.D. 5217.
Walcher, W., 1958, Electromagnetic Isotope Separators and Applications of Electromagnetically Enriched Isotopes, ed. J. Koch (North-Holland, Amsterdam).
Ward, A. G., 1950, Helv. Phys. Acta **23**, 27.
Whitton, J. L., 1969, Proc. Roy. Soc. **A311**, 63.
Wien, W., 1902, Ann. Physik **8**, 260.
Wilson, R. G., 1967, Proc. Int. Conf. on Applications of Ion Beams, 105 (C.E.N. Grenoble, France).
Wilson, R. G. and D. M. Jamba, 1967, J. Appl. Phys. **38**, 1967.
Wittkower, A. B., R. P. Bastide, N. B. Brooks and P. H. Rose, 1963, Phys. Letters **3**, 336.
Wollnick, H., 1967, Nucl. Inst. and Methods **53**, 197.
Young, R. D. and H. E. Clark, 1966, Appl. Phys. Letters **9**, 265.

| APPENDIX **1** | OPERATIONAL PROCEDURES FOR HEAVY ION PRODUCTION* |

The choice of source feed materials for heavy ion production in a particular case depends upon a large number of factors. These include the design of the source unit itself, the resolving power of the instrument, as well as the experimental requirements for beam purity, intensity and duration. The data which is listed below is largely based on operational experience, mainly with polyisotopic elements, obtained over a period of several years on electromagnetic isotope separators.

The principal sources of information which have been used are the recommended procedures and data sheets of the Oak Ridge and A.E.R.E. Harwell separation groups, but we have also drawn on other reports when these seemed appropriate.

The O.R.N.L. recommendations are particularly useful since they are derived from an unparalleled experience in the use of production separators with a wide range of elements. We are particularly grateful to L. Love and his colleagues at Oak Ridge for making this information available to us. In certain implantation experiments the beam requirements are somewhat different to those of isotope separation. For example, tin sulphide is preferred for a number of reasons as a separator source feed material, it results however in severe corrosion of copper source components and requires precise temperature control of the source oven to obtain a stable beam. For routine ion doping experiments, the easily obtainable liquid tin tetrachloride is generally much simpler to use. This, in common with several volatile chlorides (e.g. $SiCl_4$, $BCl_3$, $PCl_3$) can be introduced into the discharge from an external reservoir and the flow can be readily controlled with a needle valve.

* See also section 4.3.7 – Ion source design and operation.

454

In the light of this, we have emphasised the procedures which we have found most suitable at A.E R.E. Harwell on isotope separators which have been modified for implantation studies. Thus, although many of the recommended techniques are particularly suitable for this type of machine, and in particular for sources which operate over a temperature range of about 500–1100 °C we hope that they will provide a useful guide for other accelerator users. In the case of certain elements which have no thermally stable compounds with a suitable vapour pressure, special source techniques may be required. These include the use of sputtering probes in the discharge for the platinum elements and the use of crucibles with a very restricted orifice for very volatile materials such as mercuric fluoride. The 'in situ' preparation of volatile chloride using carbon tetrachloride vapour (see section 4.3.7.3) avoids the difficult synthesis of anhydrous chloride of polyvalent elements and is recommended in a number of cases. The technique does however require a special vapour feed inlet into the ion source oven and may present corrosion or filament deactivation problems.

The recommended source materials for each element are set bold face but in many instances other compounds may be equally suitable. Particular care must however be taken to ensure that only anhydrous compounds (particularly halides) are used in ion sources. In cases where external feed systems are used with readily hydrolysable volatile chlorides, or other halides, stringent precautions are necessary to eliminate moisture. If this is not done, the vapour feed lines and vacuum control valves may become blocked with solid hydrated reaction products, and the build up of pressure due, for example, to HCl formation may result in accidents. Finally, it is worth noting that most of these liquids and vapours are extremely corrosive and react with rubber and most plastics.

The vapour pressures listed below have been taken from a number of sources. In many cases, they are estimated or approximate and they are simply intended to provide a guide to the probable ion source operating temperature. For completeness, the appropriate physical data for each element is given even when it is not considered suitable as a source feed material.

We have noted the compounds and elements which are particularly toxic, but many of the other materials must be handled with care. In particular, most of the covalent polyhalides react vigorously with moisture to give hazardous fumes.

| Element | Symbol | At. No. | At. Wt. | Compound | Pressure mm $10^{-4}$ | $10^{-3}$ | $10^{-2}$ | | 760 | |
|---|---|---|---|---|---|---|---|---|---|---|
| | | | | | | | | | | —°C— |
| Aluminium | Al | 13(a) | 26.9815 | **Elem.** | 972 | 1082 | 1217 | [1] | 2056 | [3 |
| | | | | $AlCl_3$ | | | | | 180.2 | [3 |
| Antimony | Sb | 51 | 121.75 | **Elem.** | 425 | 475 | 533 | [1] | 1440 | [3 |
| | | | | $Sb_2O_3$ | | 399 | 450 | [2] | 1425 | [3 |
| | | | | $Sb_2S_3$ | | | | | | |
| | | | | $Sb_2I_3$ | | | | | 401.0 | [3 |
| Argon | Ar | 18 | 39.948 | **Elem.** | | | | | −185.6 | [3 |
| Arsenic | As | 33(a) | 74.9216 | **Elem.** | 204 | 237 | 247 | [1] | 610 | [3 |
| | | | | $AsH_3$ | | | | | −62.1 | [3 |
| | | | | $As_2O_3$ | | | | | 457.2 | [3 |
| | | | | $Cd_3As_2$ | | 314 | 359 | [2] | | |
| | | | | **GaAs** | | | | | | |
| | | | | $AsCl_3$ | | | | | 130.3 | [3 |
| Barium | Ba | 56 | 137.34 | **Elem.** | | 625 | 716 | [2] | | |
| | | | | | 462 | 527 | 610 | [1] | 1638 | [3 |
| | | | | **BaCl₂** | | | 970 | [2] | | |
| | | | | $BaBr_2$ | | 780 | 785 | [2] | | |
| Beryllium | Be | 4[a] | 9.01218 | **Elem.** | | 1395 | 1552 | [2] | | |
| | | | | | 997 | 1097 | 1227 | [1] | | |
| | | | | $BeCl_2$ | | 176 | 209 | [2] | 487 | [3 |
| | | | | $BeBr_2$ | | 174 | 207 | [2] | 474 | [3 |
| | | | | $BeF_2$ | | 562 | 623 | [2] | | |
| | | | | $BeI_2$ | | 160 | 194 | [2] | 487 | [3 |
| Bismuth | Bi | 83(a) | 208.9806 | **Elem.** | 517 | 587 | 672 | [1] | 1420 | [3 |
| | | | | $BiCl_3$ | | | | | 441 | [3 |
| | | | | $BiBr_3$ | | | | | 461 | [3 |
| | | | | $BiI_3$ | | | | | | |
| Boron | B | 5 | 10.81 | **Elem.** | 1707 | 1867 | 2027 | [1] | | |
| | | | | $BF_3$ | | | | | −110.7 | [3 |
| | | | | **BCl₃** | | | | | 12.7 | [3 |
| | | | | $B_2H_6$ | | | | | −86.5 | [3 |

| ther vapour ressures | Maximum ionisation cross section $10^{-16}$ cm$^2$ | Ionisation potential eV | Comments |
|---|---|---|---|
| mm at 100 °C [3] | 6.2 | 6.0 | Take care to avoid metallisation of insulators. |
| | 7.2 | 8.5 | Toxic. |
| mm at 163.6 °C [3] | | | |
| | 2.8 | 15.7 | No particular problems. |
| mm at 212.5 °C [3] | 5.0 | 10.5 | Toxic. GaAs decomposes conveniently and controllably in source furnace. |
| mm at 11.4 °C [3] | | | |
| | 17.3 | 5.2 | Toxic. Anhydrous halides convenient. Element can also be used, but surface oxide layer may inhibit vaporisation. Activates filaments and electrodes. |
| | 3.2 | 9.3 | Toxic (BeO). |
| 00 mm at 343 °C [3] 10 mm at 282 °C [3] 00 mm at 425 °C [2] | 17.3 | 8.0 | No particular problem. |
| | 2.6 | 8.3 | BCl$_3$ convenient, but deactivates filament (use tungsten not tantalum), avoid excess BCl$_3$ in discharge to minimise reaction with filament (see fig. 4.94). Boron halides hydrolyse readily, clean dry apparatus must be used. B$_2$H$_6$ spontaneously inflammable – not recommended. Diluted stabilised mixtures with inert gases are commercially available, but boron beam fraction low. |

| Element | Symbol | At. No. | At. Wt. | Compound | Pressure mm 10⁻⁴ | 10⁻³ | 10⁻² | | 760 | |
|---|---|---|---|---|---|---|---|---|---|---|
| | | | | | | | °C | | | |
| Bromine | Br | 35 | 79.904 | **Elem.** | | | | | 58.2 | [ |
| | | | | LiBr | 628 | 733 | 868 | [4] | | |
| | | | | NaBr | 687 | 791 | 932 | [4] | | |
| | | | | **KBr** | 677 | 780 | 920 | [4] | | |
| | | | | Numerous other bromides | | | | | | |
| Cadmium | Cd | 48 | 112.40 | Elem. | 177 | 217 | 265 | [1] | 765 | [ |
| | | | | CdCl₂ | | 386 | 437 | [2] | 967 | [ |
| | | | | CdBr₂ | | 341 | 390 | [2] | | |
| | | | | CdI₂ | | 289 | 340 | [2] | 796 | [ |
| | | | | CdO | | 691 | 772 | [2] | | |
| | | | | **CdS** | | 610 | 686 | [2] | | |
| | | | | CdSe | | 588 | 661 | [2] | | |
| | | | | CdTe | | 505 | 574 | [2] | | |
| Calcium | Ca | 20 | 40.08 | **Elem.** | 456 | 522 | 597 | [1] | 1487 | [ |
| | | | | **CaCl₂** | | 860 | 970 | [2] | | |
| | | | | CaBr₂ | | 780 | 875 | [2] | | |
| | | | | CaI₂ | | 485 | 555 | [2] | | |
| Carbon | C | 6 | 12.011 | **CO** | | | | | −191.3 | [ |
| | | | | **CO₂** | | | | | −78.2 | [ |
| | | | | CS₂ | | | | | 46.5 | [ |
| | | | | CCl₄ | | | | | 76.7 | [ |
| | | | | CaCO₃ | | | | | | |
| | | | | BaCO₃ | | | | | | |
| Cerium | Ce | 58 | 140.12 | Elem. | 1377 | 1522 | 1697 | [1] | | |
| | | | | **CeCl₃** | | 740 | 812 | [2] | | |
| | | | | CeBr₃ | | 619 | 684 | [2] | | |
| | | | | CeO₂+CCl₄ | see note (b) | | | | | |
| Cesium | Cs | 55(a) | 132.9055 | Elem. | 78 | 114 | 155 | [1] | 690 | [ |
| | | | | CsF | 607 | 697 | 826 | [4] | 1251 | [ |
| | | | | **CsCl** | 642 | 730 | 864 | [4] | 1300 | [ |
| | | | | CsBr | 627 | 734 | 866 | [4] | 1300 | [ |
| | | | | CsI | 622 | 724 | 855 | [4] | 1280 | [ |
| Chlorine | Cl | 17 | 35.453 | **Elem.** | | | | | −33.8 | [ |
| | | | | **NaCl** | 741 | 850 | 996 | [4] | | |
| | | | | Numerous other chlorides | | | | | | |
| | | | | (e.g. CCl₄, PCl₃ as gaseous feed materials) | | | | | | |

| Other vapour pressures | Maximum ionisation cross section $10^{-16}$ cm$^2$ | Ionisation potential eV | Comments |
|---|---|---|---|
| 00 mm at 9.3 °C [3] | 4.6 | 11.8 | Readily ionisable as gaseous element (caution), or as solid alkali bromide. In presence of water vapour or hydro-carbons large HBr$^+$ beams may be formed. Bromine reacts with hot metallic source unit components. (Graphite preferred). |
| | 6.3 | 9.0 | No particular problems. |
| | 10.4 | 6.1 | Anhydrous halides convenient. Element can also be used but surface oxide layer may inhibit vaporisation. Some filament and electrode activation. |
| | 2.0 | 11.2 | CO (toxic) gives best yield, CO$_2$ convenient but less efficient. |
| 400 mm at 28.0 °C [3] 40 mm at  4.3 °C [3] | | | |
| | 15.9 | 6.5 | Element oxidises readily. CeCl$_3$ hygroscopic, reduced filament activation with oxide plus carbon tetrachloride. For all lanthanide elements, it is preferable to use freshly prepared, low temperature, oxides to obtain maximum reactivity with carbon tetrachloride. |
| 1 mm at 279 °C [3] 1 mm at 712 °C [3] 1 mm at 744 °C [3] 1 mm at 748 °C [3] 1 mm at 738 °C [3] | 10.8 | 3.9 | No particular problems using anhydrous halides, but activates filament and electrodes. |
| | 3.4 | 13.0 | No particular problems, intense beams readily obtained from wide variety of halides. In presence of water vapour or hydrocarbons large HCl$^+$ beams may be formed. Chlorine reacts with hot metallic source unit components. (Graphite preferred). |

| Element | Symbol | At. No. | At. Wt. | Compound | Pressure mm 10$^{-4}$ | 10$^{-3}$ | 10$^{-2}$ | | 760 | |
|---|---|---|---|---|---|---|---|---|---|---|
| | | | | | | °C | | | | |
| Chromium | Cr | 24 | 51.996 | Elem. | 1157 | 1267 | 1397 | [1] | | |
| | | | | CrCl$_3$ | | | | | | |
| | | | | CrBr$_3$ | | 509 | 562 | [2] | | |
| | | | | Cr$_2$O$_3$+CCl$_4$ | see note (b) | | | | | |
| Cobalt | Co | 27(a) | 58.9332 | Elem. | 1275 | 1382 | 1517 | [1] | | |
| | | | | CoCl$_2$ | 507 | 612 | 672 | [4] | | |
| | | | | CoBr$_2$ | | | | | | |
| Copper | Cu | 29 | 63.546 | Elem. | 1027 | 1130 | 1257 | [1] | 2595 | [3 |
| | | | | CuCl$_2$ | decomposes to Cu$_2$Cl$_2$ | | | | | |
| | | | | CuBr$_2$ | | | | | | |
| | | | | Cu$_2$Cl$_2$ | | | | | | |
| | | | | CuO+CCl$_4$ | see note (b) | | | | | |
| Dysprosium | Dy | 66 | 162.50 | Elem. | 897 | 997 | 1117 | [1] | | |
| | | | | DyF$_3$ | | 1025 | 1140 | [2] | | |
| | | | | DyCl$_3$ | | 610 | 700 | [2] | | |
| | | | | DyI$_3$ | | 587 | 660 | [2] | | |
| | | | | Dy$_2$O$_3$+CCl$_4$ | see note (b) | | | | | |
| Erbium | Er | 68 | 167.26 | Elem. | | 1135 | 1270 | [2] | | |
| | | | | | 947 | 1052 | 1177 | [1]Est. | | |
| | | | | ErF$_3$ | | 1020 | 1130 | [2] | | |
| | | | | ErCl$_3$ | | 600 | 680 | [2] | | |
| | | | | ErBr$_3$ | | 585 | 665 | [2] | | |
| | | | | ErI$_3$ | | 578 | 650 | [2] | | |
| | | | | Er$_2$O$_3$+CCl$_4$ | see note (b) | | | | | |
| Europium | Eu | 63 | 151.96 | Elem. | 466 | 532 | 611 | [1] | | |
| | | | | EuCl$_3$ | | 652 | 725 | [2] | | |
| | | | | EuCl$_2$ | | 862 | 965 | [2] | | |
| | | | | Eu$_2$O$_3$+CCl$_4$ | see note (b) | | | | | |
| Fluorine | F | 9(a) | 18.9984 | Elem. | | | | | −187.9 | [3 |
| | | | | NaF | 927 | 1049 | 1218 | [4] | | |
| | | | | Numerous other fluorides | | | | | | |
| Gadolinium | Gd | 64 | 157.25 | Elem. | 1077 | 1192 | 1327 | [1] | | |
| | | | | GdF$_3$ | | 1040 | 1155 | [2] | | |
| | | | | CdCl$_3$ | | 630 | 713 | [2] | | |
| | | | | GdBr$_3$ | | 610 | 690 | [2] | | |
| | | | | GdI$_3$ | | 580 | 655 | [2] | | |
| | | | | Gd$_2$O$_3$+CCl$_4$ | see note (b) | | | | | |

| ther vapour essures | Maximum ionisation cross section $10^{-16}$ cm$^2$ | Ionisation potential eV | Comments |
|---|---|---|---|
| $)^{-1}$ mm at 417 °C [2] | 5.1 | 6.7 | No particular problem, one form of CrCl$_3$ very hygroscopic. For internal chlorination with CCl$_4$ use freshly prepared low temperature oxide for maximum reactivity. |
| 0 mm at 770 °C [3] | 6.0 | 7.8 | No particular problems with anhydrous CoCl$_2$. |
| mm at 499 °C [2] | 3.8 | 7.7 | No particular problems with anhydrous CuCl$_2$ which decomposes controllably to Cu$_2$Cl$_2$. |
| | 13.9 | 6.8 | As cerium. |
| | 13.2 | 6.1 | As cerium. |
| | 14.6 | 5.6 | Preferably as element or oxide plus carbon tetrachloride. N.B. EuCl$_3$ decomposes to less volatile EuCl$_2$. |
| mm at 1077 °C [3] | 1.0 | 17.3 | The element and certain compounds are toxic. No particular problems with a wide variety of fluorides. |
| | 12.9 | 6.7 | As cerium. |

462                                                            APPENDIX 1

| Element | Symbol | At. No. | At. Wt. | Compound | $10^{-4}$ | $10^{-3}$ | $10^{-2}$ | | 760 | |
|---|---|---|---|---|---|---|---|---|---|---|
| | | | | | Pressure mm | | | | | |
| | | | | | °C | | | | | |
| Gallium | Ga | 31 | 69.72 | Elem. | 907 | 1007 | 1132 | [1] | | |
| | | | | $GaCl_3$ | | | | | 200 | [ |
| | | | | GaN | | 715 | 825 | [2] | | |
| | | | | $Ga_2O_3+CCl_4$ | see note (b) | | | | | |
| Germanium | Ge | 32 | 72.59 | Elem. | 1137 | 1257 | 1397 | [1] | | |
| | | | | $GeH_4$ | | | | | −88.9 | [ |
| | | | | $GeCl_4$ | | | | | 84.0 | [ |
| | | | | $GeS_2$ | sublimes | | | | | |
| Gold | Au | 79(a) | 196.9665 | Elem. | 1132 | 1252 | 1397 | [1] | | |
| Hafnium | Hf | 72 | 178.49 | Elem. | 1997 | 2177 | 2397 | [1] | | |
| | | | | $HfCl_4$ | | 94 | 121 | [2] | | |
| | | | | $HfO_2+CCl_4$ | see note (b) | | | | | |
| Helium | He | 2 | 4.00260 | Elem. | | | | | −268.6 | [ |
| Holmium | Ho | 67(a) | 164.9303 | Elem. | 947 | 1052 | 1177 | [1] | | |
| | | | | $HoCl_3$ | | 1000 | | | | |
| | | | | $Ho_2O_3+CCl_4$ | see note (b) | | | | | |
| Hydrogen | H | 1 | 1.0080 | Elem. | | | | | −252.5 | [ |
| | | | | $H_2O$ | | | | | 100 | |
| | | | | $H_2S$ | | | | | −60.4 | [ |
| Indium | In | 49 | 114.82 | Elem. | 742 | 837 | 947 | [1] | | |
| | | | | $InCl_3$ | | 224 | 256 | [2] | | |
| | | | | $InBr_3$ | | 113 | 141 | [2] | | |
| Iodine | I | 53(a) | 126.9045 | Elem. | | | | | 183 | [3 |
| | | | | KI | | 499 | 564 | [2] | | |
| | | | | AgI | 722 | 845 | 995 | [4] | | |
| | | | | Numerous other iodides | | | | | | |
| Iridium | Ir | 77 | 192.22 | Elem. | 2107 | 2287 | 2497 | [1] | | |
| Iron | Fe | 26 | 55.847 | Elem. | 1227 | 1342 | 1477 | [1] | 2735 | [3 |
| | | | | $FeCl_2$ | | 428 | 482 | [2] | | |
| | | | | $Fe_2O_3+CCl_4$ | see note (b) | | | | | |

| Other vapour pressures | Maximum ionisation cross section $10^{-16}$ cm$^2$ | Ionisation potential eV | Comments |
|---|---|---|---|
| mm at 48 °C [2] [3] | 5.9 | 6.0 | No problem as element. |
| 0 mm at 16.2 °C [3] | 5.7 | 8.1 | Element convenient but requires higher temperature source, other volatile materials may be suitable. GeCl$_4$ similar to SiCl$_4$. |
| | 5.9 | 9.2 | Element convenient but requires higher temperature source, volatile gold compounds are thermally unstable. Readily ionised in sputtering source. |
| | 10.4 | 6.8 | Only limited experience, HfO$_2$+CCl$_4$ probably best. |
| | 0.21 | 24.5 | No particular problems. |
| | 13.4 | 6.0 | As cerium. |
| 7.5 mm at 20 °C | 0.22 | 13.5 | No particular operational problems but element forms explosive mixture with air. Water vapour can also be used. |
| | 7.7 | 5.8 | No particular problems. |
| 1 mm at 38.7 °C [3]<br>1 mm at 745 °C [3]<br>1 mm at 820 °C [3] | 6.8 | 10.6 | No particular problem with volatile iodides. Dry silver iodide is particularly suitable since it decomposes controllably and gives almost pure iodine beam. |
| | 7.7 | 9.1 | No convenient thermally stable compounds. Best used in sputtering source, or very high temperature source. |
| | 6.3 | 7.8 | No particular problems. |

| Element | Symbol | At. No. | At. Wt. | Compound | Pressure mm | | | | | |
|---|---|---|---|---|---|---|---|---|---|---|
| | | | | | $10^{-4}$ | $10^{-3}$ | $10^{-2}$ | | 760 | |
| | | | | | °C | | | | | |
| Krypton | Kr | 36 | 83.80 | Elem. | | | | | −156.7 | [3 |
| Lanthanum | La | 57 | 138.9055 | Elem. | 1422 | 1562 | 1727 | [1] | | |
| | | | | $LaCl_3$ | | 751 | 825 | [2] | | |
| | | | | $LaBr_3$ | | 642 | 708 | [2] | | |
| | | | | $La_2O_3 + CCl_4$ | see note (b) | | | | | |
| Lead | Pb | 82 | 207.2 | Elem. | 547 | 625 | 715 | [1] | 1744 | [3 |
| | | | | $PbF_2$ | | 528 | 606 | [2] | 1293 | [3 |
| | | | | $PbCl_2$ | | 385 | 431 | [2] | 954 | [3 |
| | | | | PbS | | 611 | 682 | [2] | 1281 | [3 |
| Lithium | Li | 3 | 6.941 | Elem. | 404 | 467 | 537 | [1] | 1114 | [3 |
| | | | | LiF | | 679 | 757 | [2] | 1681 | [3 |
| | | | | LiCl | | 519 | 586 | [2] | 1382 | [3 |
| | | | | LiI | | 463 | 530 | [2] | 1171 | [3 |
| Lutetium | Lu | 71 | 174.97 | Elem. | 1277 | 1412 | 1572 | [1] | | |
| | | | | $LuCl_3$ | | | | | | |
| | | | | $Lu_2O_3 + CCl_4$ | see note (b) | | | | | |
| Magnesium | Mg | 12 | 24.305 | Elem. | 327 | 377 | 439 | [1] | 1107 | [3 |
| | | | | $MgF_2$ | | 1058 | 1158 | [2] | | |
| | | | | $MgCl_2$ | | ≈500 | | | | |
| | | | | $MgBr_2$ | | 482 | 552 | [2] | | |
| Manganese | Mn | 25(a) | 54.9380 | Elem. | 747 | 837 | 937 | [1] | 2151 | [3 |
| | | | | $MnCl_2$ | | | | | 1190 | [3 |
| Mercury | Hg | 80 | 200.59 | Elem. | −7 | 16 | 46 | [1] | 357 | [3 |
| | | | | $HgF_2$ | | | | | | |
| Molybdenum | Mo | 42 | 95.94 | Elem. | 2117 | 2307 | 2527 | [1] | 4804 | [3 |
| | | | | $MoCl_5$ | | 20 | 43 | [2] | | |
| | | | | $MoO_2 + CCl_4$ | see note (b) | | | | | |
| | | | | $MoO_3 + CCl_4$ | | | | | | |
| | | | | $MoO_3$ | | 569 | 611 | [2] | 1151 | [3 |
| Neodymium | Nd | 60 | 144.24 | Elem. | 1047 | 1167 | 1302 | [1] | | |
| | | | | $NdCl_3$ | | 691 | 759 | [2] | | |
| | | | | $Nd_2O_3 + CCl$ | see note (b) | | | | | |

| her vapour essures | Maximum ionisation cross section $10^{-16}$ cm$^2$ | Ionisation potential eV | Comments |
|---|---|---|---|
| | 4.1 | 13.9 | No particular problems. |
| | 16.1 | 5.6 | As cerium. |
| | 7.9 | 7.4 | No particular problems. |
| | 3.3 | 5.4 | As calcium. |
| | 10.9 | 6.2 | As cerium. |
| | 5.4 | 7.6 | If the element is used, take care to avoid metallisation of insulators. |
| mm at 736 °C [3] | 6.8 | 7.4 | No particular problems. |
| | 6.4 | 10.4 | Toxic, element and most compounds require special precautions because of high volatility, $HgF_2$ most suitable. |
| | 6.9 | 7.4 | Oxide plus carbon tetrachloride preferred, but if volatile $MoO_3$ used, careful temperature control required. $MoO_3$ can be used directly without chlorination but tends to decompose under vacuum to produce deposits of $MoO_2$. Smaller beams can conveniently be obtained in certain sources using molybdenum filament with high arc voltage, or with a volatile halide. |
| | 15.6 | 6.3 | As cerium. |

| Element | Symbol | At. No. | At. Wt. | Compound | Pressure mm $10^{-4}$ | $10^{-3}$ | $10^{-2}$ | | 760 |
|---|---|---|---|---|---|---|---|---|---|
| | | | | | | °C | | | |
| Neon | Ne | 10 | 20.179 | Elem. | | | | | −246 |
| Nickel | Ni | 28 | 58.71 | Elem. | 1262 | 1382 | 1527 | [1] | 2732 |
| | | | | $NiCl_2$ | | 491 | 545 | [2] | |
| Niobium | Nb | 41(a) | 92.9064 | Elem. | 2277 | 2447 | 2657 | [1] | |
| Nitrogen | N | 7 | 14.0067 | Elem. | | | | | −195.8 |
| Osmium | Os | 76 | 190.2 | Elem. | 2487 | 2687 | 2917 | [1] | |
| | | | | $OsO_4$ | | −70 | −52.2 | [2] | |
| | | | | $Os+O_2$ | | | | | |
| Oxygen | O | 8 | 15.9994 | Elem. | | | | | −183.1 |
| | | | | $CO_2$ | | | | | |
| Palladium | Pd | 46 | 106.4 | Elem. | 1192 | 1317 | 1462 | [1] | |
| | | | | Elem. | | 1210 | 1360 | [2] | |
| Phosphorus | P | 15(a) | 30.9738 | Elem. (Yellow) | | | | | 280.0 |
| | | | | Elem. (Violet) | | | | | 417 |
| | | | | $PH_3$ | | | | | −87.5 |
| | | | | $PCl_3$ | | | | | 74.2 |
| | | | | $PBr_3$ | | | | | 175.3 |
| Platinum | Pt | 78 | 195.09 | Elem. | 1747 | 1907 | 2097 | [1] | 4407 |
| Potassium | K | 19 | 39.102 | Elem. | 123 | 161 | 208 | [1] | 774 |
| | | | | KF | | 596 | 664 | [2] | 1502 |
| | | | | KCl | | 567 | 639 | [2] | 1407 |
| | | | | KBr | | 546 | 617 | [2] | 1383 |
| | | | | KI | | 499 | 564 | [2] | 1324 |
| Praseodymium | Pr | 59(a) | 140.9077 | Elem. | 1147 | 1277 | 1427 | [1] | |
| | | | | $PrCl_3$ | | ≈800 | | | |
| | | | | $Pr_2O_3+CCl_4$ | see note (b) | | | | |
| Rhenium | Re | 75 | 186.2 | Elem. | 2587 | 2807 | 3067 | [1] | |
| | | | | $Re_2O_7$ | | 133 | 157 | [2] | 362.4 |
| | | | | $Re+O_2$ | | | | | |
| Rhodium | Rh | 45(a) | 102.9055 | Elem. | 1707 | 1857 | 2037 | [1] | |

| her vapour ssures | Maximum ionisation cross section $10^{-16}$ cm$^2$ | Ionisation potential eV | Comments |
|---|---|---|---|
| | 0.82 | 21.5 | No particular problems. |
| | 5.5 | 7.6 | No particular problem with anhydrous $NiCl_2$. |
| | 7.7 | 6.9 | Probably best as oxide plus $CCl_4$, or element in sputtering source. |
| | 1.5 | 14.5 | No particular problem. |
| mm at 6.1 °C [2] | 8.1 | 8.7 | Toxic. Use as oxide, or preferably as element plus oxygen, or as element in sputtering source. |
| | 1.3 | 13.6 | No particular problem. $CO_2$ convenient and minimises source and filament oxidation. |
| | 6.1 | 8.3 | As gold. |
| mm at 76.6 °C [3] mm at 237 °C [3] mm at 10.2 °C [3] mm at 7.8 °C [3] | 4.5 | 10.9 | $PCl_3$ convenient, some filament deactivation if excess vapour used. Phosphorus halides hydrolyse readily, clean dry apparatus must be used. Phosphine (toxic) can also be used but large $PH^+$ fraction. |
| | 6.6 | 8.9 | As iridium. |
| | 7.2 | 4.3 | As cesium. |
| | 16.0 | 5.8 | As cerium. |
| | 9.1 | 7.9 | Probably best as rhenium plus oxygen, or for smaller beams use rhenium filament with high arc voltage, or as element in sputtering source. |
| | 6.2 | 7.7 | As iridium. |

| Element | Sym-bol | At. No. | At. Wt. | Compound | Pressure mm 10$^{-4}$ | 10$^{-3}$ | 10$^{-2}$ | | 760 | |
|---|---|---|---|---|---|---|---|---|---|---|
| | | | | | | —— °C —— | | | | |
| Rubidium | Rb | 37 | 85.4678 | Elem. | 94 | 129 | 173 | [2] | 579 | ▮ |
| | | | | RbF | | 573 | 642 | [2] | 1408 | ▮ |
| | | | | **RbCl** | | 555 | 627 | [2] | 1381 | ▮ |
| | | | | RbBr | | 520 | 586 | [2] | 1352 | ▮ |
| | | | | RbI | | 483 | 547 | [2] | 1304 | ▮ |
| Ruthenium | Ru | 44 | 101.07 | Elem. | 1987 | 2147 | 2347 | [1] | | |
| | | | | RuF$_5$ | | 31 | 50 | [2] | | |
| Samarium | Sm | 62 | 150.4 | Elem. | 580 | 653 | 742 | [1] | | |
| | | | | **SmCl$_3$** | | 677 | 757 | [2] | | |
| | | | | SmCl$_2$ | | 858 | 967 | [2] | | |
| | | | | **Sm$_2$O$_3$+CCl$_4$** | see note (b) | | | | | |
| Scandium | Sc | 21(a) | 44.9559 | Elem. | 1107 | 1232 | 1377 | [1] | | |
| | | | | **ScCl$_3$** | 662 | 717 | 787 | [4] | | |
| | | | | **Sc$_2$O$_3$+CCl$_4$** | see note (b) | | | | | |
| Selenium | Se | 34 | 78.96 | Elem. | 164 | 199 | 243 | [1] | 680 | [ |
| | | | | H$_2$Se | | | | | | |
| | | | | SeO$_2$ | | | | | 317.0 | [ |
| | | | | **CdSe** | | 588 | 661 | [2] | | |
| | | | | ZnSe | | | | | | |
| | | | | PbSe | | | | | | |
| Silicon | Si | 14 | 28.086 | Elem. | 1337 | 1472 | 1632 | [1] | 2287 | [ |
| | | | | SiH$_4$ | | | | | −111.5 | [ |
| | | | | **SiCl$_4$** | | | | | 56.8 | [ |
| | | | | SiS$_2$ | | | | | | |
| Silver | Ag | 47 | 107.868 | **Elem.** | 832 | 922 | 1027 | [1] | 2212 | [ |
| | | | | **AgCl** | | 593 | 678 | [2] | 1564 | [ |
| | | | | AgBr | | 558 | 632 | [2] | | |
| Sodium | Na | 11(a) | 22.9898 | Elem. | 193 | 235 | 289 | [1] | 892 | [ |
| | | | | NaF | 927 | 1049 | 1218 | [4] | 1704 | [. |
| | | | | **NaCl** | 741 | 850 | 996 | [4] | 1465 | [ |
| | | | | NaBr | 687 | 791 | 932 | [4] | 1392 | [ |
| | | | | NaI | 657 | 753 | 885 | [4] | 1304 | [ |
| Strontium | Sr | 38 | 87.62 | **Elem.** | 404 | 465 | 537 | [1] | | |
| | | | | SrF$_2$ | | 1195 | 1315 | [2] | | |
| | | | | **SrCl$_2$** | | 856 | 965 | [2] | | |
| | | | | SrBr$_2$ | | 820 | 925 | [2] | | |
| | | | | SrI$_2$ | | 650 | 735 | [2] | | |

| er vapour ssures | Maximum ionisation cross section $10^{-16}$ cm$^2$ | Ionisation potential eV | Comments |
|---|---|---|---|
| | 8.4 | 4.2 | As cesium. |
| | 6.7 | 7.7 | Difficult to prepare and run suitable compounds, probably best as element in sputtering source. |
| | 14.9 | 6.6 | As cerium. N.B. SmCl$_3$ partially decomposes to SmCl$_2$. |
| | 9.5 | 6.7 | As calcium, but higher temperature source if used as element. |
| | 5.0 | 9.7 | Toxic. Cadmium selenide convenient source material. |
| ) mm at 5.4 °C [3] | 5.4 | 8.1 | No particular problems, similar precautions to PCl$_3$ (see phosphorus). |
| | 5.1 | 7.5 | No particular problems, (store halides in dark, AgI decomposes in ion source). |
| nm at 439 °C [3] nm at 1077 °C [3] nm at 865 °C [3] nm at 806 °C [3] nm at 767 °C [3] | 4.0 | 5.1 | As cesium. |
| | 12.9 | 5.7 | As calcium. |

| Element | Symbol | At. No. | At. Wt. | Compound | Pressure mm $10^{-4}$ $10^{-3}$ $10^{-2}$ °C | | | 760 |
|---|---|---|---|---|---|---|---|---|
| Sulphur | S | 16 | 32.06 | Elem. | 55 | 80 | 109 [1] | 444.6 |
| | | | | $H_2S$ | | | | −60.4 |
| | | | | $SO_2$ | | | | −10.0 |
| | | | | $CS_2$ | | | | 46.5 |
| | | | | COS | | | | −49.9 |
| | | | | $S_2Cl_2$ | | | | 138.0 |
| | | | | $Sb_2S_5$ | | | | |
| | | | | CdS | | 610 | 686 [2] | |
| | | | | ZnS | | 774 | 863 [2] | |
| Tantalum | Ta | 73 | 180.9479 | Elem. | 2587 | 2807 | 3057 [1] | |
| | | | | Elem. | | (3714) | (4158) [2] | |
| | | | | $TaF_5$ | | 18 | 35 [2] | 230 |
| | | | | $TaCl_5$ | | 39 | 60 [2] | |
| | | | | $TaBr_5$ | | 90 | 115 [2] | |
| | | | | $TaI_5$ | | 116 | 154 [2] | |
| | | | | $Ta_2O_5+CCl_4$ | see note (b) | | | |
| Tellurium | Te | 52 | 127.60 | Elem. | 280 | 323 | 374 [1] | |
| | | | | $TeF_6$ | | | | −38.6 |
| | | | | $TeCl_2$ | | | | |
| | | | | $TeCl_4$ | | | | 392 |
| | | | | $TeO_2$ | | 590 | 651 [2] | |
| | | | | CdTe | | 505 | 574 [2] | |
| | | | | ZnTe | | | | |
| Terbium | Tb | 65 | 158.9254 | Elem. | 1147 | 1277 | 1427 [1] | |
| | | | | $TbCl_3$ | | (<1000) | | |
| | | | | $Tb_2O_3+CCl_4$ | see note (b) | | | |
| Thallium | Tl | 81 | 204.37 | Elem. | 463 | 530 | 609 [1] | 1457 |
| | | | | TlF | | 81 | 106 [2] | |
| | | | | TlCl | | 270 | 324 [2] | 807 |
| | | | | TlBr | | 282 | 330 [2] | 819 |
| | | | | TlI | | 283 | 336 [2] | 823 |
| Thorium | Th | 90(a) | 232.0381 | Elem. | 1977 | 2167 | 2407 [1] | |
| | | | | $ThCl_4$ | | (<1000) | | |
| | | | | $ThO_2+CCl_4$ | see note (b) | | | |
| Thulium | Tm | 69(a) | 168.9342 | Elem. | 680 | 757 | 847 [1] | |
| | | | | $TmCl_3$ | | (<1000) | | |
| | | | | $Tm_2O_3+CCl_4$ | see note (b) | | | |

| er vapour ssures | Maximum ionisation cross section $10^{-16}$ cm$^2$ | Ionisation potential eV | Comments |
|---|---|---|---|
| | 3.9 | 10.3 | No particular problems. $SO_2$ convenient source feed gas. If ZnS is used $Zn^{2+}$ may contaminate $S^+$. All sulphur compounds tend to corrode copper source components. |
| mm at 10.4 °C [3] | | | |
| m at 15.7 °C [3] | | | |
| | 9.6 | 7.9 | Oxide plus carbon tetrachloride preferred. Smaller beams can be obtained from Ta filament (see molybdenum). |
| | 7.1 | 9.0 | Toxic. Cadmium telluride preferred if machine resolution is adequate to separate $Cd^+$ and $Te^+$. |
| ) mm at 240 °C [2] ) mm at 306 °C [2] | | | |
| | 12.7 | 6.7 | As cerium. |
| | 7.8 | 6.1 | Toxic. No particular problems. |
| | 16.7 | 7.0 | No particular problems except filament activation and alloying. Rhenium filaments preferred. |
| | 13.0 | 6.2 | As cerium. |

| Element | Symbol | At. No. | At. Wt. | Compound | Pressure mm $10^{-4}$ | $10^{-3}$ | $10^{-2}$ | | 760 |
|---|---|---|---|---|---|---|---|---|---|
| | | | | | °C | | | | |
| Tin | Sn | 50 | 118.69 | Elem. | 997 | 1107 | 1247 | [1] | |
| | | | | SnCl$_4$ | | | | | 113 |
| | | | | SnS | | 534 | 602 | [2] | |
| | | | | SnSe | | | | | |
| | | | | SnH$_4$ | | | | | −25.3 |
| Titanium | Ti | 22 | 47.90 | Elem. | 1442 | 1577 | 1737 | [1] | |
| | | | | TiCl$_4$ | | | | | |
| | | | | TiI$_4$ | | | | | |
| | | | | TiO$_2$+CCl$_4$ | see note (b) | | | | |
| Tungsten | W | 74 | 183.85 | Elem. | 2757 | 2977 | 3227 | [1] | |
| | | | | WF$_6$ | | | | | 17.3 |
| | | | | WO$_3$ | | 1028 | 1093 | [2] | |
| | | | | WO$_3$+CCl$_4$ | see note (b) | | | | |
| Uranium | U | 92 | 238.029 | Elem. | 1582 | 1737 | 1927 | [1] | |
| | | | | UCl$_4$ | (<1000) | | | | |
| | | | | UO$_2$+CCl$_4$ | see note (b) | | | | |
| Vanadium | V | 23 | 50.9414 | Elem. | 1547 | 1687 | 1847 | [1] | |
| | | | | VF$_3$ | | 687 | 765 | [2] | |
| | | | | VOCl$_3$ | | | | | |
| Xenon | Xe | 54 | 131.30 | Elem. | | | | | −108 |
| Ytterbium | Yb | 70 | 173.04 | Elem. | 417 | 482 | 557 | [1] | |
| | | | | Elem. | | 565 | 627 | [2] | |
| | | | | YbCl$_3$ | | 614 | 700 | [2] | |
| | | | | Yb$_2$O$_3$+CCl$_4$ | see note (b) | | | | |
| Yttrium | Y | 39(a) | 88.9059 | Elem. | 1332 | 1467 | 1632 | [1] | |
| | | | | YCl$_3$ | | 613 | 698 | [2] | |
| | | | | Y$_2$O$_3$+CCl$_4$ | see note (b) | | | | |
| Zinc | Zn | 30 | 65.37 | Elem. | 247 | 292 | 344 | [1] | |
| | | | | ZnS | | 774 | 863 | [2] | |
| | | | | ZnSe | | | | | |
| | | | | ZnTe | | | | | |
| | | | | ZnCl$_2$ | | 273 | 317 | [2] | |
| Zirconium | Zr | 40 | 91.22 | Elem. | 1987 | 2177 | 2397 | [1] | |
| | | | | ZrF$_4$ | | 481 | 531 | [2] | |
| | | | | ZrCl$_4$ | | 93 | 119 | [2] | |
| | | | | ZrO$_2$+CCl$_4$ | see note (b) | | | | |

| er vapour ssures | Maximum ionisation cross section $10^{-16}$ cm$^2$ | Ionisation potential eV | Comments |
|---|---|---|---|
| mm at 3.5 °C [2] | 7.7 | 7.3 | No particular problems. Precautions for SnCl$_4$ similar to PCl$_3$. |
| mm at 20 °C [2] mm at 191 °C [2] | 8.7 | 6.8 | No particular problems. Precautions for TiCl$_4$ similar to PCl$_3$. |
| | 9.2 | 8.1 | As molybdenum. |
| | 15.9 | 6.1 | Oxide plus CCl$_4$ preferred, freshly prepared low temperature oxide should be used (UO$_2$), UCl$_4$ preparation difficult (reflux U$_3$O$_8$ with hexachloropropene for several hours). |
| mm at 15 °C [2] | 7.9 | 6.7 | No particular problems. |
| | 6.2 | 12.1 | No particular problems. |
| | 12.8 | 7.1 | As cerium. |
| | 11.9 | 6.5 | As cerium. |
| | 4.9 | 9.4 | No particular problems. |
| | 10.9 | 6.9 | No particular problems. Use freshly prepared low temperature oxide with CCl$_4$. |

Recommended source materials are set in boldface but see also comments in the introduction above.

(a) Mono-isotopic element.

(b) When oxides are used with $CCl_4$ for internal chlorination, precise temperature control is not normally required. The source temperature should be raised well above the corresponding chloride operating temperature.

[1] Honig, R. E., 1962, Vapor Pressure Data for the Solid and Liquid Elements, R.C.A Review **23**, 4.

[2] Oak Ridge National Laboratory Data.

[3] Stull, D. R., 1947, Vapor Pressure of Pure Substances – Inorganic Compounds, Industrial and Engineering Chemistry **39**, 4.

[4] Quill, L. L., 1950, The Chemistry and Metallurgy of Miscellaneous Materials –Thermodynamics (McGraw Hill, New York).

# NATURAL ABUNDANCE OF THE ISOTOPES

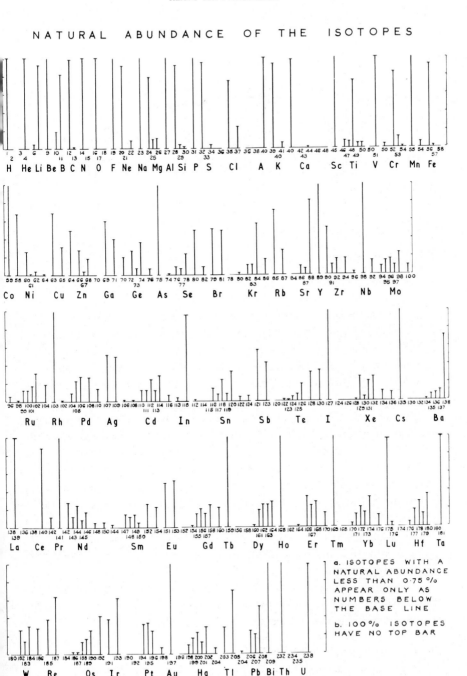

a. ISOTOPES WITH A
NATURAL ABUNDANCE
LESS THAN 0·75 %
APPEAR ONLY AS
NUMBERS BELOW
THE BASE LINE

b. 100 % ISOTOPES
HAVE NO TOP BAR

# 5 | APPLICATIONS OF ION IMPLANTATION TO SEMICONDUCTORS

## 5.1. Introduction

The application of ion implantation techniques to the doping of semi-conductors is discussed in this chapter.

The technique is described in the context of present day semiconductor fabrication methods particularly the Silicon Planar Technology. Emphasis is placed on the complementary role of ion implantation and thermal diffusion as well as on the separate applications of ion implantation doping.

A short historical introduction is followed by sections on the controlled spatial doping of semiconductors by thermal diffusion and ion implantation to give an overall picture of the available doping methods. Masking and thermal annealing techniques for ion implantation are discussed at length.

The electrical behaviour of ion implanted layers is discussed in the following section under three general headings, i) electrical properties of ion implanted layers, ii) the junction properties of the layer to the adjoining substrate and iii) the important results observed for the main dopants in silicon.

The last part of the chapter deals with the clearly identifiable applications of ion implantation in semiconductor device fabrication using published examples to illustrate the particular methods.

The principle purpose of this chapter is to provide sufficient information on ion implantation for the semiconductor device designer to readily evaluate the technique and incorporate it in his designs.

### 5.1.1. HISTORICAL INTRODUCTION

One of the main problems that has been encountered in the fabrication

476

of semiconductor devices, is the accurate and reproducible fabrication of one or more related p-n junctions to form a diode or transistor. Initially the 'grown junction method' and later the 'alloy junction method' were used and satisfactory single devices were fabricated. Research in the late 1950's on diffusion (Tannenbaum and Thomas, 1956), and masking using thin silicon dioxide layers (Frosch and Derick, 1957) and the stabilization of silicon surfaces (Atalla et al., 1959) led to a precise method of controlling the geometry of the devices and electrical state of the surface of the surrounding silicon. Precision photo-engraving techniques were used to prepare the diffusion 'windows' in the silicon dioxide layer to give the diffusant access to the silicon in the defined places. The Planar process, as it is known, was first described by Hoerni (1960) and exploited commercially by the Fairchild Transistor Corp. to produce transistors.

The next significant development in the planar process was the controlled growth of doped epitaxial layers on single crystal silicon (Theuerer, 1961). By combining these layers with diffusion, individual devices could be isolated from one another by a p-n junction and more than one device incorporated on the same piece of silicon. The integrated circuit had arrived.

The research into the stabilization of silicon surfaces using silicon dioxide layers has continued and it has resulted in a very clear understanding of the factors affecting the silicon-silicon dioxide interface (Hofstein, 1967; Kooi, 1967; Goetzberger and Sze, 1969). The influence of migrating impurities (e.g. sodium ions) in the oxide has been investigated and methods have been devised to minimise their effect. Very stable oxides with low surface state densities can now be produced and successful metal-oxide-silicon transistors (MOST's) can be made either singly (Hofstein and Heiman, 1963) or in large numbers in integrated circuits of considerable complexity. The search for the most suitable dielectric material for the MOS transistors is still being actively pursued.

Shockley (1954) discussed the use of ion beams for doping semiconductors in a farseeing patent. He described in some detail a method for producing the narrow base region of a transistor. Mono-energetic ions, of an impurity of the opposite conductivity type to the semiconductor were embedded in it to form a thin conducting layer. He anticipated that the radiation damage produced by the high energy ions would require annealing and he suggested that this could be done at low temperatures.

The sensitivity of very pure single crystal semiconductors to ionizing radiation was realised from the early days and it has given rise to a considerable amount of research into radiation damage effects.

During the 1950's there were two reports by Ohl (1952) and Cussins (1955), of heavy ion bombardment being used to modify the properties of semiconductors. The main conclusion of this work was that the principle effect of the bombardment was to damage the surface of the semiconductor.

The first published reports of ion implantation being used experimentally for doping a semiconductor appeared simultaneously with the increasing interest in solid state particle detectors (Rourke, Sheffield and White, 1961). Alväger and Hansen (1962) successfully doped a high resistivity silicon crystal using phosphorus ions, annealed it at 600 °C, and produced a working nuclear particle detector.

The importance of ion implantation in the formation of tailored junction profiles for solar cells was appreciated by King and co-workers at the Ion Physics Corporation (King et al., 1962; King and Burrill, 1964). At the same time McCaldin and Widmer (1963) were implanting caesium ions in semiconductors and producing n-type layers and consequently n-p junctions. Gibbons, Moll and Meyer (1965) studied the implantation of rare earth elements into semiconductors in an attempt to produce electroluminescent materials.

The increasing interest by 1964 is shown by a number of papers on p-n junction formation presented by King et al. (1965), Manchester et al. (1965) and Medved et al. (1965) at the Aarhus Conference on Electromagnetic Isotope Separators and Their Applications.

The research into ion implantation for the doping of silicon to produce specialized devices increased rapidly both in the United States and in Europe from 1965 onwards. The abrupt nature of the ion implanted junction (see section 5.5.4) has been used in the fabrication of avalanche photodetectors (Sherwell et al., 1967), impatt diodes (Ying et al., 1968) and a shallow junction having the approximate characteristics of a Schottky diode (Isofilm, 1969). The precise control of the depth profile offered by ion implantation has been used to make the narrow base regions in microwave planar p-n-p and n-p-n transistors (King et al., 1965; Kerr and Large, 1967; Ikeda, 1969; Fujinuma et al., 1969; Beale, 1970) and varactor diodes which incorporate a number of integrated profiles (Brook and Whitehead, 1968).

The most striking application of ion implantation, so far, in device technology has been in the automatic alignment of the gate electrode of a metal oxide silicon transistor to the source and drain regions. The technique was first described by Bower and Dill (1966) (Bower et al., 1968) while Shannon, Stephen and Freeman (1969a) have given the details of a MOST with a cut-off frequency of 1.8 GHz, and they also stressed the importance of applying ion implantation to MOS arrays (Shannon et al., 1969b). Glotin et al. (1967) have given details of similar work in France, while Burt (1969) has described the application of implantation to an MOS tetrode structure.

These initial research successes have stimulated the research and development laboratories of a number of semiconductor manufacturers to examine in what manner ion implantation can assist in their development and production problems.

Future exploitation of ion implantation will depend very critically on its economic viability and the availability of suitable implantation equipment rather than any aspect of its technical novelty.

## 5.2. Controlled spatial doping of semiconductors

### 5.2.1. INTRODUCTION

The fabrication of silicon diodes, transistors and integrated circuits by silicon planar technology requires very precise control over the spatial doping of the silicon wafer to form the devices. This controlled doping is achieved by the solid state diffusion of impurities in conjunction with the epitaxial growth of suitably doped layers. The simplest form of integrated circuit transistor requires at least five separate diffusions into one epitaxial layer.

Ion implantation is a third method now available for controllably doping semiconductors and in this chapter its role as a complementary or alternative technique will be described.

The current epitaxial and diffusion techniques will be briefly reviewed to refresh the reader with the various parameters involved. Ion implantation is described in more detail, special emphasis being given to the electrical activity of implanted atoms and the influence of the temperature during implantation and the subsequent annealing treatment.

Two other subsidiary but important aspects, the selective masking of the silicon wafer against the high energy ion beam, and the effects on passivating layers of the passage of the high energy ions, will be discussed.

The advantage and limitations of ion implantation in relation to the fabrication of semiconductor devices by planar technology will be outlined and an assessment is made of its future role in the technology.

### 5.2.2. THE SILICON PLANAR TECHNOLOGY

Planar technology is the collective name for a series of processes used to fabricate semiconductor devices extending from simple diodes to complex

circuits incorporating many thousands of MOS transistors on a single chip
of silicon.

The basic concept of this technology is the spatial distribution of impurities in a plane single crystal silicon wafer to form the complex configuration of conducting regions and p-n junctions required to make a circuit. The
distribution in depth is achieved by the controlled growth of epitaxial layers,
the temperature and duration used for diffusing the impurities. High definition photolithography is used to engrave masks to laterally control the

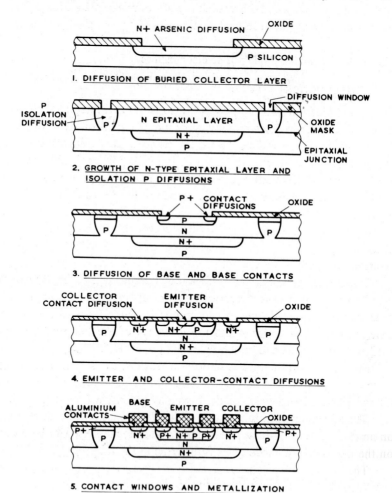

Fig. 5.1. A simplified processing sequence for fabricating a n-p-n bipolar transistor in an
integrated circuit.

Fig. 5.2. A scanning electron micrograph of a typical bipolar transistor. The collector contact is at the top, the emitter in the centre and the base contact at the bottom of the picture. The transistor is approximately 120 microns square.

position of the impurities. A metal layer, commonly aluminium, is engraved to form the series of interconnections required to make the circuit. The metal leads are insulated from the silicon by an oxide layer which also controls the charge state and the electrical leakage at the silicon surface. The processing sequence is best illustrated in fig. 5.1 which shows the basic steps needed to fabricate an n-p-n bipolar transistor while fig. 5.2 is a scanning electron micrograph of a typical bipolar transistor.

### 5.2.3. EPITAXIAL SILICON LAYERS

The ability to grow controllably near perfect single crystal silicon layers on an existing single crystal substrate has had a very pronounced influence on the fabrication of silicon planar devices.

The present day bipolar transistor and digital or linear integrated circuits cannot be made successfully without using an epitaxial layer. It is possible to grow both n and p layers on either type of doped substrate material. As examples, a discrete transistor is made in an n layer on an $n^+$

substrate to reduce the collector series resistance while a transistor made in an n layer on a p substrate is isolated by the p-n junction between the layers. In an integrated circuit there is a p type diffusion through the n layer around the transistor which isolates it from the remainder of the circuit (Warner, 1965).

The resistivity of an epitaxial layer can be controlled between degeneracy and over 50 $\Omega$ cm, the usual value being between 0.1 and 20 $\Omega$ cm. The thickness of the layer can be made as thin as 0.2 microns or well over 20 microns. It is not usual to mask the substrate with silicon dioxide for example during the growth of an epitaxial layer as the silicon that is deposited on the oxide would have to be removed by chemical etching or lapping. If the epitaxial layer is grown on a substrate that has previously been diffused with, for example antimony, for a buried $n^+$ region then any crystal defects introduced by the diffusion will be replicated in the layer.

Epitaxial silicon layers can also be grown on single crystal substrates other than silicon. A considerable amount of research has been done on silicon layers grown on sapphire and spinel structures. These layers, although single crystal in structure, contain a larger number of defects which apart from reducing minority carrier lifetime also make the depth of a thermally diffused profile irregular. Ion implantation has a role in the doping of this class of material and this will be described later.

If the silicon is deposited on an amorphous surface such as silicon dioxide, a polycrystalline layer is produced. This type of layer has found two applications. One application is to thicken an oxide coated single crystal wafer in which dielectrically isolated single crystal regions are later formed by lapping away the majority of the rear of the wafer. Here the deposited layer assumes the role of the substrate.

A more recent application is the use of deposited silicon to replace the metal in the gate electrode of a metal-oxide silicon transistor. The silicon gate technology was first described at an IEEE meeting by Sarace (1968). This technology is discussed in more detail in the section 5.5.5.2 on the alignment of the gate electrode of a MOST.

Although the emphasis in this section has been on silicon epitaxial layers, the technique is of considerable importance in the epitaxial growth of single crystal layers of compound semiconductors and certain ferromagnetic oxides. Epitaxial layers of GaAs and GaP have improved electrical properties when compared with boat grown or pulled crystals. The active regions of the Gunn diode and the electroluminescent diode lamp are in epitaxial layers of GaAs and GaP respectively.

5.2.4. THERMAL DIFFUSION OF IMPURITIES

Impurity atoms can be controllably introduced into silicon (Tannenbaum and Thomas, 1956; Warner, 1965) and other semiconductors by thermally activated diffusion at elevated temperatures. The impurity concentrations required to form device structures in a silicon substrate vary over a very wide concentration range from less $10^{15}$ to over $10^{21}$ impurities per $cm^3$.

Although many atoms will diffuse readily into silicon the important ones for doping are those in Group III (acceptors) and Group V (donors). All these atoms with the exception of nitrogen can be diffused into silicon to produce electrical activity.

The elementary theory of diffusion states that the quantity of material diffusing per unit time is proportional to the concentration gradient and that the direction is towards the lower concentration; that is,

$$J = -D \frac{\partial N}{\partial x} \quad \text{(Fick's first law)}$$

where $J$ = flux density, $N$ is the impurity concentration and $D$ the diffusion constant.

But

$$\frac{\partial N}{\partial t} = -\frac{\partial J}{\partial x} \quad \text{(from the transport equation)}$$

and hence by substitution, we obtain the diffusion equation:

$$\frac{\partial N}{\partial t} = D \frac{\partial^2 N}{\partial x^2}$$

the contribution of the ionized impurities being small enough to be neglected.

The diffusion constant $D$ is characteristic of each diffusant and it is strongly dependent on temperature:

$$D = D_0 \exp \left( - \frac{E_a}{kT} \right)$$

where $E_a$ is the activation energy. Curves of $D$ against temperature for a wide range of dopants and other elements (e.g. gold) can be found in Runyan (1965) and Grove (1967).

For most diffusion problems two general solutions to the above equations are required. These are for diffusion from an infinite surface source and from a limited surface source.

It is usual to deposit on the surface of the silicon a glassy layer containing a large percentage of the impurity to be diffused (e.g. boron or phosphorus in a glassy layer). If the diffusion is continued with this layer present or in an atmosphere of the impurity then they act as infinite sources of the dopant and the concentration $N$ at depth $x$ after time $t$ is given by:

$$N(x, t) = N_0 \left( 1 - \mathrm{erf} \frac{x}{2\sqrt{Dt}} \right)$$

where $N_0$ is the surface concentration (atoms/cm$^3$) and it is assumed to be constant throughout the diffusion. This diffusion procedure is used for transistor emitters and for isolation diffusion in epitaxial layers.

If, however, the glassy layer is removed chemically after deposition and a diffusion then continued only those impurities which were introduced during the first diffusion are available for further diffusion. This is the limited surface source case and the concentration at depth $x$ after time $t$ is given by:

$$N(x, t) = \frac{S}{\sqrt{\pi Dt}} \exp \left( - \frac{x^2}{4Dt} \right)$$

where $S$ is the number of impurity atoms per cm$^2$. If the second diffusion is carried out at a different temperature and assuming the first diffusion to be shallow, then

$$N(x, t) = \frac{2N_0}{\pi} \sqrt{\frac{Dt}{D't'}} \exp \left( - \frac{x^2}{4D't'} \right)$$

$D'$ and $t'$ are the diffusion constant and time relating to the second diffusion, and $N_0$ is the surface concentration achieved during the initial glassy step. This diffusion procedure results in more lightly doped layers and it is used for transistor base regions and diffused resistors.

There are many other cases with different boundary conditions that have been solved and the reader is referred to Runyan (1965) for the mathematical treatment of a number of cases applicable to semiconductor problems. In practice, it is customary to calculate the impurity concentration and then to verify it by examining diffused wafers using staining or other techniques. The sensitivity of the diffusion process to small changes in temperature makes this approach essential for close control of junction depths.

An examination of the diffusion constants of the Group III and V elements into silicon shows that aluminium and phosphorus are the fastest to be followed by boron, gallium, indium, arsenic and lastly antimony. This difference is exploited in the fabrication of integrated circuit bipolar transistors, for example, antimony is used as the 'buried $n^+$ layer' below the collector as it must not move during the subsequent long diffusion cycles such as the boron diffusion for the isolation which takes many hours to drive in.

Interstitial impurities such as copper, gold and the alkali metals diffuse extremely rapidly in silicon. This property is used in 'gold doping' of transistors to reduce minority carrier lifetime. However in devices (e.g. semiconductor nuclear particle detectors) where a long minority carrier lifetime is required the rapid diffusion of copper and gold is extremely undesirable.

The temperatures and times of diffusion cycles used in fabricating silicon planar devices depend on the purpose of each diffusion. The temperatures vary between 900 °C and 1250 °C while the times are from minutes to many hours. In general shallow diffusions for the emitters of bipolar transistors tend to be done at the lower temperatures for short times while deep diffusions (for example integrated circuit isolation regions) are done at high temperatures for up to 20 hours. These differences in conditions make it necessary to perform the diffusions in a sequence fixed by times and temperatures.

In the discussion of doping by diffusion it has been assumed that there are no solubility problems associated with the diffusant. Trumbore (1960) has shown that arsenic, boron and phosphorus are very soluble in silicon and that concentrations in excess of $10^{20}$ impurity atoms/cm$^3$ are possible without precipitation. Aluminium, gallium and antimony are not as soluble while interstitial impurities such as gold and iron have a solubility less than $10^{17}$ atoms/cm$^3$. The solubility of the impurity must be considered when calculating the parameters for any highly doped region whether by diffusion or ion implantation.

Thermal diffusion can be used to dope compound semiconductors but the process is not as straightforward as it is with silicon. There is the problem of the loss of stoichiometry due to the sublimation of one or more of the constituents of the semiconductor due to the high temperature used for the diffusion. This can be overcome by heating in an atmosphere with a high pressure of the volatile components, for example GaAs requires a high arsenic vapour pressure to prevent arsenic escaping from the surface.

### 5.2.5. DIFFUSION MASKS

The discussion on diffusion so far has dealt only with the factors controlling the depth and concentration of the impurity in the silicon. In order to fabricate a semiconductor device it is usually necessary to restrict the impurities to defined areas. The work of Frosch and Derick (1957) showed

that a silicon dioxide layer of adequate thickness, (a few thousand angstroms), on a silicon wafer is able to shield the silicon surface from the
diffusing impurity for a sufficiently long time for diffusion to occur in the
exposed silicon. This is due to the diffusion rate of impurities such as boron
and phosphorus being much slower in silicon dioxide than silicon, while
gallium, indium and antimony diffuse as fast in oxide as in silicon.

If apertures or windows are engraved in the silicon dioxide layer prior
to diffusion, then the impurity will only diffuse into the areas of exposed
silicon. As diffusion is a three dimensional process the impurity diffuses
sideways under the oxide layer almost as much as it diffuses normal to the
surface. As a consequence the junction region where it emerges at the silicon
surface is protected by the oxide coating (fig. 5.16a).

Silicon nitride is being investigated as an alternative masking layer to
silicon dioxide particularly for some of the more difficult dopants such as
gallium and zinc (for gallium arsenide).

The properties of a successful diffusion mask can be summarized as
follows:

(i) it must be opaque to the diffusant at high temperatures,

(ii) it must protect the silicon surface and not introduce any impurities
of its own,

(iii) it must adhere to the surface during temperature cycling,

(iv) it must be possible to photoengrave apertures in it, and

(v) it must be readily removable from the silicon surface without causing damage.

Silicon dioxide and silicon nitride possess these properties when used
as masks against boron and phosphorus diffusions.

5.2.6. SURFACE PASSIVATION

The electrical properties of semiconductor devices are strongly influenced
by the electrical conditions prevailing at the semiconductor surface. It is
essential for most devices, particularly those requiring low leakage currents,
to control the charge state at the surface. This is most satisfactorily done
by passivating the surface with a thin layer of very pure thermal silicon
dioxide (Atalla et al., 1959). A layer of phosphorus glass ($SiO_2$ with a few
percent of $P_2O_5$) or silicon nitride can be used to protect the oxide layer by
preventing impurities such as sodium ions diffusing into the oxide and introducing charge centres.

The silicon oxide on a silicon device therefore has two roles, one as a
diffusion mask and the other as a passivating layer.

## 5.2.7. Metallic interconnections

All planar devices require metal interconnections to make contact with the diffused areas. The metal most commonly used is aluminium, as it makes good ohmic contacts with both $n^+$ and $p^+$ silicon provided the impurity level is over $2 \times 10^{18}$ impurities/cm$^3$. The method of interconnecting is to evaporate a layer 1 to 1.5 microns thick over the silicon wafer after apertures have been opened in the oxide layer by etching to give access to the $n^+$ and $p^+$ regions to be contacted. The aluminium layer is photoengraved to form the interconnection pattern. The silicon wafer is then annealed (sintered) at about 500 °C for 10 to 30 minutes to ensure formation of the ohmic contacts. The wafer is then ready for testing, dicing, bonding and encapsulation. An example is shown in fig. 5.2.

## 5.3. Doping by ion implantation

### 5.3.1. Introduction

The two principle methods for the controlled doping of semiconductors, epitaxy and diffusion have been described. In this section the third and more recent method, ion implantation, will be described in depth so that its position and importance from all aspects can be assessed in relation to the other methods.

The technique of implantation will be described and its potentialities and disadvantages discussed in relation to the silicon planar technology. The discussion is concerned mainly with silicon microelectronic devices but many of the details apply equally well to large area devices (e.g. nuclear particle detectors) and to other materials such as germanium and compound semiconductors.

The basic problems in applying ion implantation will be described first and in the following sections the electrical properties of implanted layers and junctions will be discussed. Lastly the areas in which implantation can be applied to advantage will be reviewed with the aid of practical examples.

### 5.3.2. Choice of dopants

In a suitably designed ion accelerator, ion beams of nearly all the elements in the periodic table can be produced, mass analysed and implanted into the target. It is no longer necessary with implantation to consider the diffusion characteristics of the impurity element but its range and distribution when implanted into the semiconductor.

One important aspect of a mass analysed ion beam that is not always fully appreciated is that only one ion species and in some cases a particular isotope of the element (e.g. $^{11}$B) is introduced into the semiconductor. It is therefore quite simple in practice to change from one ion to another ion (e.g. boron to phosphorus) without any contamination taking place. This is in direct contrast to a diffusion furnace in which all the impurities present in the furnace, or contaminating the semiconductor surface, will diffuse into the crystal along with the wanted dopant. The effect is particularly serious if interstitial diffusers such as copper and gold are present.

Despite the wide range of dopants being available to the experimenter much of the published work on silicon has concentrated on the group III and V dopants.

The main emphasis has been on boron and phosphorus due to the wide knowledge of these two elements as impurities in silicon combined with their high substitutional solubility. At the normal implantation energies of 40 to 200 keV they have ion ranges which are directly applicable to devices (see section 5.5). Gallium, antimony and arsenic have been studied fairly extensively while aluminium, indium and nitrogen have received far less study. The electrical behaviour of all these impurities is discussed in section 5.4.

Ion implantation makes it feasible to introduce controllably a wide range of impurities into silicon and other semiconductors for the purpose of investigating their physical and electrical properties.

### 5.3.3. TEMPERATURES INVOLVED DURING IMPLANTATION

*5.3.3.1. During Implantation.*   In this section the influence of the substrate temperature, *during* implantation on the subsequent electrical behaviour of the implanted impurities is discussed.

As the ion energy and dose are independent of the substrate temperature, the same distribution of impurity atoms should result from different implantation temperatures, assuming no enhanced diffusion takes place. The substrate temperature during implantation has a pronounced effect on the subsequent electrical conduction observed after an annealing treatment. The implanted ions produce considerable lattice damage and it has been shown (Chapter 2) that the amount of damage is strongly dependent on the temperature during implantation and the ion dose. An amorphous structure is produced in silicon and other semiconductor materials if the individual damage zones become so dense that they overlap. This amorphous structure is stable until annealed at a critical temperature when it recrystallizes epitaxially on the existing crystal structure which is normally a single crystal.

The critical temperature is a characteristic of the semiconductor and it is 650 °C for silicon, 400 °C for germanium and 250 °C for GaAs. It is usual for a fraction of the implanted impurities to be incorporated in substitutional sites in the lattice during the epitaxial regrowth and become electrically active.

If the amorphous structure is prevented from forming by the substrate temperature being elevated during implantation, then the implanted region remains a single crystal but with a complex network of defects, which is stable on annealing to temperatures of about 1000 °C. The absence of an amorphous layer allows the number of channeled ions (Chapter 2) to increase despite the decrease in the channel cross-section due to increased lattice vibration.

An important practical example of the influence of the implantation temperature on the subsequent electrical activity is seen for the case of boron implanted into silicon at 77 °K. An order of magnitude increase in the electrical conductivity is found for the 77 °K implantation when compared with a room temperature implantation both being annealed at 550 °C (figs. 4.59 and 5.50).

The high energy ions deposit energy in the substrate as they come to rest. The magnitude of this energy can be calculated from the accelerating potential, the total ion dose and the charge on the ion, which is usually singly charged but in some cases may be doubly or triply ionized. The energy ultimately appears as heat which raises the temperature of the substrate and this can be appreciable unless good thermal contact is made between the substrate and a heat sink (Chapter 4).

The usual practice is to mount the wafer by clamping it at the edges with thin metal washers or clips to a metal holder. As a result, when in a vacuum, the wafer is thermally out of contact with holder except at the edges and high spots, and the heat from the beam can then only be lost by radiation and some conduction. This situation produces a steep rise in the wafer temperature which can be several hundreds of degrees centigrade. A steep lateral thermal gradient can be established across the wafer and this has been observed when the implantation produced an amorphous layer. The amorphous region is more pronounced near the cooler parts of the wafer for example the washers. This non-uniformity is clearly visible to the naked eye.

In any implantation where the subsequent electrical activity depends on the formation of an amorphous layer during implantation, serious non-uniformity of the electrical activity will result. This will be directly related to the non-uniformity of the amorphous layer and the pattern of the thermal gradients.

These thermal gradients can be avoided by cementing the wafer to the substrate holder using an electrically and thermally conducting 'cement' such as silver 'dag' or a gallium-indium alloy. The rate of heating due to the ion beam can be reduced by decreasing the beam current and extending the implantation time. This procedure has been adapted for boron implantations at 77 °K, when the wafers were cemented directly to the cryostat.

Great care must be taken to remove any cement by chemical cleaning before annealing to avoid the diffusion of unwanted impurities (e.g. silver) into the wafer.

In many cases when an increase in the substrate temperature during implantation is not important, the silicon wafer can be held simply by the metal washers mentioned earlier. Care must be exercised when a metal mask is on the silicon surface to avoid overheating thus causing melting or alloying. The details of the temperature rise of a wafer are given in Chapter 4.

*5.3.3.2. Annealing after Implantation.* In the previous section the need for annealing the implanted region was mentioned. The damage produced in a diamond type lattice by the bombardment of heavy ions is discussed in Chapter 3.

The effect of this lattice damage is to:

(i) reduce the number of electrically active implanted atoms as they will not be in substitutional sites,

(ii) introduce deep levels which act as recombination-generation centres and

(iii) scatter the carriers in the lattice reducing their mobility.

The lattice structure of silicon, germanium and compound semiconductors damaged by ion implantation has to be restored to a monocrystalline form, incorporating the impurity atoms, by suitable annealing treatments. The variation of the electrical properties with annealing temperature is discussed in section 5.4.4. This section is concerned with the temperature range and some of the techniques involved with annealing.

If a silicon wafer, for example, has been implanted with an ion dose sufficient to make the surface layer amorphous it is necessary to anneal the wafer above 650 °C to allow the amorphous layer to epitaxially regrow onto the substrate. For lower doses which do not make the silicon amorphous, the annealing temperature can be below 650 °C. Minority carrier lifetime measurements (section 5.4.3.4) have shown that annealing above 650 °C is essential for nearly complete recovery of the minority carrier lifetime. Experience has shown that it is necessary to anneal all wafers implanted at room temperature or below to obtain a contribution by the implanted dopant to the electrical conductivity.

The annealing temperatures used in practice are commonly between 300 °C and 900 °C depending on the ion, its energy and dose, the substrate and the presence of any metal layers (aluminium, silicon or molybdenum) on the substrate. If a low melting point metal such as aluminium is present, as in the ion implanted MOS transistor (section 5.5.5), then the maximum annealing temperature is determined by the melting point (660 °C) or the eutectic temperature (577 °C) with silicon, and the reaction between the silicon dioxide and the metal. For a wafer with aluminium interconnections on it, experiments have shown the maximum temperature to be 550 °C and a temperature of 500 °C is normally used in practice as it is also suitable for sintering the contacts.

For the higher annealing temperatures some thermal diffusion occurs and in some applications this is the required result.

The damage in ion implanted silicon can be annealed at temperatures well below the diffusion temperatures for boron and phosphorus. The fast diffusing impurities such as iron, copper, gold and alkali metals (Runyan, 1965) will diffuse rapidly into silicon at these annealing temperatures. Care must be taken to remove all traces of the metals particularly if they have been used in contact with the wafer for masking or as conducting cement for securing the wafer to a heat sink. If contamination from fast diffusing impurities is likely to affect the device performance (e.g. charge collection in a solid state radiation detector) then the annealing temperature must be kept as low as possible consistent with an adequate electrical activity.

The annealing environment for ion implanted silicon is not critical as successful results have been obtained using either a high vacuum or a flow of dry nitrogen. When using a high vacuum, it must be remembered that many substrate materials, for example the III–V compounds and impurities such as phosphorus, arsenic antimony and zinc have high vapour pressures at low temperatures and can evaporate very readily (see table 5.1). This evaporation is detrimental and it can be avoided by protecting the semiconductor surface with a thin layer of $SiO_2$, deposited by RF sputtering or low temperature chemical vapour deposition (section 5.3.5.3.1). In the case of silicon and germanium it is only necessary to protect the implanted surface but for GaAs the whole surface of the crystal must be protected. Ion implantation through a thin passivating layer (about 1000 Å of $SiO_2$) is a sound technique provided the ion range is sufficient to penetrate the layer. This is practical for boron and phosphorus but not for the heavier elements except for high ion energies.

TABLE 5.1

Temperatures of various elements for a vapour pressure of $10^{-6}$ torr (Honig, 1962)

| Element | Temp. °C |
|---------|----------|
| B  | 1467 |
| Al | 812  |
| P  | 88   |
| S  | 17   |
| Zn | 177  |
| Ga | 732  |
| As | 150  |
| Se | 107  |
| Cd | 119  |
| In | 597  |
| Sn | 807  |
| Sb | 345  |
| Te | 209  |

In summary, it is possible using ion implantation for doping to process a semiconductor without employing the high temperatures needed for thermal diffusion. This avoids some of the undesirable side effects of diffusion such as the propagation of lattice damage, the redistribution of impurity profiles by further diffusion, the introduction of unwanted impurities and the growth of thermal oxide with its attendant vacancy generation.

5.3.4. DISTRIBUTION OF IMPLANTED IONS

*5.3.4.1. Introduction.* A theory for the depth distribution of implanted ions in amorphous solids has been postulated by Lindhard, Scharff and Schiøtt (LSS theory, Lindhard et al., 1963) and it is discussed in Chapter 2. The channelling of ions in single crystal substrates is more complex and there is no comparable theory for calculating the channelled ranges. At present no microelectronic devices are designed using channelled profiles, though useful junctions have been made for nuclear particle detectors (see section 5.5.7) and varactor diodes (section 5.5.3.1).

This section is concerned with the control that can be exercised over the depth and lateral position of implanted ions from the aspect of device fabrication. The ion ranges are calculated using the LSS theory and the various problems associated with masking against the ion beam are discussed.

Johnson and Gibbons (1969) published, (i) a computer program for calculating the ion range, standard deviation and energy losses for any particular ion-substrate pair and (ii) a series of tables listing these details for the most frequently used ion-substrate pairs for an energy range from 10 to 1000 keV. Smith (1971) has extended the program and this has been used to plot the projected range $R_p$ and standard deviation $\Delta R_p$ against ion energy for the group III (fig. 5.3) and V (fig. 5.4) dopants in silicon and the universal tables in appendix 3. Johnson and Gibbons (1969) in an addendum to their published tables state that the mathematical approximation used in the solution of the integral equations in the LSS theory leads to a non-realistic result for the standard deviation in the projected range for ions

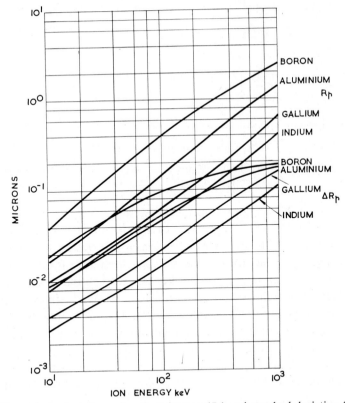

Fig. 5.3. Theoretical calculations of projected range $(R_p)$ and standard deviation $(\Delta R_p)$ of group III dopants in silicon (Smith, 1971).

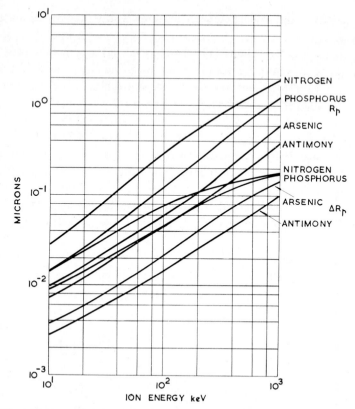

Fig. 5.4. Theoretical calculations of projected range $(R_p)$ and standard deviation $(\Delta R_p)$ of group V dopants in silicon (Smith, 1971).

heavier than the substrate atoms. These authors have recommended that the standard deviation for the projected range be obtained by first comparing the magnitudes of the standard deviations for the total range and projected range and using the larger value. This has been done for all the values of standard deviation data reproduced in this chapter and in appendix 3.

It must be emphasised that these range and standard deviation calculations are to an accuracy of about 20 % and no allowance has been made for channelling, enhanced diffusion and sputtering.

Apart from being able to predict the ion range, one of the strong features of ion implantation is that the ion dose and therefore the surface concentration can be measured accurately and repeated reproducibly.

5.3.4.2. *Ion Depth Distribution in an Amorphous Substrate.* From a knowledge of the surface concentration, $C_s$, the projected range $R_p$ and the standard deviation $\Delta R_p$ for a given ion-substrate pair it is possible to calculate some important profile parameters. Figure 5.5 shows a typical Gaussian distribution of implanted atoms with the principle parameters indicated.

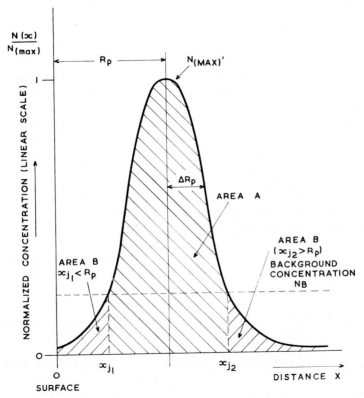

Fig. 5.5. A typical theoretical Gaussian ion distribution with depth. $R_p$ is the projected range, $\Delta R_p$ the standard deviation and $N_B$ the background doping concentration. $X_{j_1}$ and $X_{j_2}$ are the positions of p-n junctions if type conversion is produced. (See text for explanation of areas $A$ and $B$.)

The concentration $N(x)$ of implanted atoms at depth $x$ from the *surface* is given by:

$$N(x) = N_{(max)}e^{-\frac{1}{2}X^2} \tag{5.3.1}$$

and where $X = (x - R_p)/(\Delta R_p)$ and $N_{(max)}$ is the concentration at the peak of the profile corresponding to $x = R_p$.

The surface concentration $C_s$ (atoms/cm$^2$) can be related to the maximum concentration (atoms/cm$^3$) $N_{(max)}$ by integration. The number of implanted impurity atoms $C$ in a square centimeter of the implanted region between the limits $X_1 = (R_p - x_{j_1})/(\Delta R_p)$ and $X_2 = (x_{j_2} - R_p)/(\Delta R_p)$ is

$$C = \int_{X_1}^{X_2} N(x)\,dx$$

$$= N_{(max)} \Delta R_p \int_{X_1}^{X_2} e^{-\frac{1}{2}X^2}\,dX$$

by substituting for $N(x)$ from eq. (5.3.1) and putting $dx = \Delta R_p dX$.
As the Gaussian curve is symmetrical $X_1 = X_2 = X$ and hence

$$C = N_{(max)} 2\Delta R_p \int_0^X e^{-\frac{1}{2}X^2}\,dX$$

$$= N_{(max)} \sqrt{2\pi}\Delta R_p \left[ \sqrt{\frac{2}{\pi}} \int_0^X e^{-\frac{1}{2}X^2}\,dX \right] \tag{5.3.2}$$

The integral $\sqrt{2/\pi} \int_0^X e^{-\frac{1}{2}X^2}\,dX$ is the probability integral (error function) and is the normalized area under the Gaussian curve from $X_{j_1}$ to $X_{j_2}$. As $X$ tends to infinity the integral tends to unity and therefore

$C_s$ = total number of impurity atoms/cm$^2$ for the whole implantation
$\quad = N_{(max)} \Delta R_p \sqrt{2\pi}.$

Hence

$$N_{(max)} = \frac{C_s}{\sqrt{2\pi}\Delta R_p} \sim \frac{0.4 C_s}{\Delta R_p}. \tag{5.3.3}$$

Substituting eq. (5.3.3) in (5.3.2) gives the total number of implanted atoms $C_A$ from $X_{j_1}$ to $X_{j_2}$

$$C_A = C_s \left[ \sqrt{\frac{2}{\pi}} \int_0^X e^{-\frac{1}{2}X^2}\,dX \right] = C_s \cdot \text{area } A \tag{5.3.4}$$

where area $A$ is the normalized area under the Gaussian curve from $X_{j_1}$ to $X_{j_2}$. Similarly using the complementary error function the total number of implanted atoms $C_B$ from $-\infty$ to $X_{j_1}$ or $X_{j_2}$ to $\infty$ is,

$$C_B = C_s \left[ \sqrt{\frac{2}{\pi}} \int_x^{\infty} e^{-\frac{1}{2}X^2}\,dX \right] = C_s \cdot \text{area } B \tag{5.3.5}$$

where area $B$ is the normalized area under the Gaussian curve from $-\infty$ to $X_{j_1}$ or $X_{j_2}$ to $\infty$.

The functions are tabulated in appendix 2 with $N(x)/N_{(\max)}$ as the independent variable and $X$, area $A$ and area $B$ as the dependent variables.

To illustrate the use of the tables consider an acceptor implant into a donor substrate. $R_p$, $\Delta R_p$, $C_s$ and the substrate doping level $N_B$ are known and from this data some of the important parameters of the implant can be easily calculated.

The usual procedure is to calculate $N_{(\max)}$ from $C_s$ and $\Delta R_p$ and then find the ratio $N_B/N_{(\max)}$ and the corresponding value of $X$ from the tables. The junction depth $x_j$ can be calculated using eq. (5.3.1):

$$x_j = X\Delta R_p + R_p \qquad \text{for } x_j > R_p$$

and

$$x_j = R_p - X\Delta R_p \qquad \text{for } x_j < R_p.$$

It is frequently necessary to calculate the number of impurity atoms per unit area in a layer between two limits such as $x_{j_1}$ and $x_{j_2}$ (fig. 5.5). The number of impurity atoms is the normalized integral under the Gaussian curve between these limits multiplied by the total surface dose per unit area (eqs. (5.3.4) and (5.3.5)). The integrals can be evaluated using the values for area $A$ or $B$ corresponding to $N/N_{(\max)}$ and $X$ in the tables in the appendix 2.

Some of the common solutions encountered in practice are listed below. $C_s$ is the total surface ion dose in ions per unit area.

| Limits of integration | Conditions | No. of impurity atoms in the layer |
|---|---|---|
| $-\infty$ to $+x_{j_2}$ | $x_{j_2} > R_p$ | $C_1 = C_s\,(1 - \text{area } B_2)$ |
| $-\infty$ to $+x_{j_1}$ | $x_{j_1} < R_p$ | $C_2 = C_s\,(\text{area } B_1)$ |
| $x_{j_1}$ to $\infty$ | $x_{j_1} < R_p$ | $C_3 = C_s\,(1 - \text{area } B_1)$ |
| $x_{j_2}$ to $+\infty$ | $x_{j_2} > R_p$ | $C_4 = C_s\,(\text{area } B_2)$ |
| $x_{j_1}$ to $x_{j_2}$ | $R_p - x_{j_1} = x_{j_2} - R_p$ | $C_5 = C_s\,(\text{area } A)$ |
| $x_{j_1}$ to $x_{j_2}$ | $x_{j_1} < R_p,\, x_{j_2} > R_p$ | $C_6 = C_s\,(1 - \text{area } B_1 - \text{area } B_2)$ |
| $x_{j_1}$ to $x_{j_2}$ | $x_{j_2} > x_{j_1} > R_p$ | $C_7 = C_s\,(\text{area } B_2 - \text{area } B_1)$ |

Note: Area $B_1$ and area $B_2$ correspond to $x_{j_1}$ and $x_{j_2}$ respectively.

This method of approach can be applied to the solution of similar problems involving Gaussian profiles. The accuracy of these calculations is dependent on the accuracy of the LSS and other data. The LSS data for $\Delta R_p$ can only be relied on to 20 % in some cases.

A number of practical examples to illustrate the use of the table in appendix 2 are given below.

*Example 1* – For a simple implant the positions of the junction(s) and the number of impurity atoms/cm$^2$ in the layer can be found. In certain cases, boron in silicon is one, the profiles emerge at the surface and this can be interpreted as the backscattering of a small fraction of the implanted ions.

*Example 2* – The extent of the solubility limit of the impurity on either side of $N_{(max)}$ can be calculated by using a value $N_{sol}$ (Trumbore, 1960) to find $X_{j_1}$ and $X_{j_2}$ from $N_{sol}/N_{(max)}$. This region of the implant is where precipitation can affect the number of electrically active centres $N$ and carrier mobility $\mu$ (see fig. 5.60 for a practical example).

*Example 3* – For an implantation, beyond a shallow diffusion, such as the base of a bipolar transistor or the buried layer in an avalanche diode, two junctions $x_{j_1}$ and $x_{j_2}$ will be formed, one with the shallow diffusion and the other with the background doping of the substrate. The profile of the diffusion must be known to find $x_{j_1}$ and a series of successive approximations may be necessary to finally fix $x_{j_1}$. The number of impurities in the layer can be found by finding the appropriate values of 'area $B$' for $x_{j_1}$ and $x_{j_2}$, and using solutions $C_5$, $C_6$ or $C_7$ as indicated by the position of $R_p$.

*Example 4* – The number of implanted atoms which cross an interface at depth $x_j$ is given by $C_3$ $(x_j < R_p)$ or $C_4$ $(x_j > R_p)$. This is important for implantation of a semiconductor through an oxide layer as the electrical activity depends on the number of ions which have crossed the interface into the semiconductor. The profile of the ions in the semiconductor will not normally be an extension of the profile in the oxide layer but it will be deeper if the oxide has a greater stopping power than the semiconductor. Metal-oxide silicon transistors and high value resistors are made by this technique.

*Example 5* – The stopping power of a masking material can be measured by its attenuation of the ion beam. The factors $C_3$ or $C_4$ are a direct measure of this attenuation as they can be found from $R_p$, $\Delta R_p$ and the mask thickness. The tables in appendix 2 have been extended to large values of $X$ to accommodate large attenuation factors. The problems of masking are discussed in the next section.

*Example 6* – The number of ions implanted beneath the sloping edge of a mask can be found (approximately) by dividing the area beneath the mask into a number of discrete regions, calculating the number of ions in each allowing for the increasing oxide thickness. The total number of ions is found by summing the results for discrete regions. The method can be used to find the extra parallel conductance added to an implanted resistor by imperfect masking.

*Example 7* – The composite profile of a series of implants can be found by dividing the implanted region into a number of discrete levels and calculating the contribution of each implantation in each level. The composite profile is found by summing all the contributions in each level. The method is tedious if a level profile is required as a number of iterations are necessary.

*Example 8* – When an implanted layer is profiled by anodic or ionic stripping it is convenient to estimate the number of impurity atoms removed at each stage of the stripping. This can be readily calculated from factors $C_6$ and $C_7$ while $C_3$ or $C_4$ will give the number of impurity atoms remaining in the implantation.

To conclude this section, ion implantation can be used as a limited surface source for diffusion. Seidel and MacRae (1970) have shown that a time dependent factor can be included in the Gaussian distribution by replacing $\Delta R_p$ with $\sqrt{\Delta R_p^2 + 2Dt}$ where $D$ is the position independent diffusion constant and $t$ is the time. The impurity concentration at depth $x$ from the surface after time $t$ is given by

$$N(x, t) = \frac{N_{(max)}}{\left(1 + \dfrac{4Dt}{2\Delta R_p^2}\right)^{\frac{1}{2}}} \exp\left(\frac{x - R_p}{2\Delta R_p^2 + 4Dt}\right)$$

$\Delta R_p$ and $Dt$ must be in the same units of length. The solution assumes no reflection of the diffusing impurity at the silicon surface.

### 5.3.5. MASKING AGAINST AN ION BEAM

*5.3.5.1. Introduction.* The great majority of the applications of ion implantation require that the implanted ions be confined to particular areas of the substrate. It is therefore necessary to selectively shield regions of the substrate surface with a layer which will completely absorb the ion beam. In the silicon planar technology, this technique is known as diffusion masking and it is usual to use an oxide layer for this purpose.

There are two general methods of masking against an ion beam which is scanned to cover the whole of the substrate. The mask may be out of contact or in contact with the substrate. Each method has its main applications and advantages and these will be discussed in detail in the next two sections.

*5.3.5.2. Out-of-Contact Masks.* Any mask which is not in intimate contact with the substrate is treated here as an 'out-of-contact' mask (fig. 5.6). This

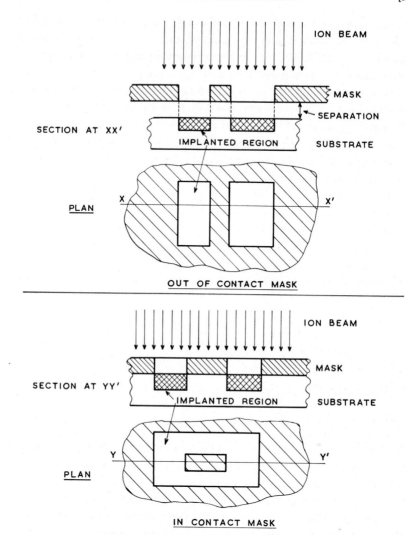

Fig. 5.6. The upper diagram shows the principle of the 'out of contact' mask and the lower figure the 'in-contact' mask. An isolated masked area is possible with the in-contact mask as it is supported by the substrate.

type of masking consists of a suitably shaped opaque sheet immediately in front of the substrate so as to absorb the ion beam in selected areas. It is a simple quick and easy method to use with some important advantages over in-contact masks.

The limited precision of registration of an out-of-contact mask makes it suitable for macroapplications rather than microapplications such as integrated circuits. By the use of mechanical registration methods it is possible to locate an out-of-contact mask to the same tolerances as masks for thin film circuitry ($\pm 2$ microns). Due to the self supporting nature of the mask the spacings in the mask are restricted to apertures as no isolated areas are possible.

Apart from its simplicity the advantages of the out-of-contact mask are: the wide choice of materials, the ability for stopping high energy ions as the mask thickness is large and the minimal contamination of the substrate.

The criteria for selecting a masking material can be listed as follows:

(i)    it must be robust and thick enough to stop the ions,

(ii)   it must be flat over the substrate area so that the variation in spacing between it and the substrate is kept to a minimum,

(iii)  it must be possible to engrave or punch apertures in it with relative ease leaving clean holes,

(iv)   the edges of the mask must not sputter rapidly as this can contaminate the substrate,

(v)    it should be a good electrical and thermal conductor especially if large ion doses are involved,

(vi)   it should have a high melting point and low vapour pressure and

(vii)  it should not oxidise or corrode readily as volatile oxides may be deposited on the substrate.

Metal foils or thin sheets fit these criteria very well and they are preferred to ceramics or glasses.

Table 5.2 lists some metals which are suitable for masking and figs. 5.7, 5.8, and 5.9 and 5.10 give the projected range $R_p$ and standard deviation of boron and phosphorus in these metals. Refer to section 5.3.4.1 for an explanation of the derivation of these standard deviations.

Some of these metals such as aluminium, copper and gold are not sufficiently rigid in thin foil form ($\sim 25$ microns thick) to be used as masks as they will not withstand the engraving and mounting steps. They are quite satisfactory when 250 microns (0.01 inch) thick or when alloyed with small quantities of other metals (e.g. beryllium, copper) to give improved mechanical properties.

Alloys with very volatile components such as brass (copper-zinc) must be avoided as the zinc has a high vapour pressure and can readily evaporate and contaminate the semiconductor.

The metals for masks which have to withstand intense ion beams for long periods must be chosen particularly to have a high melting point, low

TABLE 5.2

Metals for out-of-contact masking

| Metal | Melting point °C | Temp. °C for V.P. * $10^{-6}$ torr | Sputtering factor ** atoms/ion | Relative stopping power (silicon = 1) at 1 MeV Boron | Phosphorus |
|---|---|---|---|---|---|
| Aluminium | 660 | 812 | 0.6 | 1.46 | 1.15 |
| Silicon | 1412 | 1147 | <1 | 1 | 1 |
| Iron | 1534 | 1032 | 3 | 2.23 | 2.51 |
| Nickel | 1453 | 1072 | 3.8 | 2.48 | 2.84 |
| Steel | >1400 | 1032 | ~3 | as iron | |
| Copper | 1083 | 852 | 9 | 2.34 | 2.69 |
| Molybdenum | 2620 | 1822 | 1.5 | 2.08 | 2.55 |
| Palladium | 1552 | 992 | 8 | 2.40 | 2.87 |
| Tantalum | 3000 | 2233 | 3 | 2.34 | 3.23 |
| Tungsten | 3380 | 2407 | 5 | 2.69 | 3.72 |
| Platinum | 1769 | 1496 | 14 | 2.89 | 4.04 |
| Gold | 1063 | 947 | 52 | 2.60 | 3.64 |

\*   Honig (1962).
\*\* Nelson (1968).

vapour pressure and a low sputtering ratio. The three refractory metals, molybdenum, tantalum and tungsten in table 5.2 fulfil these requirements. They can be obtained in thin foil form and they are not difficult to engrave. Commercially produced molybdenum masks to fine tolerances can be readily purchased.

The high sputtering ratio of copper, palladium, platinum and gold detract from their use except when small ion doses are to be used.

Out-of-contact masks for ion implantation have been extensively used in three main areas:

(i)   selective masking of a substrate for control purposes,

(ii)  implanting van der Pauw patterns,

(iii) implanting small areas of a wafer to form isolated test devices, and

(iv)  implanting selected areas of large semiconductor devices such as nuclear particle detectors.

Lastly it must be emphasised that any out-of contact mask should be earthed to the target assembly to prevent electrostatic charging and the possibility of a discharge between the mask and the substrate.

*5.3.5.3. In-contact Masks.*   An in-contact mask is one in which the masking material is in direct and continuous contact with the substrate to be masked (fig. 5.6). Suitable apertures of any shape can be engraved in the mask by selective chemical etching or other methods such as ion sputtering.

The very sophisticated photolithographic techniques developed over the past few years for the silicon planar technology can be used without any modifications for engraving ion implantation masks. The compatibility between these technologies has been responsible for the acceptance of ion implantation into the silicon planar technology especially for MOS arrays.

Some of the important criteria for choosing a material for an in-contact mask are:

(i)   it must be opaque to the ion beam,

(ii)  it must be compatible with the materials of the substrate,

(iii) it must adhere to but not contaminate the semiconductor surface,

(iv)  it must be easy to apply, to engrave and later easy to remove after ion implantation and lastly

(v)   it must not sputter readily.

In-contact masking materials for semiconductors can be considered under two broad headings, insulators and metals which in turn determine how and when they are used.

*5.3.5.3.1. Insulators as masks.*   Insulating layers are used in semiconductor fabrication as masks against thermal diffusion of boron and phosphorus and as passivating layers on the semiconductor surface. The most common insulator is silicon dioxide (amorphous quartz, $SiO_2$) followed by silicon nitride ($Si_3N_4$) and to a lesser extent alumina ($Al_2O_3$).

Some of the important properties of these insulators are listed in table 5.3 while figs. 5.11, 5.12, 5.13 and 5.14 give the ranges of boron and phosphorus in these materials. Photoresist has been included, with silicon added for comparison. Refer to section 5.3.4.1 for an explanation of the derivation of these values of standard deviations.

TABLE 5.3

Insulators for in-contact masking

| Material | Relative stopping power (silicon = 1) at 40 keV | | Ease of engraving | High temp. stability |
|---|---|---|---|---|
| | Boron | Phosphorus | | |
| Silicon | 1 | 1 | two step process | good |
| Silicon dioxide | 1.25 | 1.26 | easy | good |
| Silicon nitride | 1.62 | 1.63 | two step process | good |
| Alumina | 1.83 | 1.86 | two step process | good |
| Photoresist | 0.77 | 0.73 | easy | none |

It will be seen that the stopping power of each material is very similar for both ions and taking into account the difficulty of engraving, there is no particular merit for using $Si_3N_4$ or $Al_2O_3$ in preference to $SiO_2$. The thickness of $SiO_2$ required to ensure the complete masking of a 40 keV boron ion beam is about one micron. This is the same thickness that is commonly used in integrated circuits when conductors must be insulated from the silicon substrate.

The behaviour of an insulator during ion implantation is important as the charge from the implanted ions is deposited in the surface of the insulator. In the case of oxide layers which are thin enough for the ions to penetrate, the charge is dissipated by conduction in the oxide during the implantation. However for thicker oxide layers the charge which is deposited in the upper surface must either move laterally to a sink or rupture the oxide. Experiments have shown that for a conducting substrate covered with thick oxide, no region of oxide must be further than about 500 microns from a conducting path to the substrate. An exposed area of silicon, a metal lead or a thin oxide region will act as a conducting path for the charge. Most planar devices and high density integrated circuits satisfy this dimensional requirement (see section 5.5.8).

One feature in favour of using insulators as masks is that they can remain on the semiconductor surface during the annealing stage after the implantation. In some devices, for example the ion implanted MOS transistor, it is left permanently on the device (Shannon et al., 1969b). For high annealing temperatures (over 700 °C) it is necessary to protect the semiconductor surface with an insulating layer against chemical attack as well as preventing the loss of the substrate material for example arsenic from GaAs or the loss of a dopant such as zinc from GaAs or possibly phosphorus from silicon. The layer itself must not contain a dopant for example boron in the glass or free aluminium in alumina, which may diffuse into the substrate.

An effect, which is very important for shallow doped layers, is the redistribution of impurities at a silicon-silicon dioxide interface due to the higher solubility of the impurity in the oxide. The effect is aggravated by the growth of thermal oxide as this consumes silicon and any impurity is incorporated in the resultant oxide (Grove, 1967). If an oxide is grown on a lightly doped area the redistribution of the impurity must be given special attention.

In summary, the insulating layer on the semiconductor surface serves to protect and passivate it as well as providing a convenient means of masking and registration on the surface that can remain in situ during intermediate high temperature treatments.

The properties of the three most common masking materials will now be discussed in detail.

(i) *Silicon Dioxide* – Silicon dioxide is the most common masking, insulating and passivating material at present in use in the silicon planar technology including ion implantation.

A silicon dioxide layer can be prepared by at least five methods,

(a) Thermal oxidation of a silicon surface,

(b) Chemical vapour deposition,

(c) Ion sputtering,

(d) Anodic oxidation and

(e) Massive ion implantation of oxygen.

The *thermal oxidation* of silicon is carried out at a high temperature (over 900 °C) in a stream of wet or dry oxygen in a very clean furnace (Grove, 1967; Kooi, 1967). Silicon is consumed during the oxidation process and an oxide layer of thickness $t_{ox}$ requires 0.45 $t_{ox}$ of silicon. The reaction rate is about ten times faster with wet oxygen than dry oxygen and therefore wet oxygen is used for thick oxide layers (over one micron). Thin oxide layers, as used for the gate insulator in the metal oxide semiconductor transistor, are usually grown using dry oxygen and are followed by a layer of phosphorus glass ($SiO_2$ with a few percent phosphorus) or silicon nitride to provide a barrier against the diffusion of impurities such as sodium.

The *chemical vapour deposition* (C.V.D.) method (Goldsmith and Kern, 1967) of growing silicon dioxide is very effective for making thick layers rapidly. A dilute mixture of silane ($SiH_4$) in a carrier gas (argon) is reacted with oxygen at a temperature above 300 °C in a suitable reaction chamber. Silicon dioxide is formed and it is deposited on all the heated surfaces. The semiconductor substrates are suitably placed to receive a uniform deposit. Any substrate can be coated provided it can tolerate the reaction temperature without being degraded. It is usual in the case of silicon to grow this oxide on existing thermal oxide as the latter has improved surface state characteristics. The chemically deposited oxide, as formed, is not as dense as thermally grown oxide but it can be densified by heating to over 800 °C. It may not always be possible to carry out this densification procedure as the substrate properties may be affected by the high temperature.

Silicon dioxide can be sputtered from a pure quartz surface and deposited on nearby substrates by using a radio frequency generated plasma (Davidge and Maissel, 1966) or a beam of high energy argon ions (Burrill et al., 1967). Thick layers can be deposited in this manner though not as fast as by chemical vapour deposition. An advantage of sputtering is that the substrate can be maintained near room temperature during deposition. This is important for materials such as gallium arsenide which contain a volatile component, arsenic in this example. Sputtered layers should be annealed at about 500 °C or higher to release occluded gas atoms and allow densification to occur.

Silicon can be oxidized by making it the anode in an electrolytic cell. The oxide thickness is a function of the applied voltage and the thicknesses obtained in practice vary from several hundred to over a thousand angstroms. Anodic films have not found many applications in the semiconductor technology and this may be due to their larger impurity content (Atalla et al., 1959). The main application for anodic oxidation in ion implantation is the controlled stripping of an implanted silicon surface (Johansson et al., 1970; Wilkins, 1968). Successive layers of approximately equal thickness can be removed for obtaining data on the profile of the implanted atoms.

A silicon surface can also be oxidised by making it the anode in a gas discharge (Ligenza, 1965) or by bombarding it with a massive dose of high energy oxygen ions (Freeman et al., 1970). No use is made at present of these two methods in the semiconductor technology.

Thermally produced silicon dioxide has the highest density of all the silicon dioxides discussed in this section. The range data for ions in $SiO_2$ (figs. 5.11 and 5.13) is for thermal $SiO_2$ and when considering the stopping power of the other oxides a correction for the lower density may be necessary.

All the above silicon oxides are readily soluble in buffered hydrofluoric acid. A suitable formula is (Warner, 1965): '(i) Mix 8 parts by weight of ammonium fluoride ($NH_4F$) to 15 parts by weight of deionized water. Stir until dissolved and filter extraneous material. (ii) Add 40 % hydrofluoric acid (HF) to the ammonium fluoride solution in the volume ratio of 4:1 $NH_4F$ to HF and stir. (iii) Store in clean tightly capped polyethylene bottles'. Electronic grade chemicals should be used.

The etch rate at room temperature is about 2 angstrom/sec for the thermal oxide and over 10 angstrom/sec for the other oxides. Kodak KPR and KMER negative photoresists and Shipley AZ1350 positive photoresist can be used for selectively engraving the oxides.

(ii) *Silicon Nitride* – Silicon nitride is normally not used directly on a silicon surface due to the presence of a large number of surface states at the interface. A thin silicon dioxide layer about 500 angstroms thick is sandwiched between the nitride and the silicon. The combined oxide nitride layer is a good insulator with the nitride protecting the oxide from the diffusing impurities such as sodium ions.

Silicon nitride is usually deposited from the vapour phase by decomposing silane $(SiH_4)$ in the presence of ammonia $(NH_3)$ or nitrogen. The energy for the reaction is provided by a high temperature environment or a radio frequency discharge. Silicon nitride can also be sputtered in an r.f. discharge.

Silicon nitride is soluble in hot phosphoric acid and it is necessary to use a deposited silicon dioxide mask to selectively engrave the nitride. The oxide can be engraved by the normal photolithographic tenchiques described in (i) above.

(iii) *Alumina* – Alumina is being studied as an alternative to silicon nitride as a passivating layer on silicon dioxide (Nigh, 1969).

It is more difficult to deposit alumina than silicon nitride but it is readily soluble in hot phosphoric acid at 180 °C (about 120 angstroms per min). It has been included here for completeness as there may be applications where alumina is a preferred insulator.

(iv) *Photoresist Resins* – The photoresist resins such as Kodak KPR or KTFR and Shipley AZ1350 can be used as masks against ion implantation. They are organic compounds containing carbon, hydrogen, oxygen and small quantities of other elements and when calculating the ion ranges, the carbon is considered to be the main element present. The density is also less than that of silicon dioxide. The ion ranges are thus greater than for the oxides and metals.

The action of the ions on the photoresist is to further polymerization and hardening of the resist with the result that it is difficult to remove if the ion dose is large (over $10^{14}/cm^2$). If the resist is thick compared with the ion range then polymerization occurs predominantly in the surface layers allowing the lower layers to dissolve in the stripping solvent.

The edges of photoresist layers tend to be very shelving and thus a sharp demarcation to the implantation is not posssible.

The use of a photoresist layer to define the contents of a complex 'read only' memory circuit has been described in the literature (Dill et al., 1970). In this case the photoresist is not removed until it is burnt off during the low temperature annealing stage.

*5.3.5.3.2. Metals as masks.* Metal layers are used extensively in the fabrication of microelectronic devices for contacting and interconnecting. The most commonly used metal is evaporated aluminium as it adheres well and makes good ohmic contacts to $p^+$ and $n^+$ silicon. Other metals such as

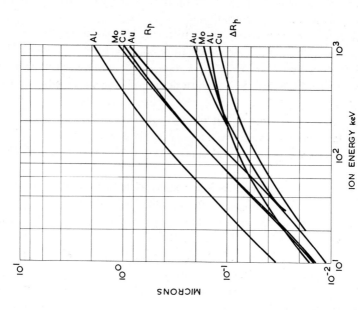
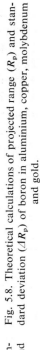

Fig. 5.8. Theoretical calculations of projected range ($R_p$) and standard deviation ($\Delta R_p$) of boron in aluminium, copper, molybdenum and gold.

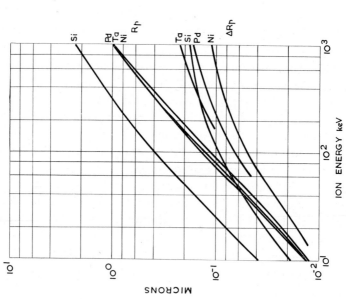

Fig. 5.7. Theoretical calculations of projected range ($R_p$) and standard deviation ($\Delta R_p$) of boron in silicon, nickel, palladium and tantalum.

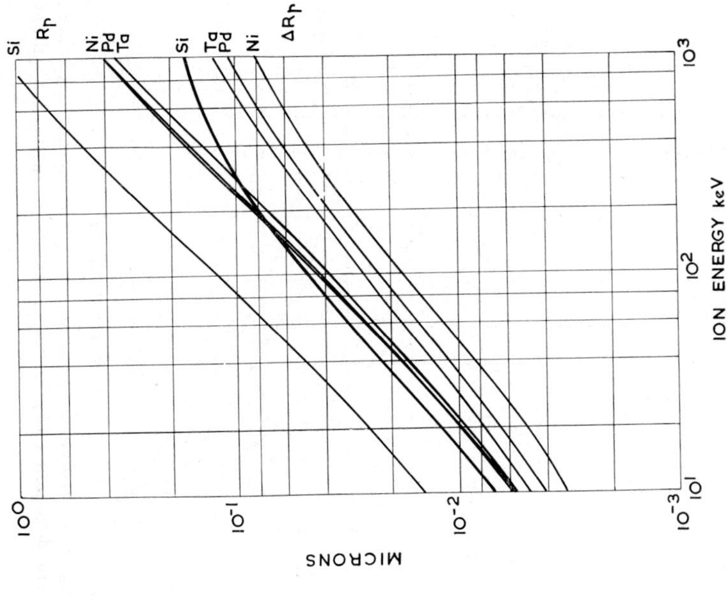

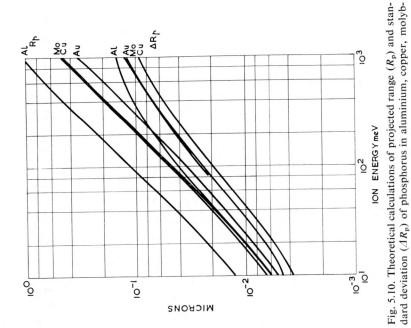

Fig. 5.10. Theoretical calculations of projected range ($R_p$) and standard deviation ($\Delta R_p$) of phosphorus in aluminium, copper, molybdenum and gold.

Fig. 5.9. Theoretical calculations of projected range ($R_p$) and standard deviation ($\Delta R_p$) of phosphorus in silicon, nickel, palladium and tantalum.

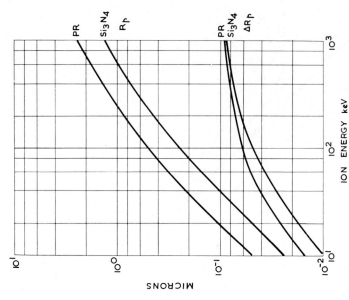

Fig. 5.12. Theoretical calculations of projected range ($R_p$) and standard deviation ($\Delta R_p$) of boron in silicon nitride and photoresist.

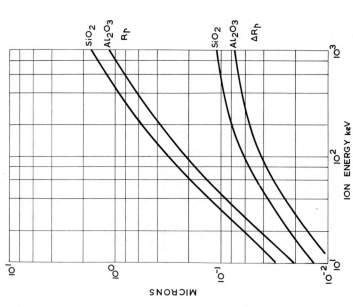

Fig. 5.11. Theoretical calculations of projected range ($R_p$) and standard deviation ($\Delta R_p$) of boron in silicon dioxide and alumina.

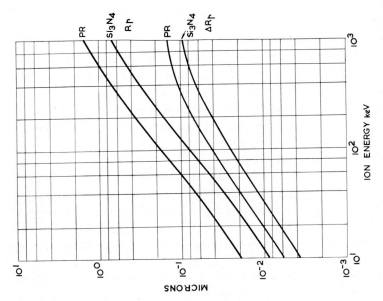

Fig. 5.14. Theoretical calculations of projected range $(R_p)$ and standard deviation $(\Delta R_p)$ of phosphorus in silicon nitride and photoresist.

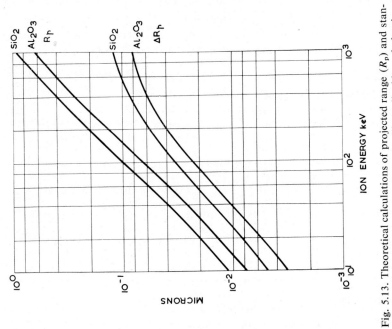

Fig. 5.13. Theoretical calculations of projected range $(R_p)$ and standard deviation $(\Delta R_p)$ of phosphorus in silicon dioxide and alumina.

titanium, molybdenum, platinum and gold are used for example in the 'beam lead' process. The use of metal layers for masking against ion implantation is an extension of an existing technique and not a new one requiring development.

The properties of semiconductors are very sensitive particularly at their surfaces to metallic impurities. It is essential, when using a metal mask directly in contact with a semiconductor surface, to remove all traces of the metal from the surface before any further high temperature processing. If a monolayer of the metal is left on the surface this is equivalent to about $10^{15}$ atoms/cm$^2$ and when diffused into the semiconductor can result in a high level of doping or trapping centres.

A good practice is to use a thin insulating layer of silicon dioxide on the semiconductor and place the metal mask on this layer. The metal can be removed after implantation and the insulator partly etched or completely removed to reduce the remaining contamination. The need for an insulating layer between the semiconductor and the metal mask restricts the implantation to ions which are sufficiently energetic to penetrate the insulator and dope the semiconductor to a sufficient depth. The range of boron and phosphorus ions in silicon dioxide are given in fig. 5.11 and fig. 5.13 respectively.

Table 5.4 shows the relative stopping powers of metals suitable for ion implantation masking, when compared with silicon for ion energies of 40 keV. There is little difference in the masking capabilities of the heavier metals and the choice of metal will depend on other factors such as the ease

TABLE 5.4

Metals for in-contact masking

| Metal | Stopping power * (Silicon = 1) 40 keV Boron | 40 keV Phosphorus | Deposition technique | Engraving | Adhesion on SiO$_2$ |
|---|---|---|---|---|---|
| Silicon | 1 | 1 | C.V.D. ** | Oxide mask reqd. | good |
| Aluminium | 1.21 | 1.28 | evap. | easy | good |
| Nickel | 3.01 | 2.73 | evap. | easy | satis. |
| Molybdenum | 2.87 | 2.43 | sputtering | satis. | good |
| Palladium | 3.26 | 2.71 | evap. | easy | poor |
| Gold | 4.21 | 3.27 | evap. | easy | poor |

\* Stopping Power $= \dfrac{\text{Projected Range in Silicon}}{\text{Projected Range in Metal}}$.

\*\* C.V.D., Chemical Vapour Deposition (Section 5.2.3).

of evaporation, engraving, adhesion and suitability in the fabrication sequence. The ranges of boron and phosphorus in these metals are given in figs. 5.7 to 5.10.

The properties of these metals will now be discussed in detail.

(i) *Aluminium* – This is the most frequently used metal for masking against ion implantation as it is the main metal used in the silicon planar technology. In certain cases, for example the MOS array, the aluminium serves first as a mask and then as the interconnecting layer.

Aluminium is evaporated onto the substrate using tungsten filaments or electron beam heating methods. The metal can be deposited up to 1.5 microns thick without difficulty and degradation of surface texture. Very thick evaporated layers tend to become rough and this makes the photoengraving difficult as the photoresist layer is not uniform and pinholes form in it after etching.

The photoengraving of aluminium can be readily performed using a positive photoresist (Shipley AZ1350) which withstands acid etchants and can be easily removed with acetone without damaging the aluminium layer. A suitable acid etchant for aluminium is: (Osborne, 1968)

Nitric acid (70 % w/w)　　　　　　　4.5 ml
Orthophosphoric acid (S.G. 1.75)　　112 ml
Glacial acetic acid　　　　　　　　　22.5 ml
Water (deionized)　　　　　　　　　　7.5 ml
(Use Electronic or analar grade chemicals).

The etching rate is slow at room temperature (1000 Å/5 minutes) but the rate can be increased by heating the solution. No gas is evolved during the etching and unwanted masking by gas bubbles is avoided.

After the engraving of the aluminium, the wafer is heated to sinter the aluminium which is in contact with the silicon to produce an ohmic contact. A suitable temperature for sintering is 500 °C for 20 to 30 minutes in an inert nitrogen atmosphere or vacuum.

(ii) *Nickel* – Nickel layers can be produced by evaporation (Holland, 1956) or sputtering. It does not adhere to silicon as well as aluminium and as its surface passivates readily, it is more difficult to engrave. Acid ferric chloride solution can be used to etch nickel in the presence of photoresist masks.

(iii) *Molybdenum* – Molybdenum is normally deposited by sputtering techniques as the temperature required for evaporation is very high. Thin molybdenum layers can be readily engraved with the same photoresists and acid etchant recommended for aluminium. Molybdenum has been investigated as an alternative metal to aluminium for contacting silicon (Hooper et al., 1965), *and* it does not react with silicon dioxide at temperatures up to 900 °C.

(iv) *Palladium and Gold* – These two metals evaporate and sputter easily and thin layers can be readily formed. Their adherence to silicon dioxide is poor and they only adhere to silicon if a eutectic (Au–Si) or a silicide (Pd–Si) is formed. The metal layers can be etched in a solution of iodine in potassium iodide using the normal photoresists. The etchant does not attack the eutectic or silicide.

The adherence to silicon dioxide can be improved by using a thin chromium layer between the oxide and the metal. The addition of the chromium complicates the evaporation and etching procedures but it may be acceptable in certain circumstances.

*5.3.5.3.3. Thick metal layers as masks.* All the evaporated or sputtered layers described in the previous section are limited in thickness to less than 0.5 microns with the exception of aluminium in which thicker layers are possible. These thicknesses are adequate for normal implantation energies but for masking against ion energies of the order of 1 MeV and over, thicker layers are needed.

A technique for meeting this requirement is to increase the evaporated layer thickness by electrodeposition of the same metal used for the evaporated layer or a more suitable one chosen for its electrochemical properties. There are no difficulties in electrodepositing onto engraved evaporated layers provided all the individual areas are interconnected to allow the plating current to flow to all parts of the mask. The beam lead process (Lepselter, 1966) is an example of the use of electrodeposition for producing thick leads on a semiconductor device.

There are three disadvantages to using electrodeposition:

(i)   contamination of the wafer by exposing it to the electrolyte,

(ii)  the loss of definition at the edges of the mask and an increasing surface roughness, and

(iii) many electrodeposits are in a state of high mechanical stress.

It is essential to use an initial layer which is very well bonded to the semiconductor substrate. Great care must be taken to ensure that the electrodeposit is free of stress and that it grows uniformly with a fine surface texture. For example, a simple evaporated gold layer is inadequately bonded for electrodeposition and an intermediate chromium layer is necessary to provide the bond.

The choice of the metal to be deposited depends on the factors mentioned above and on the metal used for the initial layer. The electropotential of the metal ion in the electrolyte must be less electro-positive (less 'noble') than the metal of the evaporated layer to prevent electrochemical exchange occurring as the substrate is placed in the electrolyte. Table 5.5 lists some of the metals which can be used for electrodeposited layers. Aluminium cannot be electrodeposited or act as substrate because of its large electro-negative potential $(-1.33V)$ and its natural protective oxide layer.

TABLE 5.5

Metals suitable for electrodeposited masks

| Metal | Electropotential Volts | Electrolyte | Remarks |
|---|---|---|---|
| Nickel | −0.24 | Buffered NiSO₄ | Deposits tend to be stressed especially if 'brighteners' are used. |
| Copper | +0.34 | Acid CuSO₄ and complex alkaline solutions | Deposits not stressed but tend to a coarse texture, less so with alkaline solutions. |
| Palladium | +0.0 | | Proprietary solutions available. |
| Platinum | +1.20 | | |
| Gold | +1.43 | Cyanide | Deposit not stressed. |

*5.3.5.4. Criterion for Determining the Thickness of a Mask.* The primary function of the masking material is to protect the substrate from the high energy ion beam. The L.S.S. theory predicts that, statistically, some ions must penetrate the mask and enter the substrate, but the number decreases very rapidly with increasing mask thickness.

In fig. 5.15, the shaded area represents the number of ions which have penetrated the mask. The attenuation of the ion beam by the mask is represented by the ratio of the shaded area to the total number of atoms entering at the surface of the mask. This ratio is area $B$ discussed in section 5.3.4.2 and it can be readily found from the ratio,

$$X = \frac{t_m - R_p}{\Delta R_p} \qquad (5.3.6)$$

where $t_m$ = mask thickness in the same units as $R_p$ and $\Delta R_p$. Area $B$ is found from $X$ by using the Gaussian table in appendix 2.

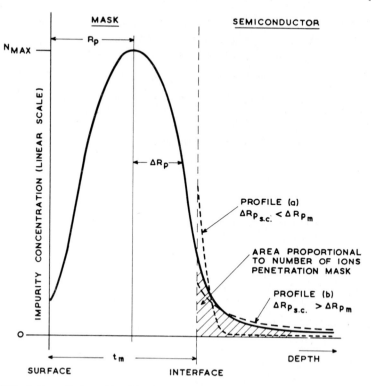

Fig. 5.15. A theoretical profile of ions in a mask on a semiconductor for the conditions where the ions penetrate the mask. $R_{p_m}$ is the projected range of the ions in the mask and $R_{p_{sc}}$ in the semiconductor.

The correct choice of the thickness $t_m$, of a mask, depends on several factors:

(i)   masking material
(ii)  ion type
(iii) ion energy
(iv)  dose (ions/cm$^2$) and
(v)   doping level of the substrate (impurities/cm$^3$).

The first three factors define $R_p$ and $\Delta R_p$, and with $t_m$, determine the attenuation factor of the mask. The ion dose and the substrate doping level must be considered together as they are inversely related, namely the mask thickness must be increased if either the dose is increased or the substrate doping is decreased.

TABLE 5.6

Relation of mask thickness to attenuation

| $\frac{t_m}{R_\nu}$ | Ratio $R_p/\Delta R_p$ | | | | | | |
|---|---|---|---|---|---|---|---|
| | 2 | 2.5 | 3 | 3.5 | 4 | 4.5 | 5 |
| | attn. | attn. | attn. | attn. | attn. | attn. | attn. |
| 1 | 0.5 | 0.5 | 0.5 | 0.5 | 0.5 | 0.5 | 0.5 |
| 1.5 | 0.158 | 0.106 | $0.669 \times 10^{-1}$ | $0.400 \times 10^{-1}$ | $0.227 \times 10^{-1}$ | $0.123 \times 10^{-1}$ | $0.620 \times 10^{-2}$ |
| 2.0 | $0.227 \times 10^{-1}$ | $0.620 \times 10^{-2}$ | $0.135 \times 10^{-2}$ | $0.234 \times 10^{-3}$ | $0.316 \times 10^{-4}$ | $0.339 \times 10^{-5}$ | $0.300 \times 10^{-6}$ |
| 2.5 | $0.135 \times 10^{-2}$ | $0.901 \times 10^{-4}$ | $0.339 \times 10^{-5}$ | $0.73 \times 10^{-7}$ | $0.85 \times 10^{-9}$ | $0.575 \times 10^{-11}$ | $0.314 \times 10^{-13}$ |
| 3.0 | $0.316 \times 10^{-4}$ | $0.300 \times 10^{-6}$ | $0.85 \times 10^{-9}$ | $0.11 \times 10^{-11}$ | $0.5 \times 10^{-15}$ | $<10^{-17}$ | $<10^{-17}$ |
| 3.5 | $0.300 \times 10^{-6}$ | $0.212 \times 10^{-9}$ | $0.26 \times 10^{-13}$ | $<10^{-17}$ | $<10^{-17}$ | $<10^{-17}$ | $<10^{-17}$ |
| 4.0 | $0.85 \times 10^{-9}$ | $0.314 \times 10^{-13}$ | $<10^{-17}$ | $<10^{-17}$ | $<10^{-17}$ | $<10^{-17}$ | $<10^{-17}$ |

It is instructive to calculate the attenuation factor for the general case. From eq. (5.3.6):

$$X = \frac{t_m - R_p}{\Delta R_p} = \frac{R_p}{\Delta R_p}\left[\frac{t_m}{R_p} - 1\right].$$

By assigning values to $t_m/R_p$ and $R_p/\Delta R_p$, the attenuation factor (area $B$) can be found from the Gaussian tables. Table 5.6 lists these values. The effect of increasing $t_m$ with respect to $R_p$ and the ratio $R_p/\Delta R_p$ is clearly shown in these calculations. In practice, with ion energies up to 100 keV it is possible to use ratios of $t_m/R_p$ between 3 and 4 as values of $t_m$ up to one micron are easily achieved in the common masking materials.

The following table 5.7 lists the ratio $R_p/\Delta R_p$ for boron and phosphorus ions implanted at 10, 40 and 1000 keV into several masking materials.

TABLE 5.7

Ratio of projected range to standard deviation for boron and phosphorus ions implanted into various masking materials ($R_p$ in microns)

| Masking material | | Boron | | | Phosphorus | | |
|---|---|---|---|---|---|---|---|
| | | 10 keV | 40 keV | 1000 keV | 10 keV | 40 keV | 1000 keV |
| Aluminium | $R_p$ | 0.0328 | 0.1324 | 0.5909 | 0.0125 | 0.0428 | 1.0277 |
| | $R_p/\Delta R_p$ | 2.12 | 3.2 | 14.0 | 2.66 | 3.08 | 7.99 |
| Silicon | $R_p$ | 0.0383 | 0.161 | 2.323 | 0.0144 | 0.0490 | 1.1762 |
| | $R_p/\Delta R_p$ | 2.03 | 2.99 | 12.9 | 2.62 | 2.99 | 7.65 |
| Silicon Dioxide | $R_p$ | 0.0304 | 0.1288 | 1.7135 | 0.0111 | 0.0388 | 0.9278 |
| | $R_p/\Delta R_p$ | 2.41 | 3.54 | 15.4 | 3.17 | 3.56 | 9.34 |
| Silicon Nitride | $R_p$ | 0.0235 | 0.0994 | 1.3243 | 0.0086 | 0.0301 | 0.7161 |
| | $R_p/\Delta R_p$ | 2.42 | 3.51 | 15.2 | 3.07 | 3.50 | 9.14 |
| Alumina | $R_p$ | 0.0207 | 0.0879 | 1.1657 | 0.0075 | 0.0264 | 0.6328 |
| | $R_p/\Delta R_p$ | 2.41 | 3.54 | 15.5 | 3.13 | 3.52 | 9.35 |
| Nickel | $R_p$ | 0.0133 | 0.0532 | 0.9369 | 0.0055 | 0.0179 | 0.4139 |
| | $R_p/\Delta R_p$ | 1.18 | 2.01 | 8.41 | 1.67 | 2.13 | 5.0 |
| Molybdenum | $R_p$ | 0.0147 | 0.0561 | 1.1183 | 0.0065 | 0.0202 | 0.4598 |
| | $\Delta R_p/\Delta R_p$ | 1.77 | 1.54 | 6.49 | – | 1.70 | 3.95 |
| Gold | $R_p$ | 0.0111 | 0.0381 | 0.8931 | 0.0051 | 0.0150 | 0.3227 |
| | $R_p/\Delta R_p$ | 0.8 | 1.49 | 4.19 | – | – | 3.68 |

(From Johnson and Gibbons, 1969).

It is very difficult to predict the profile in the semiconductor, of the ions which have passed through the mask. If the semiconductor has a smaller stopping power than the masking material then the ions will penetrate deeper into the semiconductor than if they had continued in the mask (fig. 5.15 profile (a)). The reverse situation is also true (profile (b) in the same figure). The area under each of these two profiles is the same as the area under the projection of the mask profile because the same number of ions are involved assuming no reflection of the ions at the interface.

A working value for the attenuation factor can be obtained by assuming that the majority of the implanted ions which pass through the mask are deposited in the first 100 Å of the semiconductor. As an example, silicon will have $5 \times 10^{16}$ atoms/cm$^2$ in this 100 Å layer and if the impurity level is $10^{12}$ impurities/cm$^3$ then there are $10^6$ impurities/cm$^2$ in this layer. If the number of implanted ions is to be equal to or less than the number of impurities then for a surface ion dose of $10^{14}$ ions/cm$^2$ an attenuation factor of $10^{-8}$ or greater is necessary. For 40 keV boron ions into silicon dioxide, reference to tables shows that a mask 3 times the projected range in thickness is necessary.

If a higher impurity level is present then the ion dose can be increased or the attenuation reduced by using a thinner mask or a different masking material. An intrinsic semiconductor (e.g. $n_i = 10^{10}$ carriers/cm$^3$ for silicon) requires the use of masks with attenuation factors of $10^{-10}$ or greater for ion doses of $10^{14}$ ions/cm$^2$ or over.

*5.3.5.5. Effect of the Mask Edge on the Implanted Region.* In this section, the effect of the shape of the mask edge on the impurity profile at the edge of the implanted region is considered. Many of the problems in semiconductor technology are concerned with deviations from the perfect geometrical structure such as the curvature which occurs at the edge of a diffused planar junction (Grove, 1967).

Figure 5.16 shows the edge of a typical thermal diffusion (a) and three configurations of ion implantation masks. The ideal mask (b) is difficult to attain in practice by chemical etching as usually an edge similar to (c) is obtained. The configuration in (d) is not uncommon with aluminium layers when severe undercutting occurs below the photoresist mask.

The sharp discontinuity in the implanted region produced by the ideal mask (b) is equivalent to a junction curvature with an infinitely sharp radius. In a $p^+n$ (or $n^+p$) abrupt junction this curvature should produce a concentration of the electric field in the depletion region at the corner and

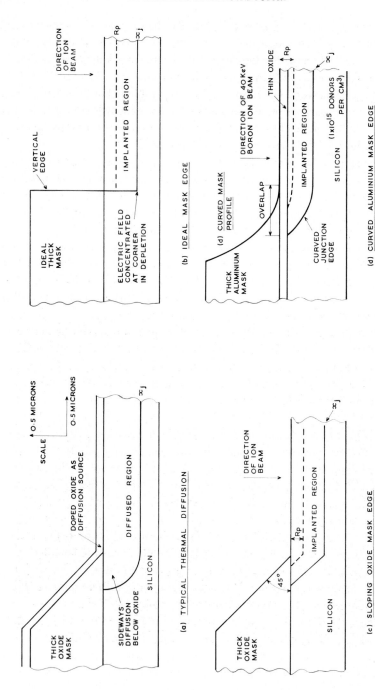

Fig. 5.16. Cross sections of a typical diffusion mask and three implantation masks.

result in premature voltage breakdown. The breakdown voltages observed in practice are not as low as indicated by extrapolating the published results (Grove, 1967) and are of the same order as a diffused junction 1 to 1.5 microns deep. A typical figure for the breakdown voltage is 30 to 40 volts for a 40 keV boron implant ($5 \times 10^{15}$ ions/cm$^2$) into a silicon substrate with a doping concentration of $10^{15}$ donors/cm$^3$, after annealing at 500 °C.

The lack of agreement with the theoretical calculations is probably due to the steep impurity step being slightly graded by the sloping edge of the mask. Figure 5.16(c) shows how this sloping edge of the mask reduces the doping level in a graded manner towards the surface and smooths off the corner with a curvature that is determined by the projected range $R_p$ and the oxide thickness. The curved effect is more pronounced if the projected ion range is an appreciable fraction of the oxide thickness.

The penetration of the ions through the mask, ensures that the junction in the semiconductor comes to the surface below the oxide which affords it some protection. There are disadvantages to the penetration of ions through

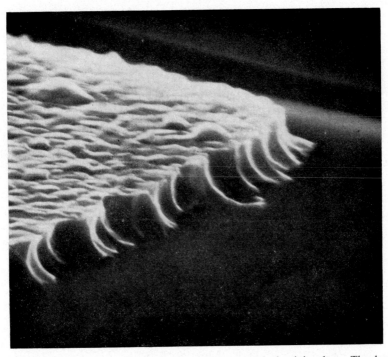

Fig. 5.17. The chemically engraved edge of a 1.5 micron thick aluminium layer. The sloping nature of the edge is clearly shown.

the mask in the fabrication of two devices. In the metal oxide silicon transistor with an aligned gate, the implanted regions will extend under the gate metal and increase the overlap capacitance (see fig. 5.16(d) and section 5.5.5.3). In making high value implanted resistors it is necessary to use narrow resistor strips (5 to 10 microns) to achieve a high resistance. If the width of these strips is increased by the implantation penetrating below the mask the accuracy of the resistor will be partly determined by the slope of the mask edge. The magnitude of the effect can be estimated from a knowledge of the dimensions of the structure by calculating the number of implanted atoms beneath the mask per unit length and comparing the number of ions with the number in the unmasked area.

The sloping edge is a common feature of the chemically engraved mask whether an insulator or a metal, due to undercutting below the photoresist mask. Figure 5.17 shows the sloping edge of an engraved aluminium mask. The influence of a sloping edge can be reduced by using a masking material with a high stopping power, for example gold. The mask need only be a

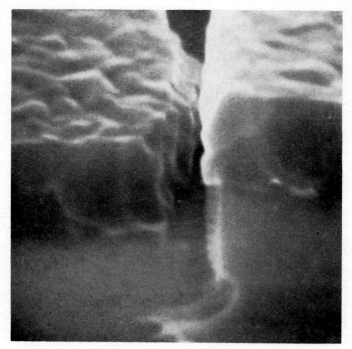

Fig. 5.18. A thin crack in a 1.5 micron thick aluminium gate electrode of a MOST as it passes over an oxide step.

fraction of the thickness of a conventional mask (e.g. aluminium) and therefore the contribution of the sloping edge will be reduced in comparison with the rest of the structure. A vertical edge can be produced by the use of ion etching in the place of chemical etching (Bauer et al., 1970) eliminating the undercutting and permitting a finer geometry to be used.

*5.3.5.6. Defects in the Masking Material.*   In the discussion of masking materials, the assumption has been made that the mask material is structurally perfect. It is not uncommon to find defects such as porosity, pinholes and cracks in evaporated and chemically vapour deposited layers. The ion beam will penetrate any defect in the mask and produce doping in unwanted areas. A thin crack in the gate metal of a metal oxide silicon transistor can lead to a complete short circuiting of the device. An example of a crack in an aluminium gate layer is shown in fig. 5.18 and a complete fracture in fig. 5.19.

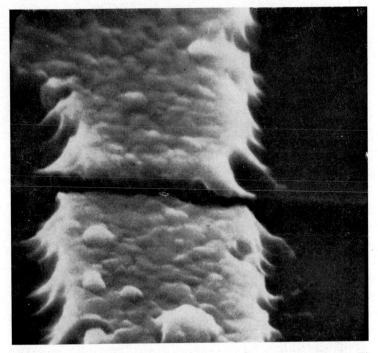

Fig. 5.19. A complete fracture in an aluminium layer passing over an oxide edge. The serrated and sloping nature of the edge is also clearly shown. The layer is about 6.5 microns wide.

Any laterial irregularities in the mask edge will be reproduced precisely in the implanted junction edge and give rise to premature breakdown due to high electric fields at these points. Figure 5.20 shows the irregular edge of an aluminium gate electrode used as a mask and the corresponding uneven implanted junction edge.

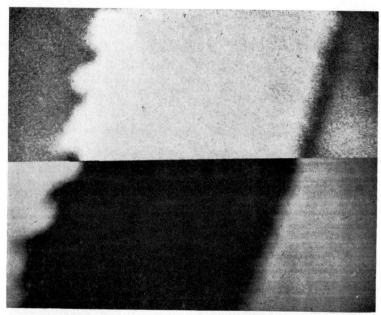

Fig. 5.20. A scanning electron micrograph of the edges of a three micron wide ion implanted MOST gate electrode (section 5.5.5.3). The upper part of the micrograph is the secondary electron picture while the lower half is the electron beam induced conductivity picture which locates the edge of the implanted junction.

Observations made by scanning electron microscopy have shown that most chemically etched edges defined by photographic masks are irregular. Oxide and polycrystalline silicon edges are less irregular than those of aluminium and the thicker the layer the more pronounced the effect. The irregularity is believed to be due to the edge of the photographic mask which is made up of clusters of grains of silver decreasing in density towards the edge of the exposed area. The outlines of these grains are reproduced faithfully when exposing the photoresist. There is no equivalent clustering in photoresist and if it is exposed directly by an electron beam the irregularities do not occur.

*5.3.5.7. Mask Registration for Ion Implantation.* In diffusion, there is no difficulty in registering one mask with the preceding one as each step leaves visible evidence of its execution on the surface of the wafer.

With ion implantation no similar evidence is left on the wafer. When a metal mask, for example, is removed after implantation it may be possible to see the implanted areas by their different reflectance due to amorphicity. After an annealing cycle the amorphicity may have disappeared losing the position of the implanted areas completely. In those cases, when the amorphous regions remain, it has been found very difficult to align a mask with any accuracy to an amorphous pattern on a wafer. This is largely due to the lack of contrast between the amorphous region and the neighbouring single crystal silicon when viewed with a microscope with vertical illumination.

It is therefore necessary to mark the semiconductor wafer surface in such a way that the register is retained throughout all the processing steps. If a thin oxide layer is used to protect and passivate the semiconductor surface then this layer can be partly or wholly engraved in suitable places, for example, with lines between the individual devices to provide a permanent register mark and later be used for scribing. It is also possible to use an aluminium layer twice by engraving one aperture in the layer to be followed by a larger one for a second implantation (fig. 5.48). Alternatively one aluminium layer can be used after another by evaporating the second layer over the first one and engraving the composite layer with the next mask (fig. 5.48).

## 5.4. Electrical properties of ion implanted layers

### 5.4.1. INTRODUCTION

The earlier part of this chapter has dealt with the techniques for controllably introducing a number of impurities into a semiconductor surface by ion implantation. The main function of these impurities is to dope the semiconductor with acceptor or donor centres in a spatially defined manner in order to form p-n structures such as simple resistors, diodes and transistors. These devices are combined in a complex manner to form integrated circuits.

The electrical behaviour of the implanted impurity should be fully understood before it is considered as part of a device structure. In this section, the electrical information required about ion implanted layers is outlined, methods of measurement are discussed and lastly some results for the common ion implanted dopants are given.

The electrical characteristics are considered under two broad headings, (i) electrical properties within the implanted layer and (ii) the electrical properties of the layer with the substrate.

Consider first the electrical behaviour of the implanted layer. The theoretical distribution of the implanted impurity atoms is a Gaussian distribution, but the distribution of electrically active centres is not necessarily the same due to the effects of precipitation, and interstitial impurity atoms which are not necessarily electrically active but reduce carrier mobility due to various scattering mechanisms. These effects are non-uniformly distributed in depth and become severe for the high impurity concentrations. It is usual to measure sheet resistance for rapid assessments such as annealing behaviour while the Hall mobility and carrier concentration are used in detailed studies.

Secondly, the properties of the interface between the layer and the substrate are of equal importance to the layer properties especially if a p-n junction is formed. The complete current-voltage characteristics and the influence of residual damage on the minority carrier lifetime must be known.

The temperatures used during implantation and annealing have a very pronounced effect on the electrical properties of the implanted layers. It is necessary to study these layers prepared under various conditions and especially those conditions pertaining to particular applications where restrictions are imposed (see also sect. 5.3.3.1).

### 5.4.2. METHODS OF MEASURING THE ELECTRICAL PROPERTIES OF ION IMPLANTED LAYERS

*5.4.2.1. General.* The electrical assessment of an ion implanted layer requires the measurement of i) carrier type ii) sheet resistance and iii) profiles in depth of carrier concentration and carrier mobility. The measurements are required for various implantation and annealing conditions.

The common methods employed for making these measurements will now be discussed.

*5.4.2.2. Determination of Carrier Type.* It is very convenient to be able to rapidly determine the carrier type of a semiconductor without having to prepare a sample for a Hall effect measurement.

This test can be made by using the thermoelectric properties of the semiconductor. Normally the semiconductor is in thermal equilibrium but if this is disturbed by heating a small region, the electrons in an n semiconductor, for example, will attain higher velocities and diffuse away from the region making it positive with respect to the surrounding area at the lower temperature.

Two probes made of the same metal, but at different temperatures are pressed onto the surface of the wafer or implanted layer. The probes are connected to a sensitive voltmeter. If the semiconductor is n type, the hot probe will be positive with respect to the cold probe. The opposite polarity holds for a p type semiconductor. The measurement is essentially qualitative but the voltage increases for higher doping levels in the semiconductor. Difficulty may be experienced in obtaining a clear response if the dopant concentration in the semiconductor or the implanted region is less than about $10^{14}$ impurities/cm$^3$.

A convenient form of the apparatus is a pair of gold tipped probes, one at room temperature and the other attached to the tip of a low voltage miniature soldering iron by a suitable electrical insulator to provide isolation. A microvoltmeter must be used to detect the voltage between the probes. It should be battery operated, or have good isolation from the a.c. mains to reduce spurious responses.

The semiconductor must be clean and free of oxide and metals in the area to be measured. The room temperature probe is pressed onto the semiconductor surface and then the hot probe is touched onto the surface nearby and the polarity of the resultant voltage noted. The voltage will decrease after the initial contact as the semiconductor chills the hot probe. The spatial resolution depends on the size of the probes and the method used for positioning. Hand held probes can be used to test to about 1 mm resolution.

*5.4.2.3. Determination of the Sheet Resistance of Ion Implanted Layers.* One of the general problems encountered in the fabrication of semiconductor devices is the measurement of the electrical conductivity of the material at various stages during processing. It is essential that the measurement be non-destructive, speedy and not involve any elaborate preparation such as the fabrication of diodes, resistors or Hall effect patterns. The four point probe technique (Valdes, 1954; Smits, 1958; v.d. Pauw, 1958; Logan, 1961; Severin, 1971) for measuring sheet resistivity has these advantages and its use is directly applicable to ion implanted layers.

The two point and three point probes used for measuring sheet resistance are not as suitable as the four point probe for ion implanted layers. The two point probe or spreading resistance probe depends on measuring the resistance adjacent to a fine point. It requires calibration against known materials and a correction if a p-n junction is present (Research Triangle Institute, 1965). The three point probe is used for measuring a lightly doped layer on a more heavily doped substrate of the same conductivity type. It depends on measuring the reverse breakdown voltage at the test probe. It has to be experimentally calibrated. The layer must be thicker than the depletion width at the breakdown voltage and is therefore not particularly suitable for thin high resistivity layers (Camenzind, 1968).

The four point probe comprises four equally spaced metal probes which are pressed onto the semiconductor surface. The probes are usually set 'in-line' but they can be at the corners of a square, the latter arrangement allowing a smaller geometry (fig. 5.21).

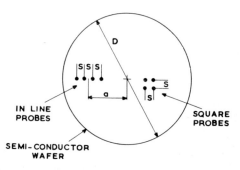

Fig. 5.21. The layout of 'in-line' and 'square' four point probe geometries. '$S$' is the probe spacing, and '$a$' the displacement from the centre of the wafer diameter '$D$'.

The *sheet* resistivity of an infinite thin layer in which the current flow is parallel to the surface is (Smits, 1958):

$$\rho_s = \frac{\pi}{\log_e 2} \frac{V}{I} \quad \text{for 'in line' probes} \tag{5.4.1}$$

with the current $I$ flowing between the outer probes and the voltage $V$ measured across the inner probes, and

$$\rho_s = \frac{\pi}{2\log_e 2} \frac{V}{I} \quad \text{for 'square' probes} \tag{5.4.2}$$

with the current $I$ flowing between adjacent probes and the voltage $V$ measured across the other two probes. If the ratio of the layer thickness $x$ to the probe spacing $S$ is less than 0.4, then the average resistivity $\rho$ is given by:

$$\rho = \rho_s x$$

$$= \frac{\pi}{\log_e 2} \cdot \frac{V}{I} \cdot x \qquad \text{for 'in line' probes}$$

$$= 4.532 \cdot \frac{V}{I} \cdot x \qquad \text{(ohm·cm).} \qquad (5.4.3)$$

This formula can be used for calculating the resistivity of wafers and diffused and implanted layers as they meet the requirement $x/S < 0.4$ for $S = 1$ mm. Many ion implanted layers are made into a substrate of the opposite conductivity type and generally the junction depth $x_j$ is not known. In these cases only the sheet resistivity can be calculated.

The concept of sheet resistivity is very important in the study and application of thin conducting layers. Consider a square of side $L$ cut from a conducting layer of thickness $x$ and resistivity $\rho$. The resistance between opposite edges of this square is given by:

$$\rho_s = \rho \cdot \frac{\text{length } (L)}{\text{cross section } (L \times x)}$$

$$= \frac{\rho}{x}.$$

$\rho_s$ is known as the sheet resistivity or sheet resistance and it has the dimensions of ohms. It is not affected by the geometrical shape of the layer and as shown above it is readily measured by the four point probe. It is a very useful parameter to use when designing diffused or ion implanted resistors as the resistance can be readily calculated from a knowledge of the sheet resistance and the number of 'squares' in the area of the device.

Bader and Kalbitzer (1970a) have described a method of measuring implanted layers in high resistivity silicon of the same conductivity type as the implanted dopant. The four point probe measures all the electrically active species in the wafer and none of the dopant atoms are excluded by being beyond a p-n junction. The use of high resistivity silicon requires special care with surface preparation to reduce leakage currents.

An estimate of the number of implanted atoms $C_s$ (per $cm^2$) can be made for certain implantations in the following manner.

The average conductivity $\sigma$ of the layer is given by:

$$\sigma = q\mu N$$

where $q$ = electronic charge, $\mu$ = average mobility, $N$ = average carrier concentration (atoms/cm$^3$). In an implanted layer the carrier concentration $N(x)$ is a function of depth and the mobility a function $N(x)$. The surface dose $C_s$ is $C_s = \int_0^{x_j} N(x)dx$ for a Gaussian distribution provided $(x_j - R_p)/\Delta R_p \geq 3$. This condition is met for large implantation doses into lightly doped substrates.

In highly doped layers $(N > 10^{19}/\text{cm}^3)$ the mobility tends to saturate at 65 or 48 cm$^2$/volt·sec for electrons and holes respectively. The mobilities may be lower in the presence of damage.

The average conductivity can be written assuming a constant mobility $\mu_c$ over the depth $x_j$ as:

$$\sigma = \frac{1}{\rho} = q\mu_c \frac{C_s}{x_j}$$

where $\rho$ is the resistivity. The *sheet resistivity*

$$\rho_s = \frac{\rho}{x_j} = \frac{1}{q\mu_c C_s}$$

$$= 4.532 \frac{V}{I}.$$

(5.4.4)

This relation can be used to obtain an estimate of $C_s$ by using the mobility data of Irvin (1962) and knowledge of the effect of damage on mobility. The interdependence of $N(x)$ and $\mu$ in the equation for conductivity is a severe limitation of the four point probe technique as an analytical method. Its use is restricted to the type of measurement where the sheet resistance is adequate and the minimum preparation is essential. A common application is in the study of the annealing of implanted layers.

Figure 5.22 is a typical circuit for a four point probe measuring assembly. The current for the probes is provided by a preset constant current supply and the voltage measured by a digital voltmeter with a floating high impedance input. A switching circuit is included for switching the current to the voltage or current probes and for reversing the direction of current flow.

The wafer, which must be free of gross contamination and any insulating layers, is placed on an insulated stage which can be positioned by a pair of orthogonal micrometers below the probes. After the probes have been lowered onto the wafer, the constant current is first passed through the voltage probes to 'form' the contacts. The current is then switched to the current probes and the voltmeter connected to the voltage probes. The first obser-

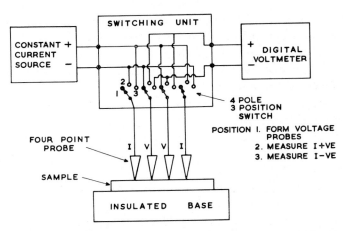

Fig. 5.22. A typical circuit for a four point probe measuring assembly. The functions of the three position switch are shown.

vation is made and then the current is reversed in direction for the second observation. The readings should agree to an accuracy better than that required of the measurement. The probes are held in contact for all this time. The micrometers can be used for moving the probes by known intervals over the surface of the wafer when measuring the variation of sheet resistivity.

There are several possible sources of error when making four point probe measurements;

(i)   surface effects
(ii)  magnitude of the measuring current
(iii) probe loading
(iv)  geometrical factors
(v)   temperature.

(i) *Surface effects* – All the calculations for the four point probe assume that there is no surface conduction and that any carriers injected at the current probes or by photons will recombine and not modulate the conductivity. The surface should have a high recombination rate which can be achieved by lapping or sandblasting etc. This is not possible for implanted layers in highly polished wafers. Bader and Kalbitzer (1970b) advocate a mild oxidising treatment by either exposing to the atmosphere for 2 hours (n layers) or a hot dip in $H_2O_2$ (p layers). Measuring immediately after a hydrofluoric acid dip should be avoided. The measurements should be made in the dark to prevent the generation of carriers by the incident light and the resultant conductivity modulation of the layer. The effect is more pronounced in high resistivity layers in which carrier lifetimes are usually long.

(ii) *Magnitude of the measuring current* – In the discussion of the four point probe measurements, so far, it has been assumed that the measuring current flows entirely in the layer and that the potential between the voltage probes is directly proportional to the current.

The potential gradient along the implanted layer will cause currents to flow in any parallel conducting path such as a thin surface layer. These currents can be comparable to the current in the implanted layer particularly for high resistivities. As an example, an n layer can form on n⁻ or p⁻ substrates due to cleaning in HF and completely obscure the voltage reading.

Many ion implanted layers are electrically isolated from the substrate by the p-n junction formed between the implantation and the substrate.

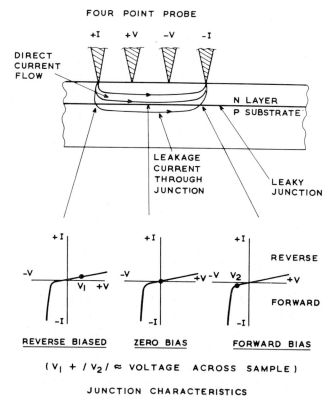

FOUR POINT PROBE

Fig. 5.23. The current flow during a four point probe measurement on an n layer with a leaky p-n junction to the substrate. The junction characteristics at three points on the leakage path are shown.

Patrick (1965) has studied the dependence of the measured sheet resistivity on the value of the constant current for n or p epitaxial layers. The potential gradient in the layer biases the n-p junction into forward bias under the negative current probe and reverse bias under the other probe (fig. 5.23). The current flowing between the probes will be divided between the layer and the path through the p-n junction. This division will depend critically on the $I/V$ characteristics of the junction (see section 5.4.3.3) and it will be negligible in the case of the ideal p-n junction. In practice many measurements are made on layers which have imperfect isolation p-n junctions or localized imperfections both of which will produce variations in the current division. A particular example is the study of the annealing behaviour in ion implanted layers, where the influence of the leaky junctions is noticeable. At the low annealing temperatures the sheet resistance is high and the junction properties poor due to damage (see section 5.4.4) and both these can combine to produce very variable observations.

Experience has shown that constant currents between 10 microamps and 1 milliamp can be used to measure ion implanted layers, the larger current being required for the lower resistance layers (less than 1000 ohms/ square). The current setting must be kept low so that the isolating junction is operating near its zero bias point. Measurements should be made over a range of currents to ascertain that the sheet resistance remains unchanged. Heating in the layer or at probes by the current must be avoided as this will affect the resistance and the junction properties by increasing the leakage current.

(iii) *Probe loadings* – The probes have to be applied with some mechanical pressure onto the semiconductor surface to help make good contact. The usual loading for probes for use on silicon is 200 grams and fig. 5.24 shows the damage that can result from this type of contact particularly if the probe '*skids*' slightly. The depth of the damage is greater than many ion implanted layers and it can cause poor contacting and excessive p-n junction leakage. The probe damage can render part of a wafer useless for subsequent device fabrication and consequently the four point probe *cannot* be considered as non-destructive when applied to thin layers. The damage produced by a probe on a lapped surface (e.g. the back of a wafer) is of no consequence and is therefore non-destructive.

For reproducible sheet resistance measurements on thin diffused or ion implanted layers it is recommended (Tong and Dupnock, 1969) that a broad probe tip of 125 microns radius be used with a probe loading of 40 grams or less. The probes must be lowered as slowly as possible to prevent them

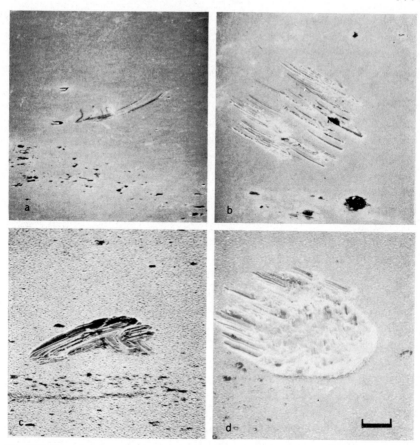

Fig. 5.24. Scanning electron micrographs of indentations made by the single operation of a four point resistivity probe. (a) silicon, 20 gm loading 100 microns probe radius; (b) silicon, 200 gm loading 40 micron probe radius; (c) aluminium 20 gm loading 100 micron probe radius; and (d) aluminium 200 gm loading 40 microns probe radius. The bar in (d) represents 10 microns.

skidding across the surface of the semiconductor. Satisfactory results with silicon have been obtained at Harwell using a 20 gram germanium probe set.

(iv) *Geometrical factors* – It has been assumed in the derivation of eqs. (5.4.1) to (5.4.3) that the four probes were placed on an infinite layer, that is, they were far from the edge of the layer. In practice, measurements are made on layers with finite dimensions and when verifying the uniformity of ion implantation, measurements are made over the whole area of the layer. Valdes (1954), Smits (1958) and Logan (1961) have calculated correction factors which apply to the above examples.

Table 5.8 (Smits, 1958) gives the geometrical factor, $C(D/S)$, for sheet resistivity for the four point probe placed in the centre of a circular sample of diameter $D$.

TABLE 5.8
Geometrical factor $C(D/S)$ for different ratios of $D/S$
$S$ = Probe spacing; $D$ = Wafer diameter.

| $D/S$ | Geometrical factor $C(D/S)$ | % Error if $C$ not applied | Remarks |
|---|---|---|---|
| 3 | 2.2662 | 50.0 | |
| 4 | 2.9289 | 35.4 | |
| 5 | 3.3625 | 25.8 | |
| 7.5 | 3.9273 | 13.4 | |
| 10 | 4.1716 | 8.0 | |
| 15 | 4.3646 | 3.71 | |
| 20 | 4.4364 | 2.12 | |
| 25 | 4.4724 | 1.32 | $1''$ wafer $S = 1$ mm |
| 31.75 | 4.4938 | 0.86 | $1\frac{1}{4}''$ wafer $S = 1$ mm |
| 40 | 4.5080 | 0.54 | $1''$ wafer $S = 0.63$ mm |
| 50.8 | 4.5172 | 0.34 | $2''$ wafer $S = 1$ mm |
| $\infty$ | 4.5326 | 0.0 | |

Sheet resistivity $\rho_s$ is given by

$$\rho_s = \frac{V}{I} \cdot C\left(\frac{D}{S}\right) = \frac{V}{I} \cdot \frac{\pi}{\left[\ln 2 + \ln\left(\frac{(D/S)^2+3}{(D/S)^2-3}\right)\right]} \tag{5.4.5}$$

where $V$ and $I$ are the voltage and current associated with the four point probe and $S$ is the probe spacing and $D$ the wafer diameter (same units).

If the four point probe is now moved off centre with the probes along a radius by a distance 'a' from the centre of the circle to the mid point between the voltage probes (fig. 5.21), then the geometrical factor is given by (Logan, 1961):

$$C\left(\frac{D}{S}, \frac{a}{D}\right) =$$

$$\frac{\pi}{\ln 2 + \frac{1}{2}\ln \dfrac{\left[1 - \left(\frac{2a}{D} + \frac{S}{D}\right)\left(\frac{2a}{D} - \frac{3S}{D}\right)\right]\left[1 - \left(\frac{2a}{D} - \frac{S}{D}\right)\left(\frac{2a}{D} + \frac{3S}{D}\right)\right]}{\left[1 - \left(\frac{2a}{D} - \frac{S}{D}\right)\left(\frac{2a}{D} - \frac{3S}{D}\right)\right]\left[1 - \left(\frac{2a}{D} + \frac{S}{D}\right)\left(\frac{2a}{D} + \frac{3S}{D}\right)\right]}} \tag{5.4.6}$$

TABLE 5.9

Geometrical factor $C(D/S, a/D)$ for 1″, 1¼″ and 2″ wafers with $S = 1.0$ mm

$D$ = wafer diameter, $a$ = displacement from centre of the wafer, $S$ = probe spacing (same units)

| $\dfrac{a}{D}$ | Diameter 1″ ($D/S = 25.4$) | | Diameter 1¼″ ($D/S = 31.75$) | | Diameter 2″ ($D/S = 50.8$) | |
|---|---|---|---|---|---|---|
| | $C(D/S, a/D)$ | % Error if $C$ not applied | $C(D/S, a/D)$ | % Error if $C$ not applied | $C(D/S, a/D)$ | % Error if $C$ not applied |
| 0 | 4.4724 | 0 | 4.4938 | 0 | 4.5172 | 0 |
| 0.1 | 4.4673 | 0.11 | 4.4905 | 0.07 | 4.5159 | 0.03 |
| 0.15 | 4.4600 | 0.28 | 4.4858 | 0.18 | 4.5141 | 0.07 |
| 0.2 | 4.4475 | 0.56 | 4.4777 | 0.36 | 4.5109 | 0.14 |
| 0.25 | 4.4260 | 1.04 | 4.4639 | 0.67 | 4.5054 | 0.26 |
| 0.3 | 4.3867 | 1.92 | 4.4385 | 1.23 | 4.4954 | 0.48 |
| 0.35 | 4.3036 | 3.78 | 4.3848 | 2.43 | 4.4742 | 0.95 |
| 0.40 | 4.0747 | 8.89 | 4.2370 | 5.72 | 4.4158 | 2.25 |
| 0.41 | 3.9797 | 11.0 | 4.1757 | 7.08 | 4.3917 | 2.78 |
| 0.42 | 3.8475 | 14.0 | 4.0907 | 8.97 | 4.3582 | 3.52 |
| 0.43 | 3.6544 | 18.3 | 3.9673 | 11.7 | 4.3097 | 4.59 |
| 0.44 | 3.3534 | 25.0 | 3.7772 | 16.0 | 4.2355 | 6.24 |
| 0.45 | — | | 3.4583 | 23.0 | 4.1131 | 8.95 |

Table 5.9 lists the correction factor for $1''$, $1\frac{1}{4}''$ and $2''$ wafers when probed by standard 1 mm spaced probe.

The table 5.9 shows the advantage in using large $D/S$ ratios when measuring conducting layers especially for measuring the uniformity of implanted layers.

(iv) *Temperature* – The conducting layer should be maintained at a constant temperature for accurate results as the sheet resistance changes with temperature. This is caused by the change in the fraction of ionized centres and their effect, due to phonon scattering, on the carrier mobility (Runyan, 1965).

### 5.4.2.4. *Measurement of Carrier Concentration and Mobility by the Hall Effect*

*5.4.2.4.1. Introduction.* The four point probe measurement, discussed in the previous section, yields the sheet resistance of a conducting layer which is inversely proportional to the product of carrier concentration and mobility. For a fuller understanding of the electrical behaviour of the ion implanted layer the carrier concentration and mobility must be measured separately and their distribution in depth must also be known. This knowledge permits a comparison of the distribution of electrically active centres with other physical measurements which can produce profiles of the implanted atoms, such as radioactive tracer and Rutherford back-scattering techniques combined with stripping of the layer (Chapter 2). From the variation of the carrier concentration with temperature, the activation energy of the electrically active centres can be calculated and the existence of other centres, due possibly to lattice damage, can be detected by the presence of additional activation energies. The carrier concentration, mobility and type can be determined by observing the Hall effect in the implanted layer.

The Hall effect has been extensively employed in the electrical studies of ion-implanted layers, but it does not appear to have been used as extensively in the study of diffused and epitaxial layers. The effect has been used for evaluating the bulk properties of semiconductors and making certain galvomagnetic devices (e.g. Hall multipliers). The subject has been reviewed by Beer (1963, 1966) and Putley (1960, 1968). The study of radiation damage produced by high energy particles in semiconductors has been reported by Billington and Crawford (1961) and the Research Triangle Institute (1964).

The first papers reporting the application of the Hall effect to isolated implanted layers appeared in the proceedings of the Grenoble conference on Ion Implantation (Gibson et al., 1967; Galaktionova et al., 1967; Marsh et al., 1967). Since this time, the majority of papers on the electrical behaviour of implanted layers have included Hall effect measurements and Johansson et al. (1970) have recently reviewed the whole technique.

*5.4.2.4.2. Measurement of the Hall effect in thin layers.* There are four different configurations for measuring the Hall effect in thin conducting layers. They all depend on the layer being homogeneous, isolated from the substrate and with contact areas small compared with the lateral dimensions of the configuration (fig. 5.25(a)). The bridge shaped sample (fig. 5.25(b)) was used by Gibson et al. (1967) and Clark and Manchester (1968). It is the direct thin layer equivalent of the semiconductor Hall bar with contacts attached at its ends and on its sides. Van der Pauw (1958) published a method for measuring the specific resistivity and Hall effect of thin conducting discs of arbitrary shape (fig. 5.26(a)). Most of the Hall effect configurations (fig. 5.26(b) and (c)) now in use are based on the results of this paper. Grosvalet et al. (1966) and Glotin (1968) described the use of the 'Corbino disc' (fig. 5.28) for Hall effect measurements. Buehler and Pearson

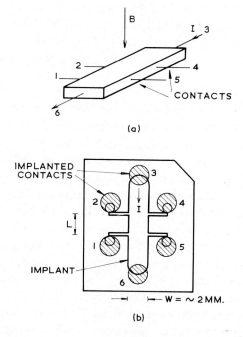

(a)

(b)

Fig. 5.25. (a) Semiconductor Hall bar. $I$ is the measuring current and $B$ the magnetic field. The Hall voltage is measured between contacts 1 and 5 (or 2 and 4) and a voltage proportional to conductivity between 1 and 2 (or 4 and 5); (b) Bridge shaped Hall sample used by Gibson et al. (1967). The magnetic field is normal to the sample.

or

$$C_S = \frac{r}{R_{HS} q} \qquad (5.4.10)$$

and

$$\mu_e = \frac{R_{HS}}{R_S} \cdot \frac{1}{r} . \qquad (5.4.11)$$

Mayer et al. (1967a) have shown that calculating $C_S$ from $R_S$ and $\mu_e$ leads to a lower value of $C_S$ than the actual number of carriers per unit area in the implanted layer. This is due to the strong depth dependence of the carrier concentration $N$ in the implanted layer. The mobility at any point is a function of the local impurity concentration and hence the depth. Petritz (1958) has shown that for a number of colaminar conducting layers with no circulating currents, the overall effective Hall mobility is weighted in favour of the layers with a higher mobility. In the calculation of $C_S$ there is an adverse weighing in favour of the lightly doped layers which can produce values of $C_S$ 20 % low in well annealed samples (Johansson and Mayer, 1970). It is most pronounced if there are strong asymmetries in the impurity distribution (e.g. channelled 'tails').

Each Hall configuration will now be considered in turn and expressions given for $R_S$ and $R_{HS}$.

(a) *Hall bar and bridge* (fig. 5.25(a) and (b)). A constant current $I$ is passed from contact 3 along the bar to contact 6. The voltage difference across the sample (1 to 5 or 2 to 4) is measured with and without the magnetic field and the difference, $V_H$, in these two observations is used to calculate the sheet Hall coefficient. The voltage $V_R$ is measured *along* the sample (1 to 2 or 4 to 5) without the magnetic field. Several readings can be obtained by different combinations of the directions of the current and the magnetic field. The polarity of the voltage across 1 to 5 or 2 to 4 depends on the relative directions of the current and the magnetic field and the type of carrier (hole or electron). Fleming's left-hand rule can be used to determine this polarity.

The sheet resistivity $R_S$ is given by:

$$R_S = \frac{V_R}{I} \left( \frac{W}{L} \right) \text{ (ohms)} \qquad (5.4.12)$$

and the Hall coefficient by:

$$R_{HS} = \frac{V_H}{I} \cdot \frac{10^8}{B} \left( \frac{cm^2}{coulomb} \right) \qquad (5.4.13)$$

where $W$ = width of the Hall pattern (cms), $L$ = distance between contacts 1 and 2 or 4 and 5 (cms), $I$ = constant current (amps), $B$ = magnetic field strength gauss. Hence

$$C_S = \frac{1}{R_S q \mu_D} = \frac{I}{V_R} \cdot \frac{L}{W} \cdot \frac{1}{q \mu_D} \text{ (carriers/cm}^2\text{)} \tag{5.4.14(a)}$$

or

$$C_S = \frac{r}{R_{HS} q} = \frac{I}{V_H} \cdot \frac{B}{10^8} \cdot \frac{r}{q} \text{ (carriers/cm}^2\text{)} \tag{5.4.14(b)}$$

and

$$\mu_e = \frac{R_{HS}}{R_S} \cdot \frac{1}{r} = \frac{V_H}{V_R} \cdot \frac{L}{W} \cdot \frac{10^8}{B} \cdot \frac{1}{r} \tag{5.4.15}$$

(b) *van der Pauw configuration* (fig. 5.26(a), (b), and (c)). In the van der Pauw configuration the voltage across A to C is measured with and without the magnetic field $B$ and the constant current $I$ flowing between B and D. $V_H$, the difference between these measurements, is used to calculate the Hall coefficient. The voltage $V_R$ is measured from A to D without the magnetic field and the current flowing from B to C. By different combinations of the magnetic field and the current direction and using all the measurement positions, 8 values of $V_R$ and 16 of $V_H$ can be made and an average value calculated (Johansson et al., 1970).

For a *symmetrical* configuration $R_S$ is given by (v.d. Pauw, 1958):

$$R_S = \frac{\pi}{\ln 2} \cdot \frac{V_R}{I} \text{ (ohms/square)} \tag{5.4.16}$$

and $R_{HS}$ by

$$R_{HS} = \frac{V_H}{I} \cdot \frac{10^8}{B} \left(\frac{\text{cm}^2}{\text{coulomb}}\right) \tag{5.4.17}$$

$C_S$ and $\mu_D$ can be calculated from eqs. (5.4.16) and (5.4.17) as described in (a).

(c) *Non-peripheral square four point probe.* Buehler and Pearson (1966) have considered two geometries where a square array of points is centred on a Hall plate as shown in fig. 5.27. Part (a) of the figure shows a circular plate and (b) a circular hole in an infinite sheet. They give the following relationships for the sheet resistivity $R_S$ and sheet Hall coefficient

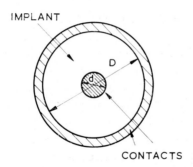

IMPLANT

D

CONTACTS

Fig. 5.28. Configuration of the Corbino disc. The current flows from the inner circle, diameter $d$, to the annulus inner diameter $D$. The magnetic field is normal to the disc.

$R_{HS}$ for the two cases as follows:

(i)  $R_S = \dfrac{V_R}{I} \dfrac{\pi}{\ln 2} \dfrac{1}{C_\rho}$ (ohms/square)                    (5.4.18)

where $V_R$ is the voltage measured between A and D for a constant current $I$ amps between B and C and with the magnetic field $B$. $C_\rho$ is the resistivity correction factor and is given by:

$$C_\rho = \left\{ \ln \sqrt{2} \cdot \left[ 2 + \left(\frac{d}{S}\right)^2 \right] \left[ 4 + \left(\frac{d}{S}\right)^4 \right]^{-\frac{1}{2}} \right.$$

$$\left. + \tan^2 \beta \ln \sqrt{2} \cdot \left[ 2 + \left(\frac{d}{S}\right)^2 \right]^{-1} \times \left[ 4 + \left(\frac{d}{S}\right)^4 \right]^{\frac{1}{2}} \right\} \Big/ \ln 2$$

where $\tan \beta$ = Hall angle = $R_{HS} B / R_S$

(ii)  $R_{HS} = \dfrac{V_H}{I} \cdot \dfrac{10^8}{B} \cdot \dfrac{1}{C_H} \left( \dfrac{cm^2}{coulomb} \right)$                    (5.4.19)

where $V_H$ is the voltage measured between A and C for a constant current between B and D with the magnetic field $B$. The polarity of $V_H$ is determined by the direction of the current, magnetic field and the type of carrier. $C_H$ is the Hall coefficient correction factor and it is given by:

$$C_H = \frac{4}{\pi} \tan^{-1} \left( \frac{\sqrt{2}S}{d} \right)^2$$                    (5.4.20(a))

for circular plates and

$$C_H = \frac{4}{\pi} \tan^{-1} \left( \frac{d}{\sqrt{2}S} \right)^2$$                    (5.4.20(b))

for circular holes where $d$ = diameter of the hole or plate and $S$ = probe spacing.

An examination of the correction factors shows that $V_R$ and particularly $V_H$ *decrease* with an increasing ratio of $d/S$, showing that it is preferable to keep the probes close to the edge of the plate or hole, i.e. approaching van der Pauw's configuration, for maximum sensitivity.

The method could be useful for measuring small discs in which $d$ is no more than 2 or 3 times $S$. In practice for $S$ = 1 mm the discs would be up to 3 mm in diameter. The correction factors, from Buehler and Pearson (1966), are given in table 5.10. $C_S$, the surface carrier density and the mobility $\mu$ can be calculated from eqs. (5.4.18) and (5.4.19).

TABLE 5.10
Correction factors for non-peripheral four point probe (equations 5.4.18 to 5.4.20(a) and (b))

| $d/S$ | $C_H$ | $C_\rho$ for $\beta$ | | | |
|---|---|---|---|---|---|
| | | 0° | 30° | 45° | 60° |
| $\sqrt{2}$ | 1.0 | 1.0 | 1.0 | 1.0 | 1.0 |
| 2 | 0.59 | 0.924 | 0.950 | 1.0 | 1.152 |
| 3 | 0.279 | 0.755 | 0.836 | 1.0 | 1.491 |
| 4 | 0.158 | 0.659 | 0.773 | 1.0 | 1.682 |
| $\infty$ | 0 | 0.5 | 0.667 | 1.0 | 2.0 |

(d) *Corbino disc.* Grosvalet et al. (1966) and Glotin (1968) have used the Corbino disc configuration (fig. 5.28) for measuring sheet resistivity and the Hall mobility.

Measurements are made by passing a constant current from the centre electrode to the outer electrode and simultaneously measuring the voltage between the two electrodes. When a magnetic field is applied normal to the disc the carriers move in a spiral path and the measured resistance between the electrodes increases.

Assuming that the sample is homogeneous and that the current lines are parallel to the surface then the resistance $R_0$ between the electrodes for no magnetic field is given by:

$$R_0 = R_S \cdot \frac{1}{2\pi} \ln \frac{D}{d}$$ (5.4.21)

or

$$R_S = \frac{V}{I} \frac{2\pi}{\ln D/d} .$$

If a magnetic field $B$ is applied with a current $I$ flowing through the disc then the resistance varies with the magnetic field $B$ and mobility $\mu$ as

$$R_H = R_0 \left( 1 + \left( \frac{\mu B}{10^8} \right)^2 \right) \tag{5.4.22}$$

where $\mu$ is in cm$^2$/volt·sec and $B$ is in gauss. By rearrangement the mobility is given by:

$$\mu = \frac{10^8}{B} \sqrt{\frac{R_H - R_0}{R_0}} \tag{5.4.23}$$

and the surface carrier density $C_S$ by

$$C_S = \frac{1}{R_S q \mu} \text{ (carriers/cm}^2\text{)} \tag{5.4.24}$$

where

$$R_S = \frac{2\pi}{\ln D/d} \cdot R_0 \text{ (ohms/square)}.$$

It is instructive to calculate the percentage change in $R_0$ due to the application of the magnetic field. The percentage change is given by:

$$\% \text{ change} = \left( \frac{R_H - R_0}{R_0} \right) \times 100 = \left( \frac{\mu B}{10^8} \right)^2 \times 100.$$

The change is only 1% for $\mu = 100$ and $B = 10^4$ gauss. To measure the mobility, $\mu$, to an accuracy of about 5%, the difference $(R_H - R_0)$ due to the magnetic field must be measured to a 10% accuracy. The method is more suitable for larger values of $\mu$, say about 1000 cm$^2$/volt·sec, which occur only in lightly doped ion implanted layers.

It is essential that both the contacts to the Corbino disc have a high conductivity to ensure a uniform radial current flow and a minimum of contact resistance to the implanted layer. Glotin (1968) used pre-diffused n$^+$ contacts in p silicon to form the contacts to a phosphorus implantation. Tungsten probes were used to make contact to the n$^+$ diffused regions.

Lastly, the Corbino disc gives no indication of the type of carrier in the layer.

*5.4.2.4.3. Sheet resistivity and Hall mobility measurements combined with layer stripping.* In the description of the technique for measuring sheet resistivity and sheet Hall coefficient it was assumed that the carrier concentration $N$ and mobility $\mu$ were constant through the thickness of the conducting layer. In an ion implanted layer $N$ is strongly dependent on the depth and $\mu$ is a function of the total impurity concentration $(N_A + N_D)$ (Irvin, 1962; Caughey and Thomas, 1967; Baron et al., 1969). The carrier mobility is also influenced by collisions with imperfections in the crystal lattice such as damage clusters and precipitates.

To obtain a picture of $N$ and $\mu$ in depth it is necessary to strip layers off the implanted region and make measurements after each step. Mayer et al. (1967a) have shown that the carrier concentration $N_i$ and mobility $\mu_i$ of the $i$th layer are given by the following equations:

$$\frac{(R_{HS})_i}{(R_S)_i^2} - \frac{(R_{HS})_{i+1}}{(R_S)_{i+1}^2} = \Delta \left(\frac{R_{HS}}{R_S^2}\right)_i = q N_i \mu_i^2 d_i \qquad (5.4.25)$$

and

$$\frac{1}{(R_S)_i} - \frac{1}{(R_S)_{i+1}} = \Delta \left(\frac{1}{R_S}\right)_i = q N_i \mu_i d_i \qquad (5.4.26)$$

where $d_i$ is the thickness of the $i$th layer and $(R_{HS})_i$ and $(R_S)_i$ are the Hall sheet coefficient and sheet resistance measured after the removal of the $i$th layer. Substituting eq. (5.4.26) in (5.4.25) gives

$$\mu_i = \frac{\Delta(R_{HS}/R_S^2)_i}{\Delta(1/R_S)_i}$$

and

$$N_i = \frac{\Delta(1/R_S)_i}{q d_i \mu_i}$$

The total number of carriers per unit area $C_S$ in the whole layer is;

$$C_S = \sum_0^d N_i d_i \ (\text{carriers/cm}^2) \qquad (5.4.27)$$

Johansson et al. (1970) state that there are two possible major sources of error introduced by the layer stripping. One is the thickness, $d_i$, removed at each step which must be known accurately. Secondly as the layer is stripped, the leakage current of the p-n junction isolating the layer, increases due to the number of processing steps and as the surface of the layer approaches the junction. The leakage current flows in parallel with the current in the layer, the sum of the two being the measuring current. This constitutes a larger error for the higher sheet resistances as the layer current has to be reduced to avoid pinching off the layer (section 5.4.2.5.4). The leakage current does not decrease by the same extent.

A profile of $N$ and $\mu$ with depth is obtained by stripping the implanted layer a few hundred angstroms at a time and measuring $(R_{HS})_i$, $(R_S)_i$, and $d_i$ after each stripping. The value of $N_i$ and $\mu$ are calculated from the above equations. It is convenient to use a computer program to reduce the data if a large number of observations is involved.

There are several methods for stripping uniform layers from a silicon surface. The most commonly used method is anodic oxidation under very carefully controlled conditions of voltage and charge. The area of silicon to be stripped is surrounded by a mask and it is made the anode in an electrolytic cell. The electrolyte is an 0.1 molar solution of sodium tetraborate in boric acid and water (Mayer et al., 1967a) or an 0.04 molar solution of potassium nitrate in n-methylacetimide (Davies et al., 1964). A platinum cathode is used. It is necessary to increase the conductivity of the isolating junction between the layer and the silicon by illuminating it while the current is flowing in the cell. The anodic oxide is stripped by giving the wafer a brief dip in buffered hydrofluoric acid. The technique requires individually calibrating by using optical interferometry and weighing as the thickness removed depends on the substrate, the type of implantation (p or n), the dose and the cell conditions.

The thickness of silicon removed by each oxidation can be up to 500 angstroms which is a convenient thickness for stripping implanted layers an order of magnitude deeper. For very shallow implanted layers (less than 500 Å) anodic oxidation can still be used down to about 100 Å per step but for steps less than 100 Å it is necessary to use a 'vibratory polisher' (Whitton et al. 1969) or form a thin oxide by boiling the wafer in deionized water or stronger oxidising agents such as hydrogen peroxide or nitric acid. For deep implantations over one micron such as 3 MeV carbon or nitrogen implants, thicker layers must be stripped at each step. This can be done with etchants based on a mixture of HF, $HNO_3$ and acetic acid. The rate of reaction is controlled by reducing the HF and $HNO_3$ concentrations and working with the etchant below room temperature.

When etching an implanted Hall effect pattern it is essential to remove the layer *uniformly* and by a known amount each time. The uniformity is essential as the calculations are based on a uniform sheet thickness.

*5.4.2.4.4. Experimental techniques.* With the exception of the square four point probe on circular discs (Buehler and Pearson, 1966), the methods described in the last section for measuring the Hall effect require some preparation of the implanted layer before any measurements can be made. The purpose of this section is briefly to outline these requirements without providing a detailed solution. The method adopted depends on the facilities available to the experimenter and the type of semiconductor being investigated.

The preparation is concerned with three problems:

(i)   isolating the implanted layer from the substrate,
(ii)  defining the Hall pattern geometry and
(iii) making contact.

The majority of implantations for Hall effect measurements are made into a semiconductor wafer doped with the opposite carrier type. A p-n junction is formed which isolates the layer from the rest of the wafer. It is also possible to use a high resistivity wafer of the same conductivity type and to measure the Hall effect and sheet resistivity of the whole wafer (Blamires, 1970; Bader and Kalbitzer, 1970a). This method avoids the junction leakage problem but it increases the surface leakage problems due to the high resistivity of the semiconductor. It is claimed that a surface carrier density approaching $10^9$ per $cm^2$ can be detected (Bader and Kalbitzer, 1970a). The use of high resistivity silicon is to be preferred if the isolating junction is likely to have a high leakage current.

The leakage current in the isolating junction arises from two sources, the junction itself and across the surface where the junction is exposed. All measurements must be made in the dark to avoid hole-electron pair generation by photons. The leakage characteristics of junctions are discussed elsewhere (section 5.4.3) but it can be stated that in general the junction leakage decreases on annealing. If separate implanted contacts are used for probing, these must be fully annealed, so that their contribution to the leakage will be a minimum.

The surface leakage across the junction can be reduced by suitable washing procedures (Bader and Kalbitzer, 1970b; Johansson et al., 1970), which involve a mild oxidation treatment following a clean-up in hydrofluoric acid. Boiling deionized water or a dilute solution of hydrogen peroxide can be used to produce a slightly oxidised surface. In the case of silicon, the surface leakage can also be controlled by a suitably prepared thin oxide layer (Grove, 1967). A typical thickness is about 1000 Å. The implantation must be done through the oxide, to gain any benefit and this severely limits the type of ion and the energy that can be contemplated. It has a place in the study of implantation the lighter ions e.g. boron and phosphorus implanted at medium to high energies. It cannot be used in conjunction with layer stripping as the oxide is removed with the anodic oxide.

The Hall pattern geometry can be defined before or after implantation, the former requiring a masking procedure and the latter the etching of the pattern in the semiconductor. The masking procedure requires two masks, one to define the contact areas and the second the implantation area. Gibson et al. (1967) used two out-of-contact tantalum masks which could be accurately registered to a reference mark on the silicon wafer. This procedure is necessary as the first implantation is not visible after annealing and it is impossible to align the second mask to it (fig 5.25b).

In-contact masking can be used by applying planar technology (section 5.2.2) methods to the engraving of the masks. Glotin (1968) used this method in preparing Corbino discs. He used thermally diffused phosphorus to make the $n^+$ contacts for his phosphorus implantations. Several patterns can be made on one wafer, implanted and then divided up (by scribing) to provide a number of similar samples for a range of experiments such as annealing at different temperatures.

The other method for defining the Hall geometry is to use a method for removing the silicon around the Hall pattern (Johansson et al., 1970; Blamires, 1970; Clark and Manchester, 1968). The wafer is first ion implanted and annealed and then a photoresist image of the pattern is formed on the implanted layer. A suitable etchant for silicon is used to dissolve away all the unmasked silicon to a depth greater than the isolation junction. The wafer is then cleaned as outlined earlier and is then ready for measurements.

It is pertinent, while discussing etching, to mention that an implanted surface, if highly disordered or amorphous, is attacked by buffered hydrofluoric acid and hydrogen peroxide mixtures, which have little effect on undamaged silicon (Gibbons in Johansson et al., 1970). This author (J.S.) has observed the same preferential attack of damaged silicon by buffered hydrofluoric acid, and a mixture of potassium hydroxide, water and iso-propyl alcohol.

The etching of the Hall geometry forms a very large mesa and exposes the isolation junction part of the way down the wall of the mesa. If the wall is not normal but slopes at an angle $\alpha$ then the electric field across the exposed junction will be reduced by sin $\alpha$. This reduction in electric field helps reduce the surface leakage current and in other circumstances allows the application of larger reverse voltages.

The normal method of making contact to the Hall effect pattern is to press metal probes on to the contact areas. It is essential that a low resistance non-rectifying contact is formed between the probe and the semiconductor. The contact pressure must be high enough to ensure a low resistance but at the same time it must not physically punch through the layer as this severely damages the isolation junction and increases the leakage current.

The contact areas of the Hall pattern have two main roles to fulfil. They must have a high conductivity with a high surface doping level to assist the formation of the 'ohmic' contact with the probe and they must provide a uniform current flow into the Hall pattern. This uniform current is particularly important for the Hall bar (fig. 5.25(b)) and the Corbino disc (fig. 5.28). The junction under the contact area must possess a low leakage, as already mentioned, but, as important, it must maintain this characteristic throughout many cycles of probing and layer stripping. If metallic contacts (e.g. aluminium) are made to the Hall pattern, these must be removed or protected before any anodic stripping is attempted. They must also be removed before annealing above 500 °C (for aluminium).

*5.4.2.4.5. Summary.* Four configurations for measuring the carrier concentration and mobility in their conducting layers have been described. Their main features are summarized in table 5.11.

TABLE 5.11

Summary of methods for measuring the Hall effect in thin conducting layers

| Method | Sheet resistivity $\Omega/\square$ | Sheet Hall coefficient $cm^2/Coulomb$ | No. of probes | Remarks |
|---|---|---|---|---|
| a) Hall bar | $\dfrac{V_R}{I} \cdot \left(\dfrac{W}{L}\right)$ | $\dfrac{V_H}{I} \cdot \dfrac{10^8}{B}$ | 6 | $V_R$ separate from $V_H$. $V_R$ sensitive to geometry |
| b) v. d. Pauw | $\dfrac{V_R}{I} \cdot \dfrac{\pi}{\ln 2}$ | $\dfrac{V_H}{I} \cdot \dfrac{10^8}{B}$ | 4 | $V_R$ and $V_H$ independent of geometry |
| c) Square four pt. probe on disc or hole | $\dfrac{V_R}{I} \cdot \dfrac{\pi}{\ln 2} \cdot \dfrac{1}{C_\rho}$ | $\dfrac{V_H}{I} \cdot \dfrac{10^8}{B} \cdot \dfrac{1}{C_H}$ | 4 | Points must be located symmetrically. $C_\rho$ and $C_H$ sensitive to geometry |
| d) Corbino disc | $\dfrac{V_R}{I} \cdot \dfrac{2\pi}{\ln \dfrac{D^*}{d}}$ | $\mu = \dfrac{10^8}{B} \sqrt{\dfrac{\Delta R_0}{R_0}}$ | 2 | Sensitive to geometry. Indirect measurement of $\mu$ |
| In-line four pt. probe | $\dfrac{V_R}{I} \cdot \dfrac{\pi}{\ln 2}$ | – | 4 | Included for comparison |
| Square four pt. probe | $\dfrac{V_R}{I} \cdot \dfrac{\pi}{2\ln 2}$ | – | 4 | |

The method using van der Pauw pattern is considered to be the most suitable one of the four methods described in this section. It requires no more physical preparation than the Hall bridge or the Corbino disc but most important, the voltage readings are independent of the geometry and are as sensitive as those from the Hall bridge. The square four point probe aligned on a disc is strongly geometry sensitive but it has been included as it can be useful for small area substrates.

### 5.4.2.5. *Measurement of Sheet Resistivity using Resistor Structures*

*5.4.2.5.1. Introduction.* The four point probe method of measuring sheet resistivity of conducting layers is only suitable when the lateral dimensions of the layer are large compared with the spacing of the probes. Correction factors must be applied to the observations when the probes approach the edge of the layer as described in section 5.4.2.3.

When ion implantation is being used to fabricate semiconductor devices in a silicon wafer, it is essential that the electrical behaviour of the implanted regions is monitored as the processing proceeds on the wafer. A knowledge of the sheet resistivity is normally adequate for monitoring and it can be measured by using resistor structures designed for this purpose. The uniformity of the implantation can be measured to a higher spatial resolution than is possible with the four point probe and no correction factors for the lateral position on the wafer are necessary. It is customary in the design of integrated circuits to incorporate monitoring resistors in the boundary region around the circuit, for example between the bonding pads. If possible the resistor should have separate pads for probing. An alternative approach, adopted on wafers with high density circuits, is to replace some of the circuit 'chips' with a special test chip incorporating resistors, diodes and transistors of varying geometries. This method has the disadvantage that extensive uniformity measurements cannot be made but the processing can be followed. In addition to the resistivity measurement, the separate resistors allow the voltage depletion of the layer and the current-voltage characteristics of the p-n junction to the substrate to be measured.

There are several references to the study of resistors and their application in integrated circuits (see section 5.5.2.2). The use of monitoring resistors has been reported by MacDougall et al. (1969), Dill et al. (1971a), Nicholas et al. (1972) and Stephen and Grimshaw (1971).

*5.4.2.5.2. Principle of the implanted resistor.* Consider the linear resistor structure, fig. 5.29(a) formed from a doped semiconductor layer, of sheet resistivity $R_s$ (ohms/square), with highly conducting contacts at either end.

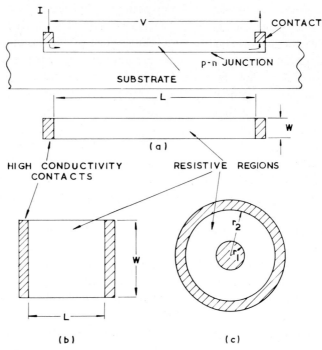

Fig. 5.29. Three implanted resistor geometries. (a) linear resistor of length $L$ and width $W$, (b) square resistors with $L/W = 1$ and (c) Corbino disc configuration.

The resistor is isolated from the remainder of the semiconductor substrate by the p-n junction. The resistance $R$ between the ends is given by

$$R = R_s \frac{L}{W} \text{ (ohms)} \tag{5.4.28}$$

where $L$ = length of the conducting path, $W$ = width of the conducting path (in the same dimensional units).

Conversely the sheet resistivity is given by

$$R_s = R \Big/ \left(\frac{L}{W}\right)$$

To measure the variation of sheet resistivity of a doped layer by the resistor method it is necessary to repeat the resistor pattern many times over the doped area spaced far enough apart to achieve the required resolution. The length $L$ and the width $W$ of the resistors can be readily defined by photoengraving techniques. Although the engraving is very accurate small variations of the order of a few microns can occur from resistor to resistor. As the variations are determined by the etching etc. and not by the magnitudes of $L$ and $W$, their effect can be minimised by making the dimensions of the resistor large in comparison with the variations. For example, the smaller dimension, usually $W$, should be made 100 microns to make the variations of the order of a few per cent.

It is necessary to consider geometries for resistors which will reduce the effects of errors due to the photoengraving. As the change in resistance is directly proportional to $L$ and inversely to $W$, then for a square resistor (fig. 5.29(b)) the changes in $L$ and $W$ will compensate and the measured resistance will not change to the same extent. The Corbino disc configuration (fig. 5.29(c)) can also be used for measuring the sheet resistivity with reduced geometrical dependance. The sheet resistivity is given by:

$$R_s = \frac{2\pi R}{\log r_2/r_1} \tag{5.4.29}$$

where $R$ = measured resistance between the inner and outer contacts, $r_1$ = inner radius of resistive region and $r_2$ = outer radius of resistive region, (in the same units), and as $r_2 > r_1$, $\log r_2/r_1$ is positive.

In the limit as $r_2$ approaches $r_1$

$$R_s = \frac{2\pi r_1 R}{\Delta r}$$

where $\Delta r = r_2 - r_1$ which is equivalent to a resistor of length $\Delta r$ and width $2\pi r_1$. As the outer radius of the resistor increases the value of $R_s$ increases as $\log r_2/r_1$ and not linearly as with the conventional geometry. It can be shown that if the outer radius increases by $\Delta r_2$ then the fractional change in resistance is

$$\frac{\Delta R}{R} = \frac{\Delta r_2/r_2}{\log r_2/r_1}$$

and therefore the linear change in dimensions is reduced by the logarithmic factor in the denominator. For this to be effective $\log r_2/r_1 > e$ indicating that a large ratio of $r_2/r_1$ must be used to decrease the effect of changes in $r_1$ and $r_2$.

The contacts to the resistor structure must be highly conducting and provide a low resistance connection to the resistive layer. It is essential to establish a uniform current density at the interface between the contact and the layer. As it is usual to make the voltage measurement across the same pair of probes which are used for injecting the current, the combined series resistance of the probes and the contacts must be at least two orders of magnitude smaller than the resistor being measured. Additional contacts can be provided for separate voltage probes to eliminate the contact resistance problem. If the use of four probes is considered then the v.d. Pauw configuration (section 5.4.3.4.2) should be used and the maximum data extracted from the measurement.

*5.4.2.5.3. Methods of fabrication.* The resistors shown in fig. 5.29((a) to (c)) are highly simplified and take no amount of delinearation, contacting and surface protection problems. Two practical resistors are shown in fig. 5.30(a) and (b). For sheet resistivity measurements a large number of these resistors are formed at the same time on the semiconductor wafer using conventional photoengraving methods. The resolution of the resistor measurements depends on the resistor dimensions and the space that is needed for other circuitry between them.

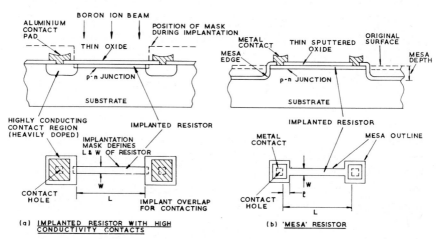

Fig. 5.30. Two practical structures for ion implanted resistors. (a) Resistor defined during implantation and having high conductivity contacts, (b) mesa resistor formed by etching the resistor pattern from a sheet implantation.

Figure 5.30(a) represents a very typical resistor structure in an ion-implanted integrated circuit. The surface is passivated with a thin silicon dioxide or nitride layer and the dopant ions are implanted through this layer with an aluminium or thick oxide layer defining the area of the resistor. It is possible to make pairs of resistors, one of which has the oxide removed to allow direct implantation into the silicon, for the study of the influence of the oxide thickness on the subsequent sheet resistivity.

The problem of making contacts to the implanted region, particularly if it is lightly doped is easily solved by using highly doped contacts made by diffusion or implantation at an earlier fabrication stage. The surface doping of the contacts must exceed $2 \times 10^{18}$ impurities/cm$^3$ to ensure an ohmic contact with the aluminium pad.

An alternative method to defining the resistor patterns by masking is to demarcate a sheet implantation in a wafer into a large number of separate resistors by mesa etching of the silicon (fig. 5.30(b)). This technique has been reported by Johansson et al. (1970) and Blamires (1970) for defining Hall effect patterns. The polished wafer can be ion implanted, annealed and assessed with a lightly loaded four point probe before the resistors are etched out. This permits more flexibility in the processing as there are no fabrication steps involved before the ion implantation. The wafer surface, however, cannot be passivated by the use of thermally grown oxide after the etching as the temperature involved is too high and some of the implanted silicon is consumed in forming the oxide.

The first step in making the resistors on a wafer is to define the resistor pattern by photolithography and remove by etching sufficient silicon around the resistors to expose the p-n junction. For the majority of low energy implantations an etched depth of 1 to 1.5 microns is adequate. The thermal probe can be used to check that the implanted layer has been completely penetrated. A silicon dioxide layer can be deposited by a suitable low temperature method (section 5.3.5.3.1) over the whole wafer to protect the exposed p-n junctions against gross contamination during subsequent processing and handling. Figure 5.30(b) shows the ends of the resistor enlarged in area to accommodate the metal contact. This contact cannot be placed at the edge of the mesa due to the strong possibility of short-circuiting the p-n junction and therefore it is placed away from the edge by a small distance $l$. The length of the resistor includes the distance $l$ at each end as the metal contact provides the high conducting contact. Some inaccuracy is introduced into the resistor calculations by it being wider than $W$ over the distance $l$. This small error is the same for all the resistors on one wafer and therefore the sheet resistivity calculated for each resistor can be compared for statistical analysis.

There are two main difficulties with the mesa structure; surface passivation and ensuring a high conductivity ohmic contact between the metal and the implanted layer over the whole contact area. The surface leakage current depends on the number of charged states formed at the silicon-oxide interface and in the oxide itself. It has been found possible to achieve sufficiently low leakage currents with radio frequency sputtered oxide layers to permit the use of currents of the order of one microamp for measuring the resistors. The ohmic contacts present a more difficult problem as aluminium when sintered into silicon dopes the re-growth region at the interface to about $2 \times 10^{18}$ acceptor/cm$^3$. This doped interface forms a good contact to p-type implants even if their surface doping is less than $10^{18}$ acceptors/cm$^3$ provided the aluminium does not penetrate the implanted layer. For an n-type implant it is possible to form a poor p-n junction and introduce considerable series resistance. If the surface doping level of the implant exceeds $10^{18}$ donors/cm$^3$ then a junction between two degenerate regions is formed and electron tunnelling occurs reducing the series resistance. Unless suitable contacting methods are available lightly doped n-type layers cannot be measured by this method. The problems are further aggravated with compound semiconductors.

The contributions of series resistance at the contacts can be eliminated in the measurement by using separate contacts for the voltage measurement as shown in the Hall bar (fig. 5.25). Four probes are required and this detracts from the advantages of the simple resistor.

In summary, a comparison of the two methods for measuring the sheet resistivity and junction properties of implanted layers shows that the mesa etching technique offers flexibility in processing prior to implantation whereas the implanted resistor with diffused contacts represents a more practical structure. The additional processing required for the diffused–implanted resistor provides good surface passivation and adequate contacting to n and p layers.

*5.4.2.5.4. Measurements of ion implanted resistors.* The primary purpose of an array of monitor resistors is to measure the sheet resistivity over the whole area of the implantation. Each resistor has to be probed separately and measured in a controlled manner. The normal procedure is to use a constant current source to produce a voltage drop across the resistor and to measure it with a digital voltmeter. A typical arrangement is shown in fig. 5.31. The digital voltmeter must have a high impedance input to avoid shunting the resistor. The data collection can be eased by recording the result directly on punched paper tape and using computer processing to perform the analysis and prepare graphs, such as histograms and distribution maps.

Before proceeding with the probing of a wafer, it is essential to check the current-voltage characteristics of a few resistors. Fig. 5.31 shows three typical current-voltage curves. Curve (a) is for a linear resistor and the

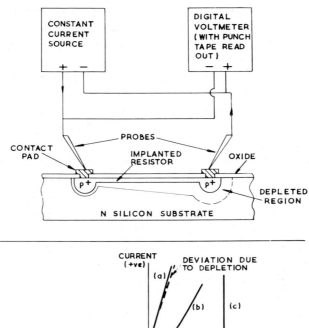

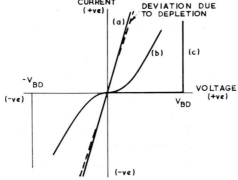

CURRENT-VOLTAGE CHARACTERISTICS OF
RESISTOR  STRUCTURES
(a) LINEAR  RESISTOR
(b) NON OHMIC  CONTACTS
(c) OPEN-CIRCUIT  RESISTOR

Fig. 5.31. Circuit for measuring implanted resistors using two probes. The current-voltage characteristics of three possible conditions are shown, (a) a linear resistor, (b) a resistor with non-ohmic contacts and (c) open circuit resistor.

slope of the line is the resistor value. If the curve deviates as shown by the dotted line, then depletion of the resistor is occurring and the resistance increases due to the reduction in the number of mobile carriers. Curve (b) is for a resistor with non-ohmic contacts which produce marked non-linearity. The 'junctions' formed at either end of the resistor are back to back and therefore a symmetrical characteristic about the origin is produced. No reliable resistance measurements can be made with this type of resistor. If the resistor is open circuit but the end contacts are good then curve (c) is observed due to the two diffused or implanted junctions being back to back. The breakdown voltages in either direction should be very similar.

The ion implanted (or diffused) resistor depends on a p-n junction to isolate it from the substrate (fig. 5.29). If a reverse voltage is applied across this junction then an equal number of immobile charge centres will be un-covered on either side of the junction. The number of mobile carriers will be reduced in the resistor structure and the value will increase. It is seen, therefore, that the implanted resistor is behaving like the channel of a junc-tion field effect transistor (FET) with the substrate acting as the gate elec-trode. There are two important differences between the resistor and FET characteristics. These are the relative doping levels of the channel and the gate and the geometry of the gate. The doping in the resistor structure is always greater than the substrate by virtue of the method of doping, while the reverse situation normally holds for the FET. The equivalent of the gate geometrically surrounds the resistor channel on three sides and therefore the depletion of carriers occurs from the base and the sides of the channel. This effect can have a very strong influence on narrow width, lightly doped resistors.

A method for calculating the effect of depletion on an implanted resistor will now be considered. Figure 5.32 shows the profile, electric field distribu-tion and the potential distribution across a typical p-n junction with a Gaussian distribution of impurity centres in the p region. The number of immobile charges uncovered on either side of the junction $x_j$ are equal, hence

No. charges in n region = No. charges in depleted region of implant.

$$N_{\mathrm{D}}(x_n - x_j) = \int_{x_p}^{x_j} N_{\mathrm{A}}(x)\,\mathrm{d}x$$

neglecting the compensation effect of the acceptor and donor centres on opposite sides of the junction. The integral $\int_{x_p}^{x_j} N_{\mathrm{A}}(x)\,\mathrm{d}x$ can be calculated with the help of the tables in Appendix 2. The electric field at the junction is

$$E_{\max} = \frac{q N_{\mathrm{D}}(x_m - x_j)}{K_s \varepsilon_0} = \frac{q}{K_s \varepsilon_0} \int_{x_p}^{x_j} N_{\mathrm{A}}(x)\,\mathrm{d}x$$

where $q$ = charge on the electron and $K_s$ = dielectric constant of silicon, $\varepsilon_0$ = permittivity of free space.

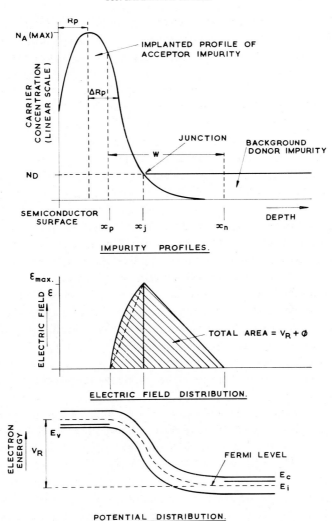

Fig. 5.32. A diagram illustrating a method for calculating the effect of depletion on an implanted resistor. The impurity profile, background doping, electric field distribution and potential distribution across the junction are given.

The potential difference from $x_p$ to $x_n$ i.e. over the depletion width $w$ is:

$$(V_R + \phi) = \int_{x_p}^{x_j} \varepsilon(x)\,\mathrm{d}x + \int_{x_j}^{x_n} \varepsilon(x)\,\mathrm{d}x$$

$$= \frac{q}{2K_s \varepsilon_0} \left[ \int_{x_p}^{x_j} N_A(x)\,\mathrm{d}x(x_j - x_p) + N_D(x_n - x_j)^2 \right]$$

$\phi$ is the built in voltage and it is dependent on the doping levels in the p and n regions. The first term is an approximation assuming that the area under the electric field curve for the p region is given by the area of a triangle with height $E_{max}$ and base $(x_j - x_p)$. The resistivity of the layer is given by:

$$R_S = \frac{1}{q \cdot \mu \cdot C_s \cdot (\text{area under implanted profile to } x_j - \text{area of depleted region from } x_p \text{ to } x_j)}$$

$C_s$ = total surface impurity concentration and $\mu$ = mobility.

A suggested procedure for calculating the effect of depletion on the resistivity of the layer is to assume a range of $E_{max}$ values and calculate, with known values of $x_j$ and $N_D$, $(x_n - x_j)$ and $(x_j - x_p)$ and then $(V_R + \phi)$. The increased sheet resistivity can then be calculated and plotted against $V_R$. The depletion width $w$ is dependent on the impurity concentrations and the applied voltage and the lower the doping level the greater the depletion and the changes in resistivity. A resistor with a low value of $N_A$ (or $N_D$) is readily depleted and it shows a strong dependence on the applied voltage.

It is essential to avoid depleting a resistor during measurement if an accurate value of sheet resistivity is required. (The problems of implanted resistors as devices are discussed in section 5.5.2.2.) The depletion of a resistor can arise during measurement by,

(i) the voltage gradient developed along the resistor due to the current flow and

(ii) reverse biasing the whole structure with respect to the substrate.
It is usual to allow the substrate to float with respect to the resistor and under these conditions the substrate will assume the potential of the positive end of a p-type resistor (or negative end of a n-type resistor) and the remainder of the structure will be reverse biased (fig. 5.31). When measuring high value resistors it is good practice to initially measure the resistance at various currents to establish the correct value to use to avoid depletion and yet be larger than the leakage current in the isolating p-n junction. A sensitive transistor curve tracer is very useful for making these initial current measurements as a deviation from linearity can be readily seen. A number of resistors in a matrix can be readily scanned. The curve tracer also indicates any non-ohmic behaviour at the contacts.

The lightly doped resistor can also be influenced by a space charge at or immediately above the surface of the resistor as the image charge induced in the silicon can either deplete or enhance the number of mobile carriers and alter the sheet resistivity. The space charge can be produced by fast

states at the silicon surface, an interface surface charge between the silicon
and an oxide layer and a space charge trapped in a passivating layer such
as some form of ionic contamination (sodium ions) or radiation damage.
All measurements must be made in the dark to avoid conductivity modulation of the resistor and increase junction leakage currents produced by the
hole-electron pairs generated by photons.

### 5.4.2.6. Profile Measurements

*5.4.2.6.1. Introduction.* The four point probe and Hall effect methods
described in the earlier section allow the measurement of the carrier surface
concentration $N_s$ and mobility $\mu$ in the whole ion implanted (or diffused)
layer, but they do not yield any information on the distribution, with depth,
of the impurity centres responsible for the mobile carriers.

Several techniques for obtaining the depth profile of electrically active
impurities will now be discussed.

*5.4.2.6.2. Differential resistivity measurements by layer stripping.*
Consider an ion implanted (or diffused) layer in a substrate of the opposite
carrier type. The profile of the implanted impurities can be obtained by
successively making Hall effect measurements and stripping thin sections
from the surface of the semiconductor. From the measurements, the surface
carrier concentration $N_s$ and the mobility $\mu$ are obtained after each step and
from a knowledge of the thickness of each section removed the carrier concentration, per $cm^3$, can be calculated. The profile of the concentration
can then be built up step by step. The thickness of silicon removed at each
step is usually between 0.02 and 0.05 microns and thus upwards of ten
successive stripping operations are required for an implantation 0.2 microns
in depth. The electrical measurements become increasingly difficult as the
junction between the layer and the substrate is approached. The measuring
current has to be reduced due to the increase in the sheet resistivity of the
test area and then the junction leakage current becomes an appreciable
fraction of the measuring current. The leakage current may also be increased
due to the stripping procedure and the frequent contacting (see section
5.4.2.4.2).

Bader and Kalbitzer (1970) have described a method of profiling ion
implanted layers in which the substrate has the same conductivity type as
the implanted impurity (i.e. there is no type conversion). A silicon wafer
with a low doping level (high resistivity) is used and the doping level is increased by the addition of the implanted impurities. A Hall effect measurement on the whole wafer, which can be regarded as a 'thick layer', yields

information on $N_s$ and $\mu$. The wafer is controllably stripped, a thin section at a time, $N_s$ and $\mu$ being measured after each removal as before. It is possible to follow an implantation profile down to the doping level of the substrate. The published curves for boron and phosphorus profiles are over a range of $10^4$ in doping level using silicon crystals doped initially to $10^{14}$ impurities per $cm^3$.

A square array of four tungsten probes was used for the Hall measurements. Corrections had to be applied to the readings as the probes were not at the edge of the implantation. In order to avoid inversion layers on the high resistivity material it was found necessary to obtain a weak concentration of majority carriers by chemical treatment of the silicon wafer after each stripping operation. Wafers with n-type impurities were rinsed with de-ionized water and exposed to the atmosphere for two hours, while p-type wafers were given a two minute dip in hot hydrogen peroxide (see 5.4.2.3).

For information on stripping techniques the reader is referred to the work of Davies et al. (1964), Mayer et al. (1967a) and Wilkins (1968) on the anodic stripping of silicon and Whitton et al. (1969) for the details of the vibratory polishing method. This latter method is applicable to a wide range of semiconductors as it does not depend on the formation of an oxidised layer.

*5.4.2.6.3. The bevel and stain technique.* Gibbons (1967, 1968) and Kleinfelder (1967) have described the use of the bevel and stain technique to produce a profile of electrically active impurities. The principle is illustrated in fig. 5.33. A number of strips of silicon of the same carrier type and crystal orientation but different impurity concentrations are mounted side by side in the target chamber of the ion implantation machine and implanted simultaneously with the same dose of impurity ions. The strips are annealed together and the depth of the junctions between the implantation and substrate in each strip located by bevelling and staining. Each strip gives one, or sometimes two points on the impurity concentration curve corresponding to the initial impurity concentration of the strip. It is stated that as a very wide range of initial resistivities are available the profile of electrically active impurities can be followed over at least six decades in concentration, which is greater than by other methods.

Significant errors can occur when the concentration of implanted impurities exceeds $2 \times 10^{19}/cm^3$ due to staining difficulties and the redistribution of the background impurities due to the vacancies created by the implantation. One disadvantage of the method is that the measurements are not made in the same piece of silicon and therefore great care is necessary

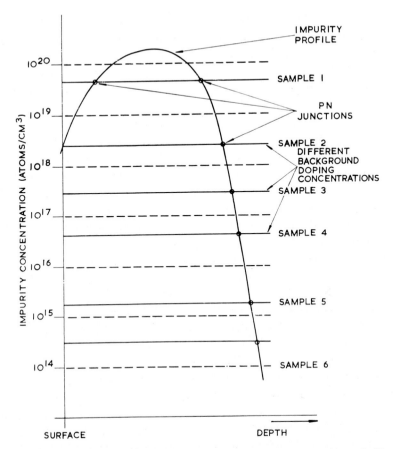

Fig. 5.33. The bevel and stain technique for impurity profiling. A number of silicon strips differing in resistivity are implanted together and then beveled and stained to locate the depths of the p-n junctions. The profile is constructed from the junction depths (Gibbons, 1967; Kleinfelder, 1967).

to ensure that the crystals apart from the impurity concentration are initially as similar as possible. The orientation of the crystals must be the same to better than 0.5 degrees especially if channelling is being studied and the dislocation densities should also be similar.

The success of this method depends on the preferential staining of the p silicon with respect to the n silicon to locate the p-n junction. The staining technique was first described in a paper by Fuller and Ditzenburger (1956) while Holmes (1962) has described a number of etchants and Smith (1964) has outlined some of the practical aspects of bevelling and staining. To obtain an accurate measurement of a shallow junction it is necessary to bevel the silicon with the p-n junction at a shallow angle $\theta$ usually less than 2 degrees, to extend the stained area over a longer distance. The degree of extension is proportional to $1/\theta$ (in radians). A precision measuring microscope incorporating an interferometer is necessary for measuring the length of the stained area from the silicon surface to the junction. A common method for polishing at a small angle to the silicon surface is to use a rotating hardened steel cylinder with a fine slurry of grinding paste to form a shallow groove in the silicon. The success of the method is strongly dependent on the success in obtaining a clean groove with a sharp edge and a clear stain. Under favourable conditions it is possible to measure a junction depth to an accuracy of better than 0.1 microns.

An alternative method to staining for junction depth measurement is to use a scanning electron microscope. The physical details of the bevel are given by the normal secondary electron picture while the p-n junction is located using the beam induced conductivity mode. By combining the two pictures the junction depth can be readily measured. Due to the high magnification of the instrument conventional mesa etching (see fig. 5.30b) is adequate for junctions 0.5 microns or more deep while bevelling is necessary for the very shallow junctions. The beam induced conductivity mode is limited to a resolution of about 0.1 microns. It is necessary to make two electrical contacts to the silicon to detect the beam induced currents and this can be a serious disadvantage.

5.4.2.6.4. *Differential capacitance-voltage measurements.* A planar p-n junction behaves as a parallel plate capacitor in which the plate separation is a function of the applied voltage. As the junction is reverse biased, the space charge volume of the junction increases exposing more fixed charge centres, and the capacitance decreases as the edges of the space charge layer are moved further apart. This phenomenon can be applied to the measurement of the spatial distribution of fixed charge centres in a semiconductor.

Consider an abrupt $p^+$-n junction of area $A$ as shown in fig. 5.34. In this example the background doping concentration $N_D$ has been increased in the vicinity of the junction by a buried n layer. As the junction is reverse biased the lightly doped n side is depleted to a depth $w$ from the junction $x_j$ while in comparison the highly doped $p^+$ side is hardly depleted due to the local high concentration of acceptor centres near $x_j$.

If the voltage across the junction is changed by $dV$ and $dC$ is the corresponding change in the charge per unit area then the capacitance $C$ per unit area is:

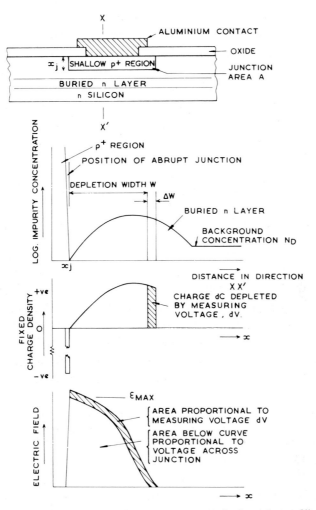

Fig. 5.34. Differential capacitance-voltage measurements for impurity profiling. The impurity concentration, fixed charge density and electric field are shown through an abrupt reversed biased p⁺-n junction (at $XX'$).

$$Q = \frac{dC}{dV}$$

$$= \frac{\varepsilon\varepsilon_0}{w} \qquad\qquad (5.4.30)$$

where $\varepsilon_0$ = dielectric constant of free space ($8.86 \times 10^{-14}$ farad/cm), $\varepsilon$ = dielectric constant (11.7 for silicon), $w$ = depletion depth (cm).

If it is assumed that the edges of the space charge layer are sharp (i.e. similar to an actual metal plate) then it can be shown that the impurity concentration $N(w)$ at the edge of the space charge layer in the lightly doped side is:

$$N(w) = \frac{1}{q\varepsilon\varepsilon_0} \frac{C^3}{\mathrm{d}C/\mathrm{d}V} \qquad (C \text{ per unit area})$$

or

$$N(w) = \frac{1}{q\varepsilon\varepsilon_0 A^2} \frac{C^3}{\mathrm{d}C/\mathrm{d}V} \tag{5.4.31}$$

and

$$w = \frac{\varepsilon\varepsilon_0 A}{C} \tag{5.4.32}$$

where $C$ is the capacitance of the junction of area $A$. $w$ is measured from $x_j$.

The method has been described by Hilibrand and Gold (1960), Grove (1967) and Kennedy et al. (1968), while Kennedy and O'Brien (1969) have suggested that the technique measures the majority carrier concentration $n(w)$ at the edge of the space charge layer and not $N(w)$. The following equation is given by Kennedy and O'Brien (1969) for correcting the capacitance voltage measurements to give $N(w)$;

$$N(w) = n(w) - \frac{kT}{q} \frac{\varepsilon\varepsilon_0}{q} \frac{\mathrm{d}}{\mathrm{d}w} \left( \frac{1}{n(w)} \frac{\mathrm{d}n(w)}{\mathrm{d}w} \right) \tag{5.4.33}$$

The above equations are also valid for a reverse biased Schottky barrier structure in which the metal electrode is equivalent to the highly doped side of the junction.

An examination of the above equations shows that the voltage and the capacitance must be measured precisely to obtain accurate values of $N(w)$. The usual method of measurement is to use a sensitive precision radio frequency bridge which will allow a d.c. bias to be applied to the p-n junction and permit the conductance component of the reactance to be balanced out over a wide range of capacitances. Differential capacitance measurements are usually made over a range of d.c. bias voltages from a few tenths of a volt forward bias to the junction breakdown voltage. It is essential that the measuring radio frequency (r.f.) voltage is small to ensure small signal conditions, that is less than $kT/q$ ($\sim 25 \text{ mV}$ at room temperature).

Copeland (1969) and Spiwak (1969) have jointly described a method in which a p-n junction or Schottky barrier is driven by a high frequency (5 MHz) alternating current. The junction voltage at the fundamental frequency is proportional to the depth ($w$) and the second harmonic voltage to $1/n(w)$. The two r.f. signals are detected and plotted on an X-Y recorder. The depth is changed by altering the d.c. bias. Miller (1972) has used a feedback method to control the motion of the low field edge of the space charge layer.

The differential capacitance method has two restrictions, a limit on the total number of electrically active centres that can be exposed in the depleted region before the junction avalanches and a decrease in the resolution due to the Debye length increasing in lightly doped semiconductors. Table 5.12 summarises the two restrictions in terms of carrier doping density in the substrate.

TABLE 5.12
The values of critical junction electric field, maximum applied voltage, maximum depletion depth, equivalent impurity concentration/cm$^2$ and Debye length for doping density on the lightly doped side of a p-n junction

| Junction doping density $N$ impurities/cm$^3$ | Critical field at junction volts/cm | Max. applied voltage volts | Max. depletion depth microns | Equivalent impurities per cm$^2$ $N_s$ | Debye length $L_D$ microns |
|---|---|---|---|---|---|
| $10^{12}$ | $0.8 \times 10^5$ | 18900 | 4720 | $4.7 \times 10^{11}$ | 4 |
| $10^{13}$ | $1 \times 10^5$ | 3250 | 650 | $6.5 \times 10^{11}$ | 1.3 |
| $10^{14}$ | $2 \times 10^5$ | 1300 | 130 | $1.3 \times 10^{12}$ | 0.4 |
| $10^{15}$ | $3 \times 10^5$ | 290 | 19.4 | $1.9 \times 10^{12}$ | 0.13 |
| $10^{16}$ | $4.5 \times 10^5$ | 56 | 2.5 | $2.5 \times 10^{12}$ | 0.041 |
| $10^{17}$ | $6 \times 10^5$ | 11.7 | 0.39 | $3.9 \times 10^{12}$ | 0.013 |
| $10^{18}$ | $1.3 \times 10^6$ | 5.5 | 0.084 | $8.4 \times 10^{12}$ | 0.004 |

where

Maximum applied voltage $= V_M = \dfrac{\varepsilon\varepsilon_0(E_{crit})^2}{2qN}$

Maximum depletion width $= w_M = \dfrac{2V_{max}}{E_{crit}}$

Maximum impurity conc/cm$^2 = N_s = Nw_M$

The electric field $E_{max}$ at the p-n junction is related to the doping density in the lightly doped substrate by;

$$E_{max} = \frac{q}{\varepsilon\varepsilon_0} \int_0^w N(w)\,dw \qquad (5.4.34)$$

where $N(w)$ is the doping concentration at the depth $w$ and is a known function of $w$. If $N(w)$ has a constant value $N$, then $E_{max}$ is given by

$$E_{max} = \frac{q}{\varepsilon\varepsilon_0} Nw \qquad (5.4.35)$$

When the electric field at the junction exceeds a critical value, $E_{crit}$, the junction breaks down due to ionization. The value of $E_{crit}$ is a function of the doping density in the silicon and it increases from less than $10^5$ volts/cm to over $10^6$ volts/cm for a six order increase in the doping density (table 5.12) (Miller, 1957; Grove, 1967). The lightly doped side of the junction will breakdown first and this will in turn determine the number of charge centres that can be uncovered before $E_{crit}$ is reached at the junction.

It is seen from table 5.12 that the maximum number of uncovered charge centres increases with the doping density at the junction and does not exceed $8 \times 10^{12}$ centres/cm$^2$. This is the approximate upper limit of the number of charge centres per unit area that can be profiled by the differential capacitance method.

The extrinsic Debye length $L_D$ is given by

$$L_D = \sqrt{\frac{kT}{q} \frac{\varepsilon\varepsilon_0}{qN}}$$

where $k$ = Boltzmann's constant and it determines the distance over which the small variations of the potentials due to fixed charges smooth themselves out. $L_D$ defines the uncertainty in the position of the edge of the space charge layer and the transition from neutrality to depletion takes place in about 6 times $L_D$. The amplitude of the measuring signal should be less than $kT/q$ ($\sim 25$ mV at 20 °C) to ensure that the modulation $w$ of the edge of the space charge layer does not exceed $6L_D$.

The effect of the perimeter of the junction (or the Schottky barrier) has been neglected. Copeland (1970) has published a numerical calculation for correcting a profile due to the edge of a Schottky barrier. The true carrier concentration $N(w)$ and depletion depth $w$ are obtained by correcting the measured values in the following manner,

$$w = \frac{w_0}{(1 - Kw_0)/r} \qquad (w_0 = w_{observed})$$

$$N(w) = \frac{N_{observed}}{(1 + Kw/r)^3}$$

where $r$ = radius of the barrier, $K$ = constant which is about 1.5 for silicon.

For a circular p-n junction of radius $r$ where the junction has a finite depth $x_j$ the area of the junction will be increased by the contribution from the edges of the junction which is

    (i)   $2\pi r \, x_j$ for a straight sided junction such as an ion implanted one or
    (ii)   $\pi^2 r \, x_j$ for a curved junction edge of radius $x_j$ (fig. 5.16).

The area of the junction $\pi r^2$ used in calculating the profile should be increased by (i) or (ii). If $r$ is made large with respect to $x_j$ then these contributions are considerably reduced and may be neglected if $r > 100 \, x_j$ which can be achieved by making $r$ over 100 microns for a one micron deep implantation.

Another factor which influences the choice of junction area is the capacitance at maximum depletion. It is essential to resolve the capacitance to four significant figures in order to obtain accurate values of $dC/dV$ for the calculation of $n(w)$. Experience has shown that it is necessary to use a Schottky barrier or a p-n junction with a diameter of 250 microns or over to obtain a capacitance of 0.5 pf for example with a depletion depth of 10 microns. The error in the capacitance due to edge effects is less than 2.5 % with a junction of these dimensions.

In the discussion, so far, of the differential capacitance measurement, attention has been concentrated on the capacitive component of the reactance. It is necessary to consider the current flowing through the junction and its effect on the measurement. The current in a p-n junction arises from three mechanisms (see section 5.4.3)

    (i)    diffusion of carriers across the space charge layer,
    (ii)   recombination-generation centres and
    (iii) surface leakage currents.

In a silicon p-n junction under reverse bias (ii) and possibly (iii) predominate, while (i) and (ii) operate under forward bias, the current increasing exponentially with applied voltage. The reverse current in a good p-n junction is normally very low but it can be large if there is a high concentration of recombination-generation centres produced by damage due to ion implantation. The electrical $Q$ of the junction is given by

$$Q = \omega \, C \, r_p$$

and it decreases as the parallel resistance across the junction $r_p$ decreases. A typical radio frequency bridge (Boonton 75D) designed for making small signal capacitance measurements on semiconductors can operate with a minimum $Q$ of 3. This corresponds to a parallel resistance of 1 Megohm for a 0.5 pF capacitance giving an equivalent leakage current of $10^{-4}$ amps at 100 volts.

A fully passivated p-n junction has much better electrical characteristics than a simple Schottky barrier formed by evaporating a metal dot on the semiconductor. Despite the degraded electrical characteristics, the Schottky barrier is preferred for differential capacitance measurements for two reasons;

(i)  The preparation is simpler and involves no high temperature processing of the semiconductor for example diffusion or annealing and

(ii)  the reference point for the depth measurement is the semiconductor surface and not a junction distance $x_j$ below the surface.

The common metals used for Schottky barrier formation on silicon are aluminium, gold, platinum, palladium, rhodium, chromium and titanium. It is extremely important that the semiconductor surface be free of contamination such as oxides before the metal is evaporated or sputtered to form the barrier. It is convenient to laterally define the individual devices by using a perforated out of contact mask. This is preferable and faster than etching the barriers out of a continuous metal layer.

The method used at Harwell is to take approximately fifty individual capacitance-voltage measurements using an r.f. bridge. The accumulated data is fed into a computer and the program computes $n(w)$ against $w$, corrects for stray capacitance and plots $n(w)$ on linear or logarithmic capacitance axes against $w$ on a linear axis. Figure 5.56 shows a profile of a buried 3 MeV nitrogen implantation in n-type silicon computed as described and measured using a gold barrier. The measurement of ion implanted profiles by the differential capacitance method have been reported by Bower et al. (1966), Brook and Whitehead (1968) and Moline (1969) for silicon while Mayer et al. (1967b) have studied Zn and Te implantations in GaAs. Schottky barriers were used in all these experiments.

In conclusion the differential capacitance method is limited by the total number of impurity centres in the semiconductor that can be uncovered by depletion before avalanche breakdown. Measurements can be made over four or five decades of impurity concentration if lightly doped substrates are used. The Schottky barrier is prefered to the p-n junction despite its poorer electrical properties as it can be readily deposited on the semiconductor. The reference position for the profile is the semiconductor surface, and not the depth of a p-n junction which is not known so accurately.

5.4.3. Methods of measuring the electrical properties of ion implanted
junctions

*5.4.3.1. Introduction.* The great majority of semiconductor devices depend
for their operation on the electrical properties of p-n junctions. Diodes,
bipolar transistors, field effect transistors both bulk and surface types, and
integrated circuits are all assemblies of p-n junctions arranged to regulate
the flow of current in the semiconductor to perform a specified function.
Despite the importance of the p-n junction, the electrical investigations into
ion implanted layers have concentrated mainly on the conduction phenomena
in the layer and there has been little published work on the junction formed
between the layer and the underlying substrate. An understanding of the
p-n junction properties is important in ion implanted silicon but it is even
more so for compound semiconductors where thermal diffusion is not as
effective as it is in silicon. Ion implantation is being vigorously investigated
as an alternative technique for type conversion with good p-n junction
properties.

The conduction in an ion implanted layer depends on the behaviour of
the majority carriers, while the p-n junction characteristics are very strongly
related to minority carrier behaviour and the density of recombination–
generation (R–G) centres in the junction especially the space charge layer.

*5.4.3.2. p-n Junction I/V Characteristics.* The original theory of the p-n
junction was first proposed by Shockley (1948) for germanium and later it
was expanded by Sah, Noyce and Shockley (1957) to include recombination–
generation (R–G) centres to explain the behaviour of silicon p-n junctions.
A great number of papers, reviews and books have dealt with the many aspects of the theory and practice of the p-n junction and the books by the
following authors are recommended for further reading, Shockley (1950),
Jonscher (1960), Gray et al. (1964), Warner and (1965), Gibbons (1966),
Grove (1967), v.d. Ziel (1968) and Sze (1969).

The current which flows in a p-n junction due to the application of an
external voltage can arise by three basic mechanisms (fig. 5.35),

(i)　the diffusion of carriers across the space charge layer,

(ii)　the recombination–generation of carriers at centres in the space
charge layer; and

(iii)　surface leakage currents (not shown in fig. 5.35).

For a complete study of the $I/V$ characteristics of a p-n junction
measurements must be made down to very low currents ($< 10^{-12}$ amps) if

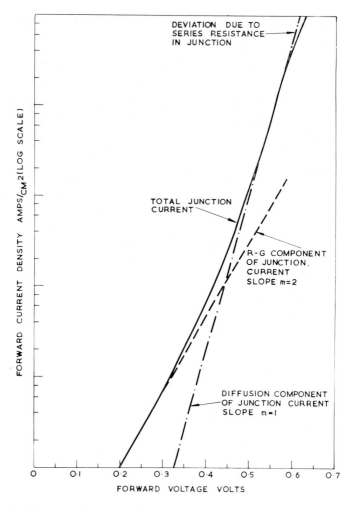

Fig. 5.35. Theoretical forward characteristic of a p-n junction showing the diffusion and recombination-generation components and the deviation due to extrinsic series resistance. The total junction current characteristic is the sum of the two components less the effect of series resistance.

the small but important differences between an ion implanted and a diffused junction are to be seen. Conducting paths in parallel with the p-n junction which may arise from gross contamination and crystal damage etc. caused by poor semiconductor processing must be avoided.

(i) *Diffusion Current Component.* Consider an abrupt p-n junction (fig. 5.34) in which the current is carried by the diffusion of carriers across the space charge layer. The junction diffusion current is given by;

$$I_D = I_{(\text{diff. electrons})} + I_{(\text{diff. holes})}$$

$$= Aqn_i^2 \left( \frac{D_h}{L_h N_d} + \frac{D_e}{L_e N_a} \right) (e^{qV/kT} - 1) \tag{5.4.36}$$

where $A$ = junction area $(\text{cm}^2)$, $q$ = electronic charge $(1.6 \times 10^{-19}$ coulombs), $n_i$ = intrinsic carrier concentration $(\text{cm}^{-3})$, $D_h$ = diff. const for holes in the n region $(\text{cm}^2/\text{sec})$, $D_e$ = diff. const for electrons in the p region $(\text{cm}^2/\text{sec})$, $L_h$ = diff. length for holes in the n region (cm), $L_e$ = diff. length for electrons in the p region (cm), $N_a$ = acceptor concentration $(\text{cm}^{-3})$, $N_d$ = donor concentration $(\text{cm}^{-3})$, $k$ = Boltzmann's constant $(1.38 \times 10^{-23}$ joules/°K), $T$ = temperature (°K), $V$ = applied voltage $(+V$ for forward bias and $-V$ for reverse bias). For the case of a $p^+$-n junction in which $N_a \gg N_d$ the current is carried by the diffusion of holes and the equation becomes:

$$I_{\text{diff}} = Aqn_i^2 \frac{D_h}{L_h N_d} (e^{qV/kT} - 1) \tag{5.4.37}$$

The saturation current $I_s$ which flows when the junction is reversed biased with $V$ several times greater than $kT/q$ is:

$$I_s = \frac{Aqn_i^2 D_h}{N_d L_h} \tag{5.4.38}$$

$I_s$ is shown in fig. 5.36 as the ideal characteristic. The above equations are of the form originally given by Shockley (1950) and they fit the experimental observations on germanium p-n junctions for low current densities. It was observed that silicon p-n junctions did not obey these equations and that another conduction mechanism was involved with the diffusion.

(ii) *R–G Current Component.* Sah, Noyce and Shockley (1957) proposed a model in which a number of R–G centres, or traps, with a single energy level in the forbidden band are uniformly distributed throughout the semiconductor. These centres may be crystal lattice defects, interstitial or substitutional impurity atoms or damage introduced by ion implantation.

These centres contribute to the junction current by allowing electrons and holes to recombine or be generated in the space charge layer. The equation for the R–G current can be written as:

$$I_{\text{R-G}} = \frac{qn_i wA}{2\tau_0} (e^{qV/2kT} - 1) \tag{5.4.39}$$

where $w$ = width of the space charge layer, $\tau_0$ = carrier lifetime in the space charge layer. $\tau_0$ is related to the number $N_t$ of R–G centres per unit volume by

$$\tau_0 = \frac{1}{\sigma v_{th} N_t}$$

where $\sigma$ = capture cross section for holes or electrons = $\sim 10^{-15}$ cm$^2$ and $v_{th}$ = thermal velocity of the carriers ($10^7$ cm/sec). The factor 2 in the exponential term is due to the contribution of carriers from both sides of the space charge layer.

For an abrupt p-n junction, the width $w$ of the space charge layer is related to the applied reverse voltage $V$ by

$$w = \sqrt{\frac{2\varepsilon\varepsilon_0}{qN} V}$$

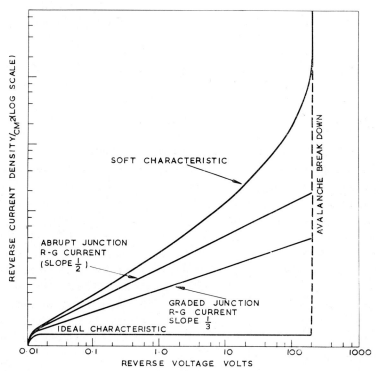

Fig. 5.36. Theoretical reverse characteristic of an p-n junction showing an ideal characteristic, the R–G components for an abrupt and graded junctions and a 'soft' characteristic. The avalanche breakdown voltage is indicated.

where $N$ is the sum of the impurity concentration and R–G centres on the lightly doped side of the junction. But $I_{R-G}$ is proportional to $w$ and assuming a uniform distribution of $N_t$, $I_{R-G} \propto (V+V_1)^{\frac{1}{2}}$ for an abrupt junction and $I_{R-G} \propto (V+V_1)^{\frac{1}{3}}$ for a graded junction (fig. 5.36). $V_1$ is the 'built in' voltage.

The total junction current $I_j$ is the sum of the diffusion and R–G currents $I_j = I_D + I_{R-G}$. For an abrupt p$^+$-n junction $I_j$ is given by:

$$I_j = Aqn_i^2 \frac{D_h}{L_h N_d} (e^{qV/kT} - 1) + \frac{qn_i WA}{2\tau_0}(e^{qV/2kT} - 1) \qquad (5.4.40)$$

(diffusion component) + (R–G component)

The ratio of the two currents is proportional to $n_i$ the intrinsic carrier concentration. Figure 5.35 shows the two currents plotted separately and combined to give the typical shape of a forward characteristic of a silicon p-n junction. If $n_i$ is large as for germanium $I_d$ dominates while for silicon $n_i$ is smaller and $I_{R-G}$ is larger at low forward voltages. In GaAs $n_i$ is very small and the conduction is carried by the $I_{R-G}$ current component.

For $V$ greater than several $kT/q$ the forward current of the p-n junction can be written as,

$$I = I_s e^{qV/mkT}$$

where $I_s$ is a current dependent on the relative contributions of $I_D$ and $I_{R-G}$ while '$m$' has a value 1 for a diffusion current and 2 for a R–G current. As $V$ increases $m$ can change from 2 to 1 (fig. 5.35), but in some junctions the diffusion and R–G contributions are comparable over a wide range of forward currents giving a value of $m$ intermediate between 1 and 2.

In a practical p-n junction diode there will be a series resistance $R_s$ associated with the metal-semiconductor contacts and the base region of the diode away from the junction particularly if normally doped silicon is used and not n on n$^+$ (or p on p$^+$) epitaxial material. The equation for a junction diode with series resistance is

$$V_{Diode} = \frac{mkT}{q} \log_e (I_j - I_s) + I_j R_s \qquad (5.4.41)$$

The series resistance will cause the $I/V$ to deviate from the exponential relation as shown in fig. 5.35. The value of $R_s$ can be readily obtained by measuring $\Delta V$, the deviation from the extrapolated exponential curve at a known current $I_j$. $R_s$ is given by

$$R_s = \frac{\Delta V}{I_j}$$

The value of $R_s$ is a constant unless conductivity modulation of the diode substrate occurs due to the minority carriers which are injected into the base region.

(iii) *Surface Leakage Currents.* The two conduction mechanisms in a p-n junction discussed so far have been entirely concerned with the bulk properties of the semiconductor. The geometrical nature of the p-n junction necessitates that the junction must be exposed at its periphery to another medium such as the atmosphere as with an etched mesa or more desirably a passivating insulating layer. The voltage applied across the junction appears directly across the exposed areas and as it is increased the depleted region spreads along the surface allowing a current to flow in any conducting path. The electric field strengths can be very high, in excess of $10^5$ volts/cm, resulting in surface currents large compared with those flowing in the bulk of the semiconductor. The conducting paths on the surface can result from contamination such as moisture, the remnants of cleaning agents or defects in the crystal caused by faulty processing.

The surface of the semiconductor can be protected from the effects of contamination by an insulating layer which protects and passivates the surface by controlling the concentration of impurities present. Silicon p-n junctions are most satisfactorily passivated by thin layers of $SiO_2$ thermally grown on the surface. Other semiconductors such as germanium or GaAs do not have a naturally occuring stable oxide and deposited layers of $SiO_2$ or $Si_3N_4$ are frequently used to passivate these materials. During the growth of thermal $SiO_2$ layers sodium ions are frequently incorporated as an impurity in the oxide. The sodium ions can move in the oxide and they can form a region of positive charge centres close to the silicon–silicon dioxide interface and induce an equal number of negative charges in the silicon. It is possible, in the case of a $p^+$-n junction, to form a field induced n layer above the $p^+$ region due to sodium ions in the oxide. The resultant field induced junction produces a large increase in the leakage current. Grove and Fitzgerald (1965) and Grove (1967) have studied this structure and they conclude that the field induced junction breaks down at a much lower voltage than the metallurgical junction because it is over a region of high doping concentration. The breakdown mechanism is either by tunnelling or avalanching and in some cases the current can completely mask the normal junction current.

For an accurate study of the bulk p-n junction it is essential that the surface effects be reduced to a minimum by the careful control of processing and junction geometry. A field plate over the oxide layer at the junction periphery can be used to control the surface states in this region. The gate controlled diode is described by Grove (1967) and it is frequently used in the study of p-n junction characteristics. The use of ion implantation through an oxide layer can leave a positive charge in the oxide which behaves in a similar manner to the sodium ions, though it anneals out at about 500 °C (section 5.5.8).

*Junction Breakdown Voltage* ($V_{BD}$)   In a reverse biased p-n junction the electric field is a maximum at the transition from p to n material. When this electric field exceeds a critical value, dependent on the semiconductor and its doping level, there is a sudden increase in the junction current. This current increase may be due to one or more of three mechanisms;

(i)   local heating causing an increase in leakage current, further dissipation and finally a run-away condition;

(ii)   avalanche multiplication of carriers and

(iii)  tunnelling.

Local heating can usually be identified using a transistor curve tracer as the heating frequently has a second or so time constant associated with it. On partial removal of the applied voltage, the dissipation decreases and there is a steady recovery of the $I/V$ characteristic to its original value assuming that there has been no permanent damage to the junction.

A measurement of the junction breakdown voltage $V_{BD}$ due to avalanche or tunnelling processes is a valuable check on surface conditions at the perimeter of the junction. Assuming that the edge effects do not influence $V_{BD}$ then it is determined by the critical electric field in the bulk of the semiconductor. The critical electric field increases with impurity doping and therefore in an abrupt p-n junction the doping level on the lightly doped side determines the breakdown voltage. The inversion layers due to surface effects lower $V_{BD}$ as their effective doping level tends to be lower than that in the crystal. Crystal defects which give rise to microplasmas also lower $V_{BD}$ as the electric field is much higher in these small regions. Sze and Gibbons (1966) and Sze (1969) have studied the effect of the radius of curvature at the edge of shallow diffused junctions. They have shown that $V_{BD}$ decreases rapidly as the radius of curvature is decreased and for a typical doping level of $10^{15}$ impurities/cm$^3$ they obtained the values given in table 5.13.

It has been observed experimentally for shallow ion implanted junctions that this relationship is not valid as the values of $V_{BD}$ are higher than those in table 5.13. The reason for this discrepancy is not known but it is probably associated with the sloping character of the mask edge (see fig. 5.17).

TABLE 5.13

Junction breakdown voltage
$V_{BD}$ as a function of junction curvature
(after Sze and Gibbons, 1966)

| Radius microns | $V_{BD}$ Volts |
|---|---|
| $\infty$ | 330 |
| 10 | 200 |
| 1 | 80 |
| 0.1 | 25 |

*5.4.3.3. Measurement of I/V Characteristics.* It was shown in the previous section that a knowledge of the $I/V$ characteristics of a p-n junction gives considerable information on the conduction mechanisms in the junction. A common and quick method of displaying the $I/V$ characteristics is to use a transistor curve tracer (such as Tektronix type 576) which gives information on the forward and reverse characteristics, the breakdown voltage and the presence of large surface currents. The $I$ and $V$ signals are displayed on linear axes and the current sensitivity is limited to $10^{-9}$ amps. A linear display of $I$ and $V$ is only satisfactory for a cursory examination, while for a rigorous study it is necessary to produce graphs of log $I_F$ against $V_F$ and log $I_R$ against log $V_R$ over a very wide range of currents and voltages. These logarithmic graphs give information on:

(i) the relation between the forward voltage and current and confirm if it has an exponential character,

(ii) the value of '$m$' which gives a measure of the relative contributions of the diffusion and R–G currents,

(iii) deviation from the exponential character due to a series resistance in the junction or in some cases a parallel leakage path,

(iv) the shape of the reverse characteristic and the sharpness of the avalanche breakdown and

(v) the distribution of R–G centres as a function of applied reverse voltage. This can be important for ion implanted junctions as the damage distribution is not uniform but is determined by the ion, substrate implantation conditions and the annealing treatment.

Conventionally, semilogarithmic or logarithmic p-n junction characteristics are often manually plotted. This is an accurate but very tedious procedure and an alternative semi-automatic plotting method has been described by Grimshawand Stephen (1971) which is based on a very sensitive logarithmic direct current amplifier with a virtual earth input (Harwell type 2256). Log $I$/linear $V$ or log $I$/log $V$ curves can be plotted with speed and accuracy. The circuit layout is given in fig. 5.37. Current from the D.C. source is passed through the junction to the input of the logarithmic D.C. amplifier. As this point is a virtual earth, there is no need for a series resistor to measure the currents. To obtain a forward characteristic the logarithmic amplifier output is connected to the Y input and the voltage across the junction to the X input of an X-Y pen recorder. For the reverse characteristic the voltage across the junction is applied via a 100 k$\Omega$ resistor to a simple D.C. logarithmic amplifier which drives the X input of the

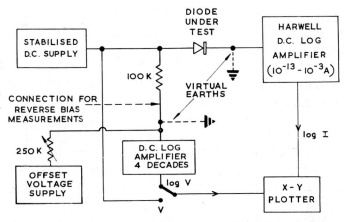

Fig. 5.37. Block diagram of the circuit used for the semi-automative plotting of forward and reverse current-voltage characteristics of p-n junctions.

recorder. The built-in potential of the junction can be simulated by applying an additional current to the simple logarithmic amplifier to offset the pen on the voltage axis.

Once set up it takes less than a minute to obtain a forward characteristic and a little longer for the reverse characteristic over the full range of currents ($10^{-12}$ to $10^{-3}$ amps) and voltage (0 to 100 volts). In taking the reverse characteristic particularly the voltage must be applied slowly to minimise errors due to displacement currents. In terms of plotting accuracy the characteristics should be comparable to any obtained manually except that a small error will be introduced by the offset voltage of the input of the logarithmic amplifier, which is about 10 millivolts. For comparative purposes the reproducibility is of the order of 0.2 %. Figure 5.38 shows the forward $I/V$ characteristics of five similar $p^{+}$-n junction diodes formed in

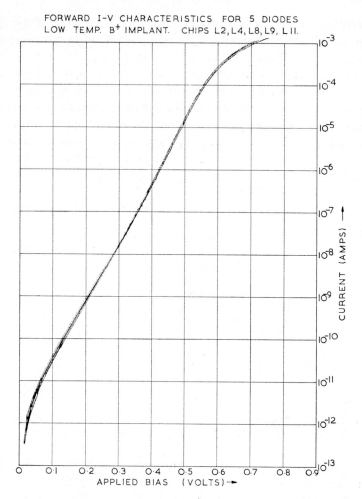

Fig. 5.38. The forward current-voltage characteristics of five diodes made by implanting 40 keV boron ions ($5 \times 10^{15}$/cm²) at 77 °K and annealing at 550 °C. The diodes were from the same silicon wafer and were mounted and bonded (Stephen and Grimshaw, 1971).

the same n type silicon wafer by implanting 40 keV boron ions at 77 °K and annealing at 550 °C. The curves show a high degree of uniformity, the change of slope where the diffusion currents exceeds the R–G currents at $V$ equals 0.4 volts and the deviation from the exponential relationship at high currents due to the series resistance.

A comparison of the forward characteristics of $p^{+}$-n diodes formed by the implantation of $5 \times 10^{15}$ ions/cm² of boron (40 keV), boron plus neon

(40 keV and 80 keV $(3 \times 10^{14}/\text{cm}^2)$), aluminium (100 keV), gallium (200 keV) and indium (400 keV) at room temperature into silicon and annealed at 500 to 550 °C was made by Stephen and Grimshaw (1971). The characteristics are shown in fig. 5.39. A boron diffused diode of the same area was included for comparison. The B, B + Ne and Ga diodes had exponential characteristics while the Al and In diodes could not be described by this simple relationship. The forward current of the diodes at low bias voltages (0.2 volts) where the $I_{R-G}$ current component is dominant increased with the

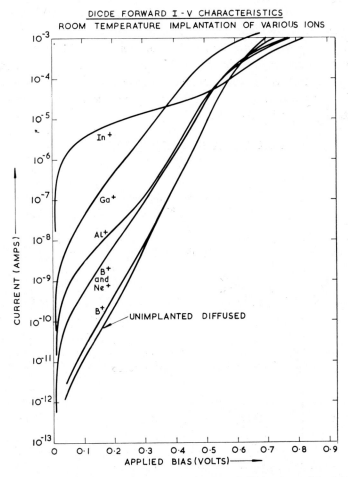

Fig. 5.39. Forward current-voltage characteristics of diodes formed by implanting the group III dopants into silicon at room temperature and annealing at 500 to 550 °C (Stephen and Grimshaw, 1971).

$Z$ number of the ions indicating that the number of defects in the junctions was directly associated with the damage produced by the heavy ions. A higher annealing temperature (650 °C) would reduce the amount of damage and number of defect centres allowing the $I/V$ characteristic to approach the exponential form.

Figure 5.40 compares the reverse characteristics of two gallium implanted diodes (G4 and G5) with two boron diffused diodes, protected by thick oxide on the same silicon chip. The diffused diodes had slightly larger leakage currents than unimplanted diffused diodes, while the gallium diodes deviated considerably from $I_{R-G} \propto (V + V_1)^{\frac{1}{2}}$ and indicated that the damage extended into the silicon below the implantation (see section 5.4.3.2(ii)). The forward characteristic did not indicate the presence of this damage to the same extent (fig. 5.39). Measurements of reverse leakage currents have been used to calculate the effective carrier lifetime and the density of defects centres (per unit volume) in the space charge region of a $p^+$-n junction (see section 5.4.3.4).

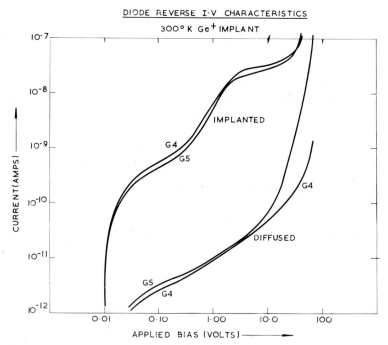

Fig. 5.40. The reverse current-voltage characteristics of two diodes implanted with gallium at room temperature and annealed at 500 °C compared with boron diffused diodes on the same piece of silicon.

*5.4.3.4. Minority Carrier Lifetime.*    The junction between the ion implanted region and the substrate is normally regarded as the lower boundary of the implantation. In general this is true but the possibility of the electrical properties of the semiconductor below the junction being influenced by the implantation must be borne in mind. The effects will be more pronounced for large doses of heavy ions. The defects produced in the semiconductor by the ion implantation are a maximum in the implanted layer and their peak concentration is closer to the surface than the peak impurity concentration (Crowder and Title, 1970). Vacancies, interstitials and impurities can diffuse into the substrate from the implanted region during the implantation itself or the subsequent thermal annealing.

This movement of defects can result in a modification to the electrical properties of the semiconductor by forming centres which allow easier recombination to occur. These centres also give rise to the R–G current component observed in the p-n junction. The most sensitive parameter to defects is the minority carrier lifetime $\tau$ which can be written in a simplified form as

$$\tau = \frac{1}{\sigma v_{th} N_t}$$

where, $\sigma$ = electron or hole capture cross section (about $10^{-15}$ cm$^2$), $v_{th}$ = thermal velocity of the carriers ($10^7$ cm/sec), $N_t$ = defect centre concentration per cm$^3$. As an example for $\tau = 10^{-6}$ seconds, $N_t$ has a value of $10^{14}$ centres/cm$^3$. This number of defects can readily be formed below a heavy implant which has peak impurity and defect concentrations exceeding $10^{19}$/cm$^3$.

The minority carrier lifetime in the lightly doped side of a p-n junction can be measured by the injection of minority carriers into this region and observing their decay time. A direct method is to forward bias the junction and inject into the lightly doped side for a time sufficient to establish a steady state distribution of the carriers. The forward current $I_F$ is then rapidly terminated and the stored carriers extracted by a defined reverse current $I_R$. Initially the junction voltage does not change but after a time $t$ when the minority carrier concentration approaches zero, the voltage suddenly reverses. It has been shown by Sze (1969) that the carrier lifetime $\tau$ is related to $t$ by the relationship:

$$\text{erf}\sqrt{\frac{t}{\tau}} = \frac{1}{1+I_R/I_F} \tag{5.4.42}$$

If $I_F = 5\,I_R$ then $\tau = t$ assuming that the width of the lightly doped region

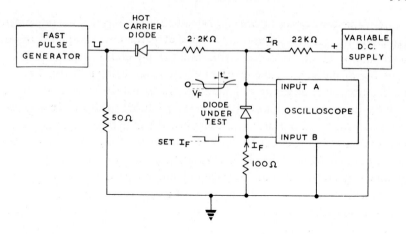

CIRCUIT FOR MINORITY CARRIER LIFETIME MEASUREMENTS.

Fig. 5.41. Block diagram of the circuit for measuring the minority carrier lifetime by the charge injection method.

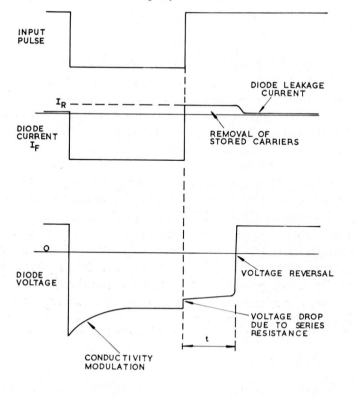

Fig. 5.42. Minority carrier lifetime measurement: switching waveforms.

is very much greater than the minority carrier diffusion length. This is the case for a $p^+$-n junction where the back contact is many diffusion lengths away from the junction. For a junction on n-$n^+$ (or p-$p^+$) epitaxial material this is not valid, as the width of the n layer is comparable with the diffusion length and n-$n^+$ interface will act as a sink for the carriers.

Figure 5.41 shows a circuit which can be used for minority carrier lifetime measurement by the injection method. The negative pulse (fig. 5.42) from the fast pulse generator turns on the forward current $I_t$ in the test diode. Input B of the oscilloscope is used to measure the voltage across the 100 ohm resistor which is proportional to $I_t$. At the end of the pulse, $I_F$ is terminated rapidly and $I_R$ as defined by the 22 k$\Omega$ flows into the diode. The hot carrier diode switches off very fast ($< 10$ ns) and isolates the pulse generator output from the test circuit. The voltage across the test diode is maintained by the minority carriers diffusing towards the junction and reverses in polarity when the carrier density approaches zero. Input A of the oscilloscope is used to observe the diode voltage. It must have a low input capacitance to avoid loading the circuit. A low capacity attenuator probe can be successfully used for connecting to the test circuit. $I_F$ is adjusted so that $I_F = 5I_R$ when $t = \tau$ and it is normally limited to low injection levels between 1 and 5 mA. It can be varied to check the consistency of $t$.

It is possible to observe the time variation of the conductivity of the lightly doped region of the junction (the substrate) during the forward current pulse (fig. 5.43). The injected minority carriers in the base region due to the forward current pulse cause the conductivity to increase and this can be seen as a decrease with time of the forward voltage. This conductivity

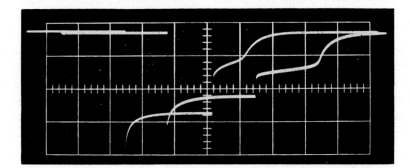

Fig. 5.43. Oscilloscope traces of the voltage across a boron implanted diode (left hand trace) and a boron diffused diode for a constant forward current of 11 mA. The vertical scale is 0.5 volts per cm and the horizontal scale 2 microseconds per cm. Refer to the lower trace of fig. 5.42 (Stephen and Grimshaw, 1971).

change can be used as a guide to the required duration of the forward current pulse to achieve a quasi stable spatial distribution of minority carriers before the lifetime measurement. At the end of the forward current pulse a small steep step is observed in the voltage waveform which is caused by cessation of the forward current in the ohmic series resistance of the diode. An ideal structure would not exhibit this effect. It is necessary to observe the voltage across the diode only for this measurement and the inputs A and B of the oscilloscope must be used in the differential mode, allowance being made for the loss of gain due to the low capacity probe on input A.

Various techniques such as Rutherford backscattering measurements, infrared absorption spectra and transmission microscopy have been used to detect defects in semiconductors. One of the more sensitive methods is to measure the minority carrier lifetime in the damaged region of the semiconductor. This can be done by the carrier injection method just described or the measurement of reverse leakage current (section 5.4.3.2). An impurity profile calculated from a capacitance-voltage plot (section 5.4.2.6.4) will also give information on the density and position of defect centres which are electrically active. These electrical measurements require a $p^+$-n (or $n^+$-p) junction structure with the defects located in the lightly doped side of the junction so that they are included in the depleted region.

Davies and Roosild (1970) used the carrier injection method to measure the minority carrier lifetimes in the n region of a shallow phosphorus diffused $n^+$-p junction in silicon implanted at 20 °C with low doses of 1 or 2 MeV carbon ions. Figure 5.44 shows some of their published results. The carrier lifetime decreased as the dose and the energy were increased as greater numbers of defects were generated in the silicon. Annealing experiments up to 650 °C showed that the lifetime recovered almost to its unirradiated value. This result was significant as it showed that approximately the same annealing temperatures were required to anneal out damage in low as well as high ion dose implantations (see section 5.4.4). Similar results were reported by Pickar and Dalton (1970), who measured the increase in the reverse leakage current due to the recombination-generation centres produced by defects. The relation between the minority carrier lifetime and the leakage current is given by Grove (1967) as (see eq. (5.4.39))

$$I_{R-G} = \frac{qn_i wA}{2\tau_0}$$

where $\tau_0$ = carrier lifetime (secs), $w$ = width of the depleted region (cm), and $A$ = junction area (cm$^2$). The effective lifetime in the depleted region

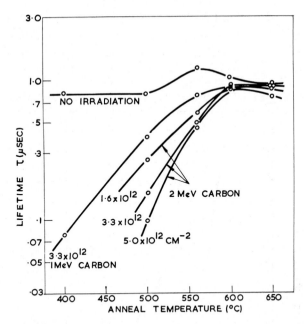

Fig. 5.44. Minority carrier lifetime in carbon implanted and unimplanted phosphorus diffused diodes in 5 ohm·cm silicon (Davies and Roosild, 1970).

can be defined as

$$\tau_0 = \frac{1}{\sigma N_t v_{\text{th}}}$$

(where $\sigma$ = capture cross section for minority and majority carriers $(\text{cm}^2)$, $v_{\text{th}}$ = thermal velocity $(\sim 10^7 \text{ cm/sec})$, $N_t$ = density of traps) and assuming that the energy level of the traps (defects) are within $1\,kT$ of the intrinsic Fermi level. Pickar and Dalton (1970) bombarded $p^+$-n shallow boron diffused junctions with helium, carbon and silicon ions with energies of 300 and 600 keV. They found that the leakage current increased with dose and energy and for 300 keV carbon ions a seven order increase in $I_{\text{R–G}}$ was observed for a 5 order increase in the dose.

The value of $\sigma \cdot N_t$ for a 600 keV $(1 \times 10^{12})$ carbon ions/cm$^2$ was about $10^{-3}$, equivalent to $\tau = 10^{-4}$ seconds which is longer than the values found by Davies and Roosild (1970). The conclusions drawn by Pickar and Dalton were that the experimental results fitted a single level Shockley–Read–Hall model with the dominant defect level within $1\,kT$ of the intrinsic Fermi level, and the change of leakage current with dose from a non-linear one for helium and carbon to a linear change for silicon indicated an important difference between defects produced by light ions and heavier ions. Annealing experiments on the carbon implantations showed that the defects were substantially but not completely annealed at 750 °C as shown in fig. 5.45.

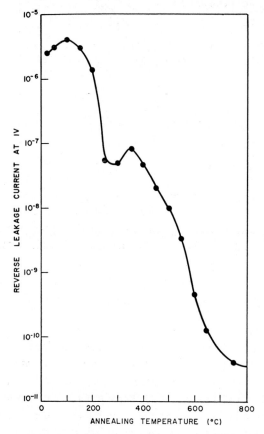

Fig. 5.45. Variation of reverse leakage current with temperature for a boron diffused diode implanted with 300 keV carbon ions ($10^{15}/cm^2$). The initial leakage current was $3 \times 10^{-12}$ amps. Negative annealing peaks occurred at 100 and 350 °C (Pickar and Dalton, 1970).

Recently Davies and Roosild (1971b) have measured the defect energy levels in low dose implanted silicon by observing the thermally stimulated current (T.S.C.) produced as a reverse biased p-n junction was warmed up from liquid nitrogen temperature. The traps in the depletion region of the junction were filled by illumination at 77 °K and then the junction was heated at about 1 °K/second to room temperature. The current increased as the traps released their carriers and the heating curves comprised a series of peaks corresponding to the various trapping levels. They were able to identify six energy levels between 0.25 and 0.40 eV for 3 MeV $3 \times 10^{12}$ oxygen ions/cm$^2$ in the p region of a shallow phosphorus diffused n$^+$-p planar diode. Integration of the areas under the peaks gave quantitative data on the number of defects present and a value of $8.3 \times 10^{11}$/cm$^2$ was obtained for the 0.38 eV and 0.40 eV peaks after a 400 °C anneal. It was possible to follow the changes in defect energy levels and their concentration (per cm$^2$) as the irradiated junction was annealed at various temperatures up to 450 °C.

These measurements on minority carrier lifetime are a direct and very sensitive method of assessing the effects of ion implantation and subsequent annealing processes on the electrical properties of the semiconductor lattice. The results are especially valuable as they are applicable without further interpretation to the study of practical ion implanted devices.

*5.4.3.5. Noise in Ion Implanted p-n Junctions.* The noise in a p-n junction due to the flow of current arises from the conduction mechanisms in the junction (van der Ziel, 1959; van der Ziel, 1968; Sze, 1969). Firstly there is the thermal noise which occurs in the ohmic parts of the junction and to this must be added the diffusion noise caused by the minority carriers moving in either direction across the space charge layer. The recombination-generation of carriers at trapping centres gives rise to noise similar to shot noise and it is independent of frequency at medium and low frequencies but becomes dependent at high frequencies.

The total rms noise current generated in forward bias in a p-n junction is given by Sze (1969), as;

$$\langle i_n^2 \rangle = 2qI_s B(e^{qV/kT} + 1)$$

(where $I_s$ = saturation current (see eq. (5.4.38)). $B$ = bandwidth in cycles/sec) and shows the direct dependence of $\langle i_n^2 \rangle$ on $I_s$. $I_s$ is in turn dependent on the number of R–G centres in the junction.

Stephen and Grimshaw (1971) have observed a partial dependence of the noise voltage on $I_s$ for a number of ion implanted diodes produced by implanting boron, gallium and indium under various conditions. The annealing temperature was limited to 500 °C. The diodes were placed in 'series' with the emitter of a low noise BFY90 n-p-n transistor and measured with a Hewlett-Packard type 4470A transistor noise analyser. Table 5.14 lists the square of the noise voltage, $\langle V_n^2 \rangle$, and the saturation currents.

TABLE 5.14

Noise voltages and saturation currents of ion implanted diodes

| Device details | $\langle V_n^2 \rangle$ volts $\times 10^{-18}$ | $I_s$ amps |
|---|---|---|
| Diffused ref. diode | 1.4 | $10^{-13}$ to $10^{-12}$ |
| Room temp. Boron implant | 3.3 | $3 \times 10^{-12}$ |
| 77 °K Boron implant | 1.2 | $2 \times 10^{-12}$ |
| 450 °C Boron implant | 6.8 | $1.5 \times 10^{-12}$ |
| Boron and Neon implant | 25 | $3.5 \times 10^{-11}$ |
| Gallium implant | 53 to 175 | $5 \times 10^{-10}$ to $2 \times 10^{-8}$ |
| Indium implant | 53 to 144 | about $10^{-6}$ |

Flicker or '$1/f$' noise can occur in a p-n junction due to surface recombination effects. It has a '$1/f$' power spectrum below 1 kHz and the magnitude of its contribution will depend on the degree of domination of the surface conditions at the exposed part of the junction. A junction with a large surface leakage current component will in general exhibit a large '$1/f$' noise component.

The measurement of noise is not a normal diagnostic tool used in the study of ion implanted devices. Noise measurements can only be meaningfully performed on p-n junctions fabricated to a very high standard of perfection and cleanliness. It is essential to study the noise generated in ion implanted p-n junctions intended for use in low noise devices such as bipolar and bulk field effect transistors. It is possible to adjust fabrication schedules such as annealing temperatures to obtain optimum performance. The noise performance of many devices is the final criterion by which they are judged for a particular application.

*5.4.3.6. Methods of Fabricating Ion Implanted Junctions.*   A number of electrical methods for measuring p-n junctions have been discussed and to complete this section some of the various methods that can be used for fabricating p-n junctions will be described. Some of these methods have already been mentioned when describing the closely related topic of forming ion implanted resistors (section 5.4.2.5).

It has been shown that the electrical behaviour of a p-n junction is determined by the bulk semiconductor properties provided that it is not dominated by surface leakage effects. The very nature of semiconductor devices necessitates that at some point the junction must be exposed at the surface. In some cases the exposed junction can be left unprotected but normally it is necessary to exercise control over the semiconductor surface to prevent it altering with the environment and time. It is also essential in practical devices such as integrated circuits to run metallic conductors over the surface of the circuits to make interconnections and form bonding pads. The conductors must be insulated from the semiconductor.

The general principle involved in the stabilization of a semiconductor surface is to chemically bond to it an appropriate solid insulating film that is well defined in composition and structure. The interface between the semiconductor and the film are determined by the properties of the film which can be made reproducible under carefully controlled preparation conditions. The silicon dioxide on silicon system is the most studied and widely used example of the stabilization of a semiconductor surface. The subject of surface stabilization is beyond the scope of this chapter and for the details of its theory and practice reference should be made to Atalla et al. (1959), Grove (1967), Hofstein (1967), Kooi (1967) and Sze (1969).

Figure 5.46(a) shows a simple p-n junction formed by implanting an impurity, in this case an acceptor, directly into the silicon with an in-contact mask. The junction is exposed at the edge of the mask and it is not protected in any way. If a metal mask is used, it has to be removed to avoid short circuiting the junction and any contamination left on the surface in the form of metallic or other impurities will give rise to a high leakage current. Figure 5.46(b) is an improved version of the simple implanted structure as the junction edge is now protected by the thin oxide layer. It is necessary to implant the silicon through this oxide layer and this requirement places a restriction on the ion energy. High ion energies are sometimes required when implanting through thin layers. This type of junction structure can be readily made by following the sequence of steps shown in fig. 5.47. A passivation layer of oxide is deposited or grown on the cleaned silicon surface to protect it during the subsequent processing and life of the junction. A metal layer is evaporated onto the surface of the oxide and apertures engraved in the metal to define the lateral extent of the implantation. After implantation, a contact window is engraved in the exposed oxide layer *before* removal of the metal layer. The position of the aperture in the metal is used to register the position of the contact window. The metal is removed and the silicon wafer thoroughly cleaned and washed in deionized water. The wafer is annealed at the required temperature to remove the damage in the implantation and then it is etched for a few seconds in buffered hydrofluoric acid to clear the contact windows prior to contact metallization. The contact conductors and pads are engraved and the wafer is sintered for a few

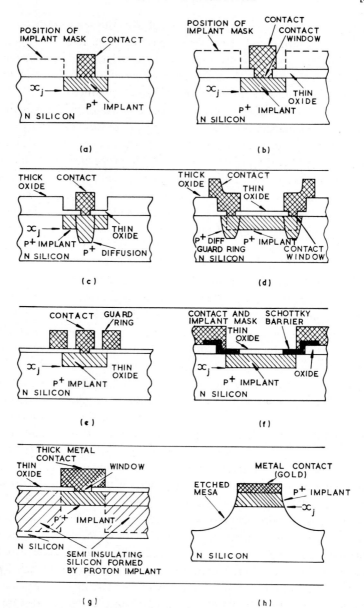

Fig. 5.46. Various configurations of ion implanted diodes. (a) simple unprotected structure, (b) simple passivated diode, (c) passivated diode with diffused contact region, (d) extension of (c) with a diffused guard ring, (e) as (b) but with a field plate, (f) combined implanted–Schottky barrier diode, (g) simple diode isolated by proton implantation and (h) mesa diode.

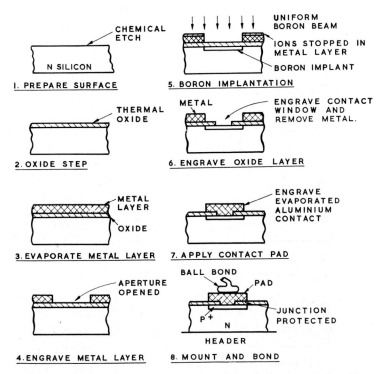

Fig. 5.47. The principal steps in the fabrication of a simple p⁺n junction diode with the junction edge protected below the passivating layer (see Fig. 5.46 (b)).

minutes to form a solid–solid very shallow diffused contact between the metal and the silicon. A 30 minute sinter at 500 °C in a flow of dry nitrogen gas is normally used for aluminium contacts onto silicon. Figure 5.48 shows the sequence for a more complex structure in which the metal mask has been used twice, once to define a buried boron implanted layer and second a phosphorus implanted layer with a larger area which forms the junction. This type of structure with a buried layer in a lightly doped substrate can be used to avoid avalanche breakdown due to the high electric fields which occur at the edge of a sharp junction (table 5.13) (Sze, 1969). The difficulty of obtaining a good ohmic contact to an implanted layer has already been discussed in the section dealing with ion implanted resistors. Figure 5.46(c) shows an implanted junction with a highly doped central region which can be produced by thermal diffusion of the appropriate impurity or implanta-tion. This can be achieved by making a structure as shown in fig. 5.46(c) and using a large ion dose and annealing at a high temperature to ensure good

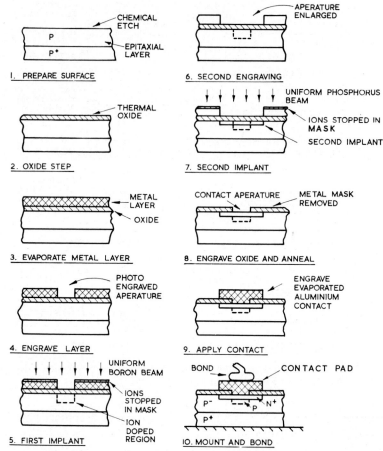

Fig. 5.48. The principal steps in the fabrication of a double implanted $n^+$-p-$p^-$-$p^+$ diode.

electrical activity. A second metal mask can be applied and implantation apertures cut in it using the existing contact windows to provide the registration. The normal contacting procedure follows the annealing of the second implantation. A more developed form of this structure is shown in fig. 5.46(d). In this example, the central diffused region has been replaced by a diffused annulus with the implanted region in the centre of the annulus. Contact to the junction is made to the diffused region through an annular window by a circular metal contact. This metal contact is shown in fig. 5.46(d) as acting as a gate control electrode for controlling the surface charge in the vicinity of the exposed junction (Grove, 1967). The use of the diffused annular ring also avoids the sharp corners of the implantation which cause the localized

high electric fields mentioned earlier. Figure 5.46(e) shows the same gate
control electrode applied to the structure, fig. 5.46(b) to form a particularly
useful device in which the surface currents can be controlled allowing the
bulk junction currents to be studied with reduced interference (Kawamoto
and Oldham, 1970 and Grove, 1967).

Lepselter et al. (1969) have employed Schottky barrier techniques to
make the source and drain contacts of a MOST with ion implantation to
produce the self aligned gate structure (see section 5.5.5). Figure 5.46(f)
shows how a diode structure can be made using PtSi for the Schottky
barrier as described by Lepselter et al. (1969). Other metals such as palladium or aluminium can be used. The method is confined to n-type substrates
as it is difficult to make good Schottky barriers on p type substrates. As the
two junctions are electrically in parallel the forward $I/V$ characteristic will
be dominated by the Schottky barrier as it conducts at a lower forward
voltage than an implanted (or diffused) junction while the reverse characteristic depends on both junctions. It is necessary to incorporate a field relieving
electrode at the edge of the Schottky barrier to prevent premature breakdown. This has been shown in fig. 5.46(f) by extending the Schottky barrier
metal up on to the oxide layer. This approach is interesting as it offers a
method for contacting to an implantation with a very thin metal layer which
in turn can be contacted by a thick metal layer for bonding.

The structure in fig. 5.46(g) is similar to fig. 5.46(b) except that the
lateral extent of the implantation has been confined by a proton implantation. The sequence for making such a structure is to firstly implant the whole
area of the wafer with the impurity through the oxide layer in the normal
manner and anneal out the damage. The contact window is then cut in the
oxide and a very thick (over 1.5 microns) metal contact pad formed over
the window. The contact is sintered to form an ohmic contact with the implanted silicon at this stage. The structure is then proton irradiated from
above to make all the unmasked silicon semi-insulating and form a guard
ring around the masked area. A very similar technique has been used by
Lindley et al. (1969) for fabricating a guard ring around a Schottky barrier
photo-diode on GaAs. Silicon $p^+$-n junctions guarded in this manner show
an increase in the breakdown voltage and the leakage current while the forward characteristic is not affected. The influence of the semi-insulating silicon
surrounding the junction prevents high electric fields occurring at the edges
of the implant thus increasing the overall breakdown voltage but the 'leaky'
insulating properties of the proton bombarded silicon increase the leakage
current.

Figure 5.46(h) shows a mesa structure formed by etching away the implanted layer after annealing to expose the junction. It is common to use a metal contact, in this case boron doped gold, to form a mask for the etching and later to serve as the contact. Wax or photoresist masking can also be used. If a non selective etchant is used to etch the silicon then a smooth curved wall to the mesa can be obtained and the electric field at the surface will be reduced as the potential will be applied over a greater distance. The etched mesa is not normally protected with a passivating layer. Kerr and Large (1967) have made experimental high voltage $p^+n$ diodes by implanting boron into 60 $\Omega$ cm n type silicon and then contour etching from both sides of the wafer. They obtained very high breakdown voltages consistent with the values predicted for 60 $\Omega$ cm silicon. Ying et al. (1968) have described an Impatt oscillator diode made by forming a mesa in an ion implanted layer.

### 5.4.4. ELECTRICAL RESULTS FOR VARIOUS ION IMPLANTED DOPANTS IN SILICON

*5.4.4.1. General.* The electrical measurements on implanted layers give an overall picture of the behaviour of the implanted impurity atoms. The lattice disorder produced during the implantation interacts with the impurities and the mobile charge carriers in a complex manner which makes the interpretation of the result very difficult. It is not possible to determine what lattice site or sites the impurity atoms are occupying other than to deduce that they are probably not substitutional and capable of contributing to the electrical activity in the normal manner. It is normally necessary to anneal an implanted layer before reliable electrical measurements are possible and therefore the state of the implant is disturbed before the measurement.

The primary purpose for studying the electrical properties of implanted dopants in silicon or any semiconductor is to exploit the results for the fabrication of solid state devices. As ion implantation is only used at one or two steps in the fabrication cycle of a device, it is necessary to adjust the implantation conditions to be compatible with an existing technology. The sole criterion of success is the production of satisfactory reproducible electrical characteristics preferably at no greater complication and expense than any other step in the cycle. A very well known example of the adaptation of implantation to an existing process is its use in the MOS planar technology to form the self-aligned gate Metal Oxide Silicon Transistor (MOST) (Bower et al., 1968).

The electrical measurements using the methods described in the earlier sections, yield information on the distribution in depth of the electrically active impurities, the carrier mobility, the minority carrier lifetime and the characteristics of any junctions that may be formed with the substrate. The results are strongly influenced by the implantation damage, the impurities already in the substrate and the thermal annealing and as a consequence the great majority of results are presented as a function of annealing temperature or after a given annealing treatment. The number of published results on implanted layers far exceed those on the junction characteristics and the effects on minority carrier lifetime. The wide range of dopants possible by implantation has been used by Meyer (1968) to study a number of elements implanted in to high resistivity silicon and germanium to form $n^+$ and $p^+$ contacts for nuclear particle detectors. He measured the $I/V$ characteristics and found acceptor behaviour from Group III (B, Al, Ga, In and Tl) and Group II (Be, Mg, Ca, Zn, Cd, Ba and Hg) elements in $n^-$ silicon and donor behaviour from Group V (N, P, As, Sb and Bi) and Group VI (O, S, Se and Te) elements in $p^-$ silicon as well as certain triple donors and acceptors and transition elements. He selected boron and tellurium for making $p^+$ and $n^+$ contacts to $n^-$ silicon for particle detectors. When considering normal device resistivities (0.1 to 10 ohm-cm), the selection of dopants is restricted to Groups III and V elements as it is necessary to obtain high electrical activity for reversing the conductivity type and making degenerate regions for ohmic contacts. Boron, phosphorus and arsenic are the most suitable elements as they are very soluble in silicon (Trumbore, 1960) allowing a high electrical conductivity to be obtained. Their behaviour in silicon is well understood from work on bulk, epitaxial and diffused silicon. One of the fortunate features of implantation is that these elements are readily ionized allowing intense ion beams to be produced for large scale implantation.

*5.4.4.2. Boron in Silicon.* The early efforts at implanting chemical dopants in silicon were aimed at making junctions for particle detectors (Rourke et al., 1961; Martin et al., 1964) and solar cells (King et al., 1965). Acceptor activity was observed in the case of boron implantations and the need for annealing was appreciated. The first detailed studies of ion implanted silicon were published by Pavlov et al. (1966b) and Kellett et al. (1966) on the conductivity of the layers and on the application to devices such as diodes by Manchester (1966) and bipolar and field effect transistors by Leith et al. (1967). Various aspects of boron implantations were discussed in a number of papers presented at the 1967 International Conference on the Application of Ion Beams to Semiconductor Technology (Glotin ed., 1967).

The annealing characteristics of boron implanted at room temperature has three stages, a rapid recovery between 300 and 400 °C, a reverse annealing effect between 400 and 600 °C and lastly a steady anneal above 600 °C as indicated by a steady decrease in sheet resistivity to 900 °C. This behaviour is characteristic of boron and it has been reported by Pavlov et al. (1966b), Blamires et al. (1967), Clark and Manchester (1968), Baron et al. (1969),

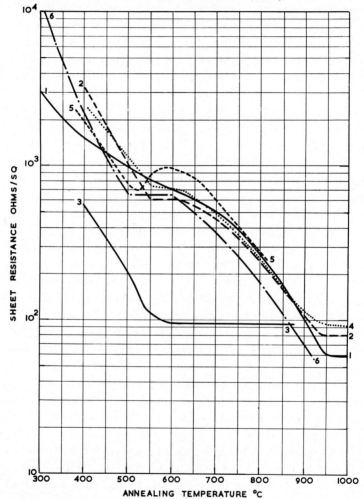

Fig. 5.49. The variation of sheet resistivity with annealing temperature for a number of boron implantations. Curve 1, 150 keV $1.5 \times 10^{15}$ at 20 °C (Seidel and MacRae, 1969); Curve 2, 200 keV $1 \times 10^{15}$, 20 °C (Davies, 1969a); Curve 3, 200 keV, $1 \times 10^{15}$, 77 °K (Davies, 1969a); Curve 4, 130 keV, $1 \times 10^{15}$ 300 °K (Blamires, 1970); Curve 5, 60 keV, $1 \times 10^{15}$, R.T. (Shannon et al., 1970a) and Curve 6, 300 keV, $1 \times 10^{15}$ R.T. (Kellett et al., 1966).

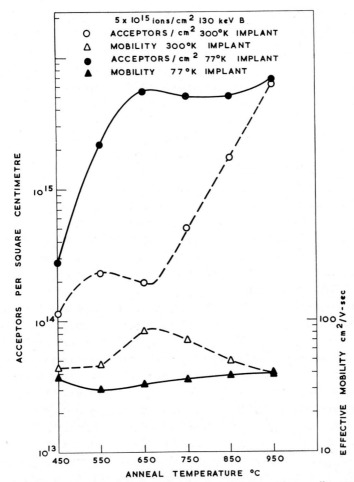

Fig. 5.50. The variation of carrier concentration and mobility with annealing temperature for 130 keV $5 \times 10^{15}/cm^2$ boron implantations at 77 °K and 300 °K (Blamires, 1970).

Gusev and Titov (1969), Stumpfi and Kalbitzer (1971) for a 7 keV implant and Lecrosnier and Pelous (1970) for a 1 MeV implantation. Figure 5.49 shows the closely similar annealing results of implantations by Kellett et al. (1966), Davies (1969a), Seidel and MacRae (1969), Shannon et al. (1970a) and Blamires (1970) while fig. 5.50 from Blamires (1970) shows the acceptor and mobility changes with annealing temperature for a 130 keV $5 \times 10^{15}$ ions/cm² implant at 300 °K. Less than 10 % of the atoms implanted at 300 °K are active between 450 and 650 °C and full activity does not appear until 950 °C. Baron et al. (1969) have studied the annealing of 50 keV boron ions implanted into $\langle 111 \rangle$ silicon at 500 °C. After implantation the electrical activity was 4 % for a dose of $10^{14}$ ions/cm$^2$ which was low when compared with 20 % for a similar room temperature implant. The activity did not increase until the annealing temperature reached 650 °C after which temperature it rose steadily until nearly 100 % activity was achieved at 900 °C. The carrier mobility at 500 °C was 250 cm$^2$/V·sec and it decreased to about 110 cm$^2$/V·sec at 900 °C.

The sites of the electrically inactive atoms are not clearly known but North and Gibson (1970) have shown a close correlation between the acceptor concentration by electrical measurements and the fraction of substitutional boron atoms by proton channelling. The substitutional fraction decreases from about 30 % at room temperature to about 5 % at 700 °C and then increases to over 90 % at 1000 °C as the boron becomes substitutional. The interstitial boron atoms up to 500 °C appear to lie along $\langle 110 \rangle$ atomic rows but not midway between the $\langle 110 \rangle$ sites. This movement of boron atoms and their interaction on the carrier mobility can account for the reverse annealing behaviour (see ch. 3).

The electrical resistivity is very dependent on the implantation temperature, dose and energy as these influence the degree of damage in the lattice. Davies (1969) implanted 200 keV boron ions at 77 °K and obtained a completely different annealing characteristic which had a low sheet resistivity at 550 °C (fig. 5.49(3)). Lecrosnier and Pelous (1970) have observed the same annealing behaviour for a 1 MeV 77 °K implant. Hart and Marsh (1969) and Blamires (1970) have observed the same effect by measuring the acceptor concentration per cm$^2$ against annealing temperature. There is a significant annealing between 500 and 600 °C after which 85 % of the acceptors are active. Figure 5.50 from (Blamires, 1970) shows the acceptor and mobility changes with temperature and above 600 °C are nearly constant resulting in a constant minimum sheet resistivity. It is essential to have 'amorphousness' present to produce the full electrical activity on annealing and Blamires (1970) has suggested that at 77 °K a dose of $10^{15}$ ions/cm$^3$ is a threshold value for energies over 100 keV while it is more than adequate at lower energies such as 20 keV when the damage is greater. A dose of $1 \times 10^{14}$ ions/cm$^2$ is inadequate to form amorphousness at 77 °K and no significant annealing is seen (Hart and Marsh, 1969). The damage produced by the boron at 77 °K can be nearly simulated by implanting neon into a room temperature boron implant and then annealing. Figure 5.51 (Blamires, 1970) shows how a dose of 80 keV neon ions ($10^{15}$/cm$^2$) has nearly the same effect on the annealing characteristic as a low temperature implant.

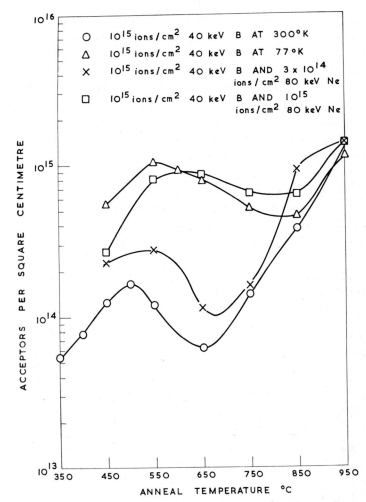

Fig. 5.51. The variation of acceptor concentration (per cm$^2$) with annealing temperature, for a 40 keV $1 \times 10^{15}$ boron implantation, is used to compare the effect on the electrical activity of substrate temperature and neon implantation at room temperature.

The ionisation energies of implanted boron atoms have been measured by Tetel'baum (1967) as 0.084 eV annealed at 500 °C and 0.052 eV at 700 °C, while Shannon et al. (1970a) found a value of 0.065 eV for a 60 keV implant annealed at 450 °C. Seidel and MacRae (1969) found good agreement with the 0.045 eV boron level at 700 °C while the results for 300 and 500 °C anneals suggest the presence of deep levels with higher activation energies.

The depth profiles of implanted boron have been studied by the successive layer removal technique, staining, capacitance/voltage and by the analysis of the emitted secondary ions during sputtering (Pistryak et al., 1970) and heavy ion X-ray excitation (Cairns et al., 1970). The last two methods are not as sensitive as the electrical methods and at present require very large boron doses. Some experimental ranges for a wide range of ion energies are given in table 5.15. The agreement with the LSS theory (Johnson and Gibbons, 1969) is good for the energies between 50 and 200 keV. Figure 5.52 (Blamires et al., 1968) compares four 40 keV $(1 \times 10^{15}$ ions/cm$^2$) boron

TABLE 5.15
Experimental values of Boron ion ranges in silicon
(The calculated ranges are taken from Johnson and Gibbons (1969))

| Energy eV | Projected range microns cal. | measured | Conditions | $T_A$ °C | Reference |
|---|---|---|---|---|---|
| 6 | 0.023 | 0.03 | $10^{14}$/cm$^2$ $\langle 111 \rangle$ | 420 | Bader and Kalbitzer (1970b) |
| 30 | 0.12 | 0.30 | $10^{18}$/cm$^2$ $\langle 111 \rangle$ | R.T. | Pistryak et al. (1970) |
| 30 | 0.12 | 0.33 | $10^{18}$/cm$^2$ $\langle 110 \rangle$ | R.T. | |
| 40 | 0.161 | 0.22 | $10^{15}$/cm$^2$ $\langle 111 \rangle$ | 1000 | Blamires et al. (1967) |
| 40 | 0.161 | 0.19 | $6.3 \times 10^{14}$/cm$^2$ 14° off $\langle 111 \rangle$ | 800 | Zorin et al. (1968) |
| 50 | 0.202 | 0.20 | $3 \times 10^{15}$ $\langle 111 \rangle$ | 625 | Gibbons (1967) |
| 100 | 0.398 | 0.43 | $10^{18}$/cm$^2$ $\langle 111 \rangle$ | R.T. | Pistryak et al. (1970) |
| 100 | 0.398 | 0.49 | $10^{18}$/cm$^2$ $\langle 110 \rangle$ | R.T. | |
| 100 | 0.398 | 0.35 | $2 \times 10^{13}$ | 600 | Large et al. (1967) |
| 150 | 0.57 | 0.35 | $10^{15}$/cm$^2$ 7° $\langle 111 \rangle$ | IMP 600 | Seidel and MacRae (1969) |
| 240 | 0.84 | 0.68 | $10^{15}$/cm$^2$ 3° $\langle 111 \rangle$ | 900 | Fairfield and Crowder (1968) |
| 150 | 0.57 | 0.52 | | | |
| 200 | 0.73 | 0.65 | | | |
| 400 | 1.24 | 1.05 | $10^{12}$/cm$^2$ 7° $\langle 111 \rangle$ | 650 | Davies (1969b) |
| 700 | 1.84 | 1.5 | 78 °K | | |
| 1000 | 2.32 | 1.9 | | | |
| 1700 | 3.4 | 2.6 | | | |
| 1250 | 2.75 | 1.8 | $10^{13}$/cm$^2$ off $\langle 111 \rangle$ | 700 | Buchanan et al. (1967) |
| 2000 | 3.8 | 2.8 | $10^{13}$/cm$^2$ off $\langle 111 \rangle$ | 700 | |
| 500 | 1.46 | 1.1 | | | |
| 1000 | 2.32 | 1.7 | $10^{15}$/cm$^2$ 7° $\langle 111 \rangle$ | 700 | Lecrosnier and Pelous (1970) |
| 1500 | 3.10 | 2.3 | | | |
| 40 | 0.161 | 0.167 | $1 \times 10^{17}$ | R.T. | Cairns et al. (1970) |
| 100 | 0.398 | 0.330 | $1 \times 10^{17}$ | R.T. | |
| 300 | 1.00 | 1.06 | $1 \times 10^{17}$ | R.T. | |

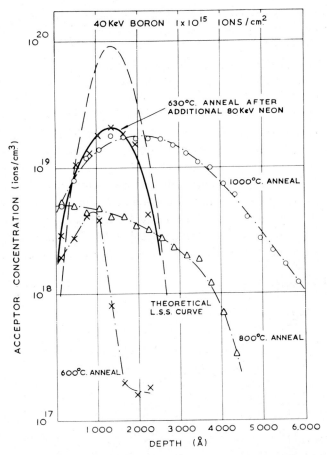

Fig. 5.52. The profiles obtained by Hall effect measurements and layer stripping on 40 keV boron implantations ($1 \times 10^{15}$/cm²) annealed at 600, 800 and 1000 °C. The profile of a combined boron and neon implantation annealed at 630 °C is included.

profiles with the theoretical LSS profile. The 600 °C profile bore no resemblance to the LSS profile and its peak concentration was closer to the silicon surface, at a distance corresponding to the position of the damage peak reported by Crowder and Title (1970) which showed that the damage on partly recovering had permitted some boron to become substitutional and electrically active. The profile for the 80 keV ($1 \times 10^{16}$ ions/cm²) neon implant into an existing 40 keV ($1 \times 10^{15}$ ions/cm²) boron layer and annealed at 630 °C agreed closely with the LSS profile and showed the increase in electrical activity seen in the annealing curves (fig. 5.51). The two profiles for implants annealed at 800 and 1000 °C exhibit high electrical activity and broadening due to thermal diffusion. The channelling of boron is less than that observed for the heavier elements phosphorus and arsenic (Dearnaley et al., 1971) and it can probably be reduced to an insignificant level by mis-orientation (Fairfield and Crowder, 1968). Bader and Kalbitzer (1971) have shown a channelled region for boron extending beyond 0.15 microns for a 6 keV implant ($R_p$ = 0.023 microns) annealed at 300 °C. The channelled region was not affected by annealing at 900 °C. Boron does not exhibit the deep enhanced diffusion effects which have been found for the other group III elements (refer to sections 5.4.4.3 to 5.4.4.5).

Figure 5.53 shows the variations of sheet resistivity against boron ion dose as measured by Davies and Roosild (1971a) (Curve C, 200 keV implants at 77 °K and annealed at 650 °C), MacDougall et al. (1969) (Curve D 55 keV implants through oxide and annealed at 900 °C, Dill et al. (1971a) (Curve E 100 keV implants in ⟨100⟩ Si through 0.12 micron oxide-nitride

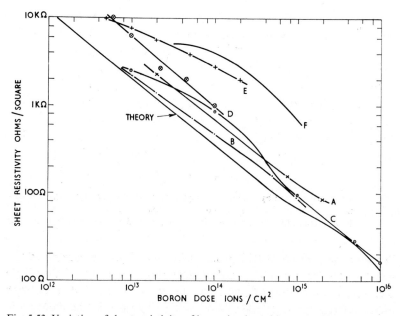

Fig. 5.53. Variation of sheet resistivity of boron implanted layers in silicon with boron ion dose. The curves are; A, 40 keV ⟨111⟩ 900 °C anneal (Stephen, 1973); B, 80 keV ⟨111⟩ 900 °C anneal (Stephen, 1973); C, 200 keV −195 °C anneal 650 °C (Davies and Roosild, 1971(b)); D, 55 keV through 0.1 to 0.2 microns of oxide 950 °C anneal (MacDougall et al., 1969); E, 100 keV ⟨100⟩ 545 °C anneal (Dill et al., 1971); F, 60 keV through 0.2 microns oxide 500 °C anneal (Shannon and Ford, 1970). A theoretical curve for 80 keV boron assuming 100 % electrical activity and Irvin's (1962) carrier mobility values is shown.

layer and annealed at 545 °C), and Shannon and Ford (1970) (Curve F 60 keV implants through 0.2 micron oxide and annealed at 500 °C). The curves A and B are unpublished results from Harwell and they are for 40 and 80 keV boron implants respectively into $\langle 111 \rangle$ silicon annealed at 900 °C. The theoretical curve was calculated for 80 keV boron ions from eq. (5.4.4.),

$$R_S = \frac{1}{q\mu C_S} \quad \text{(ohms/square)}.$$

100 % electrical activity was assumed. The values of mobility were taken from Irvin (1962) assuming an average doping concentration over 0.5 microns corresponding to

$$X = \frac{R_p - x_j}{A R_p} = 2.1$$

which includes over 98 % of the ions. A minimum value of $\mu = 48 \text{ cm}^2/\text{V} \cdot \text{sec}$ was used.

The lack of agreement between the curves A and B and the theoretical curve is probably due to the mobility being lower at the peaks of the implants where the concentration of dopant atoms is greatest. The lower resistivity of the 80 keV implant is the result of the profile being deeper and broader ($AR_p$ is larger) and the concentration at the peak being lower. Curve C shows the pronounced annealing of boron implanted at 77 °K and the very low resistivities that can be achieved for very high doses. The approach to the theoretical value at $10^{13}$ ions/cm$^2$ is possibly due to most of the boron atoms being electrically active. Curve D shows clearly the result of implanting through an oxide layer (0.1 to 0.2 microns) as a large proportion of the boron is stopped in the oxide and the effective dose is reduced by approximately a factor of four in this example. Curves E and F show the combined result of implanting through an insulating layer and limiting the annealing temperature. They are characteristic of the curves used to determine the sheet resistivity for implanted resistors and interconnections associated with aluminium gate MOST's. If precise resistivities are required from implantations, through insulating layers and at low temperatures, it is essential that the sheet resistivity/dose curve be obtained experimentally for the specific conditions.

It was observed at an early stage that the p$^+$-n junctions formed between boron implanted layers and the underlying substrate were very similar to conventional diffused junctions. Blamires et al. (1967) reported on abrupt $p^+$-n junctions made by implanting boron and annealing at 500 and 700 °C. The forward characteristics were exponential and there was evidence of recombination–generation centres in the space charge regions, while the breakdown characteristics were sharp. Kerr and Large (1967) found that there was no significant difference between boron diffused diodes and $p^+$-n, mesa diodes formed from 50 to 100 keV boron implanted layers. The minority carrier lifetime increased with annealing temperature and it was about 1 microsecond after a 700 °C anneal. A high voltage mesa diode was fabricated from a 100 keV boron implant in 60 ohm·cm silicon. The breakdown voltage was over 1500 volts. MacDougall et al. (1969), in a study of im-

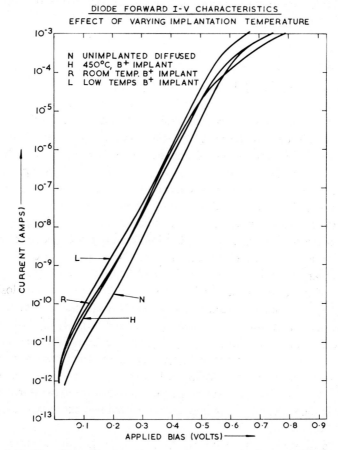

Fig. 5.54. The effect of implantation temperature on the forward $I/V$ characteristics of 40 keV ($5 \times 10^{15}/\mathrm{cm}^2$) boron implanted diodes. A diffused diode of similar geometry and in the same material is included for comparison (Stephen and Grimshaw, 1971).

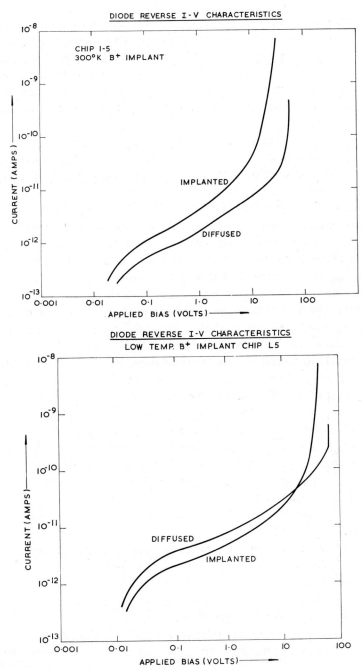

Fig. 5.55. The reverse current-voltage characteristics of boron implanted and diffused diodes. The boron diodes were implanted with 40 keV ($5 \times 10^{15}$/cm$^2$) boron ions at 77 °K and 300 °K (Stephen and Grimshaw, 1971).

planted boron resistors, which were annealed at 950 °C, found that the leakage current of the isolating junction between the resistor and the substrate was low and depended primarily on the surface treatment. This cleanliness aspect has been emphasised by Glotin et al. (1971) who found that the lowest leakage currents were obtained when the wafers were carefully cleaned before and after implantation and annealed in the high vacuum chamber.

Stephen and Grimshaw (1971) have studied boron diffused and implanted $p^+$-n diodes on the same chip to ensure that they were subjected to the same processing conditions. Figure 5.54 shows the forward characteristics of a diffused diode and three diodes implanted at 77 °K, 300 °K and 450 °C with 40 keV boron ions ($5 \times 10^{15}$ ions/cm²) and annealed at 550 °C. The curves were very similar and exponential with the two slopes associated with the R–G and diffusion components of the forward current. The currents of the implanted diodes was larger than the diffused diode which indicated that, possibly, the number of R–G centres was larger in the space charge region. The reverse characteristic (fig. 5.55) showed a similar trend with a softer approach to the breakdown voltage $V_{BD}$ than for the diffused diode. It was concluded that the boron implanted diodes had satisfactory properties which were highly reproducible although the annealing temperature was restricted to 550 °C. The work of Bower et al. (1968) and later Dill et al. (1971a), MacRae (1971), Shannon and Ford (1970) and most recently Pickar et al. (1971) have confirmed that the leakage currents associated with boron implantations were low enough for implantation to be used for MOS devices and silicon diode array camera targets. Shannon and Ford (1970) found that the leakage current was essentially constant for boron ion doses from $2 \times 10^{14}$ to $6 \times 10^{15}$ ions/cm² (annealed at 500 °C) compared with a four orders of magnitude increase for phosphorus implants under the same conditions.

The electrical properties of boron implantations may be summarized as follows:

(i)   the annealing characteristics of the layers show a pronounced recovery at 500 °C, followed by some reverse annealing, leading to 100 % electrical activity at 900 °C,

(ii)   a more rapid recovery is obtained by implanting the boron at 77 °K or with neon at room temperatures,

(iii)   the experimental ranges agree reasonably well with the LSS theory and the deep supertail associated with the other group III dopants is not present,

(iv)  ion channelling is less than that observed for phosphorus and

(v)  the $p^+$-n junction characteristics are very similar to comparable diffused junctions for annealing temperatures over 500 °C.

The general electrical behaviour of implanted boron layers make them suitable for incorporation in silicon devices. The whole subject is discussed in detail in section 5.5.

*5.4.4.3. Aluminium in Silicon.*    Aluminium implantations in silicon have been studied by several groups. Vasil'ev et al. (1968) implanted 100 keV aluminium ions into 1 ohm·cm n-type silicon at room temperature and found that the sheet resistivity of the p layers decreased on annealing up to 700 °C and was then constant at about 1 k ohm/square to 1000 °C. The electrical activity decreased when the dose was increased from $6 \times 10^{13}$ to $6 \times 10^{14}$ ions/cm$^2$ and they suggested that complexes with oxygen atoms formed and reduced the aluminium to an electrically inactive state. Clark and Manchester (1968) have shown that a 60 keV, $1 \times 10^{14}$ Al ions/cm$^2$ implant required annealing to 950 °C for 10 minutes to obtain electrical activity of half of the implanted atoms. The carrier mobility was that of bulk silicon with an equivalent doping level and they concluded that the layers had bulk properties. Itoh et al. (1968) implanted 10 keV aluminium ions into 5 ohm cm n-type silicon at room temperature. For a dose of over $10^{14}$ ions/cm$^2$ the surface was amorphous and the major part recovered on annealing at 800 °C (20 min). A 10 keV, $1 \times 10^{15}$ ions/cm$^2$, implant was profiled and the depths of the layer found to be 0.58 microns. If the first 0.08 microns of silicon, which contained the majority of the damage, was removed before annealing then the depth was reduced to 0.26 microns. The authors concluded that channelling could not be clearly established and that the depth of the layers could be the result of diffusion enhanced by crystal defects.

Baron et al. (1969) (also Mayer and Marsh, 1969) have reported comprehensive results for room temperature and hot implantations of aluminium into silicon. 30 keV aluminium ions were implanted with doses from $10^{13}$ to $10^{15}$ ions/cm$^2$ at 23 °C and 500 °C into 100 ohm·cm $\langle 111 \rangle$ n-type silicon. The value of $N_s$ (carriers/cm$^2$) and mobility were measured at different annealing temperatures. They identified three recovery stages during the annealing of a 23 °C implant ($10^{13}$ ions/cm$^2$);

(i)   300–400 °C, disorder annealed in the tail and $\mu$ rose and then decreased as $N_s$ increased

(ii)  400–650 °C, $\mu$ nearly constant as $N_s$ increased further due possibly to interstitial aluminium becoming substitutional and

(iii) 650–900 °C $\mu$ increased as aluminium diffused but $N_s$ did not approach the implanted dose.

The 500 °C, implants showed a steady increase in $N_s$ and a decrease in $\mu$ except for the lightest dose where it was nearly constant. Depth distribution of free carrier density showed that for the 23 °C $10^{14}$ ions/cm$^2$ implant the junction was 0.9 microns deep after a 700 °C anneal. Their results are summarized in table 5.16.

TABLE 5.16
Variation of electrical activity and mobility for aluminium implantation
(from Baron et al., 1969)

| Ion energy 30 keV | | | | |
|---|---|---|---|---|
| Dose ions/cm$^2$ | Implant temp. °C | Anneal temp. °C | % electrical activity | Mobility cm$^2$/V·sec |
| $10^{13}$ | 23 | 500 | 16 | 250 |
| | 23 | 800 | 55 | 280 |
| | 500 | – | 30 | 250 |
| | 500 | 800 | 55 | 260 |
| $10^{14}$ | 23 | 500 | 2.9 | 180 |
| | 23 | 900 | 30 | 180 |
| | 500 | – | 0.3 | 250 |
| | 500 | 800 | 40 | 180 |
| $10^{15}$ | 23 | 500 | 1.4 | 100 |
| | 23 | 900 | 6 | 130 |
| | 500 | – | 0.06 | 290 |
| | 500 | 800 | 4 | 160 |

The peak carrier concentration found for a $10^{15}$ ions/cm$^2$ implant annealed at 900 °C was $2 \times 10^{18}$ acceptors/cm$^3$ which is about an order of magnitude lower than the reported solid solubility (Trumbore, 1960).

Bader and Kalbitzer (1970a) implanted 20 keV aluminium ions ($10^{14}$/cm$^2$) into silicon at room temperature and found the electrical activity to be smaller than the dose by approximately a factor 10 after a 420 °C anneal. A profile of this implant confirmed the reduced electrical activity and showed a 'supertail' extending up to 0.3 microns and similar in shape to a phosphorus implant of the same energy and dose.

The general conclusions on aluminium are that the resultant electrical activity is low especially for high dose implants and that deep supertails are produced which extend up to ten times the depth predicted by the LSS theory (Lindhard et al., 1963). Baron et al. (1969) consider that the strong effect of the enhanced diffusion masks the loss of electrical activity due to solubility effects. The use of implanted aluminium for working solid state devices has not been reported but Stephen and Grimshaw (1971) have made experimental aluminium implanted diodes and resistors, which were annealed at 500 °C. The sheet resistivity was high and the $I/V$ characteristics of the diodes were very poor (fig. 5.39). Baron et al. (1969) have suggested a possible replacement of silicon by aluminium at a Si–SiO$_2$ interface causing it to act as a sink for aluminium atoms. This phenomenon would have a very serious effect on the properties of an aluminium implanted layer and the passivating oxide protecting it.

The experimental work reported at the present time (1971) indicates that aluminium is an unsuitable dopant for implanting into silicon to make devices. Aluminium is not used to dope silicon by thermal diffusion and its main application in the silicon planar technology is in the formation of ohmic contacts between an aluminium layer and degenerate p or n silicon. During the sintering operation a eutectic forms and on recrystallizing the aluminium dopes the silicon.

*5.4.4.4. Gallium in Silicon.* Gallium implantations in silicon have been studied by Marsh et al. (1967), and later results from the same experiments reported by Baron et al. (1969) and Mayer and Marsh (1969). They implanted 30 keV gallium ions into n-type 100 ohm·cm silicon at 23 and 500 °C and the results from these papers are summarized in table 5.17. For both implantation temperatures, the electrical activity of the gallium impurity increased rapidly with annealing between 400 and 600 °C and then slowly, levelling off towards 900 °C. Marsh et al. (1967) implanted layers with ion

TABLE 5.17

Variation of electrical activity and mobility for gallium implantations
(from Marsh et al. 1967, and Baron et al., 1969)

| Ion energy 30 keV | | | | |
|---|---|---|---|---|
| Dose ions/cm$^2$ | Implant temp. °C | Anneal temp. °C | % electrical activity | Mobility cm$^2$/V·sec |
| $6 \times 10^{12}$ | 500 | – | 13 | 220 |
|  | 500 | 800 | 27 | 220 |
| $6 \times 10^{13}$ | 500 | – | 1.6 | 190 |
|  | 500 | 800 | 30 | 190 |
| $1 \times 10^{14}$ | 23 | 500 | 1.6 | 240 |
|  | 23 | 800 | 14 | 190 |
| $5 \times 10^{15}$ | 500 | – | 0.08 | 160 |
|  | 500 | 800 | 3.2 | 70 |

doses from $4 \times 10^{12}$ to $2 \times 10^{16}$ ions/cm$^2$ and after annealing at 800 °C they found that the number of electrically active carriers saturated at $10^{14}$ carriers/cm$^2$. They concluded that this saturation effect was due to the strong influence of the solid solubility of gallium in silicon ($1.4 \times 10^{19}$ atoms/cm$^3$, Trumbore, 1960). More recently Johansson and Mayer (1970) have reported obtaining an average concentration of carriers nearly equal to the solubility value for a 40 keV, $5 \times 10^{13}$ ions/cm$^2$, room temperature implant. The electrical activity was about 50 %. Their results for a 350 °C implantation temperature were similar to those of Marsh et al. (1967).

Bulthuis (1968) has measured, by sectioning and staining, the junction depth of gallium layers in ⟨110⟩ n-type silicon implanted at different temperatures as high as 500 °C. He used 60 keV ions and a dose of $6 \times 10^{15}$ ions/cm$^2$. For a 7 ohm·cm substrate, the junction depth for the room temperature implant was 1.4 microns, at 200 °C 2.1 microns and between 200 and 300 °C it increased to 3 microns and then remained unchanged to 500 °C. Similar behaviour was observed in 0.3 ohm·cm material but the junctions depths were shallower which indicated that the concentration of gallium atoms was of the same order as the background n-type doping, and several orders of magnitude less than in the peak of the profile. Bulthuis did not give the details of the implantation conditions and it is possible that the temperatures during implantation were higher due to the wafer being heated by the ion beam particularly as a large dose was used. As the expected channelled depth for 60 keV gallium ions was about one micron, another explanation was necessary to understand the results. Bulthuis considered that substitutional diffusion was unlikely, as the temperatures were too low, and that interstitial diffusion was probably occurring despite the large number of vacancies and interstitials created by the implantation.

Stephen and Grimshaw (1971) made resistors and diodes by implanting at room temperature 200 keV gallium ions through a 0.13 micron oxide-nitride layer into ⟨100⟩ n-type silicon. The dose was $5 \times 10^{15}$ ions/cm$^2$ and the annealing temperature 500 °C. A low sheet resistivity of 470 ohms/square was obtained but the diode characteristics were poor which indicated the presence of residual lattice damage in the space charge region of the p$^+$-n junction (figs. 5.39, 5.40). The sheet resistivity is over an order of magnitude lower than 10 kΩ/square reported by Marsh et al. (1967) for a 30 keV $5 \times 10^{15}$ ions/cm$^2$ hot implant or the 7 kΩ/square by Oosthoek et al. (1970) for a 50 keV $1 \times 10^{15}$ ions/cm$^2$ hot implant. The discrepancy may be due to the use of a high energy and to the majority of the implant damage being confined to the oxide-nitride insulating layer.

Gallium implanted resistors have been studied by Oosthoek et al. (1970) and they are discussed in section 5.5.2.2. Howes and Knill (1970) have made junctions on lithium drifted silicon detectors by implanting gallium between 10 and 40 keV with doses of $10^{13}$ to $5 \times 10^{14}$ ions/cm². The ion beam was aligned 8° off the $\langle 111 \rangle$ axis. The junction depths were measured by the differing absorption of collimated alpha particles (from 5.49 MeV $^{241}$ Am) at various angles to the junction plane. It was found that the junction depths varied between 0.2 and 0.6 microns, which was shallower than those observed by Bulthuis (1968). The implanted detectors were given a 'clean-up' drift before the junction depth measurement was made and this was possibly responsible for the shallower junctions. The authors stated the gallium implanted detectors were more readily compensated than equivalent boron implanted detectors.

Gallium is a more promising dopant than aluminium. The electrical activity of low dose implantations is usable but care must be taken to minimise the effects of enhanced interstitial diffusion if shallow junctions are required. The p⁺-n junction properties are poor at low annealing temperatures (500 °C) but these may well improve at higher temperatures.

*5.4.4.5. Indium in Silicon.* The primary interest in indium implantations in silicon, as reflected by the literature, has been in the location of the indium atoms and their redistribution on annealing. Roughan et al. (1968) implanted 50 and 60 keV In ions into silicon and annealed at 950 °C. They found that only a small fraction of the indium was electrically active and the profile in depth showed evidence of a 'channelling' tail. Annealing experiments between 1000 and 1200 °C produced junction depths deeper than calculated from diffusion data (4.48 microns compared with 3.74 microns for a 55 keV $1.6 \times 10^{13}$ In⁺/cm² implant in 4Ω cm silicon annealed at 1200 °C for 3 h). Some similar work by Bulthuis (1968) in which he used neutron activation analysis combined with layer stripping showed that the indium penetrated deep into the lattice. 56 keV indium ions which were implanted at 450 °C penetrated to three microns. He confirmed this depth by sectioning and staining (using the method of Gibbons (1967), see section 5.4.2.6.3) a similar 56 keV implant annealed at 550 °C. A room temperature 56 keV implant only penetrated to 1.5 microns. Bulthuis (1968) suggested that interstitial diffusion of the indium atoms was the most likely explanation despite the large number of silicon interstitials produced during the implantation. The reduced penetration of the room temperature implant was possibly due to the formation of stable complexes such as clusters which required temperatures above 550 °C to be dispersed by annealing. It is likely that the deeper junctions observed by Roughan et al. (1968) were due to the initial penetration of the indium ions which would be added to the diffusion depth.

The above observations are supported by the helium channelling work of Eriksson et al. (1969). They found roughly equal numbers of substitutional and interstitial atoms for a hot implant (350–525 °C) with the interstitial components increasing on annealing to 800 °C. A room temperature implant annealed between 600 and 750 °C showed that about 75 % of the indium atoms were on interstitial sites. They concluded that there was a movement to interstitial and precipitation sites and this mechanism suggested the possibility of enhanced diffusion effects.

The electrical properties of indium implants have been studied by Johansson et al. (1970) who implanted 40 keV $3 \times 10^{14}$ In$^+$/cm$^2$ at room temperature and measured the number of carriers/cm$^2$, $N_S$, and the mobility at a number of annealing temperatures up to 900 °C. $N_S$ increased from 350 to 400 °C, decreased slightly at 500 °C and then increased sharply at 600 °C then gradually to 900 °C. The mobility increased from 150 cm$^2$/volt·sec to a nearly constant 250 cm$^2$/volt·sec with a small dip at 500 °C. The percentage of electrically active indium after a 900 °C anneal was 0.7 % and the corresponding sheet resistance was about 10 k ohm/square, which was over an order of magnitude higher than for a corresponding boron implantation. Pashley (1971) has reported the energy level of implanted indium to be 160 meV.

Stephen and Grimshaw (1971) made p$^+$-n diodes by implanting 400 keV $5 \times 10^{15}$ indium ions/cm$^2$ through a 0.13 microns oxide-nitride layer and annealed them at 500 °C. The junctions had rectifying characteristics but the leakage currents were five to six orders of magnitude greater than comparable boron implanted diodes (see section 5.4.4.2). The forward characteristic was no longer exponential and a meaningful value of $I_S$ and $m$ (section 5.4.3.2) could not be determined. These results from the p$^+$-n junctions suggested that a very considerable number of recombination-generation centres were present in the space-charge region and were responsible for the large currents. It is possible that the R–G centres are interstitial indium atoms, or lattice damage, which can be severe for a large ion dose, or a combination of the two types of defects.

The behaviour of implanted indium in silicon is very complex involving enhanced diffusion effects and a movement to interstitial sites with the result that the usual high temperature anneal does not guarantee a high electrical activity. The residual defects in the implanted region disturb the $p^+$-n junction characteristics with the result that they are unsuitable for devices. Experience so far with indium would suggest that it is not a suitable acceptor dopant unless the damage can be kept low by using low doses combined with very high annealing temperatures.

*5.4.4.6. Nitrogen in Silicon.* Nitrogen is the lightest element in Group V of the periodic table and in common with the other elements in the group it should behave as a donor impurity when occupying substitutional sites in the silicon lattice. Nitrogen normally exists in the molecular form, $N_2$, which has a binding energy of 9.8 eV. It is necessary to dissociate the molecular nitrogen into atomic nitrogen before diffusion so that the single atoms can diffuse substitutionally. The temperatures (900 to 1300 °C) used for thermal diffusion of normal impurities into silicon are too low to produce the atomic form of nitrogen and it therefore enters the lattice as a molecule and moves interstitially. No electrical activity is produced and in fact, nitrogen is frequently used as an 'inert' atmosphere when growing, diffusing or annealing silicon crystals.

Ion implantation is a method of introducing nitrogen atoms into silicon as ionized nitrogen atoms are readily produced in the source of an ion-implanting machine. Pavlov et al., (1966a) implanted 57 keV nitrogen ions into 1 ohm·cm p-type silicon at room temperature with doses ranging from $3 \times 10^{14}$ to $3 \times 10^{16}$ ions/cm$^2$ and annealed at temperatures up to 1100 °C. They found no type inversion at 500 °C but at 700 °C and above n layers were detected. For an 800 °C annealing temperature, the average concentration of carriers was from 1 to $2.5 \times 10^{12}$ carriers/cm$^2$ for a 100 : 1 change in the ion dose. They could not produce the same electrical effect by implanting neon ions under the same conditions suggesting that damage alone was not the only cause of observed effects. They concluded that the nitrogen was acting as a donor and that the solid solubility of the nitrogen was the factor limiting the electrical activity. The n-p junction formed by a nitrogen implant annealed at 800 °C exhibited rectifying properties.

Kleinfelder (1967) with Gibbons (1967) have obtained the profiles of 30 and 50 keV nitrogen implantations into silicon at 625 °C, using the bevel and stain method described in section 5.4.2.6.3. Donor activity was observed and the profiles were fitted to a Gaussian curve over 5 decades and matched the profiles computed from the LSS theory (Lindhard et al., 1963). Electrical activity of over $10^{20}$ donors/cm$^3$ was found and it was concluded that each nitrogen atom was contributing one electron. The ionization energy of the donor level was 0.142 eV (compared with 0.044 eV for phosphorus). Different results were obtained by Roughan et al. (1968) for 40 keV nitrogen implants into 100 ohm·cm p-type silicon at room temperature and annealed at 1200 °C. From Hall effect measurements they found that 1 % of the implanted nitrogen atoms were electrically active after a 900 °C anneal, 10 % after 1100 °C and nearly 100 % after a 10 minute 1200 °C anneal. The ionization energy decreased from a large value at 900 °C to 0.033 to 0.044 eV at 1200 °C. They considered that this change was possibly associated with a movement of the nitrogen from interstitial to substitutional sites taking place at the higher temperatures, and that the 0.142 eV energy found by Kleinfelder (1967) was characteristic of the interstitial nitrogen. Lastly, Roughan et al. (1968) measured the thermal diffusion constant of nitrogen atoms in silicon at 1100 °C as $7.6 \times 10^{-13}$ cm$^2$/sec which is very similar to phosphorus $(8 \times 10^{-13}$ cm$^2$/sec$)$ and greater than boron $(5 \times 10^{-13}$ cm$^2$/sec$)$ (Runyan, 1965).

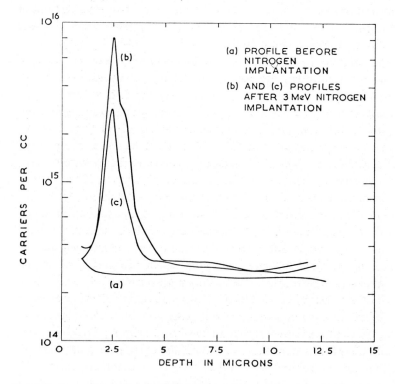

Fig. 5.56. The profiles of a 3 MeV nitrogen implantation into n-type silicon and annealed at 900 °C (Stephen et al., 1971).

As nitrogen can be very readily introduced into the source units of high energy van der Graaff type accelerators it has been possible to study nitrogen implants at energies over 1 MeV. Buchanan et al. (1967) obtained the profile of a 2 MeV nitrogen implant $(4 \times 10^{14}$ ions/cm$^2)$, annealed at 900 °C, by the bevel and stain technique (section 5.4.2.6.3). An n-type layer about two microns deep was found with a tail extending towards the surface. The projected range for 2 MeV nitrogen ions in silicon is 2.7 microns. The maximum carrier concentration was $10^{19}$ donors/cm$^3$ and a simple integration under their curve gave an electrical activity of about 50 %. Stephen et al. (1971) have studied a 3 MeV nitrogen implant annealed at 900 °C by the C–V technique (section 5.4.2.6.4) and fig. 5.56 shows two typical profiles and the background doping level. The peaks are a 2.5 microns $(R_p = 3.3$ microns) and only 0.48 to 1.3 % of the nitrogen atoms were electrically active; a result which was supported by sheet resistivity measurements made on three higher dose implants. Figure 5.57 is a scanning electron micrograph of a buried 3 MeV nitrogen layer $(1 \times 10^{15}$ ions/cm$^2)$ which is revealed

Fig. 5.57. The edge of mesa etched in p silicon implanted with 3 MeV nitrogen ions $(1 \times 10^{15}$/cm$^2)$ and annealed at 800 °C. The implanted region can be seen as a dark band extending from 1.9 to 3.3 microns.

as a dark band on the edge of a mesa etched in a p-type wafer ($5 \times 10^{15}$ impurities/cm$^3$) after a 900 °C anneal. The band extended from 1.9 to 3.3 microns and was made visible by the combined effects of a different etching rates for the damage and voltage contrast affecting the secondary electron emission in the microscope. The contrast was enhanced by slightly heating the mesa during observation. A beam induced conductivity measurement showed that the electrical junctions were at 2.2 and 6 microns (3.8 V bias) and were further apart and less well defined with no bias, which suggested graded rather than abrupt junctions. The $I/V$ characteristics showed two leaky back to back junctions with the upper one having a lower $V_{BD}$.

The ratios of the measured range to the theoretical range for the last two experimental results are 0.74 (2 MeV) and 0.76 (3 MeV). The ratios are close to that given for boron, 0.7, by Crowder and Title (1970) and suggests that the peaks were in the highly damaged regions of the implants. There is a large discrepancy in the electrical activity measured by bevel and stain methods (Kleinfelder, 1967; Buchanan et al. 1967) and direct electrical methods (Pavlov et al., 1966a; Roughan et al., 1968 (4 point probe), and Stephen et al., 1971 (C–V)). It is possible that the bevel and stain method may under certain circumstances, such as low electrical activity, preferentially stain damage regions rather than the n-type regions (fig. 5.57). The position of the stain will indicate the position of the damage, irrespective of the substrate doping level, and not the location of the p-n junction. The high values of doping level inferred from the bevel and stain profile cannot always be relied on to be accurate.

Schwuttke and Brack (1969) have studied the annealing properties of amorphous layers produced by 1.5 to 2 MeV nitrogen implants using doses from $10^{15}$ to $10^{16}$ ions/cm$^2$. By using Scanning Oscillating X-ray Techniques (SOT), they have identified two annealing stages one below 700 °C and the other above 700 °C. The annealing rate was fast in the first stage and was associated with type conversion in the buried layer due to the annealing of isolated amorphous islands. No change in the depth (1.7 microns) or width (0.5 microns) of the buried layer was observed until the annealing temperature was raised above 700 °C. In the second stage the annealing was slow and associated with the recrystallization of the amorphous buried layer which provided vacancies which in turn allowed the n layer to extend towards the surface. Resistivity measurements after a 1200°C anneal showed high electrical activity and a junction depth of 6 microns, which was in general agreement with Roughan et al. (1968). A thin layer 0.5 microns thick at a depth of 2 microns was found in the annealed n region. X-ray analysis showed that it was comprised of a high density $(>10^9/cm^2)$ of submicron size $\alpha$-Si$_3$N$_4$ precipitates and lead to the conclusion that chemical compound formation was possible and that implantation could produce a spatially continuous buried insulating film. The presence of Si$_3$N$_4$ has been confirmed by Dexter (1970) using infra-red techniques on heavily bombarded samples annealed at high temperatures. He also found an ionization energy of 0.0458 eV for a low dose implant annealed at 760 °C, which agreed with the value given by Roughan et al. (1968).

Blamires et al. (1967), Kleinfelder (1967) and Gibbons (1967) have reported making silicon nitride layers on silicon by implanting large doses (over $10^{17}/cm^2$) of low energy nitrogen ions into silicon. More recently Freeman et al. (1970) have published more details of results obtained with nitride (and oxide) layers produced by implanting $2 \times 10^{18}$ N$^+$ ions into polished silicon (section 5.5.8). Infra-red and transmission microscopy measurements confirmed that nitride was produced and optical and electrical measurements showed that the thickness was about twice the projected ion range. For example, a 25 keV implant ($R_p = 0.068$ microns) was 0.12 microns thick. A good stable insulator was formed but the silicon below the nitride was highly damaged even after annealing to 900 °C. It was concluded that the technique could not be used for making silicon nitride layers similar to those made by thermal methods and suitable for device applications. The damage in the silicon was too high in the region where a low surface state density was required. (The same general remarks apply to SiO$_2$ made by O$^+$ implantation (section 5.5.8).)

In summary, nitrogen behaves as a donor when implanted in silicon and annealed above 700 °C. A high activity can be obtained by annealing to 1200 °C which negates the normal objectives of low processing temperatures. This silicon nitride layers can be formed on the surface of, or buried in, silicon by nitrogen implantation. No silicon devices incorporating implanted nitrogen layers have been reported in the literature.

*5.4.4.7. Phosphorus in Silicon.* When considering semiconductor device fabrication phosphorus is the most important of the group V dopants in silicon. Unlike nitrogen, it has a high solid solubility and despite the mass of the phosphorus ion its range in silicon is large enough for it to be used for practical device fabrication with ion energies below 200 keV. The early work on the electrical characteristics of implanted phosphorus by Manchester (1965, 1966 and 1967), Kellett et al. (1966), Gusev et al. (1966),

Kleinfelder (1967), Gibbons (1967) and Blamires et al. (1967) established that it acted as a donor, the electrical activity was high, the p-n
junctions were good and the ions penetrated further into the silicon lattice
than predicted by the LSS theory (Lindhard et al., 1963) as profiles with
long tails were found.

The annealing behaviour of phosphorus implantations can be divided
into two regions; (i) from room temperature to 500 °C and (ii) above 500 °C.
In the first region, Glotin (1968) using electrical conductivity measurements
has identified three main annealing stages for a 20 keV phosphorus implant
$(1 \times 10^{15}$ ions/cm$^2$) channelled into $\langle 110 \rangle$ silicon. The first stage is at 170 °C
corresponding to the annealing of $E$ centres followed by the second stage
at 330 °C, which was believed to be $A$ centres annealing in the deep part of
the profile, and finally the third stage at 470 °C which was associated with
the rearrangement of the damage clusters in the shallow part of the profile.
During this last stage n-type activity appeared in the highly damaged layers
accompanied by an increase in the conductivity. Burr and Whitehouse (1970)
observed the same effect and found it to be dose dependent as the n-type
activity appeared at lower temperatures for the higher doses. Gusev and
Titov (1969) also found a dose dependence in the lower annealing characteristics as doses below $10^{15}$ ions/cm$^2$ (30 keV P$^+$ ions) only annealed above
300 °C while the high doses annealed between 200 and 300 °C. These observations are supported by measurements of surface carrier concentration,
$N_s$, and mobility $\mu$ made by Stumpfi and Kalbitzer (1970), Shannon et al.
(1970) and Andersson and Swenson (1970) on a range of phosphorus implanted layers. Both parameters increased with annealing temperature over
the range 350 to 600 °C. Andersson and Swenson concluded that the defect
centres produced by the implantation were behaving as acceptors and were
compensating the n-type activity. As the defect centres annealed out, the
compensation effect decreased and the mobility increased as the scattering
of the carriers was reduced. The electrical conductivity of the implanted
layers annealed below 500 °C is normally too low for most device applications and the higher annealing temperatures in the second region must be
used.

The annealing behaviour in this second region has been extensively
investigated by the measurement of sheet resistivity with annealing temperature. Figure 5.58 shows the annealing behaviour of a number of phosphorus
implantations in silicon from 300 to 900 °C as reported by Shannon et al.
(1970) (Curves A and B), Davies (1970) (Curve C), Webber et al. (1969)
(Curves D and F) and Vasil'ev et al. (1968) (Curve E). Figure 5.59 (Shannon

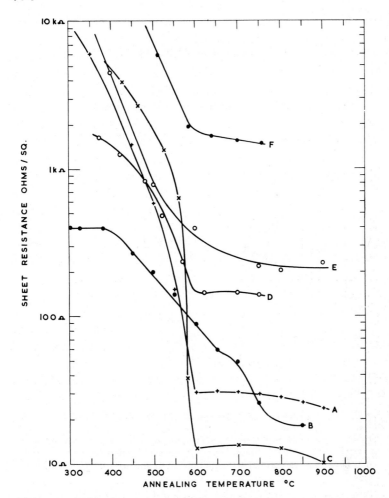

Fig. 5.58. The variation of sheet resistivity with annealing temperature for a number of phosphorus implantations. Curve A, 100 keV $6 \times 10^{15}/cm^2$ R.T. (Shannon et al., 1970); Curve B, 100 keV $6 \times 10^{15}/cm^2$ 450 °C (Shannon et al., 1970b); Curve C, 300 keV $2 \times 10^{16}/cm^2$ −195 °C (Davies, 1970); Curve D, 20+40 keV $6 \times 10^{14}/cm^2$ R.T. (calculated from Webber et al., 1969); Curve E, 180 keV $6 \times 10^{14}/cm^2$ R.T. (Vasil'ev et al., 1968); Curve F, 20+40 keV $1.1 \times 10^{13}/cm^2$ R.T. (calculated from Webber et al., 1969).

et al., 1970) extends the results to a very high dose of $5 \times 10^{16}$ phosphorus ions/cm² implanted at room temperature. Firstly the room temperature implantations, curves A, D, E and F in fig. 5.58 and fig. 5.59, all show the same general behaviour. There is a rapid decrease in sheet resistivity (increasing conductivity) between 300 and 600 °C corresponding to stages two

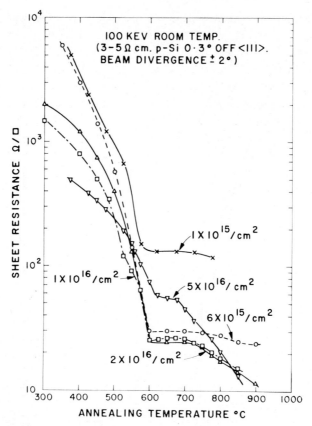

Fig. 5.59. The sheet resistance of phosphorus layers implanted at room temperature and annealed between 300 and 900 °C. Measurements were made following each successive 30 minute isochronal anneal (Shannon et al., 1970b).

and three of region i) and the start of region ii). At 600 °C there is a sharp knee in each curve and above this temperature only slight decreases in sheet resistance occur except for the highly doped layers in fig. 5.59. Shannon et al. (1970b) stated that 80 % of the phosphorus atoms in a $1 \times 10^{15}$ ions/cm$^2$ implant were active after a 600 °C anneal and the remaining 20 % after annealing at 850 °C. The curves for the very high dose implants (fig. 5.59 show a minimum sheet resistivity at 600 °C for the $1 \times 10^{16}$ ions/cm$^2$ implant with an increase in the resistivity for the $5 \times 10^{16}$ ions/cm$^2$ implant and not a further decrease as would normally be expected due to the larger dose. This phenomenon was attributed by Shannon et al. (1970) to the phosphorus in the peak of the implant exceeding the solid solubility. How-

ever on annealing at higher temperatures the solubility increased and the
sheet resistivity decreased as shown in fig. 5.60 (Shannon et al., 1970b) for
two profiles of a $2 \times 10^{16}$ ions/cm$^2$ implant annealed at 650 and 850 °C. The
lower temperature profile showed a 'flat top' with a solid solubility of $4 \times$
$10^{20}$ carriers/cm$^3$ while the 850 °C profile approximates to the theoretical
profile and has a free electron concentration of $1.5 \times 10^{21}$ electrons/cm$^3$
which is five times the solid solubility of phosphorus in silicon (Trumbore,
1960). Davies has reported similar results for 300 keV phosphorus implanted
at room temperature and $-195$ °C (Curve C, fig. 5.58). The behaviour of
the low temperature implants over the range of doses from $1 \times 10^{13}$ to
$1 \times 10^{16}$ ions/cm$^2$ was similar to that of the room temperature implants
except that the lower doses did not show the sharp levelling of the sheet
resistivity at 600 °C. The advantage found by implanting at low temperature

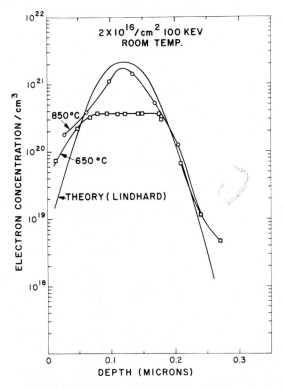

Fig. 5.60. Electrically active profiles of two 100 keV phosphorus implants annealed at 650
and 850 °C. The theoretical distribution based on the theory of Lindhard et al. (1963) is
included for comparison (Shannon et al., 1970b).

for boron was not repeated for phosphorus and Davies (1969a) attempted to lower the annealing temperatures by implanting the phosphorus into pre-damaged silicon. He used 400 keV argon ions $(1 \times 10^{15}$ ions/cm$^2$) followed by 400 keV phosphorus ions $(1 \times 10^{13}$ ions/cm$^2$) but found no reduction in the annealing temperature but instead an overall reduction in the electrical activity. The high concentration of defects produced by the phosphorus ions alone is sufficient to ensure that a large percentage of the ions go into substitutional sites at low annealing temperatures (600 °C). The additional damage produced by the argon implantation was possibly far in excess of the amount required for the annealing of the phosphorus ions and the remaining damage would have acted as carrier scattering and trapping centres. A smaller argon or neon dose may have been much more effective in altering the annealing behaviour.

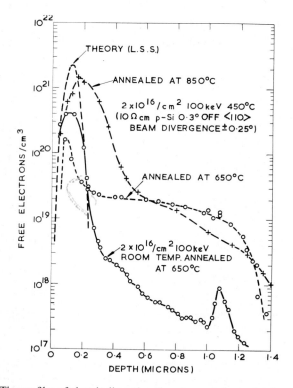

Fig. 5.61. The profiles of electrically active phosphorus in layers implanted at room temperature and 450 °C followed by annealing at 650 and 850 °C. The details of the implantations are given and the theoretical distribution is included for comparison (Shannon et al., 1970b).

When ions are implanted into silicon at elevated temperatures, continuous annealing occurs and the damage associated with the peak of the implant does not build up as it does at room temperature when the silicon can be made amorphous, Shannon et al. (1970b) and Davies (1969a) have implanted large doses of phosphorus ions ($1 \times 10^{16}$ ions/cm$^2$) into silicon held at 450 °C and 420 °C respectively. Their results are very similar, for example curve B of fig. 5.58 (Shannon et al., 1970b). The room temperature annealing curve is replaced by a curve showing a steady decrease of sheet resistivity with annealing temperature from 400 to 900 °C. The sheet resistivity at 600 °C is about three times that of the equivalent room temperature implantation indicating the desirability of using room temperature implantations for device applications. The damage produced during a high temperature implantation consists of a dense network and entanglement of dislocation loops and dipoles different in character to those produced at room temperature. This damage is more stable and higher annealing temperatures are required. During annealing the damage could act as a source

TABLE 5.18

Experimental carrier mobility values for a number of phosphorus implanted layers

| Energy keV | Dose ions/cm$^2$ | Annealing temp. °C | Effective mobility cm$^2$/v. sec. | Reference |
|---|---|---|---|---|
| 7 | $1 \times 10^{12}$ | 900 | 2450 | Stumpfi and Kalbitzer (1970) |
| 7 | $1 \times 10^{13}$ | 900 | 710 | |
| 7 | $3 \times 10^{13}$ | 900 | 410 | |
| 7 | $1 \times 10^{14}$ | 900 | 275 | |
| 7 | $8 \times 10^{14}$ | 900 | 90 | |
| 7 | $8 \times 10^{15}$ | 900 | 50 | |
| 20+40 | $1.1 \times 10^{13}$ | 760 | 520 | Webber et al. (1969) |
| 20+40 | $1.1 \times 10^{14}$ | 770 | 340 | |
| 20+40 | $6 \times 10^{14}$ | 750 | 120 | |
| 60 | $1 \times 10^{13}$ | 750 | 500 | Andersson and Swenson (1970) |
| 60 | $1 \times 10^{14}$ | 750 | 400 | |
| 60 | $1 \times 10^{15}$ | 700 | 110 | |
| 100 | $1 \times 10^{15}$ | 900 | 80 | Shannon et al. (1970b) |
| 100 | $2 \times 10^{16}$ | 900 | 50 | |
| 180 | $6 \times 10^{12}$ | 800 | 700 | Vasil'ev et al. (1968) |
| 180 | $6 \times 10^{13}$ | 800 | 390 | |
| 180 | $6 \times 10^{14}$ | 800 | 150 | |
| 400 | $1 \times 10^{13}$ | 740 | 850 | Gibson et al. (1967) |
| 400 | $1 \times 10^{15}$ | 750 | 140 | |

of defects which could increase the diffusion constant (i.e. enhanced diffusion) and allow the phosphorus to move into the bulk silicon, thus distributing the phosphorus in the peak of the implants over a larger volume. The shift of the peak of the hot implant annealed at 850 °C (fig. 5.61) is an illustration of this effect.

The carrier mobility is essentially constant after annealing above 600 °C as the increased conductivity is due solely to phosphorus ions moving from electrically inactive to substitutional sites. Table 5.18 summarizes some of the published data on carrier mobility in phosphorus layers implanted and annealed under various conditions.

The great majority of published profiles of implanted phosphorus in silicon show strong evidence of channelling by the phosphorus ions deep into the silicon lattice. Channelling is strongly dependent on the crystal

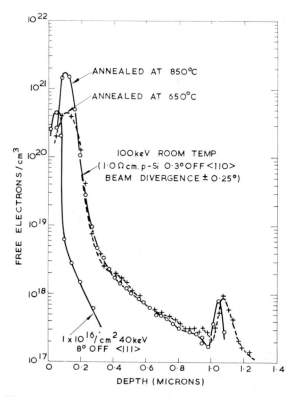

Fig. 5.62. The profiles of electrically active phosphorus implanted at room temperature into ⟨110⟩ and ⟨111⟩ oriented silicon and annealed at 650 and 850 °C. The details of the implantations are given in the figure (Shannon et al., 1970b).

orientation and of the three common orientations used for devices, the $\langle 110 \rangle$ orientation presents the widest channel followed by the $\langle 111 \rangle$ and $\langle 100 \rangle$ orientations. Figure 5.62 shows phosphorus profiles obtained by Shannon et al. (1970b) for 100 keV $P^+$ implantations into $\langle 110 \rangle$ silicon at room temperature. A room temperature 40 keV implant in $\langle 111 \rangle$ silicon oriented 8° off the ion beam direction and annealed at 650 °C is included for comparison with the channelled profiles. As the channels were directed away from the beam the number of channelled ions was considerably reduced and the profile followed the LSS profile down to $5 \times 10^{-3}$ times the peak concentration. The two profiles of 100 keV phosphorus implanted $\pm 0.3°$ along the $\langle 110 \rangle$ direction showed pronounced channelling with a small peak at 1.1 microns which suggested that this was the range of the channelled ions. There was little difference in the shape of the two profiles as a result of annealing at 650 °C and 850 °C but the carrier concentration in the main peak increased by nearly an order of magnitude (fig. 5.62). This result suggested that the channelled profile was fully active after a 650 °C anneal and that little thermal diffusion occurred due to the 850 °C anneal of 30 minutes. The number of phosphorus atoms in the channelled tail was probably limited by an amorphous layer formed just below the silicon surface by the non-channelled ions. A phosphorus ion dose of about $1 \times 10^{15}$ ions/cm$^2$ is sufficient to make the surface of a silicon crystal amorphous and 'seal off' the channels. An upper limit is therefore set to the total channelled ion dose which in this case was $4.7 \times 10^{14}$ ions/cm$^2$ and is in reasonable agreement with the dose required to make the surface amorphous.

If heavy ions are implanted into silicon at an elevated temperature the surface layer continuously anneals and the amorphous layer does not form and ions can enter the channels for the whole time of the implantation. Shannon et al. (1970b) (fig. 5.61) clearly showed this effect in the case of 100 keV phosphorus ions implanted into silicon at 450 °C with subsequent anneals at 650 and 850 °C. The number of channelled ions increased from $4.7 \times 10^{14}$ ions/cm$^2$ for the room temperature implant to about $2 \times 10^{15}$ ions/cm$^2$ for the 450 °C implant. The carrier concentration in the channelled tail was nearly constant over a micron in depth and the small peak seen earlier was not present. This observation is in agreement with a steep exponential decrease reported by Dearnaley et al. (1968) for a hot implantation and attributed to de-channelling caused by increased lattice vibration. The main peak, after annealing at 850 °C, had diffused into the bulk with an accompanying increase in the electrical activity.

The results of electrical profile measurements published by several

TABLE 5.19

Experimental ranges and electrical activity observed for a number of phosphorus implantations

| Energy keV | Dose ions/cm$^2$ | Cryst. orient. | Temp °C imp. | Temp °C anneal | Projected range theory microns | Projected range exp. microns | Channel tail depth microns | Channel tail activity donors/cm$^3$ | Reference |
|---|---|---|---|---|---|---|---|---|---|
| 20 | $2 \times 10^{13}$ | $\langle 111 \rangle$ | R.T. | 650 | 0.026 | 0.05 | 0.68 | $2 \times 10^{16}$ to $2 \times 10^{17}$ | Manchester (1966) |
| 20 | $1 \times 10^{14}$ | 15° off $\langle 111 \rangle$ | R.T. | 420 | 0.026 | 0.036 | 0.42 | | Bader and Kalbitzer (1970a) |
| 20 | $1 \times 10^{14}$ | $\langle 110 \rangle$ | R.T. | 420 | 0.026 | 0.062 | 0.67 | | |
| 20 | $1 \times 10^{14}$ | $\langle 111 \rangle$ | R.T. | 900 | 0.026 | 0.036 | 0.29 | $\sim 1 \times 10^{15}$ | Kleinfelder (1967) |
| 30 | $3 \times 10^{15}$ | $\langle 111 \rangle$ | 625 | – | 0.038 | – | 0.40 | | |
| 30 | $3.6 \times 10^{16}$ | $\langle 111 \rangle$ | R.T. | R.T. | 0.038 | $\sim 0.09$ | 0.38 | $8 \times 10^{16}$ | Golovner et al. (1968) |
| 30 | $3.6 \times 10^{16}$ | $\langle 111 \rangle$ | R.T. | 850 | 0.038 | $\sim 0.3$ | 1.05 | $5 \times 10^{18}$ | |
| 30 | $3.6 \times 10^{15}$ | $\langle 111 \rangle$ | R.T. | 600 | 0.038 | 0.05 | 0.25 | $1.5 \times 10^{17}$ | |
| 30 | $3.6 \times 10^{15}$ | $\langle 111 \rangle$ | R.T. | 800 | 0.038 | | 0.28 | $1.5 \times 10^{18}$ | Titov (1970) |

| Energy (keV) | Dose | Orientation | Implant temp | Anneal temp | Depth | | | Concentration | Reference |
|---|---|---|---|---|---|---|---|---|---|
| 100 | $2 \times 10^{16}$ | $\langle 110 \rangle$ | R.T. | 650 | 0.123 |  | 1.10 | $2$ to $9 \times 10^{17}$ | Shannon et al. (1970b) |
| 100 | $2 \times 10^{16}$ | $\langle 110 \rangle$ | R.T. | 850 | 0.123 | 0.125 | 1.25 |  |  |
| 100 | $2 \times 10^{16}$ | $\langle 110 \rangle$ | 450 | 650 | 0.123 | 0.08 | 1.36 |  |  |
| 100 | $2 \times 10^{16}$ | $\langle 110 \rangle$ | 450 | 850 | 0.123 | 0.17 | 1.40 | $1.2 \times 10^{19}$ to $1.5 \times 10^{18}$ | Goode et al. (1970) (b), (c) |
| 80 | $5 \times 10^{13}$ | $\langle 110 \rangle$ | R.T. | 700 | 0.098 | 0.14 | 1.27 | $1.5 \times 10^{19}$ |  |
| 120 | $5 \times 10^{13}$ | $\langle 110 \rangle$ | R.T. | 700 | 0.149 | 0.23 | 1.46 | $2 \times 10^{17}$ |  |
| 200 | $5 \times 10^{13}$ | $\langle 110 \rangle$ | R.T. | 700 | 0.254 | 0.39 | 1.95 | $3 \times 10^{17}$ |  |
| 200 | $1 \times 10^{14}$ | $\langle 110 \rangle$ | R.T. | 700 | 0.254 | 0.48 | 1.96 | $2\text{--}3 \times 10^{17}$ |  |
| 400 | $1 \times 10^{14}$ | $\langle 110 \rangle$ | R.T. | 700 | 0.516 | 0.72 | 2.70 | $3.4 \times 10^{17}$ |  |
| 280 | $1 \times 10^{15}$ | $\langle 111 \rangle$ | R.T. | 800 | 0.359 | 0.39 | 0.84 | $3 \times 10^{17}$ | Fairfield and Crowder (1969) |
| 100 | $1 \times 10^{15}$ | $7°$ off $\langle 111 \rangle$ | 78 °K | 650 | 0.123 | $<0.23$ | 0.90 | Exp tails from $10^{17}$ to $10^{13}$ | Davies (1970) |
| 200 | $1 \times 10^{15}$ | $7°$ off $\langle 111 \rangle$ | 78 °K | 650 | 0.254 | $\sim 0.34$ | 1.25 |  |  |
| 400 | $1 \times 10^{15}$ | $7°$ off $\langle 111 \rangle$ | 78 °K | 650 | 0.516 | $\sim 0.5$ | 1.6 |  |  |
| 700 | $1 \times 10^{15}$ | $7°$ off $\langle 111 \rangle$ | 78 °K | 650 | 0.866 | $\sim 0.75$ | 2.05 |  |  |
| 1000 | $1 \times 10^{15}$ | $7°$ off $\langle 111 \rangle$ | 78 °K | 650 | 1.18 | $\sim 1.0$ | 2.40 |  |  |
| 1500 | $1 \times 10^{15}$ | $7°$ off $\langle 111 \rangle$ | 78 °K | 650 | 1.6 | 1.35 | 2.8 |  |  |
| 2000 | $1 \times 10^{15}$ | $7°$ off $\langle 111 \rangle$ | 78 °K | 650 | 1.85 | 1.6 | 2.98 |  |  |

(a) Implants made through 0.05 micron layer of $SiO_2$.
(b) Ion beam $\pm 0.1°$ off optimum $\langle 110 \rangle$ channel.
(c) Second 200 keV implant made into float zone material.

authors are summarized in table 5.19 and illustrated in fig. 5.63. The approximate electrical activity in carriers/cm$^3$ in the channelled region is also given. There is a very considerable scatter on the projected ranges at the low ion energies which may be attributed to the different implantation, surface treatment and annealing conditions combined with the difficulties of depth measurement below 0.1 microns. The channelled results are consistent and show the $dE/dx \propto E^{\frac{1}{2}}$ dependence associated with electronic stopping (Chapter 2). The increased range of the ions in the $\langle 110 \rangle$ direction over the ions in the $\langle 111 \rangle$ direction is very evident. At the higher energies the difference between the channelled and projected ranges becomes smaller as at the higher energies the non-channelled ions lose a large fraction of their energy by electronic stopping. A thin oxide layer on the surface of the silicon crystal behaves in the same manner as an amorphous layer in preventing ion chan-

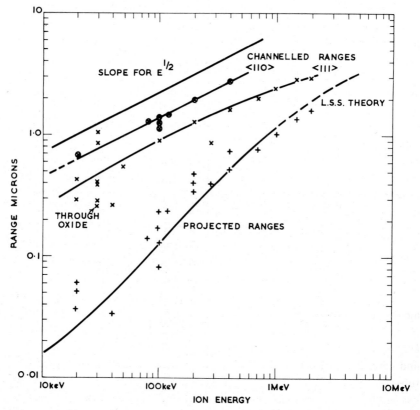

Fig. 5.63. Experimental projected and channelled range-energy relationships for the phosphorus implantations listed in table 5.17. The slope for $dE/dx \propto E^{\frac{1}{2}}$ is indicated.

nelling. Titov (1970) implanted 30 keV phosphorus ions through 0.05 microns of silicon dioxide into $\langle 111 \rangle$ silicon and annealed at 600 and 900 °C (see table 5.19 and fig. 5.63) for up to six hours. The channelled range was reduced and agreed with the range found by Shannon et al. (1970) for 40 keV phosphorus ions implanted 8° off the $\langle 111 \rangle$ direction.

The electrical activity found by the various authors in the channelled tails varied between $1 \times 10^{16}$ and $1 \times 10^{18}$ donors/cm$^3$ and it showed an increase with the ion dose, and both the implantation and annealing temperatures. The shape of the channelled portion of the phosphorus profiles usually approximated to an exponential curve but if the beam was very carefully aligned to the crystal axis ($\pm 0.3°$) then a nearly uniform doping profile was found (Shannon et al., 1970b; Goode et al., 1970). The latter authors also observed significant dechannelling in dislocation free silicon crystals which exhibited much less uniformity of the lattice parameter (Chapter 2).

Channelled phosphorus profiles have not been applied to practical silicon devices for several reasons. Although the ranges of channelled phosphorus ions in $\langle 110 \rangle$ and $\langle 111 \rangle$ silicon are known, the control in dose and position of the phosphorus ions in the tail is difficult. It depends on the ion; its energy and dose, the crystal orientation, its alignment to the ion beam, the state of the crystal and the implantation temperature. The surface of the crystal must be free of any non-crystalline layer (e.g. an oxide layer) which could disturb the direction of the ion beam before it reaches the silicon. As the doping level of the channelled region is three to four orders lower in magnitude than diffused or non-channelled implanted layers it is most readily applied to devices using lightly doped silicon. This will ensure that the channelled implant will be large enough to completely reverse the carrier type and for n$^+$-p junctions with a large concentration differential. Varactor diodes (section 5.5.3.1) have been made using channelled phosphorus ions while Dearnaley et al. (1969) have applied a channelled ion junction in making a high energy proton detector (see section 5.5.7).

Figure 5.64 shows the variation of sheet resistivity in ohms per square against phosphorus ion dose as measured by Davies and Roosild (1971a) (Curve A, 400 keV implants at $-195$ °C annealed at 650 °C), Webber et al. (1969) (Curve B, calculated from $N_s$ surface concentration (donors/cm$^2$) and mobility for a combined 20 keV plus 40 keV implant at room temperature and annealed at 750 °C), Vasil'ev et al. (1968) (Curve C, 180 keV implants at room temperature and annealed at 800 °C) and Stumpfi and Kalbitzer (1970) (Curve D, 7 keV implants at room temperature and annealed at 900 °C). The theoretical curve was calculated for 100 keV phosphorus by

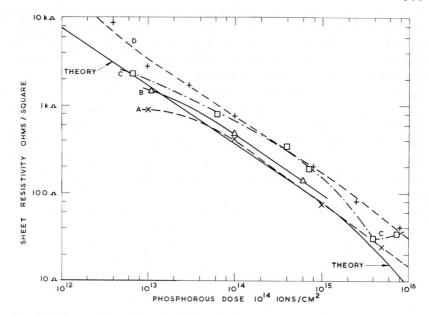

Fig. 5.64. The variation of sheet resistivity of phosphorus implanted layers in silicon with phosphorus ion dose. The curves are; Curve A, 400 keV $-195$ °C annealed at 650 °C (Davies and Roosild, 1971(b)); Curve B, 20 keV plus 40 keV at R.T. annealed at 750 °C (Webber et al., 1969); Curve C, 180 keV at R.T. annealed at 800 °C (Vasil'ev et al., 1968); Curve D, 7 keV at R.T. annealed at 900 °C (Stumpfi and Kalbitzer, 1970). A theoretical curve for 100 keV phosporus assuming 100 % electrical activity and Irvin's (1962) carrier mobility values is shown.

the same method used in the boron curve (fig. 5.53) with $X = 2.58$ and a minimum value of mobility of 65 cm$^2$/volt sec. The curves A and B show good agreement with the theory except at low doses while curve C shows good agreement at high and low doses but not in the intermediate region. The displacement of curve D to higher resistivity values is probably due to the very shallow nature of the 7 keV implant which produces high dopant concentrations per unit volume even for low ion doses. The extrapolated standard deviation at 7 keV for phosphorus is 0.004 microns (40 Å) giving $N_{(max)}$ of $10^{19}$ donors/cm$^3$ for a dose of $10^{13}$ ions/cm$^2$. Using a mobility figure of 125 cm$^2$/volt · sec corresponding to $N_{(max)}$ gives a sheet resistivity of 5 k ohms/square which is about twice the measured value. A further reduction in sheet resistivity will occur if allowance is made for the contribution from the channelled tail in which the mobility will be higher. Figure 5.64 can be used as a guide for the dose required to obtain a given sheet resistivity and the theoretical curve applies to all donor impurities.

The electrical properties of phosphorus implanted $n^+$-p junctions have been reported by several authors. Manchester (1966) measured the electrical characteristics of diodes made by implanting $20+40$ keV phosphorus ions ($5 \times 10^{14}$ ions/cm$^2$ at each energy) into 10 ohm $\cdot$ cm p-type silicon and annealing at 600 °C. The leakage durrent at $+30$ V bias was $4.2 \times 10^{-5}$ amps/cm$^2$ and the breakdown voltage was between 80 and 90 volts. Some of the diodes were not annealed at 600 °C but were briefly annealed at 370 °C during the attachment of the dice and the leads. These diodes showed a two order of magnitude increase in the leakage current and the breakdown voltage was reduced to 35 to 50 volts. The series resistance in the forward direction increased from 500 ohms to over 10 k ohms. These results showed clearly that the electrical activity was low and that there was a considerable amount of residual damage present which was producing recombination-generation centres. Glotin et al. (1971) measured the breakdown voltage for diodes produced by implanting 20 keV phosphorus ions ($1 \times 10^{15}$ ions/cm$^2$) into 2 ohm $\cdot$ cm p-type silicon and annealing at 250 and 600 °C. Their results were:

| Annealing temp. °C | Reverse breakdown voltage (V) | | |
|---|---|---|---|
| | Direction of implantation | | |
| | $\langle 110 \rangle$ | $\langle 111 \rangle$ | random |
| 250 | 105 | 75 | 65 |
| 600 | 60 | 55 | 50 |

They explain these results by using a model for the diode which is split into two parts. Firstly a 'deep' diode associated with the channelled region and having a graded impurity concentration, and secondly a 'shallow' diode associated with the 'amorphous' range and having a steep impurity gradient at its perimeter. The channelled ions in the deep diode would be active after annealing at 250 °C while the shallow diode would only be active after annealing at 600 °C. The high breakdown voltage in the $\langle 110 \rangle$ silicon after a 250 °C anneal was due to the channelled tail. The $\langle 111 \rangle$ silicon showed the effect to a lesser extent. A random implantation which inhibited the channelling and removed the deep diode reduced $V_{BD}$ even further. When the shallow diode was electrically active it broke down at its perimeter at a lower voltage than the deep diode. A channelled diode annealed at 600 °C was mesa etched (section 5.4.3.6) so that the shallow diode was no longer in contact directly with the substrate and only the deep diode was connected.

The breakdown voltage increased to the value measured at 250 °C. Glotin et al. (1971) obtained leakage currents of 2 $\mu$A/cm$^2$ after annealing in the vacuum system used for the implantation. They found that this procedure was better than removing the implantations for annealing. In forward bias, they found a value of $m$ (in $I = I_0\, e^{qV/mkT}$ (section 5.4.3.2)) between 1.3 and 1.4 which compared favourably with the value for a phosphorus diffused junction. Kleinfelder (1967) obtained a value of $m$ from 1.8 to 2.27 for mesa diodes formed from a phosphorus layer implanted at 625 °C (50 keV P$^+$ ions, $2.25 \times 10^{15}$/cm$^2$) and annealed at 625 °C. The breakdown voltage was 38 volts. Blamires et al. (1967) produced n$^+$-p diodes by implanting 40 keV phosphorus ions ($1 \times 10^{15}$/cm$^2$) into 1 ohm·cm p-type silicon. Their results can be summarized as follows:

| Annealing temp. °C | Saturation current $I_s$ amps | $m$ | Series resistance ohms | $V_{BD}$ volts |
|---|---|---|---|---|
| 500 | $4 \times 10^{-10}$ | 1.76 | 13.5 | 25 |
| 550 | $2 \times 10^{-11}$ * | 1.56 | 80 | 34 |
| 750 | $1.3 \times 10^{-10}$ | 1.56 | 13 | 28 |

\* Small geometry.

The value of $V_{BD}$ was the value expected for the 1 ohm·cm substrate while the low series resistance indicated that the implant had high electrical activity. The values of $m$ and $V_{BD}$ are in agreement with Kleinfelder (1967).

Shannon and Ford (1970) have shown that the leakage current of phosphorus implanted diodes annealed at 500 °C increased with the phosphorus ion dose whereas the leakage current of equivalent boron diodes remained unchanged. A possible explanation is that the amount of damage in the space charge region of the diode is directly related to the phosphorus ion dose. A large number of recombination–generation centres will be formed and give rise to a high leakage currents similar to those observed for the heavy elements in group III. The damage produced by light boron ions is less and does not extend into the space charge layer in the same way.

*5.4.4.8. Arsenic in Silicon.*   Arsenic, despite its importance in silicon device fabrication, has received far less attention than the other dopants in Group V. It has the highest solubility of any dopant in silicon (Trumbore, 1960) and with its lower diffusion constant it is applied in silicon technology where highly doped n layers are required which must not move during subsequent high temperature processing. Arsenic is frequently used for the buried $n^+$ layer below the collector junction of an integrated bipolar transistor (Warner, 1965).

The annealing behaviour and electrical activity of arsenic atoms implanted into silicon at 78 °K, room temperature and at elevated temperatures have been studied over a range of annealing temperatures. Davies and Roosild (1971) implanted 150 keV arsenic ions into silicon at 78 °K and then annealed at 650 °C. The sheet resistivities they obtained are given in table 5.20 and agree reasonably well with the theoretical values calculated for phosphorus and shown in fig. 5.58. A higher annealing temperature would probably reduce the measured values nearer to the theoretical values by increasing the fraction of electrically active impurity atoms. Crowder and Morehead (1969) implanted 280 keV arsenic ions into silicon at room temperature and annealed at a number of temperatures between 500 and 800 °C. The measurements were made using prepared van der Pauw configurations and conventional direct current potentiometric techniques. Above a critical arsenic dose of $3 \times 10^{14}$ ions/cm$^2$, the implant annealed rapidly between 500 and 550 °C and a large fraction of the implanted atoms was in substitutional sites electrically active and uncompensated. For ion doses less than the critical dose, higher annealing temperatures were required to obtain the same fraction of active atoms. The effect was dose dependent, the annealing temperature increasing with the ion dose. The sheet resistivities reported by Crowder and Morehead are given in table 5.20 and they agree reasonably well with the theoretical values especially for the higher ion doses. They also found 100 % electrical activity for a 260 keV $3 \times 10^{18}$ ions/cm$^2$ implant at room temperature and annealed at 600 °C.

Arsenic implantations into silicon at elevated temperatures have been reported by Mayer et al. (1967a), Baron et al. (1969), Johnson (1969) and one result by Crowder and Morehead (1969). The results which are in table 5.20 are in general agreement and they show a considerable reduction in electrical activity when compared with the room temperature implantations. Mayer et al. (1967) and Baron et al. (1969) measured the carrier concentration $N_s$ per cm$^2$ and showed that for hot arsenic implantations between $1 \times 10^{14}$ and $8 \times 10^{15}$ ions/cm$^2$ $N_s$ increased with annealing temperature from 500 °C and saturated before the annealing temperature reached 900 °C. The carrier mobility varied by about 2 : 1 over this annealing temperature range. For example, the $1 \times 10^{15}$ arsenic ions/cm$^2$ implant showed a variation in mobility from 170 to 95 cm$^2$/volt·sec. at 900 °C. The percentage of electrically active arsenic atoms at 900 °C was 70 %, 25 % and 10 % for ion doses of $1 \times 10^{14}$, $1 \times 10^{15}$ and $8 \times 10^{15}$ ions/cm². The reduced electrical activity of the hot implants has been supported by Eriksson et al. (1969) who used Rutherford backscattering to locate the arsenic atoms. Their experiments showed that when arsenic ions were implanted into a hot silicon substrate (450 °C) the percentage of substitutional arsenic atoms was about 60 % irrespective of the dose. They concluded that this phenomenon had nothing to do with solubility and therefore it might persist to much lower doses. Many arsenic atoms may become trapped on interstitial lattice sites during the implantation and are not readily released on annealing. This is a surprising result when compared with the high electrical activity of a high dose room temperature implant anneal at 600 °C (Crowder and Morehead, 1969), in view of the lattice damage produced during the implantation. The recrystallization of the damage which occurs during the annealing of a room temperature implant clearly favours the arsenic atoms being incorporated into substitutional sites.

TABLE 5.20

Experimental sheet resistivities observed for a number of arsenic implantations in silicon

| Energy keV | Dose ions/cm² | Cryst. orient. | Temperature | | Sheet resistivity | | Reference |
|---|---|---|---|---|---|---|---|
| | | | imp. °C | anneal °C | theory | meas. | |
| 10 | $1 \times 10^{14}$ | $\langle 111 \rangle$ | 600 | 600 | 330 | 1040 | Johnson (1969) |
| 10 | $1 \times 10^{14}$ | $\langle 111 \rangle$ | 600 | 600 | 330 | 1270 | Johnson (1969) |
| 10 | $1 \times 10^{14}$ | $\langle 111 \rangle$ | 600 | 600 | 330 | 1320 | Johnson (1969) |
| 20 | $1 \times 10^{14}$ | $\langle 111 \rangle$ | 500 | 900 | 330 | 560 | Baron et al. (1969) |
| 20 | $1 \times 10^{15}$ | $\langle 111 \rangle$ | 500 | 900 | 79 | 250 | Mayer et al. (1967a) |
| 20 | $8 \times 10^{15}$ | $\langle 111 \rangle$ | 500 | 900 | 11 | 111 | Mayer et al. (1967a) |
| 150 | $1 \times 10^{13}$ | 7° off $\langle 111 \rangle$ | −195 | 650 | 1700 | 670 | Davies and Roosild (1971a) |
| 150 | $1 \times 10^{14}$ | 7° off $\langle 111 \rangle$ | −195 | 650 | 330 | 490 | Davies and Roosild (1971a) |
| 150 | $1 \times 10^{15}$ | 7° off $\langle 111 \rangle$ | −195 | 650 | 79 | 100 | Davies and Roosild (1971a) |
| 150 | $5 \times 10^{15}$ | 7° off $\langle 111 \rangle$ | −195 | 650 | 19 | 49 | Davies and Roosild (1971a) |
| 280 | $3 \times 10^{12}$ | $\langle 100 \rangle$ | 30 | 800 | 3750 | 5000 | Crowder and Morehead (1969) |
| 280 | $1 \times 10^{13}$ | $\langle 100 \rangle$ | 30 | 800 | 1700 | 1200 | Crowder and Morehead (1969) |
| 280 | $3 \times 10^{13}$ | $\langle 100 \rangle$ | 30 | 800 | 810 | 560 | Crowder and Morehead (1969) |
| 280 | $1 \times 10^{14}$ | $\langle 100 \rangle$ | 30 | 800 | 330 | 350 | Crowder and Morehead (1969) |
| 280 | $3 \times 10^{14}$ | $\langle 100 \rangle$ | 30 | 550 | 175 | 290 | Crowder and Morehead (1969) |
| 280 | $1 \times 10^{15}$ | $\langle 100 \rangle$ | 30 | 550 | 79 | 75 | Crowder and Morehead (1969) |
| 280 | $3 \times 10^{15}$ | $\langle 100 \rangle$ | 30 | 800 | 30 | 30 | Crowder and Morehead (1969) |
| 280 | $3 \times 10^{15}$ | $\langle 100 \rangle$ | 30 | 600 | 30 | 29 | Crowder and Morehead (1969) |
| 260 | $2.5 \times 10^{15}$ | $\langle 100 \rangle$ | 600 | 600 | 37 | 133 | Crowder and Morehead (1969) |

The results obtained by Crowder and Morehead (1969) for 280 keV arsenic room temperature implantations from $3 \times 10^{14}$ to $3 \times 10^{15}$ ions/cm$^2$ and annealed at 550 °C showed a 1 : 1 ratio between the number of implanted arsenic atoms and the number of electrons. Below $3 \times 10^{14}$ ions/cm$^2$ a higher temperature was required to obtain a reasonable fraction of electrically active centres. Davies and Roosild (1971a) found for a 150 keV arsenic implant at 78 °K a decrease in the fraction of electrically active centres from 1.0 at a dose of $1 \times 10^{13}$ ions/cm$^2$ to 0.17 at $5 \times 10^{15}$/cm$^2$ when annealed at 650 °C.

The profiles of arsenic ions implanted in silicon at 78 °K, room temperature and elevated temperatures have been reported in several papers. Gibbons (1967) and (1968) implanted arsenic ions ($1.5 \times 10^{15}$/cm$^2$) at energies between 10 and 70 keV into $\langle 111 \rangle$ silicon at 625 °C. The bevel and stain method of junction location in p silicon of different resistivities was used for profiling (section 5.4.2.6.3). The implanted carrier concentration was roughly constant out to a depth $R$ where it decreased very rapidly. There was evidence of a buried layer at the high concentrations. The relation between the depth $R$ and the ion energy was

$$R = 0.0445(E)^{\frac{1}{2}} \text{ (microns)}$$

where $E$ = ion energy in keV. The $E^{\frac{1}{2}}$ dependence strongly indicated ion channelling. It was suggested by Gibbons (1967) that there might be less crystal damage as the majority of the ion energy was lost by electronic stopping. Fairfield and Crowder (1969) have implanted 240 and 280 keV arsenic ions into $\langle 111 \rangle$ and $\langle 110 \rangle$ silicon at room temperature and obtained electrical profiles by differential sheet resistivity measurements and the total profiles by activation analysis and layer stripping. The total profile for a 280 keV $6 \times 10^{15}$ ions/cm$^2$ implant into silicon (2° off $\langle 111 \rangle$) annealed at 750 °C agreed closely with Lindhard et al. (1963) but showed some evidence of a channelled tail. The depth of the profile, at the point where the concentration was four orders of magnitude below the peak, was about 0.6 microns which is slightly shorter than the value of 0.75 microns calculated from Gibbons(1967). The difference can be attributed partly to the 280 keV implant having been done at room temperature and not at an elevated temperature. The electrical profile was very similar including the tail except that the concentration at the peak was level at $1 \times 10^{20}$ carriers/cm$^3$ and was attributed to the assumptions made regarding the mobility. (A similar level profile for phosphorus (fig. 5.60) was reported by Shannon et al. (1970).) Annealing at 900 °C 'smeared' out the peak due to diffusion but the channelled tail was not affected. A 280 keV $3 \times 10^{15}/cm^2$ arsenic implantation into $\langle 110 \rangle$ silicon showed considerable channelling to over 1 micron without annealing and was measured by activation analysis. A similar implantation into $\langle 110 \rangle$ silicon held at 350 °C showed enhanced channelling with the tail extending to 1.4 microns. The authors confirmed the enhanced channelling rather than diffusion by implanting a $\langle 110 \rangle$ wafer with arsenic at room temperature, neutron irradiating it to activate the arsenic and then re-implanting it at 350 °C. The profile showed no measurable change from the original room temperature profile showing that the second implant did not cause the radioactive atoms of the first implant to diffuse further into the silicon. By studying the profiles in $\langle 110 \rangle$ silicon for different ion doses, a dose of $1 \times 10^{13}$ ions/cm$^2$ was found as the threshold of saturation for the fraction of ions in the channelled tail. The channelling fraction could be increased by implanting at an elevated temperature (350 °C) to prevent damage building up near the surface. If the silicon was pre-bombarded with $1 \times 10^{15}$ argon ions to make the surface amorphous, the arsenic channelled tail was reduced to an insignificant minimum. It was suggested that in practice a pre-implant of silicon or argon and a 3° to 7° misorientation should produce accurate Gaussian profiles with no channelled tail. The measured values of the peak ranges at different energies were compared with Lindhard et al. (1963) and were found to be deeper by over 20 % (see table 5.21). Davies (1970) implanted arsenic ions $(1 \times 10^{15}/cm^2)$ with energies of 100 keV, 400 keV, 1 MeV and 1.7 MeV into $\langle 111 \rangle$ silicon (off set by 7°) at 78 °K and annealed at 650 °C. The profiles were measured by the bevel and stain method using silicon of different resistivities (section 5.4.2.6.3). The range of the ions for the three higher energies agreed closely with the calculated projected ranges (Lindhard et al., 1963) (Table 5.21). The profile of the 100 keV implant did not have a peak but comprised an exponential tail extending over four decades of concentration to a depth of 0.6 microns. The profiles of the three higher energy implants had exponential tails followed by a deeper penetration which was observed in higher resistivity material $(1 \times 10^{14}$ acceptors/cm$^3)$. For example, the 1.7 MeV implant penetrated 2.55 microns while extrapolation of the exponential tail showed that it should have reached only 1.95 microns. The profile of a room temperature 1 MeV implantation was similar to the 78 °K profile except that the deep part of the tail extended further into the silicon. A pre-bombardment of the silicon with $1 \times 10^{15}/cm^2$ 400 keV argon ions did not make a significant difference to the profile of a 1 MeV arsenic implantation into this pre-damaged silicon at 78 °K. This observation is contrary to the results

TABLE 5.22

Experimental sheet resistivity observed for a num

| Energy keV | Dose ions/cm² | Cryst. orient. | Temperature implant °C | annea °C |
|---|---|---|---|---|
| 20 | 3 × 10¹⁴ | – | 500 | 800 |
| 40 | 1.5 × 10¹³ | – | −125 | 550 |
| 40 | 1.5 × 10¹² | – | −125 | 850 |
| 40 | 1.5 × 10¹³ | – | R.T. | 550 |
| 40 | 1.5 × 10¹³ | – | R.T. | 850 |
| 40 | 2 × 10¹⁴ | – | R.T. | 550 |
| 40 | 2 × 10¹⁴ | – | R.T. | 770 |
| 40 | 2 × 10¹⁴ | – | 350 | 550 |
| 40 | 2 × 10¹⁴ | – | 350 | 900 |
| 40 | 5 × 10¹⁴ | ⟨111⟩ | 23 | 500 |
| 40 | 5 × 10¹⁴ | ⟨111⟩ | 23 | 700 |
| 40 | 9 × 10¹⁴ | – | R.T. | 550 |
| 40 | 9 × 10¹⁴ | – | R.T. | 900 |
| 260 | 3 × 10¹⁵ | ⟨100⟩ | 30 | 600 |
| 260 | 3 × 10¹⁵ | ⟨100⟩ | 500 | – |
| 260 | 3 × 10¹⁵ | ⟨100⟩ | 600 | – |

[1] Marsh et al. (1967)
[2] Johannson and Mayer (1970)
[3] Baron et al. (1969)
[4] Crowder and Morehead (1969)

the room temperature implantations to 8(
number of carriers. The low dose implant
not show the sharp increase in activity a
increased steadily to level off at 800 °C. /
at −125 °C behaved in a similar manner t
implantations (Johansson and Mayer, 197
firmed by Rutherford backscattering measu
Johansson and Mayer (1970) showed that
atoms for room temperature implantations
temperature and then depending on the d
peratures. For example, a 40 keV, 2.6 × 1

TABLE 5.21

Experimental range measurements of arsenic implantations in silicon

| Energy keV | Projected ranges microns (Ref. [3]) | Measured depth microns | Reference |
|---|---|---|---|
| 70 | 0.043 | 0.056 | [1] |
| 140 | 0.08 | 0.095 | [1] |
| 220 | 0.13 | 0.148 | [1] |
| 280 | 0.155 | 0.187 | [1] |
| 400 | 0.22 | 0.274 | [1] |
| 400 | 0.22 | 0.25 | [2] |
| 500 | 0.28 | 0.332 | [1] |
| 1000 | 0.56 | 0.69 | [2] |
| 1700 | 1.0 | 1.00 | [2] |

[1] Fairfield and Crowder (1969)
[2] Davies (1970)
[3] Johnson and Gibbons (1969)

of Fairfield and Crowder (1969) discussed earlier and there is no clear explanation for this discrepancy.

The use of arsenic implantations for silicon devices has not been reported in the literature except in a paper by Drum and Miller (1971) on ion implanted n⁺ buried layers for silicon integrated circuits. They implanted 150 keV arsenic ions with doses between $1 \times 10^{15}$ and $1 \times 10^{16}$ ions/cm² into p-type silicon and then annealed at 1200 °C first in oxygen and then nitrogen for periods of 2 to 18 hours. An oxide layer was grown to protect the surface. After the diffusion the oxide was removed and an n-type epitaxial silicon layer grown on the surface. The quality of the epitaxial layer was very good and this was attributed to the arsenic not exceeding its solid solubility in the silicon and disrupting the crystal structure by the formation of precipitates which can occur with thermally diffused arsenic layers. The sheet resistivities observed for the annealed arsenic implants were from 60 to 6 ohms/square for the $1 \times 10^{15}$ and $1 \times 10^{16}$ ions/cm² doses respectively. These resistivities are lower than the theoretical values from fig. 5.64 and this is most likely due to the diffusion of the arsenic during the 1200 °C anneal, reducing the carrier concentration and increasing the mobility. The results indicate that the arsenic must be nearly 100 % electrically active and that there was little if any loss during the annealing stage.

In summary, for arsenic the annealing results show clearly that if high electrical activity is required with low annealing temperatures (550 °C), the

ions must be implanted at room
appear to be any advantage in in
than to increase the channelled fr
arsenic profiles are consistent and
to occur, and consequently implar
from the LSS theory.

### 5.4.4.9. Antimony in Silicon.　Ant

used for the regular doping of silic
diffusion constant and therefore i
excessive diffusion is a disadvantag
main dopants for $n^+$ substrate ma

One of the first published resul
1966) shows that n-type layers wer
$2 \times 10^{15}$ antimony ions/cm$^2$ into si
tion was high, $(1 \times 10^{20}/cm^3)$ with
microns of the surface.

The annealing behaviour of in
on the implantation dose and temp
the implantations at room temperatu
Room temperature implantations h
(20 keV, $1 \times 10^{15}$ ions/cm$^2$), Baron
Johannson and Mayer (1970) (40
ions/cm$^2$) and Crowder and Moreh
Table 5.22 lists their results togethe
sheet resistances calculated, where
surface concentration $N_s$ and mobi
agreement, show that the annealing
p-type activity of the substrate to b
antimony. The number of electrical
with annealing between 350 and 40(
$1 \times 10^{14}/cm^2$, where an amorphous
there was a second sharp increase in
of the implanted dose. There was an
as the number of carriers increased
et al. (1967) observed a change in
surface for a large ion dose of $1 \times 1$
of an amorphous layer. The 'milky'
was annealed above 600 °C and the

from 0 % at 550 °C to a maximum of 50 % at 650 °C and then dropped to
25 % at 750 °C. A $4.5 \times 10^{14}/cm^2$ dose implantation increased from 0 % at
450 °C to 90 % at 600 °C and above. These results agree with the observed
change in the electrical activity of the room temperature implantation annealed from 350 to 900 °C. The high activity reported by Crowder and
Morehead (1969) for a 260 keV $3 \times 10^{15}$ dose at room temperature annealed
at 600 °C does not agree with these Rutherford backscatter measurements
and this may be due to the higher energy giving a deeper implantation and
removing the implanted atoms from the vicinity of the surface. Mayer et
al. (1967) have suggested that antimony atoms may be trapped at the surface and simple etching of the surface indicated that this was a possibility.

Hot implantations were studied by Mayer et al. (1967a) (20 keV, $1 \times 10^{15}/cm^2$, 450 °C), Baron et al. (1969) (30 keV, $2 \times 10^{14}/cm^2$, 350 °C),
Johansson and Mayer (1970) (40 keV, $2 \times 10^{14}/cm^2$, 350 °C) and Crowder
and Morehead (1968) (260 keV, $3 \times 10^{15}/cm^2$, 500 and 600 °C). The annealing characteristics of the 350 °C implantations were very similar to the
low dose room temperature implantations but the 450 °C implantations
gave a high electrical activity of over 50 % without annealing. The activity
decreased by about three times after annealing at 800 °C. The loss of antimony to the surface may have accounted for a proportion of the decrease
in activity at the higher temperatures. The change in the reflectivity of the
silicon surface seen for the room temperature implantations was not observed for implantations at 450 °C (Mayer et al., 1967). The two 260 keV
implantations at 500 and 600 °C (Crowder and Morehead, 1968) gave electrical activities of 23 % and 13 % respectively which were lower than those
of Mayer et al. (1967). The high electrical activity observed for the 450 °C
implantation indicated that most of the antimony atoms were on substitutional sites. Davies et al. (1967) have shown by Rutherford backscattering
experiments that 95 % of the antimony atoms in a layer implanted at 450 °C
to a dose of $1 \times 10^{14}$ ions/cm$^2$ were substitutional while Eriksson et al.
(1969) found that the substitutional component of a $2 \times 10^{14}$ ions/cm$^2$
40 keV implantation increased with implantation temperature from 0 % at
room temperature to 90 % at 300 °C and above. The substitutional component was strongly correlated with the decrease in lattice disorder and the
increase in electrical activity of the 350 °C low dose implantations. As the
dose was increased above $1 \times 10^{15}$ ions/cm$^2$ the substitutional component
decreased due to the rapid increase in lattice disorder despite the elevated
temperature. About 50 % of the antimony atoms were substitutional after
a 40 keV $4 \times 10^{15}$ ions/cm$^2$ implant at 450 °C. These results could explain the

difference in the observations of Mayer et al. (1967) and Crowder and More-head (1969) for their large dose hot implantations of about $1 \times 10^{15}/\text{cm}^2$ and $3 \times 10^{15}/\text{cm}^2$ respectively. Lastly, Eriksson et al. (1969) found 90 % of the antimony atoms to be substitutional in layers implanted with less than $5 \times 10^{14}$ ions/cm$^2$, hot or at room temperature, and annealed between 550 and 850 °C.

The electrical measurements for doses below $5 \times 10^{14}$ ions/cm$^2$ do not agree with the Rutherford backscattering observations as the electrical activity increases steadily with temperature while the percentage substitutional remains constant. Johansson and Mayer (1970) have suggested that compensation centres were formed during the hot implantation and that high annealing temperatures were required to remove these centres. The same centres do not appear to be present when an amorphous layer, which was formed at room temperature, was annealed, as high electrical activity was observed. However, for low dose implantations ($1.5 \times 10^{13}$ ions/cm$^2$) they may be present as the same high annealing temperatures were required.

The solid solubility of the antimony in the layers with high substitutional fractions exceeded the solid solubility of antimony in silicon at 1100 °C ($4.5 \times 10^{19}/\text{cm}^3$, Trumbore (1960)) by over an order of magnitude. This supersaturation phenomenon was investigated by Mayer et al. (1967a) (also Marsh et al., 1967), who implanted 20 keV antimony ions into silicon above and below the dose corresponding to the solubility limit (about $5 \times 10^{14}$ ions/cm$^2$). For the high dose implantation, the activity was high and decreased as expected on annealing at 750 °C. By implanting carbon ions at 500 °C into the antimony implant the activity was increased nearly to the value immediately after the implantation. The same behaviour was observed for a second cycle. The *opposite* behaviour was found for the low dose implantation and this was attributed to the annealing of compensation centres at 750 °C and their creation by the carbon ions at 500 °C. The carbon atoms were assumed to be electrically inactive as carbon implantations into non-implanted silicon produced no activity. If a high dose room temperature antimony implantation was bombarded with carbon ions while being annealed at 500 °C, the resultant electrical activity was higher than that observed by simply annealing at 500 °C but it was lower than the activity found by implanting at 500 °C (Mayer et al., 1967).

Hall effect measurements at low temperatures on implanted antimony layers (Johansson et al., 1970) have shown that the doping was degenerate for doses as low as $10^{13}$ ions/cm$^2$.

The profile of a 20 keV $1 \times 10^{15}$ ions/cm$^2$ implantation in silicon at

500 °C and annealed at 750 °C (Mayer et al., 1967a) showed that 90 % of the electrically active atoms were within 0.06 microns of the surface with a doping concentration of $2 \times 10^{20}$ atoms/cm$^3$. The profile extended to 0.12 microns where the concentration was $1 \times 10^{18}$/cm$^3$. The mobility was low, 20 cm$^2$/volt·sec at the peak and 250 cm$^2$/volt·sec in the tail. Nelson et al. (1970) and Cairns et al. (1971) have studied the total antimony profiles by the technique of selective X-ray generation (Cairns and Nelson, 1971) and standard Hall effect measurements. Anodic stripping was used to remove successive layers. Figure 5.65 shows two typical 100 keV total antimony profiles obtained by the X-ray method and a theoretical LSS curve for 100 keV antimony in amorphous silicon. The agreement with the projected range was good while the standard deviation was larger and it increased on annealing at 650 °C. The peak of the electrical profile was deeper at 0.05 microns but the activity was less than half that of the total profile. The electrical activity per unit volume increased with depth and nearly all the antimony was active in the tail. The discrepancy between the total and electrical profiles can be understood by examining the residual damage by transmission microscopy. It has been shown (Nelson et al., 1970) that the amorphous layer recrystallizes in an epitaxial manner on to the underlying

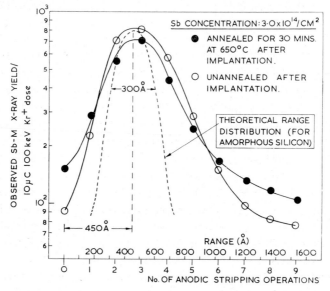

Fig. 5.65. The profile of 100 keV antimony ions implanted in silicon obtained by selective X-ray generation combined with layer stripping. The effect of annealing at 650 °C is shown and the theoretical curve is included for comparison (Nelson et al., 1970).

single crystal silicon. The regrowth was not perfect and a network of micro-twins and faults, misoriented by a few degrees from the host crystal, were observed. The presence of these defects would adversely affect the electrical behaviour by providing sites for trapping antimony atoms reducing the carrier mobility, and may be responsible for the compensation effect suspected in low dose implantations.

There are no published results of the electrical properties of the junctions formed between antimony implanted layers and the underlying silicon. The high damage concentration in antimony implantations may adversely affect the properties of the p-n junctions and it would be necessary to select the correct implantation and annealing conditions for the particular application.

Antimony implantations have not been applied to the fabrication of silicon devices in the same manner as phosphorus and arsenic though Bower and Dill (1966) have used a 15 keV antimony implant at 475 °C ($1 \times 10^{13}$ to $1 \times 10^{15}$ ions/cm$^2$) to make self aligned metal oxide silicon transistors (see section 5.5.5.3). The implantation was chosen to give a high electrical activity without degradation of the gate oxide by the aluminium gate electrode. It was made directly into the silicon as the gate oxide had been removed on either side of the channel region.

In summary, the electrical behaviour of implanted antimony is dominated by the lattice damage and the number of substitutional antimony atoms, which are in turn determined by the dose and the implantation and annealing temperatures. High electrical activity can be obtained from room temperature implantations $> 5 \times 10^{14}$/cm$^2$ by annealing above 550 °C or by implanting at 450 °C to 500 °C. High annealing temperatures are required for all low dose implantations. The measured range of the antimony ions agrees with the LSS theory. There has been little application of antimony implants to silicon devices.

## 5.5. The application of ion implantation to solid state device fabrication

### 5.5.1. INTRODUCTION

It is clear from the earlier sections in this chapter that ion implantation is very compatible with the various processing steps involved in the silicon planar technology. This section will be concentrating on silicon devices as these have the widest application at present, but other semiconductors can be considered in the same light. Some of the advantages offered by ion implantation to device fabrication are listed below in no special order;

(1) Wide range of dopants with no masking complications.

(2) Profile control independent of temperature.

(3) Precise control over ion dose.

(4) Fine control (to the limits of photolithography) over lateral dimensions.

(5) High degree of uniformity of impurity concentration over one wafer and from wafer to wafer.

(6) Low processing temperatures which can be used to reduce the diffusion of unwanted impurities (e.g. copper), restrict the propagation of dislocations etc. and therefore maintain the minority carrier lifetime.

(7) The implantation profile is independent of dislocation density and

(8) The implantation environment is clean, as there is no exposure to very high temperatures with the risk of surface contamination and corrosion.

The various applications of ion implantation to device fabrication will be considered with specific examples under the following headings:

(1) Controlled surface doping.

(2) Buried layers.

(3) Abrupt junctions.

(4) Precision alignment.

(5) Doping of imperfect material and

(6) Nuclear radiation detectors.

The behaviour of insulating layers during implantation will also be considered as many devices are made with an insulating layer protecting the surface of the semiconductor during implantation.

### 5.5.2. CONTROLLED SURFACE DOPING

Unlike thermal diffusion, the total number of ions deposited per unit area during implantation can be controlled and measured accurately over a very wide range of surface concentrations. Implantations can be made with ion doses from $10^{10}$ to $10^{18}$ ions/cm$^2$, with the errors being larger for the lower doses due to the accuracy of current integration for small charges. It is possible, therefore, to introduce a precise number of impurity atoms into the surface of a semiconductor for various purposes. This control is unique to ion implantation. The lateral position of the atoms can be defined by in-contact masks (section 5.3.5.3).

*5.5.2.1. Shallow Diffusion Source.* Roughan et al. (1968) have described the use of ion implantation as a source of impurities for thermal diffusion listing the following potential advantages;

(i)   The diffusion source can be introduced at low temperatures rather than by the normal error function deposition at elevated temperatures,

(ii)  the amounts of dopant can be more precisely controlled than thermal diffusion especially at low doping levels and

(iii) dopants which are difficult to introduce by thermal diffusion can be used.

They have reported results for nitrogen, boron and indium, the last being very difficult to predeposit while a protective oxide is grown. Indium also shows a marked tendency to accumulate in $SiO_2$ which reduces its masking capability. This problem does not occur with implantation. The electrical activity for nitrogen was 1 % after a 900 °C anneal and nearly 100 % after a 1200 °C anneal (section 5.4.4.6). The indium implants were successfully defined by a $SiO_2$ mask annealed at 950 °C and then driven in over four microns by diffusing at 1200 °C for 3 hours (section 5.4.4.5). The experimental value of electrically active centres was considerably lower than the implanted dose due to the low ionization of the indium acceptor centres at room temperature.

It can be shown that a Gaussian profile is maintained as the implanted impurities are thermally diffused assuming no reflection at the surface. If reflection occurs at the surface then the peak of the Gaussian distribution does not appreciably move towards the surface until the quantity $2Dt$ is greater than 5 times the standard deviation $\Delta R_p$ ($D$ = diffusion constant at temperature $T$ and the time $t$) assuming no enhanced diffusion effects produced by radiation damage (Seidel and MacRae, 1969). The situation is altered if the surface is covered with an oxide layer which takes up or rejects the impurity. For example, $SiO_2$ takes up boron more readily than silicon with the result that the surface will act as a sink and become depleted in boron.

An integrated circuit using complementary MOST's can be operated at a lower power than a circuit using single channel devices. The complementary circuit requires the use of p-type regions (or 'wells') in an n-type wafer for the n-channel devices (Dill et al., 1971b). MacRae (1971) has reported considerable difficulty in producing the p-type regions by diffusion methods, and that ion implantation has been used successfully to overcome the problem by implanting controlled amounts of boron. Windows in a thick oxide mask are used to define the p-type regions which are annealed after implantation during the growth of the gate oxide. The circuit is completed by normal planar processing.

*5.5.2.2. High Value Resistors.*  One of the problems in the design of many integrated circuits for example low power bipolar and MOS digital circuits and bipolar linear circuits is the incorporation of high value resistors. The maximum practical sheet resistivity obtainable by diffusion is about 500 ohms/square and to make high value resistors of over 100 k$\Omega$ requires a length-width ratio of over 100 : 1. The width is limited by photolithography to about 10 microns, thus making the resistor over a 1000 microns in length and occupying a large percentage of the circuit area. The resistor will also have a large parasitic capacitance which shunts it at high frequencies (Warner, 1965). The diffused resistor can be replaced by using a thin metal film resistor placed on an insulating layer over the circuit. This approach involves a hybrid technology and it is costly and not suitable for the mass production of cheap circuits.

Before discussing the ion implanted resistor, the factors influencing the sheet resistivity of a doped layer on a substrate of the opposite carrier type must be considered (5.4.2.5.2). The layer is isolated by the p-n junction formed between it and the substrate and its resistivity is given by the total number, $N_s$, of mobile carriers per cm$^2$ and their mobility. From eq. (5.4.8)

$$R_s = \frac{1}{q\mu N_s} \tag{5.5.1}$$

$N_s$ is a function of the difference between the acceptor $(N_A)$ and donor $(N_D)$ carrier concentrations in the layer, while $\mu$ is an inverse function of the sum of $N_A$ and $N_D$ together with the contributions from lattice vibration and lattice damage. For $R_s$ to be large both $N_s$ and $\mu$ must be small at the same time, which is contrary to normal majority carrier behaviour. The temperature coefficient of resistance (T.C.R.) of diffused resistors is a function of the change of mobility of the carrier with temperature rather than any change in $N$, if it is assumed that all the impurity centres are ionized at room temperature. The change of mobility with temperature increases with the mobility of electron and holes (Sze, 1969), and high value sheet resistivities show large T.C.R. values. Warner (1965) gives 2800 parts per 10$^6$ for a 300 ohm/square boron diffusion. A disadvantage of reducing $N_s$ to increase $R_s$ by making $N_A$ approach $N_D$ is the reduction of mobile carriers in the layer due to the depletion region of the junction spreading deep into the layer under reverse bias. A doped resistor behaves as the channel of a junction field effect transistor with the substrate acting as the gate. The $I/V$ characteristic of the resistor will be increasingly non-linear with increasing applied voltage until it 'pinches off' at a voltage $V_p$ given by

Warner (1965) for a symmetrically doped abrupt structure as;

$$V_p = \frac{qN_A t^2}{\sqrt{2\varepsilon\varepsilon_0}}$$

$$= \frac{qN_s t}{\sqrt{2\varepsilon\varepsilon_0}} \qquad (5.5.2)$$

where $t$ is the layer thickness and $N_s = N_A t$.

Thus the smaller the value of $N_s$ the lower $V_p$ and the more non-linear the resistor. Surface states, and charge trapped on the oxide above the resistor can also deplete the layer of mobile carriers and produce non-linearity. To overcome the difficulties associated with a low value of $N_s$, $\mu$ must be reduced so that $N_s$ can be *increased* by a similar fraction to obtain the same $R_s$. Ion implantation gives an accurate control over $N_s$ and by adopting certain techniques $\mu$ can be reduced.

MacDougall et al. (1969) have described making high value resistors by implanting 55 keV boron ions through oxide windows into n-type silicon and annealing at 950 °C. The implanted resistor comprised a narrow strip between two boron diffused contacts. Aluminium was used as the contact metal after annealing (see fig. 5.30). The sheet resistivity was inversely proportional to dose and varied from 0.8 to 11 k ohms/square with the T.C.R. changing from 800 to 4000 p.p.m/°C. The higher value resistors were non-linear due to 'pinch-off'. The leakage across the isolation junction was less than 2 μA ($-$ 10 V) which was considered to be low enough not to interfere with the function of the resistor. MacDougall and Manchester (1969) have explored the possibility of improving the T.C.R. by impurity compensation using implantation instead of diffusion as studied by Tufte (1966). They implanted 40 keV boron ions to give a sheet resistance of 1.84 k ohms/square and then implanted increasing amounts of 110 keV phosphorus ions. When the phosphorus dose was 88 % of the boron dose the sheet resistance had increased to 4.8 k ohm/square and the T.C.R. reduced from 2000 to 400 p.p.m/°C (fig. 5.66). The other electrical properties of the resistors showed no change due to the phosphorus compensation. Figure 5.67 shows the comparison between boron diffused and implanted resistors in the same linear integrated circuit. The smaller area occupied by the implanted resistors is very apparent. The process is stated to be compatible with high $\beta$ bipolar transistor technology.

Rosendal (1971) has studied similar boron implanted resistors to those of MacDougall et al. (1969) over a range of boron ion doses at two energies 120 and 200 keV. He annealed the resistors between 300 and 950 °C and ob-

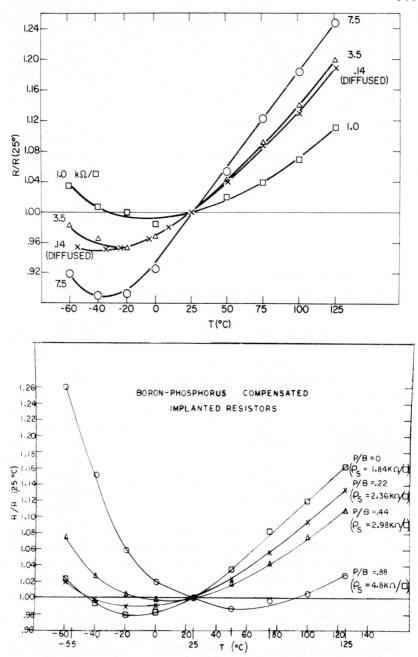

Fig. 5.66. The temperature coefficient of implanted resistors. The upper curves are for 55 keV boron implanted resistors with a diffused resistor for comparison. The lower curves are for boron-phosphorus compensated resistors. The boron to phosphorus ratio is indicated (MacDougall et al., 1969).

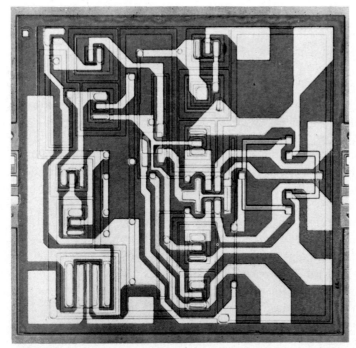

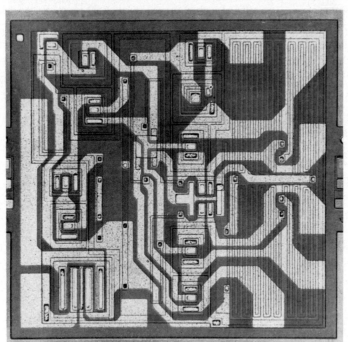

Fig. 5.67. A linear integrated circuit designed with diffused resistors (left) and implanted resistors (right). The resistors are the fine tracks below the aluminium connection pattern (MacDougall et al., 1969).

tained similar annealing characteristics to those reported earlier for boron (see fig. 5.49) including the reverse annealing effect. The sheet resistivity of a 120 keV boron implants varied from 50 ohms/square for a dose of $1 \times 10^{15}$ ions/cm$^2$ annealed at 950 °C to 50 k ohms/square for a dose of $10^{13}$ ions/cm$^2$ annealed at 300 °C. The T.C.R. was nearly constant at $-4000$ p.p.m/°C for the low dose ($1 \times 10^{13}$ ions/cm$^2$) implants but it could be controlled by annealing between 400 and 600 °C for ion doses above $10^{14}$ ions/cm$^2$. The T.C.R. changed from $-1500$ p.p.m/°C through zero to $+1500$ p.p.m/°C. He suggested that this behaviour was possibly due to the interaction of the positive temperature coefficient of the carrier concentration which decreases with annealing temperature and the negative coefficient of Hall mobility which increases, negatively, with annealing temperature. He also reported large area homogeneity in doping to an extent where the photolithography was influencing the spread in resistor values and not the implantation.

Oosthoek et al. (1970) have extended the study of high value resistors by measuring the sheet resistivity, carrier density and mobility for 50 keV, boron and gallium implantations in silicon after annealing between 400 and 800 °C. A $5 \times 10^{12}$ ions/cm$^2$ boron implant varied between 58 and 31 k ohms/square with the T.C.R. changing from over $-6000$ ppm/°C to $+2200$ ppm/°C. A $2.5 \times 10^{13}$ ion/cm$^2$ gallium implant varied between 6.8 and 4 k ohms/square with the T.C.R. changing from $-3000$ ppm/°C to $+2000$ ppm/°C with the annealing temperature changing from 600 to 800 °C. Their measurements of carrier concentration $N_s$ and mobility $\mu$ made on implants annealed to have a zero T.C.R. show a strong positive temperature coefficient for $N_s$ and a negative coefficient for $\mu$. Their conclusions, in agreement with those of Rosendal (1971), are that the low T.C.R. is the consequence of the positive coefficient of $N_s$ due to lattice disorder just balancing the negative coefficient of $\mu$ caused by lattice scattering of the carriers.

The results discussed in the previous paragraphs have depended on the damage produced by the implanted impurity atoms themselves. Nicholas and Ford (1971) have deliberately introduced damage into a 40 keV boron implanted resistor by implanting neon at 100 keV with the purpose of reducing the carrier mobility. They have reported a ten times reduction in mobility and very high sheet resistances of the order of 100 k ohm/square with an improvement in the linearity of the $I/V$ characteristics of the resistor. The characteristics of the isolation junction despite the damage produced by the neon ions has been found to be satisfactory for MOS applications. The leakage current was less than $7 \times 10^{-6}$ A/cm$^2$ and the T.C.R. was $-4300$ ppm/°C.

An alternative approach to the implanted resistor has been described by Hodges et al. (1969). They have used rhodium-silicide (RhSi) low barrier height Schottky diodes on p-type silicon as resistors. The diode is reverse biased and the leakage current is a direct function of the acceptor doping below the RhSi contact. A boron implantation of $4 \times 10^{12}$ ions/cm$^2$ at 150 keV was used to establish a maximum doping density of $10^{17}$ acceptors/cm$^3$. The reverse current does not saturate as in a normal Schottky but increases with applied voltage. A typical diode, 150 microns square, has a resistance of 20 k ohms and occupies less that 3 % of the area of a conventional diffused 20 k ohm resistor. The T.C.R. was 3000 ppm/°C.

Shannon and Ford (1970) have measured the spread in the values of boron implanted resistors on one wafer, and found 89 % of the resistors to be within $\pm 4 \%$ of mean (55 k ohm). More recently Nicholas et al. (1972) have made an assessment of resistors on a large number of wafers formed by implanting boron at 40 keV in the Harwell-Lintott industrial ion implanting machine (Freeman, 1970). Figure 5.68 shows the histograms of the values of resistors on two wafers implanted with $2 \times 10^{15}$ ions/cm$^2$. The upper histogram is for a wafer 'loosely' mounted during implantation so that its temperature could rise. It was annealed at 500 °C. The lower histogram is for a wafer in good thermal contact with a heat sink during implantation and annealed at 800 °C. The results showed the wide variation to be expected with a 500 °C anneal when heating was permitted during implantation and the small spread when care was taken to reduce these effects. The results observed for a set of wafers showed a $\pm 2.5 \%$ spread over a single wafer and a $\pm 5 \%$ spread from wafer to wafer. This spread was expected from processing tolerances and the spread in implantation dose was estimated to be $\pm 1 \%$. The measurement of a large number of van der Pauw patterns (section 5.4.2.4) will be necessary to confirm the spread in the ion dose. These results show that tight control of the ion dose is possible but that very wide spreads in resistivity can be expected with wafers 'loosely' mounted on the target plate and given a low temperature anneal which is a condition under which many wafers with MOS circuits are implanted.

The use of implanted resistors in a commercial device has been reported by Dill et al. (1971a) who have incorporated implanted boron resistors in a 64 bit MOST shift register (fig. 5.69) in place of the usual enhancement MOST loads. The purpose of using the resistors was to achieve a high operating speed of 30 MHz at the expense of increased dissipation. At 10 MHz the dissipation was 150 milliwatts compared with 50 milliwatts for a similar circuit with enhancement MOST loads operating at the same frequency.

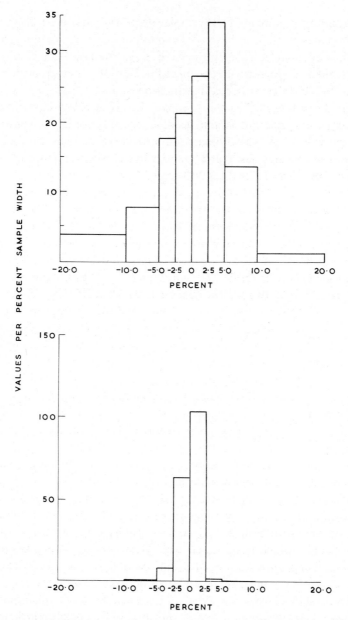

Fig. 5.68. Histograms showing the spread in sheet resistivity for two boron implantations. The upper histogram is for an implantation annealed at 500 °C and the lower one for an implantation annealed at 800 °C with precautions being taken to avoid a temperature rise during the implantation at room temperature (Nicholas et al., 1972).

Fig. 5.69. The cell of a 64 bit shift register incorporating two boron implanted resistors shown in the upper half of the photograph. The MOST gates are 5 microns long. (Dill et al., 1971a).

*5.5.2.3. Doped Channel Regions of Metal Oxide Silicon Transistors.* Ion implantation can be used to modify the conditions in the channel of a MOST transistor for several purposes.

Shannon et al. (1969a) have implanted a low dose of boron into the channel region of an n-channel u.h.f. MOST made on a $p^-$ substrate. The increased p-type doping in the channel prevents punch through from drain to source while permitting a large depleted volume to exist below the drain and reduce the output capacitance. Figure 5.70 shows the transistor in cross section. It is self aligned (section 5.5.5.3) and has an $f_{max}$ of 1.4 GHz.

Aubuchon (1969), Shannon and Ford (1970) and MacPherson (1971) have shown that the threshold voltage $V_T$ of a p-channel MOST can be shifted in a positive direction by implanting a small dose of boron ions through the gate oxide into the channel region. The change $\Delta V_T$ in the threshold voltage is $\Delta V_T = qN_s/C_0$ where $N_s$ is the ion dose/cm$^2$ and $C_0$ the gate oxide capacitance/cm$^2$, assuming all the ions are at the Si–SiO$_2$ interface. Shannon and Ford used 60 keV ions while Aubuchon used 16 keV ions with a relatively large dose of $4 \times 10^{13}$ ions/cm$^2$ and relied on the leading edge of the Gaussian distribution in the oxide to dope the channel.

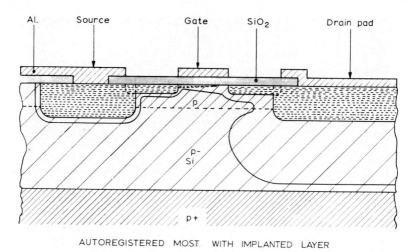

AUTOREGISTERED MOST. WITH IMPLANTED LAYER

Fig. 5.70. The schematic cross section of a u.h.f. MOST showing the position of a p implantation between the source and drain. The drain depletion is shown with its restriction in the p region (Shannon et al., 1969a).

The ions must be implanted and annealed before the gate is placed on the oxide. Aubuchon (1969) considered the process suitable for MOS circuits and sufficiently reproducible to pre-select threshold shifts up to $+5$ volts without changing the other electrical characteristics except for a doubling of the $1/f$ noise. MacPherson (1971) tried two techniques employing much deeper implantations with lower ion doses to change $V_T$. In the first, the peak of the implant was placed at the $SiO_2$–Si interface and in the second it was placed well beyond it, on the assumption that it was less sensitive to the oxide thickness. In theory, as the ion energy was increased in the second case the effects of the implant should decrease as the peak moved into the silicon. Experiment showed that the effect saturated as the actual profile was level back to the interface and contained about 20 % of the total ion dose. The same techniques have been tried on n-channel devices but field inversion problems considerably complicated the process.

MacDougall et al. (1970) have outlined the need to reduce $V_T$ to make MOS arrays compatible with the logic levels of T.T.L. circuitry. They have also pointed out that $\langle 100 \rangle$ oriented silicon which is used in normal processing for its low $V_T$ has a lower carrier mobility and lower threshold voltage under the field oxide than the $\langle 111 \rangle$ oriented silicon. By implantation it is simple to reduce the $V_T$ of a MOST on $\langle 111 \rangle$ silicon and make it compatible with T.T.L. logic voltages and retain the advantages of the higher field threshold voltage and carrier mobility. They discuss the possibility of forming depletion MOST's to act as load 'resistors' for enhancement MOST's on the same chip. The depletion load MOST is smaller than a resistor and a factor of two in speed is claimed for the same dissipation.

*5.5.2.4. Solar Cells.* Gusev et al. (1966) and Burrill et al. (1967) have described making solar cells in high resistivity p silicon by implanting phosphorus to produce shallow $n^+$ on p layers. The area of the cells was several square centimeters. Gusev et al. observed a relatively high collection efficiency of 7 % when compared with a diffused cell. Burrill et al. (1967) found that the implanted cells had the same efficiency as the best diffused solar cells over 10 %. There was an improved collection in the ultraviolet wavelengths due to the cell having a shallow, 0.25 micron junction. The difference in the results may be accounted for by the bulk recombination rates for the minority carriers in the p silicon.

5.5.3. BURIED LAYERS

The Gaussian distribution of ion implanted atoms makes it possible to form buried impurity layers in a semiconductor. The higher the ion energy, the deeper are the layers. This aspect of implantation is a unique method of doping and it has found applications in several devices.

*5.5.3.1. Varactor Diodes.* Varactor diodes are specially designed p-n junctions which have a rapid change of capacitance with applied reverse voltage. They are used for voltage controlled tuning in domestic radio and television equipment and microwave mixers and parametric amplifiers. The sensitivity of the junction is measured by

$$m = - \frac{d(\log C)}{d(\log V)}$$

where $m$ is 0.5 for an abrupt p-n junction or Schottky and 0.33 for a graded p-n junction.

By implanting tailored phosphorus impurity profiles below Schottky barriers it is as possible to obtain larger values of $m$. Brook and Whitehead (1968) obtained $m = 2.1$ for a series of summed phosphorus implants, while MacRae (1971) obtained $m = 2.5$ with a spread of less than 3 % for twelve devices. He considered the variation to be probably due to non-uniformity in the background doping of the epitaxial silicon and not the implantation. A shallow $p^+$-n diffused implanted junction can be used in place of the Schottky barrier.

The voltage 'tuning diodes' are of commercial importance and they are being manufactured by ion implantation in the U.S. (KEV, 1971).

*5.5.3.2. Impatt Diodes.*  The IMPATT (*Imp*act *A*valanche *T*ransit *T*ime) diode has a negative resistance at microwave frequencies and when placed in a resonant cavity it forms a powerful source of microwave power up to 100 GHz. The conventional avalanche diode is a single drift $p^+$-n-$n^+$ structure, but a more efficient device is the double drift diode of the form $p^+$-p-n-$n^+$. MacRae (1971) has described a device of this type in which the p drift region was implanted. An epitaxial n on $n^+$ wafer was doped by a series of boron implantations with energies between 80 and 200 keV to form a p layer in the surface of the n epitaxial layer with the same doping concentration. This degree of control is very difficult by diffusion. The $p^+$ contact was boron diffused at 875 °C. An output of 1 watt CW at 50 GHz with an efficiency of 12 % was obtained which was about twice the performance of a single drift device. Similar diodes have operated up to 100 GHz and Trapatt operation has been reported.

*5.5.3.3. Isolating Layers.*  In a bipolar integrated circuit it is frequently necessary to isolate one component from another to avoid unwanted coupling (section 5.2.3). This is normally achieved by p-type isolation diffused through an n epitaxial layer on a p substrate or by using a series of complex processing steps to introduce a dielectric between the individual regions of the layer and the substrate. An alternative method is to isolate the surface layer of a wafer by an implanted buried layer of the opposite carrier type and then to separate the surface layer into regions by further selected area implantation.

Buchanan et al. (1967), Davies (1968) and Lecrosnier and Pelous (1970) have studied (in part) high energy (over 1 MeV) implantations of boron, nitrogen, phosphorus and arsenic, in silicon. The layers of boron phosphorus and arsenic were highly conducting though the range is less than predicted by LSS theory (section 5.4.4). Nitrogen has a very poor electrical activity (1 %) and some recent work of Stephen et al. (1971) confirmed this result. Schwuttke and Brack (1969) have implanted 2 MeV nitrogen and oxygen ions in silicon and using X-ray topographs they have found a buried insulating layer of nitride or oxide. In a later paper Schwuttke et al. (1970) have reported the formation of high resistance layers in silicon by proton bombardment which were stable up to 400 °C.

There are no published reports of ion implanted layers being used for isolation in integrated circuits. There is considerable potential in the technique especially if the integrated circuits do not require deep layers. The need for ion energies above 1 MeV is at present a technical disadvantage as the majority of machines cannot accelerate ions to this energy range.

Dolan et al. (1966) produced a feasibility model of a buried gate field effect transistor. Partly isolated n regions in an n-type wafer were formed by a surrounding deep p-type diffusion. High energy boron ions were implanted to produce a p-type grid across the n region at the depth of the diffusion. The grids formed the gate, with the diffusion as the contact, and the source and drain were sited above and below the grid. Difficulty was experienced with the edges of the gate as they penetrated to the surface due to the sloping edge of the mask (fig. 5.16). Because of a high doping level in the channel, a low breakdown voltage and only limited channel modulation was observed.

5.5.3.4. *Narrow Base Bipolar Transistors*.   The bipolar transistor can be used as an efficient amplifier at frequencies well over 1 GHz. The main problems in making such a transistor are (i) obtaining narrow base widths with graded profiles and low intrinsic and extrinsic base resistances and (ii) a highly doped emitter with an abrupt emitter to base junction. One of the problems encountered when diffusing the emitter of a narrow base transistor is the 'emitter-push' (or emitter dip) effect (Grove, 1967) in which the base collector junction moves ahead of the emitter base junction during the emitter diffusion making the control of base width very difficult. Shockley (1954) appreciated at an early stage that ion beams offered an alternative method of forming the base region of a bipolar transistor (section 5.1.1). The early work on implanted bipolar transistors was reported briefly by Manchester et al. (1965), and more fully by Kellett et al. (1966) (with Leith et al., 1967) who described some initial experiments which produced a mesa transistor with a beta of 40. They listed the flow chart with the detailed steps for producing an all-implanted n-p-n or p-n-p planar transistor. A passivating layer of $SiO_2$ was sputtered on the polished wafer surface and photoresist was used for masking against the implantation. A p-n-p transistor with an 0.25 micron base width was produced but the $f_T$ was not given. Kerr and Large (1967) made p-n-p and n-p-n transistors by (i) implanting the emitter into a diffused base and (ii) implanting both the emitter and base. In the first case they found the common emitter current gain ($h_{fe}$) to be between 10 and 50 and the cut-off frequency $f_T$ limited by the base width. $f_T$ values as high as 1.5 GHz were measured. An emitter dip was observed

TABLE 5.23

Parameters of n-p-n transistors produced by combinations of diffusion and ion implantation (from Ikeda et al., 1969; Tokoyama et al., 1970)

| Combination and order | T Implant °C 900 °C anneal | Base width $W_B$ microns | $h_{fe}$ | $I_{CEO}$ | E-B junction | Remarks |
|---|---|---|---|---|---|---|
| Diff B and diff E | – | 0.3–0.5 ($f_T$ 500 MHz) | 100–200 | Small | Good | Emitter dip (0.25 $\mu$) |
| Diff B and imp E | RT-350 400–600 | 0.25 ($f_T$ 710 MHz) | – 100–200 | V. Large Small | Good Good | Emitter dip (0.2 $\mu$) |
| Imp B and diff E | RT-600 | | 10 | Large | No good (soft) | Emitter dip small |
| Imp B and imp E | RT-200 400–600 | 0.25 ($f_T$ 880 MHz) | – 60–80 | V. Large Small | No good Fair | Emitter dip (0.15 $\mu$) |
| Diff E and imp B | RT-200 600 | 0.20 0.40 | 50–60 5 | Slightly Large Large | Fair No good | Flat Emitter 'suck' in |
| Imp E and imp B | RT-600 | | – | V. Large | No good | Emitter 'suck' in |

after an 850 °C anneal for 30 minutes. This effect was removed by adopting the second approach and implanting the emitter first annealing it at 960 °C and then implanting the base through it and annealing at 600 °C. The performance of these transistors was not reported but considerable redistribution of the emitter profile occurred with a reduced concentration at the surface of $10^{19}$ donors/cm$^2$. Similar experiments were made by Gibbons (1967) and Kleinfelder (1967) on n-p-n structures. He found an emitter push for large dose phosphorus emitter implants and an increase in $h_{f_e}$ with dose. His all implanted transistors had very low $h_{f_e}$ values (5 to 8). The results of these groups were only mildly encouraging and clearly indicated a lack of understanding of the mechanisms involved in the implantation and annealing behaviour of two acceptor and donor impurities in one substrate.

Ikeda et al. (1969) and Tokoyama et al. (1970) from the same laboratory have reported on the fabrication of n-p-n transistors by diffusion, implantation and combinations (hybrids) of diffusion and implantation using phosphorus for the emitters and boron for the base regions. Their results are summarised in table 5.23. Some double implanted transistors had $f_T$'s of 1.85 GHz. These results support the earlier results of Kerr and Large (1967) and Gibbons (1967) and the authors concluded that in devices with implanted base regions, the electrical characteristics are considerably affected by residual lattice defects in the base regions. Lattice defects had little effect on the emitter base junctions in the case of implanted emitters but the phosphorus supertail (section 5.4.4.7) and the emitter dip made control of the base width difficult. The role of enhanced diffusion during hot emitter implants (emitter dip) or base implants ('suck'-in) must be carefully studied.

Fujinuma et al. (1969) have made low noise implanted base n-p-n transistors (fig. 5.71). The base region was surrounded by a boron diffused guard ring which lowered the extrinsic base resistance and helped with contacting. The base region was implanted with boron *before* the emitter was diffused from a special solid to solid diffusion source using arsenic doped germanosilicate glass which gave a very steep shallow emitter profile with a high surface doping (Abe et al., 1969). The base implantation was annealed during the temperature rise to 900 °C prior to emitter diffusion. The resultant base width was 0.05 to 0.1 microns. A multilayer metal contacting system was used to avoid the alloying effect experienced with aluminium which can penetrate the shallow layers. The transistors had a minimum gain of 8 dB at 4 GHz with a total base resistance of 20 ohms and noise figure of 4 dB. These results were better than to those of Tokoyama et al. (1970) who suggested that this was due to the use of arsenic in place of phosphorus. In this

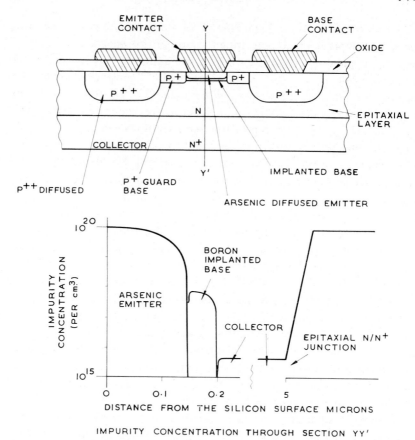

Fig. 5.71. Cross section of an ion implanted base, arsenic diffused emitter low noise microwave transistor. The doping profile of the transistor at $XX'$ is shown in the lower half of the figure (Fujinuma et al., 1969).

case, the majority of defects introduced by the low dose boron base implantation will have annealed out before diffusion commences thus avoiding enhanced diffusion effects. The use of germanosilicate glass for the arsenic emitter was reported *not* to produce emitter dip (Abe et al., 1967) as seen for phosphorus diffusion. This particular combination of implanted base and diffused emitter gives the necessary control over the base width.

Pavlov et al. (1970) discussed the problems of fabricating u.h.f. transistors by ion bombardment and emphasised the effect of the profile tails on the transistor structure. Experimental all implanted p-n-p transistors were made with $h_{fe}$'s about 50 and good breakdown voltages for the emitter-base

($-7$V) and collector base ($+60$V) junctions. Some diffused p-n-p transistors, using previously implanted boron and phosphorus, were fabricated on 5 ohm·cm p silicon. The two implants were annealed successively at 1200 °C in wet oxygen. The $h_{fe}$ was about 100 and $f_T$ 100 MHz. The junction $I/V$ characteristics were better than for normal diffused devices.

More recently, Assemat (1971) made n-p-n transistors by combining implantation and diffusion of boron and phosphorus. He obtained $f_T$'s above 1 GHz and observed a reduction of leakage current and an increase in $h_{fe}$ for transistors with implanted bases and diffused emitters over completely implanted transistors (c.f. Ikeda et al., 1969). Morizot et al. (1971) studied boron, phosphorus and arsenic profiles and confirmed the earlier reported results. A transistor with an implanted base diffused emitter and a chromium-gold contact metal system had an $h_{fe}$ of 100 and the $f_T$ was 2 GHz. Transistor action was observed for an implanted arsenic emitter and a phosphorus emitter implanted through a thin oxide layer to avoid channelling.

Beale (1970) in a recent review stated that bipolar transistors are entering a new phase. Much of the work on diffused transistors has been empirical but computer simulation of the transistor has indicated the need for a very abrupt emitter base junction. Ion implantation offered control in a less empirical manner and unorthodox ideas which were earlier discarded as impractical by diffusion are now being reconsidered. (see note added in proof).

### 5.5.4. ABRUPT JUNCTIONS

The bipolar transistor discussed in the previous section depends in part for its performance on an abrupt emitter base junction. Although the current solution for the transistor is the arsenic diffused emitter, there are some devices which have used to advantage the abrupt ion implanted junction.

*5.5.4.1. Impatt Diode Structure.* Ying et al. (1968) have made $p^+$-n-$n^+$ diodes in which the $p^+$-n junction was boron implanted (60 keV, $1 \times 10^{15}$ ions/cm$^2$ at 500 °C; annealed 800 °C). A power output of 1.4 watts at 11.7 GHz with an efficiency of 6.3 % was obtained. The diode had an improved performance over a similar diffused one (1.2 watts at 6 %) because:

(i) the junction was shallow, uniform and closer to the heat sink making the diode operated at a lower temperature

(ii) the signal to noise ratio was better and

(iii) implantation followed the surface profile of the n-$n^+$ epitaxial wafer and did not distort the n-$n^+$ interface during annealing. This condition ensured that the unswept volume of the n region was minimal and increased the efficiency.

*5.5.4.2. Avalanche Photodiodes.*  Sherwell et al. (1967) studied avalanche multiplication in photosensitive diodes made by implanting phosphorus (15 to 120 keV, $7 \times 10^{15}$ ions/cm$^2$, 800 °C anneal) into $\langle 111 \rangle$p type silicon to form abrupt n$^+$p junctions. The photosensitive area was defined by etching out a mesa structure. The junction must be free of microplasma breakdown for useful multiplication to occur and the diodes were tested specifically under avalanche conditions. A 90 % yield of useful devices was obtained with a one order of magnitude increase in the signal to noise ratio. This was to be compared with 10 % for diffused diodes. The minority carrier lifetimes were longer in the implanted diodes. They concluded that the number of microplasmas in ion implanted diodes were considerably reduced. It was believed that the lower temperatures used to make the diodes resulted in less structural damage and reduced contamination.

5.5.5. PRECISION ALIGNMENT OF THE GATE ELECTRODE OF THE METAL OXIDE
       SILICON TRANSISTOR

*5.5.5.1. Introduction.*  The metal oxide silicon transistor (or MOST) has become in the past few years one of the most important active components in silicon integrated circuits. The operating speed of a MOS array is limited by the switching speed of the individual transistors. Three factors determine the switching speed;

(i) the carrier transit time from source to drain given by,

$$t_{sd} = \frac{l}{v_d} = \frac{l^2}{\mu V_d} \qquad (5.5.3)$$

where $l$ = channel length (cm), $v_d$ = carrier velocity (cm/sec) = $\mu V_d/l$, $V_d$ = drain voltage and $\mu$ = carrier mobility. A short channel length is clearly indicated and the minimum length is reached when $v_d = \mu V_d/l$ is about $10^7$ cm/sec, the limiting carrier velocity. An advantage is gained by using electrons (n-channel) instead of holes (p-channel) as their mobility is higher. The enhancement type transistor is required in digital circuits to allow direct coupling between the stages without d.c. level shifting. Unfortunately it is very difficult to make n-channel enhancement MOST's by the conventional technology and consequently p-channel MOST's are used for digital circuits.

(ii) the stray capacitance associated with the source, drain and gate electrodes which can be reduced to a minimum by the use of good circuit layout principles and the use of thick field insulators to separate the interconnection leads away from the silicon substrate, and

(iii) the feedback capacitance (Miller effect) from drain to gate due to the overlapping of the drain by the gate electrode. Peters (1968) has shown that during switching, a reverse polarity voltage spike appears at the gate via the overlap capacitance which reduces the drain current and delays the turn-on of the transistor. The opposite effect occurs during the turn-off operation. A capacitance network is formed between the overlap capacitance and the input capacitance and Bazin et al. (1968) have analysed the circuit and they have shown that if the overlap capacitance is large enough the output voltage swing is reduced so that a subsequent transistor cannot be switched.

Research to improve the switching speed of MOST's for arrays in particular has concentrated on reducing the channel length and the overlap capacitance at the same time. Enhancement p-channel devices have been used in most of this work.

*5.5.5.2. Gate Self Alignment Techniques by Thermal Diffusion.* One of the features of the MOST (fig. 5.72(a)) made by standard silicon planar technology (see Grove, 1967; Hofstein, 1967; Kooi, 1967) is the need to ensure that the gate electrode overlaps the source and drain regions so that the whole length of the channel is controlled by the gate. Modern photolithographic methods require a tolerance of two to three microns of overlap to ensure that all the gates in a MOS array will be located correctly. This overlap produces an appreciable feedback capacitance and various changes have been made to the conventional MOS processing in an effort to reduce the capacitance and the gate length at the same time. Heiman (1966) has described the use of thick (1 micron) p-type doped oxides pyrolytically deposited to form the source and drain regions (fig. 5.72(b)). The area between the two regions is the channel. Boron from the doped oxide is driven into the n-type substrate to form the source and drain and at the same time the thin gate oxide (0.13 microns) is grown. The aluminium gate electrode is placed over the channel with its overlapping edges on the thick oxide, thus reducing the overlap capacitance by the increase in oxide thickness. A reduction of eight times has been observed. Brown et al. (1968) have used refractory metals such as molybdenum and tungsten as diffusion masks against boron and phosphorus and afterwards as the gate electrode. The source and drain diffusion areas are aligned to the gate and they state that

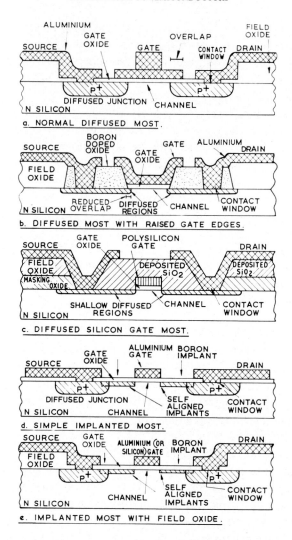

Fig. 5.72. Cross section of five typical metal oxide silicon transistor structures.

the practical limit for the overlap was about 0.1 microns. A further development of the source technique is to use vapor deposited polysilicon in place of the refractory metal (fig. 5.72(c)). This is a particularly attractive technique as the polysilicon is completely compatible with the rest of the materials in the system. An added benefit is the lower threshold voltage obtained with silicon gates due to the higher work function of the doped polysilicon.

There are difficulties with the resistance of thin polysilicon gate leads and connections to the substrate but these have been overcome and the very successful silicon gate technology has emerged (Sarace et al., 1968; Vadasz et al., 1969; Faggin and Klein, 1970). Very complex read only and random access memories and shift registers are being made using very large numbers of silicon gate transistors in a single device.

*5.5.5.3. Gate Self Alignment by Ion Implantation.* The first paper to describe the self alignment of an MOST gate by ion implantation was presented by Bower and Dill (1966) at the International Electron Devices Meeting in Washington D.C. in October 1966.

They described p and n channel MOS devices with gate lengths between 7.9 and 15.2 microns and with no oxide covering the source and drain regions on either side of the gate electrodes. In section, the transistors were similar to that shown in fig. 5.72(d) but with the gate oxide only below the gate metal. The source and drain contacts were extended to the edges of the gate by implanting 15 keV aluminium or antimony ions with doses up to $10^{15}/cm^2$ for p and n channel devices respectively. The annealing temperature was 475 °C. The transistors performed as expected, but with some parasite resistance, and the carrier mobility in the channel was 300 cm$^2$/V · sec. An important feature of these transistors was the 40 times reduction in the feedback capacitance between the gate and the drain when compared with a comparable diffused MOST. In a later paper (Bower et al., 1968) they described oxide passivated p channel MOS transistors with the gates offset from the drain by 2.5 to 23 microns. The transistors were fabricated in 10 ohm·cm $\langle 111 \rangle$ n type material with 0.14 microns of gate oxide and of 0.4 microns gate metal. The alignment or 'fill in' was done with 50 to 100 keV boron ions with doses from $1 \times 10^{13}$ to $5 \times 10^{15}$ ions/cm$^2$ and annealed up to 475 °C. The electrical properties of devices implanted with 50 to 60 keV were encouraging though lower sheet resistances were obtained with the 100 keV implantations. The drain leakage currents were typically about 10 nA and the low frequency ($< 1$ kHz) noise voltage $e_n$ was about 25 % higher in the implanted transistors in comparison with similar diffused transistors. The experiments had shown that alignment was achieved and high quality junctions were formed beneath the passivated surface. The low feedback capacitance could not be exploited in single devices mounted in a package due to the added capacitance of the leads but full advantage could be taken in integrated MOS arrays if the interconnection capacitances were reduced to a minimum.

The self aligned or 'autoregistered' MOS transistor, both p and n channel, has been extensively investigated by Shannon et al. (1969a) and (1969b),
who adopted a different approach to the fabrication technique. The basic
steps they used are shown in fig. 5.73 and are described below:

1. The source and drain regions were diffused by normal thermal diffusion methods using oxide masking and they were spaced further apart than
the length of the gate electrode.

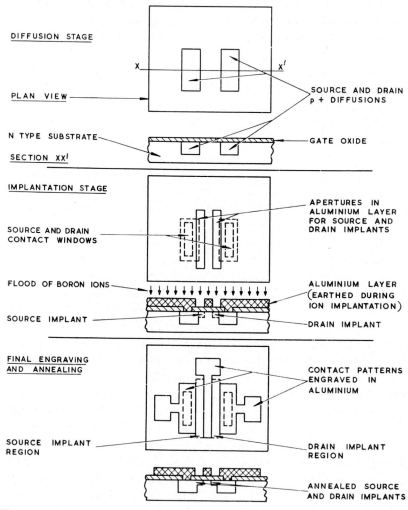

Fig. 5.73. The three basic steps in the fabrication of a simple ion implanted MOST (fig.
5.72(d)) (Shannon et al., 1969b).

2. The masking oxide was removed and replaced by the stable gate oxide, 2000 Å thick or less, which was grown over the whole silicon wafer. Contact windows, aligned to the diffused source and drain regions, were engraved in the thin oxide;

3. An aluminium layer, sufficiently thick to act as a mask was evaporated over the whole wafer. Two apertures were engraved in the aluminium so as to overlap the edges of the source and drain diffusions but not the contact windows. The strip of aluminium between the two apertures was later to be the gate electrode but at this stage it was connected to the rest of the aluminium layer. The width of the strip was the gate length;

4. The wafer was implanted with boron (or phosphorus) ions with energies between 60 and 100 keV. These energies were sufficient for the ions to penetrate through the oxide and into the silicon below and dope it. This operation extended the source and drain to the edges of the gate electrode. The aluminium layer was earthed during implantation to prevent any charging up and rupture of the oxide layer.

5. The aluminium was engraved a second time with another pattern to define the conductors and contact pads, and

6. The wafer was annealed at 500 °C for 30 minutes to produce electrical activity in the implanted regions and sinter the aluminium in the contact windows. A sheet resistivity of 2.5 to 3 k ohms/square was obtained for boron $(6 \times 10^{15}$ ions/cm$^2)$ and 0.6 k ohms/square for a similar dose of phosphorus ions.

A scanning electron micrograph of the actual transistor is shown in fig. 5.74. It is a dual transistor which has a common source with dual gates and two drains. The implanted regions are showing up as dark areas by voltage contrast. The initial separation between the diffused source and drain was 37 microns and the gate length defined by the aluminium strip was about 6 microns. Careful examination of the nearer gate reveals a 'crank' in the lead due to the slight misalignment of the second engraving mask (step 5).

Successful transistors were made by this method and the results showed that ion implantation doping could be used to extend the source and drain up to the gate and produce self-alignment. The gate oxide in the apertures was not destroyed by the passage of the ions and there was no evidence of extra gate shorts or excessive surface leakage currents. The results of stability tests of MOST's passivated with phosphorus glass were no different to those obtained with conventional devices. Damage does not occur in the gate oxide proper as this was covered by the gate aluminium. During implantation

Fig. 5.74. A scanning electron micrograph of an ion implantated MOST. Two transistors with a common source in the centre, two strapped 7 microns long gate electrodes and drains at either side are illustrated. The implanted areas are shown up by voltage contrast and appear as the dark areas on either side of the gates (Shannon et al., 1969b).

the ion dose used for self alignment was not critical, the main requirement being the need for a low sheet resistivity in the implanted regions.

The technique was extended to a n-channel depletion type ultra high frequency (u.h.f.) MOST, which is shown in cross section in fig. 5.70 and in the scanning electron micrograph fig. 5.75. The circular drain was surrounded by a circular gate of length 3 microns and width 230 microns (i.e. circumference) with the source on the outside. The implanted area is darker due to voltage contrast. A n channel transistor gave 18 dB gain at 100 MHz and an $f_{max}$ of 800 MHz. The design was improved by using a high resistivity p-type epitaxial substrate (15 ohm·cm layer) and implanting the whole region between the source and drain with 100 keV $6 \times 10^{15}$ boron ions/cm$^2$ to increase the doping and prevent the depleted region from the drain extending to the source (fig. 5.70). The boron implantation was annealed during the growth of the gate oxide. A phosphorus implantation annealed

Fig. 5.75. A scanning electron micrograph of a n-channel u.h.f. ion implanted MOST. The drain is in the centre surrounded by the gate electrode which is 3 microns long. The source surrounds the whole structure. The implanted regions appear darker on either side of the gate by voltage contrast (Shannon et al., 1969a).

at 500 °C was used for the self alignment. The large depleted volume below the drain reduced the parasitic drain capacitance. These transistors had a gain of 8 dB at 1 GHz and $f_{max}$ of 1.4 GHz. The carrier mobility in the channel was 340 cm$^2$/volt·sec and 275 cm$^2$/volt·sec for the boron doped channel. The threshold voltage was 1.5 volts and the extrinsic channel resistance about 3 ohms. The overlap capacitance was 0.040 pF in a TO-18 package. Figure 5.20 is a scanning electron micrograph of the 3 micron long gate of the u.h.f. MOST with the ion implanted regions showing up (lower part of figure) by beam induced conductivity on the p-n junctions on either side of the gate. The overlap between the gate metal and the implantations was estimated to be less than 0.25 microns and this was confirmed by the overlap capacitance measurements.

Recently Josephy (1970), Nishimatsu et al. (1971) and Dill et al. (1971b) have replaced the aluminium gate with a silicon gate in both p and n channel

MOST's. Apart from the advantages of a lower threshold voltage, the poly-silicon gate allowed higher annealing temperatures to be used with a corresponding decrease in the extrinsic resistances of the transistor. Nishimatsu et al. (1971) implanted 100 keV $1 \times 10^{15}/cm^2$ boron ions, annealed at 900 °C and obtained sheet resistances below 100 ohms/square. The MOST's annealed at 900 °C had $1/f$ noise voltage 1.5 times lower than MOST's annealed at 400 °C. Josephy has made similar MOST's but with gate lengths of about one micron and he estimated that operating speeds upto 100 MHz could be obtained in logic circuits incorporating MOST's of this type. Dill et al. (1971b) have described the case of the self-aligned silicon gate for p and n channel MOST's in complementary MOS (C-MOS) circuitry. Implantation was also used to dope the p wells for the n channel transistors. The advantages gained by using the silicon gate combined with implantation were, full electrical activity of the n and p implanted regions, an increase of three times in the packing density when compared with present C-MOS technology and operating speeds upto 15 MHz.

Alternative techniques have been tried to replace the diffused source and drain contact regions. MacDougall and Manchester (1969) produced an all-implanted self-aligned p channel MOST with a molybdenum gate over a thin oxide-nitride layer (0.07 + 0.06 microns) with thick oxide (0.5 microns) mask elsewhere (fig. 5.76). A 50 keV boron implantation of $3 \times 10^{14}$ ions/cm$^2$ was used and after annealing at 800 °C and 450 °C in hydrogen gave a sheet resistance for the implanted regions of 400 ohms/square. The molybdenum did not attack the oxide-nitride layer at 800 °C. The $g_m$ was 480

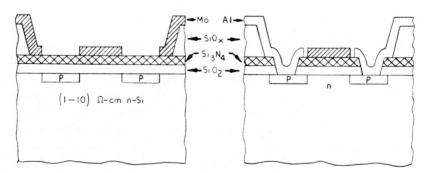

Fig. 5.76. An all implanted MOS transistor. The left hand figure shows the transistor after implantation and annealing with the molybdenum as a mask. The right hand figure shows the completed device with contact windows cut through the $SiO_2/Si_3N_4$ to the source and drain. The molybdenum except that of the gate has been removed and replaced by aluminium connections (MacDougall and Manchester, 1969).

micro-mhos with a threshold voltage of −3.2 volts. A second metal layer (aluminium) was required to make ohmic contacts to the source and drain regions.

Lepselter et al. (1969) have combined Schottky barrier techniques with MOS technology and ion implantation to produce a p channel MOST, with Schottky barriers for the source and drain contacts. Figure 5.77 shows the basic device in plan and cross section. After the growth of 0.1 microns of oxide on ⟨100⟩ n-type silicon, the source and drain contact windows were opened and platinum silicide films formed in them. A titanium-platinum-gold layer one micron thick was deposited to form the source, drain and gate electrodes. The source and drain were extended to the edge of the gate

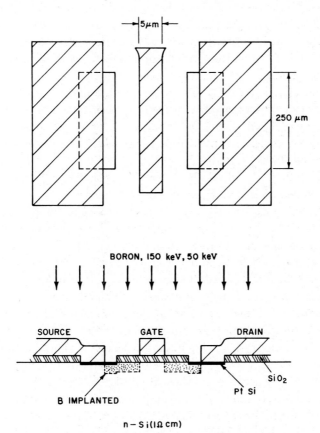

Fig. 5.77. The plan and cross section of an ion implanted MOST with Schottky barriers for the source and drain electrodes. The gate oxide was 0.1 microns thick. The shallow Schottky barrier effectively overlaps the deeper boron implantation (Lepselter et al., 1969).

by implanting 150 keV $(1.5 \times 10^{14}/\text{cm}^2)$ and 50 keV $(1 \times 10^{14}/\text{cm}^2)$ boron ions through the oxide into the silicon with the gate electrode forming a very effective mask. The structure was heated to 350 °C to reduce the radiation damage and produce a p layer with a doping density of $10^{18}/\text{cm}^3$ extending to a depth of 0.4 microns. This electrical activity was adequate to ensure essentially ohmic contacts between the PtSi and the implantation. The transistor had a normal threshold voltage and $g_m$. The experiments demonstrated the feasibility of combining Schottky barriers with p-type ion implanted layers (see also section 5.5.2.2).

From the research on single ion implanted MOST's, it was generally appreciated that the main future for the ion implanted MOST was in large MOS arrays. The electrical characteristics of the MOST, particularly the one with the self aligned gate, are such that the maximum benefit can only be realised when it is in a system where the interconnection capacitances are kept to a minimum and one transistor is directly coupled to another. The p channel enhancement transistor is used in the greater majority of arrays at present as it is basically simple in structure and the output from one transistor can switch the gate electrodes of other transistors. The first large MOS array using self aligned gates was made by Moyer, Bower and others and reported at the GOMAC conference in 1969 (Moyer et al., 1969 and later by Dill et al., 1971a, b).

The methods described above for making single MOS transistors have to be modified when making MOS arrays. In an array it is necessary to use two layers of oxide, a thin gate oxide for the active region of the transistors and a thick oxide to insulate the interconnecting leads and allow them to pass over the silicon substrate without inducing MOS action and forming short-circuit paths. The thick oxide ensures that the capacitance of the leads to the substrate is minimised. It also serves as an implantation mask making it unnecessary to cover the whole wafer with an aluminium layer. The final gate electrode and interconnection patterns are engraved before implantation and consequently the implanted regions are defined by the combination of the two masking systems. Figures 5.72(e) and 5.78 show diagrammatically a typical MOS array transistor. It should be noted that a linear geometry employed as easy access to all three electrodes is essential.

The method of fabricating an ion implanted MOS array is basically the same as that used for conventional MOS arrays. The steps in the fabrication of a typical p channel MOS array transistor are described below (refer to fig. 5.78). The substrate for these circuits is an n-type wafer, 3 to 5 ohm · cm.

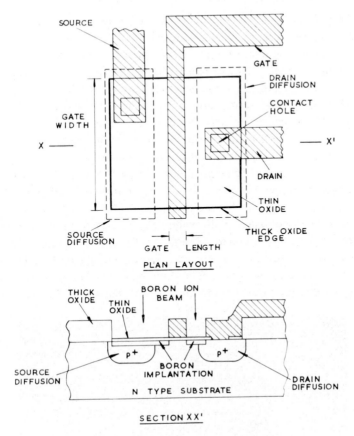

SECTION XX'

TYPICAL p CHANNEL ION IMPLANTED
MOS TRANSISTOR WITH THICK OXIDE
AND ALUMINIUM MASKING

Fig. 5.78. The plan and cross section of a p channel ion implanted MOS transistor with thick oxide and aluminium masking.

1. All the source, drain and interconnection regions are thermally diffused with boron using conventional oxide masking.

2. The thick field oxide is produced by increasing the thickness of the masking oxide or by growing a fresh layer of oxide. This can be done by first growing thermal oxide and then depositing $SiO_2$ by chemical vapour deposition (section 5.3.5.3.1).

3. Apertures are engraved in the thick oxide to give access to all the areas for the transistors resistors and contacting requirements.

4. The stable gate oxide is grown in these apertures and contact holes engraved to give access to the diffused areas.

5. A thick aluminium layer is evaporated over the whole wafer and engraved with the gate and interconnection patterns.

6. The wafer is implanted with boron ions with energies of 40 to 100 keV and a dose usually exceeding $1 \times 10^{15}/cm^2$. The implantation conditions are not critical for the self-alignment requirement but a specific sheet resistivity may be needed for any implanted resistors. The thick oxide and aluminium act jointly as masks during the implantation.

7. The wafer is annealed at about 500 to 545 °C for 30 minutes to partly anneal the boron implantation and to sinter the aluminium–diffused silicon contacts.

The wafer is complete at this stage. Figure 5.79 is an electron micrograph of a MOS array transistor. The diffused source and drain regions can be clearly seen on either side of the gate electrode in the thick oxide depression. The gate is 5 microns long. The thinning of the gate as it passes over the thick oxide step is due to non-uniformity of the aluminium evaporation.

Fig. 5.79. Scanning electron micrograph of a typical transistor from a MOS array. The diffusions for the source and drain can be seen on either side of the gate electrode which is about 5 microns long.

The above description gives only a bare outline of the fabrication procedure. Considerable modifications will be necessary if special techniques such as oxide-nitride passivation, silicon gate and methods for avoiding large steps in the metallization over oxide edges are to be incorporated.

Recently Maloney (1971) and later Sigournay (1971) have given the details of a new high yield all ion implantation process suitable for fabricating MOS arrays. The process is novel and is not an extension of accepted MOS diffusion practice. Figure 5.80 shows a typical MOST made by this process. The major steps in the fabrication sequence are:

(i)    grow the gate dielectric (oxide-nitride) over the area of the whole silicon wafer (3 to 5 ohm·cm n-type $\langle 100 \rangle$),

(ii)    engrave the contact windows in the gate dielectric,

(iii)    boron implant (40 keV, $5 \times 10^{15}$ ions/cm$^2$) the source drain and interconnection regions using the first in-contact metal mask. After removing the mask, anneal at 900 °C to give a sheet resistivity of 70 ohms per square,

(iv)    grow the thick field silicon dioxide by chemical vapour deposition and engrave in it the windows for the devices,

(v)    evaporate a thick layer of aluminium over the wafer and engrave the windows for the second implantation, the transistor gate lengths and widths being defined by the separation between the windows (compare with the method described earlier by Shannon et al. (1969b) in fig. 5.73),

(vi)    implant boron (40 keV $5 \times 10^{15}$ ions/cm$^2$),

(vii)    engrave the interconnections and contact pad pattern in the remainder of the aluminium layer, and

(viii)    anneal at 500 °C to give a sheet resistivity of 3.8 k ohm per square for the second implant. The contacts are sintered at the same time.

The initial evaluation was carried out on a test chip with resistors, capacitors, diodes, MOST's and lateral bipolar transistors and a delay line oscillator with 54 MOST inverter stages with an MOS bistable feedback element. Each oscillator stage comprised a MOST with a gate area of 5 microns (length) by 10 microns and a resistor load. Five wafers each with 168 test chips were processed in one batch. 96 % of all the oscillators worked at frequencies between 2.04 and 2.73 MHz and 90 % of them were within $\pm 4.8$ % of the mean frequency of 2.4 MHz. The propagation delay for each inverter stage was 3.9 nanoseconds which compares with the carrier transit time of 2.5 nanoseconds. The test resistors showed a $\pm 3$ % variation across a wafer and $\pm 5$ % from wafer to wafer. The resistor variations correlated closely with the capacitor variations across a wafer (both devices on the same

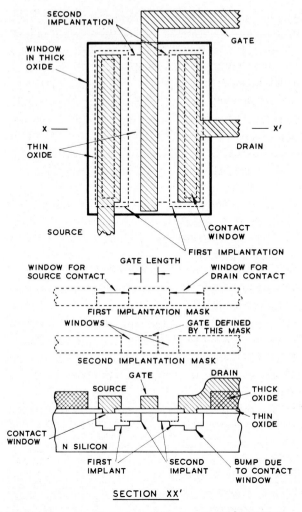

Fig. 5.80. The plan and cross section of an all implanted p channel MOS transistor. The positions of the masks for the two implantations are indicated (Maloney, 1971).

chip) indicating that the gate dielectric thickness was changing from place to place and influencing the surface concentration of the implanted boron in the silicon. The resistor values were corrected for the change in dielectric thickness and a variation of better than $\pm 1\%$ was obtained. The sheet resistivity at the overlap of the first and second implants was found to be 500 ohms per square, showing that lattice damage introduced into the first implant by the second one was sufficient to increase the sheet resistivity by nearly an order of magnitude. The decrease in electrical activity is possibly due to a combination of a loss of carrier mobility and increase in the fraction of interstitial boron from both the first and second implants.

The $p^+$-n diodes formed by this process were very similar to diffused diodes and the breakdown voltages were over 80 volts.

There are some important points to be noted about this process. These are:

(i)  no junction is exposed to contamination at any stage during the fabrication,

(ii)  all the gate dielectric and field oxide is covered during implantation except for the areas being doped,

(iii)  apart from the growth of the gate dielectric all processing temperatures involving doping are below 900 °C,

(iv)  the yield and uniformity are high.

The direct benefit obtained by implanting the gate of a MOST to achieve self-alignment may be summarised as:

(i)  minimal overlap capacitance,

(ii)  short gate lengths, giving with (i) an increase in operating speed,

(iii)  easing of photolithographic tolerances as precise overlap alignments are not necessary,

(iv)  n and p channel devices can be produced on the same chip if necessary, and

(v)  a high yield.

To these benefits must be added (i) the advantages of threshold control for making depletion load MOST's and making the operating voltages of p-channel enhancement MOST's fall within transistor-transistor-logic levels, (ii) high value resistors which can be incorporated on the same chip as enhancement MOST's and (iii) the use of controlled surface doping to produce wells or 'tubs' for complementary MOS circuits (MacRae, 1971; Dill et al., 1971).

It is relevant as a conclusion to this section on the self-aligned MOST to briefly describe some of the other ion implanted MOS devices which have appeared in the literature. (The experimental devices which have already been discussed are not included.)

Burt (1969) has produced an n channel tetrode structure in which a second gate in the channel is used to isolate the drain from the input gate electrode. Power gains of 20 dB at 4 MHz and 10 dB at 1 GHz with noise figures of 3 dB and 5.5 dB respectively were measured.

The application of ion implantation to complex MOS devices has been reported by Bower (1969) and later by Dill, Bower and Toombs (1971). Among the devices they have made are:

(a)  a dual 64 bit shift register which can operate up to 30 MHz.

(b)  10 channel and 50 channel multiplexers, the former circuit working from 100 Hz to 10 MHz.

(c)  a 9 bit digital to analogue converter with implanted resistors for defining the output current. The resistors were matched to within 1 % and the accuracy was limited by the geometry of the resistor rather than the implantation resistivity,

(d)  2048 bit read-only memory with transistor-transistor-logic compatibility and an access time of 100 ns. The data in the store is defined by a special photoresist mask which prevents implantation whereever an 'O' state was to be stored.

(e)  an all implanted MOS operational amplifier with a gain of 1,000 times and the capability of operating at 4 °K.

5.5.6. DOPING OF IMPERFECT SEMICONDUCTOR MATERIALS

The production of uniform junctions by thermal diffusion depends on the use of material with a high degree of crystalline perfection. The presence of a dislocation or grain boundary produces a region of enhanced diffusion as the diffusivity in such a defect is higher than in the crystal lattice. Many semiconductor crystals contain a large number of defects such as grain boundaries, dislocations, twinning, vacancy clusters etc., which will distort the diffusion profile and produce non-planar junctions with poor electrical characteristics. The range distribution of implanted ions depend on the ion type, its energy and the elements of the substrate. The defects enumerated above have no effect on the range distribution but gross defects such as voids, cracks or inclusions will distort the distribution. Silicon on sapphire is a typical single crystal material which has a high defect content. Figure 5.81 shows, by bevel and staining, the comparison between a phosphorus diffused and a phosphorus implanted junction in silicon on sapphire (Lawson, 1970). The uneven edge of the diffusion is very clear. The vertical edge of the implanted junction should be as uniform as that of the planar junction and in those cases where the ion penetrates completely through the layer to the sapphire good vertical junctions should be feasible. If the annealing temperature is high enough to permit diffusion then both the planar and vertical junctions will be degraded.

The single crystals of many III–V and II–VI semiconductors contain a

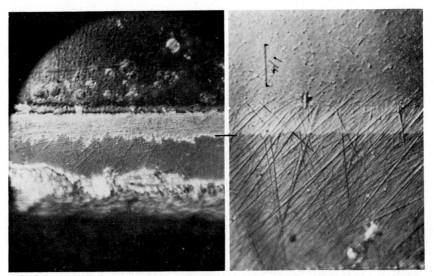

Fig. 5.81. Photographs of beveled and stained p-n junctions in silicon on sapphire. The left hand junction was phosphorus diffused and the right hand one was ion implanted with phosphorus. The horizontal marker indicates the line of the junctions. The uneven profile of the diffused junction is very apparent (Lawson, 1970).

large number of defects and this makes implantation a suitable method for introducing impurities without interference from the defects. The same situation exists in many thin semiconducting films, produced by various deposition techniques, which may either be monocrystalline, polycrystalline or amorphous. There are no published papers describing this application of ion implantation except for the special case of thin insulating layers which are considered in section 5.5.8.

### 5.5.7. NUCLEAR RADIATION DETECTORS

The first practical device to be made by ion implantation appears to have been a nuclear particle detector, reported by Alväger and Hansen in 1962. Phosphorus ions of 10 keV energy were implanted into 9000 ohm · cm. silicon, and this was then annealed at 600 °C to produce a detector with reasonably good energy resolution. The advantages of ion implantation in producing a uniform, shallow junction by a low-temperature technique were immediately recognised, since diffused contacts had proved less controllable and the high temperatures led to a degradation of the carrier lifetime of the bulk material.

Subsequent work by Martin et al. (1964, 1966) failed to produce devices with characteristics better than the established surface-barrier technique, and it is only recently that the reasons for this have become more apparent, as a result of better understanding of ion channelling and the annealing of radiation damage. Martin et al. (1964) implanted $B^+$ ions at 500 keV at a variety of temperatures ranging from 100 to 300 °K. The low-temperature implantations were more successful, but even so the leakage currents, after annealing at 300 °C for 4 hours in argon, were around 10 $\mu$A at 50 volts reverse bias. This is far greater than the leakage typical in surface-barrier detectors, and the energy resolution was correspondingly poorer (about 100 keV for 5 MeV alphas). There was evidence of non-uniform response over the detector area, which was attributed to recombination at damage centres, but could have been due to a non-uniform window thickness.

There were three major differences in the work of Meyer and Haushahn (1967) and Kalbitzer et al. (1967a): very low ion energies were used, between 2 and 10 keV; the beams were aligned with a channelling direction and low annealing temperatures, of 300 °C to 400 °C were employed. Meyer implanted p-type contacts with $10^{14}$ $B^+$ ions/cm$^2$ and n$^+$ contacts with $10^{14}$ $P^+$ or Te$^+$ ions/cm$^2$, in high-resistivity n-type silicon. Annealing was performed for 10 minutes at 300 °C. The n$^+$ rear contact showed very little injection on full depletion, which is a valuable feature for transmission detectors, which are used for d$E$/d$x$ measurements. A room temperature resolution of 16.7 keV was achieved for 5.5 MeV alpha-particles, while at 77 °K the resolution for $^{207}$Bi electrons (of about 1 MeV) was 6.6 keV. These excellent results compare with the best reported for surface-barrier detectors.

Kalbitzer et al. (1967a) used even lower ion energies, and annealing temperatures of 400 °C. Very good reverse bias current-voltage characteristics were observed, but the results were less satisfactory if non-channelled implants were used. Meyer and Haushahn (1967) also report this and both groups stressed the simplicity and reproductivity of the technique which they had applied in the production of several hundred detectors.

Meyer (1969a, b, c) has made a detailed study of the window-thickness (or dead-layer thickness) on the surface of diodes implanted under varied conditions. This window thickness was found to be strongly voltage-dependent and it was also influenced by the type of ion implanted and the substrate resistivity. Under certain conditions extremely thick windows up to 10 $\mu$m were found and were attributed to penetrating tails in the distribution profile, due to channelling and, perhaps, anomalous diffusion. These tails are of very low concentration ($\sim 10^{12}$ ions/cm$^3$) and are not readily observable by other means. However, adequately thin windows ($< 0.02$ $\mu$m) could be achieved without difficulty, and this is important in the use of the detectors for accurate energy measurement and with radiation of low penetrating power (e.g. X-rays).

Meyer (1968) and Kalbitzer et al. (1967a) consider that better performance results from channelled implants because the junction is then formed in the region where channelled ions came to rest and the density of damage in the crystal is low. Moreover, this low-density damage is readily annealed at low temperatures (Eriksson et al., 1968), which involve less degradation of carrier lifetime due to the diffusion of unwanted trapping impurities (such as copper). Necessarily, if channelling is employed, low ion energies must be used to provide a thin entrance window.

Other work on ion-implanted silicon detectors has exploited the uniformity and high sheet resistivity of implanted contacts in the preparation of position-sensitive detectors, for determining the point at which a charged-particle enters a detector. Several types of such detectors have been described, for both arc and two-dimensional position sensing (see for example Owen and Awcock, 1968), but most employ a resistive electrode layer as a charge divider. The charges released by ionization on the depletion layer are swept to the corresponding point on this electrode layer and currents then flow so as to restore the circuit potentials. From the relative magnitude or duration of these currents the entry point of the particle can be determined.

Evaporated metal films have generally been used since uniform layers with the required sheet resistivity of the order of $10^3$ ohms per square cannot be realized by diffusion. Few metals, however, give sufficiently stable resistive layers, and it is not possible to use such contacts for a two-dimensional position sensing detector, owing to injection problems. For these reasons, ion implanted contacts were attempted and were found to be so successful that they are now preferred to those prepared by other techniques. Three groups have published results in this field (Kalbitzer et al., 1967b; Owen and Awcock, 1968; Laegsgaard et al., 1968) and position resolutions of about $\frac{1}{2}$ per cent of the width of the detector have been attained (typically 0.2 mm in 4 cm). Naturally, the linearity of the response is dependent upon the uniformity of sheet resistivity, and tests with a 4-point probe indicate (fig. 5.82) that the uniformity of doping is nearer to 1–2 %, for a sheet resistivity of $10^4$ ohms per square. This remarkable result, which the authors feel may yet be improved by better control of beam stability, shows a very considerable advantage of ion implantation over diffusion. This control is desirable in integrated circuits where compact high-value resistors are required. The use of ion implantation should lead to better reproducibility and uniformity of resistance values over a large-area slice and furthermore, to a

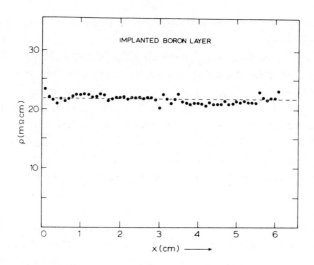

Fig. 5.82. The sheet resistivity of a boron implanted layer along a position sensitive detector, showing a high degree of doping uniformity (Laegsgaard et al., 1968).

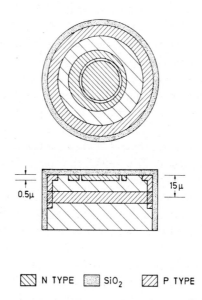

Fig. 5.83. The diagram of a particle detector for the measurement of energetic heavy ions. It is an $n^+$-$n^-$-p-$n^-$ structure with the buried p layer being produced by a 22 MeV boron implantation (Martin, 1969).

reduction in the area occupied by large-value resistors. In addition the use of higher resistor values leads in turn to a reduction in the power consumption of an integrated circuit, and an avoidance of heating problems.

An interesting type of particle detector for the identification and energy measurement of energetic heavy ions has recently been reported by Martin (1969). This consists of two p-n junctions, forming an $n^+$-$n^-$-p-n structure, (fig. 5.83) and capable of measuring the energy, $E$, and specific energy loss, $dE/dx$, for heavy ions, and hence their atomic number. A buried p-type layer was produced by the implantation of 11 or 22 MeV $B^+$ ions, from a heavy-ion linear accelerator. This is surely the most energetic ion implantation yet attempted. The mean range of 11 MeV $B^+$ ions was estimated to be about 13 $\mu$m, and doses of about $10^{14}$ ions per cm$^2$ were used. Surface $n^+$ and $p^+$ layers were formed by implantation of 200 keV $B^+$ and 250 keV $P^+$ ions. Annealing was carried out at 600 °C for 15 minutes. Unfortunately, the leakage current in these diodes was high and the large capacitance of the surface diode coupled with the high sheet resistance ($\sim 10^3$ ohms/sq) of the implanted p-layer resulted in a long charge collection time constant. Martin (1969) suggests several ways of improving the design, including smaller area devices, channelled implantations at lower energies and reduced annealing temperatures.

In addition to the above work on ion-implanted silicon detectors, the technique has been employed to produce good detectors in germanium. This material is preferred to silicon in cases where its greater stopping power, due to the higher atomic number, is an advantage. Owing to the lower energy gap of germanium, it is necessary to operate the detectors at or near liquid nitrogen temperature. Germanium detectors have been used for gamma spectrometry and for the detection of medium-energy protons (30–60 MeV). Gunnersen et al. (1967) reported measurements of the sheet conductivity and thickness of ion implanted layers in germanium, as a function of annealing. $Ga^+$ and $Sb^+$ ions were used, at 80 keV energy and annealing temperatures of 400 °C to 800 °C were investigated. This work showed that relatively low temperature annealing would result in a good sheet conductivity with $Ga^+$ ions and that window thicknesses of about 0.25 $\mu$m could be achieved.

Dearnaley, Hardacre and Rogers (1969) described $Ga^+$ implanted germanium detectors for proton detection. In this work low ion energies, about 12 keV, were used and the ions were intentionally channelled into the germanium, at room temperature. Doses of $10^{14}$ ions/cm$^2$ were normally used. Lithium ion-drift was carried out to compensate the germanium and so obtain the large depletion layer thickness (5–9 mm) needed to stop energetic protons. Lithium diffusion was performed at 400 °C over a period of 20 minutes, and this sufficed to anneal the previously implanted gallium layer. In this way, there was no additional thermal processing of the material. Lithium was drifted up to, and into, the gallium layer, to form an $n^+$-i-$p^+$ structure which could be almost totally depleted, at low voltages. The resulting diodes had extremely low leakage currents, sometimes as low as $10^{-11}$ A/cm$^2$, at operating bias voltages of 500 V, and the energy resolution of the detectors was very good (fig. 5.84) in some cases about 0.06 per cent, for 50 MeV protons. Window thicknesses were between 0.5 and 1.0 $\mu$m, despite the low ion energies and presumably this is due to the deep penetration of channelled Ga$^+$. We recall that the electronic stopping cross-section of channelled Ga$^+$ ions is very low (see fig. 2.34). Successful thin $p^+$ electrodes on germanium have been produced up to 40 cm$^2$ in area, and this is important in the fabrication of large-volume gamma-ray spectrometers.

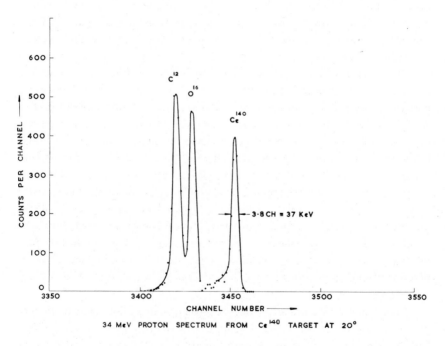

34 MeV PROTON SPECTRUM FROM Ce$^{140}$ TARGET AT 20°

Fig. 5.84. Proton energy spectrum observed with an ion implanted germanium detector by Dearnaley et al. (1969). An overall percentage resolution of 0.11 % was obtained for 34 MeV protons on a $^{140}$Ce target at an observation angle of 20°.

These authors have stressed, as did those using ion implantation in silicon radiation detectors, the high degree of uniformity and reliability of the technique and the advantage that relatively low temperatures are employed. It is particularly important to minimise the degradation of carrier lifetime, due to the diffusion of deep-trapping impurities, in detectors with such large dimensions.

5.5.8. ION IMPLANTATION OF INSULATING LAYERS

It has already been stressed that for many device applications, notably the MOST, it is necessary to implant ions through or into an insulating layer. It is therefore important to consider the behaviour of these layers during the ion bombardment and the changes in their electrical properties due to the bombardment and subsequent thermal annealing.

When a high energy ion beam passes through or is stopped in an insulator it produces ionization in the layer and this can make it sufficiently conducting to prevent the surface charging up to a high voltage. The dielectric strength of the insulator is therefore not exceeded and no permanent damage is done by the insulator being ruptured. The charge state of the surface will be determined by the number of secondary electrons ejected from the surface. Experiments (Stephen, 1973) with high energy (3 MeV) protons and nitrogen ions have shown that silicon oxide layers up to one micron and oxide-nitride layers up to 0.1 micron became conducting during the ion bombardment and returned to their normal insulating state as soon as the ion flux was removed. In the case of thick oxide layers where the ions are stopped in the surface layer of the oxide, the whole oxide layer is not 'ionized' and the deposited charge must leak away laterally to the nearest 'sink' such as a thin oxide window, contact hole or metal conductor. The exact mechanism is not understood at present, but experience has shown that with the short distances between the 'sinks' in complex MOS circuits no damage is produced. The charging of the oxide can be minimised by flooding the substrate with low energy electrons from a hot filament placed nearby (Bower, 1970). The need for such precautions will depend on the layout of the target chamber. The use of electron flooding will complicate the measurement of the ion dose during implantation.

There are no published reports of the dielectric layers on semiconductor devices being permanently damaged during implantation. The experience at Harwell with a wide range of device wafers including MOS arrays of differing surface geometries, has shown that the insulating layers were electrically intact after implantation. There has been one instance of a device being damaged during implantation and no satisfactory explanation has been produced for this case and it did not recur on further wafers.

The electrical properties of implanted insulating layers are considerably changed by the radiation damage produced by the implantation. The behaviour of insulators, mainly silicon dioxide, has been studied by several groups, Shannon and Ford (1970), Glotin et al. (1971), and Nishimatsu et al. (1971) and Fritzche et al. (1971) who studied the influence of implantation damage on the mobility of sodium ions on $SiO_2$. The overall results are in general agreement. The technique for observing the damage is to measure the capacitance–voltage (C–V) characteristic of an MOS capacitor incorporating the implanted dielectric and observe the shift in the flat band voltage $V_{FB}$. $V_{FB}$ suffers a pronounced shift in a negative direction from its normal value of a few volts to, in some cases, many tens of volts. This negative shift indicates the presence of a positive charge located near the silicon-silicon dioxide interface and it behaves electrically in a similar manner to the radiation damage produced by electrons, X-rays, gamma rays or neutrons. The implantation damage, like radiation damage can be annealed out by heating above about 300 °C, which is normally achieved during the annealing of the implantation which was responsible for the damage.

Nishimatsu et al. (1971) have found that the etch rate of boron implanted (100 keV $10^{15}$ ions/cm$^2$) unannealed $SiO_2$ in buffered HF is about 1.7 times faster than that of unimplanted $SiO_2$. The etch rate is gradually reduced for annealing temperatures up to 600 °C and rapidly from 600 to 800 °C. Infra-red absorption spectra of implanted $SiO_2$ show a breaking or weakening of the Si–O bonds which confirms the faster etch rate. Shannon and Ford (1970) have found that $SiO_2$ layers stabilized with phosphorus glass show complete recovery of the C–V characteristic on annealing up to 500 °C, while Glotin et al. (1971) have reported that a 20 keV phosphorus implant ($10^{14}$ ions/cm$^2$) on a 0.5 micron thick $SiO_2$ layer produced the same effect. Ahn et al. (1970) have produced phosphosilicate glass layers on thin thermal $SiO_2$ layer by bombarding with phosphorus ions to a dose of $5 \times 10^{16}$ ions/cm$^2$ with ion energies from 30 to 100 keV. The oxide is annealed at 400 °C and was shown by C–V measurements to act as a trap for mobile carriers. These results indicate that the phosphorus glass layers used for mobile charge trapping in silicon technology are able to trap the charge centres produced by implantation. Fritzche et al. (1971) implanted sodium ions into thermal $SiO_2$ on silicon and found them to be less mobile than sodium ions introduced as contaminants. This suggested that the implanted sodium atoms were trapped in the lattice disorder. If lattice disorder was introduced by a nitrogen implantation and contaminant sodium ions drifted at 200 °C into the disorder region, then successive drift treatments gave reduced $V_{FB}$ shifts indicating that an increasing number of sodium ions were being trapped in the damage.

Freeman et al. (1970) have shown that silicon dioxide and nitride layers can be formed in polished silicon wafers by implanting very large doses of oxygen or nitrogen respectively. Good stable layers with high dielectric strengths were produced but the C–V characteristic showed no depletion or accumulation effects with changing polarizing voltage indicating a high defect density at the Si–SiO$_2$ interface. The silicon below the oxide exhibited no electron channelling pattern which suggested that it had largely lost its monocrystalline structure. Thermal annealing to high temperatures produced no improvement. As a result of the damage to the silicon this technique of forming insulating layers cannot be considered as an alternative to the conventional thermal methods used in the silicon planar technology.

In conclusion, insulating layers on silicon can be implanted with heavy ions without permanent damage being produced and the charge centres in the layers associated with the implantation damage can be removed by annealing above 300 °C. The recovery is improved if the insulating layer has a layer of phosphosilicate covering it to act as a getter for the defects.

### 5.6. Compound semiconductors

#### 5.6.1. INTRODUCTION

The increasing use of compound semiconductors for device electronics such as electroluminescent diodes, Gunn diodes, infrared detectors, microwave transistors, has emphasized the difficulties associated with the technology of fabricating with compound semiconductors. This technology is not as advanced as that of silicon due to several serious problems;

(i)  the materials have two or three components and the choice of a dopant is not always clear.

(ii)  the basic crystal material is difficult to prepare in a form free of gross defects such as inclusions, grain boundaries etc. (epitaxy is playing an important role here in producing high quality materials).

(iii) diffusion is difficult as one of the components of the semiconductor is usually very volatile.

(iv)  suitable materials for diffusion masks are difficult to find, and

(v)  there is no natural passivating layer as exists in the case of silicon.

In the light of these problems, ion implantation potentially has a good opportunity of making a significant contribution to compound semiconductor technology, but there are some difficulties to overcome, (Allen, 1970);

(i)   the choice of dopant is not clear.

(ii)  experimental range data is sparse,

(iii) the projected ranges are small as calculated by the LSS theory as most substrates and dopants have high $Z$ numbers (Lindhard et al., 1963),

(iv)  annealing of radiation damage is difficult,

(v)   the electrical activity of chemical dopants and residual damage are difficult to separate, and

(vi)  it is difficult to make ohmic contacts to the thin implanted layers.

The choice of dopant is not restricted by the implantation process but by the desired electrical activity required in the semiconductor. Compound semiconductors are more susceptible to radiation damage than silicon. As an example a dose of $10^{13}$ 100 keV protons/cm$^2$ reduced the carrier concentration in a piece of n-type gallium arsenide from $2 \times 10^{16}$ carrier/cm$^3$ to less than $10^{11}$/cm$^3$ by introducing traps. The damage was stable at 500 °C but annealed out at 700 °C (Allen, 1970).

Table 5.24 attempts to summarize the present position of the ion implantation of compound semiconductors and for further reference, the review by Allen (1970) should be consulted.

From the device aspect, the main activity has been concerned with the formation of n and p doped layers and the formation of p-n junctions. This is discussed in the next section.

### 5.6.2. DEVICE FABRICATION BY ION IMPLANTATION IN COMPOUND SEMICONDUCTORS

One of the main problems with compound semiconductors is making ohmic contacts. This is particularly important in the case of n-type GaAs because of several important microwave applications. Ion implanted contacts to n-type GaAs have been studied by Hunsperger et al. (1968) and later Dunlap et al. (1969). They used sulphur, tin and argon ions implanted at room temperature and found that a high resistance ohmic contact was formed provided the implant was not annealed above 200 °C. They concluded that the damage rather than the electrical activity of the particular ion in the GaAs was responsible. Allen (1970) states that if this method can be extended to other materials, it should be a useful aid in the assessment of implanted layers.

TABLE 5.24

Details of ion implantation into compound semiconductors

| Ion | Experimental details and results | Reference |
|---|---|---|
| | (i) Gallium arsenide | |
| H | $10^{13}$ $H^+$ ions/cm$^2$ implanted in n and p GaAs reduced carrier concentration to $< 10^{11}$/cm$^3$. | Foyt et al. (1969a) |
| S | n-p diodes formed in p substrate by 400 and 800 keV implants. | Leith et al. (1967) |
| | 30 keV implant produced n-type activity before annealing but no activity afterwards. | Dunlap et al. (1969) |
| | Another implant produced n layer ($\rho \sim 10^4$ ohms/square) which was removed on annealing to 700 °C and reappeared again after annealing above 800 °C. | Hunsperger and Marsh (1970(b)) |
| Ar | 30 keV $1.1 \times 10^{16}$ Ar$^+$/cm$^2$ implant without annealing produced high resistance ohmic contact to n type GaAs | Dunlap et al. (1969) |
| Zn | 80 keV implant formed p-n diode with abrupt junction. Light emitted at $\lambda = 0.92$ microns. | Manchester (1967) Roughan et al. (1968) |
| | Implant formed p-n junction with large series resistance. | Schroeder and Dieselman (1967) |
| | 20 keV implant formed p layer after 400 °C anneal with $\rho = 10^3$ ohms/square. | Mayer et al. (1967b) |
| | 70 keV implant formed p layer with $\rho = 598$ ohms/square and $\mu = 45.8$ cm$^2$/volt · sec. | Hunsperger et al. (1968) |
| | 20 keV implant formed p layer with $\rho = 10^4$ ohms/square and $\mu = 10$ cm$^2$/volt · sec. | Dunlap et al. (1969) |
| | 20 keV implant formed p layer with 1 % electrical activity after 600 °C anneal with $\mu = 10$ cm$^2$/volt · sec. | Hunsperger and Marsh (1970(b)) |
| | 84 keV implant formed p layer after 600 °C anneal. | Hunsperger and Marsh (1970(b)) |
| Se | 400 keV implant at 500 °C formed n layer with 50 % electrical activity after 800 °C anneal. | Foyt et al. (1969b) |
| Cd | 20 keV implant formed p layer, with $\rho = 2k\Omega$/square and $\mu = 40$ cm$^2$/V · sec. | Hunsperger et al. (1968) |
| | 20 keV implant produced p-i-n diodes. | Hunsperger and Marsh (1970(a)) |
| | 60 keV implant annealed at 800 °C formed p layer with 100 % elect. activity for dose $< 10^{14}$ ions/cm$^2$. | Hunsperger and Marsh (1970(b)) |
| Te | 20 keV implants at 400 °C formed n layer. A C-V plot indicated a graded junction. | Mayer et al. (1967b) |
| Sn | 35 keV implant at 400 °C gave n layer after 600 °C anneal. | Hunsperger et al. (1968) |
| | (ii) Semi insulating gallium arsenide | |
| C | 70 keV implant formed p layer with 2 % elect. activity and $\mu = 250$ cm$^2$/V · sec. | Sansbury and Gibbons (1970) |

TABLE 5.24 (continued)

| | (ii) *Semi-insulating gallium arsenide* (contd.) | *Reference* |
|---|---|---|
| Si | $10+30+50$ keV multiple implant formed n layer with $\mu = 2600$ cm$^2$/V·sec. | Sansbury and Gibbons (1969) |
| | 70 keV implant, formed n layer after 700 °C anneal with 10 % elect. activity and $\mu = 2700$ cm$^2$/V·sec. | Sansbury and Gibbons (1970) |
| S | 70 keV implant formed n layer with low electrical activity and $\mu = 3600$ cm$^2$/ V·sec. | Sansbury and Gibbons (1970) |
| Cd | Cr doped GaAs when implanted with $5 \times 10^{14}$ ions/cm$^2$ and annealed at 800 °C gave 20 % electrical activity with $\mu = 20$ cm$^2$/V·sec. | Shifrin et al. (1969(b)) |
| Zn Se Kr Xe | These four ions are reported to produce conducting layers in $10^8$ ohm-cm GaAs. The energy was $< 50$ keV and dose about $6 \times 10^{13}$ ions/cm$^2$. | Zelevinskaya et al. (1970) |
| | (iii) *Gallium phosphide* | |
| O | 100 keV oxygen was implanted in Zn doped GaP. Cathodoluminescence was observed at 1.8 eV. | Lacey et al. (1969) |
| Bi | Isoelectronic traps produced in layer after 800 °C anneal. Photoluminescence increases with fraction Bi substitutional. | Merz et al. (1970) |
| | (iv) *Indium arsenide* | |
| S | 800 keV implant formed n layers with up to $4 \times 10^{19}$ S atoms/cm$^3$ after 550 °C anneal. p-n junction formed. | McNally and King (1968) |
| Zn | 400 keV implant gave 2 % doping efficiency after 450 °C anneal used for infrared detector. | McNally (1970) |
| | (v) *Indium antimonide* | |
| H | Proton implant produced n layer with $N_s = 10^{13}$ carriers/cm$^2$ and $\mu = 15,000$ cm$^2$/V·sec used for photo detector. | Foyt et al. (1970) |
| S | 800 keV implant formed n layers with up to $8 \times 10^{17}$ S atoms/cm$^3$ after 300 °C anneal, p-n junction formed. | McNally and King (1968) |
| Zn | 400 keV implant gave 2 % doping efficiency after 450 °C anneal. Used for infrared detector array. | McNally (1970) |
| | (vi) *Zinc oxide* | |
| H | 10 keV, $2 \times 10^{15}$ H$^+$/cm$^2$ implant with no anneal gave n type activity with $\rho = 10^3$ ohms/square. | Shifrin et al. (1969(a)) |
| | $1 \times 10^{14}$ H$^+$/cm$^2$ implant gave n activity. | Shifrin et al. (1969(b)) |
| | (vii) *Zinc selenide* | |
| Li | 37 and 27 keV implant formed p layer after 400 °C anneal p-n junction exhibited electroluminescence. | Park and Chung (1971) |

TABLE 5.24 (continued)

|  | (viii) *Zinc telluride* | *Reference* |
|---|---|---|
| H | $5 \times 10^{14}$ H$^+$/cm$^2$ converted p material with $2 \times 10^{18}$ carriers/cm$^3$ to semi-insulating form. Depth 1 micron per 100 keV. | Donnelly et al. (1970) |
| F | 400 keV, $1 \times 10^{15}$ F$^+$/cm$^2$ implant at 77 °K produced n layer in p material after 550 °C anneal. | Hou et al. (1969) |
| O | Luminescent centres observed for a concentration of about $10^{18}$ implanted oxygen atoms/cm$^3$. | Merz et al. (1970) |
| Cl | A multienergy implant (30-120 keV) of $1.6 \times 10^{15}$ Cl$^+$/cm$^2$ at room temp. into p material gave n activity after 520 °C anneal. Diode formed and electroluminescence observed. | Marine (1970) |

|  | (ix) *Cadmium sulphide* |  |
|---|---|---|
| B | 70 keV $5 \times 10^{14}$ B$^+$/cm$^2$; formed n layer. | Shifrin et al. (1969(a)) |
| F | 80 keV $5 \times 10^{14}$ F$^+$/cm$^2$, gave slight n activity. | Shifrin et al. (1969(a)) |
| Al | 80 keV $6 \times 10^{14}$ Al$^+$/cm$^2$, gave n activity with $\rho_s = 10^8$ ohms/square. | Shifrin et al. (1969(a)) |
| P | Acceptor centres formed in n CdS. Activation energy 0.13 eV; some evidence of large amounts of residual damage. | Hou and Marley (1970) |
| Ga | 100 keV $2 \times 10^{14}$ Ga$^+$/cm$^2$; gave n activity. | Shifrin (1969(a)) |
| Bi | 25 keV room temp. implant produced p layers about 3000 Å deep with 1 to 5 ohm·cm resistivity and $\mu = 3$ cm$^2$/V·sec. | Chernow et al. (1968) Eldridge et al. (1970) |

|  | (x) *Cadmium telluride* |  |
|---|---|---|
| P | p layers reported. | Kachurin et al. (1969) |
| As | 400 keV hot implant (500 °C) produced p layer, doping efficiency $<1\%$ and $\mu = 3$ cm$^2$/V·sec. | Donnelly et al. (1968) |

|  | (xi) *Silicon carbide* |  |
|---|---|---|
| B | Implant fails to exhibit acceptor activity. | Hunsperger et al. (1968) Dunlap et al. (1969) Marsh and Dunlap (1970) |
| N | Donor activity observed after the implant annealed at over 1000 °C. Diodes with good $I/V$ characteristics up to 400 °C formed in α-SiC. | Dunlap et al. (1969) Marsh and Dunlap (1970) |
| Al | 500 °C implant into n β-SiC produced p type activity, $\rho = 900$ ohms/square, $\mu = 400$ cm$^2$/V·sec. Later work shows no p-type activity. | Hunsperger et al. (1968) Dunlap et al. (1969) Marsh and Dunlap (1970) |

TABLE 5.24 (continued)

---

(xi) *Silicon carbide* (contd.)                                              *Reference*

| | | |
|---|---|---|
| P | 400 keV implant gave $2 \times 10^{20}$ donors/cm³ junction depth 0.6 microns diode formed. | Leith et al. (1967) |
| | Donor activity observed after high temp. anneal ($> 1000$ °C). Diodes with good $I/V$ properties. | Dunlap et al. (1969)<br>Marsh and Dunlap (1970) |
| Ga | Implant failed to exhibit acceptor activity. | Hunsperger et al. (1968)<br>Marsh and Dunlap (1970) |
| Sb | Implant formed n layer in p α-SiC after 1700 °C anneal. Poor diode characteristics; improved results in later paper. | Hunsperger et al. (1968) |
| Bi | 50 keV implant at 500 °C gave n layer in p α-SiC after 1000 °C anneal. | Hunsperger et al. (1968)<br>Marsh and Dunlap (1970) |
| He<br>Ge<br>Kr | These three ions were implanted between 160 and 300 keV to a dose of $10^{14}$ ions/cm² and annealed up to 1400 °C.<br>Bright yellow luminescence in n epitaxial layers weak green luminescence in p epitaxial layers. | Brander et al. (1970) |

---

Lindley et al. (1969) have used proton bombardment to produce a semi-insulating guard ring around a platinum Schottky barrier photodiode made on n-type GaAs. These devices operated with a gain of 100 and showed gain-bandwidth products greater than 50 GHz.

GaAs, CdS and ZnO have been investigated by Shifrin et al. (1969a) and (1969b) for piezoelectric crystals in acoustic microwave devices. The requirement for these devices is a thin conducting layer on an insulating substrate to allow interaction between the shallow surface wave and the electric field. Ion implantation was being tried to produce these layers particularly in ZnO as it offers the highest performance according to computer calculations.

Foyt, Lindley and Donnelly (1970) have made a photovoltaic detector in InSb by a proton implantation. The p-type InSb was covered with 0.15 microns of $SiO_2$ and masked with a thick layer of photoresist. Gold contacts were later made through windows in the oxide layer to the n regions produced by the proton bombardment. Good $I/V$ characteristics were obtained and the $C/V$ characteristic suggested a slightly graded profile. The peak detectivity of $10^{11}$ cm Hz$^{\frac{1}{2}}$ watt$^{-1}$ was measured at 4.9 microns. The quantum efficiency at this wavelength was 35 %.

The high resistivity layer produced in ZnTe by proton implantation has been exploited by Donnelly et al. (1970) to make a metal-insulator-semiconductor structure in p-type ZnTe. Gold was used for the metal contacts. When the p-ZnTe was biased positive with respect to the gold contact on the insulating ZnTe, electrons were injected into the p-region and they produced electroluminescence at 300 °K and 77 °K in the green and red with quantum efficiencies up to 0.3 %.

Ion implanted planar mosaic photovoltaic infrared detectors in InAs and InSb have been described by McNally and King (1968) and McNally (1970). Sulphur and zinc ions were used to produce n layers in the p-type InAs and InSb. Surface layers of $SiO_2$, $Al_2O_3$ or MgO have been tried for surface passivation and anti-reflection coatings. The detectivities $(D^*)$ (500 °K, $2\pi$) were;

$$5 \times 10^8 \text{ cm Hz}^{\frac{1}{2}} \text{ watt}^{-1} \text{ for InAs}$$

and

$$5 \times 10^9 \text{ cm Hz}^{\frac{1}{2}} \text{ watt}^{-1} \text{ for InSb}$$

The fact that all the devices described above were made by the planar technology, emphasis that there are various features of this technology which are directly applicable to compound semiconductors, when ion implantation is used for doping. There is no restriction on the choice of dopant as it can be implanted at room temperature and the masking is not exposed to the chemical attack, which frequently occurs at diffusion temperatures. The annealing of the implantation can be done at temperatures lower than diffusion temperatures and for shorter times, which helps in avoiding the degradation of the semiconductor surface due to the loss of one of the components which can occur at elevated temperatures. Many experimenters have found it beneficial to use a thin layer of $SiO_2$ on the semiconductor to protect it during implantation and annealing. This has the disadvantage that higher ion energies are required to penetrate the $SiO_2$ layer. The layer can be readily removed by buffered hydrofluoric acid (section 5.3.5.3.1) which does not attack compound semiconductors such as gallium arsenide and gallium phosphide. Many compound semiconductors are not available in the form of large flat circular polished wafers as is the case for silicon. In many cases pieces of material are small, irregular in shape and they vary considerably in thickness. This problem with the material preparation makes the use of high definition contact photolithography very difficult and it is suggested that projection photolithography (Osborne, 1968) or electron beam fabrication techniques (Bauer et al., 1970) could be used as alternative methods for generating the masking patterns.

## Note added in proof (May 1973)

INTRODUCTION

The use of ion implantation for the production of silicon semiconductor devices has been receiving increasing attention over the past two years particularly in the United States. The research has shown that implantation is a controllable process for doping silicon and this fact together with the emergence of ion implanting equipment capable of handling very large numbers of wafers has caused manufacturers seriously to consider the process for use in production. Increasing demands are being made on the silicon integrated circuit technology to produce more complex circuits capable of performing a very large number of logical functions. The pocket electronic calculator using a single MOS chip is a good example of increased function density and complexity. To meet these demands requires the use of smaller transistors and larger chip areas with the consequent reduction in yield. It is therefore imperative to increase the yield by employing methods which reduce the defect density and it is in this area that ion implantation has an important role to play. These general remarks apply equally to MOS and bipolar integrated circuits. It must also be emphasised that any solution to the yield problem must also be economically viable.

The MOS integrated circuit is receiving most of the attention at present but there is considerable interest in bipolar devices particularly due to the very encouraging results that have been obtained with arsenic implanted emitters.

MOS INTEGRATED CIRCUITS

Dill et al. (1972) and Swanson and Meindl (1972) have described the use of boron implantation for MOS transistor threshold control and the doping of p-type wells in n-type substrates for complementary MOS circuitry (C-MOS), while MacPherson (1972) has made calculations of threshold shifts in ion implanted devices. The results agree well with published data. Crawford (1972) described the fabrication of depletion load MOS transistors for use in p-MOS circuitry and the resultant advantages of increased speed, lower power consumption and greater tolerance to power supply fluctuations. The technique has been used in the manufacture of a random access memory (RAM) and the microprocessor and random logic chips for the Hewlett Packard h/p 35 pocket scientific calculator.

At a recent I.E.E.E. Electron Devices Meeting in Washington (Dec. 1972) papers from three prominent U.S. semiconductor manufacturers concentrated on the threshold control of p and n channel MOS transistors (Moline et al. (1972), Sigmon (1972) and Dingwall (1972)) and for a channel stop below the field oxide in a p-channel 1024 bit RAM (Moline et al. (1972)). The actual ion energy and dose required in practice to alter the MOS transistor threshold is dependent on several critical processing parameters;

(i) the thickness of the insulator (e.g. $SiO_2$) through which the implantation is made,

(ii) the annealing temperature and time and

(iii) any regrowth or continued growth of the gate oxide which can lead to a depletion of boron for example at the $SiO_2$–Si interface.

A typical boron dose required to alter a threshold voltage by one or two volts is between 1 and $2 \times 10^{12}$ ions/$cm^2$ with the ion energy in the 30 to 100 keV range. In the case of phosphorus the doses are lower ($< 10^{12}$ ions/$cm^2$) and the energy higher because of the shorter range of the phosphorus ions.

The influence of an oxide layer on silicon when implanting boron ions has been studied by Bauer et al. (1973) and they concluded that the stopping power of $SiO_2$ can be up to 20 % lower than that predicted from the L.S.S. theory. This result has been confirmed by Chu et al. (1972) using Rutherford backscattering techniques. Reddi and Yu (1972) have measured the attenuation of high energy boron and phosphorus ions (300 to 600 keV) in $SiO_2$, $Si_3N_4$, aluminium and gold layers.

BIPOLAR TRANSISTORS

The work in the bipolar field has been concentrated on the use of ion implantation for doping the emitter and base regions of microwave n-p-n transistors. Scavuzzo et al. (1972) and Reddi and Yu (1972) have described the fabrication of transistors with a wide range of $h_{fe}$ (20 to over 1000) and with $f_T$ from 1 to 8 GHz using two or three implantations. Reddi and Yu (1972) gave a noise figure of 2.7 dB at 4 GHz.

The general technique for fabricating the transistor is as follows. The emitter is formed by implanting a large dose of arsenic (1 to $2 \times 10^{16}$ ions/$cm^2$ at over 100 keV) and then driving it in at a temperature between 900°C and 1100 °C. Arsenic is chosen for four reasons:

(i) it is free of the material dependant exponential tail (see Chapter 2) which interferes with the base doping and is associated with phosphorus,

(ii) the diffusion constant of arsenic is enhanced at high concentrations and this results in a flat topped very abrupt profile after diffusion,

(iii) the arsenic can be diffused deeper than the damaged region near the surface allowing long lifetime material to be present in the base and

(iv) as arsenic is very soluble in silicon it is possible to obtain very high emitter concentrations.

A deep boron implantation (200 to 275 keV 2 to $5 \times 10^{12}$ ions/cm$^2$) is used to form the intrinsic base region and control the Gummel number (active centres/cm$^2$ in the base) and hence the $h_{fe}$. The base width and therefore $f_T$ is a function of the ion energy. A third boron implant (about 40 to 50 keV and 1 to $4 \times 10^{14}$ ions/cm$^2$) is used by Scavuzzo et al. (1972) to reduce the extrinsic base resistance by increasing the doping between the buried intrinsic base layer and the base contact areas.

In addition to this basic work on the bipolar transistor structure, implantation has also been used for arsenic doping buried layers prior to epitaxy and for phosphorus doped isolation walls in integrated circuits.

## DAMAGE GETTERING

The damage produced by ion implantation has been studied by Seidel et al. (1972) as a means of gettering metallic impurities such as copper and gold in silicon. The damage was produced by implanting silicon ions (100 keV, $10^{16}$ ions/cm$^2$) and then annealing at 900° C. High concentrations of copper and gold were found by Rutherford backscattering measurements in wafers deliberately contaminated with the metals. The gettering of copper was aided by the presence of oxygen during the annealing. A comparison with the conventional phosphorus glass gettering showed that

(i) the gold atoms were on non-substitutional sites in the damaged layer but were on substitutional sites in a phosphorus doped layer and the gold profile was an error function similar to the phosphorus profile.

(ii) the doped layer was more effective than the damage layer as measured by the reduction in the leakage current of gold doped $p^+$-n diodes.

Hseih et al. (1973) have reported a considerable improvement in the performance of silicon photodiode vidicon targets by a reduction in the leakage current of individual diodes by gettering the impurities with damaged layers produced by P, As or Ar implantation.

## SUMMARY

Important applications are being found for ion implantation in the production of MOS and bipolar silicon transistors. MOS circuits are already in limited production using implantation to produce depletion load MOS transistors and this use is likely to expand rapidly as experience and demand dictate. The encouraging results with bipolar transistors indicate that a similar expansion is to be expected shortly in this field.

## References

The references for chapter 5 are listed below in alphabetical order. A comprehensive list of references to all aspects of Ion Implantation can be found in the Bibliography on Ion Implantation compiled by Morgan and Greenhalgh (1971).

Abe, T., K. Sato, M. Konaka and A. Miyazaki, 1969, Extended Abstract 184, Electrochem. Soc. Fall Meeting Detroit.

Ahn, J., B. M. Gordon and W. A. Pliskin, 1970, I.B.M. Tech. Disclosure Bulletin 13, 7, 1798.

Allen, R. M., 1970, Proc. European Conf. Ion Implantation (Peter Peregrinus Ltd.) p. 127.

Alväger, T. and N. J. Hansen, 1962, Rev. Sci. Instrum. 33, 567.

Andersson, A. and G. Swenson, 1970, Proc. European Conf. Ion Implantation (Peter Peregrinus Ltd.) p. 65.

Assemat, J. L., 1971, Proc. 2nd Int. Conf. Ion Implantation, Garmisch (Springer-Verlag) p. 351.

Atalla, M. M., E. Tannenbaum and E. J. Schreibner, 1959, Bell Syst. Tech. J. 38, 749.

Aubuchon, K. G., 1969, Proc. Int. Conf. Properties and Use of M.I.S. structures, Grenoble, p. 575.

Bader, R. and S. Kalbitzer, 1970a, App. Phys. Letters 16, 1, 13.

Bader, R. and S. Kalbitzer, 1970b, Radiation Effects 6, No. 3 and 4, 211.

Baron, R., G. A. Shifrin, O. J. Marsh and J. W. Mayer, 1969, J. Appl. Phys. 40, 9, 3702.

Bauer, L. O., R. W. Bower and E. Wolf, 1970, I.E.E.E. Trans. Electron Devices ED-17, No. 6, 446.

Bazin, B., E. H. Crawford and L. J. Levin, 1968, Proc. I.E.E.E. 56, 6, 1088.

Beale, J. R. A., 1970, Proc. European Conf. Ion Implantation, Reading (Peter Peregrinus Ltd.) p. 81.

Beer, A. C., 1963, Supp. 4 Solid State Physics Series (Academic Press).

Beer, A. C., 1966, Solid State Electronics 9, 339.

Billington, D. S. and J. H. Crawford Jr., 1961, Radiation Damage in Solids (Oxford University Press) p. 312.

Blamires, N. G., 1970, Proc. European Conf. on Ion Implantation (Peter Peregrinus Ltd.) p. 52.

Blamires, N. G., 1971, private communication (AERE, Harwell).

Blamires, N. G., D. N. Osborne, R. B. Owen and J. Stephen, 1967, Proc. Int. Conf. on Application of Ion Beams to Semiconductor Technology, Grenoble, p. 669.

Blamires, N. G., M. D. Matthews and R. S. Nelson, 1968, Physics Letters 28A, 3, 178.

Bower, R. W., 1969, Nerem Record 120.

Bower, R. W., 1970, U.S. Patent. 3, 507, 709.

Bower, R. W., R. Baron, J. W. Mayer and O. J. Marsh, 1966, App. Phys. Letters 9, 5, 203.

Bower, R. W. and H. G. Dill, 1966, Paper presented at Int. Electron Devices Meeting 28/10/66, Washington.

Bower, R. W., H. G. Dill, K. G. Aubuchon and S. A. Thompson, 1968, I.E.E.E. Trans. Electron Devices ED-15, 10, 757.

Brander, R. W., M. P. Callaghan and A. Todkill, 1970, Proc. European Conf. Ion Implantation, Reading (Peter Peregrinus Ltd.) p. 135.
Brook, P. and C. S. Whitehead, 1968, Electronics Letters 4, 335.
Brown, D. M., W. E. Engeler, M. Garfinkel and P. V. Gray, 1968, Solid State Electronics 11, 1105.
Buchanan, B., R. Dolan and S. Roosild, 1967, Proc. Int. Conf. Application of Ion Beams to Semiconductor Technology, Grenoble, p. 649.
Buehler, M. G. and G. L. Pearson, 1966, Solid State Electronics 9, 395.
Bulthuis, K., 1968, Physics Letters 27A, 4, 193.
Bulthuis, K. and R. Tree, 1969, Physics Letters 28A, 558.
Burr, P. and J. E. Whitehouse, 1970, Proc. European Conf. Ion Implantation (Peter Peregrinus Ltd.) p. 61.
Burrill, J. T., W. J. King, S. Harrison and P. McNally, 1967, I.E.E.E. Trans. Electron Devices ED-14, No. 1, 10.
Burt, D. J., 1969, Proc. Int. Conf. Prop. and Use of M.I.S. Structures, Grenoble, p. 605.
Burt, D., 1971, private communication (Hirst Research Centre).
Cairns, J. A., D. F. Holloway and R. S. Nelson, 1970, Proc. European Conf. Ion Implantation (Peter Peregrinus Ltd.) p. 203.
Cairns, J. A. and R. S. Nelson, 1971, Rad. Effects 7, 163.
Cairns, J. A., D. F. Holloway and R. S. Nelson, 1971, Rad. Effects 7, 167.
Camenzind, H. R., 1968, Circuit Design for Integrated Electronics (Addison Wesley) (276 pages).
Caughey, D. M. and R. E. Thomas, 1967, Proc. I.E.E.E. Letters 55, 2192.
Chernow, F., G. Eldridge, G. Ruse and L. Wahlin, 1968, App. Phys. Letters 12, 10, 339.
Clark, A. H. and K. E. Manchester, 1968, Trans. Met. Soc. A.I.M.E. 242, 1173.
Copeland, J. A., 1969, I.E.E.E. Trans. on Electron Devices ED-16, 445.
Copeland, J. A., 1970, I.E.E.E. Trans. on Electron Devices ED-17, 5, 404.
Crowder, B. L. and F. F. Morehead, Jr., 1969, App. Phys. Letters 14, 10, 313.
Crowder, B. L. and R. S. Title, 1970, Rad. Effects 6, 63.
Cussins, W. D., 1955, Proc. Phys. Soc. 68, 213.
Davidge, P. D. and L. I. Maissel, 1966, J. App. Phys. 37, 574.
Davies, D. E., 1968, Appl. Phys. Letters 13, 243.
Davies, D. E., 1969a, App. Phy. Letters 14, 227.
Davies, D. E., 1969b, Can. Jour. Phys. 47, 1750.
Davies, D. E., 1970, Solid State Electronics 13, 229.
Davies, D. E. and S. A. Roosild, 1970, App. Phy. Letters 17, 107.
Davies, D. E. and S. A. Roosild, 1971a, Solid State Electronics 14, 975.
Davies, D. E. and S. A. Roosild, 1971b, Proc. 2nd Int. Conf. Ion Implantation, Garmisch (Springer-Verlag) p. 23.
Davies, J. A., G. C. Ball, F. Brown and B. Domeij, 1964, Can. Jour. Physics 42, 1070.
Davies, J. A., J. Denhartog, L. Eriksson and J. W. Mayer, 1967, Can. J. Phys. 45, 4053.
Dearnaley, G., 1969, Reports on Progress in Physics 32, 4, 405.
Dearnaley, G., J. H. Freeman, G. A. Gard and M. A. Wilkins, 1968, Can. Jour. Phys. 46, 587.
Dearnaley, G., A. G. Hardacre and B. D. Rogers, 1969, Nucl. Instr. and Methods 71, 86.
Dexter, J. R., 1970, Thesis Virginia Polytechnic Institute Blacksberg (No. 70-9911).
Dill, H. G., R. W. Bower and T. M. Toombs, 1971a, Rad. Effects 7, Nos. 1 and 2, 45.
Dill, H. G., T. N. Toombs and L. O. Bauer, 1971b, Proc. 2nd Int. Conf. Ion Implantation, Garmisch (Springer-Verlag) p. 315.
Dolan, R., S. Roosild and B. Buchanan, 1966, Proc. 2nd Conf. Microelectronics Meeting of the I.N.E.A. (R. Oldenbourg, München) p. 207.

Donnelly, J. P., A. G. Foyt, E. D. Hinkley, W. T. Lindley and J. O. Dimmock, 1968, App. Phy. Letters **12**, 303.

Donnelly, J. P., A. G. Foyt, W. T. Lindley and G. W. Iseler, 1970, Solid State Electronics **13**, 755.

Drum, C. M. and P. Miller, 1971, Paper 17.7 Int. Electron Devices Conf. Washington Oct. 1971.

Dunlap, H. L., R. G. Hunsperger and O. J. Marsh, 1969, Development of Ion Implantation techniques for Microelectronics NAS 12-124, N70-17314.

Eldridge, G., P. K. Govind, D. A. Nieman and F. Chernow, 1970, Proc. European Conf. Ion Implantation, Reading (Peter Peregrinus Ltd.) p. 143.

Eriksson, L., J. A. Davies and J. W. Mayer, 1968, Radiation Effects in Semiconductors, Ed. F. Vook, p. 398 (Plenum Press, N.Y.).

Eriksson, L., J. A. Davies, N. G. E. Johansson and J. W. Mayer, 1969, Jour. App. Phys. **40**, 2, 842.

Faggin, F. and T. Klein, 1970, Solid State Electronics **13**, 1125.

Fairfield, J. M. and B. L. Crowder, 1969, Trans. Met. Soc. A.I.M.E. **245**, 469.

Ford, R. A., K. H. Nicholas and M. G. Vry, 1971, 5th Annual Conf. Solid State Devices, Lancaster (Sept. 1971).

Foyt, A. G., W. T. Lindley, C. M. Wolfe and J. P. Donnelly, 1969a, Solid State Electronics **12**, 209.

Foyt, A. G., J. P. Donnelly and W. T. Lindley, 1969b, App. Phys. Letters **14**, 372.

Foyt, A. G., W. T. Lindley and J. P. Donnelly, 1970, App. Phys. Letters **16**, 335.

Freeman, J. H., 1970, Proc. European Conf. on Ion Implantation, Reading (Peter Peregrinus Ltd.) p. 1.

Freeman, J. H., G. A. Gard, D. J. Mazey, J. Stephen and F. B. Whiting, 1970, Proc. European Conf. Ion Implantation (Peter Peregrinus Ltd.) p. 74.

Fritzche, C., A. Goetzberger, A. Axmann, W. Rothemund and G. Sixt, 1971, Rad. Effects **7**, Nos. 1-2, 87.

Frosch, C. J. and L. Derick, 1957, J. Electrochem. Soc. **104**, 547.

Fujinuma, K., T. Sakamoto, T. Abe, K. Sato and Y. Ohmura, 1969, Proc. 1st Conf. Solid State Devices Supp. J. Jap. Soc. App. Phy. **39**, 71.

Fuller, C. S. and J. A. Ditzenberger, 1954, J. App. Phys. **25**, 1439.

Fuller, C. S. and J. A. Ditzenberger, 1956, J. App. Phys. **27**, 544.

Galaktionova, I. A., V. M. Gusev, V. G. Naumenko and V. V. Titov, 1967, Proc. Int. Conf. Application of Ion Beams to Semiconductor Technology, Grenoble, p. 503.

Gibbons, J. F., 1966, Semiconductor Electronics (McGraw-Hill).

Gibbons, J. F., 1967, Proc. Int. Conf. Application of Ion Beams to Semiconductor Technology, Grenoble, p. 561.

Gibbons, J. F., J. L. Moll and N. I. Meyer, 1965, Nucl. Instr. and Methods **38**, 165.

Gibbons, J. F., 1968, Proc. IEEE **56**, 3, p. 305.

Gibson, W. M., F. W. Martin, R. Stensgaard, F. Palmgren-Jensen, N. I. Meyer, G. Galster, A. Johansen and J. S. Olsen, 1967, Proc. Int. Conf. Application of Ion Beams to Semiconductor Technology, Grenoble, p. 449.

Glotin, P., 1967, Proc. Int. Conference Application of Ion Beams to Semiconductor Technology, Edition Ophyrs, Grenoble.

Glotin, P., 1968, Can. Jour. Physics **46**, 705.

Glotin, P., J. Grapa and A. Monfret, 1967, Proc. Int. Conf. Applications of Ion Beams to Semiconductor Technology, Grenoble, p. 619.

Glotin, P., J. Bernard and A. Monfret, 1971, Rad. Effects **7**, Nos. 1-2, 65.

Goetzberger, A. and S. M. Sze, 1969, Applied Solid State Science Vol. 1, 153.

Goldsmith, N. and W. Kern, 1967, RCA Review **28**, 153.

Golovner, T. M., V. V. Zadde, A. K. Zaitseva, M. M. Koltun and A. P. Landsman, 1968, Soviet Physics 2, 5, 598.

Goode, P. D., M. A. Wilkins and G. Dearnaley, 1970, Rad. Effects 6, 3 and 4, 237.

Gray, P. E., D. de Witt, A. R. Boothroyd and J. F. Gibbons, 1964, Physical Electronics and Circuit Models of Transistors (S.E.E.C. Vol. 2) (Wiley).

Grimshaw, J. A. and D. N. Osborne, 1971, Solid State Electronics 14, 603.

Grimshaw, J. A. and J. Stephen, 1971, A.E.R.E. Report R6567.

Grove, A. S., 1967, Physics and Technology of Semiconductor Devices (Wiley).

Grove, A. S. and O. J. Fitzgerald, 1965, IEEE Trans. Electron Devices ED-12, p. 619.

Grosvalet, J., C. Jund, C. Motsch and R. Poirier, 1966, Surface Science 5, 49.

Gunnersen, E. M., A. J. Hitchcock and G. G. George, 1967, Proc. Int. Conf. Applications of Ion Beams to Semiconductor Technology, Grenoble, p. 487.

Gusev, V. M., V. V. Titov, M. I. Guseva and V. I. Kurimiyi, 1966, Soviet Physics–Solid State 7, No. 7, 1673.

Gusev, V. M. and V. V. Titov, 1969, Soviet Physics – Semiconductors 3, 1, 1.

Hart, R. R. and O. J. Marsh, 1969, App. Phys. Letters 15, 7, 206.

Heiman, F. P., 1966, Field Effect Transistors, Eds. J. T. Wallmark and H. Johnson (Prentice-Hall) p. 187.

Hilibrand, J. and R. D. Gold, 1960, R.C.A. Review 21, 245.

Hodges, D. A., M. P. Lepselter, R. W. MacDonald, A. U. MacRae and H. A. Waggener, 1969, I.E.E.E. Jour. Solid State Circuits SC-4, No. 5, 280.

Hoerni, J. A., 1960, I.R.E. Electron Devices Meeting, Washington D.C.

Hofstein, S. R. and F. P. Heiman, 1963, Proc. I.E.E.E. 51, 9, 1190.

Hofstein, S. R., 1967, Solid State Electronics 10, 657.

Holland, L., 1956, Deposition of Thin Films (Chapman and Hall).

Holmes, P. J., 1962, Ed. The Electrochemistry of Semiconductors (Academic Press).

Honig, R. E., 1962, R.C.A. Review 23, 574.

Hooper, R. C., J. A. Cunningham and J. G. Harper, 1965, Solid State Electronics 8, 833.

Hou, S. L., K. Beck and J. A. Marley, 1969, App. Phys. Letters 14, 151.

Hou, S. L. and J. A. Marley, 1970, App. Phy. Letters 16, 467.

Howes, J. H. and G. Knill, 1970, Proc. European Conf. Ion Implantation, Reading (Peter Peregrinus Ltd.) p. 97.

Hunsperger, R. G., H. L. Dunlap and O. J. Marsh, 1968, Development of Ion Implantation techniques for Microelectronics NAS 12-124 and N69-24439.

Hunsperger, R. G. and O. J. Marsh, 1970a, Metallurgical Transactions 1, 603.

Hunsperger, R. G. and O. J. Marsh, 1970b, Rad. Effects 6, No. 3 and 4, 263.

Ikeda, T., T. Tokoyama and T. Tsuchimuto, 1969, Extended Abstract 170, Electrochem. Soc. Fall Meeting, Detroit (1969).

Irvin, J. C., 1962, Bell Syst. Tech. J. 41, 387.

Isofilm International Inc., 1969, (Chatsworth Calif.) News article: Electrotechnology (Oct.) p. 20.

Itoh, T., T. Inada and K. Kanekawa, 1968, App. Phys. Letters 12, 8, 244.

Johansson, N. G. E. and J. W. Mayer, 1970, Solid State Electronics 13, 123.

Johansson, N. G. E., J. W. Mayer and O. J. Marsh, 1970, Solid State Electronics 13, 317.

Johnson, W. S., 1969, Thesis SU-SEL-69-014, Stanford University.

Johnson, W. S. and J. F. Gibbons, 1969, Projected Range Statistics in Semiconductors (Stanford University Bookstore).

Jonscher, A. K., 1960, Principles of Semiconductor Device Operation (Bell and Sons Ltd.).

Josephy, R., 1970, private communication (G.E.C. Hirst Research Centre).

Kachurin, G. A., V. M. Zelevinskaya and L. S. Smirvov, 1969, Sovjet Physics–Semi-conductors 2, 1527.

Kalbitzer, S., R. Bader, H. Herzer and K. Bethge, 1967a, Z. Physik 203, 117.

Kalbitzer, S., R. Bader, W. Melzer and W. Stumpfi, 1967b, Nucl. Inst. and Methods **54**, 323.

Kawamoto, H. and W. G. Oldham, 1970, I.E.E.E. Trans. Nucl. Sci. **NS-17**, 2, 26.

Kellett, C. M., W. J. King, F. A. Leith, et al., 1966, High Energy Implantation of Materials (U.S.) Air Force Contract AF19(628)-4970 AD635267.

Kennedy, D. P., P. C. Murley and W. Kleinfelder, 1968, I.B.M. Jour. Research and Development **12**, 399.

Kennedy, D. P. and R. R. O'Brien, 1969, I.B.M., Jour. of Research and Development **13**, 212.

Kerr, J. A. and L. N. Large, 1967, Proc. Int. Conf. on Applications of Ion Beams to Semiconductor Technology, Grenoble, p. 601.

Kerr, J. A., 1971, private communication (Mullard Research Laboratories).

KEV Electronics Corp., 1971, KEVICAP VHF tuner diode data sheet.

King, W. J., J. T. Burrill and D. Bumiller, 1962–1965, p-n Junction Formation Techniques, U.S. Air Force Contracts AF33(657)-10505, AF33(615)1097 and AF33(615)-2292.

King, W. J. and J. T. Burrill, 1964, Proc. 4th Photovoltaic Specialists Conf. Cleveland, Ohio.

King, W. J., J. T. Burrill, S. Harrison, F. W. Martin and C. M. Kellett, 1965, Nucl. Inst. and Methods **38**, 178.

Kleinfelder, W. J., 1967, Technical Report K701-1, Stanford Electronics Laboratories, California.

Kooi, E., 1967, The Surface Properties of Oxidised Silicon (Cleaver-Hume Press Ltd., Philips Technical Library).

Lacey, S. D., L. N. Large and D. R. Wright, 1969, Electronics Letters **5**, 10, 203.

Laegsgaard, E., F. W. Martin and W. M. Gibson, 1968, Nucl. Inst. and Methods **54**, 223.

Large, L. N., H. Hill and M. P. Ball, 1967, Int. Jour. Electronics **22**, 153.

Lawson, R. W., 1970, private communication (Post Office Research Station).

Lecrosnier, D. P. and G. P. Pelous, 1970, Proc. European Conf. Ion Implantation, Reading (Peter Peregrinus Ltd.) p. 102.

Leith, F. A., W. J. King, P. McNally, E. Davies and C. M. Kellett, 1967, High Energy Ion Implantation of Materials, AFCRL-67-0123 AD 651313.

Lepselter, M. P., 1966, Bell Syst. Tech. Jour. **45**, 2, 223.

Lepselter, M. P., A. U. MacRae and R. W. MacDonald, 1969, Proc. I.E.E.E. **57**, 5, 812.

Ligenza, J. R., 1965, J. App. Physics **36**, 2703.

Lindhard, J., M. Scharff and H. E. Schiøtt, 1963, Kgl. Danske Videnskab Selskab, Mat.-fys. Medd. **33**, 14.

Lindley, W. T., R. J. Phelan Jr., C. M. Wolfe and A. G. Foyt, 1969, App. Phys. Letters **14**, 6, 197.

Logan, M. A., 1961, Bell Syst. Tech. J. **40**, 3, 885.

McCaldin, J. O. and A. E. Widmer, 1963, J. Phys. and Chem. Solids **24**, 1073.

MacDougall, J. D. and K. E. Manchester, 1969, Paper presented at the National Electronics Conf. Chicago (Dec. 1969).

MacDougall, J. D., K. E. Manchester and P. E. Roughan, 1969, Proc. I.E.E.E. **57**, 9, 1538.

MacDougall, J., K. Manchester and R. B. Palmer, 1970, Electronics **43**, 13, 86.

McNally, P. J., 1970, Rad. Effects **6**, 1 and 2, 149.

McNally, P. J. and W. J. King, 1968, paper presented at 16th Nat. Infrared Information Symposium, Ft. Monmouth.

MacPherson, M. R., 1971, App. Phys. Lett. **18**, 11, 502.

MacRae, A. U., 1971, Rad. Effects **7**, 59.

Maloney, C. R. C., 1971, private communication (Smiths Industries Ltd.).

Manchester, K. E., C. B. Silbey and G. Alton, 1965, Nucl. Inst. and Methods **38**, 169.

Manchester, K. E., 1966, S.C.P. and Solid State Technology, September, 48.

Manchester, K. E., 1967, Electronics **40**, 3, 116.

Marine, J., 1970, Proc. European Conf. Ion Implantation Reading (Peter Peregrinus Ltd.) p. 153.

Marsh, O. J., J. W. Mayer and G. A. Shifrin, 1967, Proc. Int. Conf. Application of Ion Beams to Semiconductor Technology, Grenoble, 513.

Marsh, O. J. and H. L. Dunlap, 1970, Rad. Effects **6**, 3 and 4, 301.

Martin, F. W., W. J. King and S. Harrison, 1964, Trans. I.E.E.E. Nucl. Sci. **NS-11**, No. 3, 280.

Martin, F. W., S. Harrison and W. J. King, 1966, Trans. I.E.E.E. on Nucl. Sci. **NS-13**, 22.

Martin, F. W., 1969, Nucl. Instr. and Methods **72**, 223.

Mayer, J. W., O. J. Marsh, G. A. Shifrin and R. Baron, 1967a, Can. J. Phys. **45**, 4073.

Mayer, J. W., O. J. Marsh, R. Mankarious and R. Bower, 1967b, Jour. App. Phys. **38**, 4, 1975.

Mayer, J. W., L. Eriksson, S. T. Picraux and J. A. Davies, 1968, Can. Jour. Phys. **46**, 663.

Mayer, J. W. and O. J. Marsh, 1969, Ion Implantation in Semiconductors, 239, Applied Solid State Science, Ed. R. Wolfe (Academic Press, New York).

Medved, D. B., J. Perel, H. L. Daley and G. P. Rolik, 1965, Nucl. Inst. and Methods **38**, 175.

Merz, J. L., L. C. Feldman and E. A. Sadowski, 1970, Rad. Effects **6**, 3 and 4, 285.

Meyer, O., 1968, Trans. I.E.E.E. Nucl. Sci. **NS-15**, No. 3, 232.

Meyer, O., 1969a, Proc. Meeting on Special Techniques and Materials for Semiconductor Detectors, Ispra, Euratom publication EUR 4269e, p. 161.

Meyer, O., 1969b, Nucl. Inst. and Meth. **70**, 279.

Meyer, O., 1969c, Nucl. Inst. and Meth. **70**, 285.

Meyer, O. and G. Haushahn, 1967, Nucl. Inst. and Meth. **56**, 177.

Miller, S. L., 1957, Phys. Rev. **105**, 1246.

Miller, S. L., 1972, IEEE Trans. Electron Devices **ED-19**, 10, 1103.

Mitchell, I. V. and D. A. Marsden, 1969, The Electrochem. Soc. Detroit Extended Abstracts (Abstract 164).

Moline, R. A., 1969, Electrochem. Soc. Detroit Meeting (October 1969) (Abstract 165).

Morgan, R. and K. R. Greenhalgh, 1971, Ion Implantation, Bibliography, Information Services, Harwell.

Morizot, M., A. Dubee and A. Cornette, 1971, Proc. Second Int. Conf. Ion Implantation, Garmisch (Springer-Verlag) p. 345.

Moyer, N. E., R. W. Bower and H. G. Dill, 1969, Proc. 1969 GOMAC Conference.

Nelson, R. S., 1968, The Observations of Atomic Collisions in Crystalline Solids (North-Holland) p. 221.

Nelson, R. S., J. A. Cairns and N. Blamires, 1970, Rad. Effects **6**, 1 and 2, 131.

Nicholas, K. H. and R. A. Ford, 1971, Proc. Second Int. Conf. Ion Implantation, Garmisch (Springer-Verlag) p. 357.

Nicholas, K. H., B. J. Goldsmith, J. H. Freeman, G. A. Gard, J. Stephen and B. J. Smith, 1972, J. Phys. **E5**, 309.

Nigh, H. E., 1969, Int. Conf. Properties and Uses on MIS Structures, Grenoble, p. 77.

Nishimatsu, S., N. Natsuaki, T. Warabisako and T. Tokoyama, 1971, Supp. Jour. Jap. Soc. App. Phys. **40**, 29.

North, J. C. and W. M. Gibson, 1970, App. Phys. Letters **16**, 3, 126.

Ohl, R. S., 1952, Bell Syst. Tech. J. **31**, 104.

Oosthoek, D. P., J. A. den Boer and W. K. Hofker, 1970, Proc. European Conf. Ion Implantation, Reading (Peter Peregrinus Ltd.) p. 88.

Osborne, D. N., 1968, AERE Harwell, R5631.

Owen, R. B. and M. L. Awcock, 1968, Trans. I.E.E.E. **NS-15**, No. 3, 290.

Park, Y. S. and C. H. Chung, 1971, App. Phys. Letters **18**, 3, 99.

Pashley, R. D., 1971, Rad. Effects **11**, 1, 1.

Patrick, W. J., 1965, Solid State Electronics **9**, 203.

van der Pauw, L. J., 1958, Philips Research Reports **13**, 1.

Pavlov, P. V., E. I. Zorin, D. I. Tetel'baum and Yu. S. Popov, 196a, Soviet Physics **10**, 6, 786.

Pavlov, P. V., E. I. Zorin, D. I. Tetel'baum and E. K. Granitsyna, 196b, Soviet Physics Solid State **7**, 10, 2386.

Pavlov, P. V., V. K. Vasil'yev, Ye. I. Zorin, I. Tetel'baum, V. S. Tulovchikov and T. Yu. Chigirinskaya, 1970, Izvestiya vuz S.S.S.R., Radioelektronika **13**, 4, 493.

Peters, D. W., 1968, Proc. I.E.E.E. **56**, 89 (correspondence).

Petritz, R. L., 1958, Phys. Rev. **110**, 1254.

Pickar, K. and J. V. Dalton, 1970, Rad. Effects **6**, 1 and 2, 89.

Pickar, K. A., J. V. Dalton, H. D. Seidel and J. R. Matthews, 1971, App. Phys. Letters **19**, 2, 43.

Pistryak, V. M., A. K. Gnap, V. F. Kozlov, R. I. Garber, A. I. Fedorenko and Ya. M. Fogel, 1970, Soviet Physics–Solid State **12**, 4, 1005.

Putley, E. H., 1960, The Hall Effect and Related Phenomena (Butterworths, London, Reprinted by Dover, 1968).

Research Triangle Institute, 1964, Integrated Silicon Device Technology Vol. V, 119, AD 605558.

Research Triangle Institute, 1965, Integrated Silicon Device Technology Vol. IX, 81, AD 624520.

Rosendal, K., 1971, Rad. Effects **7**, 1 and 2, 95.

Roughan, P. E., J. D. MacDougall, A. H. Clark, K. E. Manchester and F. W. Anderson, 1968, Paper presented to Electrochem. Soc. Boston Meeting (May 1968).

Rourke, F. M., J. C. Sheffield and F. A. White, 1961, Rev. Sci. Instrum. **32**, 455.

Runyan, W. R., 1965, Silicon Semiconductor Technology (McGraw-Hill) p. 164.

Sah, C. T., R. N. Noyce and W. Shockley, 1957, Proc. I.R.E. **45**, 1228.

Sansbury, J. D. and J. F. Gibbons, 1969, App. Phys. Lett. **14**, 10, 311.

Sansbury, J. D. and J. F. Gibbons, 1970, Rad. Effects **6**, 3 and 4, 269.

Sarace, J. C., R. E. Kerwin, D. L. Klein and R. Edwards, 1968, Solid State Electronics **11**, p. 653.

Schroeder, J. B. and H. D. Dieselman, 1967, Proc. I.E.E.E. **55**, 125.

Schwuttke, G. H. and K. Brack, 1969, Trans. of the Met. Soc. of A.I.M.E. **245**, 475.

Schwuttke, G. H., K. Brack, E. F. Gorey, A. Kahan and L. F. Lowe, 1970, Rad. Effects **6**, 103.

Sebillotte, Ph., P. Siffert and A. Coche, 1969, Nuclear Science Symposium, San Francisco (29/10/69).

Seidel, T. E. and A. U. MacRae, 1969, Trans. Met. Soc. A.I.M.E. **245**, 491.

Seidel, T. E. and D. L. Scharfetter, 1970, Proc. I.E.E.E. **58**, 7, 1135.

Seidel, T. E., R. E. Davis and D. E. Iglesias, 1971, Proc. I.E.E.E. **59**, 8, 1222.

Severin, P. J., 1971, Philips Research Reports **26**, 279.

Shannon, J. M. and R. A. Ford, 1970, Int. Conf. Ion Implantation, Thousand Oaks, Paper J1 (not published).

Shannon, J. M., J. Stephen and J. H. Freeman, 1969a, Electronics **42**, 3, 96.

Shannon, J. M., J. Stephen and J. H. Freeman, 1969b, Proc. Int. Conf. Prop. and Use of M.I.S. Structures, Grenoble, p. 593.

Shannon, J. M., R. Tree and G. A. Gard, 1970a, Can. Jour. Phys. **48**, 229.

Shannon, J. M., R. A. Ford and G. A. Gard, 1970b, Rad. Effects **6**, 3 and 4, 217.

Sherwell, R. J., J. A. Raines and L. N. Large, 1967, Proc. Int. Conf. on Application of Ion Beams to Semiconductors Technology, Grenoble, p. 641.

Shifrin, G. A., K. R. Zanio, D. M. Jamba, W. R. Jones, O. J. Marsh and R. G. Wilson, 1969a, Selective Doping of Piezoelectric Crystals by Ion Implantation. Semiannual Report N00014-69-C-0171, AD 693154.

Shifrin, G. A., D. M. Jamba, W. R. Jones, O. J. Marsh, M. T. Wauk and R. G. Wilson, 1969b, Selective Doping for Piezoelectric Crystals by Ion Implantation. Tech. Report 2, NR 251-001, AD 702778.

Shockley, W., 1949, Bell Syst. Tech. Jour. 28, 435.

Shockley, W., 1950, Electrons and Holes in Semiconductors (van Nostrand).

Shockley, W., 1954, U.S. Patent 2, 787, 564.

Sigournay, N., 1971, I.E.E. Colloquium MOS Integrated Circuits, 13 Dec. 1971.

Smith, B. J., 1971, A.E.R.E. Report R6660.

Smith, A. M., 1964, Integrated Silicon Device Technology Vol. IV – Diffusion, Research Triangle Inst. AD 603716.

Smits, F. M., 1958, Bell Syst. Tech. Jour. 37, 711.

Spiwak, R. R., 1969, I.E.E.E. Trans. Inst. and Measurements IM-18, 3, 197.

Stephen, J., 1973, private communication.

Stephen, J., B. J. Smith, G. W. Hinder, D. C. Marshall and E. M. Wittam, 1971, Proc. 2nd Int. Conf. Ion Implantation, Garmisch (Springer-Verlag) p. 489.

Stephen, J. and G. A. Grimshaw, 1971, Rad. Effects 7, 1 and 2, 73.

Stumpfi, W. and S. Kalbitzer, 1970, Rad. Effects 6, 3 and 4, 205.

Sze, S. M. and G. Gibbons, 1966, Solid State Electronics 9, 831.

Sze, S. M., 1969, Physics of Semiconductor Devices (Wiley Interscience).

Tannenbaum, M. and D. E. Thomas, 1956, Bell Syst. Tech. Jour. 35, 1.

Tetel'baum, D. I., 1967, Soviet Physics Semiconductors 1, 5, 593.

Theuerer, H. C., 1961, J. Electrochem. Soc. 108, 649.

Titov, V. V., 1970, Phys. Stat. Sol. (a) 2, 203.

Tokoyama, T., T. Ikeda and T. Tsuchimoto, 1970, Proc. 4th Microelectronics Congress, Munich (Oldenbourg Verlag) p. 36.

Tong, A. H. and A. Dupnock, 1969, Electrochem. Soc., Extended Abstracts (Detroit 5/10/69) Abstract 185.

Trumbore, F. A., 1960, Bell Syst. Tech. Jour. 39, 205.

Tufte, O. N., 1966, Diffused Resistor Temperature Coefficient Improvement, U.S. Gov. Report (RADC-TR-66-156) AD 633945.

Vadasz, L. L., A. S. Grove, T. A. Rowe and G. E. Moore, 1969, I.E.E.E. Spectrum 6, 10, 28.

Valdes, L. B., 1954, Proc. I.R.E. 42, 420.

Vasil'ev, V. K., E. I. Zorin, P. V. Pavlov and D. I. Tetel'baum, 1968, Soviet Physics – Solid State 9, 7, 1503.

Warner, R. M., 1965, Integrated Circuits, Design Principles and Fabrication (McGraw-Hill).

Webber, R. F., R. S. Thorn and L. N. Large, 1969, Int. J. Electronics 26, 2, 163.

Whitton, J. L., G. Carter, J. H. Freeman and G. A. Gard, 1969, J. Mat. Science 4, 208.

Wilkins, M. A., 1968, A.E.R.E. Harwell, R.5875.

Ying, R. S., R. G. Mankarious, D. L. English, R. W. Bower and L. E. Coerver, 1968, I.E.E.E. J. Solid State Circuits SC-3, 3, 225.

Zelevinskaya, V. M., G. A. Kachurin, N. B. Pridachin and L. S. Smirnov, 1970, Soviet Physics – Semiconductors 4, 258.

van der Ziel, A., 1959, Fluctuation Phenomena in Semiconductors (Butterworths Scientific Publications).

van der Ziel, A., 1968, Solid State Physical Electronics (Prentice-Hall).

Zorin, E. I., P. V. Pavlov, D. I. Tetel'baum and Yu. N. Shutor, 1968, Soviet Physics – Semiconductors 1, 8, 1051.


## References added in proof

Bauer, L. O., M. R. MacPherson, A. T. Robinson and H. G. Dill, 1973, Solid State Electronics, 16, p. 289.
Chu, W. K., B. L. Crowder, J. W. Mayer and J. F. Ziegler, 1972, Paper III–6, Int. Conf. on Ion Implantation in Semiconductors and other Materials, Yorktown Heights, New York, Dec. 1972.
Crawford, B., 1972, Electronics 45, p. 85.
Dill, H. G., R. M. Finnila, A. M. Leupp and T. N. Toombs, 1972, Solid State Technology 15, p. 27.
Dingwall, A. G. F., 1972, Paper 19.3, I.E.E.E. Electron Devices Meeting Washington December 1972.
Hseih, C. M., J. R. Matthews, H. D. Seidel, K. A. Pickar and C. M. Drum,1973, Applied Phys. Letters 22, p. 238.
MacPherson, M. R., 1972, Solid State Electronics 15, p. 1319.
Moline, R. A., G. W. Reutlinger, R. R. Buckley, A. U. MacRae and S. E. Haszko, 1972, Paper 19.1, I.E.E.E. Electron Devices Meeting Washington December 1972.
Reddi, V. G. K. and A. Y. C. Yu, 1972, Solid State Technology 15, p. 35.
Scavuzzo, R. J., R. S. Payne, K. H. Olson, J. M. Nacci and R. A. Moline, 1972, Paper 8.3, I.E.E.E. Electron Devices Meeting Washington December 1972.
Seidel, T. E., R. L. Meek and J. M. Poate, Papers IV–8 and IV–9, Int. Conf. on Ion Implantation in Semiconductors and other Materials, Yorktown Heights, New York, Dec. 1972.
Sigmon, T. W., 1972, Paper 19.2, I.E.E.E. Electron Devices Meeting Washington December 1972.
Swanson, R. M. and J. D. Meindl, 1972, I.E.E.E. Jour. of Solid State Circuits, SC-7, 2, p. 146.

# 6

# APPLICATIONS OF ION IMPLANTATION OUTSIDE THE SEMICONDUCTOR FIELD

The technique of ion implantation is rapidly finding application in many fields other than semiconductors. An attempt to review these applications together with some speculation as to the future development has recently been made by Thompson (1970). However, in this book we will limit our discussion to those other applications which have been taken to the point where our understanding and knowledge are sufficiently firm for us to make some definite statements. In several other cases, for example in regard to potentially important work in ion implanted catalysts, or use of the release of implanted gaseous ions as a measure of temperature, insufficient material has been published so far to allow scientific discussion.

## 6.1. Ion implantation in superconductors

Apart from the obvious application to semiconductors it has been suggested that ion implantation might be a useful technique for studying superconductivity. As yet, very little work has been published and it is therefore only possible to present a superficial review.

Work at Harwell (Bett and Howlett, 1970), has been directed towards the implantation of $Sn^+$ into Nb. The aim being to produce a thin surface layer of $Nb_3Sn$ precipitates. $Nb_3Sn$ is an important superconductor, with a high critical field ($> 200$ kOe), a high current density ($> 10^6$ A cm$^{-2}$) and a high critical temperature (18.5 °K). It is thought that a large number of fine $Nb_3Sn$ precipitates within the surface Nb could screen the underlying Nb from the penetration of the magnetic field by pinning the flux vortices at the surface so enhancing its superconducting properties. Alternatively, such a fine distribution of precipitates could itself carry a superconducting current by a tunnelling phenomenon. Lengths of 3 mm × 10 $\mu$m Nb tape were bombarded with ~ 100 keV $Sn^+$ ions to doses such that a mean concentration of 25 at. % was implanted to a depth of about 1000 Å. After implantation the tapes were heated to 950 °C in argon at 360 torr for times up to 20 min. The change in critical current with increase in magnetic field is illustrated in the typical example shown in fig. 6.1, which clearly shows a significant increase compared with non-implanted material.

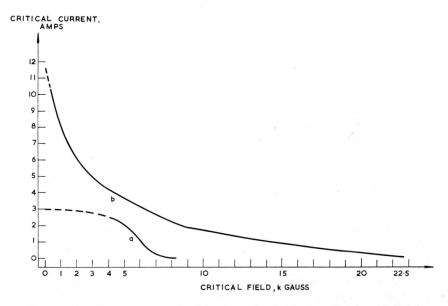

Fig. 6.1. The effect on superconductivity due to ion implanting Nb tape with $Sn^+$ ions. Curve (a) The critical current $v$ critical field for pure Nb tape. Curve (b) The critical current $v$ critical field for $Sb^+$ implanted Nb tape (from Bett and Howlett, 1970).

Other work by Chang and Rose-Innes (1970) has used ion implantation to study the mechanisms controlling the critical currents in very pure type II superconductors. By implanting $Mo^+$ into a Ni–Mo alloy they discovered that it was possible to suppress or enhance the superconductivity, and that the superconducting current was controlled almost entirely by that part of the surface parallel to the applied magnetic field. Figure 6.2 shows the measurement of critical current at 2.7 °K for an implanted $Nb_{77}$–$Mo_{23}$ alloy together with a sample which has been implanted with 80 keV $Mo^+$

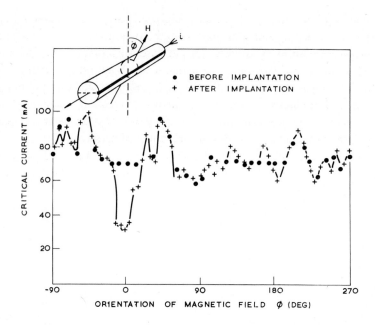

Fig. 6.2. Effect of an implanted $Mo^+$ stripe on critical current of a cylinder of $Nb_{77}$–$Mo_{23}$ as a function of the direction of the magnetic field. ($T = 2.7K$, $H = 0.6H_{c1}$) (from Chang and Rose-Innes, 1970).

ions so as to increase the surface Mo component from 23 to 26 atomic per cent. The critical current is unaltered except for the fact that there is a pronounced dip when the applied magnetic field is in a direction where it is tangential to the surface, e.g. at $\phi = 0$. It is interesting to note that although the critical current is reduced when the field is oriented at an angle $\phi = 0$, the critical current is *not* reduced when the field is rotated 180° so that it is again tangential to the implanted strip but in the opposite direction. Consideration of the direction of the Lorentz force on the fluxons shows that the critical current is only reduced when the direction of applied magnetic field is such that the fluxons are driven from the implanted strip towards the opposite side.

### 6.2. Hyperfine interactions

A branch of physics of growing importance concerns the measurement of hyperfine interactions in the gamma-decay of radioactive nuclei embedded in solids. These interactions involve the product of the magnetic field, electric field, or field gradients and the appropriate electric or magnetic multipole of the ion. As such, the measurement involves a nuclear parameter and one that is determined by its environment in the lattice, and if either of these is known the other may be determined. Therefore, depending upon one's motivation, the same experiment may be a contribution to nuclear physics or to solid state physics.

Ion implantation has been used as a versatile means of incorporating radioactive ions into a variety of materials. There are two principal ways of achieving this: one is by the conventional means of a beam of isotopically-separated ions; the other is by allowing the radioactive products of a nuclear reaction to recoil into a solid sample by virtue of the energy released in the reaction. Much greater penetration is feasible in the latter case, but the concentrations of active atoms are low.

Of the various hyperfine interactions that might be measured, the magnetic dipole interaction between the nuclear magnetic moment, $\mu$, and the magnetic field present at the nucleus has received most attention. By means of a low-temperature nuclear orientation technique, Grace et al. (1959) were able to demonstrate that this magnetic field may be extraordinarily high, $10^5$ to $10^6$ Oe being commonly present in iron.

There are, broadly speaking, three techniques for the observation of hyperfine magnetic interactions. The *nuclear orientation* method is carried out at extremely low temperatures of around 0.01 °K, attained by adiabatic demagnetization. At these temperatures the nuclear dipole moment becomes oriented by the applied field and the measurement consists in detecting the ratio of gamma-rays emitted at 0° and 90° to the axis of polarization, and fitting the variation of this ratio with $1/T$ to the theoretical prediction. The second method is by measurement of *perturbed angular correlations*. The angular distribution of gamma rays from a nuclear reaction, measured with respect to the beam direction, involves a series of Legendre polynomials, the maximum order of which is governed by the spin and parity of the initial and final nuclear states. An external perturbation, such as an applied magnetic field, has the effect of introducing a Larmor precession of the nuclear spin by an extent $\theta = \omega t$ involving the strength of the hyperfine interaction. There are various techniques of making such measurements, but one is illustrated in fig. 6.3. An ion beam strikes a target placed, at room temperature, in an applied magnetic field. Charged reaction products are detected in an annular detector situated along the beam axis, and two gamma-ray detectors, at 0° and 90° to this axis, measure the gamma-ray yield. In other cases, a gamma-ray detector is moved around to plot out the angular cor-

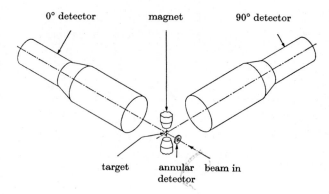

Fig. 6.3. Schematic drawing of a typical experimental arrangement used for particle-gamma ray perturbed angular correlation measurements (from Fossan and Poletti, 1968).

relation. Application of the magnetic field may shift the curve showing the number of gamma-rays measured in coincidence with charged reaction products through $\theta = \omega t$, integrated over the lifetime of the decaying state. In other experiments the yield is measured as a function of time after a short pulse of ions has generated the radioactive species. The third method of measurement utilises the *Mössbauer effect*. Under suitable circumstances the recoil of a nucleus as a result of gamma emission is absorbed by the whole crystal, and the emitted gamma-ray line is exceedingly well-defined in energy due to the extremely low recoil energy taken up by the lattice. A magnetic dipole interaction will produce a Zeeman splitting of this line which can be measured by the conventional Mössbauer technique using an absorber foil, moved at different speeds relative to the emitter, and a gamma-ray detector behind.

Since the hyperfine field is due to the probability of finding an unpaired electron at the point in question, it will not be surprising that the lattice field differs at different sites in a crystal, and thus the net magnitude of a hyperfine interaction will be determined by the lattice site occupancy of the gamma-emitting species. There is therefore interest in studying the lattice site location of the implanted atoms used in such experiments (Feldman and Murnick, 1970; Alexander et al., 1970), since the results are meaningful only if they can be shown to correspond to a particular lattice field. Thus, the annealing of $Xe^+$ implanted iron crystals has been shown to result in a decrease in the fraction of emitting ions occupying a high-field substitutional lattice site (de Waard and Drentje, 1969) and this is correlated with measurements, by the Rutherford back-scattering method (see section 2.19), of the substitutional fraction of $Xe^+$ ions in Fe, before and after annealing (Feldman and Murnick, 1970). It is very likely that during annealing at 450 °C much of the xenon accumulates into small gas bubbles and so is no longer subject to the lattice magnetic field but a much weaker macroscopic field. The back-scattering technique is therefore very useful in confirming the substitutional occupancy of implanted ions as a prerequisite to interpreting hyperfine interaction data. Conversely, hyperfine interaction measurements can provide information regarding the influence of ion implantation on lattice site location and the microscopic environment of the implanted ions.

A further interesting piece of information has come from hyperfine magnetic interaction measurements. Herskind et al. (1968) showed, by the time-dependent perturbed angular correlation method, that during the slowing down of an ion in a ferromagnetic lattice there is a strong transient field of 20 to 50 MOe. They ascribed this field to the capture, by the ion, of polarized electrons. Lindhard (1969) has put forward an alternative explanation, in which the scattering of free polarized electrons by the screened Coulomb field of the ion increases their probability density at the nucleus by a velocity-dependent factor, which can be of the order 300. Lindhard's model leads to a transient field $B(v)$ which increases with decreasing ion velocity $v$ and can reach several mega-Oersteds at low velocities. Lindhard further predicts that the effect will be diminished if the ions are channelled into the ferromagnetic crystal, since then they are steered into regions of low electron density: it remains to be seen whether this is borne out by experiment.

Readers interested in pursuing the interrelation of ion implantation and hyperfine interaction measurements in more detail are referred to the proceedings of a discussion meeting on the topic organized by the Royal Society of London (Kurti et al., 1969).

### 6.3. Ion implanted phosphors

Many different transparent materials with wide band-gaps are employed as phosphors in cathode ray tubes and other electron beam devices. The luminescent efficiency of inorganic phosphors is small, however, unless a suitable 'activator' is incorporated. Generally activators are rare earth or transition metal ions and the transport of energy from the ionization event to an activator centre, for example in the form of excitons, results in radiative recombination at a wavelength characteristic of the activator species, its state of ionization and local configuration.

Phosphors are generally applied as a thin layer of a fine powder mixed with binder, and the activator is introduced during the initial preparation of the powder by the techniques of phosphor chemistry. There are some instances, however, when it may be advantageous to be able to activate a uniform, shallow layer of pre-deposited phosphor or a localized area of phosphor. Then ion implantation of the activator may have some useful applications.

First, however, it is necessary to cope with the radiation damage which implantation brings about and the likelihood that this will so increase the probability of radiationless recombination to render the phosphor useless. There are two approaches to this problem: one is to anneal the specimen after implantation, or alternatively to implant at an elevated temperature as described in earlier chapters; the other is to find a phosphor which is not subject to severe damage. Strongly ionic materials, as we have seen earlier, have a considerable power of restoring their short-range order after atomic displacement.

Dearnaley, Goode and Turner (1971) have investigated a number of systems in order to find efficient combinations of phosphor and implanted activator. Some of the most promising materials proved to be alumina, silica and alumino-silicates. When implanted to concentrations approaching 1 per cent with activators such as terbium, europium, samarium or dysprosium useful and grain-free luminescent screens were obtained, showing respectively a green, orange-red, rose-red and yellowish luminescence, with either electron or ion excitation. The cathodoluminescent spectrum of terbium-implanted silica is shown in fig. 6.4. The luminescent efficiency was similar in fused silica, anodic silicon dioxide, and thermally-grown silicon dioxide. In the case of the amorphous phosphors, alumina and silica, there was no significant visible discolouration as a result of implantation, and annealing was unnecessary. Furthermore, unlike any commercially-available phosphors tested, these screens withstand subsequent ion bombardment indefinitely, at least until sputtering away of the surface may be assumed to have occurred. Therefore they prove useful in making visible the location and focus of an ion beam.

In other cases, standard phosphors such as calcium alumino-silicate were bombarded, and it was observed that the luminescence due to the conventional $Ce^{2+}$ activator was soon quenched. The mechanisms by which ion bombardment produces a deterioration in the luminescence of phosphors has been studied in ZnS:Ag, $Zn_2 SiO_4$: Mn and $MgWO_4$ by Hanle and Rau (1952) and in ZnS:Ag and ZnO:Zn by van Wijngaarden and Hastings (1967).

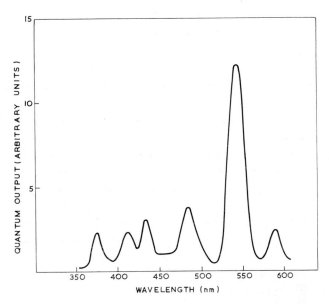

Fig. 6.4. Electron-excited luminescence spectrum of $SiO_2$ implanted with $10^{15}$ cm$^2$ 250 keV terbium ions. The electron energy was 15 keV.

These authors propose a model in which damage centres allow radiationless transitions to occur in competition with light emission from activator sites. If the probability of a transition via an activator centre is $\alpha$ and via a defect site is $\delta$ and if $A$ and $D$ are the numbers of activator atoms and defects per unit volume, one has as an expression for the luminescent efficiency, $\eta$

$$\eta = \frac{\eta_0}{1+(\delta D/\alpha A)}$$

in which $\eta_0$ is the initial efficiency. This formula holds, of course, only for a uniformly damaged phosphor, but agrees with the form of Hanle and Rau's (1952) results.

Dearnaley, Goode and Turner (1971) have observed that, after damaging a $Ca_2Al_2SiO_9$:Ce phosphor until its luminescent efficiency was near zero, it was possible to impart a green luminescence by implantation of terbium ions. It would appear that the product $\alpha A$ in the above expression, for Tb sites, must be significantly larger than the corresponding values for Ce so that, although $\delta D$ must be increased by the implantation a good luminescent efficiency $\eta$ is attained.

Most common phosphors, such as zinc sulphide, calcium tungstate, yttrium orthovanadate, etc. are rapidly discoloured by room temperature ion implantation, even in the absence of an organic binder. This is probably due to the formation of colour centres, such as F-centres, and it is likely that these can be eliminated by annealing the material or exposing it to ultra-violet irradiation.

In colour television receivers the most widely-adopted design makes use of a shadow-mask tube in which tiny dots of phosphor are deposited in groups of three, excited separately by electron beams from three guns directed through a shadow-mask. In principle, it would be feasible to produce such a screen by the implantation of suitable activators into the appropriately-masked areas of un-activated phosphor.

In this and the earlier applications mentioned, it is fortuitous that the penetration of typical activator ions, of energies 200–500 keV, into phosphors is comparable with that of electrons of energy 5–10 keV. The production of phosphors which will withstand ion bombardment is also of value in the ion implantation field itself.

### 6.4. Chemical effects of ion implantation

Although many of the consequences of ion implantation are ultimately related to the chemical fate of the dopant species certain areas of work can be more directly considered as ion bombardment chemistry. These include, in particular, the chemical transformation of surfaces with very high doses of reactive ions and the use of energetic ions in 'hot atom' chemistry. The work which has been carried out in the latter field has up to now been largely subordinate to the well established and extensive studies of the chemical effects of nuclear transformations in solids and as such is beyond the scope of this book. We simply note that the results to-date of doping with energetic radioactive ions are generally similar to those obtained using the more conventional neutron irradiation methods. It is apparent however that ion implantation is a very versatile technique and that it offers important advantages in certain 'hot atom' studies. The choice of dopant and host is much less restrictive. The ion dose and energy can be precisely controlled and in single crystal bombardments channeling may be exploited to minimise the radiation damage. Oxidising or reducing ions can also be implanted to modify the chemical environment in the solid.

The general techniques of doping with radioactive ions are described in section 4.6 and typical experimental results can be found in the following references, Ascoli and Cacace (1965); Andersen and Sorensen (1965), Kasrai (1967).

### 6.4.1. ION SURFACE CHEMISTRY

The degree of chemical conversion of a surface which may be attained by bombardment with reactive ions is generally restricted both by sputtering and by the form of the dopant penetration profile.

With most ions at sufficiently high bombardment doses an equilibrium is established in which the loss by sputtering of previously implanted dopant is precisely matched by the rate of arrival of new ions. The only significant exceptions to this rule occur with very light ions such as boron or carbon which even at energies of tens of kilovolts have very low self-sputtering coefficients and can thus be directly deposited on suitable targets to produce well defined layers of any required thickness. In the more general case the equilibrium dopant concentration is inversely proportional to the sputtering rate with the frequent result that the required high doping level for stoichiometric conversion of the surface to a new compound cannot be attained.

In cases where the sputtering rate is low and does not provide a limitation on the doping level the Gaussian form of the penetration profile will generally result in a situation in which precise chemical conversion can only be achieved over a very limited region of the total ion range. Whilst this effect could be minimised by careful energy programming of the beam such a technique would clearly be of limited application and in general it appears that the use of reactive ion beams will be most appropriate when only relatively imprecise or minor changes in stoichiometry are required.

There is however one particular area in which these restrictions do not apply and in which the technique has proved to be quite successful for the production of well defined compound layers. This is in the conversion of certain elemental surfaces to oxides and nitrides using very high doses of oxygen and nitrogen ions. In these particular cases the sputtering coefficient of the compounds is very low and the use of a large ion dose to ensure chemical conversion over the whole range is unimportant since the excess of the gaseous dopant deposited at the peak of the concentration profile can diffuse away.

The very high retention of implanted oxygen and nitrogen in certain metallic foils was established in isotope separators many years ago and was extensively exploited as a convenient method of preparing isotopic targets of these elements (Smith, 1956). The implanted surfaces showed clear visual evidence of chemical conversion. Trillat (1961) has described the production of certain oxides and nitrides by the ion bombardment of metals and compounds with oxygen and nitrogen beams. A more detailed characterisation of such an ion interaction compound was made by Watanabe and Tooi (1966) who described the formation of well defined silicon dioxide layers by

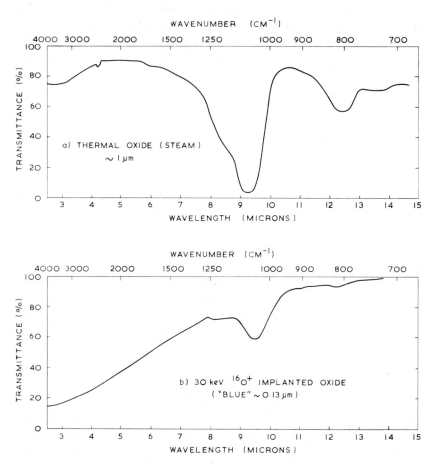

Fig. 6.5. Infrared spectra of a 30 keV $^{16}O^+$ implanted layer, compared with a 1 $\mu$m thermal $SiO_2$ grown in steam at 1150 °C. The thermal oxide shows the strong absorption band at 9.2 $\mu$m with a weaker band at 12.5 $\mu$m and a distinct shoulder in the 8–9 $\mu$m region, characteristic of amorphous $SiO_2$. The absorption band at 3400 $cm^{-1}$ and 1630 $cm^{-1}$ shows the presence of water in the film. By comparison the implanted oxide shows a weaker absorption at 9.5 $\mu$m and 12.5 $\mu$m and a gradual strong absorption down to 3 $\mu$m. Pavlov and Shitova (1967) reported a large absorption at 10.0 $\mu$m, characteristic of silicon monoxide, in addition to the $SiO_2$ absorption at 9.2 $\mu$m which suggests that their substrates were subjected to insufficient dose of $O^+$ ions to convert the layer entirely to $SiO_2$. No indication of the dose given to their samples was reported.

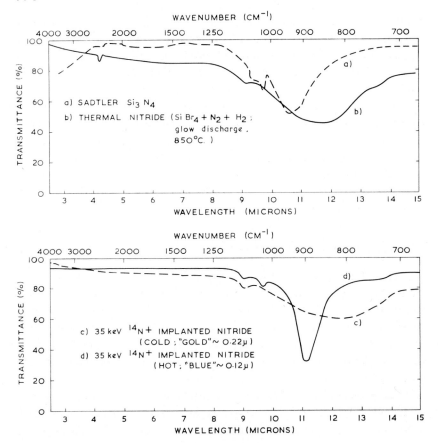

Fig. 6.6. Infrared spectra of a 35 keV $^{14}N^+$ implanted nitride. The 'blue' implanted film (hot) shows an intense absorption band at 11.15 $\mu$m and a weak one at 9.7 $\mu$m; a further weak band at 9.1 $\mu$m is due to oxygen in the silicon substrate. The spectrum is typical of crystalline $Si_3N_4$ in that the bands are sharp, but the actual wavelengths do not tie up with any published spectra. The spectrum is not dissimilar to the Sadtler spectrum shown in fig. 1(a) (which is assumed to be the usual mixture of $\alpha$- and $\beta$-form). The 'gold' implanted film (cold) has a very broad absorption band centred at about 12.3 $\mu$m and this is more typical of amorphous $Si_3N_4$.

oxygen ion bombardment of single crystal silicon surfaces. Although good electrical insulating behaviour was obtained, the results of capacitance measurements were less satisfactory and indicated the presence of a damaged or doped region below the oxide skin. Similar results were obtained by Freeman et al. (1970, 1967) who also produced equally well defined layers of both crystalline and amorphous silicon nitrides in this way. The compounds were characterised by infra-red, electron microscopy and electrical measurements. The infra-red spectra of both the oxides and nitrides are compared with those of more conventionally produced compounds in figs. 6.5; 6.6. The very sharp spectrum of the crystalline ('hot') modification of the nitride is particularly notable. It was found that the thickness of the layers was sensitive to the ion energy but quite insensitive to the total ion dose (up to five fold excess) once equilibrium had been attained. Pavlov and Shitova (1967) have also studied oxygen implanted silicon and noted a large absorption in the infra-red spectrum at 10 $\mu$m characteristic of SiO in addition to the $SiO_2$ absorption at 9.2 $\mu$m. No indication of the ion dose was reported. Batarin et al. (1969) have performed similar experiments with a number of metallic conductors including silicon.

## 6.5. Implantation of magnetic materials

There have been no published reports of the use of ion implantation to modify the magnetic properties of conventional thin films of the iron–cobalt–nickel alloy system. The very thin character of such films would appear to lend themselves to compositional and structural modifications. There have, however, been a number of recent papers from the Bell Laboratories on the use of ion implantation to control the properties of magnetic bubble garnets (Brown, 1971, Wolfe et al., 1971 and Wolfe and North, 1972).

The magnetic material used for magnetic bubble research is a thin sheet of a single-crystal material in which the easy axis of magnetisation is perpendicular to the plane of the sheet. Under equilibrium conditions half the magnetic domains will have their north poles directed up and the other half down (fig. 6.7). If an external normal magnetic field is applied then a number of the domains can be made to shrink into nearly perfect cylinders several microns in diameter known as 'bubbles'. The bubbles can be moved about by several techniques but the most popular is by means of a thin nickel–iron alloy pattern on the surface of the bubble material which sets up magnetic poles to selectively repel or attract and hence control the magnetic bubbles. The thin Ni–Fe alloy film obtains its energy from a rotating magnetic field in the plane of this film. In a practical device the bubbles are about 7.5 microns in diameter and they have domain wall velocities between 25 cm/sec and 2500 cm/sec in a field strength of one oersted.

The early bubble materials were rare-earth orthoferrites (e.g. yttrium, samarium or terbium samarium orthoferrite (Bobeck, 1967), while later

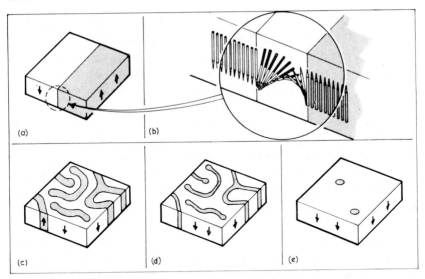

Fig. 6.7. Each unpaired electron in a bubble material is like a tiny magnet. The north poles of such 'atomic magnets' all point up as shown in the grey region (a) of a plate of single crystal bubble material whose easy axis of magnetisation is normal to the plate. This region is called a magnetic domain. Domain walls (b) separate domains where the north poles point down (grey regions) from domains where north poles point up. By dividing these regions into alternating sections, magnetised upward and downward (c), the magnetostatic energy can be greatly reduced. A downward directed bias field applied to (c) causes the grey strips to narrow at (d). At a critical value of field, the single-walled strip domain becomes cylindrical (e), resulting in a 'bubble domain'. (After Bell Labs. Record **49** (8) 240; Copyright 1971, Bell Telephone Laboratories, Inc. Reprinted by permission, Editor, Bell Laboratories Record.)

work has been concentrated on single-crystal films of magnetic garnets of the general composition $(A_{3-x}B_x)(Fe_{5-y}Ga_y)O_{12}$. A and B are rare earths or yttrium (Wolfe et al., 1971). These garnet films possess a unique easy axis of magnetisation which can be made normal to the plane of the film by growth-induced anisotropy during the epitaxial growth or inducing a tensile stress by a lattice mismatch with the gallium garnet substrate.

Wolfe et al. (1971) grew a 3.5 micron film of $(Tb_{2.5}Er_{0.5})(Fe_5)O_{12}$ garnet on a (111) $Sm_3Ga_5O_{12}$ substrate by chemical vapour deposition. The stress in the film was so high and combined with a positive magnetostriction resulted in magnetisation being in the plane of the film and consequently no domains were observed. A part of the film was implanted with $1 \times 10^{17}$ 300 keV hydrogen ions/cm$^2$. Domains formed in the implanted region, which was visibly darker, and bubbles could be formed with a normal bias field of about 20 Oe. Annealing experiments showed that the implantation effects persisted upto 1000 °C. A 300 keV helium implant $(1 \times 10^{17}/\text{cm}^2)$ showed darkening of the garnet film but no normal anisotropy while a 300 keV lithium implant $(5 \times 10^{15} \text{ ions/cm}^2)$ showed perpendicular magnetisation in a very thin surface layer. A similar effect was observed in a 4.5 micron film of $(\text{Eu}_{1.75}\text{Er}_{1.25})$ $(\text{Fe}_{4.1}\text{Al}_{0.9})\text{O}_{12}$ grown by liquid phase epitaxy on a $\text{Gd}_3\text{Ga}_4\text{O}_{12}$ substrate with a (100) orientation. On the other hand, the reverse effect was seen in a $(\text{Y}_3\text{Fe}_{2.5})(\text{Ga}_{0.5})\text{O}_{12}$ film grown by chemical vapour deposition on a (111) substrate of $\text{Dy}_{0.75}\text{Gd}_{2.25}\text{Ga}_{0.5}\text{O}_{12}$. Initially this film exhibited normal domains due to a combination of tensile stress and negative magnetostriction but after implantation with 150 keV hydrogen ions $(1 \times 10^{17}/\text{cm}^2)$ the magnetisation was parallel to the surface as expected for a negative magnetostrictive material in lateral compression.

These experiments showed that ion implantation produced lattice expansion and that this was chemical in nature rather than being related to the lattice damage. The implanted layer becomes laterally compressed and this allows the easy axis of magnetisation to be perpendicular or parallel to the surface depending on the magnetostriction constants being positive or negative respectively. The magnetostriction constant is dependent on the chemical composition of the garnet. The use of implantation offers the possibility of an extension of the range of materials suitable for bubble applications.

Recently a new form of bubble, known as the 'hard' bubble has been observed. It has a different magnetic spin arrangement in the domain wall. A large number of Bloch-to-Néel transitions are postulated around the bubble circumference with a left hand or right hand orientation resulting in a locked- in spin arrangement. A normal bubble is assumed to have no Bloch-to-Néel transitions. It has been shown that the locked-in arrangements can be unwound by a domain wall at the bottom of the bubble leaving only two Bloch transitions and a bubble which does not behave in a hard manner, and can be considered as a normal bubble. Multilayer magnetic garnet films (Bobeck et al., 1972) and ion implantation (Wolfe and North, 1972) into single layer films is being investigated as a means for providing 'bottoms' and 'lids' respectively with their magnetisation parallel to the film surface to suppress hard bubbles in the films. As already discussed, (Wolfe et al., 1971) ion implantation expands the lattice and in a negative magnetostriction garnet this results in the magnetisation being parallel to the surface in the implanted volume. Wolfe and North (1972) have implanted a (111) liquid phase epitaxy film of $(\text{Y Gd Tm})_3(\text{Fe Ga})_5\text{O}_{12}$, which exhibited hard bubbles, with $2 \times 10^{16}$ 100 keV hydrogen ions/cm$^2$ and produced regions in which only 'normal' bubbles would form, hence confirming the formation of a 'lid' (fig. 6.8). They found that hydrogen ion energies between 25 keV and 300 keV with doses from $10^{16}$ to $10^{17}$ ions/cm$^2$ were effective in eliminating hard bubbles. Other garnet compositions, with negative values of magnetostriction in the (111) direction normal to the film, which gave a similar result were $(Y \; Gd \; Yb)_3(Fe \; Ga)_5O_{12}$, $(Gd \; Er)_3(Fe \; Ga)_5O_{12}$ and

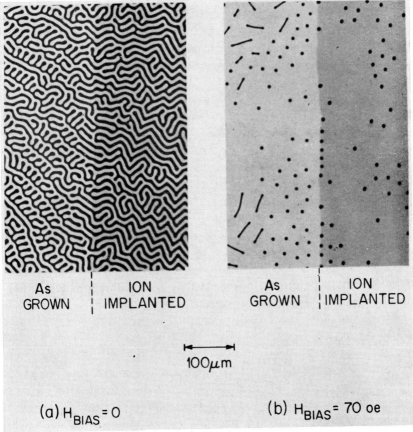

Fig. 6.8. The effect of hydrogen ion implantation in a magnetic garnet film of composition $(YGdTm)_3(FeGa)_5O_{12}$. (a) Rapidly demagnetised state. The unimplanted area on the left shows the comb-like pattern of magnetic domains, typical of materials in which hard bubbles form. This pattern is absent from the implanted area on the right. (b) Bias field of 70 Oe applied. Many of the domains in the unimplanted area are still strips. At higher fields they form hard bubbles with collapse fields up to 105 Oe. All of the bubbles in the implanted part have normal characteristics with collapse fields near 75 Oe. (After Wolfe and North, 1972; Copyright, 1972, American Telephone and Telegraph Company, reprinted by permission.)

(Y Eu Yb)$_3$(Fe Ga)$_5$O$_{12}$. Preliminary isochronal annealing experiments for one film showed that the implantation was still effective after a 1000 °C anneal though in another film, hard bubbles appeared after a 300 °C anneal.

Wolfe and North (1972) also found that a 300 keV helium ion implant with a dose over $10^{16}$ ions/cm$^2$ suppressed hard bubbles. A large increase in the lattice constant was also observed. These results contradict the earlier findings of Wolfe et al. (1971) in which no effect with helium was observed. This raises some doubts as to whether the effect of the implantation is purely chemical or depends on the lattice damage. Implantation is considered to be an economic method when compared with multilayer epitaxial growth techniques, for modifying the properties of garnet films to prevent hard bubble formation.

North and Wolfe (1972) have proposed that the boundaries of implanted regions on hard bubble garnets could be used as 'rails' for guiding the bubbles as the rail pattern could be predetermined by photoresist masking during ion implantation. They have also pointed out that the in-plane magnetisation in the implanted regions can be rotated by an external rotating magnetic field. The bubbles in the underlying garnet are attracted to the magnetic poles in the layer and as a result the bubbles follow the poles as the external field is rotated. This raises the possibility of using a drive pattern in the garnet itself, rather than in a permalloy overlay pattern, for the propagation of the bubbles.

This research on the implantation of thin film magnetic garnets is at an early stage and there is a possibility that in the future it may be instrumental in realising high density magnetic storage elements.

### 6.6. Refractive index modification

Since ion implantation can be used to induce controllable changes in the optical properties of materials and at depths comparable with, or greater than, optical or near infra-red wavelengths, the possibility exists of producing a number of interesting optical components. The fine lateral dimensions achievable with electron-beam or photolithographic mask fabrication allow, in principle, the construction of buried optical waveguides, or a variety of optical couplers as described by Brown (1971); see figure 6.9.

In the first experiments of this kind, Schineller et al. (1968) at Wheeler Laboratories produced a buried optical guide by the proton irradiation of fused silica. An increase in the refractive index towards the end of the proton range provided the propagating channel. Two closely-spaced guides were

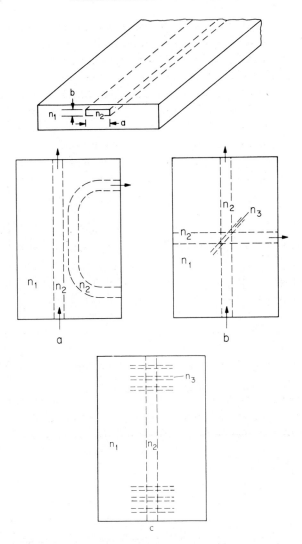

Fig. 6.9. Geometry of a simple rectangular laser light guides: $n_2$ must exceed $n_1$ (after Brown, 1971); (a), (b) Two possible types of directional couplers, on the left a hybrid type, on the right a junction type (from Brown, 1971 after Miller, 1969;) (c) A possible design of optical wave-guide resonator; (from Brown, 1971 after Miller, 1969).

shown to act as an optical waveguide coupler. The degree of attenuation in the structures was not reported and may have been high.

The work was followed up by Bayly and Townsend (1970), who implanted several different types of ion, including $A^+$ and $C^+$, into silica and

measured the resulting refractive index changes by the method of ellipso-
metry: the refractive index always increased with dose, the maximum value
observed being 1.560 after $5 \times 10^{16}$ $He^+$ ions/cm$^2$, a rise of 7 % over the
unirradiated value of 1.458.

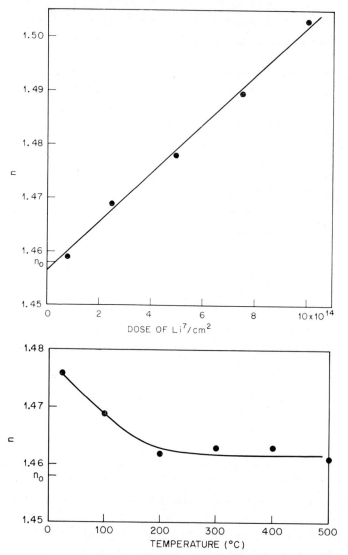

Fig. 6.10. (a) Index of refraction of fused $SiO_2$ as a function of $Li^+$ ion dose (from Standley
et al., 1972). (b) Annealing of the refractive index change brought about by $5 \times 10^{14}$ $Li^+$
ions/cm$^2$ in $SiO_2$ (from Standley et al., 1972).

Standley, Gibson and Rodgers (1972) reported the effects of annealing $Li^+$ implanted silica (see fig. 6.10). Annealing brought about a rapid fall of the refractive index between room temperature and 200 °C. Following a dose of about $10^{15}$ $Li^7$ ions/cm$^2$ at energies up to 200 keV the residual refractive index after a 300 °C anneal was 1.468. Standley et al. also measured the optical loss in a 1 $\mu$m guide produced in this way and observed that the value immediately after implantation, 1.8 dB/cm, fell to below 0.2 dB/cm as a result of annealing.

A debate has arisen with regard to the extent to which the refractive index changes are brought about by atomic displacement or alternatively by the polarizability of the implanted ions. There is little doubt that the initial effects are dominated by displacements, but the residual modification after annealing could be partly due to the nature of the ion species implanted.

Since a variety of ions ($H^+$, $He^+$, $Li^+$, $C^+$, $O^+$, $A^+$, $Tl^+$ and $Bi^+$) produce broadly comparable initial effects (Standley et al,. 1972; Bayly and Townsend, 1970) it seems likely that chemical doping is of minor importance. However, it is interesting to compare the effects of ion bombardment with the results of Primak (1958), obtained with the neutron irradiation of silica and quartz. Primak found a rapid increase of refractive index in fused silica, saturating at a value of 1.47 after $5 \times 10^{19}$ neutrons/cm$^2$: this fell steadily towards the initial value of 1.458 after a 1000 °C anneal. The refractive index changes and annealing behaviour followed the density, which Primak also measured. The maximum density rise in silica was nearly 3 %, and it is interesting to note that neutron-irradiated quartz exhibits a decrease in density, finally arriving at the same value (2.253 g/cm$^3$) as that of irradiated silica. Weissman and Nakajima (1963) have demonstrated that both products of irradiation contain some 20 % $\alpha$-quartz (the normal stable form), which suggests that an equilibrium between order–disorder processes is taking place. X-ray diffraction studies by Lukesh (1955) indicate that the deformation produced neutron irradiation, or strictly by the recoiling silicon and oxygen atoms, takes the form mainly of a bending and twisting of the relatively stable Si–O tetrahedral structures with respect to each other. Primak noted that by far the major contribution to the refractive index of silica is made by the oxygen ions, and attributed the change in the index brought about by irradiation to an alteration of the polarizability of the oxygen ions by the compacting of the structure.

Bayly (priv. comm.) has pointed out that the changes in refractive index caused by ion bombardment of silica often exceed the value of 1.5 %

observed by Primak in neutron irradiation. Furthermore, the annealing behaviour appears to be quite different (fig. 6.10). Moreover, the ion-induced effects show no maximum, but appear to continue rising with increasing dose.

It appears to us possible that the implantation of foreign species, and particularly glass-forming ions, even in concentrations below 1 %, could influence the stability of the deformed Si–O tetrahedra and so modify the annealing behaviour as well as the final refractive index change. It is well-known that the crystallization or de-vitrification of fused silica, especially at elevated temperatures, is strongly influenced by trace impurities such as sodium. Bayly has developed a technique of stripping the surface of implanted $SiO_2$ and thus determining the profile of both the refractive index and the absorption. In this way it should be possible to resolve the question of the chemical influence of the implanted ions, which will lie at a greater depth than the damage they induce.

In practical devices, the interesting feature is the small refractive index changes necessary to produce an optical guide. Brown (1971) has discussed this point, and considers a buried rectangular guide (as shown in fig. 6.9) with a ratio of dimensions $a:b$ of 2:1. At a wavelength of 6000 Å, taking $a = 10 \ \mu m$, a change of 1 % in refractive index will confine the energy of the lowest mode almost entirely to the modified region marked $\eta_2$. Even smaller changes will allow the lowest mode to be propagated, other modes being beyond their cut-off. Goell et al. (1972) have achieved the fabrication of a semi-circular waveguide 5–12 $\mu m$ in width and a few mm long, in a piece of fused silica masked by electron-beam resist (PMMA) and exposed to $Li^+$ ions so as to introduce a uniform doping of $2.2 \times 10^{19}$ $Li^+$ ions/$cm^3$ over a region 1 $\mu m$ in depth. The estimated loss was 10 dB/cm, reduced to 2 dB/cm after a 1 hour anneal at 200 °C.

Silica is not the only attractive material for optical waveguide construction: the feasibility of producing injection lasers in GaAs has led Garmire et al. (1972) to investigate guides formed by proton irradiation in polished GaAs crystals, at an energy of 300 keV, with the eventual aim of producing active semiconductor integrated optics. In the case of this experiment the proton bombardment, to a dose of $10^{15}$/$cm^2$, brings about a compensation of the GaAs (see Ch. 5, section 5.6) and so reduces the free-carrier concentration and thereby the plasma contribution to the refractive index. The observed guiding behaviour, using focused 1.15 $\mu m$ radiation from a He–Ne laser, was consistent with the plasma model. It was possible to measure the mode profile (figure 6.11), and to demonstrate that the maximum transmission occurs at the depth (3 $\mu m$) corresponding to the proton range. The

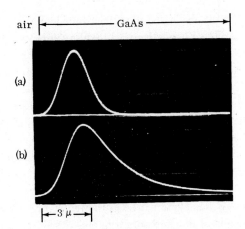

Fig. 6.11. A measurement of the optical mode profile in a buried light-guide produced by the proton bombardment of a GaAs single crystal for two different proton doses (after Garmire et al., 1972).

attenuation measured was reduced very considerably by annealing the crystals to 500 °C, under an $SiO_2$ encapsulation.

The attenuation values achieved in these experiments so far do not hold out much hope for the fabrication of extended optical waveguides for communication purposes, though in principle the idea of producing a buried guide, below the region in which scattering by surface cracks occurs, is an attractive one. It is more likely that the first applications of ion-implanted optical waveguides will be in the form of small couplers, as described by Brown (1971) or in the production of integrated optical components (Garmire et al., 1972). Precise masking will be required in each case, and the energies of the ions are sufficiently high that the thickness of the masking layer must be large: this in turn means that difficulties may be encountered with the mask-edges. If there proves to be any merit in implanting the heavier ion species, from the point of view of their polarizability or glass-forming properties, their energies will need to be well into the MeV range in order to achieve a penetration of at least 1 $\mu$m. This could prove an interesting application for heavy ion beams now increasingly available from Van de Graaff, Tandem accelerators and cyclotrons.

## 6.7. Corrosion studies in ion-implanted metals

Relatively small amounts of certain impurities can influence strongly the corrosion properties of metals. Normally the beneficial elements are incorporated by alloying, as in stainless steel, but there are occasions when, for reasons of cheapness, strength, machinability, nuclear properties, etc. such an admixture is ruled out by the requirements for the bulk properties of the material. Surface protection is then provided by such means as painting, electroplating, spray coating, etc. Problems can arise due to interfacial corrosion beneath such coatings, and it is sometimes difficult to maintain tight tolerances on the dimensions. Ion implantation is a versatile technique that avoids these difficulties, and so interest has been stimulated in applying it to corrosion inhibition in the technologically-important metals and alloys, particularly now that more intense beams of ions are available.

A second impetus has stemmed from the realization that it is possible to introduce, by means of ion implantation, controlled amounts of different substances into a metal without the need for developing the appropriate metallurgy and without the varied grain structure in the resulting specimens. It is feasible to compare the effects of adding many different species, separately or in combination, upon the subsequent corrosion or electrochemical behaviour of a metal surface. In this way it begins to seem possible to discover more readily what are the desirable properties of an alloying constituent, even though it could prove simpler in practice to introduce it by means other than implantation. This semi-empirical approach is necessary in corrosion science because of the great complexity of the mechanisms of corrosion, involving as they do charge transport by ions and electrons, grain boundary phenomena, mechanical effects due to surface stresses, and so on.

As yet a further tool in the study of corrosion science, the technique of ion backscattering, now greatly refined for purposes we have discussed in the semiconductor field, has proved to be a most useful means of investigating surface corrosion. The combination of ion implantation and ion backscattering, to detect for example the distribution of implanted material in the oxide, or to measure oxide thickness, seems capable of yielding valuable new information on corrosion behaviour, particularly in the lighter metals.

That ion bombardment, even with inert gas ions, can strongly affect the oxidation of certain metals has been known for well over a decade. Trillat and Haymann (1961) reported that argon ion bombardment will inhibit the normally rapid tarnishing of a freshly-polished uranium surface. Presumably this effect is due to some damaging mechanism, not yet fully understood, rather than any chemical influence of the implanted argon (Thompson, 1970).

More recently, Crowder and Tan (1971) have reported that the implantation of boron ions, to doses of around $10^{16}/\text{cm}^2$, will reduce the atmospheric tarnishing of copper. This work was possibly the first published example of the modification of metallic corrosion by ion implantation rather than bombardment. It was followed up by a study, begun at Harwell, of the oxidation of single-crystal copper, using the ion back-scattering technique to examine the surface (Rickards, 1972): the observation was made that implanted $Al^+$ ions will also reduce the subsequent thermal oxidation of crystalline copper.

Other work at Harwell has concentrated upon the corrosion of titanium and stainless steels, both materials which are often employed because of their combination of strength and corrosion resistance. The oxidation of titanium follows different rate equations depending upon temperature and time: below about 400 °C the oxidation is logarithmic, but between 400 and 600 °C a transition to a parabolic oxidation takes place, while above about 800 °C a linear rate of oxidation commences after a time (Kofstad, 1966). Miller (1972) chose to examine the influence of ion implantation in

Fig. 6.12. Example of an oxidised ion-implanted Ti foil specimen, in which $2 \times 10^{16}$ $Bi^+$ ions (over the circular central area) have enhanced the take-up of oxygen by 95 % compared with the unimplanted area, and so produced an oxide film showing a different interference colour.

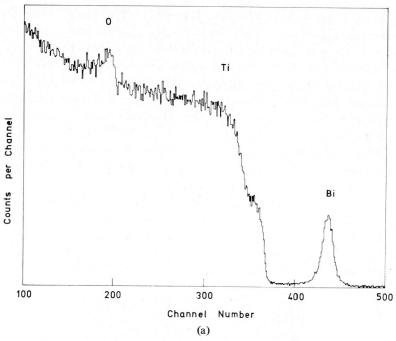

(a)

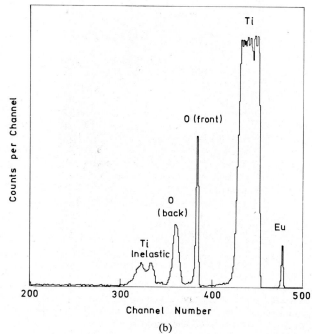

(b)

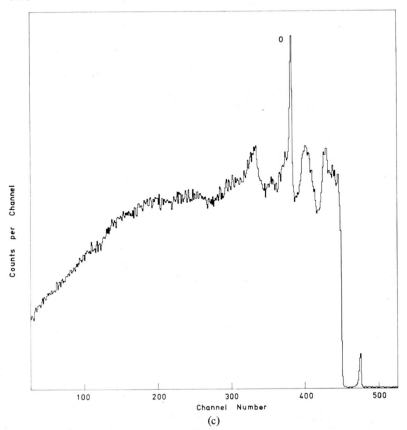

Counts per Channel

100   200   300   400   500

Channel Number

(c)

Fig. 6.13 (a) Back-scattering spectrum of 1.5 MeV $^4He^+$ ions from a $Bi^+$-implanted titanium specimen, which had received $5 \times 10^{16}$ ions/cm$^2$: note the symmetrical Bi distribution and the non-uniform Ti: O stoichiometry in the oxide, revealed in this spectrum (from Dearnaley et al., 1972); (b) Back-scattering spectrum of 4.0 MeV protons at 164° from a 3 $\mu$m thick $Eu^+$-implanted titanium foil, showing protons scattered from the oxide films on each face of the metal (from Dearnaley et al., 1972); (c) A back-scattering spectrum obtained under the same conditions, but in a thick-specimen of titanium, showing the effect of resonances in the elastic scattering cross-section of $^{48}Ti$.

the intermediate region, at 600 °C, where it is believed (Markali, 1962) that grain boundary diffusion of oxygen ions may be the dominant mechanism in polycrystalline material. Smeltzer et al. (1961) proposed a phenomenological theory to describe the oxidation kinetics of Ti, Zr and Hf at these intermediate temperatures. In this theory it is assumed that the oxygen migrates simultaneously by lattice diffusion and along low-resistance grain-boundary paths within which the number of available oxygen sites decreases exponentially with time. The theory provides excellent fits to experimental data, and suggests that, for the short times of oxidation employed by Miller (about 15 minutes), grain-boundary migration is predominant. By contrast, the oxidation of multicomponent stainless steels is highly complex, and is known to involve the formation of protective layers of $FeCr_2O_4$ and spinel, $Ni_xFe_{3-x}O_4$. The migration of both anions and cations and their inter-actions are all involved and, over prolonged periods, the porosity of the scale formed and its adherence to the substrate are important. However, for short oxidation times and relatively low temperatures, it is generally thought that grain boundary diffusion again plays a prominent role in the oxidation kinetics. It is therefore interesting to compare the behaviour in stainless steel and the simpler case of titanium.

Polished specimens were implanted with doses of $10^{15}$ to $10^{17}$ ions per $cm^2$ of about ten different species of metal, choosing the energies so as to give a theoretical ion range of around 1000 Å. Specimens were then oxidised in dry $O_2$ at 600 °C (for titanium) and 800 °C (for 18/8/1 stainless steel). In order to test the influence of radiation damage, samples were implanted with argon ions, and subsequently oxidised, both with and without vacuum annealing.

The simplest method of estimating the thickness of the oxide film is by its optical interference colour, and this proved a valuable qualitative guide, fig. 6.12. Ion back-scattering provided a better quantitative measure of the oxygen take-up, but an obvious difficulty arises due to the low mass of oxygen: normally its scattering cross-section is low, and the peak observed in the spectrum is superimposed upon an intense continuum of scattering from the substrate metal. Dearnaley et al. (1972) took advantage of the strong and broad $O^{16}$ (p, p) elastic scattering resonance around 4 MeV (Laubenstein et al., 1954) which provides a cross-section some forty times the Rutherford value, in order to increase the sensitivity of this technique. Furthermore, by the use of thin foil specimens (3 to 6 $\mu$m in thickness) it is possible to achieve a complete separation of the peaks due to scattering from metal or oxygen nuclei (figure 6.13), and this overcomes the difficulty arising from the presence of strong resonances in the $Ti^{48}$ (p, p) cross-section between 3.5 and 4.0 MeV: in measurements on a thick substrate the $O^{16}$ peak is superimposed on a sharply-varying Ti continuum (fig. 6.14). It would have been feasible to make a direct comparison of the oxygen take-up on the implanted face of such a thin foil and the unimplanted rear face, by comparison of the counts in the two, separated peaks observable due to

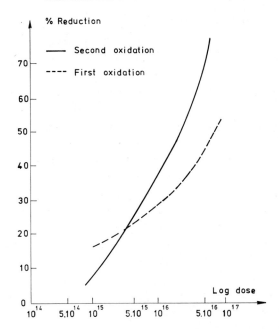

Fig. 6.14. Percentage reduction in thermal oxidation of titanium brought about by implantation with 135 keV $Ca^+$ ions, as a function of dose. The solid curve shows the result of polishing away the oxide first grown, and re-oxidizing (from Miller, 1972).

scattering at the entry and exit faces of the foil, but for the fact that the surface finish of the foils available differed visibly and this itself was found to influence the oxidation rate.

The results of these experiments can be summarized as follows: their implications in terms of the theories of corrosion mechanisms have yet to be worked out. Argon implantations produced only a slight increase of approximately 10 % in a 1000 Å oxide grown on titanium: annealing in vacuum eliminates this effect which is presumably a damage phenomenon. Other (metal) species varied widely in their effects, from a 75 % reduction in the oxidation of Ti (by $Ca^+$ ions) to a 52 % reduction of the oxidation of stainless steel by $Bi^+$ ions. The effects observed increased with ion dose (fig. 6.15). The most interesting observation was that those ions which enhance the oxidation of Ti will inhibit the oxidation of 18/8/1 stainless steel: no exception to this rule was found. Thus $Ca^+$ ions ($5 \times 10^{16}/cm^2$) would enhance the oxidation of stainless steel by 100 %. There appeared, tentatively, to be a correlation between the effectiveness of a given ion and its electronegativity or its position in the electro-chemical series. There was

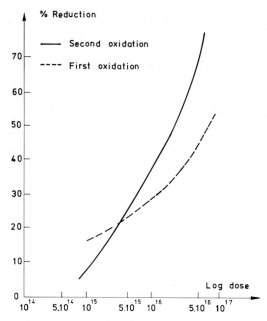

Fig. 6.15. Percentage enhancement of thermal oxidation of titanium brought about by implantation with 270 keV $Y^+$ ions, as a function of dose. As in fig. 6.14, the upper curve shows the result of polishing away the oxide first grown, and the re-oxidizing (from Miller, 1972).

TABLE 6.1

Metal ion species arranged in order of their effectiveness in inhibiting the thermal oxidation of titanium at 600 °C (from Miller, 1972).

| Relative effectiveness of ion | Electro-negativity | Position in electrochemical series | | Ionisation potential | | | Ionic Crystal Radius (Å) |
|---|---|---|---|---|---|---|---|
| | | | | First | Second | Third | |
| Calcium | 1.0 | $Ca^{2+}-2e^-$ | −2.76 V | 6.1 | 11.87 | 51.2 | 0.99 |
| Europium | 1.1 | Not quoted | | 5.67 | 11.24 | N-q | 0.95 |
| Cerium | 1.1 | $Ce^{3+}-3e^-$ | −2.34 V | 5.6 | 12.3 | 20.0 | 1.03 |
| Yttrium | 1.3 | $Y^{3+}-3e^-$ | −2.37 V | 6.38 | 12.23 | 20.5 | 0.89 |
| Zinc | 1.5 | $Zn^{2+}-2e^-$ | −0.76 V | 9.39 | 17.96 | 37.7 | 0.74 |
| Aluminium | 1.5 | $Al^{3+}-3e^-$ | −1.706 V | 5.98 | 18.82 | 28.4 | 0.51 |
| Indium | 1.7 | $In^{3+}-3e^-$ | −0.34 V | 5.78 | 18.86 | 28.03 | 0.81 |
| Nickel | 1.7 | $Ni^{2+}-2e^-$ | −0.23 V | 7.63 | 18.15 | 35.16 | 0.69 |
| Bismuth | 1.9 | Not quoted | | 7.29 | 16.68 | 25.6 | 0.96 |

certainly no correlation with ion size, though possibly one could be drawn with the second ionization potential (table 6.1).

A similar correlation has come to light in the electrical behaviour of anodic oxide films grown on ion-implanted titanium (Suffield, 1972). In this work, polished specimens of pure Ti were implanted with similar doses of the same species of ions as those used in the thermal oxidation studies. After anodizing to an oxide thickness of about 1000 Å, gold electrodes were deposited by vacuum evaporation, and the electrical conductivity of the film was measured. Three regions of conductivity were identified: initially the film would usually display a very low conduction, but might undergo 'forming' to an intermediate state, in which it could exhibit negative resistance and memory phenomena (for a review of these effects in oxide and other films, see Dearnaley et al., 1970). Ultimately, all the films would break down to a terminal state of high conductivity. Different implanted ions, which were shown by ion back-scattering (Suffield and Dearnaley, 1972) to be dispersed throughout the anodic oxide, appeared to control relative stability of the three conductivity regions. Argon ion implantation gave no effect. $Ca^+$ ions (electronegativity 1.0) produced a stable high-impedance film that rarely 'formed', but $Bi^+$ (electronegativity 1.8) led to films which very readily broke down to the low-impedance state. Since the conduction in these films is localized, and electrical 'forming' is believed to involve the migration of oxygen ions along grain-boundary paths (Dearnaley et al., 1970), it is not surprising that a correlation should exist between the electrical effects and the corrosion behaviour, the latter being essentially a charge-transport phenomenon. Ions, such as $Bi^+$, that increase the tendency for conduction in anodic $TiO_2$ are the very ones that will enhance the thermal oxidation of titanium at 600 °C.

The fact that yttrium, specifically, will convey good resistance to high temperature oxidation of austenitic stainless steels has been known for several years (Antill and Peakall, 1967). Antill et al. (1972) have carried out lengthy tests of the effectiveness of ion-implanted yttrium on the corrosion resistance of a 20 % Cr/25 % Ni/Nb-stabilized stainless steel in $CO_2$ at up to 850 °C. Polished samples of the steel were subjected to multiple implantations at four ion energies up to 900 keV, the total dose implanted being $3.5 \times 10^{15}$ ions/cm². From Lindhard, Scharff and Schiøtt data the yttrium concentration was estimated to be 0.2 % down to a depth of 0.25 $\mu$m. In all temperature ranges this yttrium implantation brought about a substantial reduction in oxidation, amounting at 850 °C to a factor of 5 or more, though diminishing somewhat with time over a 4000 h test. The degree of oxidation inhibition exceeded that in samples of similar steel but alloyed with 0.4 % yttrium, and yet the improvement was maintained for weight gains corresponding to oxide thicknesses of around 3 $\mu$m, i.e. at least an order of magnitude greater than the implantation depth. Besides inhibiting the oxidation process, the yttrium implantation was observed to produce a significant reduction in the degree of oxide spalling that results from the thermal cycling of the specimens during frequent weight-gain measurements at room temperature. In many practical applications this improved oxide adhesion is very important, as loss of the oxide scale would deplete yttrium present only near the surface of the metal.

These studies, and similar ones in progress, reveal that ion implantation is a most useful tool in the study of corrosion science, and that ion backscattering can supplement the more conventional techniques of measurement (such as weight gain) particularly if advantage is taken of the high oxygen cross-section at certain proton energies. The doses of ions required to produce significant effects ($\sim 10^{16}$ ions/cm$^2$) place high demands upon the ion intensities, but we have seen in chapter 4 that industrial implantation machines now exist capable of producing the milliampere intensities which are appropriate for research into corrosion. Further work aimed at still higher beam intensities should extend the scope of this work still more. It will be surprising if this relatively novel area of ion implantation research is not the subject of considerable growth.

### 6.8. Low-friction and wear resistant metal surfaces

Friction and wear of metal surfaces are normally minimised by lubrication. Sometimes, however, conventional fluid lubricants may be inadequate. Under extreme pressure conditions the fluid may be absent from contacting surfaces and friction, heating, adhesion and wear increase abruptly. In space or vacuum conditions or at high temperatures fluid lubricants are unsuitable and solid or 'dry' lubricants are preferred owing to their low vapour pressure. These solid lubricants are usually applied as thin coatings, often containing some form of binder, but their adherence is frequently inadequate and failure occurs due to peeling of the coating or accumulation of soft wear debris.

Ion implantation offers an attractive solution to these problems by providing a means of incorporating into the surface of a machined metal component suitably-chosen foreign atoms to lessen friction or improve wear resistance. There is in this case no coating that might peel away, and the physical dimensions are not changed during the implantation process.

That moderate doses of certain ions can have a strong effect on the frictional properties of a steel surface has now been demonstrated. Hartley and Dearnaley (1972), in work carried out jointly at the University of Sussex and at Harwell, have studied the effects of implanting a number of selected ion species into polished or ground specimens of a case-hardening steel, EN 352. The frictional force necessary to drag a loaded tungsten carbide ball across the steel surface was measured with the equipment shown schematically in figure 6.16 under dry conditions. With a 4 mm diameter ball, bearing pressures as high as $10^5$ g · $cm^{-2}$ were reached – sufficient to score the surface of the metal visibly. Doses of between $10^{16}$ and $10^{17}$ ions/$cm^2$ of $Mo^+$, $Sn^+$, $Pb^+$, $In^+$, $Ag^+$ and $Kr^+$ were implanted into rectangular areas of the steel, and the frictional behaviour was measured along tracks which traversed implanted and unimplanted regions. No significant effects were observed when $Kr^+$ was implanted, but increases or decreases of up to 50 % in the frictional force were observed after implantation with metal species (fig. 6.17). Clearly, therefore, the effect is associated with

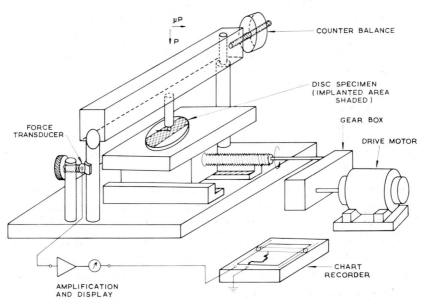

Fig. 6.16. Schematic drawing of the apparatus used by Hartley and Dearnaley (1972) to measure changes in the coefficient of friction in ion-implanted metals. The arm at the top can be loaded to a predetermined extent, and the transducer (left) allows the frictional force to be recorded as the specimen is traversed so that a test-ball is tracked across it in a line that traverses an implanted band (cross-hatched).

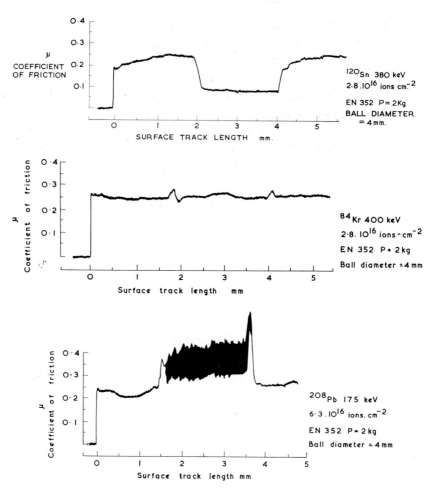

Fig. 6.17. Changes in coefficient of friction brought about in EN 352 steel by ion implantation, measured with the equipment shown in fig. 6.16. as a result of: (a) $2.8 \times 10^{16}$ $^{120}Sn^+$ ions/cm$^2$ at 380 keV; load: 2 kg; ball diameter 4 mm; (b) $2.8 \times 10^{16}$ Kr$^+$ ions/cm$^2$ at 400 keV; load: 2 kg; ball diameter 4 mm; (c) $6.3 \times 10^{16}$ Pb$^+$ ions/cm$^2$ at 175 keV; load: 2 kg; ball diameter 4 mm (after Hartley and Dearnaley, 1972).

incorporation of foreign species, in concentrations of a few atomic per cent, and it is not the result of a damage process or mere surface contamination. Care was taken to minimise any risk of hydrocarbon build-up in these implantations.

The effects observed increased with ion dose and it was further found that a combination of elements, such as Mo$^+$ followed by twice the dose of

$S^+$ ions, gave a greater decrease in friction than $Mo^+$ alone. In this experiment the ion energies were adjusted so as to provide similar mean ranges, but the presence of $MoS_2$ molecules in the implanted layer has not been proved.

Large fluctuations in frictional force (fig. 6.17c) observed with $Ag^+$ or $Pb^+$ implantations appear to be due to local adhesion and a consequent stick-slip motion of the ball. The effect was lessened by using a 25 mm test ball, but the frictional force was still high, and there was evidence of the transfer of material from the steel to the tungsten carbide.

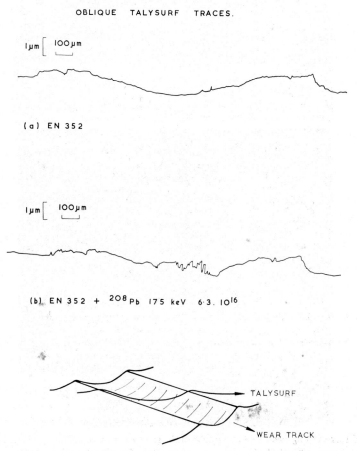

Fig. 6.18. The results of an obliquely-inclined trace taken with a 'Talysurf' recorder showing the topography of the wear track in unimplanted and $Pb^+$-implanted EN 352 steel, revealing considerable irregularities in the latter case (from Hartley and Dearnaley, 1972).

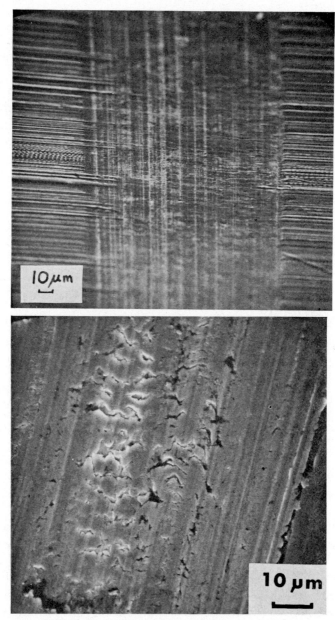

Fig. 6.19. Scanning electron micrographs of the grooves produced in ion-implanted EN 352 steel (a) after $2.8 \times 10^{16}$ Mo$^+$ ions/cm$^2$ at 400 keV, (b) after $6.3 \times 10^{16}$ Pb$^+$ ions/cm$^2$ at 175 keV. Lateral cracks responsible for the stick-slip adhesion apparent in the frictional trace (fig. 6.17c) are clearly visible in the lower photograph (from Hartley and Dearnaley, 1972).

Examination of the profile of the surface after these tests was made by a 'Talysurf' mechanical profiling instrument (Rank, Taylor-Hobson Ltd.) (fig. 6.18) and by a scanning electron microscope (fig. 6.19). In cases in which the friction is reduced, the groove shows a uniform profile along its length; high friction seems to be associated with irregular grooves.

A striking feature of these experiments was the way in which the friction changes would persist for several operations of re-tracking the ball along the same path, despite the small amount of material implanted to a shallow depth in the surface. It would appear that an important function of the lubricant is to lessen adhesion and consequent galling.

Previous work by Harris and Warwick (1969) and others on the use of thin coatings of soft metals such as lead, silver or gold in friction tests over long periods in ultrahigh vacuum has demonstrated their excellent performance. Lead is particularly good in this application, and so it is interesting to contrast the poor results obtained with both implanted $Pb^+$ and $Ag^+$ in the tests described above. Almost certainly the explanation lies in the fact that these tests were made in air: lead oxidises rapidly in air, particularly when the temperature rises, as it does at points of contact on a bearing surface. It is quite possible that the lead could melt and so relieve the friction. However, if oxide is formed this no longer has the low shear strength that is required. Tests of the behaviour of lead-implanted steel surfaces in ultrahigh vacuum are an obvious next step.

We have dwelt so far on low-friction surfaces, and have given little consideration to wear. Wear is a more complex phenomenon, and one must distinguish between conditions of mild and severe wear: in the former case there is little or no welding, and the wear debris consists of small particles, mostly of oxide. Corrosion is an important factor in mild wear, while fatigue may contribute to severe wear. High wear-resistance is not necessarily associated with low friction: soft materials may provide good lubricants but may not wear well. The best combination of material properties for wear resistance, under dry lubrication, is likely to be a very hard smooth surface into which small particles ('microreservoirs') of soft metal or lamellar solid (such as $MoS_2$) have been incorporated. Ion implantation, combined with appropriate thermal treatment, is a good way to produce such precipitates, particularly of rather insoluble species. In the case of solid lubricant coatings it is often recommended that the component surface is roughened by etching or sandblasting, so as to improve adhesion. Ion implantation renders this quite unnecessary. Optimum coating thicknesses for wear resistance (under high load conditions) are surprisingly thin (Hopkins and Campbell, 1969) and values of only 2 $\mu$m have been quoted for a molybdenum disulphide compound. Probably this is due to the need to conduct heat away from the contact surface, and this is difficult in a coating which includes $MoS_2$ and resin binder, superimposed on an oxide film. An ion implanted surface, on the other hand, is in excellent thermal contact with the substrate. It is difficult to apply very thin coatings by spraying methods, etc. with any uniformity, but ion implantation can be highly uniform, and provide a controlled thickness.

Much more work is needed in order to establish the value of ion implantation for the production of low-friction and wear-resistant surfaces, but the early indications are promising. Because areas of severe wear are often limited, the process need be applied only to small regions of a bearing or shaft and the costs need not therefore be very great.

## 6.9. The use of ion implantation in the study of mechanical properties and void formation in irradiated metals

### 6.9.1. GENERAL

During the lifetime of nuclear reactors the inert gas He is produced in copious amounts as a consequence of nuclear transmutation. The most significant implications of He production are in the stainless steel fuel cladding and other stainless steel structural components in the vicinity of the reactor core. In thermal reactors He is generally created from slow neutron reactions with the boron constituents of the steel; whereas in the modern generation of fast reactors the majority of He is produced as a consequence of (n, $\alpha$) reactions between fast neutrons and the iron or nickel constituents of steel. Helium is highly insoluble in metals and rapidly collects together to form gas bubbles, as described in chapter 3. It is such gas bubbbles which are thought to have important implications on the physical and mechanical integrity of reactor components. However, He production is generally accompanied by radiation damage – which itself can influence the physical and mechanical properties of metals. However, if we can independently simulate the production of He and radiation damage, using for instance, particle accelerators we may be able to separately identify the independent effects of He and of radiation damage. In this section we will therefore briefly discuss the techniques which have been developed to study the effects of He in reactor steels using ion implantation.

### 6.9.2. HELIUM IMPLANTATION TECHNIQUES

Unlike the implantation of semiconductors, we are generally interested in producing uniform implantation, to atomic concentrations up to say $10^{-4}$, to depths of up to over 100 $\mu$m. Only in such relatively large quantities of material can we be confident that spurious effects due to the free surfaces etc. can be kept to a minimum. For this reason it is necessary to use significantly higher energies than we have previously discussed in this book. For instance it is common practice to use He beams from large cyclotrons having energies ranging typically between 10–50 MeV. The technique of producing uniform implantations has been discussed in detail by Worth (1969) and by King (1969). In this book it will suffice simply to point out that such uniform implantations can easily be achieved by the use of a programmed energy degrader, either in the form of a tapered wedge or a tapered wheel. From our knowledge of the stopping power of He ions in solids we can simply compute the rate at which the energe degrader should be moved in order to produce a uniform concentration of implanted atoms.

### 6.9.3. HELIUM EMBRITTLEMENT

The first study to assess the effect of helium implantation on the mechanical properties of steel was carried out by Higgins and Roberts (1965). However, as an example of He embrittlement we will discuss the effect of He implantation on the ductility of 304 stainless steel at higher temperatures carried out by King (1969) in the Oak Ridge Isochronous Cyclotron. Tensile specimens of 304 steel were uniformly implanted with He to an atomic concentration of $0.83 \times 10^{-6}$ and annealed for 1 hour at 929 °C. These specimens were then subjected to standard tensile tests at temperatures from 500 °C to 900 °C, together with control specimens, at a strain rate of 0.026 $\text{min}^{-1}$. No significant changes were found in the yield strength or ultimate tensile strength, however the effects of He on the total strain of 304 steel are illustrated in fig. 6.20. Within the limits of the experimental uncertainty, the results demonstrate that although $0.83 \times 10^{-6}$ at. conc. He has essentially no effect on the tensile ductility at 500 and 600°C, severe reductions of ductility occur over the 700°–900 °C test temperature range. Experiments on similar samples irradiated in a thermal reactor to produce comparable He concentrations show qualitatively similar reductions in ductility. However, in the reactor case substantial radiation damage is produced within the specimens, and we are therefore led to conclude that the loss in ductility can be ascribed to the presence of He.

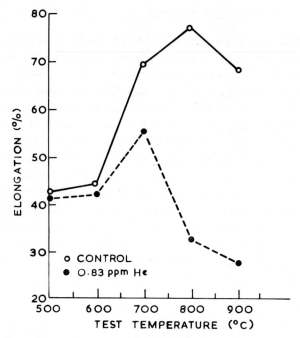

Fig. 6.20. Ductility of 304 steel annealed 1 h at 925 °C and tensile tested at 0.026$^{-1}$ strain rate (from King, 1969).

This book is not the place to discuss the possible mechanisms of the embrittlement of steels and the above experiment will suffice as an example of how cyclotrons have helped our understanding of the problems.

6.9.4. VOID FORMATION

*6.9.4.1. Introduction.* In 1966 Cawthorne and Fulton working at the UKAEA Fast Reactor Establishment at Dounreay discovered a new phenomenon of radiation damage – now know as 'void formation'. Upon examination, in the electron microscope, of stainless steel fuel cladding irradiated at about 500 °C to a high fast neutron dose, they observed a large number of small (100 Å) cavities. This observation immediately resulted in the instigation of major research programmes in those countries already engaged in the design and construction of fast breeder reactors. Such a phenomenon had naturally not been allowed for in reactor design, and its late arrival on the scene gave rise to significant concern. Cawthorne and Fulton (1967) pointed out that such cavities could not be accounted for in terms of equilibrium

He gas bubbles, as although significant He is produced during fast neutron irradiation, it was insufficient to account for the total cavitation. Such cavities were therefore, essentially empty and consequently named 'voids'. However, it was considered that the agglomeration of He, which was created during transmutation, perhaps played an important role in nucleating such voids. The significance of void formation to the reactor designer and his initial concern due to lack of knowledge can readily be understood. The initial observation suggested that even at doses where every atom had been displaced only about 10–20 times (10–20 dpa) the volume increase as a consequence of void formation, was about 1 %. Fast reactor steels were being designed to withstand a damage dose of up to 100 dpa and there was no data to even indicate the magnitude of the volume increase at such a dose. Furthermore, it would take many years for such doses to be achieved in existing reactors. Fortunately, radiation damage and the effects of He were being studied using particle accelerators in a variety of establishments. Such damage doses are readily achieved during ion bombardment within just a few hours irradiation and it was therefore a natural extension of such work to embark on a series of experiments specifically designed to study void formation and to isolate the possible effects of He.

*6.9.4.2. Ion beam simulation studies.* It is a relatively simple matter to produce voids during ion bombardment of steels and a typical transmission electron micrograph is shown in fig. 6.21 (see also fig. 3.15). The difficulty is to ensure that the simulation is as close as possible to that pertaining during fast neutron irradiation. Early reactor data showed a denudation of voids in the vicinity of grain-boundaries to a distance of up to 0.5 $\mu$m in some cases. At high damage levels this denudation is accompanied by a region of enhanced void growth immediately adjacent to it, and only at distances of the order of 1 $\mu$m is the void formation typical of bulk material. Such effects are thought to result from the influence of the grain boundaries as sinks for migrating defects. Similar effects are also to be expected near free surfaces and furthermore, because the dislocation densities near a free surface are generally lower than in the bulk, it has been common practice in simulation experiments to produce a uniform damage region about 2 $\mu$m wide situated at least 1 $\mu$m below the surface. As it is important to keep the complicating effects of impurity atoms to a minimum, it is highly desirable to use so-called 'self ions' as the damaging species. For instance in the study of steels it has been common practice to use Ni ions, however Cr or Fe ions can also be used. In order to satisfy these conditions a significant amount

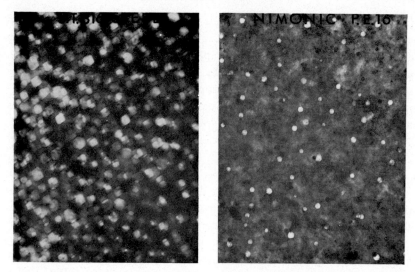

Fig. 6.21. Electron micrographs of typical void distributions in 316 stainless steel and Nimonic P.E.16 irradiated at 525 °C to 100 dpa.

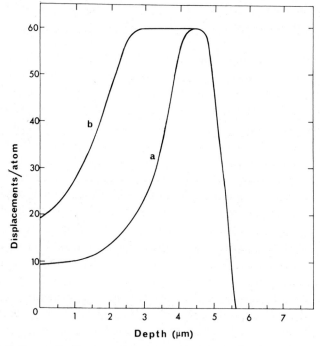

Fig. 6.22. Depth distributions of damage for 48 MeV $Ni^{6+}$ ions into steels–rocked and unrocked.

of work has been carried out at Harwell using 45 MeV $Ni^{6+}$ ions from the Variable Energy Cyclotron (VEC), and in order to provide a uniformly damaged region it has been common practice to utilise a programmed rocking of the target (see for example Nelson et al. (1970), Worth et al. (1971)). In the remainder of this section we will briefly report the main features of the void simulation work, with specific reference to the effects of implanted He. Fig. 6.22 shows a typical damage distribution using 45 MeV Ni ions in steel together with the corresponding distribution produced by a programmed rocking of the target. It is important to relate the total damage produced during ion bombardment to that produced during neutron irradiation. It is therefore of some importance to check that irradiations to the same calculated damage dose in the reactor and accelerator do in fact correspond. We must, of course, be careful to choose a system such that any dose rate effects can be neglected, or at least accounted for. In this context it is possible to choose corresponding temperatures for nickel, which at the peak swelling temperatures show little or no effect of dose rate on the magnitude of the swelling. In order to eliminate the variability which might result from different helium concentrations, it was decided to perform these check irradiations with nickel samples, taken from the same parent material, each of which having been previously annealed to the same temperature and implanted uniformly with identical concentrations of He ($10^{-5}$ atom/atom). A selection of samples were then irradiated in either the Dounreay fast reactor (DFR) or the VEC at corresponding temperatures to equal total damage doses. Fig. 6.23 shows two electron micrographs of voids which illustrate the results of this experiment. A computation of swelling from a knowledge of the void density, the void size and the foil thickness in the two cases gives answers which within the experimental error, agree remarkably well. This result, therefore gives us confidence both in the simulation technique and in the model for calculating damage.

Due to the restricted depth of the irradiated volume during ion bombardment we are limited to the use of the transmission electron microscope for the examination of samples. The volume swelling ($\Delta V/V$) is calculated from the void size and density. Fig. 6.24 shows a sequence of electron micrographs showing the change in void microstructure in 316 steel as a function of displacement dose at 525 °C. Fig. 6.25 shows this same data in graphical form whereas fig. 6.26 shows the temperature dependence of void swelling at 40 dpa (equivalent to $8 \times 10^{22}$ neutrons/cm$^2$) (Mazey et al., 1971).

The effect of He on void swelling has been studied for a variety of materials. For instance in 316 stainless steel voids are found whether or

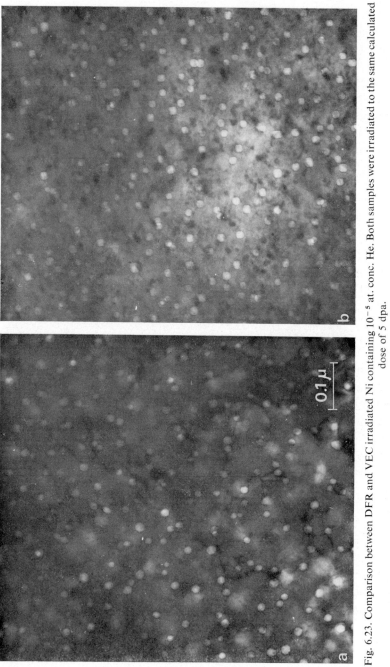

Fig. 6.23. Comparison between DFR and VEC irradiated Ni containing $10^{-5}$ at. conc. He. Both samples were irradiated to the same calculated dose of 5 dpa.

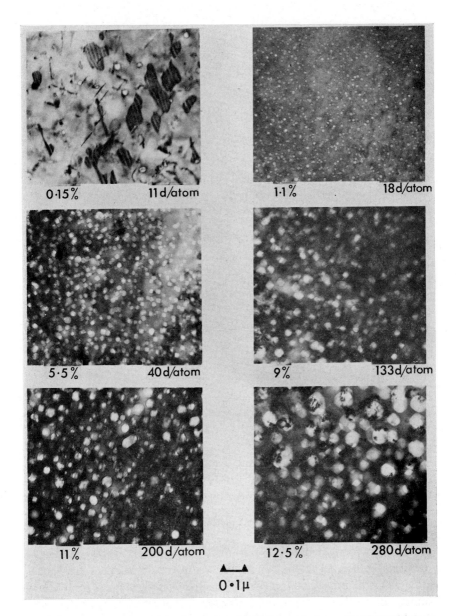

Fig. 6.24. Micrographs showing the increase in voidage with dose – 20 MeV $C^{++}$ bombardment of 316 steel at 525 °C.

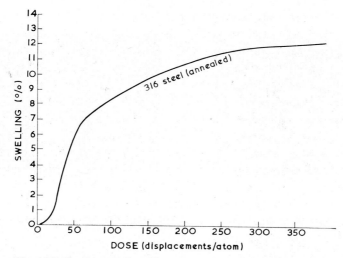

Fig. 6.25. Graph of the dose dependence of swelling in 316 steel at 525 °C.

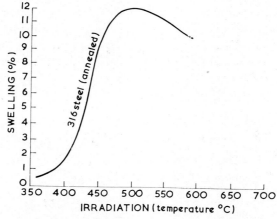

Fig. 6.26. Temperature dependence of void swelling in 316 steel at 40 dpa. The curve has been moved by 100 °C to lower temperatures to account for the increased dose rate compared with reactor irradiation.

not He was implanted prior to irradiation, however the void density is significantly affected at the concentrations in excess of about $10^{-5}$ atom/atoms. On the other hand in EN 58B stainless steel, voids are only found subsequent to the implantation. This difference is thought to be due to the Ti additions in the case of EN 58B as follows. Both 316 and EN 58B have significant amounts of oxygen within their matrix, and as such these gas atoms can agglomerate to form void nuclei. However, only in the case of 316 steel is the oxygen considered to be in solution and free to agglomerate, for in the case of the EN 58B the Ti additions are thought to be potent trapping sites for any free oxygen.

Data such as illustrated in figs. 6.24, 6.25 and 6.26 above, together with that from reactor irradiations, can be used to assess the relevance of void swelling in fast reactor design. For instance, due to the non-uniform damage rates throughout a reactor, some key components are expected to suffer distortion as a consequence of differential swelling. Such distortions must be minimised either by operational or engineering modifications or by finding alternative materials. The choice of other materials is restricted to those which are both compatible with liquid sodium at temperatures up to about 700 °C, and at the same time, exhibit the correct mechanical behaviour. For some time scientists within the UKAEA have been studying a selection of nickel based alloys, and in particular a nimonic alloy called PE16. After a suitable heat treatment this material contains within its structure a fine dispersion of very small precipitates, called gamma prime precipitates ($\gamma'$). For instance, after heating to 750 °C for four hours the $\gamma'$ precipitates grow to about 100 Å diameter and have an average separation of just over 500 Å. It was decided to perform a series of accelerator irradiations on this material. The results turned out to be extremely encouraging. For irradiation at 525 °C to 200 dpa (about five years in DFR) the total swelling was less than about 0.4 %. Fig. 6.27 shows the swelling plotted as a function of dose (Hudson et al., 1971).

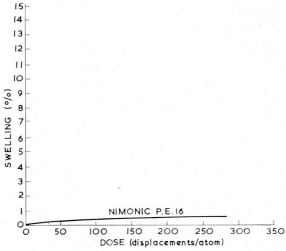

Fig. 6.27. Dose dependence of swelling in Nimonic P.E.16 at 525 °C.

The main role of the accelerator studies has been to produce advance data to damage doses which will not be achieved in the reactor for many years. However, the final tests will, of course, have to be made using reactors.

## References

Alexander, R. B., J. M. Poate and D. V. Morgan, 1970, Proc. Conf. on Hyperfine Inter- actions detected by Nuclear Radiations, Rehovoth (to be published).

Andersen, T. and G. Sorensen, 1965, Nucl. Inst. and Methods **38**, 204.

Antill, J. E., M. Bennett, G. Dearnaley, F. Fern, P. D. Goode and J. F. Turner, 1972, Proc. Conf. on Ion Implantation, Yorktown Heights, N.Y. (to be published).

Antill, J. E. and K. A. Peakall, 1967, J. Iron and Steel Institute **205**, 1136.

Ascoli, S. and F. Cacace, 1965, Nucl. Inst. and Methods **38**, 198.

Balarin, M., G. Otto, I. Storbeck, M. Sckenti and H. Wagner, 1969, Thin Solid Films **4**, 255.

Bayly, A. R., 1973 (to be published).

Bayly, A. R. and P. D. Townsend, 1970, Proc. Conf. on Ion Implantation, Reading, p. 120 (Peter Peregrinus, Stevenage, 1970).

Bett, R. and B. W. Howlett, 1970, private communication, to be published.

Bobeck, A. H., 1967, Bell Syst. Tech. Jour. **46**, 1901.

Bobeck, A. H., S. L. Blank and H. J. Levinstein, 1972, Bell Syst. Tech. Jour. **51**, 6, 1431.

Brown, W. L., 1971, 2nd Ion Implantation conference Garmisch (Springer-Verlag) 430.

Cawthorne, C. and E. J. Fulton, 1967, Nature **216**, 575.

Chang, C. C. and A. C. Rose-Innes, 1970, Proc. XII Int. Conf. on Low Temp. Physics, Kyoto.

Crowder, B. L. and S. I. Tan, 1971, IBM Technical Disclosure Bulletin **14**, no. 1, 198.

Dearnaley, G., P. D. Goode, W. S. Miller and J. F. Turner, 1972, Proc. Conf. on Ion Implantation, Yorktown Heights, N.Y. (to be published).

Dearnaley, G., P. D. Goode and J. F. Turner, 1971, to be published.

Dearnaley, G., A. M. Stoneham aud D. V. Morgan, 1970, Rep. Prog. Phys. **33**, 1129.

De Waard, H. and S. A. Drentje, 1969, Proc. Roy. Soc. **A311**, 139.

Feldman, L. C. and D. E. Murnick, 1970, Proc. Conf. on Hyperfine Interactions detected by Nuclear Radiations, Rehovoth (to be published).

Fossan, D. B. and A. R. Poletti, 1968, Hyperfine Structure and Nuclear Radiations, Ed. Matthias, E. and D. Shirley (North-Holland, Amsterdam).

Freeman, J. H., G. A. Gard, D. J. Mazey, J. H. Stephen and F. B. Whiting, 1970, Proc. Conf. Ion Implantation, 74, (Reading, England), also, 1967, Proc. Int. Conf. Application of Ion Beam to Semiconductor Technology 75 and 669 (Grenoble, France 1967).

Garmire, E., H. Stoll, A. Yariv and R. G. Hunsperger, 1972, Appl. Phys. Lett. **21**, 87.

Goell, J. E., R. D. Standley, W. M. Gibson and J. W. Rodgers, 1972, Appl. Phys. **21**, 72.

Grace, M. A., C. E. Johnson, N. Kurti, R. G. Scurlock and R. T. Taylor, 1959, Phil. Mag. **4**, 948.

Hanle, W. and K. H. Rau, 1952, Z. Physik **133**, 297.

Harris, C. L. and M. G. Warwick, 1969, Proc. I. Mech. Eng. **183**, 31.

Hartley, N. E. W. and G. Dearnaley, 1972, Proc. 3rd International Conf. on Ion Implanta- tion, Yorktown Heights, N.Y. (to be published).

Herskind, B., R. R. Borchers, J. D. Bronson, D. E. Murnick, L. Grodzins and R. Kalish, 1968, Hyperfine Structure and Nuclear Radiations (North-Holland, Amsterdam).

Higgins, P. R. B. and A. C. Roberts, 1965, Nature **20b**, 1249.

Hopkins, V. and M. Campbell, 1969, Lubrication Engineering **23**, 288.

Hudson, J. A., D. J. Mazey, R. S. Nelson, 1971, Proc. European Conf. on Voids formed by the Irradiation of Reactor Materials, Reading (B.N.E.S.).

Kasrai, M., 1967, Thesis, The State of $S^{35}$ in alkali chloride crystals, Cambridge, England, also Freeman, J. H., M. Kasrai and A. G. Maddock, 1967, Chemical Communications 979.

King, R. T., 1969, Proc. Uses of Cyclotron in Chemistry, Metallurgy and Biology, (Butterworths, London) p. 294.

Kofstad, P., High Temperature Oxidation of Metals (Wiley, New York, 1966).

Kurti, N., N. J. Stone, G. Dearnaley and J. H. Freeman (Ed.), 1969, Proc. Discussion Meeting on Ion Implantation and Hyperfine Interactions, Proc. Roy. Soc. A311, 1.

Laubenstein, R. A., M. J. W. Laubenstein, L. J. Koester and R. C. Mobley, 1951, Phys. Rev. 84, 12.

Lindhard, J., 1969, Proc. Roy. Soc. A311, 11.

Lukesh, J. S., 1955, Phys. Rev. 97, 345.

Markali, J., 1962, Proc. 5th International Congress on Electron Microscopy (Academic Press, New York, 1962).

Mazey, D. J., J. A. Hudson and R. S. Nelson, 1971, J. Nucl. Materials 41, 257.

Miller, W. S., 1972, M. Sc. Dissertation, Brighton Polytechnic.

Nelson, R. S., D. J. Mazey and J. A. Hudson, 1970, J. Nucl. Materials 37, 1.

North, J. C. and R. Wolfe, 1972, Abstract submitted to 3rd Ion Implantation Conference, Yorktown Heights.

Pavlov, P. V. and E. V. Shitova, 1967, Sov. Phys. Doklady 12, 11.

Primak, W., 1958, Phys. Rev. 110, 1240.

Rickards, G., 1972, private communication.

Schineller, E. R., R. Flam and D. Wilmot, 1968, J. Opt. Soc. Amer. 58, 1171.

Smeltzer, W. W., R. R. Haering and J. S. Kirkaldy, 1961, Acta Met. 9, 880.

Smith, M. L., 1956, 100, Electromagnetically enriched isotopes and mass spectrometry (Butterworths, London).

Standley, R. D., W. M. Gibson and J. W. Rodgers, 1972, Applied Optics, 11, 1313.

Suffield, N. W., 1972, AERE Report R-7395.

Suffield, N. W. and G. Dearnaley, 1972, Proc. Conf. on Ion Implantation, Yorktown Heights, N.Y. (to be published).

Thompson, M. W., 1970, Proc. European Conf. on Ion Implantation, Reading (Peter Peregrinus Ltd., Stevenage England) p. 109.

Trillat, J. J., 1961, Le bombardement ionique, Théories et applications, Editions due C.R.N.S., Paris, France.

Trillat, J. J. and Haymann, 1961, in Trillat (1961).

Turner, J. F., W. Temple and G. Dearnaley, 1972, Proc. Conf. on Ion Implantation, Yorktown Heights, N.Y. (to be published).

Van Wijngaarden, A. and L. Hastings, 1967, Can. J. Phys. 45, 2239; 3803; 4039.

Watanabe, M. and A. Tooi, 1966, Japan J. App. Phys. 5, 737.

Weissmann, S. and K. Nakajima, 1963, J. Appl. Phys. 34, 3152.

Wolfe R., J. C. North, R. L. Barnes, M. Robinson and H. J. Levinstein, 1971, App. Phys. Letters, 19, 298.

Wolfe, R. and J. C. North, 1972, Bell Syst. Tech. Jour., 51, 1436.

Worth, J. H., Proc. Uses of Cyclotrons in Chemistry, Metallurgy and Biology (Butterworths, London, 1969) p. 283.

Worth, J. H., P. A. Clark and J. A. Hudson, 1971, J. of B.N.E.S. Oct., p. 329.

# TABLES FOR
# CALCULATING GAUSSIAN
# PROFILE PARAMETERS

In this appendix, the function

$$N(x) = N_{(\text{max})} e^{-\frac{1}{2}X^2}$$

where

$$X = \frac{x - R_p}{\Delta R_p}$$

and

$$N_{(\text{max})} = \frac{\text{surface concentration, } C_s}{\sqrt{2\pi}\Delta R_p}$$

has been tabulated with the ratio $N(x)/N(\text{max})$ as the independent variable with $X$, and the two areas $A$ and $B$ under the curve as dependent variables (see fig. A2.1). This form of presentation is more convenient than the usual Gaussian tables in reference books.

The use of the tables, with examples, is described in section 5.3.4.2 (Chap. 5). The table is a direct copy of a computer print-out in order to avoid the introduction of transcription errors. The representation of the powers of ten used in the tables are:

.... E 01 ≡ X $10^1$

.... E 00 ≡ X $10^0$

.... E-01 ≡ X $10^{-1}$

etc.

and for example,

0.96936 E-01

$= 0.96936 \times 10^{-1}$

$= 0.096936$

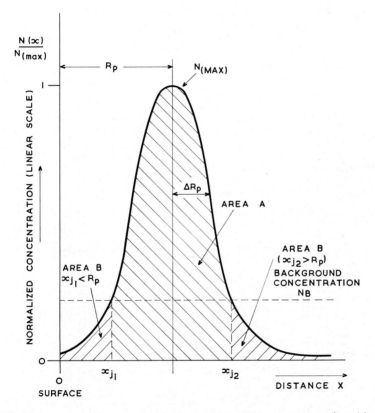

Fig. A2.1. A theoretical Gaussian distribution with depth. $R_p$ is the projected ion range, $\Delta R_p$ the standard deviation and $N_B$ the background doping concentration. The areas $A$ and $B$ are given in the accompanying table for the corresponding values of $N(x)/N_{(max)}$.

TABLE A2.1.

The tabulation of $X$, Area $A$ and Area $B$ (see fig. A2.1) against $N(x)/N_{(max)}$.

| N/N(MAX) | X | AREA(A) | AREA(B) |
|---|---|---|---|
| 0.10000E 01 | 0.0 | 0.0 | 0.50000E 00 |
| 0.99000E 00 | 0.14178 | 0.11274 | 0.44363E 00 |
| 0.98000E 00 | 0.20101 | 0.15931 | 0.42034E 00 |
| 0.97000E 00 | 0.24682 | 0.19495 | 0.40253E 00 |
| 0.96000E 00 | 0.28573 | 0.22492 | 0.38754E 00 |
| 0.95000E 00 | 0.32029 | 0.25125 | 0.37437E 00 |
| 0.94000E 00 | 0.35178 | 0.27500 | 0.36250E 00 |
| 0.93000E 00 | 0.38097 | 0.29678 | 0.35161E 00 |
| 0.92000E 00 | 0.40837 | 0.31700 | 0.34150E 00 |
| 0.91000E 00 | 0.43431 | 0.33593 | 0.33203E 00 |
| 0.90000E 00 | 0.45904 | 0.35380 | 0.32310E 00 |
| 0.89000E 00 | 0.48277 | 0.37074 | 0.31463E 00 |
| 0.88000E 00 | 0.50563 | 0.38689 | 0.30656E 00 |
| 0.87000E 00 | 0.52775 | 0.40233 | 0.29884E 00 |
| 0.86000E 00 | 0.54922 | 0.41715 | 0.29143E 00 |
| 0.85000E 00 | 0.57012 | 0.43140 | 0.28430E 00 |
| 0.84000E 00 | 0.59051 | 0.44515 | 0.27742E 00 |
| 0.83000E 00 | 0.61046 | 0.45844 | 0.27078E 00 |
| 0.82000E 00 | 0.63000 | 0.47131 | 0.26435E 00 |
| 0.81000E 00 | 0.64919 | 0.48378 | 0.25811E 00 |
| 0.80000E 00 | 0.66805 | 0.49590 | 0.25205E 00 |
| 0.79000E 00 | 0.68662 | 0.50768 | 0.24616E 00 |
| 0.78000E 00 | 0.70493 | 0.51914 | 0.24043E 00 |
| 0.77000E 00 | 0.72300 | 0.53032 | 0.23484E 00 |
| 0.76000E 00 | 0.74086 | 0.54122 | 0.22939E 00 |
| 0.75000E 00 | 0.75853 | 0.55186 | 0.22407E 00 |
| 0.74000E 00 | 0.77602 | 0.56226 | 0.21887E 00 |
| 0.73000E 00 | 0.79336 | 0.57243 | 0.21378E 00 |
| 0.72000E 00 | 0.81056 | 0.58238 | 0.20881E 00 |
| 0.71000E 00 | 0.82764 | 0.59212 | 0.20394E 00 |
| 0.70000E 00 | 0.84460 | 0.60167 | 0.19917E 00 |
| 0.69000E 00 | 0.86147 | 0.61102 | 0.19449E 00 |
| 0.68000E 00 | 0.87825 | 0.62019 | 0.18990E 00 |
| 0.67000E 00 | 0.89496 | 0.62919 | 0.18540E 00 |
| 0.66000E 00 | 0.91161 | 0.63803 | 0.18099E 00 |
| 0.65000E 00 | 0.92821 | 0.64670 | 0.17665E 00 |
| 0.64000E 00 | 0.94476 | 0.65522 | 0.17239E 00 |
| 0.63000E 00 | 0.96129 | 0.66359 | 0.16820E 00 |
| 0.62000E 00 | 0.97779 | 0.67182 | 0.16409E 00 |
| 0.61000E 00 | 0.99428 | 0.67991 | 0.16004E 00 |
| 0.60000E 00 | 1.01077 | 0.68787 | 0.15606E 00 |
| 0.59000E 00 | 1.02726 | 0.69570 | 0.15215E 00 |
| 0.58000E 00 | 1.04377 | 0.70341 | 0.14830E 00 |
| 0.57000E 00 | 1.06030 | 0.71099 | 0.14450E 00 |
| 0.56000E 00 | 1.07686 | 0.71846 | 0.14077E 00 |

# TABLE A2.1.

(continued)

| N/N(MAX) | X | AREA(A) | AREA(B) |
|---|---|---|---|
| 0.55000E 00 | 1.09347 | 0.72581 | 0.13709E 00 |
| 0.54000E 00 | 1.11012 | 0.73305 | 0.13347E 00 |
| 0.53000E 00 | 1.12683 | 0.74019 | 0.12991E 00 |
| 0.52000E 00 | 1.14361 | 0.74722 | 0.12639E 00 |
| 0.51000E 00 | 1.16047 | 0.75414 | 0.12293E 00 |
| 0.50000E 00 | 1.17741 | 0.76097 | 0.11952E 00 |
| 0.49000E 00 | 1.19445 | 0.76770 | 0.11615E 00 |
| 0.48000E 00 | 1.21159 | 0.77433 | 0.11284E 00 |
| 0.47000E 00 | 1.22884 | 0.78087 | 0.10957E 00 |
| 0.46000E 00 | 1.24622 | 0.78732 | 0.10634E 00 |
| 0.45000E 00 | 1.26373 | 0.79367 | 0.10316E 00 |
| 0.44000E 00 | 1.28139 | 0.79994 | 0.10003E 00 |
| 0.43000E 00 | 1.29921 | 0.80613 | 0.96936E-01 |
| 0.42000E 00 | 1.31719 | 0.81223 | 0.93887E-01 |
| 0.41000E 00 | 1.33536 | 0.81824 | 0.90879E-01 |
| 0.40000E 00 | 1.35373 | 0.82418 | 0.87911E-01 |
| 0.39000E 00 | 1.37230 | 0.83003 | 0.84984E-01 |
| 0.38000E 00 | 1.39110 | 0.83581 | 0.82097E-01 |
| 0.37000E 00 | 1.41014 | 0.84150 | 0.79249E-01 |
| 0.36000E 00 | 1.42944 | 0.84712 | 0.76439E-01 |
| 0.35000E 00 | 1.44901 | 0.85267 | 0.73667E-01 |
| 0.34000E 00 | 1.46888 | 0.85814 | 0.70932E-01 |
| 0.33000E 00 | 1.48907 | 0.86353 | 0.68235E-01 |
| 0.32000E 00 | 1.50959 | 0.86885 | 0.65574E-01 |
| 0.31000E 00 | 1.53048 | 0.87410 | 0.62949E-01 |
| 0.30000E 00 | 1.55176 | 0.87928 | 0.60360E-01 |
| 0.29000E 00 | 1.57345 | 0.88439 | 0.57807E-01 |
| 0.28000E 00 | 1.59560 | 0.88942 | 0.55289E-01 |
| 0.27000E 00 | 1.61823 | 0.89439 | 0.52807E-01 |
| 0.26000E 00 | 1.64139 | 0.89928 | 0.50359E-01 |
| 0.25000E 00 | 1.66511 | 0.90411 | 0.47945E-01 |
| 0.24000E 00 | 1.68945 | 0.90887 | 0.45567E-01 |
| 0.23000E 00 | 1.71445 | 0.91355 | 0.43223E-01 |
| 0.22000E 00 | 1.74019 | 0.91817 | 0.40913E-01 |
| 0.21000E 00 | 1.76672 | 0.92272 | 0.38638E-01 |
| 0.20000E 00 | 1.79412 | 0.92721 | 0.36397E-01 |
| 0.19000E 00 | 1.82249 | 0.93162 | 0.34190E-01 |
| 0.18000E 00 | 1.85192 | 0.93596 | 0.32019E-01 |
| 0.17000E 00 | 1.88253 | 0.94024 | 0.29882E-01 |
| 0.16000E 00 | 1.91446 | 0.94444 | 0.27781E-01 |
| 0.15000E 00 | 1.94788 | 0.94857 | 0.25715E-01 |
| 0.14000E 00 | 1.98298 | 0.95263 | 0.23685E-01 |
| 0.13000E 00 | 2.02001 | 0.95662 | 0.21691E-01 |
| 0.12000E 00 | 2.05925 | 0.96053 | 0.19735E-01 |
| 0.11000E 00 | 2.10108 | 0.96437 | 0.17817E-01 |

TABLE A2.1.

(continued)

| N/N(MAX) | X | AREA(A) | AREA(B) |
|---|---|---|---|
| 0.10000E 00 | 2.14597 | 0.96812 | 0.15938E-01 |
| 0.98000E-01 | 2.15536 | 0.96887 | 0.15567E-01 |
| 0.96000E-01 | 2.16491 | 0.96961 | 0.15197E-01 |
| 0.94000E-01 | 2.17461 | 0.97034 | 0.14830E-01 |
| 0.92000E-01 | 2.18448 | 0.97107 | 0.14464E-01 |
| 0.90000E-01 | 2.19451 | 0.97180 | 0.14099E-01 |
| 0.88000E-01 | 2.20473 | 0.97253 | 0.13737E-01 |
| 0.86000E-01 | 2.21513 | 0.97325 | 0.13375E-01 |
| 0.84000E-01 | 2.22573 | 0.97397 | 0.13016E-01 |
| 0.82000E-01 | 2.23653 | 0.97468 | 0.12658E-01 |
| 0.80000E-01 | 2.24754 | 0.97539 | 0.12303E-01 |
| 0.78000E-01 | 2.25878 | 0.97610 | 0.11948E-01 |
| 0.76000E-01 | 2.27025 | 0.97681 | 0.11596E-01 |
| 0.74000E-01 | 2.28197 | 0.97751 | 0.11246E-01 |
| 0.72000E-01 | 2.29394 | 0.97821 | 0.10897E-01 |
| 0.70000E-01 | 2.30619 | 0.97890 | 0.10550E-01 |
| 0.68000E-01 | 2.31873 | 0.97959 | 0.10205E-01 |
| 0.66000E-01 | 2.33157 | 0.98028 | 0.98618E-02 |
| 0.64000E-01 | 2.34473 | 0.98096 | 0.95205E-02 |
| 0.62000E-01 | 2.35823 | 0.98164 | 0.91812E-02 |
| 0.60000E-01 | 2.37209 | 0.98231 | 0.88438E-02 |
| 0.58000E-01 | 2.38634 | 0.98298 | 0.85085E-02 |
| 0.56000E-01 | 2.40100 | 0.98365 | 0.81751E-02 |
| 0.54000E-01 | 2.41610 | 0.98431 | 0.78439E-02 |
| 0.52000E-01 | 2.43167 | 0.98497 | 0.75147E-02 |
| 0.50000E-01 | 2.44775 | 0.98562 | 0.71876E-02 |
| 0.48000E-01 | 2.46437 | 0.98627 | 0.68628E-02 |
| 0.46000E-01 | 2.48158 | 0.98692 | 0.65401E-02 |
| 0.44000E-01 | 2.49943 | 0.98756 | 0.62197E-02 |
| 0.42000E-01 | 2.51797 | 0.98820 | 0.59017E-02 |
| 0.40000E-01 | 2.53727 | 0.98883 | 0.55860E-02 |
| 0.38000E-01 | 2.55741 | 0.98945 | 0.52728E-02 |
| 0.36000E-01 | 2.57846 | 0.99008 | 0.49620E-02 |
| 0.34000E-01 | 2.60054 | 0.99069 | 0.46539E-02 |
| 0.32000E-01 | 2.62375 | 0.99130 | 0.43484E-02 |
| 0.30000E-01 | 2.64823 | 0.99191 | 0.40457E-02 |
| 0.28000E-01 | 2.67415 | 0.99251 | 0.37459E-02 |
| 0.26000E-01 | 2.70172 | 0.99310 | 0.34490E-02 |
| 0.24000E-01 | 2.73119 | 0.99369 | 0.31553E-02 |
| 0.22000E-01 | 2.76287 | 0.99427 | 0.28648E-02 |
| 0.20000E-01 | 2.79715 | 0.99484 | 0.25778E-02 |
| 0.18000E-01 | 2.83457 | 0.99541 | 0.22944E-02 |
| 0.16000E-01 | 2.87582 | 0.99597 | 0.20149E-02 |
| 0.14000E-01 | 2.92188 | 0.99652 | 0.17396E-02 |
| 0.12000E-01 | 2.97417 | 0.99706 | 0.14689E-02 |

TABLE A2.1.

(continued)

| N/N(MAX) | X | AREA(A) | AREA(B) |
|---|---|---|---|
| 0.10000E-01 | 3.03485 | 0.99759 | 0.12033E-02 |
| 0.98000E-02 | 3.04150 | 0.99765 | 0.11770E-02 |
| 0.96000E-02 | 3.04828 | 0.99770 | 0.11508E-02 |
| 0.94000E-02 | 3.05517 | 0.99775 | 0.11246E-02 |
| 0.92000E-02 | 3.06221 | 0.99780 | 0.10986E-02 |
| 0.90000E-02 | 3.06937 | 0.99785 | 0.10725E-02 |
| 0.88000E-02 | 3.07669 | 0.99791 | 0.10466E-02 |
| 0.86000E-02 | 3.08415 | 0.99796 | 0.10207E-02 |
| 0.84000E-02 | 3.09177 | 0.99801 | 0.99483E-03 |
| 0.82000E-02 | 3.09956 | 0.99806 | 0.96906E-03 |
| 0.80000E-02 | 3.10751 | 0.99811 | 0.94335E-03 |
| 0.78000E-02 | 3.11565 | 0.99816 | 0.91771E-03 |
| 0.76000E-02 | 3.12397 | 0.99822 | 0.89213E-03 |
| 0.74000E-02 | 3.13250 | 0.99827 | 0.86662E-03 |
| 0.72000E-02 | 3.14123 | 0.99832 | 0.84119E-03 |
| 0.70000E-02 | 3.15019 | 0.99837 | 0.81582E-03 |
| 0.68000E-02 | 3.15938 | 0.99842 | 0.79053E-03 |
| 0.66000E-02 | 3.16881 | 0.99847 | 0.76532E-03 |
| 0.64000E-02 | 3.17851 | 0.99852 | 0.74017E-03 |
| 0.62000E-02 | 3.18848 | 0.99857 | 0.71511E-03 |
| 0.60000E-02 | 3.19875 | 0.99862 | 0.69013E-03 |
| 0.58000E-02 | 3.20933 | 0.99867 | 0.66522E-03 |
| 0.56000E-02 | 3.22024 | 0.99872 | 0.64041E-03 |
| 0.54000E-02 | 3.23152 | 0.99877 | 0.61567E-03 |
| 0.52000E-02 | 3.24318 | 0.99882 | 0.59102E-03 |
| 0.50000E-02 | 3.25525 | 0.99887 | 0.56647E-03 |
| 0.48000E-02 | 3.26776 | 0.99892 | 0.54200E-03 |
| 0.46000E-02 | 3.28076 | 0.99896 | 0.51764E-03 |
| 0.44000E-02 | 3.29428 | 0.99901 | 0.49337E-03 |
| 0.42000E-02 | 3.30837 | 0.99906 | 0.46920E-03 |
| 0.40000E-02 | 3.32309 | 0.99911 | 0.44513E-03 |
| 0.38000E-02 | 3.33849 | 0.99916 | 0.42118E-03 |
| 0.36000E-02 | 3.35464 | 0.99921 | 0.39733E-03 |
| 0.34000E-02 | 3.37164 | 0.99925 | 0.37361E-03 |
| 0.32000E-02 | 3.38957 | 0.99930 | 0.35001E-03 |
| 0.30000E-02 | 3.40856 | 0.99935 | 0.32653E-03 |
| 0.28000E-02 | 3.42874 | 0.99939 | 0.30319E-03 |
| 0.26000E-02 | 3.45029 | 0.99944 | 0.27999E-03 |
| 0.24000E-02 | 3.47341 | 0.99949 | 0.25695E-03 |
| 0.22000E-02 | 3.49837 | 0.99953 | 0.23406E-03 |
| 0.20000E-02 | 3.52551 | 0.99958 | 0.21133E-03 |
| 0.18000E-02 | 3.55527 | 0.99962 | 0.18880E-03 |
| 0.16000E-02 | 3.58825 | 0.99967 | 0.16646E-03 |
| 0.14000E-02 | 3.62527 | 0.99971 | 0.14433E-03 |
| 0.12000E-02 | 3.66754 | 0.99976 | 0.12245E-03 |

TABLE A2.1.

(continued)

| N/N(MAX) | X | AREA(A) | AREA(B) |
|---|---|---|---|
| 0.10000E-02 | 3.71692 | 0.99980 | 0.10083E-03 |
| 0.95000E-03 | 3.73070 | 0.99981 | 0.95476E-04 |
| 0.90000E-03 | 3.74516 | 0.99982 | 0.90139E-04 |
| 0.85000E-03 | 3.76039 | 0.99983 | 0.84824E-04 |
| 0.80000E-03 | 3.77648 | 0.99984 | 0.79530E-04 |
| 0.75000E-03 | 3.79353 | 0.99985 | 0.74260E-04 |
| 0.70000E-03 | 3.81167 | 0.99986 | 0.69014E-04 |
| 0.65000E-03 | 3.83107 | 0.99987 | 0.63794E-04 |
| 0.60000E-03 | 3.85190 | 0.99988 | 0.58602E-04 |
| 0.55000E-03 | 3.87443 | 0.99989 | 0.53438E-04 |
| 0.50000E-03 | 3.89895 | 0.99990 | 0.48306E-04 |
| 0.45000E-03 | 3.92588 | 0.99991 | 0.43207E-04 |
| 0.40000E-03 | 3.95577 | 0.99992 | 0.38145E-04 |
| 0.35000E-03 | 3.98938 | 0.99993 | 0.33123E-04 |
| 0.30000E-03 | 4.02784 | 0.99994 | 0.28146E-04 |
| 0.25000E-03 | 4.07285 | 0.99995 | 0.23221E-04 |
| 0.20000E-03 | 4.12727 | 0.99996 | 0.18354E-04 |
| 0.15000E-03 | 4.19640 | 0.99997 | 0.13560E-04 |
|  |  |  |  |
| 0.10000E-03 | 4.29193 | 0.99998 | 0.88563E-05 |
| 0.95000E-04 | 4.30387 | 0.99998 | 0.83921E-05 |
| 0.90000E-04 | 4.31641 | 0.99998 | 0.79293E-05 |
| 0.85000E-04 | 4.32963 | 0.99999 | 0.74679E-05 |
| 0.80000E-04 | 4.34361 | 0.99999 | 0.70079E-05 |
| 0.75000E-04 | 4.35845 | 0.99999 | 0.65495E-05 |
| 0.70000E-04 | 4.37425 | 0.99999 | 0.60926E-05 |
| 0.65000E-04 | 4.39116 | 0.99999 | 0.56375E-05 |
| 0.60000E-04 | 4.40935 | 0.99999 | 0.51842E-05 |
| 0.55000E-04 | 4.42904 | 0.99999 | 0.47328E-05 |
| 0.50000E-04 | 4.45050 | 0.99999 | 0.42835E-05 |
| 0.45000E-04 | 4.47411 | 0.99999 | 0.38364E-05 |
| 0.40000E-04 | 4.50036 | 0.99999 | 0.33919E-05 |
| 0.35000E-04 | 4.52994 | 0.99999 | 0.29501E-05 |
| 0.30000E-04 | 4.56384 | 0.99999 | 0.25113E-05 |
| 0.25000E-04 | 4.60361 | 1.00000 | 0.20761E-05 |
| 0.20000E-04 | 4.65183 | 1.00000 | 0.16450E-05 |
| 0.15000E-04 | 4.71327 | 1.00000 | 0.12189E-05 |

TABLE A2.1.

(continued)

| N/N(MAX) | X | AREA(A) | AREA(B) |
|---|---|---|---|
| 0.10000E-04 | 4.79853 | 1.00000 | 0.79919E-06 |
| 0.95000E-05 | 4.80920 | 1.00000 | 0.75766E-06 |
| 0.90000E-05 | 4.82043 | 1.00000 | 0.71624E-06 |
| 0.85000E-05 | 4.83228 | 1.00000 | 0.67491E-06 |
| 0.80000E-05 | 4.84481 | 1.00000 | 0.63368E-06 |
| 0.75000E-05 | 4.85811 | 1.00000 | 0.59256E-06 |
| 0.70000E-05 | 4.87229 | 1.00000 | 0.55156E-06 |
| 0.65000E-05 | 4.88748 | 1.00000 | 0.51069E-06 |
| 0.60000E-05 | 4.90383 | 1.00000 | 0.46994E-06 |
| 0.55000E-05 | 4.92154 | 1.00000 | 0.42934E-06 |
| 0.50000E-05 | 4.94086 | 1.00000 | 0.38888E-06 |
| 0.45000E-05 | 4.96214 | 1.00000 | 0.34860E-06 |
| 0.40000E-05 | 4.98582 | 1.00000 | 0.30849E-06 |
| 0.35000E-05 | 5.01253 | 1.00000 | 0.26859E-06 |
| 0.30000E-05 | 5.04319 | 1.00000 | 0.22891E-06 |
| 0.25000E-05 | 5.07922 | 1.00000 | 0.18950E-06 |
| 0.20000E-05 | 5.12296 | 1.00000 | 0.15039E-06 |
| 0.15000E-05 | 5.17881 | 1.00000 | 0.11165E-06 |
| | | | |
| 0.10000E-05 | 5.25652 | 1.00000 | 0.73403E-07 |
| 0.90000E-06 | 5.27653 | 1.00000 | 0.65827E-07 |
| 0.80000E-06 | 5.29880 | 1.00000 | 0.58282E-07 |
| 0.70000E-06 | 5.32394 | 1.00000 | 0.50771E-07 |
| 0.60000E-06 | 5.35282 | 1.00000 | 0.43297E-07 |
| 0.50000E-06 | 5.38677 | 1.00000 | 0.35867E-07 |
| 0.40000E-06 | 5.42804 | 1.00000 | 0.28488E-07 |
| 0.30000E-06 | 5.48078 | 1.00000 | 0.21173E-07 |
| 0.20000E-06 | 5.55427 | 1.00000 | 0.13939E-07 |
| | | | |
| 0.10000E-06 | 5.67769 | 1.00000 | 0.68262E-08 |
| 0.90000E-07 | 5.69622 | 1.00000 | 0.61247E-08 |
| 0.80000E-07 | 5.71686 | 1.00000 | 0.54255E-08 |
| 0.70000E-07 | 5.74017 | 1.00000 | 0.47291E-08 |
| 0.60000E-07 | 5.76696 | 1.00000 | 0.40357E-08 |
| 0.50000E-07 | 5.79849 | 1.00000 | 0.33457E-08 |
| 0.40000E-07 | 5.83685 | 1.00000 | 0.26599E-08 |
| 0.30000E-07 | 5.88593 | 1.00000 | 0.19791E-08 |
| 0.20000E-07 | 5.95442 | 1.00000 | 0.13050E-08 |

TABLE A2.1.

(continued)

| N/N(MAX) | X | AREA(A) | AREA(B) |
|---|---|---|---|
| 0.10000E-07 | 6.06971 | 1.00000 | 0.64071E-09 |
| 0.90000E-08 | 6.08704 | 1.00000 | 0.57508E-09 |
| 0.80000E-08 | 6.10636 | 1.00000 | 0.50964E-09 |
| 0.70000E-08 | 6.12819 | 1.00000 | 0.44442E-09 |
| 0.60000E-08 | 6.15329 | 1.00000 | 0.37945E-09 |
| 0.50000E-08 | 6.18285 | 1.00000 | 0.31477E-09 |
| 0.40000E-08 | 6.21884 | 1.00000 | 0.25043E-09 |
| 0.30000E-08 | 6.26493 | 1.00000 | 0.18650E-09 |
| 0.20000E-08 | 6.32932 | 1.00000 | 0.12313E-09 |
| 0.10000E-08 | 6.43790 | 1.00000 | 0.60570E-10 |
| 0.90000E-09 | 6.45424 | 1.00000 | 0.54381E-10 |
| 0.80000E-09 | 6.47247 | 1.00000 | 0.48208E-10 |
| 0.70000E-09 | 6.49306 | 1.00000 | 0.42054E-10 |
| 0.60000E-09 | 6.51676 | 1.00000 | 0.35921E-10 |
| 0.50000E-09 | 6.54468 | 1.00000 | 0.29812E-10 |
| 0.40000E-09 | 6.57869 | 1.00000 | 0.23731E-10 |
| 0.30000E-09 | 6.62227 | 1.00000 | 0.17686E-10 |
| 0.20000E-09 | 6.68322 | 1.00000 | 0.11688E-10 |
| 0.10000E-09 | 6.78614 | 1.00000 | 0.57587E-11 |
| 0.90000E-10 | 6.80165 | 1.00000 | 0.51714E-11 |
| 0.80000E-10 | 6.81894 | 1.00000 | 0.45856E-11 |
| 0.70000E-10 | 6.83850 | 1.00000 | 0.40014E-11 |
| 0.60000E-10 | 6.86100 | 1.00000 | 0.34190E-11 |
| 0.50000E-10 | 6.88752 | 1.00000 | 0.28386E-11 |
| 0.40000E-10 | 6.91985 | 1.00000 | 0.22607E-11 |
| 0.30000E-10 | 6.96130 | 1.00000 | 0.16858E-11 |
| 0.20000E-10 | 7.01930 | 1.00000 | 0.11149E-11 |
| 0.10000E-10 | 7.11736 | 1.00000 | 0.55005E-12 |
| 0.90000E-11 | 7.13215 | 1.00000 | 0.49406E-12 |
| 0.80000E-11 | 7.14865 | 1.00000 | 0.43819E-12 |
| 0.70000E-11 | 7.16730 | 1.00000 | 0.38245E-12 |
| 0.60000E-11 | 7.18878 | 1.00000 | 0.32687E-12 |
| 0.50000E-11 | 7.21409 | 1.00000 | 0.27147E-12 |
| 0.40000E-11 | 7.24496 | 1.00000 | 0.21628E-12 |
| 0.30000E-11 | 7.28456 | 1.00000 | 0.16136E-12 |
| 0.20000E-11 | 7.34001 | 1.00000 | 0.10679E-12 |

TABLE A2.1.

(continued)

| N/N(MAX) | X | AREA(A) | AREA(B) |
|---|---|---|---|
| 0.10000E-11 | 7.43384 | 1.00000 | 0.52743E-13 |
| 0.90000E-12 | 7.44800 | 1.00000 | 0.47381E-13 |
| 0.80000E-12 | 7.46380 | 1.00000 | 0.42031E-13 |
| 0.70000E-12 | 7.48167 | 1.00000 | 0.36692E-13 |
| 0.60000E-12 | 7.50225 | 1.00000 | 0.31367E-13 |
| 0.50000E-12 | 7.52651 | 1.00000 | 0.26057E-13 |
| 0.40000E-12 | 7.55610 | 1.00000 | 0.20767E-13 |
| 0.30000E-12 | 7.59408 | 1.00000 | 0.15500E-13 |
| 0.20000E-12 | 7.64728 | 1.00000 | 0.10264E-13 |
| 0.10000E-12 | 7.73739 | 1.00000 | 0.50739E-14 |
| 0.90000E-13 | 7.75100 | 1.00000 | 0.45587E-14 |
| 0.80000E-13 | 7.76618 | 1.00000 | 0.40445E-14 |
| 0.70000E-13 | 7.78335 | 1.00000 | 0.35314E-14 |
| 0.60000E-13 | 7.80313 | 1.00000 | 0.30195E-14 |
| 0.50000E-13 | 7.82646 | 1.00000 | 0.25090E-14 |
| 0.40000E-13 | 7.85492 | 1.00000 | 0.20001E-14 |
| 0.30000E-13 | 7.89146 | 1.00000 | 0.14933E-14 |
| 0.20000E-13 | 7.94268 | 1.00000 | 0.98933E-15 |
| 0.10000E-13 | 8.02947 | 1.00000 | 0.48947E-15 |
| 0.90000E-14 | 8.04258 | 1.00000 | 0.43983E-15 |
| 0.80000E-14 | 8.05721 | 1.00000 | 0.39027E-15 |
| 0.70000E-14 | 8.07377 | 1.00000 | 0.34081E-15 |
| 0.60000E-14 | 8.09284 | 1.00000 | 0.29145E-15 |
| 0.50000E-14 | 8.11534 | 1.00000 | 0.24222E-15 |
| 0.40000E-14 | 8.14279 | 1.00000 | 0.19314E-15 |
| 0.30000E-14 | 8.17804 | 1.00000 | 0.14425E-15 |
| 0.20000E-14 | 8.22747 | 1.00000 | 0.95605E-16 |
| 0.10000E-14 | 8.31129 | 1.00000 | 0.47333E-16 |
| 0.90000E-15 | 8.32396 | 1.00000 | 0.42537E-16 |
| 0.80000E-15 | 8.33810 | 1.00000 | 0.37748E-16 |
| 0.70000E-15 | 8.35409 | 1.00000 | 0.32968E-16 |
| 0.60000E-15 | 8.37253 | 1.00000 | 0.28198E-16 |
| 0.50000E-15 | 8.39427 | 1.00000 | 0.23439E-16 |
| 0.40000E-15 | 8.42082 | 1.00000 | 0.18694E-16 |
| 0.30000E-15 | 8.45491 | 1.00000 | 0.13965E-16 |
| 0.20000E-15 | 8.50273 | 1.00000 | 0.92591E-17 |
| 0.10000E-15 | 8.58386 | 1.00000 | 0.45869E-17 |

# NORMALISED PROJECTED RANGE AND STANDARD DEVIATION TABLES

The tables in this appendix are for calculating appropriate values of the projected range, $R_p$, and the standard deviation, $\Delta R_p$, of $R_p$ for ions with atomic number $(Z_1)$ from 5 to 80 into substrates with atomic number $(Z_2)$ from 10 to 80. The calculations were done by B. J. Smith (1971) using his modified form of the program by Johnson and Gibbons (1969). Refer to section 5.3.4.1 for an explanation of the deviation of these standard deviations.

There are nine tables one for each value of $Z_1$ with eight values of $Z_2$ varying from 10 to 80 in steps of 10. For the purpose of these calculations, unity density was assumed for the substrate with the following values for the atomic masses of the ion and the substrate for the various values of $Z_1$ and $Z_2$:

| $Z_1$ and $Z_2$ | $M$ (amu) |
|---|---|
| 10 | 21 |
| 20 | 44 |
| 30 | 67 |
| 40 | 92 |
| 50 | 118 |
| 60 | 144 |
| 70 | 171 |
| 80 | 199 |

The ranges are given in milligrams/cm$^2$ of the substrate and are tabulated for ion energies from 10 to 1000 keV. The program is unstable when computing $\Delta R_p$ for the higher values of $Z_2$ and where this has occurred the value has not been included.

In general, the values of $Z_1$ and $Z_2$ will fall between the calculated values and therefore the following method of interpolating first between the values of $Z_2$ and then $Z_1$ is suggested;

(1) For the given ion energy, plot $R_p$ and $\Delta R_p$ against $Z_1$ on double logarithmic paper for $Z_2 = 10, 20, 30, 40$ etc. Sufficient points must be plotted to establish the trend in $R_p$ and $\Delta R_p$.

(2) From the above curves plot $R_p$ and $\Delta R_p$ against the value of $Z_2$ for the substrate on double logarithmic paper for the required values of $Z_1$.

(3) Read off $R_p$ and $\Delta R_p$ for the particular value of $Z_2$ in terms of milligrams/cm$^2$. Convert these values to g/cm$^2$ and divide by the density of $Z_2$ (g/cm$^3$) to obtain the actual ranges in cm. A list of $Z$ numbers, densities and other data are given below for some elements, compound semiconductors and insulators.

The references are given in Chapter 5.

List of useful data on the elements, semiconductors and insulators

(i) Elements

| Element | Atomic Number $Z$ | Density g/cm$^3$ | Element | Atomic Number $Z$ | Density g/cm$^3$ |
|---|---|---|---|---|---|
| Beryllium | 4 | 1.85 | Ruthenium | 44 | 12.6 |
| Boron | 5 | 2.34 | Rhodium | 45 | 12.4 |
| Carbon Diamond | 6 | 3.51 | Palladium | 46 | 12.0 |
| Magnesium | 12 | 1.74 | Silver | 47 | 10.5 |
| Aluminium | 13 | 2.70 | Cadmium | 48 | 8.64 |
| Silicon | 14 | 2.33 | Indium | 49 | 7.30 |
| Titanium | 22 | 4.50 | Tin | 50 | 7.30 |
| Vanadium | 23 | 6.00 | Antimony | 51 | 6.68 |
| Chromium | 24 | 7.20 | Tantalum | 73 | 16.6 |
| Manganese | 25 | 7.20 | Tungsten | 74 | 19.3 |
| Iron | 26 | 7.86 | Rhenium | 75 | 20.5 |
| Cobalt | 27 | 8.90 | Osmium | 76 | 22.5 |
| Nickel | 28 | 8.90 | Iridium | 77 | 22.4 |
| Copper | 29 | 8.92 | Platinum | 78 | 21.5 |
| Zinc | 30 | 7.10 | Gold | 79 | 19.3 |
| Germanium | 32 | 5.35 | Thallium | 81 | 11.9 |
| Zirconium | 40 | 6.40 | Lead | 82 | 11.3 |
| Niobium | 41 | 8.55 | Bismuth | 83 | 9.80 |
| Molybdenum | 42 | 10.20 | Thorium | 90 | 11.2 |
|  |  |  | Uranium | 92 | 18.7 |

(ii) Semiconductors

| Semiconductor | Avg. atomic no. | Density g/cm³ | Energy gap eV | Dielectric constant |
|---|---|---|---|---|
| SiC | 10 | 3.21 | 2.4–3 | 10 |
| GaP | 23 | 4.13 | 2.3 | 10.2 |
| GaAs | 32 | 5.27 | 1.4 | 12.5 |
| GaSb | 41 | 5.61 | 0.7 | 15 |
| InP | 32 | 4.97 | 1.3 | 14 |
| InAs | 41 | 5.66 | 0.3 | 14.5 |
| InSb | 50 | 5.78 | 0.2 | 17.6 |
| ZnS | 23 | 4.80 | 3.6 | 8 |
| ZnSe | 32 | 5.42 | 2.6 | 5.8 |
| ZnTe | 41 | 5.54 | 2.2 | 18.6 |
| ZnO | 19 | 5.6 | | 9 |
| ZnSb | 41 | 6.38 | 0.53 | – |
| CdS | 32 | 4.82 | 2.4 | 5.9 |
| CdSe | 41 | 5.81 | 1.7 | 10 |
| CdTe | 50 | 6.20 | 1.4 | 11 |
| CdSb | 50 | 6.92 | 0.46 | – |
| PbS | 49 | 7.6 | 0.37 | 17 |
| PbSe | 58 | 8.15 | 0.26 | – |
| PbTe | 67 | 8.16 | 0.29 | 30 |
| $Bi_2Te_3$ | 64 | 7.86 | | 81 |
| C | 6 | 3.51 | 5.2 | 5.7 |
| Si | 14 | 2.33 | 1.1 | 11.7 |
| Ge | 32 | 5.35 | 0.67 | 16.3 |

(iii) Insulators

| Insulator | Avg. atomic no. | Density g/cm³ | Dielectric constant | Dielectric strength volts/cm |
|---|---|---|---|---|
| Silicon dioxide | 10 | 2.27 | 3.9 | $6 \times 10^6$ |
| Silicon nitride | 10 | 3.44 | 6 | $6 \times 10^6$ |
| Aluminia | 10 | 3.98 | 8.9–9.3 | $7 \times 10^6$ |
| Zirconia | 18 | 5.6 | 18 | $1–2 \times 10^6$ |

TABLE A3.1.

Normalised projected range ($R_p$) and standard deviation ($\Delta R_p$) for ions with $Z_1 = 5$ in substrates with $Z_2$ from 10 to 80 for ion energies from 10 keV to 1000 keV.

| $Z_1 = 5$ | $Z_2 = 10$ | $Z_2 = 20$ | $Z_2 = 30$ | $Z_2 = 40$ | $Z_2 = 50$ | $Z_2 = 60$ | $Z_2 = 70$ | $Z_2 = 80$ |
|---|---|---|---|---|---|---|---|---|
| Energy keV | $R_p$ $(\Delta R_p)$ mg/cm$^2$ | $R_p$ $(\Delta R_p)$ mg/cm$^2$ | $R_p$ $(\Delta R_p)$ mg/cm$^2$ | $R_p$ $(\Delta R_p)$ mg/cm$^2$ | $R_p$ $(\Delta R_p)$ mg/cm$^2$ | $R_p$ $(\Delta R_p)$ mg/cm$^2$ | $R_p$ $\Delta R_p)$ mg/cm$^2$ | $R_p$ $(\Delta R_p)$ mg/cm$^2$ |
| 10 | 0.0084 (0.0038) | 0.0111 (0.0071) | 0.0130 | 0.0149 | 0.0166 | 0.0182 | 0.0199 | 0.0216 |
| 20 | 0.0175 (0.0063) | 0.0224 (0.0114) | 0.0252 (0.0165) | 0.0281 | 0.0310 | 0.0337 | 0.0363 | 0.0389 |
| 30 | 0.0267 (0.0083) | 0.0341 (0.0154) | 0.0383 (0.0221) | 0.0424 (0.0283) | 0.0458 | 0.0492 | 0.0527 | 0.0563 |
| 40 | 0.0360 (0.0101) | 0.0459 (0.0190) | 0.0515 (0.0272) | 0.0570 (0.0360) | 0.0617 | 0.0654 | 0.0694 | 0.0737 |
| 50 | 0.0453 (0.0118) | 0.0577 (0.0222) | 0.0649 (0.0319) | 0.0717 (0.0422) | 0.0777 (0.0501) | 0.0825 | 0.0869 | 0.0916 |
| 60 | 0.0541 (0.0131) | 0.0694 (0.0250) | 0.0784 (0.0361) | 0.0866 (0.0480) | 0.0937 (0.0592) | 0.0996 (0.0480) | 0.1051 | 0.1103 |
| 70 | 0.0625 (0.0142) | 0.0812 (0.0276) | 0.0918 (0.0400) | 0.1015 (0.0533) | 0.1099 (0.0664) | 0.1167 (0.0728) | 0.1233 | 0.1296 |
| 80 | 0.0707 (0.0152) | 0.0929 (0.0299) | 0.1050 (0.0435) | 0.1165 (0.0582) | 0.1261 (0.0728) | 0.1339 (0.0843) | 0.1415 | 0.1488 |
| 90 | 0.0787 (0.0161) | 0.1044 (0.0321) | 0.1182 (0.0468) | 0.1314 (0.0628) | 0.1424 (0.0787) | 0.1512 (0.0928) | 0.1597 | 0.1680 |
| 100 | 0.0865 (0.0170) | 0.1158 (0.0340) | 0.1314 (0.0499) | 0.1463 (0.0670) | 0.1587 (0.0843) | 0.1686 (0.1001) | 0.1780 (0.1091) | 0.1873 |
| 200 | 0.1542 (0.0220) | 0.2162 (0.0466) | 0.2562 (0.0711) | 0.2914 (0.0979) | 0.3188 (0.1257) | 0.3408 (0.1527) | 0.3620 (0.1797) | 0.3820 (0.2062) |
| 300 | 0.2098 (0.0246) | 0.3017 (0.0537) | 0.3654 (0.0837) | 0.4223 (0.1169) | 0.4719 (0.1516) | 0.5090 (0.1866) | 0.5420 (0.2225) | 0.5732 (0.2588) |
| 400 | 0.2579 (0.0262) | 0.3779 (0.0583) | 0.4637 (0.0922) | 0.5424 (0.1302) | 0.6099 (0.1705) | 0.6673 (0.2107) | 0.7171 (0.2530) | 0.7616 (0.2968) |
| 500 | 0.3010 (0.0274) | 0.4468 (0.0616) | 0.5534 (0.0984) | 0.6538 (0.1400) | 0.7402 (0.1848) | 0.8124 (0.2301) | 0.8810 (0.2768) | 0.9432 (0.3259) |
| 600 | 0.3404 (0.0282) | 0.5098 (0.0641) | 0.6372 (0.1031) | 0.7565 (0.1479) | 0.8634 (0.1958) | 0.9516 (0.2453) | 1.0339 (0.2969) | 1.1135 (0.3501) |
| 700 | 0.3769 (0.0289) | 0.5685 (0.0660) | 0.7156 (0.1068) | 0.8538 (0.1541) | 0.9786 (0.2050) | 1.0847 (0.2576) | 1.1818 (0.3132) | 1.2744 (0.3711) |
| 800 | 0.4110 (0.0294) | 0.6236 (0.0676) | 0.7891 (0.1099) | 0.9464 (0.1591) | 1.0878 (0.2128) | 1.2112 (0.2679) | 1.3244 (0.3267) | 1.4310 (0.3886) |
| 900 | 0.4431 (0.0298) | 0.6758 (0.0690) | 0.8587 (0.1125) | 1.0346 (0.1633) | 1.1927 (0.2192) | 1.3310 (0.2770) | 1.4614 (0.3382) | 1.5829 (0.4033) |
| 1000 | 0.4737 (0.0302) | 0.7255 (0.0701) | 0.9251 (0.1147) | 1.1187 (0.1669) | 1.2937 (0.2247) | 1.4464 (0.2848) | 1.5924 (0.3484) | 1.7299 (0.4160) |

## TABLE A3.2.

Normalised projected range ($R_p$) and standard deviation ($\Delta R_p$) for ions with $Z_1 = 10$ in substrates with $Z_2$ from 10 to 80 for ion energies from 10 keV to 1000 keV.

| $Z_1 = 10$ $Z_2 = 10$ | $Z_2 = 20$ | $Z_2 = 30$ | $Z_2 = 40$ | $Z_2 = 50$ | $Z_2 = 60$ | $Z_2 = 70$ | $Z_2 = 80$ |
|---|---|---|---|---|---|---|---|
| Energy keV | $R_p$ ($\Delta R_p$) mg/cm$^2$ | $R_p$ ($\Delta R_p$) mg/cm$^2$ | $R_p$ ($\Delta R_p$) mg/cm$^2$ | $R_p$ ($\Delta R_p$) mg/cm$^2$ | $R_p$ ($\Delta R_p$) mg/cm$^2$ | $R_p$ ($\Delta R_p$) mg/cm$^2$ | $R_p$ ($\Delta R_p$) mg/cm$^2$ | $R_p$ ($\Delta R_p$) mg/cm$^2$ |

| Energy keV | $Z_2=10$ | $Z_2=20$ | $Z_2=30$ | $Z_2=40$ | $Z_2=50$ | $Z_2=60$ | $Z_2=70$ | $Z_2=80$ |
|---|---|---|---|---|---|---|---|---|
| 10 | 0.0042 (0.0020) | 0.0060 (0.0034) | 0.0072 | 0.0085 | 0.0097 | 0.0108 | 0.0119 | 0.0131 |
| 20 | 0.0081 (0.0036) | 0.0112 (0.0058) | 0.0133 (0.0081) | 0.0154 | 0.0172 | 0.0188 | 0.0204 | 0.0221 |
| 30 | 0.0123 (0.0050) | 0.0164 (0.0080) | 0.0193 (0.0107) | 0.0221 (0.0145) | 0.0247 (0.0190) | 0.0269 | 0.0290 | 0.0311 |
| 40 | 0.0165 (0.0064) | 0.0219 (0.0100) | 0.0253 (0.0133) | 0.0288 (0.0178) | 0.0321 (0.0226) | 0.0348 | 0.0375 | 0.0401 |
| 50 | 0.0209 (0.0076) | 0.0275 (0.0118) | 0.0317 (0.0160) | 0.0356 (0.0210) | 0.0394 (0.0265) | 0.0427 | 0.0459 | 0.0491 |
| 60 | 0.0254 (0.0088) | 0.0332 (0.0135) | 0.0381 (0.0184) | 0.0426 (0.0243) | 0.0468 (0.0303) | 0.0505 (0.0363) | 0.0542 | 0.0580 |
| 70 | 0.0298 (0.0099) | 0.0388 (0.0151) | 0.0445 (0.0208) | 0.0498 (0.0273) | 0.0543 (0.0341) | 0.0584 (0.0408) | 0.0626 | 0.0667 |
| 80 | 0.0343 (0.0108) | 0.0446 (0.0165) | 0.0510 (0.0231) | 0.0571 (0.0303) | 0.0622 (0.0378) | 0.0664 (0.0451) | 0.0709 | 0.0755 |
| 90 | 0.0388 (0.0117) | 0.0503 (0.0180) | 0.0575 (0.0252) | 0.0643 (0.0331) | 0.0701 (0.0412) | 0.0746 (0.0492) | 0.0794 | 0.0843 |
| 100 | 0.0433 (0.0126) | 0.0561 (0.0195) | 0.0640 (0.0273) | 0.0715 (0.0358) | 0.0780 (0.0446) | 0.0831 (0.0532) | 0.0880 (0.0609) | 0.0932 |
| 200 | 0.0870 (0.0187) | 0.1134 (0.0314) | 0.1293 (0.0447) | 0.1447 (0.0590) | 0.1575 (0.0734) | 0.1679 (0.0874) | 0.1778 (0.1019) | 0.1873 (0.1164) |
| 300 | 0.1263 (0.0220) | 0.1681 (0.0398) | 0.1935 (0.0576) | 0.2165 (0.0764) | 0.2369 (0.0958) | 0.2531 (0.1147) | 0.2680 (0.1339) | 0.2823 (0.1535) |
| 400 | 0.1618 (0.0241) | 0.2178 (0.0458) | 0.2552 (0.0674) | 0.2874 (0.0903) | 0.3144 (0.1136) | 0.3366 (0.1366) | 0.3578 (0.1603) | 0.3775 (0.1845) |
| 500 | 0.1945 (0.0257) | 0.2649 (0.0505) | 0.3119 (0.0750) | 0.3559 (0.1015) | 0.3910 (0.1285) | 0.4190 (0.1550) | 0.4455 (0.1824) | 0.4715 (0.2107) |
| 600 | 0.2251 (0.0268) | 0.3090 (0.0543) | 0.3661 (0.0812) | 0.4204 (0.1105) | 0.4656 (0.1408) | 0.5004 (0.1707) | 0.5325 (0.2015) | 0.5636 (0.2333) |
| 700 | 0.2538 (0.0277) | 0.3503 (0.0573) | 0.4182 (0.0865) | 0.4812 (0.1182) | 0.5371 (0.1512) | 0.5801 (0.1840) | 0.6186 (0.2180) | 0.6551 (0.2531) |
| 800 | 0.2807 (0.0284) | 0.3898 (0.0599) | 0.4681 (0.0910) | 0.5403 (0.1249) | 0.6045 (0.1602) | 0.6572 (0.1956) | 0.7031 (0.2324) | 0.7457 (0.2705) |
| 900 | 0.3062 (0.0290) | 0.4276 (0.0622) | 0.5155 (0.0949) | 0.5977 (0.1308) | 0.6697 (0.1682) | 0.7311 (0.2058) | 0.7856 (0.2451) | 0.8350 (0.2859) |
| 1000 | 0.3306 (0.0296) | 0.4639 (0.0641) | 0.5608 (0.0983) | 0.6533 (0.1359) | 0.7334 (0.1754) | 0.8015 (0.2150) | 0.8654 (0.2564) | 0.9225 (0.2997) |

TABLE A3.3.

Normalised projected range $(R_p)$ and standard deviation $(\Delta R_p)$ for ions with $Z_1 = 20$ in substrates with $Z_2$ from 10 to 80 for ion energies from 10 keV to 1000 keV.

| $Z_1=20$ Energy keV | $Z_2 = 10$ $R_p$ $(\Delta R_p)$ mg/cm$^2$ | $Z_2 = 20$ $R_p$ $(\Delta R_p)$ mg/cm$^2$ | $Z_2 = 30$ $R_p$ $(\Delta R_p)$ mg/cm$^2$ | $Z_2 = 40$ $R_p$ $(\Delta R_p)$ mg/cm$^2$ | $Z_2 = 50$ $R_p$ $(\Delta R_p)$ mg/cm$^2$ | $Z_2 = 60$ $R_p$ $(\Delta R_p)$ mg/cm$^2$ | $Z_2 = 70$ $R_p$ $(\Delta R_p)$ mg/cm$^2$ | $Z_2 = 80$ $R_p$ $(\Delta R_p)$ mg/cm$^2$ |
|---|---|---|---|---|---|---|---|---|
| 10 | 0.0026 (0.0011) | 0.0036 (0.0019) | 0.0045 (0.0026) | 0.0054 | 0.0062 | 0.0070 | 0.0077 | 0.0084 |
| 20 | 0.0045 (0.0018) | 0.0063 (0.0031) | 0.0077 (0.0043) | 0.0092 (0.0057) | 0.0105 | 0.0117 | 0.0129 | 0.0140 |
| 30 | 0.0064 (0.0025) | 0.0088 (0.0043) | 0.0107 (0.0059) | 0.0126 (0.0076) | 0.0144 (0.0094) | 0.0160 | 0.0175 | 0.0190 |
| 40 | 0.0083 (0.0032) | 0.0113 (0.0054) | 0.0137 (0.0073) | 0.0161 (0.0093) | 0.0182 (0.0114) | 0.0202 | 0.0220 | 0.0238 |
| 50 | 0.0101 (0.0039) | 0.0138 (0.0065) | 0.0167 (0.0087) | 0.0195 (0.0111) | 0.0221 (0.0134) | 0.0242 (0.0159) | 0.0264 | 0.0285 |
| 60 | 0.0120 (0.0045) | 0.0163 (0.0075) | 0.0196 (0.0100) | 0.0229 (0.0127) | 0.0259 (0.0153) | 0.0284 (0.0178) | 0.0308 | 0.0332 |
| 70 | 0.0140 (0.0052) | 0.0187 (0.0085) | 0.0225 (0.0113) | 0.0263 (0.0143) | 0.0296 (0.0172) | 0.0325 (0.0199) | 0.0352 | 0.0378 |
| 80 | 0.0160 (0.0058) | 0.0213 (0.0094) | 0.0254 (0.0125) | 0.0296 (0.0158) | 0.0334 (0.0189) | 0.0365 (0.0219) | 0.0395 (0.0258) | 0.0425 (0.0304) |
| 90 | 0.0180 (0.0065) | 0.0238 (0.0104) | 0.0283 (0.0137) | 0.0329 (0.0172) | 0.0371 (0.0207) | 0.0406 (0.0238) | 0.0439 (0.0280) | 0.0471 (0.0327) |
| 100 | 0.0201 (0.0071) | 0.0264 (0.0114) | 0.0313 (0.0149) | 0.0362 (0.0186) | 0.0407 (0.0223) | 0.0446 (0.0257) | 0.0482 (0.0302) | 0.0517 (0.0351) |
| 200 | 0.0413 (0.0129) | 0.0534 (0.0202) | 0.0621 (0.0258) | 0.0707 (0.0315) | 0.0782 (0.0372) | 0.0844 (0.0439) | 0.0906 (0.0511) | 0.0967 (0.0587) |
| 300 | 0.0634 (0.0176) | 0.0813 (0.0276) | 0.0938 (0.0350) | 0.1064 (0.0423) | 0.1174 (0.0513) | 0.1263 (0.0606) | 0.1347 (0.0703) | 0.1425 (0.0803) |
| 400 | 0.0860 (0.0217) | 0.1094 (0.0337) | 0.1261 (0.0428) | 0.1425 (0.0517) | 0.1569 (0.0638) | 0.1686 (0.0755) | 0.1797 (0.0874) | 0.1902 (0.0999) |
| 500 | 0.1085 (0.0252) | 0.1377 (0.0389) | 0.1582 (0.0493) | 0.1788 (0.0611) | 0.1965 (0.0754) | 0.2110 (0.0890) | 0.2247 (0.1032) | 0.2379 (0.1178) |
| 600 | 0.1305 (0.0281) | 0.1659 (0.0436) | 0.1901 (0.0549) | 0.2149 (0.0695) | 0.2363 (0.0861) | 0.2536 (0.1017) | 0.2699 (0.1177) | 0.2855 (0.1344) |
| 700 | 0.1513 (0.0304) | 0.1940 (0.0477) | 0.2220 (0.0600) | 0.2505 (0.0769) | 0.2759 (0.0958) | 0.2962 (0.1134) | 0.3151 (0.1314) | 0.3332 (0.1499) |
| 800 | 0.1716 (0.0324) | 0.2213 (0.0512) | 0.2539 (0.0645) | 0.2861 (0.0840) | 0.3149 (0.1045) | 0.3385 (0.1243) | 0.3603 (0.1442) | 0.3810 (0.1646) |
| 900 | 0.1916 (0.0342) | 0.2478 (0.0542) | 0.2854 (0.0687) | 0.3216 (0.0907) | 0.3536 (0.1126) | 0.3804 (0.1341) | 0.4053 (0.1561) | 0.4287 (0.1784) |
| 1000 | 0.2111 (0.0358) | 0.2731 (0.0567) | 0.3163 (0.0732) | 0.3569 (0.0969) | 0.3923 (0.1204) | 0.4217 (0.1433) | 0.4498 (0.1671) | 0.4762 (0.1913) |

TABLE A3.4.

Normalised projected range $(R_p)$ and standard deviation $(\Delta R_p)$ for ions with $Z_1 = 30$ in substrates with $Z_2$ from 10 to 80 for ion energies from 10 keV to 1000 keV.

| $Z_1 = 30$ $Z_2 = 10$ | $Z_2 = 20$ | $Z_2 = 30$ | $Z_2 = 40$ | $Z_2 = 50$ | $Z_2 = 60$ | $Z_2 = 70$ | $Z_2 = 80$ |
|---|---|---|---|---|---|---|---|
| Energy keV $R_p$ $(\Delta R_p)$ mg/cm² | $R_p$ $(\Delta R_p)$ mg/cm² | $R_p$ $(\Delta R_p)$ mg/cm² | $R_p$ $(\Delta R_p)$ mg/cm² | $R_p$ $(\Delta R_p)$ mg/cm² | $R_p$ $(\Delta R_p)$ mg/cm² | $R_p$ $(\Delta R_p)$ mg/cm² | $R_p$ $(\Delta R_p)$ mg/cm² |
| 10   0.0022 | 0.0029 | 0.0036 | 0.0043 | 0.0050 | 0.0056 | 0.0062 | 0.0068 |
| (0.0008) | (0.0014) | (0.0019) | (0.0025) | (0.0034) | | | |
| 20   0.0036 | 0.0049 | 0.0060 | 0.0071 | 0.0082 | 0.0091 | 0.0101 | 0.0110 |
| (0.0013) | (0.0022) | (0.0031) | (0.0040) | (0.0050) | | | |
| 30   0.0050 | 0.0066 | 0.0081 | 0.0096 | 0.0111 | 0.0124 | 0.0136 | 0.0148 |
| (0.0017) | (0.0030) | (0.0041) | (0.0054) | (0.0067) | | | |
| 40   0.0063 | 0.0084 | 0.0101 | 0.0120 | 0.0138 | 0.0154 | 0.0169 | 0.0184 |
| (0.0022) | (0.0037) | (0.0051) | (0.0066) | (0.0082) | (0.0097) | | |
| 50   0.0076 | 0.0100 | 0.0121 | 0.0143 | 0.0164 | 0.0183 | 0.0201 | 0.0218 |
| (0.0026) | (0.0044) | (0.0060) | (0.0078) | (0.0096) | (0.0113) | | |
| 60   0.0089 | 0.0117 | 0.0141 | 0.0166 | 0.0190 | 0.0211 | 0.0231 | 0.0251 |
| (0.0030) | (0.0051) | (0.0069) | (0.0089) | (0.0109) | (0.0128) | | |
| 70   0.0101 | 0.0134 | 0.0160 | 0.0189 | 0.0215 | 0.0238 | 0.0261 | 0.0283 |
| (0.0035) | (0.0058) | (0.0078) | (0.0100) | (0.0122) | (0.0143) | (0.0165) | |
| 80   0.0114 | 0.0150 | 0.0180 | 0.0211 | 0.0240 | 0.0266 | 0.0291 | 0.0315 |
| (0.0039) | (0.0065) | (0.0087) | (0.0111) | (0.0135) | (0.0158) | (0.0182) | (0.0206) |
| 90   0.0127 | 0.0166 | 0.0199 | 0.0233 | 0.0265 | 0.0293 | 0.0320 | 0.0346 |
| (0.0043) | (0.0071) | (0.0095) | (0.0121) | (0.0147) | (0.0172) | (0.0197) | (0.0223) |
| 100   0.0139 | 0.0183 | 0.0218 | 0.0256 | 0.0290 | 0.0320 | 0.0349 | 0.0378 |
| (0.0047) | (0.0077) | (0.0104) | (0.0132) | (0.0159) | (0.0186) | (0.0213) | (0.0241) |
| 200   0.0272 | 0.0348 | 0.0409 | 0.0474 | 0.0535 | 0.0588 | 0.0638 | 0.0687 |
| (0.0086) | (0.0138) | (0.0180) | (0.0226) | (0.0270) | (0.0312) | (0.0353) | (0.0395) |
| 300   0.0412 | 0.0522 | 0.0607 | 0.0696 | 0.0778 | 0.0850 | 0.0919 | 0.0987 |
| (0.0123) | (0.0196) | (0.0253) | (0.0311) | (0.0367) | (0.0420) | (0.0474) | (0.0527) |
| 400   0.0557 | 0.0701 | 0.0811 | 0.0926 | 0.1031 | 0.1118 | 0.1202 | 0.1286 |
| (0.0157) | (0.0249) | (0.0320) | (0.0391) | (0.0459) | (0.0520) | (0.0582) | (0.0657) |
| 500   0.0705 | 0.0883 | 0.1017 | 0.1159 | 0.1288 | 0.1395 | 0.1495 | 0.1591 |
| (0.0190) | (0.0299) | (0.0382) | (0.0466) | (0.0544) | (0.0614) | (0.0689) | (0.0785) |
| 600   0.0854 | 0.1068 | 0.1226 | 0.1394 | 0.1546 | 0.1673 | 0.1792 | 0.1906 |
| (0.0218) | (0.0345) | (0.0440) | (0.0534) | (0.0623) | (0.0702) | (0.0797) | (0.0908) |
| 700   0.1004 | 0.1254 | 0.1437 | 0.1630 | 0.1805 | 0.1952 | 0.2090 | 0.2222 |
| (0.0245) | (0.0388) | (0.0494) | (0.0598) | (0.0695) | (0.0783) | (0.0898) | (0.1023) |
| 800   0.1156 | 0.1440 | 0.1649 | 0.1868 | 0.2066 | 0.2231 | 0.2388 | 0.2538 |
| (0.0270) | (0.0426) | (0.0545) | (0.0658) | (0.0763) | (0.0859) | (0.0995) | (0.1133) |
| 900   0.1308 | 0.1626 | 0.1861 | 0.2107 | 0.2328 | 0.2511 | 0.2687 | 0.2855 |
| (0.0294) | (0.0462) | (0.0591) | (0.0714) | (0.0827) | (0.0939) | (0.1087) | (0.1238) |
| 1000   0.1461 | 0.1814 | 0.2072 | 0.2346 | 0.2590 | 0.2793 | 0.2985 | 0.3171 |
| (0.0316) | (0.0496) | (0.0633) | (0.0767) | (0.0887) | (0.1016) | (0.1174) | (0.1338) |

TABLE A3.5.

Normalised projected range ($R_p$) and standard deviation ($\Delta R_p$) for ions with $Z_1 = 40$ in substrates with $Z_2$ from 10 to 80 for ion energies from 10 keV to 1000 keV.

| $Z_1=40$ | $Z_2 = 10$ | $Z_2 = 20$ | $Z_2 = 30$ | $Z_2 = 40$ | $Z_2 = 50$ | $Z_2 = 60$ | $Z_2 = 70$ | $Z_2 = 80$ |
|---|---|---|---|---|---|---|---|---|
| Energy keV | $R_p$ ($\Delta R_p$) mg/cm² | $R_p$ ($\Delta R_p$) mg/cm² | $R_p$ ($\Delta R_p$) mg/cm² | $R_p$ ($\Delta R_p$) mg/cm² | $R_p$ ($\Delta R_p$) mg/cm² | $R_p$ ($\Delta R_p$) mg/cm² | $R_p$ ($\Delta R_p$) mg/cm² | $R_p$ ($\Delta R_p$) mg/cm² |
| 10 | 0.0020 (0.0006) | 0.0026 (0.0011) | 0.0032 (0.0015) | 0.0038 (0.0020) | 0.0044 (0.0025) | 0.0049 (0.0030) | 0.0054 | 0.0060 |
| 20 | 0.0033 (0.0010) | 0.0043 (0.0018) | 0.0051 (0.0024) | 0.0061 (0.0032) | 0.0070 (0.0040) | 0.0078 (0.0048) | 0.0087 | 0.0095 |
| 30 | 0.0044 (0.0014) | 0.0057 (0.0024) | 0.0069 (0.0033) | 0.0081 (0.0042) | 0.0094 (0.0052) | 0.0104 (0.0062) | 0.0115 (0.0073) | 0.0126 |
| 40 | 0.0055 (0.0017) | 0.0071 (0.0029) | 0.0085 (0.0040) | 0.0101 (0.0052) | 0.0116 (0.0064) | 0.0129 (0.0076) | 0.0142 (0.0088) | 0.0155 (0.0101) |
| 50 | 0.0065 (0.0020) | 0.0084 (0.0034) | 0.0101 (0.0047) | 0.0119 (0.0061) | 0.0137 (0.0075) | 0.0153 (0.0089) | 0.0168 (0.0103) | 0.0183 (0.0118) |
| 60 | 0.0076 (0.0023) | 0.0097 (0.0039) | 0.0116 (0.0054) | 0.0137 (0.0069) | 0.0157 (0.0085) | 0.0175 (0.0101) | 0.0193 (0.0117) | 0.0210 (0.0134) |
| 70 | 0.0086 (0.0026) | 0.0110 (0.0044) | 0.0131 (0.0060) | 0.0154 (0.0078) | 0.0176 (0.0095) | 0.0197 (0.0113) | 0.0217 (0.0130) | 0.0236 (0.0149) |
| 80 | 0.0096 (0.0029) | 0.0123 (0.0049) | 0.0146 (0.0067) | 0.0171 (0.0086) | 0.0196 (0.0105) | 0.0218 (0.0124) | 0.0240 (0.0143) | 0.0261 (0.0163) |
| 90 | 0.0106 (0.0032) | 0.0135 (0.0054) | 0.0160 (0.0073) | 0.0188 (0.0094) | 0.0215 (0.0114) | 0.0239 (0.0135) | 0.0263 (0.0156) | 0.0286 (0.0177) |
| 100 | 0.0116 (0.0035) | 0.0148 (0.0059) | 0.0175 (0.0079) | 0.0205 (0.0101) | 0.0234 (0.0124) | 0.0260 (0.0145) | 0.0285 (0.0168) | 0.0310 (0.0191) |
| 200 | 0.0214 (0.0063) | 0.0270 (0.0104) | 0.0317 (0.0138) | 0.0369 (0.0174) | 0.0418 (0.0210) | 0.0461 (0.0244) | 0.0503 (0.0278) | 0.0544 (0.0314) |
| 300 | 0.0315 (0.0090) | 0.0393 (0.0146) | 0.0458 (0.0192) | 0.0529 (0.0240) | 0.0598 (0.0288) | 0.0658 (0.0333) | 0.0717 (0.0378) | 0.0774 (0.0424) |
| 400 | 0.0419 (0.0117) | 0.0521 (0.0187) | 0.0601 (0.0243) | 0.0690 (0.0302) | 0.0776 (0.0360) | 0.0852 (0.0415) | 0.0926 (0.0469) | 0.0999 (0.0525) |
| 500 | 0.0526 (0.0142) | 0.0651 (0.0228) | 0.0749 (0.0295) | 0.0856 (0.0363) | 0.0957 (0.0430) | 0.1047 (0.0492) | 0.1135 (0.0554) | 0.1221 (0.0618) |
| 600 | 0.0635 (0.0167) | 0.0783 (0.0266) | 0.0899 (0.0344) | 0.1026 (0.0422) | 0.1144 (0.0497) | 0.1245 (0.0566) | 0.1345 (0.0635) | 0.1444 (0.0706) |
| 700 | 0.0746 (0.0191) | 0.0918 (0.0303) | 0.1050 (0.0391) | 0.1196 (0.0479) | 0.1332 (0.0563) | 0.1449 (0.0639) | 0.1560 (0.0714) | 0.1669 (0.0790) |
| 800 | 0.0859 (0.0213) | 0.1054 (0.0339) | 0.1203 (0.0436) | 0.1368 (0.0534) | 0.1522 (0.0626) | 0.1653 (0.0708) | 0.1779 (0.0790) | 0.1900 (0.0872) |
| 900 | 0.0972 (0.0235) | 0.1191 (0.0373) | 0.1357 (0.0479) | 0.1541 (0.0586) | 0.1713 (0.0686) | 0.1859 (0.0776) | 0.1999 (0.0864) | 0.2135 (0.0951) |
| 1000 | 0.1086 (0.0256) | 0.1330 (0.0407) | 0.1513 (0.0521) | 0.1715 (0.0636) | 0.1904 (0.0744) | 0.2065 (0.0840) | 0.2220 (0.0934) | 0.2369 (0.1028) |

TABLE A3.6.

Normalised projected range $(R_p)$ and standard deviation $(\Delta R_p)$ for ions with $Z_1 = 50$ in substrates with $Z_2$ from 10 to 80 for ion energies from 10 keV to 1000 keV.

| $Z_1 = 50$ | $Z_2 = 10$ | $Z_2 = 20$ | $Z_2 = 30$ | $Z_2 = 40$ | $Z_2 = 50$ | $Z_2 = 60$ | $Z_2 = 70$ | $Z_2 = 80$ |
|---|---|---|---|---|---|---|---|---|
| Energy keV | $R_p$ $(\Delta R_p)$ mg/cm$^2$ | $R_p$ $(\Delta R_p)$ mg/cm$^2$ | $R_p$ $(\Delta R_p)$ mg/cm$^2$ | $R_p$ $(\Delta R_p)$ mg/cm$^2$ | $R_p$ $(\Delta R_p)$ mg/cm$^2$ | $R_p$ $(\Delta R_p)$ mg/cm$^2$ | $R_p$ $(\Delta R_p)$ mg/cm$^2$ | $R_p$ $(\Delta R_p)$ mg/cm$^2$ |
| 10 | 0.0019 (0.0006) | 0.0025 (0.0009) | 0.0029 (0.0013) | 0.0035 (0.0017) | 0.0040 (0.0021) | 0.0045 (0.0026) | 0.0050 (0.0030) | 0.0054 (0.0038) |
| 20 | 0.0031 (0.0009) | 0.0039 (0.0015) | 0.0047 (0.0021) | 0.0055 (0.0027) | 0.0063 (0.0033) | 0.0070 (0.0040) | 0.0078 (0.0047) | 0.0086 (0.0054) |
| 30 | 0.0042 (0.0012) | 0.0053 (0.0020) | 0.0062 (0.0027) | 0.0073 (0.0035) | 0.0083 (0.0044) | 0.0093 (0.0052) | 0.0103 (0.0061) | 0.0113 (0.0070) |
| 40 | 0.0051 (0.0014) | 0.0065 (0.0024) | 0.0076 (0.0033) | 0.0089 (0.0043) | 0.0102 (0.0053) | 0.0114 (0.0063) | 0.0126 (0.0074) | 0.0138 (0.0085) |
| 50 | 0.0061 (0.0017) | 0.0076 (0.0029) | 0.0090 (0.0039) | 0.0105 (0.0051) | 0.0120 (0.0062) | 0.0134 (0.0074) | 0.0148 (0.0086) | 0.0161 (0.0098) |
| 60 | 0.0070 (0.0019) | 0.0087 (0.0033) | 0.0103 (0.0045) | 0.0120 (0.0058) | 0.0138 (0.0071) | 0.0153 (0.0084) | 0.0169 (0.0097) | 0.0184 (0.0111) |
| 70 | 0.0078 (0.0022) | 0.0098 (0.0037) | 0.0116 (0.0050) | 0.0135 (0.0064) | 0.0154 (0.0079) | 0.0172 (0.0094) | 0.0189 (0.0108) | 0.0206 (0.0124) |
| 80 | 0.0087 (0.0024) | 0.0109 (0.0041) | 0.0128 (0.0055) | 0.0149 (0.0071) | 0.0170 (0.0087) | 0.0190 (0.0103) | 0.0209 (0.0119) | 0.0228 (0.0136) |
| 90 | 0.0095 (0.0027) | 0.0119 (0.0044) | 0.0140 (0.0060) | 0.0163 (0.0077) | 0.0186 (0.0095) | 0.0207 (0.0112) | 0.0228 (0.0129) | 0.0249 (0.0148) |
| 100 | 0.0104 (0.0029) | 0.0130 (0.0048) | 0.0152 (0.0065) | 0.0177 (0.0083) | 0.0202 (0.0102) | 0.0224 (0.0120) | 0.0247 (0.0139) | 0.0269 (0.0159) |
| 200 | 0.0186 (0.0051) | 0.0230 (0.0084) | 0.0268 (0.0112) | 0.0310 (0.0141) | 0.0350 (0.0171) | 0.0388 (0.0200) | 0.0425 (0.0229) | 0.0461 (0.0260) |
| 300 | 0.0266 (0.0071) | 0.0328 (0.0117) | 0.0380 (0.0155) | 0.0439 (0.0195) | 0.0495 (0.0235) | 0.0546 (0.0273) | 0.0596 (0.0311) | 0.0645 (0.0350) |
| 400 | 0.0348 (0.0091) | 0.0426 (0.0148) | 0.0492 (0.0196) | 0.0566 (0.0246) | 0.0637 (0.0295) | 0.0702 (0.0342) | 0.0765 (0.0388) | 0.0827 (0.0436) |
| 500 | 0.0432 (0.0112) | 0.0526 (0.0180) | 0.0604 (0.0235) | 0.0692 (0.0294) | 0.0778 (0.0352) | 0.0855 (0.0406) | 0.0931 (0.0461) | 0.1006 (0.0516) |
| 600 | 0.0517 (0.0132) | 0.0629 (0.0211) | 0.0718 (0.0275) | 0.0820 (0.0341) | 0.0919 (0.0407) | 0.1008 (0.0468) | 0.1096 (0.0530) | 0.1183 (0.0593) |
| 700 | 0.0604 (0.0151) | 0.0733 (0.0242) | 0.0835 (0.0314) | 0.0950 (0.0388) | 0.1061 (0.0461) | 0.1162 (0.0528) | 0.1261 (0.0596) | 0.1359 (0.0666) |
| 800 | 0.0692 (0.0170) | 0.0838 (0.0272) | 0.0953 (0.0352) | 0.1083 (0.0435) | 0.1207 (0.0514) | 0.1317 (0.0587) | 0.1426 (0.0661) | 0.1536 (0.0736) |
| 900 | 0.0780 (0.0188) | 0.0945 (0.0301) | 0.1073 (0.0390) | 0.1217 (0.0480) | 0.1355 (0.0566) | 0.1475 (0.0645) | 0.1594 (0.0724) | 0.1713 (0.0804) |
| 1000 | 0.0870 (0.0206) | 0.1052 (0.0330) | 0.1193 (0.0426) | 0.1352 (0.0524) | 0.1504 (0.0617) | 0.1636 (0.0702) | 0.1765 (0.0786) | 0.1892 (0.0871) |

TABLE A3.7.

Normalised projected range $(R_p)$ and standard deviation $(\Delta R_p)$ for ions with $Z_1 = 60$ in substrates with $Z_2$ from 10 to 80 for ion energies from 10 keV to 1000 keV.

| $Z_1 = 60$ | $Z_2 = 10$ | $Z_2 = 20$ | $Z_2 = 30$ | $Z_2 = 40$ | $Z_2 = 50$ | $Z_2 = 60$ | $Z_2 = 70$ | $Z_2 = 80$ |
|---|---|---|---|---|---|---|---|---|
| Energy keV | $R_p$ $(\Delta R_p)$ mg/cm$^2$ | $R_p$ $(\Delta R_p)$ mg/cm$^2$ | $R_p$ $(\Delta R_p)$ mg/cm$^2$ | $R_p$ $(\Delta R_p)$ mg/cm$^2$ | $R_p$ $(\Delta R_p)$ mg/cm$^2$ | $R_p$ $(\Delta R_p)$ mg/cm$^2$ | $R_p$ $(\Delta R_p)$ mg/cm$^2$ | $R_p$ $(\Delta R_p)$ mg/cm$^2$ |
| 10 | 0.0019 (0.0005) | 0.0024 (0.0008) | 0.0028 (0.0012) | 0.0032 (0.0015) | 0.0037 (0.0019) | 0.0042 (0.0023) | 0.0046 (0.0026) | 0.0051 (0.0030) |
| 20 | 0.0031 (0.0008) | 0.0038 (0.0013) | 0.0044 (0.0018) | 0.0051 (0.0024) | 0.0058 (0.0029) | 0.0065 (0.0035) | 0.0072 (0.0041) | 0.0079 (0.0047) |
| 30 | 0.0040 (0.0010) | 0.0050 (0.0017) | 0.0058 (0.0024) | 0.0067 (0.0031) | 0.0077 (0.0038) | 0.0086 (0.0045) | 0.0094 (0.0053) | 0.0103 (0.0061) |
| 40 | 0.0050 (0.0013) | 0.0061 (0.0021) | 0.0071 (0.0029) | 0.0082 (0.0038) | 0.0094 (0.0046) | 0.0104 (0.0055) | 0.0115 (0.0064) | 0.0126 (0.0073) |
| 50 | 0.0058 (0.0015) | 0.0071 (0.0025) | 0.0083 (0.0034) | 0.0096 (0.0044) | 0.0109 (0.0054) | 0.0122 (0.0064) | 0.0134 (0.0074) | 0.0147 (0.0085) |
| 60 | 0.0066 (0.0017) | 0.0082 (0.0029) | 0.0095 (0.0039) | 0.0110 (0.0050) | 0.0125 (0.0061) | 0.0139 (0.0072) | 0.0153 (0.0084) | 0.0167 (0.0096) |
| 70 | 0.0074 (0.0019) | 0.0091 (0.0032) | 0.0106 (0.0043) | 0.0123 (0.0056) | 0.0140 (0.0068) | 0.0155 (0.0080) | 0.0170 (0.0093) | 0.0186 (0.0107) |
| 80 | 0.0082 (0.0021) | 0.0101 (0.0035) | 0.0117 (0.0048) | 0.0136 (0.0061) | 0.0154 (0.0075) | 0.0171 (0.0088) | 0.0188 (0.0102) | 0.0205 (0.0117) |
| 90 | 0.0090 (0.0023) | 0.0110 (0.0038) | 0.0128 (0.0052) | 0.0148 (0.0067) | 0.0168 (0.0082) | 0.0187 (0.0096) | 0.0205 (0.0111) | 0.0223 (0.0127) |
| 100 | 0.0097 (0.0025) | 0.0119 (0.0041) | 0.0138 (0.0056) | 0.0160 (0.0072) | 0.0182 (0.0088) | 0.0202 (0.0104) | 0.0222 (0.0120) | 0.0242 (0.0137) |
| 200 | 0.0170 (0.0043) | 0.0206 (0.0071) | 0.0238 (0.0094) | 0.0273 (0.0120) | 0.0309 (0.0146) | 0.0342 (0.0170) | 0.0375 (0.0196) | 0.0408 (0.0223) |
| 300 | 0.0239 (0.0060) | 0.0290 (0.0098) | 0.0333 (0.0130) | 0.0382 (0.0165) | 0.0430 (0.0199) | 0.0474 (0.0231) | 0.0518 (0.0265) | 0.0562 (0.0299) |
| 400 | 0.0308 (0.0076) | 0.0373 (0.0125) | 0.0427 (0.0165) | 0.0489 (0.0207) | 0.0549 (0.0250) | 0.0605 (0.0289) | 0.0659 (0.0330) | 0.0714 (0.0371) |
| 500 | 0.0377 (0.0092) | 0.0455 (0.0150) | 0.0520 (0.0198) | 0.0594 (0.0248) | 0.0667 (0.0298) | 0.0733 (0.0345) | 0.0799 (0.0392) | 0.0864 (0.0441) |
| 600 | 0.0447 (0.0108) | 0.0538 (0.0175) | 0.0613 (0.0230) | 0.0699 (0.0288) | 0.0784 (0.0345) | 0.0860 (0.0398) | 0.0937 (0.0452) | 0.1013 (0.0507) |
| 700 | 0.0519 (0.0124) | 0.0622 (0.0199) | 0.0706 (0.0261) | 0.0804 (0.0326) | 0.0900 (0.0390) | 0.0987 (0.0450) | 0.1074 (0.0510) | 0.1160 (0.0571) |
| 800 | 0.0591 (0.0140) | 0.0708 (0.0224) | 0.0801 (0.0292) | 0.0909 (0.0364) | 0.1016 (0.0434) | 0.1113 (0.0499) | 0.1210 (0.0566) | 0.1306 (0.0633) |
| 900 | 0.0664 (0.0155) | 0.0795 (0.0249) | 0.0898 (0.0323) | 0.1016 (0.0401) | 0.1133 (0.0477) | 0.1239 (0.0548) | 0.1345 (0.0620) | 0.1451 (0.0693) |
| 1000 | 0.0738 (0.0171) | 0.0882 (0.0273) | 0.0995 (0.0354) | 0.1125 (0.0439) | 0.1252 (0.0520) | 0.1366 (0.0596) | 0.1481 (0.0673) | 0.1597 (0.0751) |

TABLE A3.8.

Normalised projected range $(R_p)$ and standard deviation $(\Delta R_p)$ for ions with $Z_1 = 70$ in substrates with $Z_2$ from 10 to 80 for ion energies from 10 keV to 1000 keV.

| $Z_1=70$ | $Z_2 = 10$ | $Z_2 = 20$ | $Z_2 = 30$ | $Z_2 = 40$ | $Z_2 = 50$ | $Z_2 = 60$ | $Z_2 = 70$ | $Z_2 = 80$ |
|---|---|---|---|---|---|---|---|---|
| Energy keV | $R_p$ $(\Delta R_p)$ mg/cm$^2$ | $R_p$ $(\Delta R_p)$ mg/cm$^2$ | $R_p$ $(\Delta R_p)$ mg/cm$^2$ | $R_p$ $(\Delta R_p)$ mg/cm$^2$ | $R_p$ $(\Delta R_p)$ mg/cm$^2$ | $R_p$ $(\Delta R_p)$ mg/cm$^2$ | $R_p$ $(\Delta R_p)$ mg/cm$^2$ | $R_p$ $(\Delta R_p)$ mg/cm$^2$ |
| 10 | 0.0019 (0.0005) | 0.0023 (0.0008) | 0.0027 (0.0011) | 0.0031 (0.0014) | 0.0035 (0.0017) | 0.0039 (0.0020) | 0.0043 (0.0023) | 0.0047 (0.0027) |
| 20 | 0.0030 (0.0007) | 0.0037 (0.0012) | 0.0042 (0.0017) | 0.0049 (0.0021) | 0.0055 (0.0026) | 0.0062 (0.0031) | 0.0068 (0.0036) | 0.0074 (0.0042) |
| 30 | 0.0040 (0.0009) | 0.0048 (0.0016) | 0.0055 (0.0022) | 0.0064 (0.0028) | 0.0072 (0.0034) | 0.0080 (0.0041) | 0.0089 (0.0047) | 0.0097 (0.0054) |
| 40 | 0.0048 (0.0012) | 0.0059 (0.0019) | 0.0068 (0.0026) | 0.0078 (0.0034) | 0.0088 (0.0041) | 0.0097 (0.0049) | 0.0107 (0.0057) | 0.0117 (0.0065) |
| 50 | 0.0057 (0.0014) | 0.0068 (0.0022) | 0.0079 (0.0030) | 0.0091 (0.0039) | 0.0102 (0.0048) | 0.0114 (0.0057) | 0.0125 (0.0066) | 0.0136 (0.0075) |
| 60 | 0.0065 (0.0015) | 0.0078 (0.0026) | 0.0090 (0.0034) | 0.0103 (0.0044) | 0.0116 (0.0054) | 0.0129 (0.0064) | 0.0142 (0.0074) | 0.0154 (0.0085) |
| 70 | 0.0072 (0.0017) | 0.0087 (0.0029) | 0.0100 (0.0038) | 0.0115 (0.0049) | 0.0129 (0.0060) | 0.0143 (0.0071) | 0.0158 (0.0082) | 0.0172 (0.0094) |
| 80 | 0.0080 (0.0019) | 0.0096 (0.0031) | 0.0110 (0.0042) | 0.0126 (0.0054) | 0.0143 (0.0066) | 0.0158 (0.0078) | 0.0173 (0.0090) | 0.0189 (0.0103) |
| 90 | 0.0087 (0.0021) | 0.0105 (0.0034) | 0.0120 (0.0046) | 0.0138 (0.0059) | 0.0155 (0.0072) | 0.0172 (0.0085) | 0.0188 (0.0098) | 0.0205 (0.0112) |
| 100 | 0.0094 (0.0022) | 0.0113 (0.0037) | 0.0130 (0.0050) | 0.0149 (0.0064) | 0.0168 (0.0078) | 0.0185 (0.0091) | 0.0203 (0.0105) | 0.0221 (0.0120) |
| 200 | 0.0159 (0.0038) | 0.0191 (0.0062) | 0.0219 (0.0083) | 0.0250 (0.0105) | 0.0281 (0.0128) | 0.0311 (0.0150) | 0.0340 (0.0172) | 0.0370 (0.0196) |
| 300 | 0.0222 (0.0052) | 0.0266 (0.0085) | 0.0302 (0.0113) | 0.0345 (0.0143) | 0.0387 (0.0173) | 0.0426 (0.0202) | 0.0466 (0.0231) | 0.0506 (0.0262) |
| 400 | 0.0284 (0.0066) | 0.0339 (0.0108) | 0.0385 (0.0142) | 0.0437 (0.0179) | 0.0490 (0.0216) | 0.0538 (0.0251) | 0.0587 (0.0287) | 0.0636 (0.0324) |
| 500 | 0.0344 (0.0079) | 0.0411 (0.0129) | 0.0466 (0.0171) | 0.0529 (0.0215) | 0.0591 (0.0258) | 0.0649 (0.0299) | 0.0707 (0.0341) | 0.0765 (0.0384) |
| 600 | 0.0405 (0.0093) | 0.0482 (0.0150) | 0.0546 (0.0198) | 0.0619 (0.0249) | 0.0692 (0.0299) | 0.0759 (0.0346) | 0.0826 (0.0393) | 0.0893 (0.0442) |
| 700 | 0.0465 (0.0105) | 0.0554 (0.0171) | 0.0626 (0.0225) | 0.0709 (0.0282) | 0.0792 (0.0338) | 0.0868 (0.0391) | 0.0944 (0.0444) | 0.1020 (0.0498) |
| 800 | 0.0527 (0.0118) | 0.0625 (0.0191) | 0.0706 (0.0251) | 0.0799 (0.0314) | 0.0891 (0.0376) | 0.0975 (0.0434) | 0.1060 (0.0493) | 0.1146 (0.0553) |
| 900 | 0.0589 (0.0131) | 0.0698 (0.0211) | 0.0786 (0.0277) | 0.0888 (0.0346) | 0.0990 (0.0413) | 0.1083 (0.0477) | 0.1176 (0.0541) | 0.1270 (0.0606) |
| 1000 | 0.0652 (0.0144) | 0.0771 (0.0231) | 0.0866 (0.0302) | 0.0978 (0.0377) | 0.1089 (0.0450) | 0.1190 (0.0518) | 0.1292 (0.0587) | 0.1394 (0.0657) |

TABLE A3.9.

Normalised projected ranges $(R_p)$ and standard deviation $(\Delta R_p)$ for ions with $Z_1 = 80$ in substrates with $Z_2$ from 10 to 80 for ion energies from 10 keV to 1000 keV.

| $Z_1 = 80$ | $Z_2 = 10$ | $Z_2 = 20$ | $Z_2 = 30$ | $Z_2 = 40$ | $Z_2 = 50$ | $Z_2 = 60$ | $Z_2 = 70$ | $Z_2 = 80$ |
|---|---|---|---|---|---|---|---|---|
| Energy keV | $R_p$ $(\Delta R_p)$ mg/cm$^2$ | $R_p$ $(\Delta R_p)$ mg/cm$^2$ | $R_p$ $(\Delta R_p)$ mg/cm$^2$ | $R_p$ $(\Delta R_p)$ mg/cm$^2$ | $R_p$ $(\Delta R_p)$ mg/cm$^2$ | $R_p$ $(\Delta R_p)$ mg/cm$^2$ | $R_p$ $(\Delta R_p)$ mg/cm$^2$ | $R_p$ $(\Delta R_p)$ mg/cm$^2$ |
| 10 | 0.0019 (0.0004) | 0.0023 (0.0007) | 0.0026 (0.0010) | 0.0030 (0.0013) | 0.0034 (0.0015) | 0.0037 (0.0018) | 0.0041 (0.0021) | 0.0045 (0.0024) |
| 20 | 0.0030 (0.0007) | 0.0036 (0.0011) | 0.0041 (0.0015) | 0.0047 (0.0020) | 0.0053 (0.0024) | 0.0059 (0.0029) | 0.0065 (0.0033) | 0.0071 (0.0038) |
| 30 | 0.0040 (0.0009) | 0.0047 (0.0015) | 0.0054 (0.0020) | 0.0061 (0.0025) | 0.0069 (0.0031) | 0.0077 (0.0037) | 0.0084 (0.0043) | 0.0092 (0.0049) |
| 40 | 0.0048 (0.0011) | 0.0057 (0.0018) | 0.0065 (0.0024) | 0.0074 (0.0031) | 0.0084 (0.0038) | 0.0093 (0.0044) | 0.0102 (0.0052) | 0.0111 (0.0059) |
| 50 | 0.0056 (0.0012) | 0.0067 (0.0021) | 0.0076 (0.0028) | 0.0087 (0.0036) | 0.0097 (0.0044) | 0.0107 (0.0051) | 0.0118 (0.0060) | 0.0128 (0.0068) |
| 60 | 0.0064 (0.0014) | 0.0076 (0.0023) | 0.0086 (0.0031) | 0.0098 (0.0040) | 0.0110 (0.0049) | 0.0122 (0.0058) | 0.0133 (0.0067) | 0.0145 (0.0077) |
| 70 | 0.0071 (0.0016) | 0.0084 (0.0026) | 0.0096 (0.0035) | 0.0109 (0.0045) | 0.0123 (0.0055) | 0.0135 (0.0064) | 0.0148 (0.0074) | 0.0161 (0.0085) |
| 80 | 0.0078 (0.0017) | 0.0093 (0.0029) | 0.0105 (0.0038) | 0.0120 (0.0049) | 0.0134 (0.0060) | 0.0148 (0.0070) | 0.0163 (0.0081) | 0.0177 (0.0093) |
| 90 | 0.0085 (0.0019) | 0.0101 (0.0031) | 0.0115 (0.0042) | 0.0130 (0.0053) | 0.0146 (0.0065) | 0.0161 (0.0076) | 0.0176 (0.0088) | 0.0192 (0.0101) |
| 100 | 0.0092 (0.0020) | 0.0109 (0.0033) | 0.0124 (0.0045) | 0.0140 (0.0057) | 0.0158 (0.0070) | 0.0174 (0.0082) | 0.0190 (0.0095) | 0.0207 (0.0108) |
| 200 | 0.0153 (0.0034) | 0.0182 (0.0055) | 0.0206 (0.0074) | 0.0234 (0.0094) | 0.0262 (0.0114) | 0.0288 (0.0134) | 0.0315 (0.0154) | 0.0342 (0.0175) |
| 300 | 0.0211 (0.0046) | 0.0249 (0.0076) | 0.0282 (0.0100) | 0.0319 (0.0127) | 0.0357 (0.0154) | 0.0392 (0.0180) | 0.0428 (0.0206) | 0.0464 (0.0234) |
| 400 | 0.0267 (0.0058) | 0.0315 (0.0095) | 0.0355 (0.0126) | 0.0401 (0.0159) | 0.0448 (0.0191) | 0.0491 (0.0223) | 0.0536 (0.0255) | 0.0581 (0.0288) |
| 500 | 0.0322 (0.0070) | 0.0380 (0.0114) | 0.0428 (0.0150) | 0.0483 (0.0189) | 0.0538 (0.0228) | 0.0589 (0.0264) | 0.0641 (0.0302) | 0.0694 (0.0340) |
| 600 | 0.0377 (0.0082) | 0.0444 (0.0132) | 0.0499 (0.0174) | 0.0563 (0.0219) | 0.0627 (0.0263) | 0.0686 (0.0305) | 0.0745 (0.0348) | 0.0806 (0.0391) |
| 700 | 0.0431 (0.0093) | 0.0508 (0.0150) | 0.0570 (0.0198) | 0.0643 (0.0248) | 0.0715 (0.0298) | 0.0781 (0.0345) | 0.0849 (0.0392) | 0.0917 (0.0441) |
| 800 | 0.0485 (0.0104) | 0.0571 (0.0168) | 0.0641 (0.0221) | 0.0722 (0.0277) | 0.0802 (0.0332) | 0.0876 (0.0383) | 0.0952 (0.0436) | 0.1028 (0.0489) |
| 900 | 0.0540 (0.0114) | 0.0635 (0.0185) | 0.0711 (0.0243) | 0.0800 (0.0304) | 0.0889 (0.0365) | 0.0971 (0.0421) | 0.1054 (0.0478) | 0.1137 (0.0537) |
| 1000 | 0.0594 (0.0125) | 0.0698 (0.0202) | 0.0781 (0.0265) | 0.0879 (0.0332) | 0.0976 (0.0397) | 0.1065 (0.0458) | 0.1155 (0.0520) | 0.1246 (0.0583) |

The sheet resistance $\rho_s$ of an ion implanted layer can be calculated from the equation (Smith and Stephen, 1972)

$$\rho_s = \int \frac{dx}{N(x)\mu(N(x))q}$$

where

$$N(x) = \frac{C_s}{\sqrt{2\pi\Delta R_p}} \exp\left(-\tfrac{1}{2}X^2\right),$$

$$X = \frac{x - R_p}{\Delta R_p} \text{ (section 5.3.4.2)},$$

$C_s$ = surface concentration atoms/cm$^2$, $R_p$ = projected ion range and $\Delta R_p$ = standard deviation.

The mobility $\mu(N(x))$ is a function of $N(x)$ and the value for a particular doping concentration can be found from the resistivity against doping concentration curves of Irvin (1962).

These values are for bulk doped or diffusion doped silicon. Ion implanted silicon may have a lower mobility due to the influence of lattice defects and therefore a higher value of sheet resistance.

The above integral has been evaluated over the range from $x = -3\Delta R_p$ to $x = +3\Delta R_p$ which includes 99.7 % of the dopant atoms under the profile. The range was divided into 100 equal divisions and the value of $N(x)$ determined for the centre of each. $\mu(N(x))$ was interpolated from mobility values calculated from Irvin's resistivity against dose data for the appropriate dopant type. The contribution of each segment to the integral was summed and the value for the whole profile obtained. It was assumed that all the dopant atoms were electrically active.

The calculations were performed for 31 values of $C_s$ and 10 values of $\Delta R_p$ in the range 0.0025 to 0.175 microns. As the mobility values are different for n- and p-type carriers two sets of calculations were carried out. Figs. A4.1 and A4.2 show the results of the calculations for n- and p-type carriers respectively.

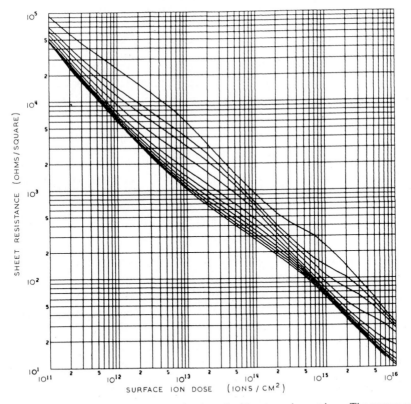

Fig. A4.1. Sheet resistance of silicon implanted with n-type dopant ions. The curves are, in descending order, for standard deviations of 0.0025, 0.0075, 0.0125, 0.0250, 0.0500, 0.0750, 0.100, 0.125, 0.150 and 0.175 microns.

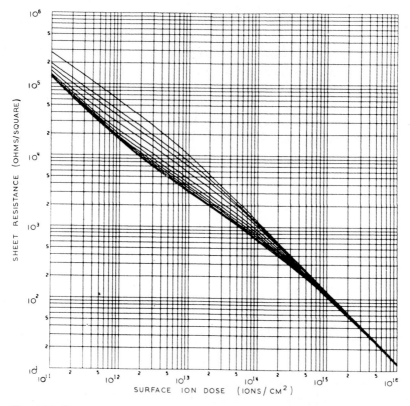

Fig. A4.2. Sheet resistance of silicon implanted with p-type dopant ions. The curves are, in descending order, for standard deviation of 0.0025, 0.0075, 0.0125, 0.0250, 0.0500, 0.0750, 0.100, 0.125, 0.150 and 0.175 microns.

## References

Irvin, J. C., 1962, Bell Syst. Tech. J. **41**, 387.
Smith, B. J. and J. Stephen, 1972, Rad. Effects, **14**, 181.

# AUTHOR INDEX

# SUBJECT INDEX